High Marks:

Regents Living Environment Made Easy

Sharon Welcher

Adjunct Instructor
Chemistry and Physics
City University of New York

Chairperson
Machon Academy High School

Teacher of High School Chemistry, Living Environment, Physics, Earth Science,
Regents Review Courses

See our Website:
http://www.HighMarksinSchool.com

High Marks Made Easy
Forest Hills, NY *(877) 600-7466*

DEDICATION

I dedicate this book to

my father, Rav Jacob Joseph Mazo, זצ"ל,

my mother, Claire Mazo, ע"ה

my husband, Dr. Marvin Welcher,

and my children

HIGH MARKS: REGENTS LIVING ENVIRONMENT MADE EASY

Printed in the United States of America.

 For information, contact **HIGH MARKS MADE EASY**, 62-27 108th Street, Forest Hills, NY 11375, (877) 600-7466.

ISBN:0-9714662-2-X

First Printing: High Marks Made Easy, October 2009
Second Printing: High Marks Made Easy, March 2010

10 9 8 7 6 5 4 3 2

All sales through
(718) 271-7466 / (877) 600-7466

INTRODUCTION

WHY IS THIS BOOK SO GOOD AND SO NECESSARY?

1 This book contains **everything you need to know** for the revised New York State Living Environment Regents.

2 This book is in **simple, clear, easy language**, explaining everything you need to get **high marks** on the Living Environment Regents and all living environment exams.

3 If you **don't understand** a topic in living environment, **read** the same topic in this **book**, and it will **help** you **understand** it. This book is your **private tutor.**

4 This book helps the students understand the concepts, apply them to answer questions and get high marks on the Regents, other state exams, tests and quizzes.

5 All questions are **Regents** and **Regents-type questions** to give you practice for the examination. Included in the book are **constructed response questions**, a type of question that is in the living environment Regents.

6 At the beginning of the Exam section is a detailed description of the Living Environment Regents and **test-taking strategies** to get higher marks.

7 The exam section contains **June**, **August** and **January Regents**.

8 If you don't understand a term, go to the **glossary**, which has definitions in simple, easy language.

9 At the end of the book is an **index**, which makes it easier for you to find what you are looking for.

10 **Answer keys** to homework questions and Regents are available to the **students** at the **teacher's request.**

With this clear and simple book, Living Environment is made **EASY**, and you can get **High Marks** on the **Living Environment Regents** and all Living Environment exams.

Good Luck!

FOREWORD

This is the first edition of **High Marks: Regents Living Environment Made Easy**. This book is based on the New York State Living Environment core curriculum. Teachers should consult the State Education website, www.NYSED.gov, for updates.

My students are the ones who gave me the idea to write a book. They realized that my review sheets were in simple, clear, easy language, while the other books were difficult for many of them to understand.

I wrote High Marks: Regents Chemistry Made Easy, which was very successful. Over 100,000 books were sold. Numerous chairmen, teachers, parents, and students informed me that the book High Marks: Regents Chemistry Made Easy was a tremendous help and benefit to the students. **People** called and thanked me for writing the book and **asked** if there were **other books** in other subjects, such as physics, biology, and math.

Because physics is such a difficult subject, I decided to take on the challenge and write a physics review book to help students get high marks on the Regents and other state exams.

I knew a tremendous number of students take **Living Environment;** it was an opportunity to help the students understand these concepts, apply them to answer questions and get high marks on the Regents, other state exams, tests and quizzes. I have taken **all** the **strategies of teaching** over the years and **incorporated** them **into** the **living environment book** to **help all students,** and not only my own.

ACKNOWLEDGEMENTS

I thank my brilliant father, Rav Jacob Joseph Mazo, זצ"ל, for teaching me how to be an excellent teacher and how to help students get high marks on exams. I thank my dear mother, Claire Mazo, ע"ה, for encouraging me to write a book. I also thank my thoughtful husband, Dr. Marvin Welcher, for typing the book and helping to proofread the book, and my children for being considerate and good.

I express my gratitude to S. Malkah Cohen for a professional job in typesetting and typing four books and helping to bring these books to publication. I thank Devorah Moskowitz for an excellent job on the computer graphics in the physics and living environment books.

I thank my students for giving me the idea to write a book. They knew my book would be in simple, clear, easy language and would help students get high marks on the Regents and all exams.

Sharon Welcher

TABLE OF CONTENTS

About the Author: Sharon Welcher

Sharon Welcher is an adjunct instructor of **chemistry** and **physics** at **City University of New York**, Science **Chairperson** of Machon Academy High School, and a proven **master teacher**, teaching **chemistry**, **physics**, **living environment,** and **earth science** in both **public** and **private** high schools in New York City.

Sharon Welcher wrote High Marks: Regents Chemistry Made Easy, which was very successful. Over 100,000 books were sold. Numerous chairmen, teachers, parents, and students informed her that the book **High Marks: Regents Chemistry Made Easy** was a tremendous help and benefit to the students.

Almost all her students **pass** the **Regents** every year, because she explains the subjects in a simple, clear, **easy to understand** manner and **concentrates** on what the New York State **Regents emphasizes**. Her ability to focus on what is important for the Regents and make it easy to understand has helped innumerable students get **excellent marks** on the **Regents** and exams.

Because physics is such a difficult subject, Sharon Welcher decided to take on the challenge and write a physics review book to help students get high marks on the Regents and other exams. She **wrote High Marks: Regents Living Environment Made Easy** to help students understand the concepts, apply them to answer questions and get high marks on the Regents, other state exams, tests and quizzes. She took all her strategies of teaching over the years and incorporated them into books to help all students, and not only her own.

Sharon Welcher is the **author** of the following books:

High Marks: Regents Chemistry Made Easy

High Marks: Regents Chemistry Made Easy - The Physical Setting

High Marks: Regents Physics Made Easy - The Physical Setting

High Marks: Regents Living Environment Made Easy

CHAPTER 1: LIVING THINGS SIMILARITIES AND DIFFERENCES

Biology is the study of life (living things). A cat, bird or a human being is living. A book, pen, pencil is nonliving. How are living things different from nonliving things?

All **living things** are **similar**:

1. All **living things** (example human, cats, birds) are **made of** one or more **cells** which is the basic unit of structure and function. (Non living things do not have cells).
2. All **living things carry on life processes** (life functions) (for example digestion-breaking down food into simpler substances (things) that can be used by the body) see below. Nonliving things do not carry on life processes.

Life Processes

1. **Nutrition** includes
   a. taking in food (or plants making their own food).
   b. breaking down large food molecules into smaller molecules so they can be used by the body (digestion).
   c. getting rid of (eliminating) undigested food.
2. **Transport:** taking materials (digested food, oxygen, etc.) into the organism (living thing) and spreading (circulating) the material (digested food, oxygen, etc.) throughout the organism.
3. **Respiration** (Cellular Respiration): producing energy from breaking down nutrients (example glucose which is a simple sugar) into simpler, smaller pieces; the energy (chemical bond energy) that is produced is stored in the form (ATP) that can be used for life processes (life activities). Glucose unites with oxygen, producing water, carbon dioxide, and energy (ATP).
4. **Excretion**: Waste materials are produced when the cell carries out these life processes. Excretion is removal (getting rid of) wastes (metabolic wastes) produced from life processes.
5. **Synthesis:** Small molecules are joined together (combined) to form larger molecules. Synthesis includes all the chemical reactions that take place when smaller molecules join together (combine) to form larger molecules.

6. **Regulation:** control and coordination of all life activities (life processes) by nerves and chemicals in the blood (hormones) in order to maintain homeostasis. Homeostasis means a stable or balanced internal environment (same or constant amounts of sugar, salt, and water in the organism).

7. **Growth**: increase in size and number of cells in the organism.

8. **Reproduction:** producing new individuals (babies, offspring) which is not needed for the survival of each individual.

9. **Locomotion**: moving from place to place.

All of these **life functions (life processes) together** are called **metabolism.** Life processes require energy (in the form of ATP). Wastes produced by these life processes are called metabolic wastes.

All living things carry on these life processes but different living things (example plants and animals) carry on these life processes in a different way (plants make their own food and animals take in food). We will explain more about these life processes later in the chapter.

Chemical composition

All living things have the same chemical composition (made up of the same things, carbon, hydrogen, oxygen, nitrogen, etc.). All living things are made of these four main elements **carbon, hydrogen, oxygen,** and **nitrogen** and many other elements in small amounts. Elements (examples: carbon, hydrogen, oxygen) combine (join together) to form molecules (example, sugar).

Organic molecules have both carbon and hydrogen. For example, carbon, hydrogen, and oxygen combine to form sugar and starch. Carbon, hydrogen, oxygen, and nitrogen combine to form protein and parts of living things (example cell wall and cell membrane, which you will learn later). Carbon, hydrogen, oxygen, and nitrogen also combine to form enzymes, needed for chemical reactions in living things, and to form DNA, which is used for heredity (example how the children, grandchildren, great-grandchildren look–eye color, height, etc.).

Inorganic molecules do not have carbon and hydrogen together but can have any element (example oxygen or hydrogen or nitrogen) combined (joined together) with other elements to form inorganic molecules (examples, salts, minerals, water). Living things are mostly made of water.

In short, all living things are similar; living things are made of cells, carry on life processes, and have the same chemical composition.

Now Do Homework Questions #1-4, page 41.

Organization

All living things (organisms) are made of one or more cells. The cell is the basic unit of structure and function of living things. Simple organisms (example ameba) have one cell; complex organisms (example, human being, dog, tree) can have billions of cells. You will learn later that inside a cell are specialized structures called organelles that carry out different functions.

A **cell** is the basic unit of structure and function of all living things.

A group of **similar cells** (similar in structure and similar in function (what the cells do)) **forms** a **tissue** (which is more complex than cells).

Examples of tissues: 1. Millions of skin cells make up a skin tissue, which covers and protects the body. 2. Cells in the body (similar cells) which clean the air before it (air) gets into the lungs form a tissue. 3. Millions of muscle cells make up a muscle tissue. One muscle cell cannot pick up a piece of paper, but a muscle tissue (made of millions of muscle cells) can pick up a baseball, basketball, soccer ball, etc.

Different types of **tissues combine** (join together) to **form** an **organ** (even more complex than tissues) which carries out a life function . Epithelial (skin) tissue, muscle tissue, nerve tissue, blood tissue etc. combine to form the stomach, an organ that digests food.

Different organs work together to **form** a **system** (even much more complex than organs) which also carries out life functions. For example, the digestive system is made up of these organs: mouth, esophagus, stomach, small intestine, which carries out the life function of nutrition (taking in food, digesting food and eliminating undigested wastes). The respiratory system is made of the following organs: nose, trachea, bronchial tubes, and lungs, which carry out breathing, taking in air, bringing oxygen to the blood and taking away carbon dioxide.

In short, similar cells combine (join together) to form tissues, tissues combine to form organs, organs combine to form organ systems (digestive system, respiratory system, etc.), and organ systems combine to make up the organism.

Cells $\xrightarrow[\text{to form}]{\text{combine}}$ Tissues $\xrightarrow[\text{to form}]{\text{combine}}$ Organs $\xrightarrow[\text{to form}]{\text{combine}}$ Organ Systems $\xrightarrow[\text{to form}]{\text{combine}}$ Organism

Another way of showing cells combine to form tissues, tissues combine to form organs, organs combine to form organ systems (digestive system, respiratory system, etc.), and organ systems combine to make up the organism is shown in the next figure:

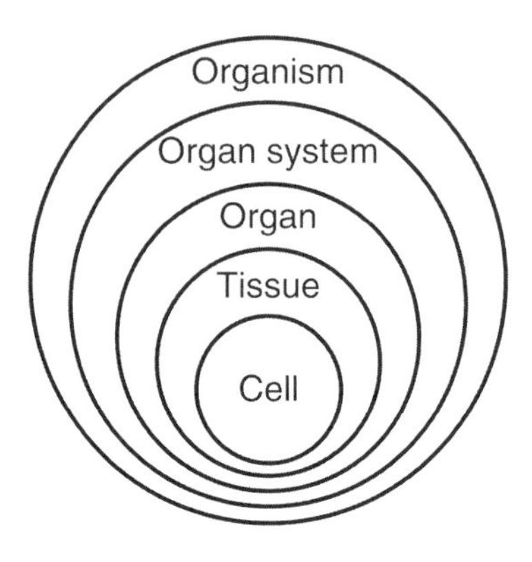

Cells combine to form tissues, tissues combine to form organs, organs combine to form organ systems, and organ systems combine to form the organism.

Now Do Homework Questions #5-8, pages 41-42.

CELLS

Living things (organisms) are made up of one or more cells. You can see cells using a compound light microscope. Each cell carries out the life processes and all the cells work together in a coordinated manner.

Look at the picture of the cell. The cytoplasm is the jellylike substance inside the cell, surrounded by the cell membrane. The cytoplasm **transports** (carries) material through the cell. Many chemical reactions take place in the cytoplasm.

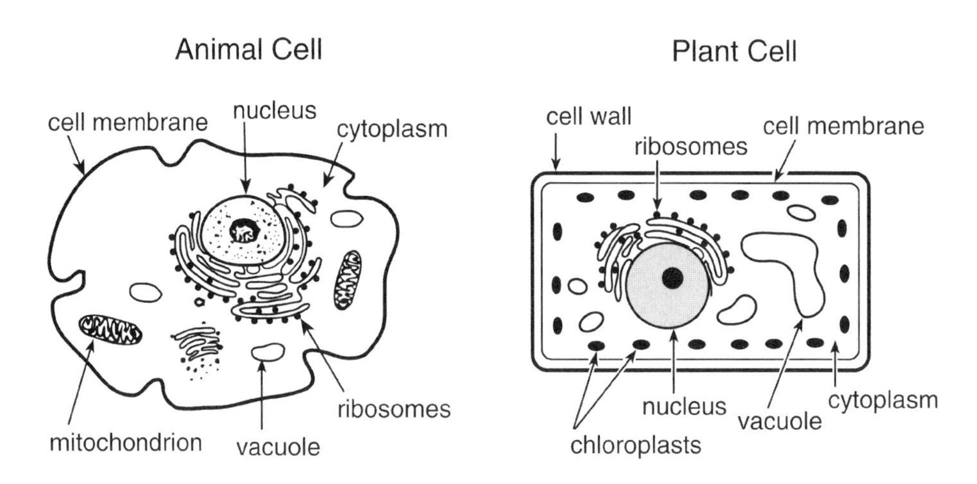

Look again at the picture of the cell. The **structures** (examples: nucleus, ribosomes, mitochondria) that are **inside the cell** are called **organelles**.

Organelles

Organelles are **structures** (examples: nucleus, ribosomes, vacuoles) that are **inside** the **cell**. Each organelle (examples: nucleus, ribosomes) carries out a specific life function (see below). All organelles together do all life functions; all life functions (examples respiration, synthesis, nutrition) together are called metabolism.

Cell membrane surrounds the cell. Cell membrane is made mostly of fats (lipids) and some proteins. The cell membrane controls (regulates) which materials (or how much of a material, example how much water) enters the cell or leaves the cell; you will learn about this later. The cell membrane lets digested food (example, simple sugar) enter the cell. The cell membrane lets wastes leave the cell (waste disposal).

Nucleus is the control center; it controls all life processes (metabolism). The nucleus stores genetic information (information storage); information in the nucleus directs protein synthesis (the synthesis of proteins (joining together of smaller molecules to form proteins (large molecules)).

Vacuoles storage sacs that are inside the cytoplasm. Some vacuoles store food and digest food; other vacuoles store water and get rid of excess (too much) water and other vacuoles store wastes. Vacuoles can store different materials, such as food, water, or waste.

Mitochondria are called the **powerhouse** of the **cell**. Mitochondria are the place where **cellular respiration** takes place. Mitochondria contain enzymes that take the energy out of food and **produce energy** in the form of **ATP**. Cells that need more energy (example muscle cells) have more mitochondria to produce more energy (in the form of ATP).

Ribosomes site (place) of protein synthesis (place where protein is made). Some ribosomes are attached to membranes; other ribosomes are floating in the cytoplasm.

Chloroplasts are **only** in **plants** (and some one celled organisms) but not in animals. **Plants** have chloroplasts (contain chlorophyll) and can **make their own food** in the presence of light (when there is light). When plants make their own food (glucose) in the presence of light, it is called **photosynthesis.**

Cell walls are found in plant cells and not in animal cells. Cell walls are outside the cell membrane and are made of a hard, nonliving material (cellulose). Cell walls support the plant.

Organelles work together: You know **organelles** are **structures** (example, nucleus) **inside the cell.** These **organelles interact** (work together) to **maintain** a **balanced internal environment (homeostasis).** Examples:

1. The nucleus and ribosomes are interrelated. The nucleus is the control center; it directs the cell what to do and tells the ribosome what protein to make. Ribosome makes proteins (protein synthesis) by joining together (synthesis) amino acids to form proteins.

2. Mitochondria and ribosomes interact. Mitochondria contain enzymes that take the energy out of food and **produce energy** in the form of **ATP**. Ribosomes use energy in the form of ATP to make protein.
3. Cell membrane and ribosomes interact. Cell membrane lets amino acids enter the cell Ribosomes use the amino acids as building blocks (synthesis) to make protein.

Organelles, cells, tissues, organs, and organ systems work together to maintain homeostasis (constant internal environment).

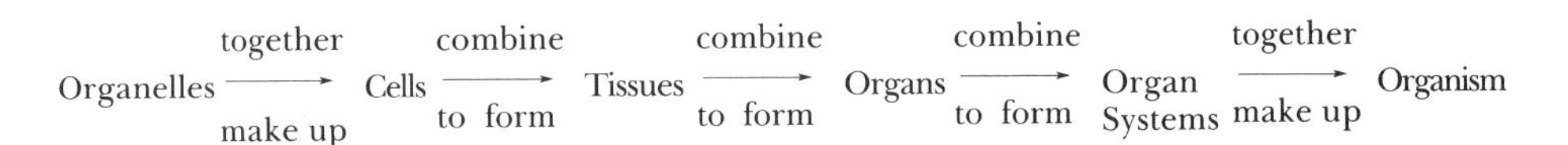

There are two bar graphs below, one bar graph of a plant cell and one bar graph of an animal cell. Look at the bar graphs (a bar graph uses bars).

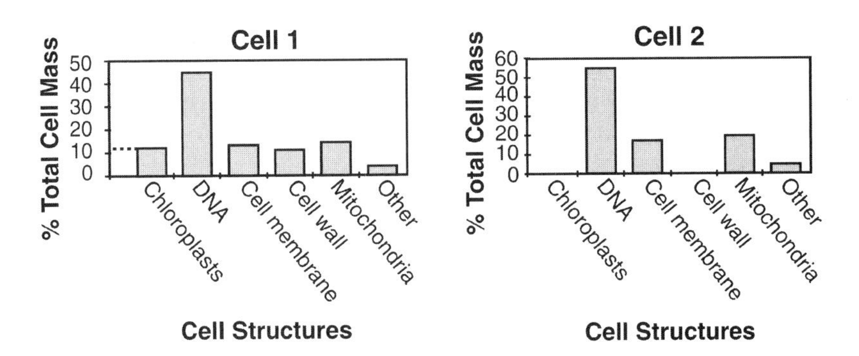

On the vertical axis is percent cell mass (example, mitochondria make up what percentage of the cell). You can tell that **cell 1** is a **plant cell** because it has **chloroplasts** and a **cell wall**. Chloroplasts and cell wall are only in plants and not in animals. Look at the top of the bar for chloroplasts; the student draws a dotted line to the vertical axis (see Cell 1). You see the dotted line is a little above 10% but less than 20%, therefore the chloroplasts are about 12% of the cell mass (material).

Look at cell 2. Cell 2 has **no** (zero) **chloroplasts** and **no cell wall** (there is no bar above the word chloroplasts and no bar above the word cell wall). **Cell 2** is an **animal cell**.

Question: The diagram below represents two cells, X and Y.

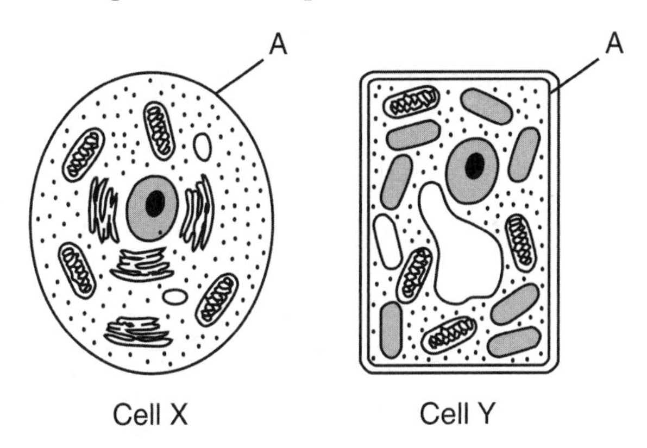

Which statement is correct concerning the structure labeled A?

(1) It aids in the removal of metabolic wastes in both cell X and cell Y.

(2) It is involved in cell communication in cell X but not in cell Y.

(3) It prevents the absorption of CO_2 in cell X and O_2 in cell Y.

(4) It represents the cell wall in cell X and the cell membrane in cell Y.

Solution: The structure labeled A is the cell membrane. The cell membrane lets wastes leave the cell which means the cell membrane helps the cell remove (get rid of) wastes both from animal cells (cell X) and plant cells (cell Y). Answer *1*

Question: The diagram represents one cell and some of its parts. Identify the organelles labeled X, Y, and Z.

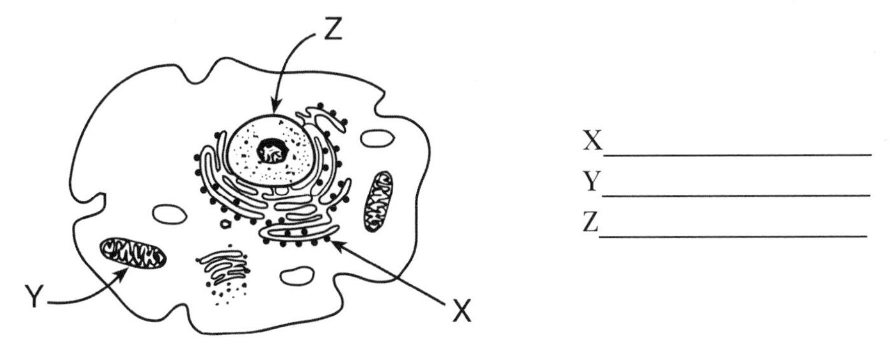

X________________

Y________________

Z________________

Solution: X ribosome

Y mitochondrion (mitochondria)

Z nucleus

Question: An organelle that releases energy for metabolic activity in a nerve cell is the

(1) chloroplast (2) ribosome (3) mitochondrion (4) vacuole

Solution: Mitochondria contain enzymes that take energy out of food and produce (release) energy in the form of ATP. Answer *3*

Now Do Homework Questions #9-23, pages 42-44.

Cell membrane

The cell membrane surrounds the cell and is made mainly of fats (lipids) and some protein.

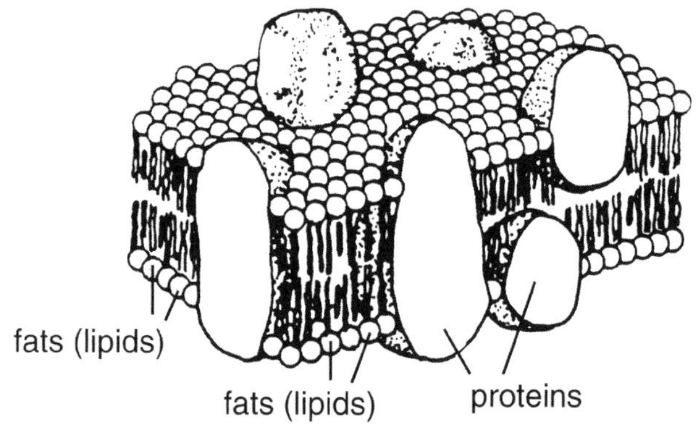

Cell Membrane

The figure below shows a cross section of a cell membrane (this view shows only one side of the membrane from the figure above). **Proteins** in the cell membrane (called receptors) **recognize** and **respond** to **chemical signals** (example insulin). The chemical signals (example insulin) attach to the proteins (receptors) on the membrane, causing the cell to respond (react).

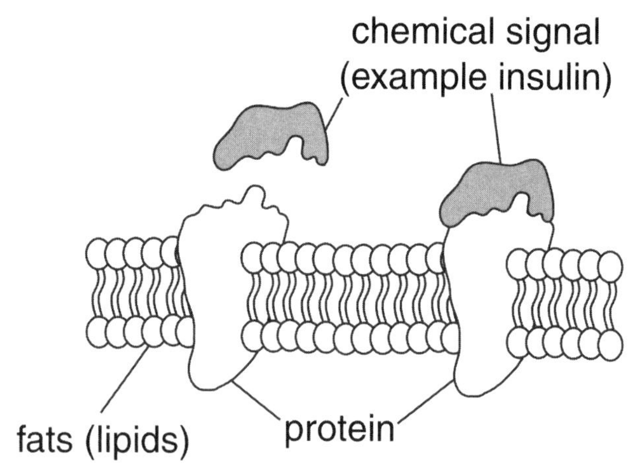

Cell Membrane (cross section)

You learned the cell membrane controls which materials (or how much of a material, example how much water) enters the cell or leaves the cell. Only small, soluble molecules such as glucose (simple sugar) and dissolved gases such as oxygen can pass through the cell membrane; large molecules (example, starch) cannot pass through the cell membrane. Starch, fats, and proteins must first be digested (broken down into simple substances), then they can go through the cell membrane. The cell membrane lets wastes leave the cell (waste disposal).

You learned the cell membrane regulates the transport (moving) of material into and out of the cell. Diffusion and active transport are important in moving material into and out of the cell.

Diffusion

If there is more concentration of a dissolved substance (example, dissolved sugar) outside the cell and a lower concentration (less) of the dissolved substance (dissolved sugar) inside the cell, some dissolved sugar will go across the membrane into (inside) the cell so that there is the same concentration (amount) of sugar inside and outside the cell.

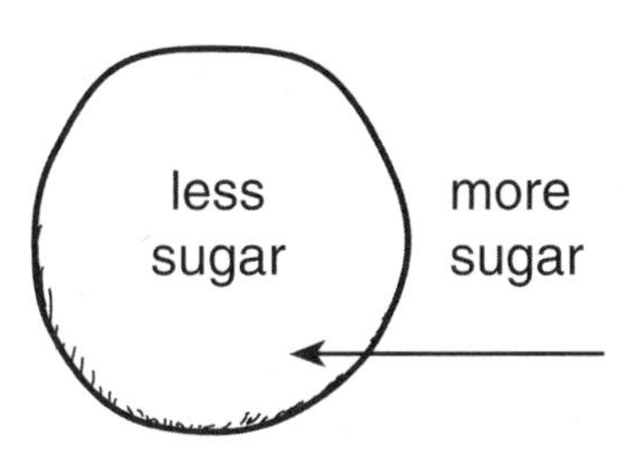

If there is more concentration of a dissolved substance (example dissolved sugar) inside the cell and a lower concentration (less) of a dissolved substance (dissolved sugar) outside the cell, some of the dissolved sugar will go across the membrane out of the cell so that there is the same concentration (amount) of dissolved sugar inside and outside the cell.

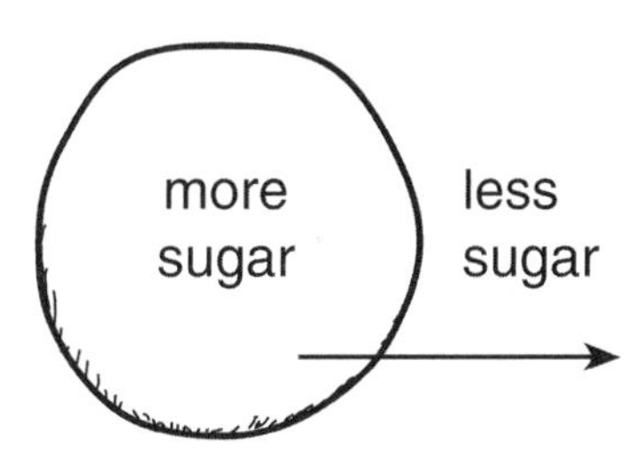

Diffusion: molecules go from an **area of higher concentration** (more concentration) across a membrane **to** an **area of lower concentration**.

Diffusion of water is called **osmosis.** When **water goes from higher concentration** of **water** across a membrane **to a lower concentration** of **water** it is called **osmosis.**

Examples of osmosis:

Example 1: cell has 94% water, surrounding area has 96% water. Water goes from higher concentration of water across a membrane to lower concentration of

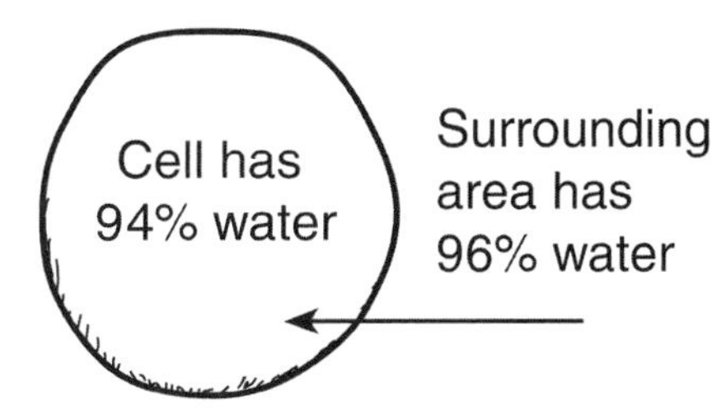

water, therefore, water will enter the cell until the concentration of water inside and outside the cell is the same.

Example 2: cell has 97% water, surrounding area has 92% water. Water goes from higher concentration of water (97% water inside the cell) to outside the cell (92%) until the concentration of water inside and outside the cell is the same.

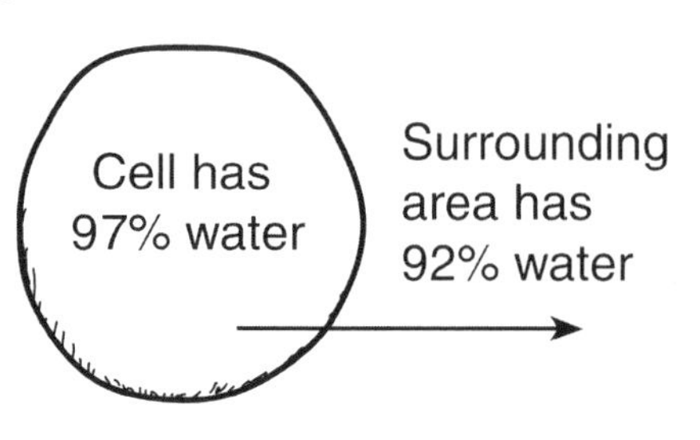

Note: A **3% salt solution** (salt in water) means it has 3% salt and **97% water**. Percent of salt (3%) and percent of water (97%) must equal 100%.An **8% sugar solution** (sugar in water) means it has 8% sugar and **92%** water. Percent of sugar (8%) and percent of water (92%) must equal 100%.

Diffusion and osmosis (diffusion of water) are called passive transport, because no energy is required for diffusion.

Example 1: Red onion cell in salt water (see Figure 1 below).

Look at Figure 2A below. **Rinse** the **onion cell with** distilled water (**pure water**, 100% water, which has no salt), which means there is a higher concentration of water outside the cell than inside the cell. You learned water goes from higher concentration of water (which is now outside the cell) to lower concentration of water (inside the cell), therefore water will enter the cell and cause the cell with its cell membrane to swell, get bigger, until the concentration of water outside and inside the cell is the same (see Figure 2B).

Original cell in salt water	cell wall, cell membrane, water enters Put onion cell in pure water and rinse the cell. Water enters cell until concentration of water inside and outside the cell is the same.	cell wall, cell membrane Result: Cell with its cell membrane got bigger. Cell swelled because water entered cell.
Figure 1	Figure 2A	Figure 2B
	Note: Cell wall is rigid and does not change its size.	

In short:

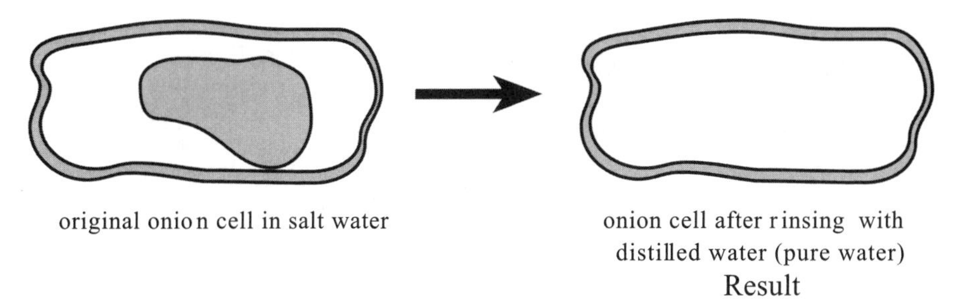

original onion cell in salt water

onion cell after rinsing with distilled water (pure water)
Result

If a cell was left in pure water, the water would keep entering the cell and the cell could burst.

Example 2: Cell A shown at right is a typical red onion cell in water on a slide viewed with a compound light microscope.

Cell A

Draw a diagram of how Cell A would most likely look after salt water has been added to the slide and label the cell membrane in your diagram.

Solution: Cell A is in water (see Figure 1 below).

Look at Figure 2A, which is the same picture as Figure 1, but is labeled. Now, put salt water on the slide (salt water is surrounding the cell). Since **salt** water has **some salt** in the **water**, obviously it has a lower concentration of water than pure water.

You learned water goes from higher concentration of water (inside the cell) to lower concentration of water (outside the cell), therefore water goes out of the cell and the **cell with its cell membrane shrinks (gets smaller)**. See figure 2B.

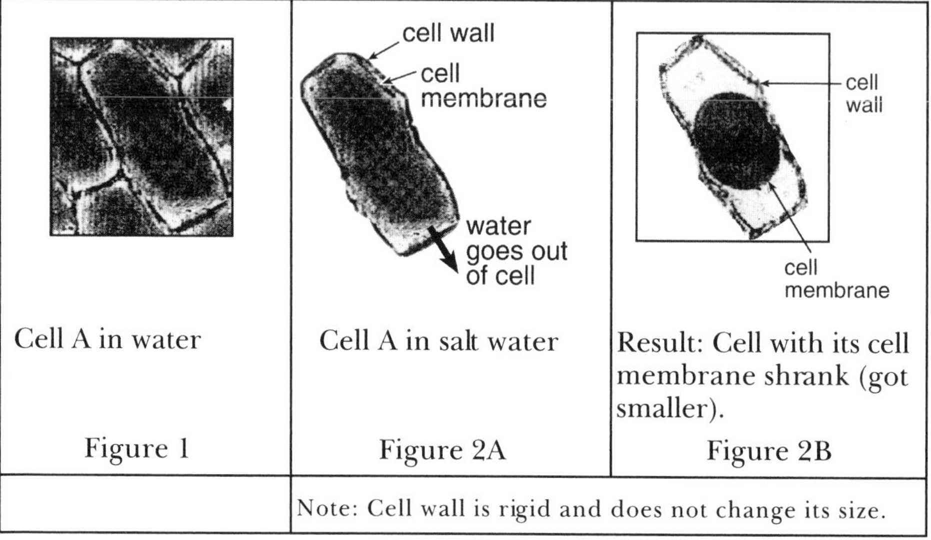

Cell A in water	Cell A in salt water	Result: Cell with its cell membrane shrank (got smaller).
Figure 1	Figure 2A	Figure 2B
	Note: Cell wall is rigid and does not change its size.	

The question asks to draw a diagram of how the cell would look in salt water. Draw the result, figure 2B.

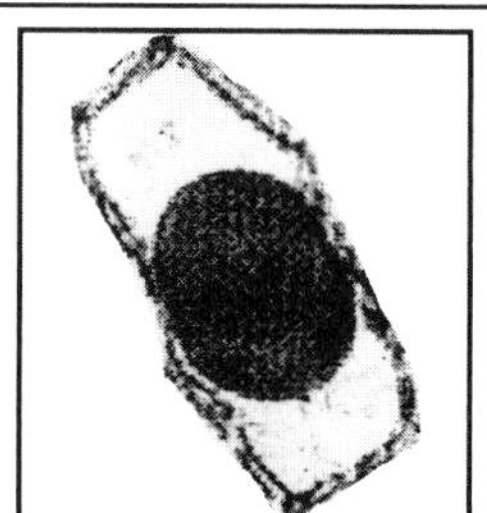

Similarly, just like in example 2 above, you see again that when a plant cell in water is put into salt water, the **cell with its cell membrane shrinks (gets smaller).**

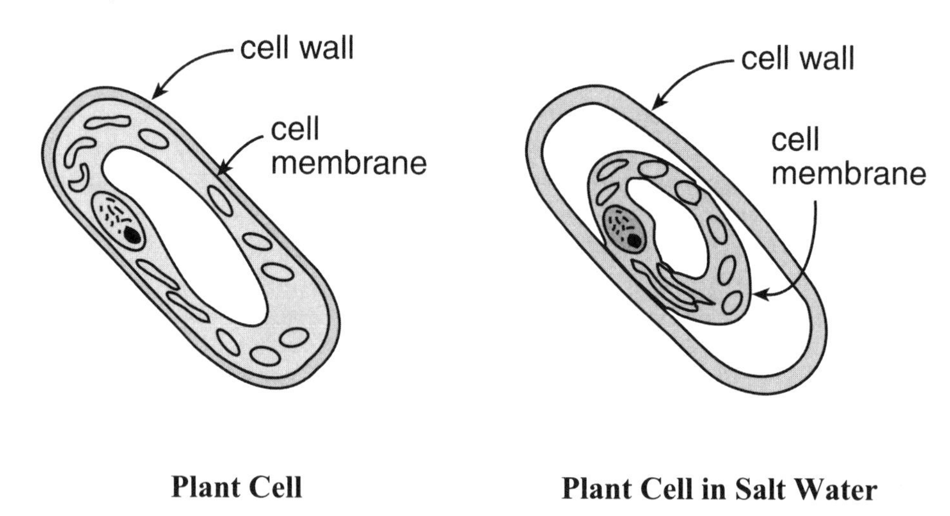

Plant Cell **Plant Cell in Salt Water**

Diffusion through a dialysis membrane in an artificial (model) cell

We can show diffusion or osmosis by making an artificial cell (model cell) using a bag made of dialysis tubing and placing it in a beaker of water. The bag represents a cell; the dialysis tubing (dialysis membrane) represents the cell membrane.

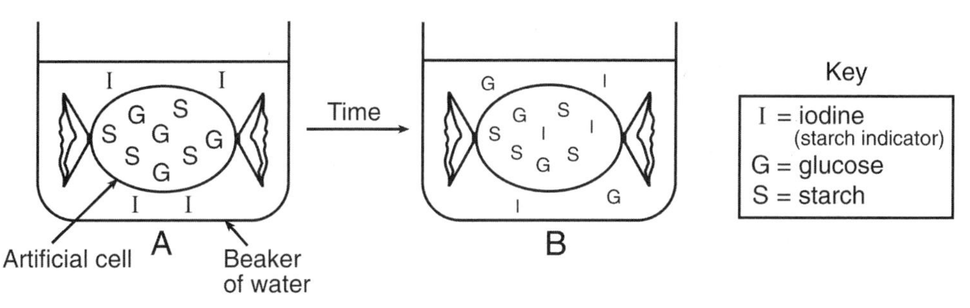

The beaker and the artificial cell contain water

Inside the bag (artificial cell), there is glucose, starch, and water. Outside the artificial cell is only water. In the figure above, glucose diffuses out of the artificial cell into the water in the beaker, going from the area of higher concentration (of glucose) to an area of lower concentration (of glucose). Glucose is a small molecule and will diffuse out of the bag (artificial cell) into the water in the beaker. Starch is a large molecule and will not diffuse out of the bag (artificial cell) into the water.

Iodine turns blue-black if starch is present. The iodine in the beaker of water did not change color because the **starch** (in the artificial cell) is a large molecule and did **not** diffuse into the beaker of **water.**

Also, the iodine is a small molecule and will diffuse from outside the bag to inside the bag. The iodine inside the bag turns blue-black because there is starch inside the bag.

Active Transport

Some cells can **use cellular energy** (energy from ATP) to **force materials to go** across a membrane **from areas of lower concentration** (less concentration) **to areas of higher concentration** (more concentration). (This is the **opposite of diffusion**; diffusion is from higher concentration to lower concentration). Since energy is required, it is called **active transport**. Desert plants use active transport to take water into their roots from desert soil, which has very little water.

Practice Questions and Solutions

Questions 1 and 2: The diagram shows the changes that occurred in a beaker after 30 minutes. The beaker contained water, food coloring, and a bag made from dialysis tubing membrane.

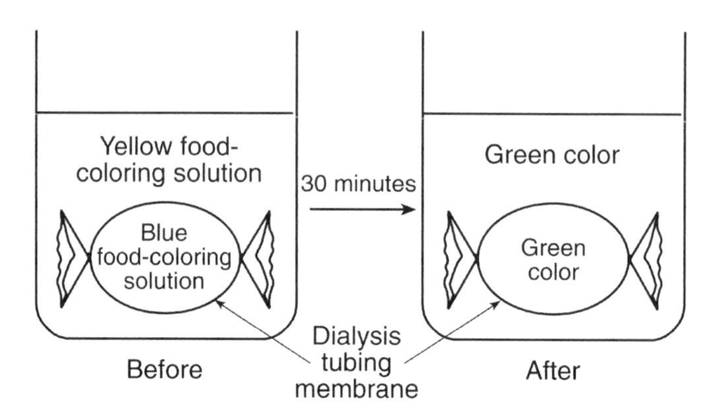

Question 1: When the colors yellow and blue are combined, they produce a green color. Which statement most likely describes the relative sizes of the yellow and blue food-coloring molecules in the diagram?

(1) the yellow food-coloring molecules are small, while the blue food-coloring molecules are large.

(2) the yellow food-coloring molecules are large, while the blue food-coloring molecules are small.

(3) both the yellow food-coloring molecules and the blue food-coloring molecules are large.

(4) both the yellow food-coloring molecules and the blue food-coloring molecules are small.

Solution 1: The water (both inside and outside the bag (made from dialysis tubing **membrane**)) became green (see diagram above at right). Blue color and yellow color mixed to form a green color.

Blue food color diffused from inside the bag to outside the bag and the **yellow food color diffused** from outside the bag to inside the bag **causing blue and yellow colors** to **mix** together and **form** a **green color** (inside and outside the bag.). Therefore, the blue and yellow colors must be small molecules to diffuse (go) through a membrane (bag made from membrane) Large molecules cannot diffuse through a membrane. Answer *4*

Question 2: Which statement best explains the changes shown?

(1) molecular movement was aided by the presence of specific carbohydrate molecules on the surface of the membrane.

(2) molecular movement was aided by the presence of specific enzyme molecules on the surface of the membrane.

(3) molecules moved across the membrane without additional energy being supplied.

(4) molecules moved across the membrane only when additional energy was supplied.

Solution 2: Blue food color diffused across the membrane from inside the dialysis bag (high concentration of blue food color) to outside the bag (low concentration of blue color). Yellow food color diffused across the membrane from outside the dialysis bag (high concentration of yellow food color) to inside the bag (low concentration of yellow color). Diffusion is when molecules go from an area of high concentration to an area of low concentration across a membrane and does not need energy, Answer *3*

Now Do Homework Questions #24-55, pages 44-52.

Receptor Molecules: Cellular Communication (Communication Between Cells)

The nervous system (example nerve cells) and the endocrine system (example pancreas) produce chemicals that help in cellular communication (communication between cells).

1. The endocrine system (example, pancreas) produces chemicals in the blood called **hormones** that regulate how the body works (make sure the body works properly). For example, the pancreas produces the hormone insulin, which regulates the amount of sugar in the blood (you will learn about this later). When the hormone (chemical signal) in the blood moves toward the cell membrane with its receptor, the hormone attaches itself to the receptor (receptor molecule) on the cell membrane (see figure 2). The receptor is a protein on the cell membrane. Each **receptor** is **specific** and can only **recognize** and **respond** to a **specific hormone** (shape of the receptor matches only the shape of the specific hormone–see figure 2). The receptor is on the cell membrane; the **receptor molecule with** the **hormone** sends a signal to that cell, telling the cell what to do (how to respond). For example: When there is too much glucose (sugar) in the blood, the pancreas produces insulin. The **hormone insulin** in the blood attaches to a receptor on liver cells, telling (communicating to) the liver cells to remove (take out) glucose (sugar) from the blood and store it in the liver. Now the blood has the right amount of glucose.

As you can see, there is **communication** between cells. The pancreas tells (communicates to) the liver cells what to do. The **pancreas produces** the hormone **insulin;** the insulin goes to the liver and **tells** the **liver** "Take glucose out of the blood."

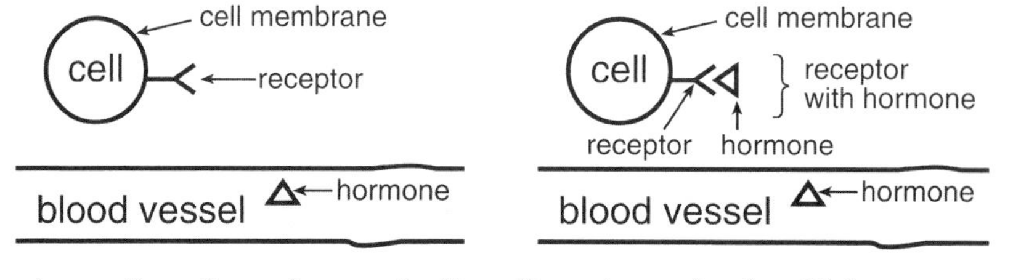

Receptor on the cell membrane of cell

Figure 1

Receptor molecule with hormone sends signal to that cell

Figure 2

You learned the cell membrane with the receptors and chemical signals (example hormones) can also be shown as below.

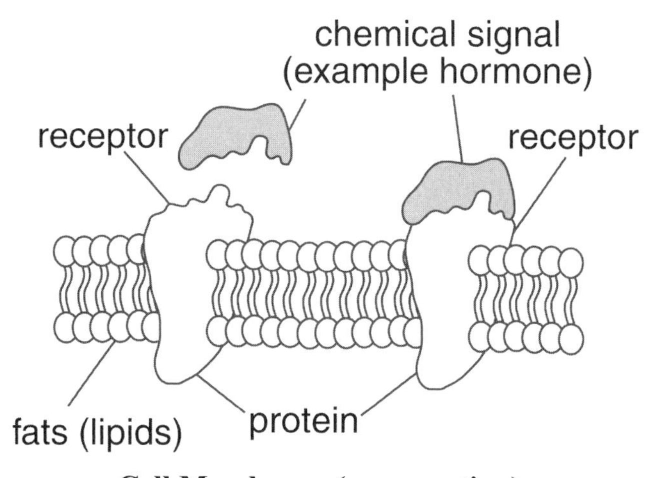

Cell Membrane (cross section)

The hormone (example insulin) attaches itself to the receptor (a protein) on the cell membrane, sending a signal to the cell telling the cell what to do (how to respond).

As you can see, there is **communication** between cells. The pancreas tells (communicates to) the liver cells what to do. The **pancreas produces** the hormone **insulin;** the insulin goes to the liver and **tells** the **liver** "Take glucose out of the blood."

2. Nerve cells produce **chemicals** called neurotransmitters (example acetylcholine) that regulate how the body responds (reacts) to the environment (example, hand jumps away when it touches a hot pot). The chemical at the end of the first nerve cell (see A in the diagram) moves

toward the receptors which are at the beginning of the next nerve cell. The chemical attaches itself (like a lock and key) to the receptors (see receptor with chemical at A in the diagram below). At A, a few receptors are enlarged to show you what is happening. Each **receptor** is **specific** and can only recognize and **respond** to a **specific chemical (nerve signal)**, which fits into the receptor. The **receptor with** the **chemical** (acetylcholine) sends a message through its nerve cell to a muscle, gland, or to another nerve cell telling (communicating to) the muscle, gland, or another nerve cell what to do (how the cell should respond). If the nerve cell dies or the chemical (neurotransmitter) does not work, the nerve cell cannot send a message to the muscle (or another nerve cell); then the muscle cannot respond.

In short, nerve cells can tell (communicate with) gland, muscle or nerve cells what to do (how to respond). There is **communication between** nerve cells and other **cells** (gland cell or muscle cell or nerve cell).

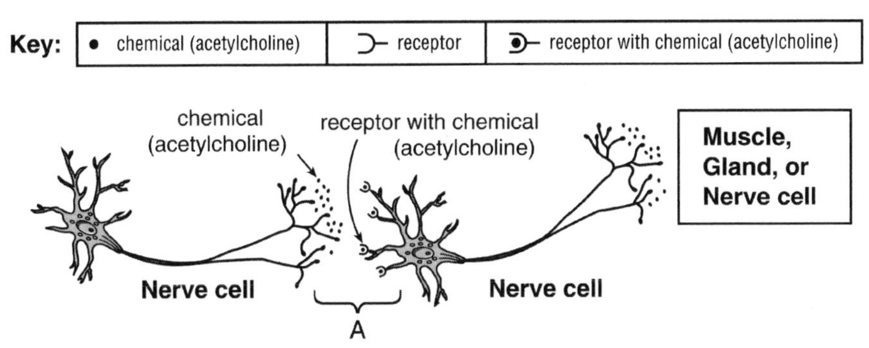

Note: Area between the two nerve cells is enlarged.

Communication Between Nerve Cells and Other Cells

Receptors can also be drawn as:

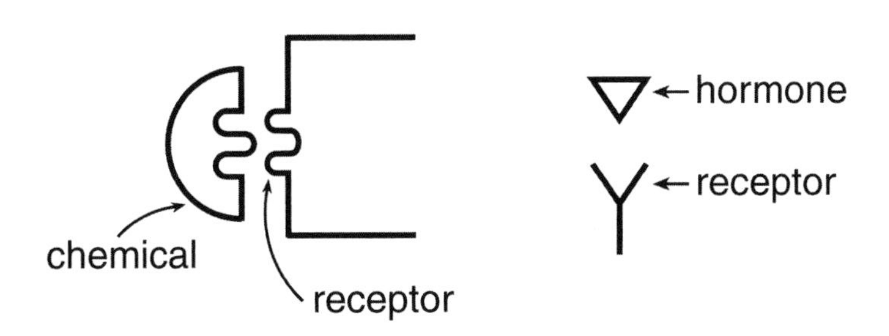

Figure 1 **Figure 2**

In short, **receptor molecules** on the cell membrane recognize and **respond only to specific** chemicals (chemical signals, **nerve signals, hormones,** hormone signals) **which have** the **exact shape** that **fits into** the **receptor** (see figures 1 and 2 above). The chemical (example hormone, nerve signal etc.) must have the exact shape that fits into the receptor so the hormone or nerve signal can work.

Examples of chemicals (example hormones) having the exact shape that fits into the receptor so the chemical can work:

Example 1: Cell B with the circular hormone (•) sends a message to Cell A. Cell A recognizes the circular hormone (•) from cell B and attaches it to the receptor on cell A. Cell A responds to the hormone (signal) (see figure below).

Note: Hormone from Cell A fits into the receptor on Cell B.

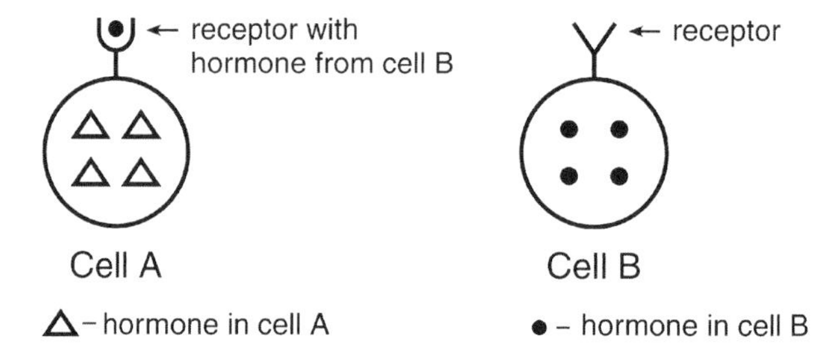

Example 2: A drug is given to block substance A. This is the shape of substance A (see Figure 1 below).

The drug must have a shape that fits into substance A (lock and key), therefore the shape of the drug must be (see Figure 2).

The result of the drug attached to substance A is (see Figure 3).

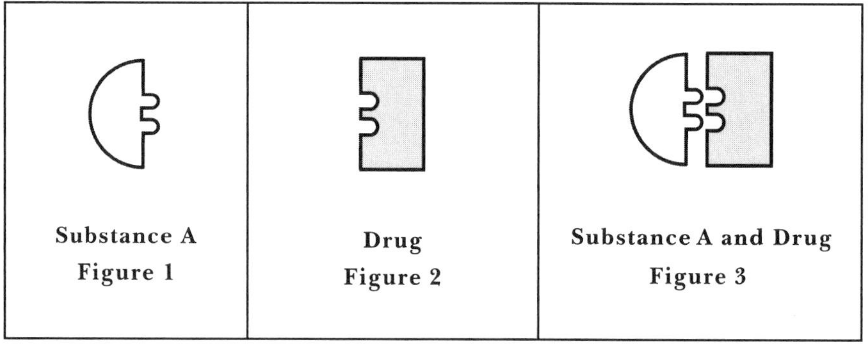

Example 3: Some cells in a female body respond to reproductive hormones while other cells do not. Only cells with receptors for these reproductive hormones respond.

Question: Identify the specialized structures in the cell membrane that are involved in communication.

Solution: Receptor molecules in the cell membrane help communication between cells (cellular communication).

Question: The diagram below represents two molecules that can interact with each other to cause a biochemical process to occur in a cell.

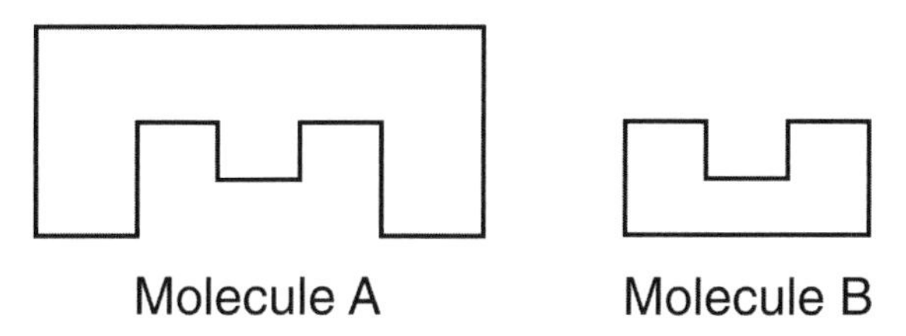

Molecules A and B most likely represent

(1) a protein and a chromosome
(2) a receptor and a hormone
(3) a carbohydrate and an amino acid
(4) an antibody and a hormone

Solution: You learned **receptor molecules** on the cell membrane recognize and **respond only to specific chemicals (example** chemical signals, **nerve signals, hormones,** hormone signals) **which have** the **exact shape** that **fits into** the **receptor** (see figures 1 and 2).

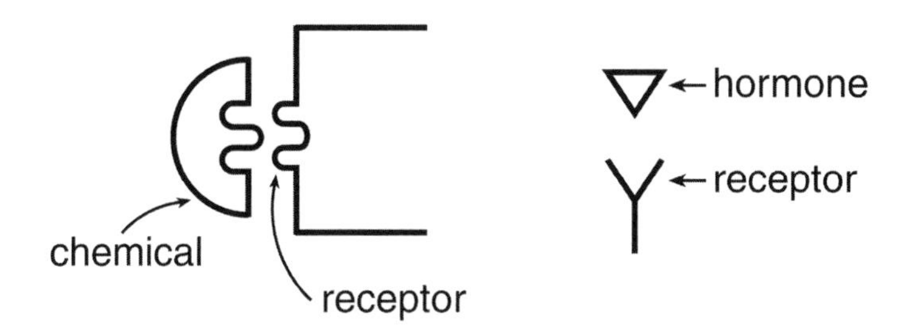

Figure 1 **Figure 2**

The chemical (example hormone, nerve signal etc.) must have the exact shape that fits into the receptor so the hormone or nerve signal can work.

Look at molecule A and molecule B in the question. Molecule B has the exact shape that fits into molecule A. One molecule can be the receptor and the other molecule can be the hormone that fits into the receptor so the hormone can do its work. From the four choices, only choice 2 **receptor** and **hormone, must fit together** like a lock and key so the hormone can do its work (do a biochemical process).

Now Do Homework Questions #56-67, pages 53-54.

SYSTEMS OF THE BODY

Let's understand the systems of our body in more detail. These systems are interrelated (systems work together).

Digestive system

When you eat a sandwich, the food goes from your mouth to the esophagus, to the stomach and small intestine (see figure below), where the digested food goes (diffuses) into the blood and then goes (diffuses) into the cells of your body. The function of the digestive system is to digest food (break down food) into smaller pieces (mechanical digestion) and into simpler substances (chemical digestion) so it can be used by the cells of the body. Starch, protein, and fat are large molecules, which need to be broken down (digested) to simpler (smaller) substances which can be used by the body. Starch is digested (broken down) into simple sugars, protein is digested (broken down) into amino acids, and fat is digested to fatty acids and glycerol. Enzymes help to break down the food chemically, forming simpler substances (examples amino acids, fatty acids), which can be used by the body.

Starch, protein, fats, sugars, amino acids, fatty acids, glycerol, and **vitamins** are **nutrients** that are used by the body to help in life processes (examples: respiration, nutrition, synthesis (such as amino acids joining together to form protein), and growth).

Food is moved down the digestive tract (food tube) by muscular contraction called peristalsis (see diagram of digestive system).

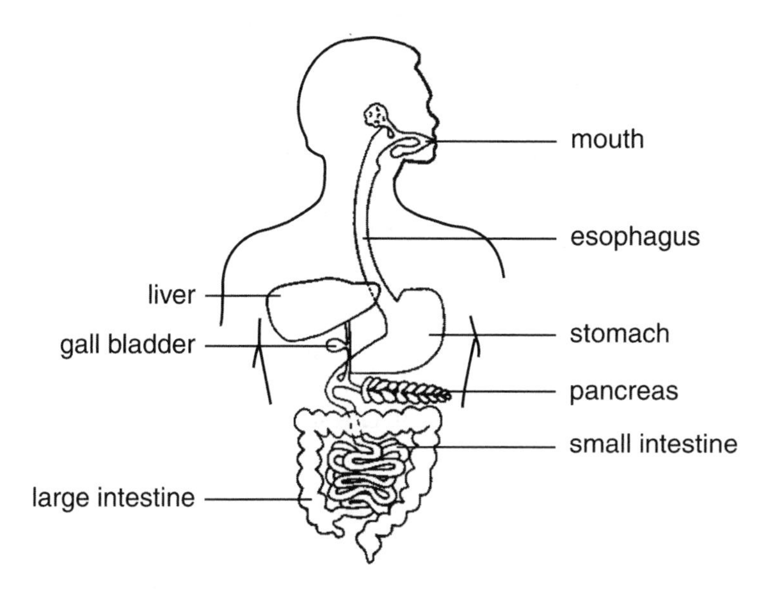

Digestive System

1. In the mouth

a. the teeth break the food mechanically (example, a sandwich broken into smaller pieces).

b. The **enzyme** (ptyalin) in saliva (the liquid in the mouth) begins the **digestion** of **starch**, breaking down starch chemically into a type of sugar (which is a simpler substance).

2. Food goes down the esophagus (tube that connects the mouth to the stomach).

3. Food goes to the stomach. The **stomach** (see digestive system diagram above) has gastric juice that has **enzymes** that begin **protein digestion.** The stomach has **hydrochloric acid**.

4. Partially digested food goes into the **small intestine** (see digestive system diagram). The small intestine is a long coiled tube. The **enzymes** in the **small intestine** help more to **break down chemically** the partially digested protein and sugar into simpler substances that the body can use.

The **liver** (see figure) **produces bile** (not an enzyme) which **goes** through a duct (tube) **into the small intestine**. Bile helps mechanically to break down fat into smaller pieces.

The pancreas (see figure) produces pancreatic juice, which goes into the small intestine. Pancreatic juice contains enzymes that digest fats and continue the digestion of starch and proteins.

Digestion of food is finished in **the small intestine. Starch** is **changed to** simple **sugar, fat** is **changed to fatty acids and glycerol, and protein** is **changed to amino acids.** The end products of digestion are simple sugar, fatty acids and glycerol, and amino acids.

The digested food goes (diffuses) from the small intestine into the blood. The lining of the **small intestine** has **villi** which increase the surface area so more digested food can go into the blood. Digested food (examples simple sugars, amino acids, and fatty acids) **goes through** the **villi** and **goes** into the **bloodstream** (which is part of the circulatory system) and then **goes** by **diffusion** to the **cells of the body** and even into parts of the cell such as the mitochondria. Some of the small molecules of simple sugar, amino acids and fatty acids combine together **(synthesis)** to form larger molecules of starch, proteins, fat, and DNA (or, you can say, sugar, amino acids and fatty acids are used as building blocks forming starch, protein and fat). Sugar molecules are the building blocks which combine together (synthesis) forming starch. Amino acids are the building blocks that combine together (synthesis) forming protein (example of a protein is an enzyme; enzymes are proteins). Fatty acids and glycerol are the building blocks (synthesis) forming fats.

5. **Undigested food** goes from the small intestine **to** the **large intestine** (see diagram of digestive system). The large intestine absorbs excess water from the undigested food and the water goes into the blood (bloodstream) by diffusion and then goes by diffusion to the cells of the body.

Undigested food (wastes) go into the rectum (lower part of the large intestine) and the undigested food (wastes) **leave** the **body (egestion, elimination)** through an opening called the anus.

Laboratory Experiment

You learned starch is a large molecule and cannot go through the cell membrane. It must be broken down into simple sugars.

A laboratory **experiment shows** that **starch cannot go through** the **membrane.** Look at the diagram. Put starch-water mixture in a test tube. Put a membrane at the open end of the test tube. Put the test tube upside down in a beaker of water.

Let's see if starch goes through the membrane in to the beaker of water. Iodine is an indicator that turns blue-black if starch is present. Put iodine (starch indicator solution) in the beaker of water.

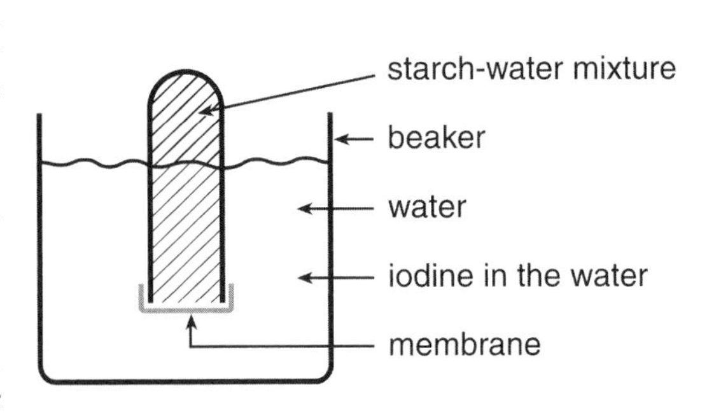

The water with the iodine did not turn blue-black, which means there is no starch in the beaker of water. There was no starch in the beaker of water because starch is a large molecule and cannot go through the membrane.

Now Do Homework Questions #68-73, pages 55-56.

Circulatory System

The **circulatory system** consists of the **heart, blood vessels** (blood vessels that go away from the heart are called arteries; blood vessels that go to the heart are called veins), **and blood** (which has plasma (the liquid part of the blood), and red blood cells, white blood cells and platelets).

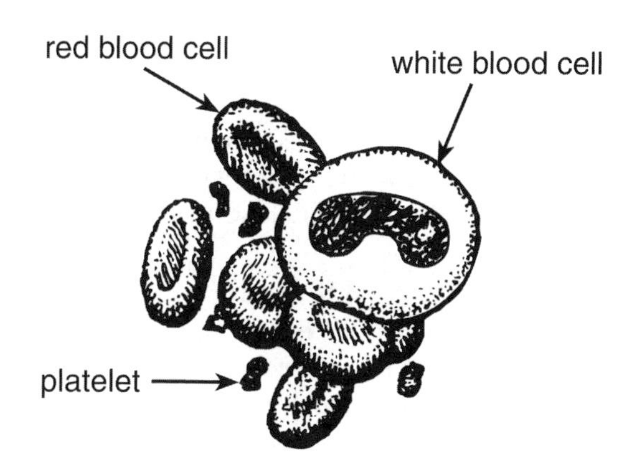

Blood Cells

The function of the circulatory system, or you can say the **blood,** is to **transport (carry)** dissolved material (examples: **glucose, oxygen, amino acids**) to all parts of the body and carry and take away **wastes** and **carbon dioxide.** The blood also carries hormones and antibodies (which fight disease).

Look at the heart diagram. Blood from the lungs which has a lot of oxygen (oxygenated blood) goes to the left atrium of the heart, then goes to the left ventricle of the heart. The left ventricle pumps the oxygenated blood through the **arteries** (**blood vessels** that go **away** from the **heart**). Then the blood goes to capillaries (very thin-walled blood vessels) where oxygen and glucose go from the blood to the cells of the body (by diffusion) and wastes and carbon dioxide go from the cells of the body to the blood (by diffusion). The blood (which has a lot of carbon dioxide and wastes) goes through blood vessels called **veins back to the** right atrium of the **heart**, then the right ventricle, then the blood goes to the pulmonary arteries to the lungs. In the lungs, carbon dioxide goes out and then leaves the body; oxygen is taken in.

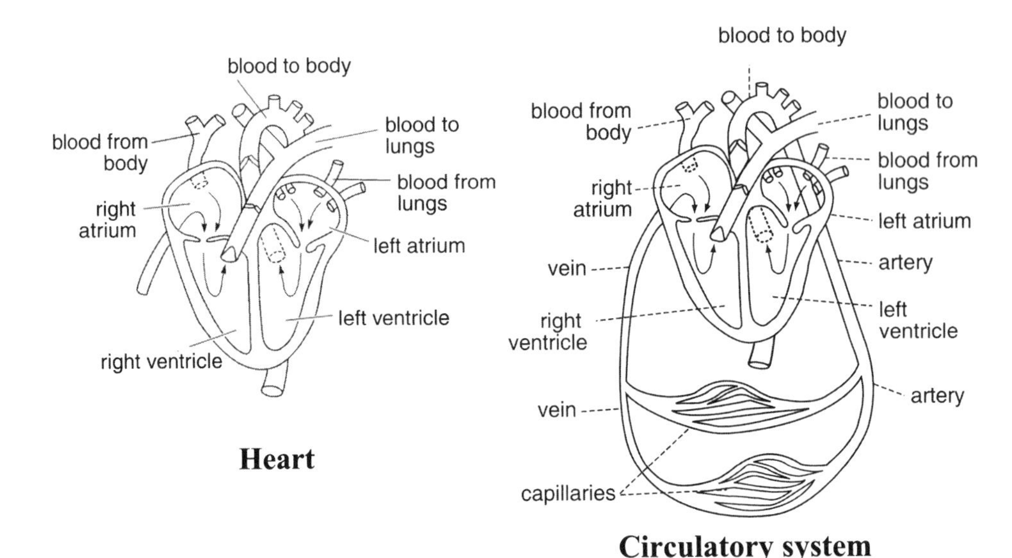

Circulatory system

The heart beats (contracts and relaxes) to pump blood. Every time the heart contracts, blood is forced into the arteries , making the arteries bulge (get a little wider). When an artery is close under the skin, we feel these bulges as a pulse.

Your **heart rate** (example 72 or 76 beats per minute) is **equal** to the **pulse** (example 72 or 76 per minute.) People have different pulse rates: A person's pulse rate depends on the person's weight, whether they are male or female, how physically fit they are, how fast their body uses food (faster or slower metabolism) and differences in genes, therefore people have different pulse rates (examples 60, 72, 80 per minute) even when resting, sitting.

During **exercise, heart rate and pulse increase.** When sitting your pulse might be 72 per minute, when walking, your pulse might increase to 92 per minute and when running your pulse might even increase more to 108 per minute.

Note: The heart beats at a rate causing the blood to flow, bringing enough oxygen and glucose to the cells and taking away carbon dioxide. During exercise, the heart beats faster, therefore the blood moves faster and brings more oxygen and glucose to the cells of the body (example to muscle cells) and removes more wastes.

Experiment

In an investigation, 14 female (girl) students in a class determined their pulse rates after performing each of three different activities. The bar graph below shows the **average pulse rate** for females **after sitting, walking, and running.**

Look at the bar graph. The vertical axis shows pulse rate. The bar for sitting is shorter than the bar for walking, which means the pulse rate while sitting is less than the pulse rate when walking. The bar for walking is shorter than the bar for running, which means the pulse rate when walking is less than the pulse rate when running.

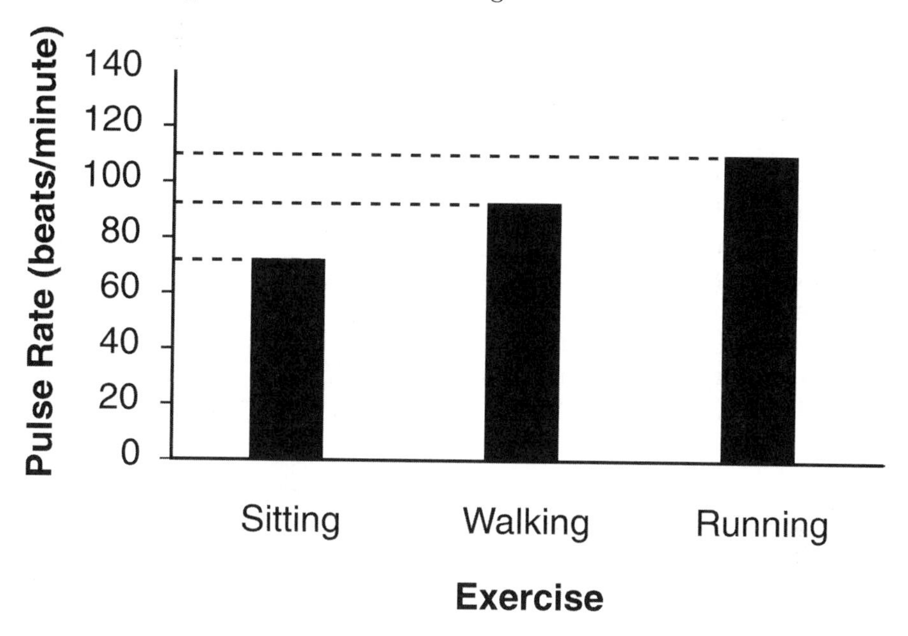

What is the pulse rate for females sitting? Look at the bar graph. There are 20 beats per minute between each two numbers on the vertical line. Look at the top of the bar for sitting and go across to the vertical line; the dotted line is more than halfway between 60 and 80 per minute, or about 72 per minute.

What is the pulse rate for females walking? Look at the bar graph. Look at the top of the bar for walking and go across to the vertical line; the dotted line is more than halfway between 80 and 100 per minute. The pulse rate for females walking is about 92 per minute. The pulse rate for females running is about 110 per minute, halfway between 100 and 120.

Respiratory System

Air goes from the nose to the trachea (windpipe), to the bronchi (bronchial tubes), and then to the lungs (see diagram of respiratory system). The oxygen in the air goes from the alveoli in the lungs (by diffusion) to the blood, which goes to all parts of the body. Carbon dioxide from the cells of the body goes into the blood, then the carbon dioxide from the blood goes into the alveoli

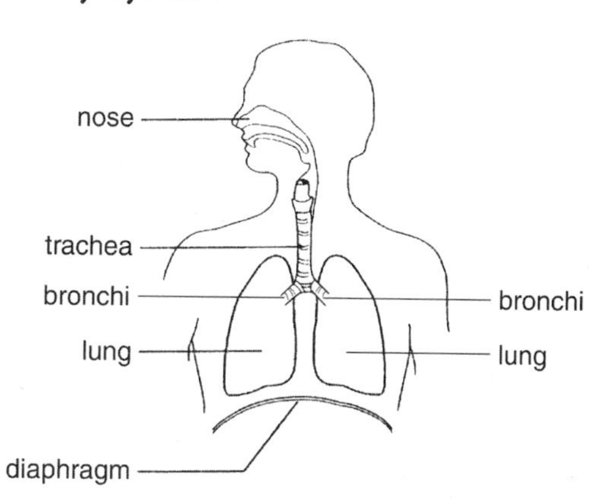

in the lungs and then the carbon dioxide goes to the bronchi (bronchial tubes) and the trachea and out the nose. In short, the function of the respiratory system is breathing, also called gas exchange (to take oxygen into the body and remove carbon dioxide).

Excretory System

The **excretory system** consists of the **lungs, kidneys** (also ureters, urinary bladder, urethra, see diagram at right), and **sweat glands** in the skin. The function of the excretory system is to remove wastes produced by the cells of the body.

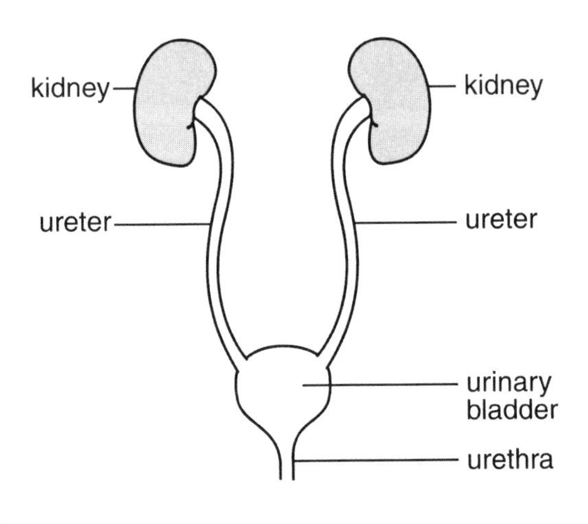

The kidneys remove wastes from the blood. The wastes (urine) go from the kidney to the ureter, to the urinary bladder, and then to the urethra and the urine goes out of the body (see diagram of excretory system).

Nervous System

The nervous system is made of the brain, spinal cord, and nerves. Nerves carry messages from the body, from your hands, feet, etc. (such as you are touching a very hot spoon, you are stepping on a sharp object, a person is stepping on your foot, etc.) to the brain or spinal cord; the brain or spinal cord sends instructions to different parts of the body, telling the body how to respond (react), example pull your hand away.

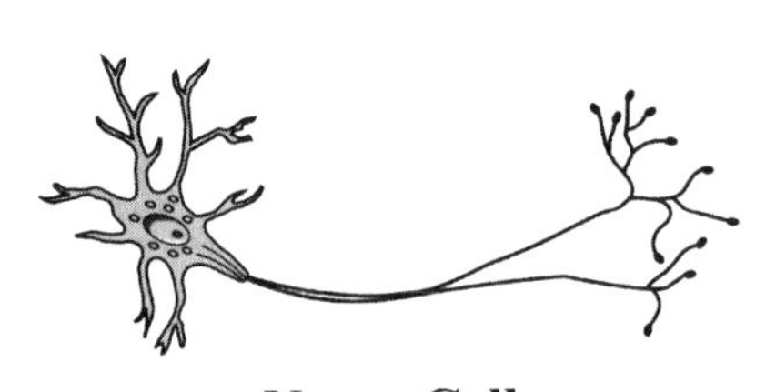

Nerve Cell

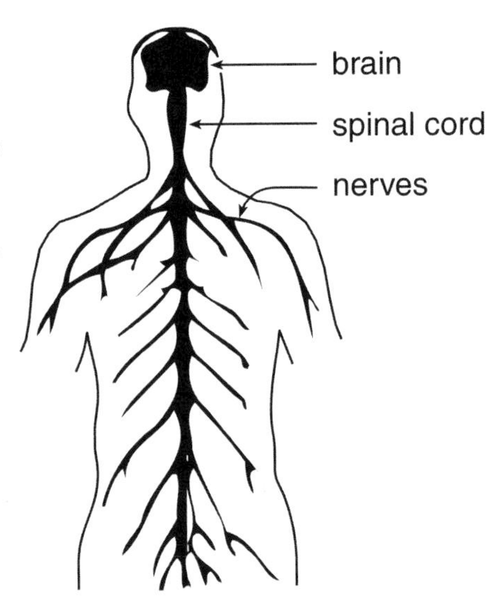

Nervous System

Endocrine System

The endocrine system is made of endocrine glands (examples: pancreas, thyroid) that produce hormones (chemicals in the blood), which regulate the body. For example, the pancreas produces the hormone insulin, which regulates the amount of glucose (sugar) in the blood.

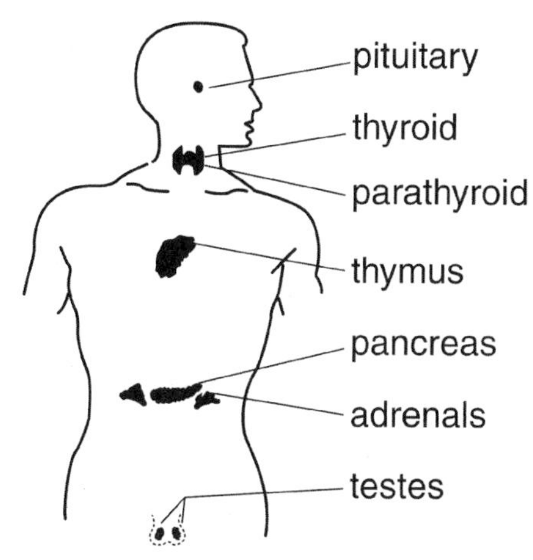

Male Endocrine System

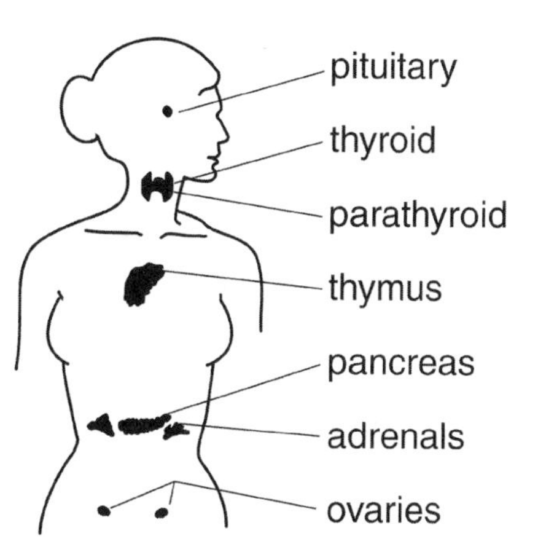

Female Endocrine System

PRACTICE QUESTIONS AND SOLUTIONS

Question: What will most likely happen to wastes containing nitrogen produced as a result of the breakdown of amino acids within liver cells of a mammal?

(1) they will be digested by enzymes in the stomach
(2) they will be removed by the excretory system
(3) they will be destroyed by specialized blood cells
(4) they will be absorbed by mitochondria in nearby cells

Solution: The excretory system removes wastes produced by the cells of the body (in this example, by liver cells) Answer *2*

Now Do Homework Questions #74-78, pages 56-57.

Reproductive System

The function of the reproductive system is to produce more organisms (example more cats, dogs, humans or apple trees) of the same type.

Male Reproductive System:

The testes produce the sperm (male gamete). The sperm go from the testes to the sperm duct and then to the urethra, which is in the penis.

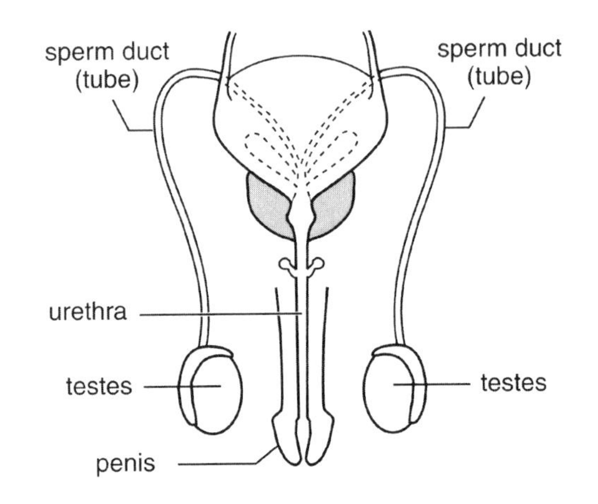

Hormone: The testes produce testosterone, male sex hormone which influences (regulates) the development of secondary sexual characteristics (examples, facial hair (a beard) and a deep voice).

Note: As you can see, the urethra is used in both the excretory system and reproductive system.

Female Reproductive System:

The ovaries produce eggs (female gamete). The egg goes from the ovary to the oviduct (fallopian tube) and then to the uterus.

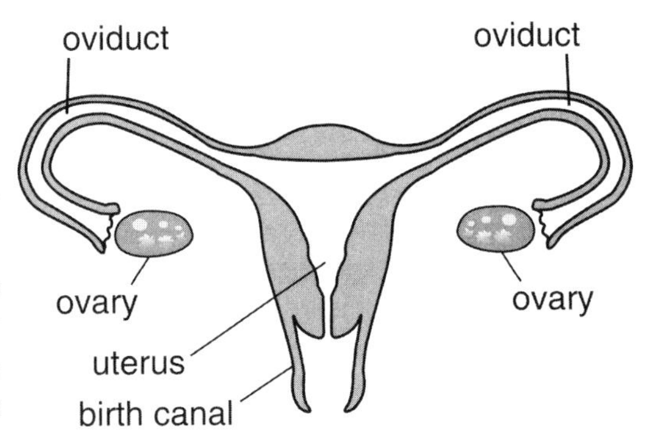

If sperm are present, egg unites with a sperm (fertilization, egg fertilized) in the oviduct. The fertilized egg goes down the oviduct to the uterus.

Hormones: Ovaries produce estrogen (a female sex hormone) which influences the development of the secondary sex characteristics (developing mammary glands) and regulates the reproductive cycle.

Ovaries produce progesterone (a female sex hormone), which regulates the menstrual cycle and prepares the uterus for a pregnancy. The hormone progesterone goes and attaches itself to the receptor on the uterus and tells (communicates to) the uterus to make a thick lining, which prepares the uterus for a pregnancy.

In short, the male reproductive system produces the sperm; the female reproductive system produces the egg. The sperm and egg unite to form a new individual.

Now Do Homework Questions #79-84, page 57.

Regulation

The environment outside the body (such as a very loud noise or very high temperature) and inside the body (such as too much or too little sugar in the blood) keep changing. The **nervous system** (brain, spinal cord, and nerves) and the **endocrine system** (glands, examples: pancreas, ovaries) **help** to **regulate**, **control** how our bodies should react (respond) to the changing environment, (too hot or too cold etc), what our bodies should do. Our reactions (responses) to the changing environment should be **coordinated.** The **nervous and endocrine system** carry out the life process of **regulation (control and coordination of life activities).** Some of the responses(reactions) of the nervous and endocrine systems are needed for homeostasis (maintaining a constant internal environment such as a constant amount of salt or sugar in the blood).

You learned, the nervous system (example nerve cells) and the endocrine system (example pancreas) produce chemicals that help in cellular communication (communication between cells). The nervous system sends signals through the nerves. The endocrine system consists of glands that produce hormones that are carried by the blood. The nerves produce chemicals (neurotransmitters) and the endocrine system (glands) produces chemicals (hormones) to help in cellular communication (communication between the cells).

Movement

The **skeletal system** (bones) supports the body. The **muscular system** (muscles) helps the animal (example human) move.

Immunity

The immune system fights organisms (example viruses, bacteria) that cause disease, protecting the body against disease. The immune system has white blood cells and antibodies that kill organisms that cause disease.

White Blood Cell

Look at the figure below. On the surface of pathogens (organisms that cause disease, example bacteria) are antigens. The **white blood cells** recognize (notice) the antigen and surround and engulf (eat) the pathogen (examples bacteria, viruses) which has the antigen on it.

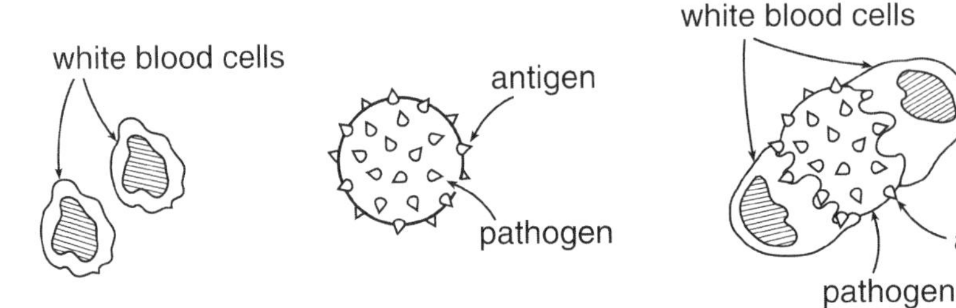

White blood cell recognizes antigen **Then white blood cell engulfs (eats) the pathogen with antigen on it**

Other white blood cells produce antibodies. Antibodies help white blood cells to fight pathogens (organisms that cause disease). Antibodies and white blood cells together kill pathogens.

Now Do Homework Questions #85-92, pages 57-58.

Interaction Among Systems (Body Systems Work Together)

Different body **systems work together** to carry out life functions (example digestion, circulation).

Examples of **systems working together :**

1. Digestive and circulatory systems: The digestive system digests food (breaks down food into simpler substances). Digested food is carried all over the body by the blood, which is part of the circulatory system.
You can also say life functions (life processes) of digestion and transport (or circulation) work together.

2. Respiratory and circulatory systems: The respiratory system (nose, trachea, lungs) takes in oxygen, which goes into the blood (circulatory system)). The blood carries the oxygen to all cells of the body. The respiratory and circulatory systems work together.

The blood (circulatory system) carries the oxygen to the cells of the body. In cellular respiration (also called respiration), glucose (simple sugar) unites with oxygen, producing energy, water, and carbon dioxide. The life function of transport (circulation) works together with the life function of respiration (celllular respiration).

3. Digestive and respiratory systems: The digestive system breaks down food into simpler substances (example: starch is broken into glucose). The respiratory system brings in oxygen. The glucose and oxygen combine (unite) in cells to give off energy (ATP).
You can also say the life functions (life processes) of digestion and respiration work together.

4. Excretory and circulatory systems: Wastes produced in the cells go into the blood (part of the circulatory system) The blood brings the wastes to the kidneys (part of the excretory system) where the wastes are removed from the blood and leave the body. You can say life functions (life processes) of excretion and transport (or circulation) work together.

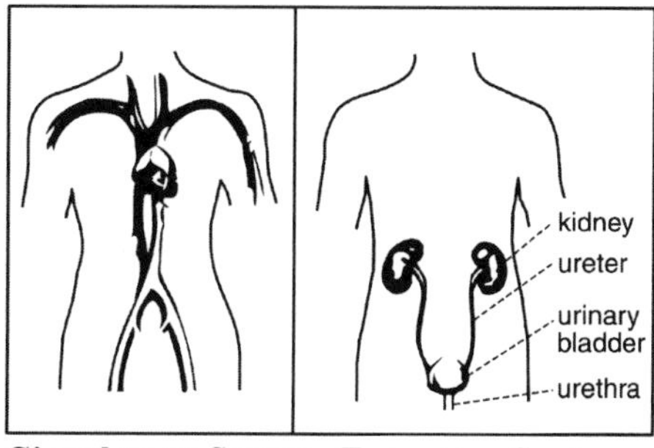

Circulatory System Excretory System

5. Endocrine and reproductivesystems: The female reproductive system has ovaries which produce eggs. An egg and sperm unites together to form a child. Hormones from the endocrine system are needed to produce eggs and to prepare the uterus for a pregnancy. You can say life functions (life processes) of regulation (hormones) and reproduction work together.

6. Skeletal and muscle systems: the skeletal system (bones) supports the body. The muscular system (muscles) helps the animal (example human) move. When a person moves, it involves both the muscles and the bones, carrying on the life function (life process) of locomotion (moving from place to place).

7. Nervous system (brain, spinal cord, and nerves) **and muscular system** (muscles): when temperature drops (animal gets very cold), messages from the brain (part of the nervous system) signal the muscles to shiver, which produces heat and warms the body. The **temperature** of the animal is kept **constant** (humans have a constant temperature of about 98.6° F). The nervous system carries out the life function of regulation (helps the body to adjust to changes in the environment such as too cold or too hot).

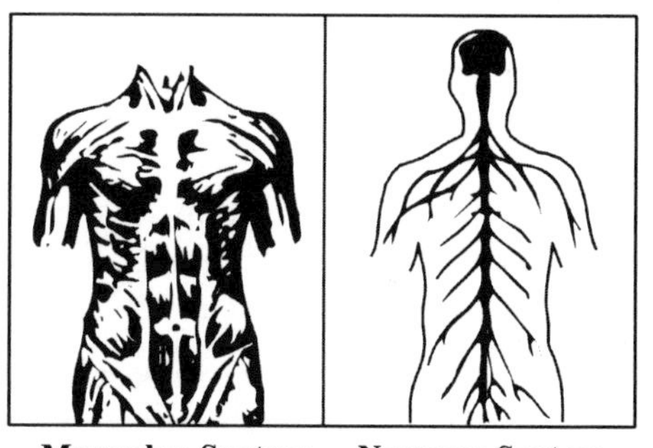
Muscular System Nervous System

8. Endocrine system (produces hormones) **and circulatory system**: The pancreas (part of the **endocrine system**) and blood (part of the **circulatory system)** work together. If the blood has too much sugar, the pancreas produces the hormone **insulin**, which attaches itself to receptors on the liver cells and tells the liver to remove (take away) sugar from the blood and store the sugar. Now the blood has the right amount of sugar.

If the blood has too little sugar, a different hormone attaches itself to a specific receptor on the liver cells and tells the liver to release (give off) the stored sugar, which goes into the blood. Now the blood has the right amount of sugar. The endocrine system carries out the life function of regulation (helps the body adjust to changes in the environment (in this case inside the body, too much or too little sugar in the blood)).

You can say life functions of regulation (hormones) and transport (or circulation) work together.

As you can see from the eight examples above, the systems of the body work together to carry out life functions (life processes).

Examples of Homeostasis (maintaining a constant internal environment)

The body (which has systems, organs, tissues, cells, and organelles) must work properly in order to **maintain homeostasis**. **Homeostasis** is a stable or **balanced internal environment (same** or **constant amount** of **sugar, salt, water)** in the cells of the organism.

1. You learned hormones from the pancreas (part of the endocrine system) keep the amount of **sugar constant** in the blood (see endocrine system, above).

2. The kidneys (part of the excretory system) also regulate the amount of water and salts (example sodium) in the body and keep them constant.

3. White blood cells and antibodies help the body to stay healthy by fighting disease and maintaining homeostasis (a healthy constant environment inside the body).

4. The heart pumps blood to all parts of the body; the blood carries (brings) oxygen and food (glucose) to the cells and takes away carbon dioxide and wastes.

In short, **homeostasis** means **constant internal environment**, constant amount of sugar, salt, water, etc, which is needed for the animal to be alive.

Homeostasis is not maintained:
(when systems, organs, tissues, cells, or organelles are not working properly):

If a system (example circulatory system, excretory system) is not working properly (you can say, a disruption in the system), homeostasis is not maintained.

1. If the **pancreas** (part of the **endocrine system**) is not working properly, then obviously the sugar in the blood would not be constant and homeostasis would not be maintained (imbalance in homeostasis). If the pancreas produces too little insulin, there is too much sugar in the blood. Diabetes is a disease where there is too much sugar in the blood because the pancreas produces too little insulin.

2. If the **kidneys** (part of the **excretory system**) are not working properly, then obviously the amount of water and salts (example sodium) would not be constant and homeostasis would not be maintained (imbalance in homeostasis or you can say, a disturbance in homeostasis). It is dangerous if there is too little sodium (salt) in the blood; it can be fatal. If there is too

much salt in the body and the body cannot keep it at a good constant level, it can cause high blood pressure or too much fluid in the body.

Note: other causes of high blood pressure are stress, genetics, and cigarette smoking. High blood pressure damages blood vessels and causes heart attacks, strokes and kidney damage.

3. If there are too few **white blood cells,** the body cannot fight diseases. Note: Blood cells are part of the **circulatory system**.

4. In a heart attack, the **heart** muscle (the heart is part of the **circulatory system)** is damaged; the heart cannot pump enough blood to all parts of the body.

In short, if any **system,** organ, tissue, cell, or organelle does **not work properly**, **homeostasis** would **not** be **maintained** (**imbalance in homeostasis**), or you can say, a **disturbance or disruption in homeostasis.**

Homeostasis is also not maintained (disturbance (disruption) of homeostasis) if there is too much or too little of a needed chemical (example insulin, estrogen).

Maintaining Homeostasis While Doing Very Strenuous Exercise

During strenuous exercise, such as running a marathon, the runner sweats (perspires) and the body loses (gives off) salt and water. The body may lose too much salt, that the body cannot replace. The runner can drink sports drinks, which contain salt, (or eat salty foods) to make up (replace) the salt the runner lost by sweating. This helps the runner to maintain homeostasis, a constant internal environment, a constant amount of salt.

Now Do Homework Questions #93-105, pages 58-59.

Diseases Caused by Organ Malfunctions Disrupt Homeostasis

Diseases caused by organ malfunctions (organs such as the stomach, heart, kidney or brain not working properly) are explained in the chart below.

Organ Part of System	Disease: Organ Malfunction	Causes	Effect
Heart part of circulatory system	Heart attack	Blood clot (blockage) in artery that brings blood to heart muscle; therefore, blood cannot go to the heart muscle. Also, heart attacks can be caused by a high-fat diet, which narrows the arteries of the heart, causing blood clots (blocking blood) in the heart.	Damaged heart muscle; cannot pump enough blood containing oxygen to body organs (example brain, part of nervous system).
Pancreas part of endocrine system	Diabetes: too much sugar in the blood (high blood sugar)	Too little insulin (undersecretion of insulin)	Too much sugar in the blood; causes poor circulation and damage to eyes.
Brain part of nervous system	Stroke	Blood clot in blood vessel in brain	Causes brain damage: affects organs or systems, can harm movement, speech, breathing, etc..
Stomach part of digestive system	Ulcers: painful sores in the stomach wall	Too much stomach acid damages the stomach wall	Painful sores in the stomach wall; causes bleeding.
Bronchi (bronchial tubes) part of respiratory system	Bronchitis: inflammation of the membrane of the bronchi	Infection or other irritant (such as cigarette smoke)	Difficulty breathing; not enough oxygen goes to cells (example brain cells), which are part of nervous system.

Organ Part of System	Disease: Organ Malfunction	Causes	Effect
Kidney part of excretory system	Kidney disease: kidney not working properly	High blood pressure, diabetes	Wastes not removed from blood; wastes can enter other systems and poison them. Kidneys not working properly can cause dangerous amount of sodium in blood, which can cause death.

In the next chapter you will learn that diseases caused by viruses, bacteria, fungi, and parasites also disrupt homeostasis and hurt the body.

Now Do Homework Questions #106-109, page 60.

All living things carry on Life Functions

You learned that **all living things** (organisms that have only one cell or many cells, such as a human being) **carry on life processes (life functions).** You learned all life processes together (example transport, excretion, respiration) are called metabolism.

In any cell (in a one-celled organism or a many-celled organism) the function of the nucleus, cytoplasm, cell membrane, etc., is similar.

Examples of one-celled organisms (organisms which have only one cell) are the ameba and paramecium. **One-celled organisms** are **also called single celled** organisms **or unicellular** organisms.

One-Celled Organisms

Ameba **Paramecium**

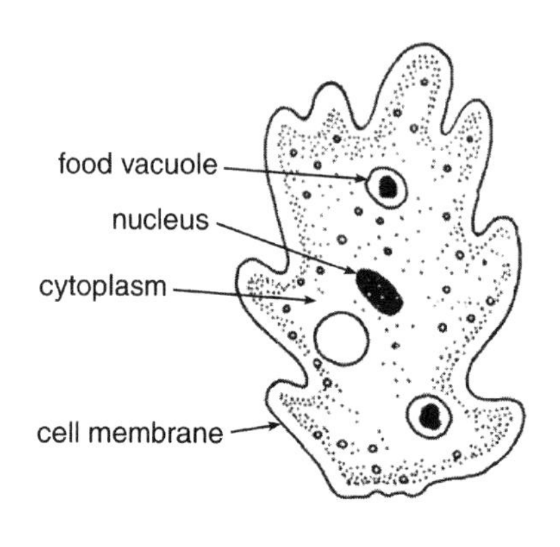

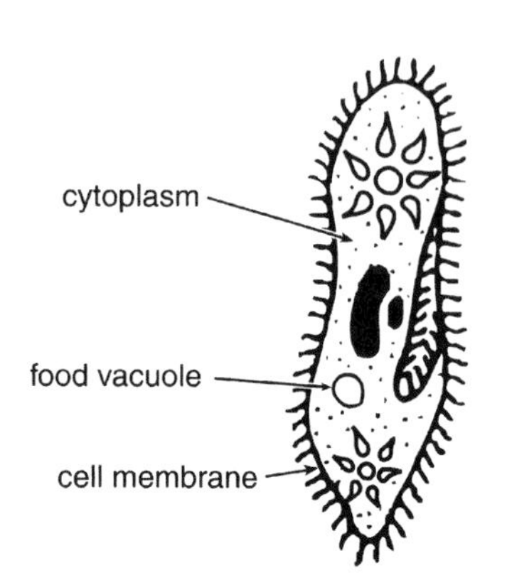

Digestion: Look at the diagram above and the chart of life functions below. In a single-celled (one-celled) organism like an ameba or a paramecium, food is digested in the food vacuole. In a multi-cellular animal (many-celled animal, example human), food is digested in the organs (mouth, stomach, small intestine) of the digestive system.

Transport: The liquid inside the organism carries oxygen, carbon dioxide, and digested food throughout the organism: In a single-celled (one-celled) organism like the ameba or paramecium, the cytoplasm carries (transports) the digested food and oxygen throughout the organism. In the multi-cellular (many-celled) organism, the circulatory system (blood) carries digested food, oxygen, etc., throughout the organism.

Taking in oxygen, giving off carbon dioxide: Look again at the diagram above and the chart below. In the one-celled organism, oxygen is taken in and carbon dioxide is given off through the cell membrane. In a multi-cellular (many-celled) organism, the organs (nose, trachea, lungs) of the respiratory system take in oxygen and give off carbon dioxide.

Excretion: In a single-celled organism (example ameba or paramecium) wastes go through the cell membrane. In a multi-cellular organism, wastes go out through the organs (kidneys, urethra, etc.) of the excretory system (see chart below).

Life Functions in a Single-Celled and Multi-Cellular Organism

Life Function	Single Cell (one-celled)	Multi-Cellular Organism (many-celled)
Nutrition (includes digestion of food)	food vacuole	digestive System
Liquid inside cell, carries food, oxygen, etc.	cytoplasm	circulatory system
Taking in oxygen, giving off carbon dioxide	cell membrane	respiratory system
Excretion (getting rid of wastes)	cell membrane	excretory system

As you can see, the organelles (example, food vacuole) inside the one-celled organism do the same function as the organ systems (example digestive system) in a many-celled organism like the human body.

You learned in a multi-cellular organism, the systems of the body must work properly in order to maintain homeostasis. Similarly, in a unicellular (single-celled, one-celled) organism, the organelles (structures inside the cell, such as a vacuole) must work properly in order to maintain homeostasis.

You realize one-celled organisms and many-celled organisms carry out the same life processes (life functions).

PRACTICE QUESTIONS AND SOLUTIONS

Question: The table below provides some information concerning organelles and organs.

Function	Organelle	Organ
gas exchange	cell membrane	lung
nutrition	food vacuole	stomach

Based on this information, which statement accurately compares organelles to organs?

(1) functions are carried out more efficiently by organs than by organelles.

(2) organs maintain homeostasis while organelles do not.

(3) organelles carry out functions similar to those of of organs,

(4) organelles function in multicellular organisms while organs function in single celled organisms.

Solution: The organelle (example food vacuole) inside the one-celled organism does the same function (nutrition, which includes digestion of food) as the stomach (organ) which is part of the digestive system in a multi-cellular organism.

The organelle (example cell membrane) in the one-celled organism does the same function (gas exchange) as the lung (organ) which is part of the respiratory system in a multi-cellular organism. Answer *3*

Now Do Homework Questions #110-116, pages 60-62.

Drawing Graphs

Let's see how we can draw graphs based on experimental data. Data (from an experiment) is written on a data table (see below). Draw the graph based on the data table.

Problem 1: The experiment used five tubes to study the effect of temperature on protein digestion (amount or how much protein is digested). The results of the experiment are shown in the data table below.

Protein Digestion at Different Temperatures

Tube #	**Temperature** (°C)	**Amount of Protein Digested** (grams)
1	5	0.5
2	10	1.0
3	20	4.0
4	37	9.5
5	85	0.0

How to draw the line graph:

1. On the x axis, put "Temperature, °C". **The thing you change** (in this case temperature) is always put on the **x axis.** This is the independent variable. Space the lines along the axis equally. "Make an appropriate scale" by spacing the numbers on the graph so that all the data fits on the graph and it is easy to read. There must be an equal number of degrees between lines (see graph). On the x axis, put 10°C between lines (every two lines) (scale on the x axis), then all the temperatures on the data table fit on the graph and it is easy to read.

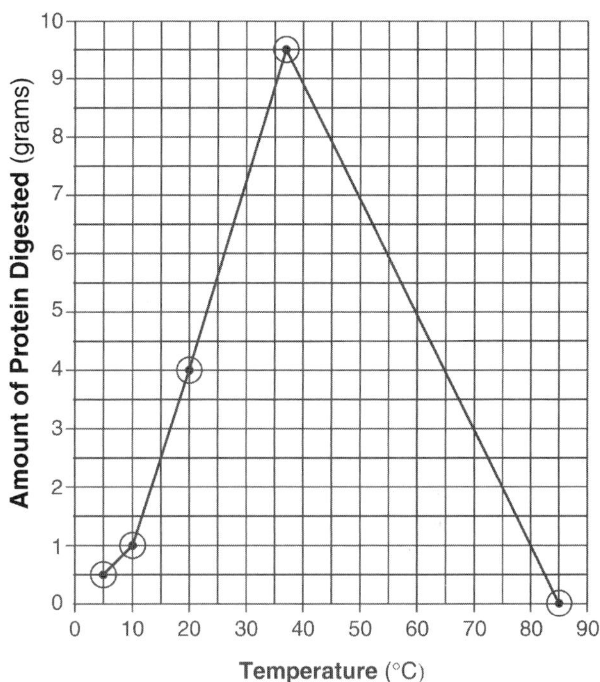

2. On the y axis, put "Amount of Protein Digested (grams)". The **result** you get (amount of protein digested) is always put on the **y axis**. This is the dependent variable. Space the lines along the axis equally. "Make an appropriate scale" by spacing the numbers on the graph so all the data fits on the graph and it is easy to read. There must be an equal number of grams of protein digested between lines (see graph.) On the y axis, put one gram of protein between lines (every two lines) (scale on the y axis), then all the grams of protein digested in the data table fits on the graph and is easy to read.

3. Plot the experimental data on the graph. Draw a circle around each point. Draw a line that connects the points. Do not continue the line past the last point.

4. Put a title on the graph which shows what the graph is about. Examples: "Effect of temperature on protein digestion" or "Protein digestion at different temperatures."

Now Do Homework Questions #117-120, pages 62-63.

NOW LET'S TRY A FEW HOMEWORK QUESTIONS

To assign fewer questions, you may want to skip the questions with underlined numbers.

1. Energy from organic molecules can be stored in ATP molecules as a direct result of the process of
   (1) cellular respiration (3) diffusion
   (2) cellular reproduction (4) digestion

2. A single-celled organism is represented in the diagram below. An activity is indicated by the arrow.

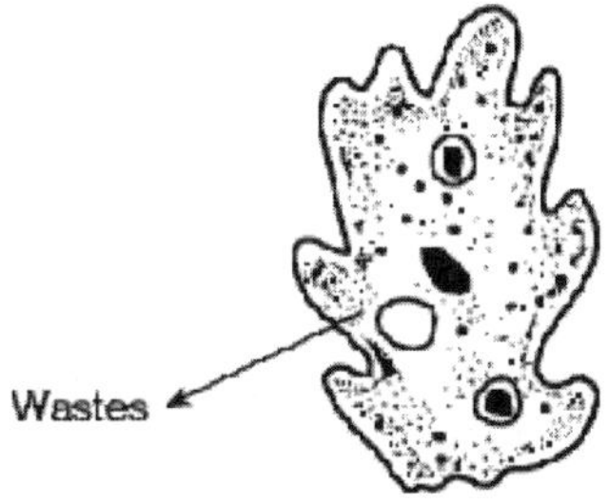

If this activity requires the use of energy, which substance would be the source of this energy?
   (1) DNA (3) a hormone
   (2) ATP (4) an antibody

3. To remain healthy, organisms must be able to obtain materials, change the materials, move the materials around, and get rid of waste. These activities directly require
   (1) energy from ATP
   (2) the replication of DNA
   (3) nutrients from inorganic sources
   (4) manipulation of altered genes

4. Living organisms must be able to obtain materials, change the materials into new forms, remove poisons, and move needed material from one place to another. Many of these activities directly require
   (1) energy released from ATP
   (2) carbohydrates formed from receptor molecules
   (3) the synthesis of DNA
   (4) the breakdown of energy-rich inorganic molecules

<u>5</u>. Which structure is best observed using a compound light microscope?
   (1) a cell (3) a DNA sequence
   (2) a virus (4) the inner surface of a mitochondrion

<u>6</u>. In a cell, all organelles work together to carry out
   (1) diffusion (3) information storage
   (2) active transport (4) metabolic processes

7. Write the structures listed below in order from least complex to most complex.
   organ
   cell
   organism
   organelle
   tissue

   Least complex: organelle
   cell
   tissue
   organ
   Most complex: organism

8. The respiratory system includes a layer of cells in the air passages that clean the air before it gets to the lungs. This layer of cells is best classified as
   (1) a tissue
   (2) an organ
   (3) an organelle
   (4) an organ system

9. Which structure is best observed using a compound light microscope?
   (1) a cell
   (2) a virus
   (3) a DNA sequence
   (4) the inner surface of a mitochondrion

Base your answers to the next three questions on the diagrams below and on your knowledge of biology. The diagrams represent two different cells and some of their parts. The diagrams are not drawn to scale.

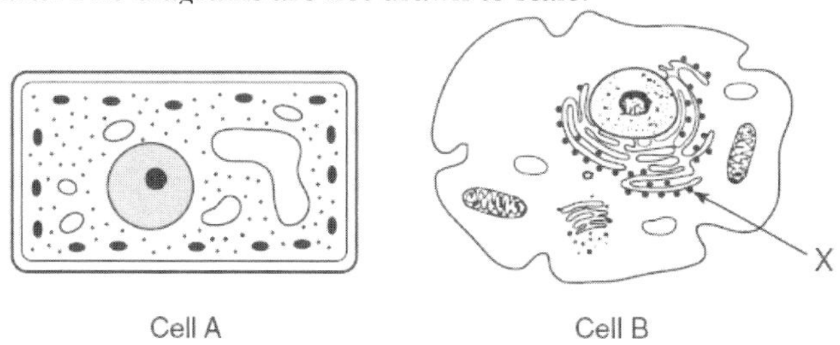

Cell A Cell B

10. Identify an organelle in cell *A* that is the site of autotrophic nutrition.

11. Identify the organelle labeled *X* in cell *B*.

12. Which statement best describes these cells?
   (1) Cell *B* lacks vacuoles while cell *A* has them.
   (2) DNA would not be found in either cell *A* or cell *B*.
   (3) Both cell *A* and cell *B* use energy released from ATP.
   (4) Both cell *A* and cell *B* produce antibiotics.

13. Which organelle is correctly paired with its specific function?
   (1) cell membrane—storage of hereditary information
   (2) chloroplast—transport of materials
   (3) ribosome—synthesis of proteins
   (4) vacuole—production of ATP

14. Just like complex organisms, cells are able to survive by coordinating various activities. Complex organisms have a variety of systems, and cells have a variety of organelles that work together for survival. Describe the roles of two organelles. In your answer be sure to include:
   - the names of two organelles and the function of each
   - an explanation of how these two organelles work together
   - the name of an organelle and the name of a system in the human body that have similar functions

Base your answers to the following two questions on the diagram of a cell below.

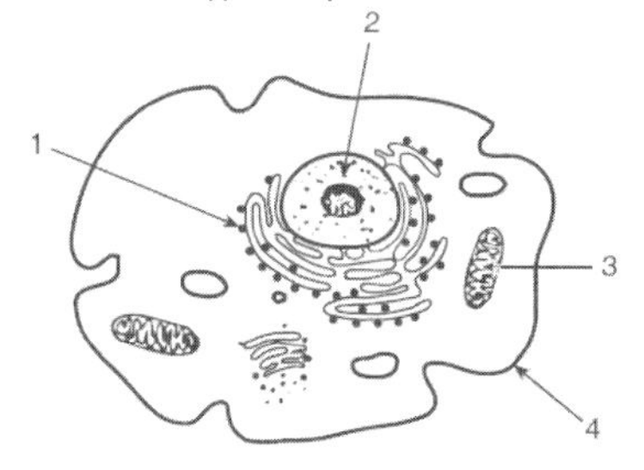

15. Choose either structure *3* or structure *4*, write the number of the structure on the line provided, and describe how it aids the process of protein synthesis.

16. The diagram below represents a sequence of events in a biological process that occurs within human cells.

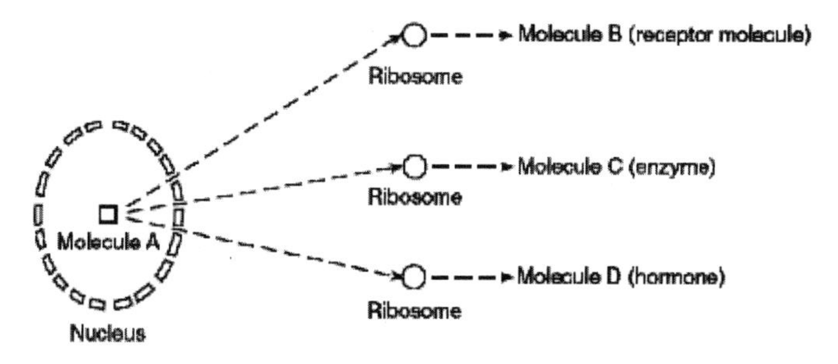

Molecule A contains the
(1) starch necessary for ribosome synthesis in the cytoplasm
(2) organic substance that is broken down into molecules *B*, *C*, and *D*
(3) proteins that form the ribosome in the cytoplasm
(4) directions for the synthesis of molecules *B*, *C*, and *D*

Base your answers to the next three questions on the diagrams below of two cells, *X* and *Y*, and on your knowledge of biology.

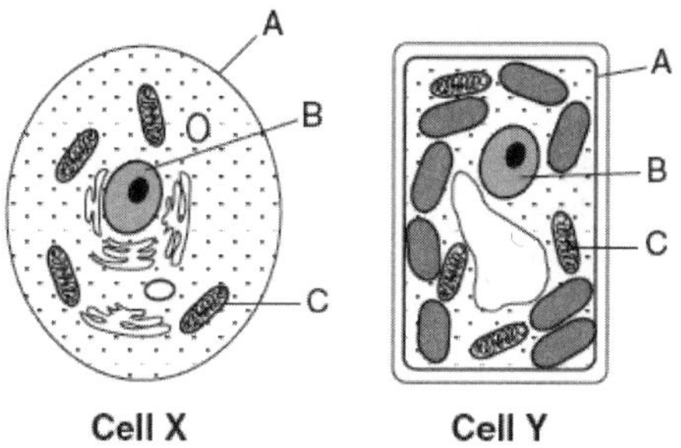

17. Select *one* lettered organelle and write the letter of that organelle in the space provided. Identify the organelle you selected.

18. State *one* function of the organelle that you identified in question 17, above.

19. Identify *one* process that is carried out in cell *Y* that is not carried out in cell *X*.

---

20. In a cell, information that controls the production of proteins must pass from the nucleus to the
(1) cell membrane
(2) chloroplasts
(3) mitochondria
(4) ribosomes

21. Data from two different cells are shown in the graphs below.

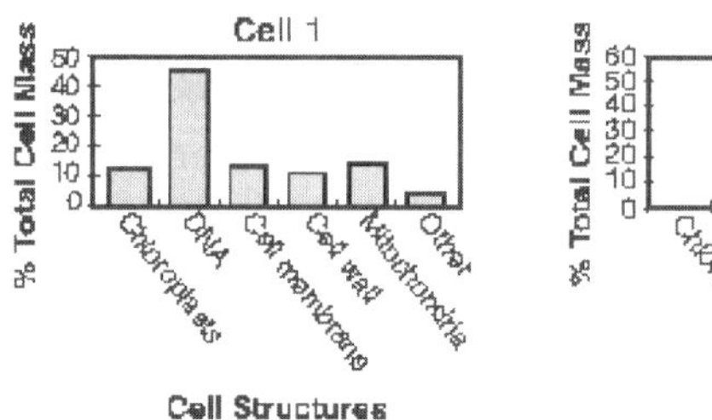

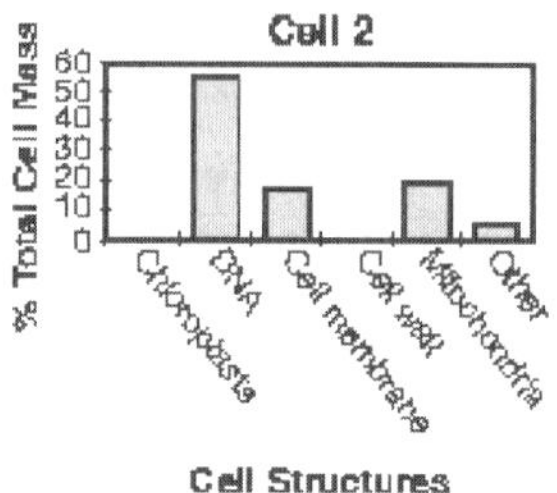

Which cell is most likely a plant cell? Support your answer.

22. The diagram below represents two cells, *X* and *Y*.

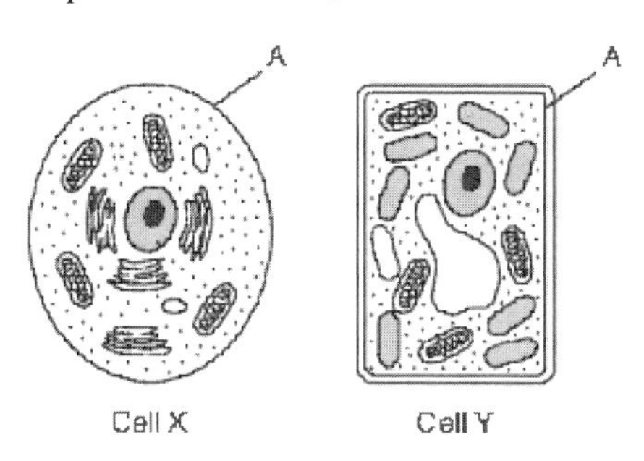

Which statement is correct concerning the structure labeled *A*?

(1) It aids in the removal of metabolic wastes in both cell *X* and cell *Y*.
(2) It is involved in cell communication in cell *X*, but not in cell *Y*.
(3) It prevents the absorption of CO_2 in cell *X* and O_2 in cell *Y*.
(4) It represents the cell wall in cell *X* and the cell membrane in cell *Y*.

23. Damage to which structure will most directly disrupt water balance within a single-celled organism?

(1) ribosome
(2) cell membrane
(3) nucleus
(4) chloroplast

24. In the *Diffusion Through a Membrane* lab, the model cell membranes allowed certain substances to pass through based on which characteristic of the diffusing substance?

(1) size
(2) shape
(3) color
(4) temperature

25. Molecule *X* moves across a cell membrane by diffusion. Which row in the chart below best indicates the relationship between the relative concentrations of molecule *X* and the use of ATP for diffusion?

Row	Movement of Molecule X	Use of ATP
(1)	high concentration → low concentration	used
(2)	high concentration → low concentration	not used
(3)	low concentration → high concentration	used
(4)	low concentration → high concentration	not used

26. Base your answers to this question on the information and diagram below and on your knowledge of biology. The diagram illustrates an investigation carried out in a laboratory activity on diffusion. The beaker and the artificial cell also contain water.

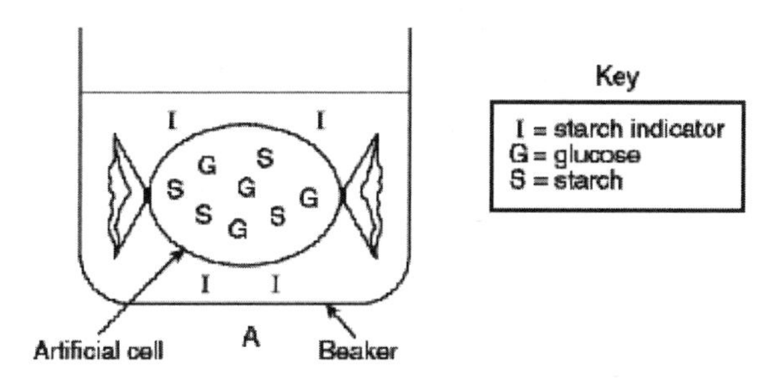

Predict what would happen over time by showing the location of molecules *I*, *G*, and *S* in diagram *B* below.

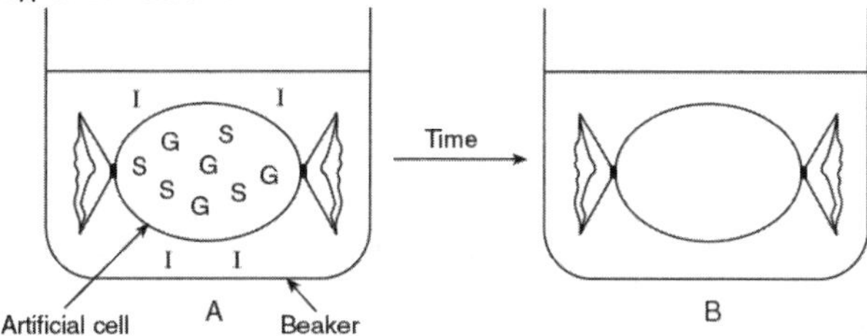

27. The diagram below represents a container of water and two different kinds of molecules, *A* and *B*, separated into two chambers by a membrane through which only water and molecule *A* can pass.

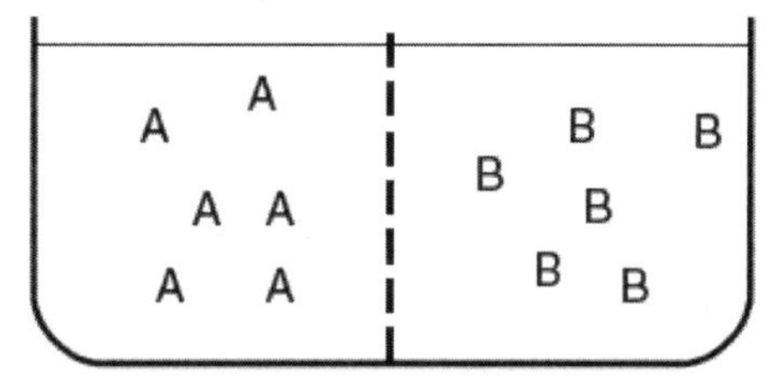

On the diagram of the container below, indicate the distribution of molecules *A* and *B* after the net movement of these molecules stops.

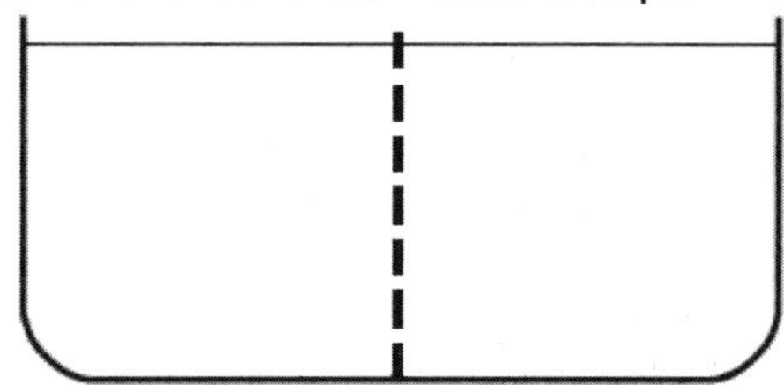

28. A red onion cell has undergone a change, as represented in the diagram below.

This change is most likely due to the cell being placed in

(1) distilled water
(2) light
(3) salt water
(4) darkness

29. A student prepared a wet-mount slide of some red onion cells and then added some salt water to the slide. The student observed the slide using a compound light microscope. Diagram *A* is typical of what the student observed after adding salt water.

Complete diagram *B* to show how the contents of the red onion cells should appear if the cell were then rinsed with distilled water for several minutes.

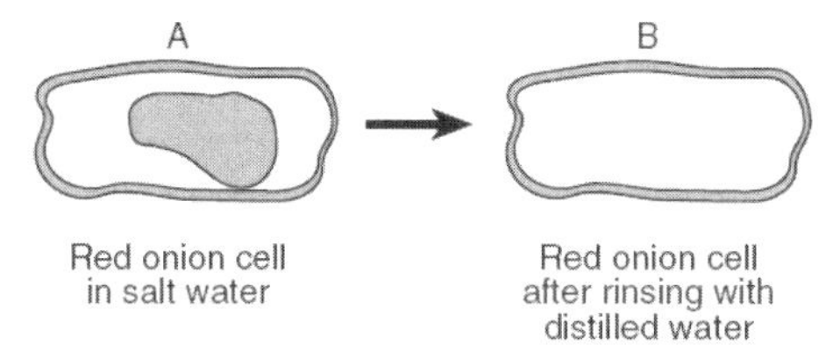

30. Molecules *A* and *B* are both organic molecules found in many cells. When tested, it is found that molecule *A* cannot pass through a cell membrane, but molecule *B* easily passes through. State *one* way the two molecules could differ, that would account for the difference in the ability to pass through the cell membrane.

31. Which statement regarding the functioning of the cell membrane of all organisms is *not* correct?

(1) The cell membrane forms a boundary that separates the cellular contents from the outside environment.
(2) The cell membrane is capable of receiving and recognizing chemical signals.
(3) The cell membrane forms a barrier that keeps all substances that might harm the cell from entering the cell.
(4) The cell membrane controls the movement of molecules into and out of the cell.

32. If vegetables become wilted, they can often be made crisp again by soaking them in water. However, they may lose a few nutrients during this process. Using the concept of diffusion and concentration, state why some nutrients would leave the plant cell.

33. The photos below show two red onion cells viewed with the high power of a compound light microscope. Describe the steps that could be used to make cell *A* resemble cell *B*, using a piece of paper towel and an eyedropper or a pipette without removing the coverslip.

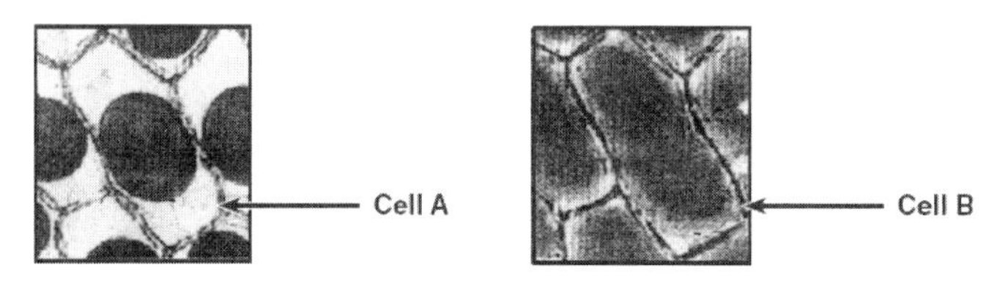

Base your answers to the next two questions on the diagram below and on your knowledge of biology. The diagram shows the changes that occurred in a beaker after 30 minutes. The beaker contained water, food coloring, and a bag made from dialysis tubing membrane.

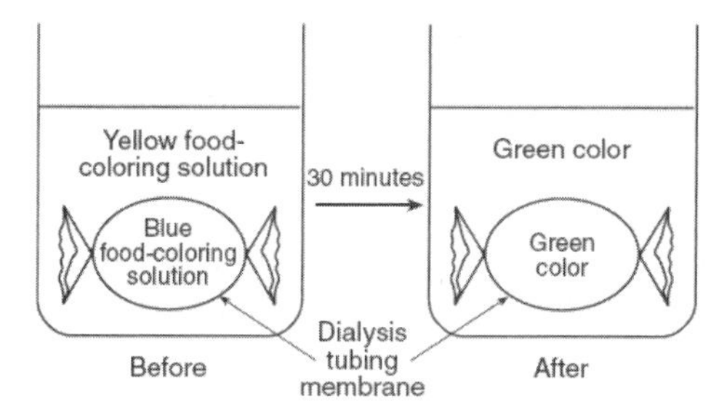

When the colors yellow and blue are combined, they produce a green color.

34. Which statement most likely describes the relative sizes of the yellow and blue food-coloring molecules in the diagram?
    (1) The yellow food-coloring molecules are small, while the blue food-coloring molecules are large.
    (2) The yellow food-coloring molecules are large, while the blue food-coloring molecules are small.
    (3) Both the yellow food-coloring molecules and the blue food-coloring molecules are large.
    (4) Both the yellow food-coloring molecules and the blue food-coloring molecules are small.

35. Which statement best explains the changes shown?
    (1) Molecular movement was aided by the presence of specific carbohydrate molecules on the surface of the membrane.
    (2) Molecular movement was aided by the presence of specific enzyme molecules on the surface of the membrane.
    (3) Molecules moved across the membrane without additional energy being supplied.
    (4) Molecules moved across the membrane only when additional energy was supplied.

---

36. Cell *A* shown below is a typical red onion cell in water on a slide viewed with a compound light microscope.

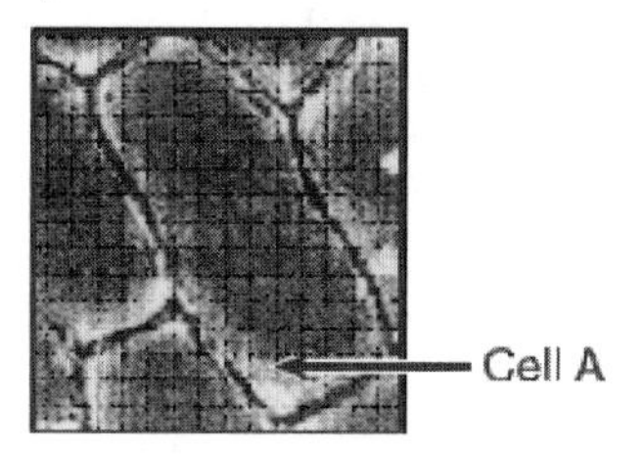

Draw a diagram of how cell A would most likely look after salt water has been added to the slide and label the cell membrane in your diagram.

37. An investigation was set up to study the movement of water through a membrane. The results are shown in the diagram below.

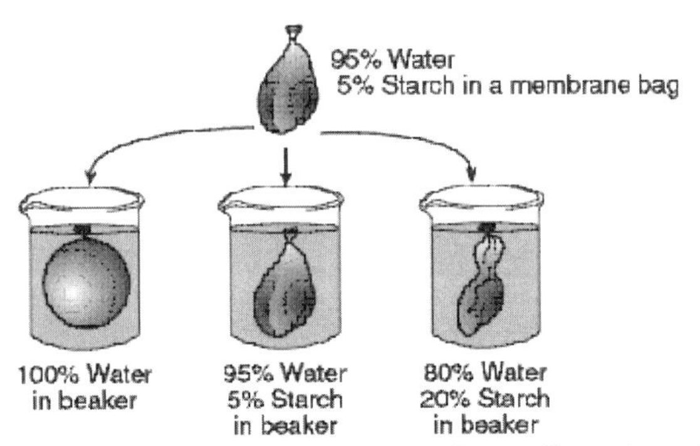

Based on these results, which statement correctly predicts what will happen to red blood cells when they are placed in a beaker containing a water solution in which the salt concentration is much higher than the salt concentration in the red blood cells?

(1) The red blood cells will absorb water and increase in size.
(2) The red blood cells will lose water and decrease in size.
(3) The red blood cells will first absorb water, then lose water and maintain their normal size.
(4) The red blood cells will first lose water, then absorb water, and finally double in size.

38. A biologist observed a plant cell in a drop of water as shown in diagram *A*. The biologist added a 10% salt solution to the slide and observed the cell as shown in diagram *B*.

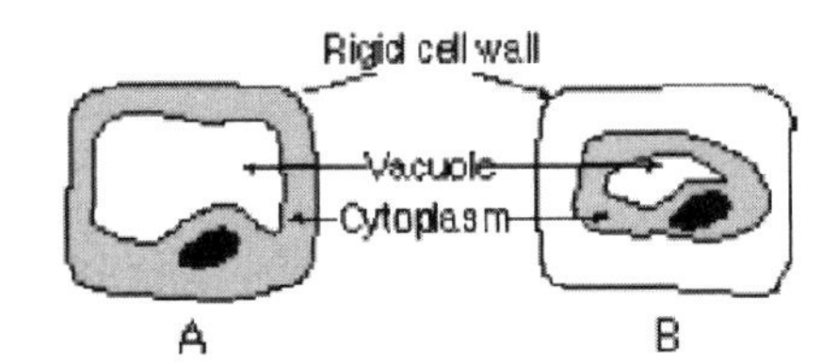

The change in appearance of the cell resulted from

(1) more salt moving out of the cell than into the cell
(2) more salt moving into the cell than out of the cell
(3) more water moving into the cell than out of the cell
(4) more water moving out of the cell than into the cell

39. Base your answers to the question on the diagram below of sugar in a beaker of water and on your knowledge of biology.

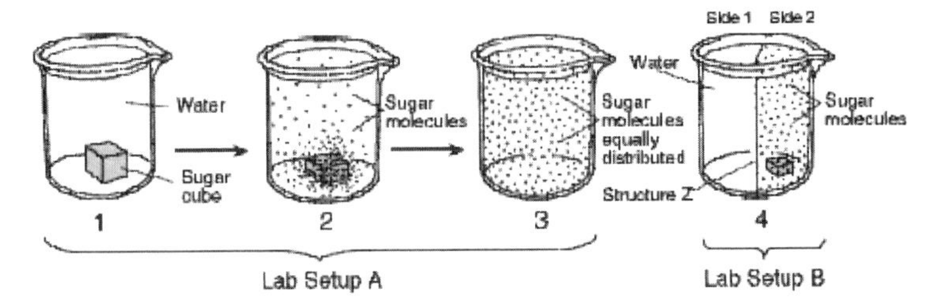

What process accounts for the change shown in lab setup *A*?

Base your answers to the next two questions on the information below and on your knowledge of biology.

Students prepared four models of cells by using dialysis tubing containing the same blue solution. Each of the model cells originally weighed 10 grams. They then placed each model cell in a beaker containing a different concentration of water. After 24 hours, they recorded the mass of the model cells as shown in the data table below.

Data Table

Concentration of Water Surrounding the Model Cell	Mass of Model Cell
100%	12 grams
90%	11 grams
80%	10 grams
70%	9 grams

40. Why did the model cell that was placed in 100% water increase in mass?

41. What was the concentration of water in the original blue solution? State evidence in support of your answer.

42. The diagram below represents a laboratory setup used by a student during an investigation of diffusion.

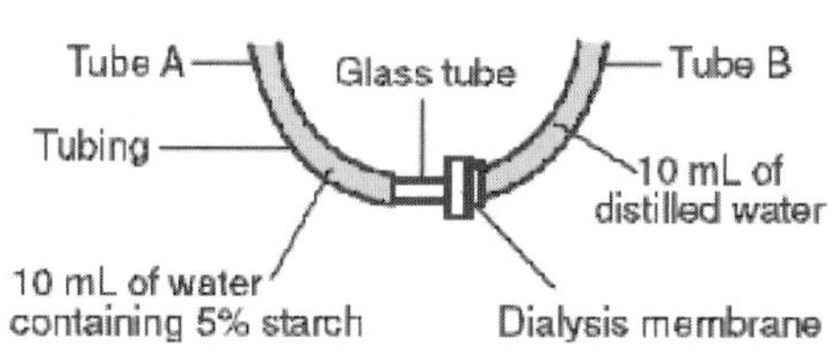

Which statement best explains why the liquid in tube *A* will rise over a period of time?

(1) The starch concentrations are equal on both sides of the membrane.
(2) The water will pass from a region of lower starch concentration to one of higher starch concentration.
(3) Water and starch volumes are the same in both tubes *A* and *B*.
(4) The fluids in both tubes *A* and *B* will change from a higher temperature to a lower temperature.

43. The diagram below shows the relative concentration of molecules inside and outside of a cell.

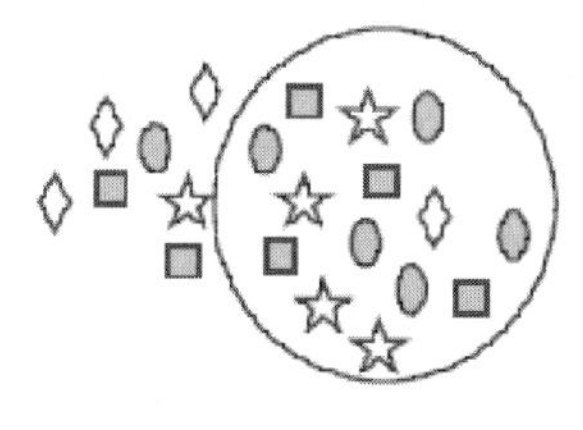

Key
= Protein
= Oxygen
= Glucose
= Carbon dioxide

Which statement best describes the general direction of diffusion across the membrane of this cell?

(1) Glucose would diffuse into the cell.
(2) Protein would diffuse out of the cell.

(3) Carbon dioxide would diffuse out of the cell.
(4) Oxygen would diffuse into the cell.

44. Which row in the chart below best describes the active transport of molecule *X* through a cell membrane?

Row	Movement of Molecule X	ATP
(1)	high concentration → low concentration	used
(2)	high concentration → low concentration	not used
(3)	low concentration → high concentration	used
(4)	low concentration → high concentration	not used

Base your answers to the next two questions on the information and table below and on your knowledge of biology.

A model of a cell is prepared and placed in a beaker of fluid as shown in the diagram below. The letters *A*, *B*, and *C* represent substances in the initial experimental setup.

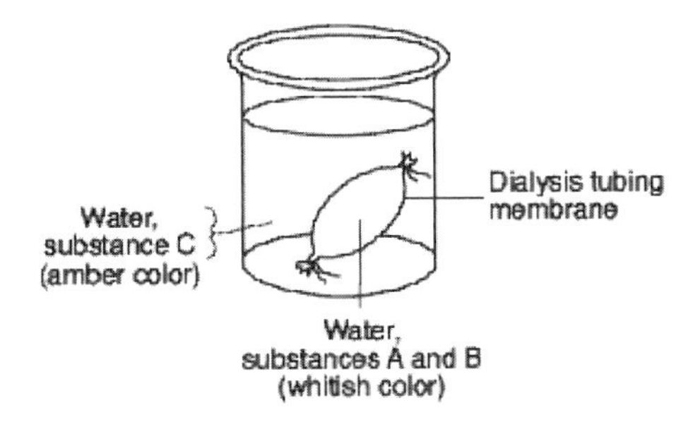

The table below summarizes the content and appearance of the cell model and beaker after 20 minutes.

Results After 20 Minutes

	Outside of Cell Model	Inside of Cell Model
Substances	water, A, C	water, A, B, C
Color	amber	blue black

45. Complete the table below to summarize a change in location of substance *C* in the experimental setup.

Name of Substance C	Direction of Movement of Substance C	Reason for the Movement of Substance C

46. Identify substance *B* and explain why it did not move out of the model cell.

47. The diagram below represents a plant cell in tap water as seen with a compound light microscope.

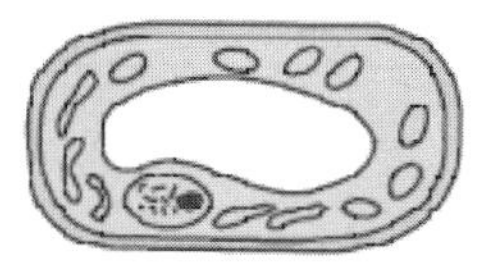

Which diagram best represents the appearance of the cell after it has been placed in a 15% salt solution for two minutes?

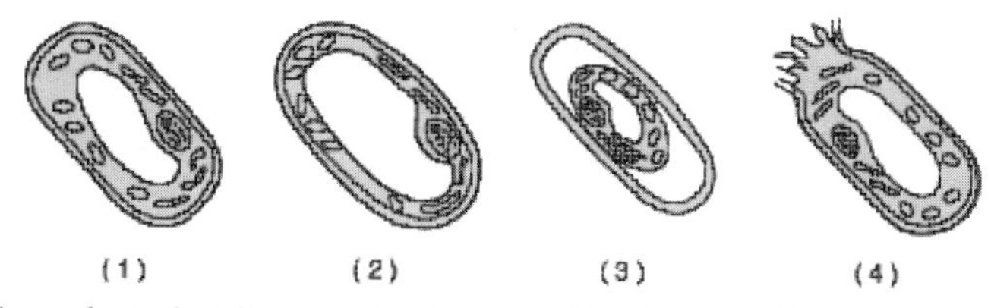

48. Elodea is a plant that lives in freshwater. The diagram below represents one Elodea leaf cell in its normal freshwater environment.

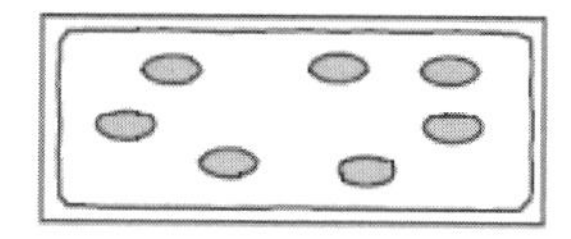

Elodea cell in freshwater

Predict how the contents of the Elodea cell would change if the cell was placed in saltwater for several minutes by completing the diagram, "Elodea cell in saltwater" below. Label the location of the cell membrane.

Elodea cell in saltwater

49. If frog eggs taken from a freshwater pond are placed in a saltwater aquarium, what will most likely happen?
   (1) Water will leave the eggs.
   (2) Salt will leave the eggs.
   (3) Water will neither enter nor leave the eggs.
   (4) The eggs will burst.

50. Arrows *A*, *B*, and *C* in the diagram below represent the processes necessary to make the energy stored in food available for muscle activity.

Food —A→ Simpler molecules —B→ Mitochondria —C→ ATP in muscle cells

The correct sequence of processes represented by *A*, *B*, and *C* is

(1) diffusion → synthesis → active transport
(2) digestion → diffusion → cellular respiration
(3) digestion → excretion → cellular respiration
(4) synthesis → active transport → excretion

Base your answers to the next two questions on the experimental setup shown below.

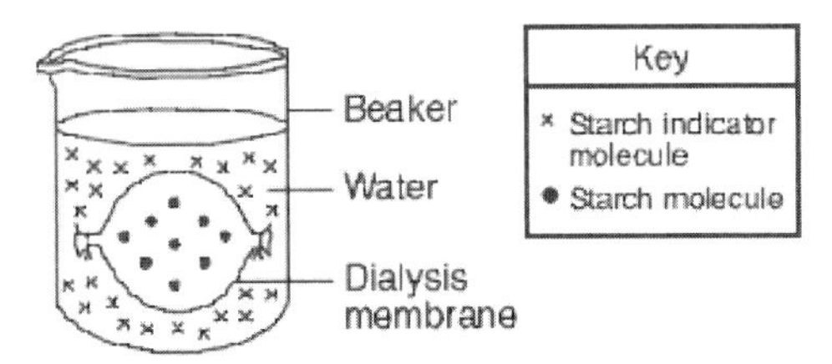

51. On the diagram below, draw in the expected locations of the molecules after a period of one hour.

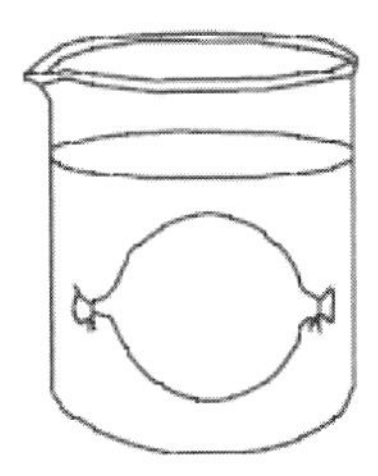

52. When starch indicator is used, what observation would indicate the presence of starch?

53. State *one* reason why some molecules can pass through a certain membrane, but other molecules can not.

54. Explain why high salt concentrations can kill organisms.`

55. The diagram below represents the distribution of some molecules inside and outside of a cell over time.

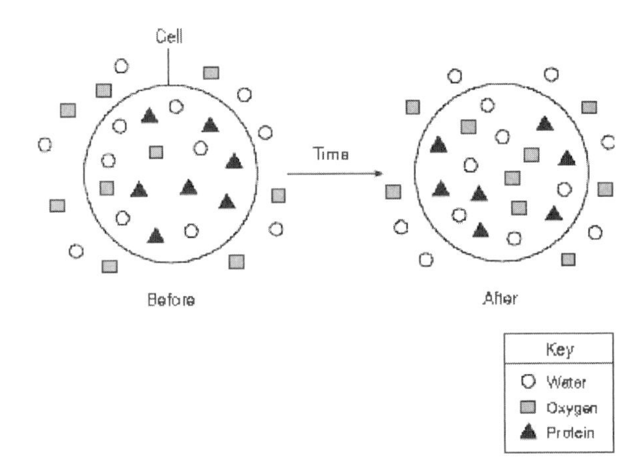

Which factor prevented the protein molecules (▲) from moving out of the cell?

(1) temperature
(2) pH
(3) molecule size
(4) molecule concentration

56. Which substances are found on cell surfaces and respond to nerve and hormone signals?

(1) starches and simple sugars
(2) subunits of DNA
(3) vitamins and minerals
(4) receptor molecules

57. After a hormone enters the bloodstream, it is transported throughout the body, but the hormone affects only certain cells. The reason only certain cells are affected is that the membranes of these cells have specific

(1) receptors (2) tissues (3) antibodies (4) carbohydrates

58. Cellular communication is illustrated in the diagram below.

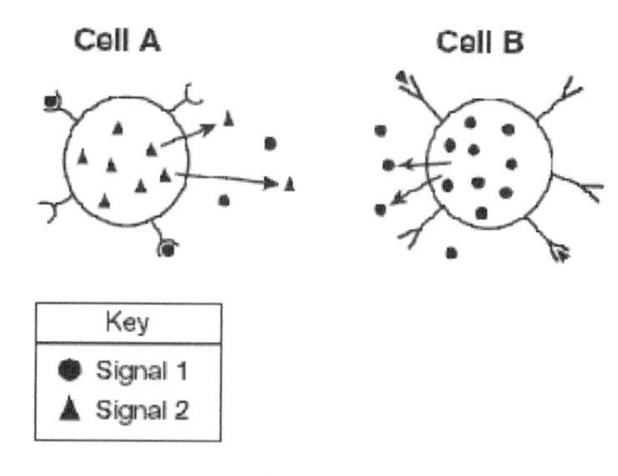

Information can be sent from

(1) cell *A* to cell *B* because cell *B* is able to recognize signal *1*
(2) cell *A* to cell *B* because cell *A* is able to recognize signal *2*
(3) cell *B* to cell *A* because cell *A* is able to recognize signal *1*
(4) cell *B* to cell *A* because cell *B* is able to recognize signal *2*

59. The diagram below represents two molecules that can interact with each other to cause a biochemical process to occur in a cell.

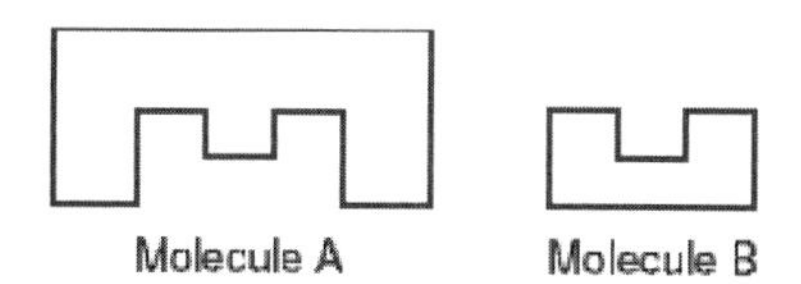

Molecules *A* and *B* most likely represent

(1) a protein and a chromosome
(2) a receptor and a hormone
(3) a carbohydrate and an amino acid
(4) an antibody and a hormone

60. A process that occurs in the human body is represented in the diagram below.

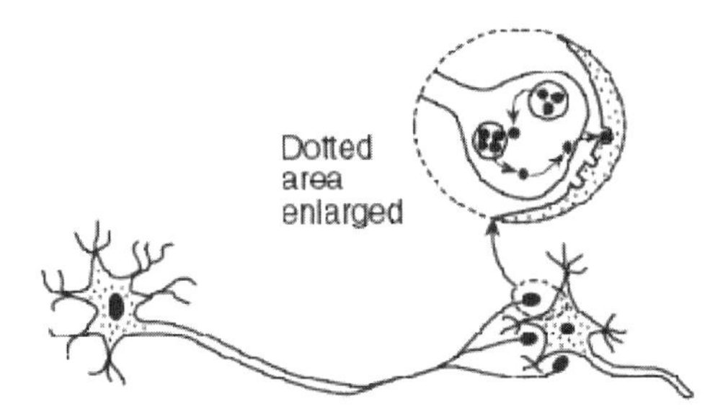

Which statement is most closely associated with the diagram?
(1) Small molecules are obtained from large molecules during digestion.
(2) Certain molecules are replicated by means of a template.
(3) Receptor molecules play an important role in communication between cells.
(4) Energy from nutrients is utilized for waste disposal.

61. Acetylcholine is a chemical secreted at the ends of nerve cells. This chemical helps to send nerve signals across synapses (spaces between nerve cells). After the signal passes across a synapse, an enzyme breaks down the acetylcholine. LSD is a drug that blocks the action of this enzyme. Describe one possible effect of LSD on the action of acetylcholine.

62. Nerve cells are essential to an animal because they directly provide
(1) communication between cells
(2) transport of nutrients to various organs
(3) regulation of reproductive rates within other cells
(4) an exchange of gases within the body

63. A protein on the surface of HIV can attach to proteins on the surface of healthy human cells. These attachment sites on the surface of the cells are known as
(1) receptor molecules
(2) genetic codes
(3) molecular bases
(4) inorganic catalysts

64. Enzyme molecules normally interact with substrate molecules. Some medicines work by blocking enzyme activity in pathogens. These medicines are effective because they
(1) are the same size as the enzyme
(2) are the same size as the substrate molecules
(3) have a shape that fits into the enzyme
(4) have a shape that fits into all cell receptors

65. Two primary agents of cellular communication are
(1) chemicals made by blood cells and simple sugars
(2) hormones and carbohydrates
(3) enzymes and starches
(4) hormones and chemicals made by nerve cells

Base your answers to the next two questions on the information below and on your knowledge of biology.

Insulin is a hormone that has an important role in the maintenance of homeostasis in humans.

66. Identify the structure in the human body that is the usual source of insulin.

67. Identify a substance in the blood, other than insulin, that could change in concentration and indicate a person is not secreting insulin in normal amounts.

68. Describe *one* example of diffusion in the human body. In your description be sure to:

- identify the place where diffusion takes place
- identify a substance that diffuses there
- identify where that substance diffuses from and where it diffuses to, at that place

---

Base your answers to the next two questions on the diagram below, which represents a sequence of events in a biological process that occurs within human cells, and on your knowledge of biology.

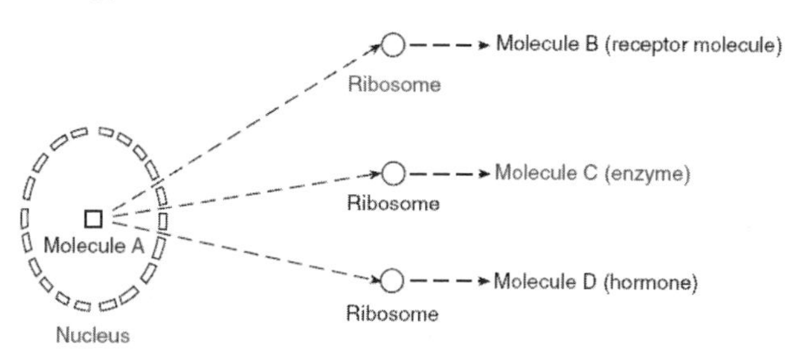

69. Molecule *A* contains the
    (1) starch necessary for ribosome synthesis in the cytoplasm
    (2) organic substance that is broken down into molecules *B*, *C*, and *D*
    (3) proteins that form the ribosome in the cytoplasm
    (4) directions for the synthesis of molecules *B*, *C*, and *D*

70. Molecules *B*, *C*, and *D* are similar in that they are usually
    (1) composed of genetic information
    (2) involved in the synthesis of antibiotics
    (3) composed of amino acids
    (4) involved in the diffusion of oxygen into the cell

---

71. Which order of metabolic processes converts nutrients consumed by an organism into cell parts?
    (1) digestion → absorption → circulation → diffusion → synthesis
    (2) absorption → circulation → digestion → diffusion → synthesis
    (3) digestion → synthesis → diffusion → circulation → absorption
    (4) synthesis → absorption → digestion → diffusion → circulation

72. Which row in the chart below contains correct information concerning synthesis?

Row	Building Blocks	Substance Synthesized Using the Building Blocks
(1)	glucose molecules	DNA
(2)	simple sugars	protein
(3)	amino acids	enzyme
(4)	molecular bases	starch

73. A laboratory setup for a demonstration is represented in the diagram below.

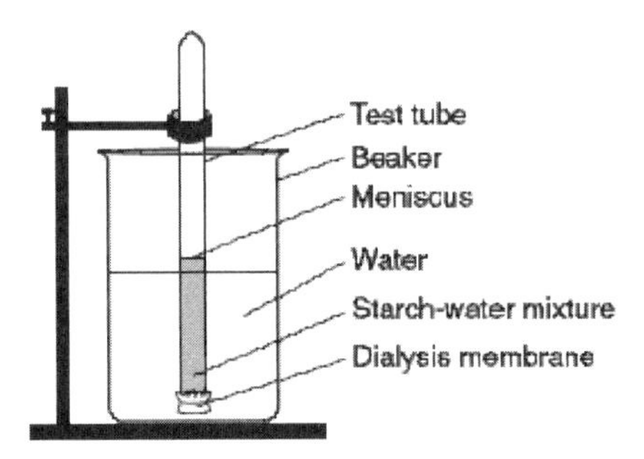

Describe how an indicator can be used to determine if starch diffuses through the membrane into the beaker. In your answer, be sure to include:

- the procedure used
- how to interpret the results

74. Carbon exists in a simple organic molecule in a leaf and in an inorganic molecule in the air humans exhale. Identify the carbon-containing molecule that humans exhale and the process that produces it.

75. Humans require multiple systems for various life functions. Two vital systems are the circulatory system and the respiratory system. Select *one* of these systems, write its name in the chart below, then identify *two* structures that are part of that system, and state how each structure you identified functions as part of the system.

System:	
Structure	Function
(1)	
(2)	

76. An increase in heart rate will most likely result in
    (1) a decrease in metabolic rate
    (2) an increase in pulse rate
    (3) an increase in cell division
    (4) a decrease in body temperature

77. Which statement accurately compares cells in the human circulatory system to cells in the human nervous system?
    (1) Cells in the circulatory system carry out the same life function for the organism as cells in the nervous system.
    (2) Cells in the circulatory system are identical in structure to cells in the nervous system.
    (3) Cells in the nervous system are different in structure from cells in the circulatory system, and they carry out different specialized functions.
    (4) Cells in the nervous system act independently, but cells in the circulatory system function together.

78. The diagram below represents three human body systems.

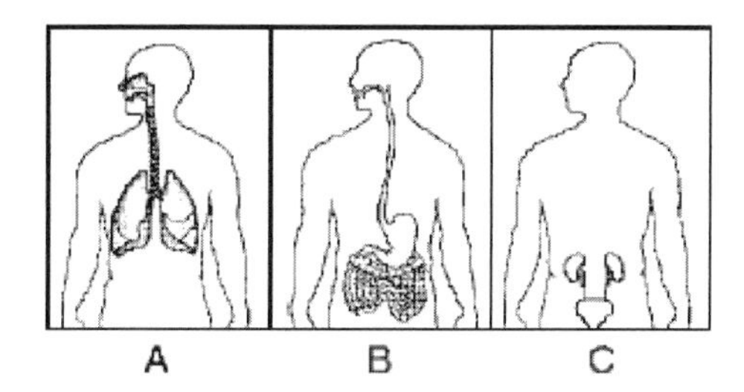

Which row in the chart below correctly shows what systems *A*, *B* and *C* provide for the human body?

Row	System A	System B	System C
(1)	blood cells	glucose	hormones
(2)	oxygen	absorption	gametes
(3)	gas exchange	nutrients	waste removal
(4)	immunity	coordination	carbon dioxide

79. Testes are adapted to produce
   (1) body cells involved in embryo formation
   (2) immature gametes that undergo mitosis
   (3) sperm cells that may be involved in fertilization
   (4) gametes with large food supplies that nourish a developing embryo

80. Which statement describes the reproductive system of a human male?
   (1) It releases sperm that can be used only in external fertilization.
   (2) It synthesizes progesterone that regulates sperm formation.
   (3) It produces gametes that transport food for embryo formation.
   (4) It shares some structures with the excretory system.

Base your answers to the next two questions on the information below and on your knowledge of biology.

The reproductive cycle in a human female is not functioning properly. An imbalance of hormones is diagnosed as the cause.

81. Identify *one* hormone directly involved in the human female reproductive system that could cause this problem.

82. Explain why some cells in a female's body respond to reproductive hormones while other cells do not.

---

83. Removal of one ovary from a human female would most likely
   (1) affect the production of eggs
   (2) make fertilization impossible
   (3) make carrying a fetus impossible
   (4) decrease her ability to provide essential nutrients to an embryo

84. State *two* ways cells of the immune system fight disease.

85. Identify the substance produced by the cells of all the endocrine glands that helps maintain homeostasis.

86. Identify *one* specific product of one of the endocrine glands and state how it aids in the maintenance of homeostasis.

87. The energy demands of a cell or an organism are met as a result of interactions between several life functions.
    - Identify *two* life functions involved in meeting the energy demands of a cell or an organism.
    - Explain how these *two* life functions interact to make energy available.

88. The diagrams below represent some of the systems that make up the human body.

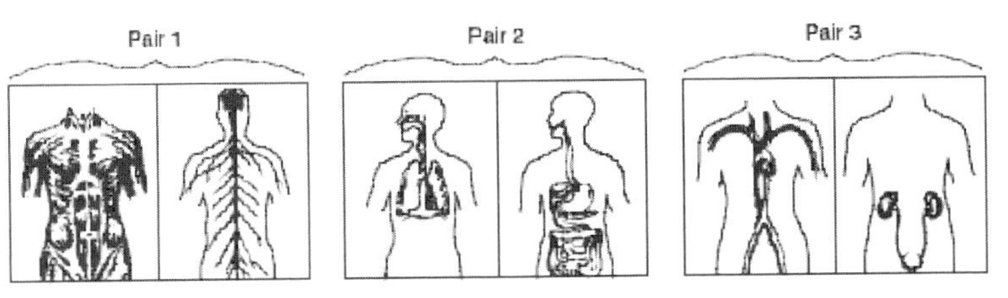

Select *one* of the pairs of systems and write its number in the space provided. For the pair selected, identify each system and state *one* function of that system. Explain how the two systems work together to help maintain homeostasis in an individual.

89. Organisms undergo constant chemical changes as they maintain an internal balance known as
    (1) interdependence
    (2) homeostasis
    (3) synthesis
    (4) recombination

90. The human reproductive system is regulated by
    (1) restriction enzymes
    (2) antigens
    (3) complex carbohydrates
    (4) hormones

Base your answers to the next two questions on the information below and on your knowledge of biology.

The reproductive cycle in a human female is not functioning properly. An imbalance of hormones is diagnosed as the cause.

91. Identify *one* hormone directly involved in the human female reproductive system that could cause this problem.

92. Removal of one ovary from a human female would most likely
    (1) affect the production of eggs
    (2) make fertilization impossible
    (3) make carrying a fetus impossible
    (4) decrease her ability to provide essential nutrients to an embryo

---

93. The human reproductive system is regulated by
    (1) restriction enzymes
    (2) antigens
    (3) complex carbohydrates
    (4) hormones

94. Identify the substance produced by the cells of *all* the endocrine glands that helps maintain homeostasis.

95. Identify *one* specific product of one of the endocrine glands and state how it aids in the maintenance of homeostasis.

96. Every single-celled organism is able to survive because it carries out
    (1) metabolic activities
    (2) autotrophic nutrition
    (3) heterotrophic nutrition
    (4) sexual reproduction

97. Contractile vacuoles maintain water balance by pumping excess water out of some single-celled pond organisms. In humans, the kidney is chiefly involved in maintaining water balance. These facts best illustrate that
    (1) tissues, organs, and organ systems work together to maintain homeostasis in all living things
    (2) interference with nerve signals disrupts cellular communication and homeostasis within organisms
    (3) a disruption in a body system may disrupt the homeostasis of a single-celled organism.
    (4) structures found in single-cell organisms can act in a manner similar to tissues and organs in multicellular organisms.

98. The following diagram of human blood below show the structures that help to maintain homeostasis in humans.

Identify the cell labeled *X*.

99. Homeostasis in unicellular organisms depends on the proper functioning of
    (1) organelles
    (2) insulin
    (3) guard cells
    (4) antibodies

100. Which statement best compares a multicellular organism to a single-celled organism?
    (1) A multicellular organism has organ systems that interact to carry out life functions, while a single-celled organism carries out life functions without using organ systems.
    (2) A single-celled organism carries out fewer life functions than each cell of a multicellular organism.
    (3) A multicellular organism always obtains energy through a process that is different from that used by a single-celled organism.
    (4) The cell of a single-celled organism is always much larger than an individual cell of a multicellular organism.

101. The energy demands of a cell or an organism are met as a result of interactions between several life functions.
    - Identify *two* life functions involved in meeting the energy demands of a cell or an organism.
    - Explain how these *two* life functions interact to make energy available.

102. The best way to reduce the symptoms of hyponatremia would be to
    (1) drink more water
    (2) eat chocolate
    (3) eat salty foods
    (4) drink cranberry juice

103. Many runners pour water on their bodies during a race. Explain how this action helps to maintain homeostasis.

104. How would running in a marathon on a warm day most likely affect urine production? Support your answer.

105. Many people today drink sport drinks containing large amounts of sodium. Describe *one* possible effect this might have on a person who is not very active.

106. In the diagram below, which structure performs a function similar to a function of the human lungs?

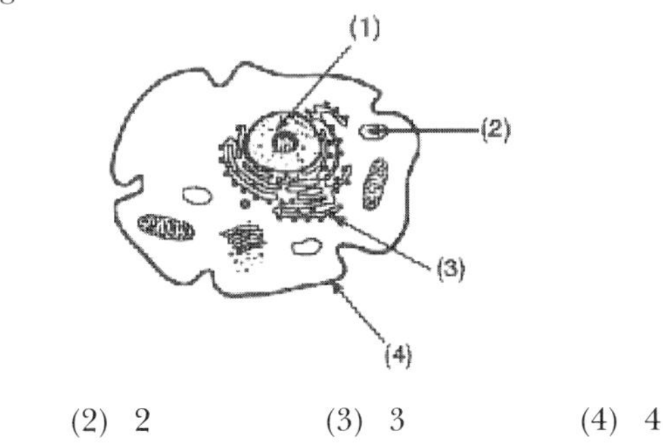

(1) 1 (2) 2 (3) 3 (4) 4

107. Which structures in diagram *I* and diagram *II* carry out a similar life function?

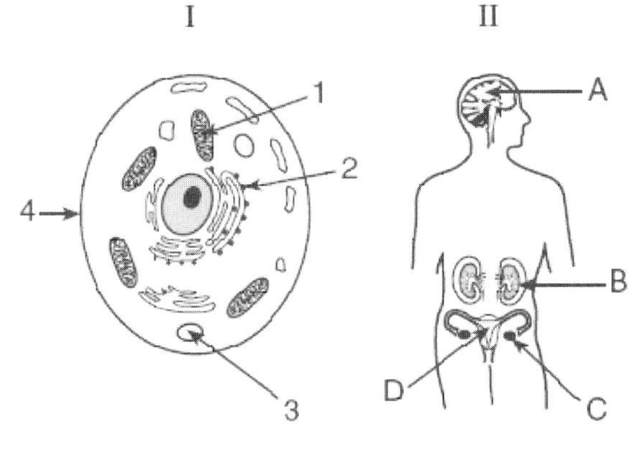

(1) *1* and *C* (2) *2* and *D* (3) *3* and *A* (4) *4* and *B*

108. Contractile vacuoles maintain water balance by pumping excess water out of some single-celled pond organisms. In humans, the kidney is chiefly involved in maintaining water balance. These facts best illustrate that
(1) tissues, organs, and organ systems work together to maintain homeostasis in all living things
(2) interference with nerve signals disrupts cellular communication and homeostasis within organisms
(3) a disruption in a body system may disrupt the homeostasis of a single-celled organism.
(4) structures found in single-cell organisms can act in a manner similar to tissues and organs in multicellular organisms.

109. Homeostasis in unicellular organisms depends on the proper functioning of
(1) organelles (2) insulin (3) guard cells (4) antibodies

110. Which statement best compares a multicellular organism to a single-celled organism?
(1) A multicellular organism has organ systems that interact to carry out life functions, while a single-celled organism carries out life functions without using organ systems.
(2) A single-celled organism carries out fewer life functions than each cell of a multicellular organism.

(3) A multicellular organism always obtains energy through a process that is different from that used by a single-celled organism.
(4) The cell of a single-celled organism is always much larger than an individual cell of a multicellular organism.

111. The graph below indicates the size of a fish population over a period of time.

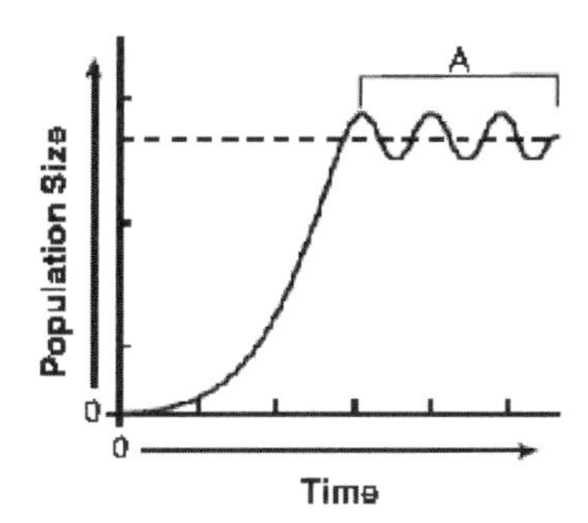

The section of the graph labeled *A* represents
(1) biodiversity within the species
(2) nutritional relationships of the species
(3) a population becoming extinct
(4) a population at equilibrium

112. Humans require organ systems to carry out life processes. Single-celled organisms do not have organ systems and yet they are able to carry out life processes. This is because
(1) human organ systems lack the organelles found in single-celled organisms
(2) a human cell is more efficient than the cell of a single-celled organism
(3) it is not necessary for single-celled organisms to maintain homeostasis
(4) organelles present in single-celled organisms act in a manner similar to organ systems

113. Systems in the human body interact to maintain homeostasis. Four of these systems are listed below.

Body Systems
circulatory
digestive
respiratory
excretory

a. Select *two* of the systems listed. Identify each system selected and state its function in helping to maintain homeostasis in the body.
b. Explain how a malfunction of *one* of the four systems listed disrupts homeostasis and how that malfunction could be prevented or treated. In your answer be sure to:
- name the system and state *one* possible malfunction of that system
- explain how the malfunction disrupts homeostasis
- describe *one* way the malfunction could be prevented or treated

114. In the diagram below, which structure performs a function similar to a function of the human lungs?

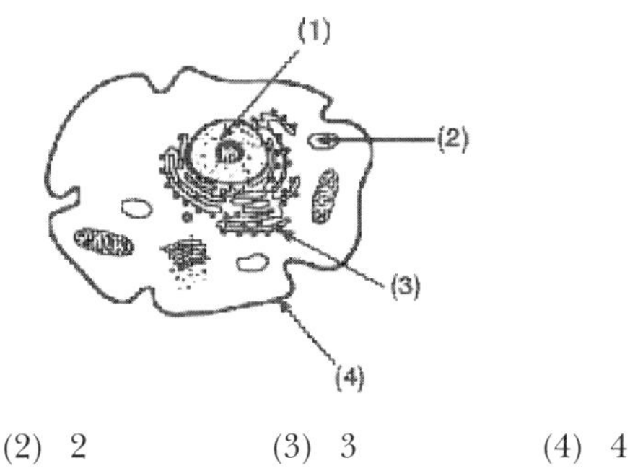

(1) 1 (2) 2 (3) 3 (4) 4

115. Select one human body system from the list below.

Body Systems
Digestive
Circulatory
Respiratory
Excretory
Nervous

Describe a malfunction that can occur in the system chosen. Your answer must include at least:

- the name of the system and a malfunction that can occur in this system
- a description of a possible cause of the malfunction identified
- an effect this malfunction may have on any other body system

116. Base your answer to the next question on the diagram below and on your knowledge of biology.

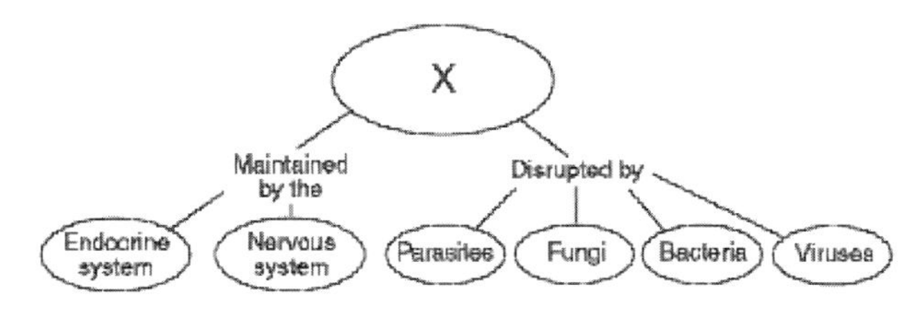

What term or phrase does letter *X* most likely represent?

Base your answers to the next four questions on the information and data table below and on your knowledge of biology.

The effect of temperature on the action of pepsin, a protein-digesting enzyme present in stomach fluid, was tested. In this investigation, 20 milliliters of stomach fluid and 10 grams of protein were placed in each of five test tubes. The tubes were then kept at different temperatures. After 24 hours, the contents of each tube were tested to determine the amount of protein that had been digested. The results are shown in the table below.

Protein Digestion at Different Temperatures

Tube #	Temperature (°C)	Amount of Protein Digested (grams)
1	5	0.5
2	10	1.0
3	20	4.0
4	37	9.5
5	85	0.0

117. The dependent variable in this investigation is the
   (1) size of the test tube
   (2) time of digestion
   (3) amount of stomach fluid
   (4) amount of protein digested

Directions for the next two questions: Using the information in the data table, construct a line graph on the grid provided, following the directions below.

118. Mark an appropriate scale on each axis.

119. Plot the data on the grid. Surround each point with a small circle and connect the points.

Example:

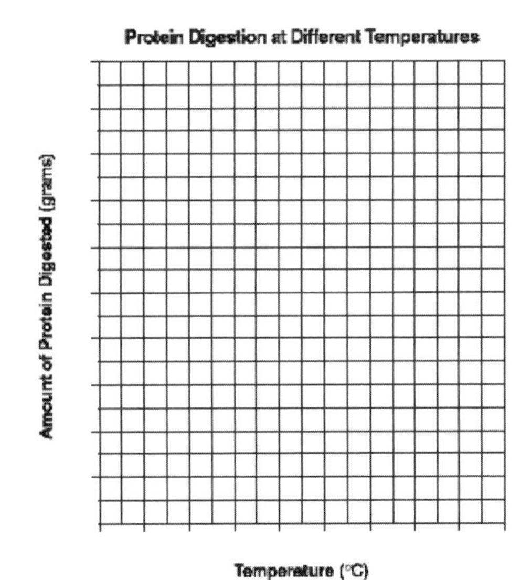

120. If a sixth test tube identical to the other tubes was kept at a temperature of 30°C for 24 hours, the amount of protein digested would most likely be
   (1) less than 1.0 gram
   (2) between 1.0 and 4.0 grams
   (3) between 4.0 and 9.0 grams
   (4) more than 9.0 grams

CHAPTER 2: HOMEOSTASIS (DYNAMIC EQUILIBRIUM)

You will learn in the chapter that biochemical processes of photosynthesis, respiration, enzymes, feedback, immune system, and regulation (by using hormones and nerves) help to maintain homeostasis.

Organisms (living things) need energy and raw materials (example, oxygen) to live (survive). **Photosynthesis** and **cellular respiration** are **biochemical processes** (see below) that produce energy; energy is needed for obtaining (getting) raw materials (example water and minerals in plants), for active transport (example water goes from areas of **less** concentration of **water to** areas of **more** concentration of **water)**, for changing small molecules to large molecules, for eliminating waste, etc.

Photosynthesis

Plants and algae carry on photosynthesis. In **photosynthesis,** in the presence of **sunlight, plants take in carbon dioxide** (CO_2) and **water** (H_2O) and **produce glucose** (a single sugar) and **oxygen** (O_2). Glucose provides energy for life processes (examples digestion, respiration, transport).

Plants and **algae** carry on **photosynthesis, making their own food** (glucose, a simple sugar); plants and algae are called **autotrophs (autotrophic nutrition)** because they make their own food.

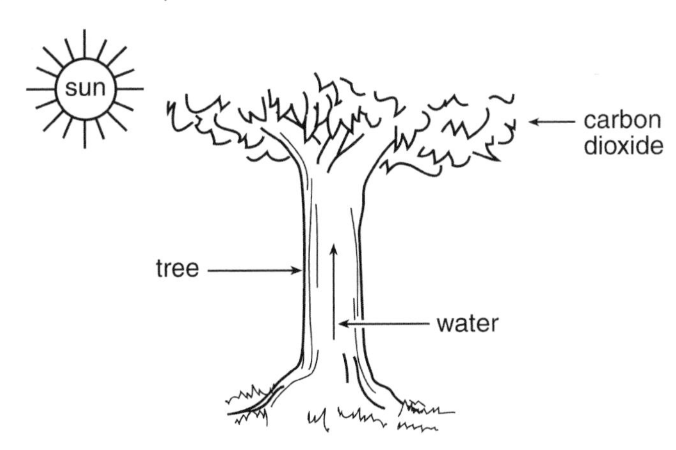

Look at the leaf diagram below. There are openings in the leaf called stomates. Carbon dioxide enters (goes into) the leaf through the stomates (openings) and oxygen goes out (gas exchange, meaning exchange of gases, carbon dioxide (gas) goes in and oxygen goes out). The guard cells that surround the openings regulate the amount of carbon dioxide going in and oxygen and water vapor going out.

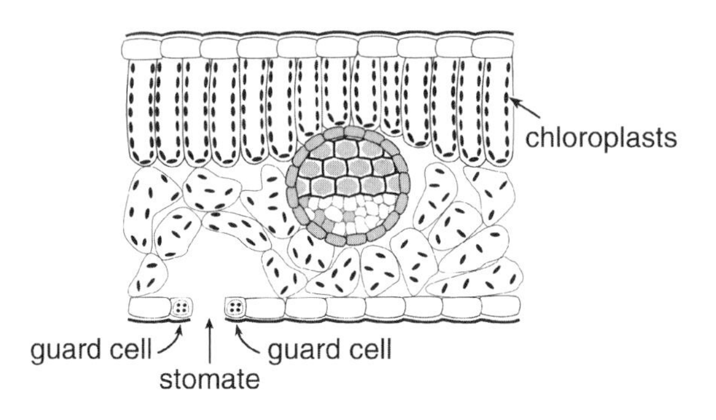

Cross section of leaf

You learned the **chloroplasts in** the cells of the plant **leaf** and in one-celled organisms such as **euglena** are the site (place) of **photosynthesis. Photosynthesis** takes place in the **chloroplasts**. The chloroplasts have a green pigment called chlorophyll. The chlorophyll takes in the **sun's (light) energy,** the roots take in water which goes up the stem and to the leaf, and the leaf takes in carbon dioxide (see figure of tree); this produces glucose (simple sugar) and oxygen (see equation below). A specific enzyme is used in photosynthesis. An enzyme (biological catalyst) affects the **rate of** a chemical **reaction,** but the enzyme is not used up in the reaction.

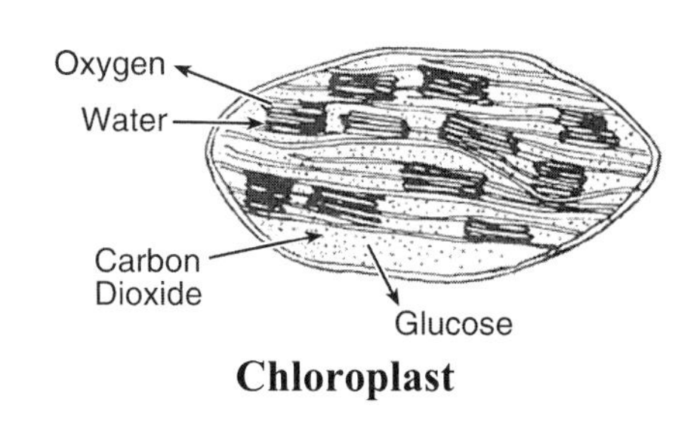

Chloroplast

Sun (light) energy	+	carbon dioxide $6\ CO_2$	+	water $6\ H_2O$	$\xrightarrow{\text{enzyme}}$	glucose $C_6H_{12}O_6$	+	oxygen $6\ O_2$

The process of photosynthesis uses solar energy (sun's energy) to combine carbon dioxide and water into glucose (which has chemical bond energy) and oxygen; oxygen is given off to the environment (see equation above). Chemical bond energy (example, chemical bond energy in glucose) provides energy for life activities (life processes), such as digestion, transport, and growth.

In photosynthesis, glucose is produced. Glucose ($C_6H_{12}O_6$) is an organic molecule because it has both C (carbon) and H (hydrogen). Water (H_2O) and carbon dioxide (CO_2) are inorganic molecules because they do not have both C and H.

Note: When there is very little sunlight (example, far down in the ocean), very little photosynthesis takes place in plants and algae. Also, the amount of photosynthesis depends on the color of the light. In the presence of red light or blue light, plants can easily carry on photosynthesis; in green light, very little photosynthesis takes place.

Note: When there are more algae or plants in a lake or ocean, more photosynthesis takes place and more glucose and oxygen are produced.

Experiments on photosynthesis

The diagram shows light shining on a plant; in the presence of light, photosynthesis takes place. In photosynthesis, the plant takes in carbon dioxide and water and gives off oxygen. (Since the plant is in the water (see figure), the plant takes in carbon dioxide which is in the water). We tested the bubbles in the upside-down test tube for oxygen. A glowing splint put into the upside-down test tube bursts into flame; this shows that oxygen is there. The plant gives off oxygen in photosynthesis.

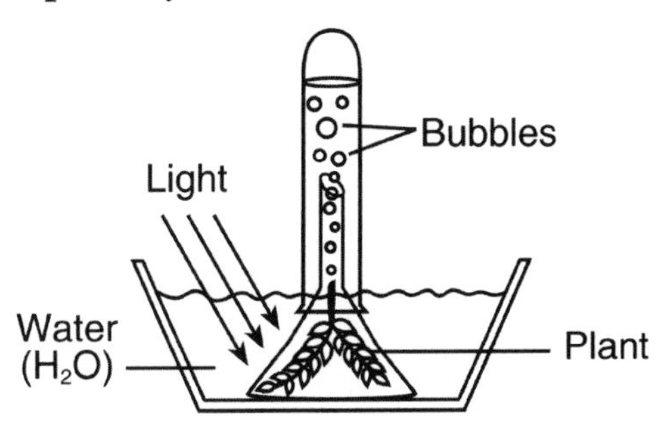

Test for oxygen: a glowing splint put into a test tube bursts into flame; this shows that oxygen is there.

An experiment can be done to show if light is needed for photosynthesis (or, you can say, if light is needed to produce glucose, or, if plants in the dark can produce glucose)(see figure at right). Cover both sides of part of a leaf on a plant with black paper so the covered part has no light. Leave the plant in sunlight for a number of days.

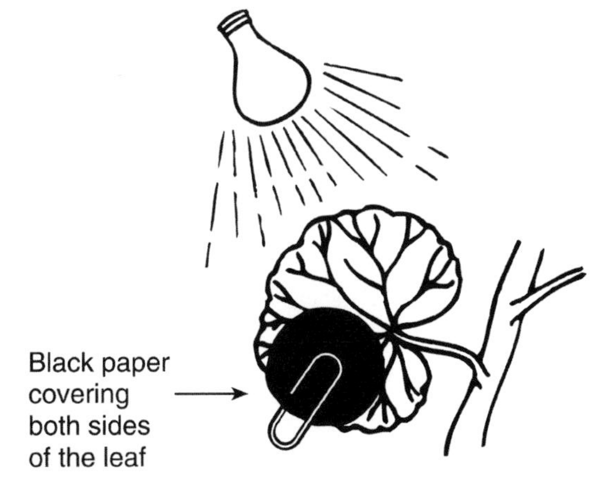

In photosynthesis, glucose (simple sugar) is produced. Sugar is changed to starch; the starch is stored (remains) in the leaf until needed. Test the covered and uncovered parts of the leaf for starch; Lugol's solution (or iodine solution) turns blue-black in the presence of starch (when there is starch). The presence of starch shows that photosynthesis takes place or glucose is produced (which is stored as starch).

In this experiment, the students observe that the covered part of the leaf does not turn blue-black with Lugol's solution, showing that starch is not present and therefore photosynthesis did not take place in the dark (no light). The uncovered part of the leaf turns blue-black with Lugol's solution, showing that photosynthesis takes place in the light and glucose is produced, which is stored as starch.

Note: Sugars and starches are both called carbohydrates.

PRACTICE QUESTIONS AND SOLUTIONS

Question: A five-year study was carried out on a population of algae in a lake. The study found that the algae population was steadily decreasing in size. Over the five-year period, this decrease most likely led to

(1) a decrease in the amount of nitrogen released into the atmosphere

(2) an increase in the amount of oxygen present in the lake

(3) an increase in the amount of water vapor present in the atmosphere

(4) a decrease in the amount of oxygen released into the lake

Solution: You learned, when there are more algae or plants in a lake, more photosynthesis takes place and more glucose and oxygen are produced. Therefore, when there are less algae (**decrease in algae**) in a lake, like in this example, less photosynthesis takes place and less oxygen (decrease in oxygen) is produced (released). Answer *4*

Question: The diagram below represents a biological process.

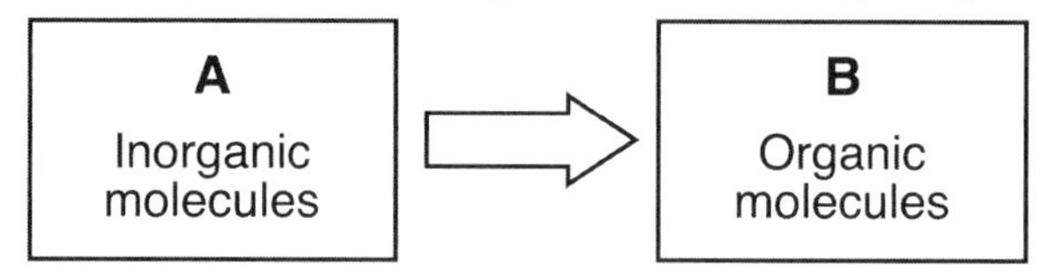

Which set of molecules is best represented by letters A and B?

(1) A: oxygen and water B: glucose

(2) A: glucose B: carbon dioxide and water

(3) A: carbon dioxide and water B: glucose

(4) A: glucose B: oxygen and water

Solution: You learned **organic** molecules have both C and H (example **glucose**, $C_6H_{12}O_6$). **Inorganic** molecules do not have both C and H (examples **carbon dioxide** CO_2 and **water** H_2O).

You learned in the process (biological) of photosynthesis, carbon dioxide and water (both inorganic molecules) produce glucose (organic molecules).

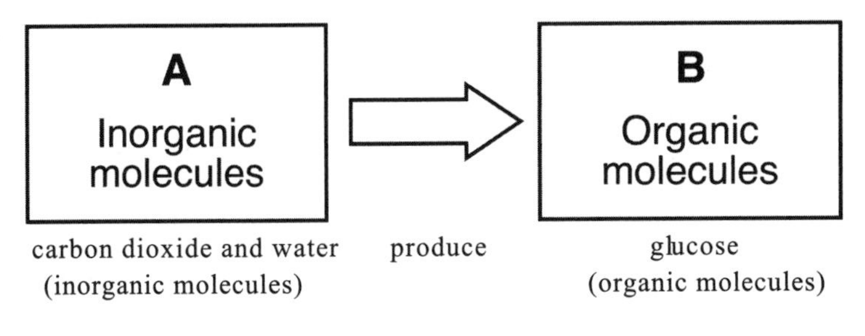

carbon dioxide and water (inorganic molecules) produce glucose (organic molecules)

Answer *3*

Now Do Homework Questions #1-22, pages 33-38.

Plants and animals use glucose

Plants use glucose to make ATP; glucose and ATP provide (give) energy for living things. Glucose and ATP are energy-rich molecules (have a great deal of energy). Plants change the sun's energy into glucose (which has chemical bond energy). Glucose is then changed into ATP (stored as ATP), which also has chemical bond energy. As you can see, there is a connection between the sun's energy and the chemical bond energy in glucose and in ATP.

Plants also **use glucose** (a small molecule) as building blocks **to build complex molecules** (compounds) such as **starch**. Glucose is stored as starch.

Similarly, **animals** eat plants and **use** the **glucose** (simple sugar) in the plants **to produce** (make) **ATP**, a usable form of energy. Animals (example humans) eat the plants, and animals eat other animals (example cows) that eat the plants which produce glucose; animals convert (change) the glucose into ATP.

The animals also use the glucose (small molecule) as building blocks to make complex molecules (large molecules).

Simple Compounds Make Complex Compounds

Simple **(small)** organic compounds (examples: glucose, amino acids, fatty acids, glycerol) can be used as building blocks to **make larger, complex** molecules (examples: starch, protein, fats, DNA). Glucose (small molecule) can be made into starch (large molecule), fatty acids (small molecules) into lipids (fats or oils, which are large molecules) and amino acids can be made into proteins (large molecules). As you can see, the amino acids are the building blocks to make protein (a large molecule). (Sugar, phosphate, and nitrogen bases can be made into DNA, which you will learn later).

Or, you can say, complex **large** (larger) molecules such as starch, fat, protein, and DNA are made **from small** molecules. Example: starch (large molecule) is made from glucose (small molecules), fat (large molecule) is made from fatty acids (small molecules), and protein (large molecule) is made from amino acids (small molecules).

Note: Fats and starches are food reserves (stored food) (can be broken down into simpler foods, which give off energy when needed by living things). Protein makes up enzymes, hormones, and parts of cells (enzymes, hormones, and parts of cells are made of proteins). Sugar, starch, fat, protein, vitamins and minerals are called **nutrients**.

Now Do Homework Questions #23-28, pages 38-39.

Cellular Respiration

In the biochemical process of photosynthesis, glucose is produced; glucose has chemical bond energy (the energy in glucose is in the chemical bond).

In **cellular respiration**, glucose unites with oxygen, forming water, carbon dioxide, and ATP (see equation below). Or, you can say **glucose** (a simple sugar) is **broken down** into **water, carbon dioxide**, and **ATP**, a usable form of energy.

$$\text{Glucose} + \text{oxygen} \xrightarrow{\text{enzymes}} \text{water} + \text{carbon dioxide} + \text{ATP}$$

Breaking the bond in glucose gives off energy (chemical bond energy); the energy is temporarily stored as ATP. Carbon dioxide and water are given off as wastes. You learned**, cellular respiration** takes place (is completed)in the **mitochondria** in **both plants and animals** all the time, both **day** and **night.**

In cellular respiration, glucose ($C_6H_{12}O_6$) and oxygen (O_2) are taken in; water (H_2O) and carbon dioxide (CO_2) go out. ATP is produced, which provides energy for life processes (life activities).

In the cells, the ATP (adenosine triphosphate, which means it has three phosphates) is broken down into ADP (adenosine diphosphate, which has two phosphates) + P (one phosphate), and **energy** is **released** (given off).

$$\text{ATP} \longrightarrow \text{ADP} + \text{P} + \text{energy}$$

The energy (given off from ATP) is used to carry on life processes, to obtain (get) materials, to transform (change) small molecules into big molecules, to pump blood, to use active transport to move materials in and out of cells, and to eliminate wastes.

Note: In the cells, energy in glucose unites ADP and P, forming ATP. This is the reverse of the above equation (see equation above).

$$ADP + P + energy \rightarrow ATP$$

ATP provides energy for living things. Muscle cells need more energy, and therefore they have more mitochondria and produce (make) more ATP.

Recycling materials: In cellular respiration, oxygen is taken in and carbon dioxide is given off. The plants take in carbon dioxide for photosynthesis and give off oxygen, which is used in cellular respiration. Oxygen and carbon dioxide are recycled.

During **strenuous exercise** the heart beats faster, causing the blood to move faster, and **more oxygen** unites with glucose to produce (make) **more ATP**. (When less oxygen unites with glucose, less ATP is produced).

A student did an experiment showing how the amount of oxygen affects ATP production in muscle cells. The data shows, when X amount of oxygen is used, the amount of ATP produced.

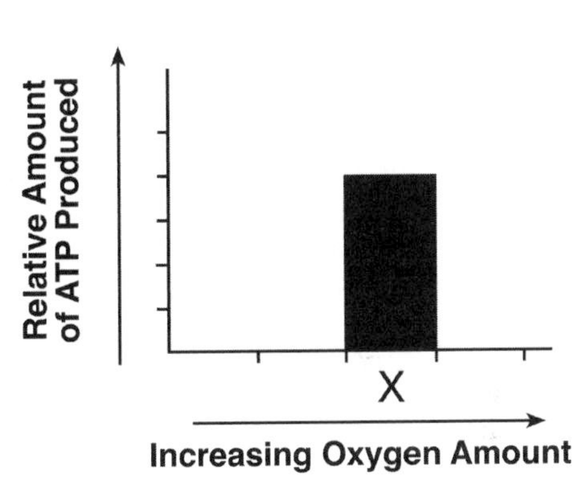

If in a new trial, there was **less oxygen**, the new bar would be placed to the left of bar X , which is less oxygen, and the bar would be shorter because less ATP would be produced.

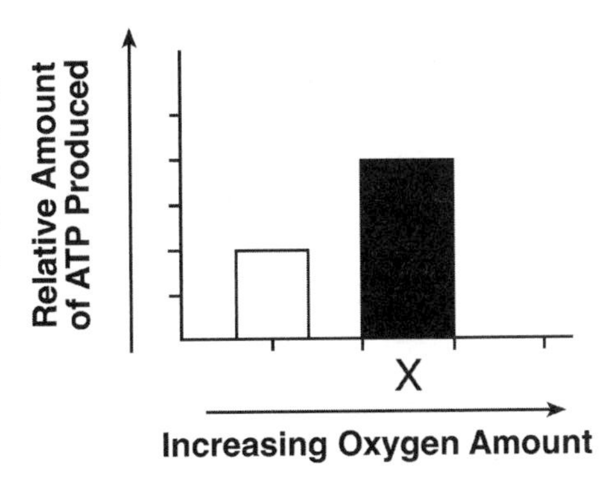

Question: Which statement best describes these cells?

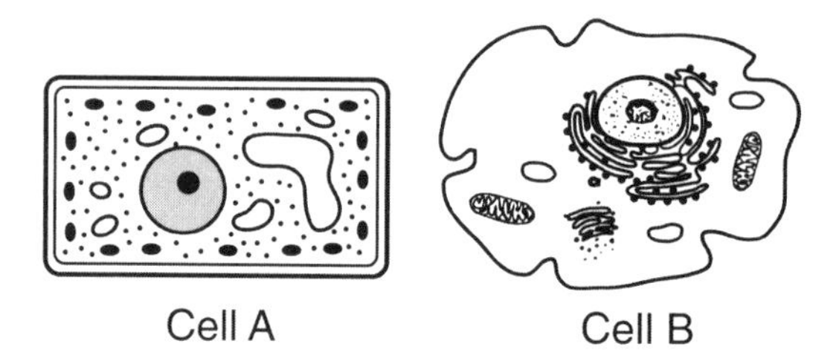

(1) Cell B lacks vacuoles while cell A has them
(2) DNA would not be found in either cell A or cell B
(3) Both cell A and cell B use energy released from ATP
(4) Both cell A and cell B produce antibiotics

Solution: Plants and animals use ATP for energy. ATP is broken down into ADP and P and energy is released (given off). You know cell A is a plant cell because it has a cell wall and chloroplasts. You know cell B is an animal cell because it does **not** have a cell wall and chloroplasts.

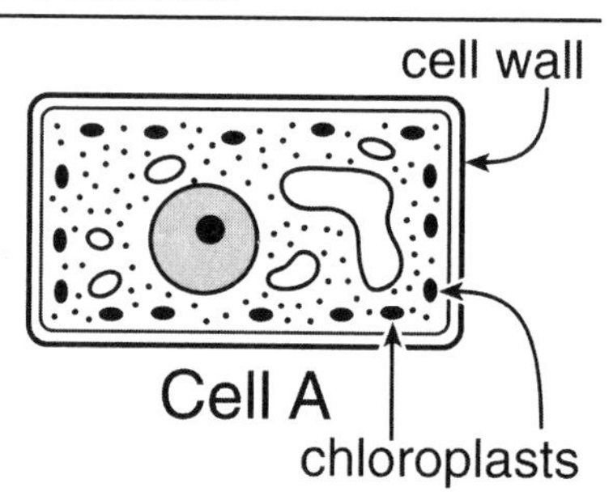

Answer *3*

Question: Which set of terms best identifies the letters in the diagram below?

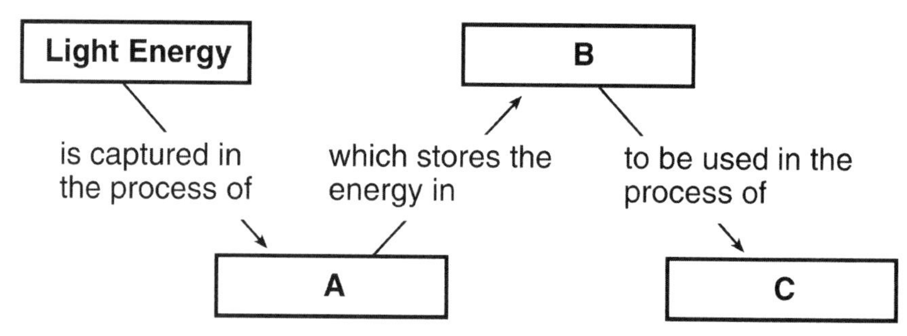

	A	B	C
(1)	photosynthesis	inorganic molecules	decomposition
(2)	respiration	organic molecules	digestion
(3)	photosynthesis	organic molecules	respiration
(4)	respiration	inorganic molecules	photosynthesis

Solution:

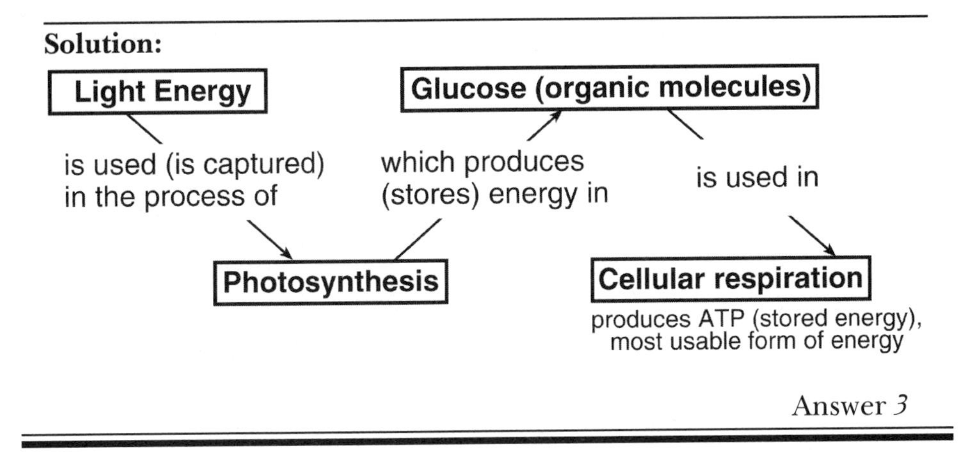
Light Energy
is used (is captured) in the process of
Photosynthesis
which produces (stores) energy in
Glucose (organic molecules)
is used in
Cellular respiration
produces ATP (stored energy), most usable form of energy

Answer *3*

Now Do Homework Questions #29-51, pages 39-45

Enzymes

An **enzyme (biological catalyst) affects** the **rate of** a chemical **reaction.** An enzyme speeds up the rate of chemical reaction, but the enzyme is not used up in the reaction; the enzyme can be used over and over again. Enzymes are proteins.

Enzymes are **needed** in **biochemical processes** (life processes, life activities) such as **photosynthesis**, **cellular respiration**, **synthesis** (small molecules join together to form larger molecules, example: amino acids join together to form protein), and **enzymes are needed to break down** large molecules into smaller molecules (examples: digestion of starch into sugar, protein broken down into amino acids). Without enzymes, these biochemical processes or reactions cannot happen (examples, photosynthesis, cellular respiration, digestion, synthesis would not happen). All **organisms** (all **living things),** which includes all **animals** (examples deer, dogs, cats, fish, wolves), all **plants** (example grasses, trees), all organisms that you can only see with a microscope (example bacteria), and all fungi need (depend on) enzymes (biological catalysts).

Each **enzyme** has a **specific shape** and a **specific function**. A specific enzyme joins together two small substances (example two simple sugars) to form a bigger substance (example a bigger sugar $C_{12}H_{22}O_{11}$, which has 12 C (carbons)). The same enzyme can break the bigger sugar into the two small simple sugars. Figure 1 shows a specific enzyme and two simple sugars. The shape of the simple sugars must fit exactly into the shape of the

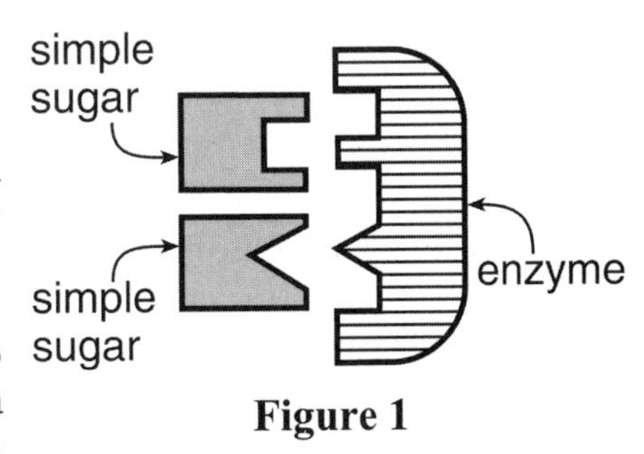
simple sugar
simple sugar
enzyme

Figure 1

enzyme; the enzyme can now do its work. It can be compared to a lock and key. The key must fit exactly into the lock.

Look at Figure 2. The two simple sugars attach themselves to the active site on the enzyme. The "active site" on the enzyme is the part of the enzyme where the sugars directly attach themselves to the enzyme (touching the enzyme). See Figure 2.

The two simple sugars join together and form the bigger sugar $C_{12}H_{22}O_{11}$.

Note: The simple sugar ($C_6H_{12}O_6$) has six C (carbon atoms); the bigger sugar ($C_{12}H_{22}O_{11}$) has twelve C (carbon atoms).

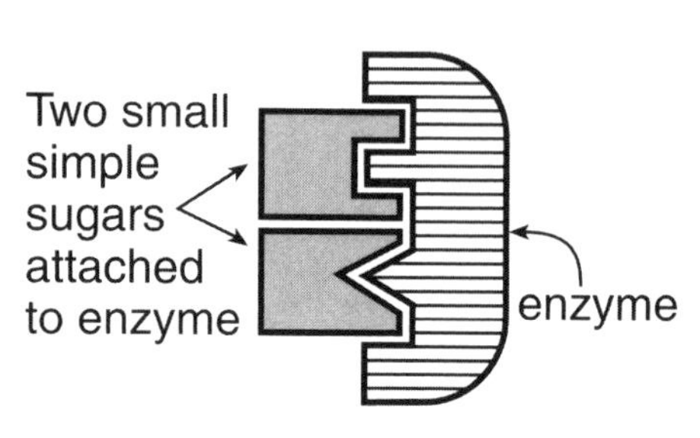

Figure 2

Look at Figure 3. Now the enzyme finished its work (function) in putting two simple sugars together and the enzyme separates from the bigger sugar. The enzyme can now put together two other simple sugars to make a bigger sugar or break down a bigger sugar to form two simple sugars.

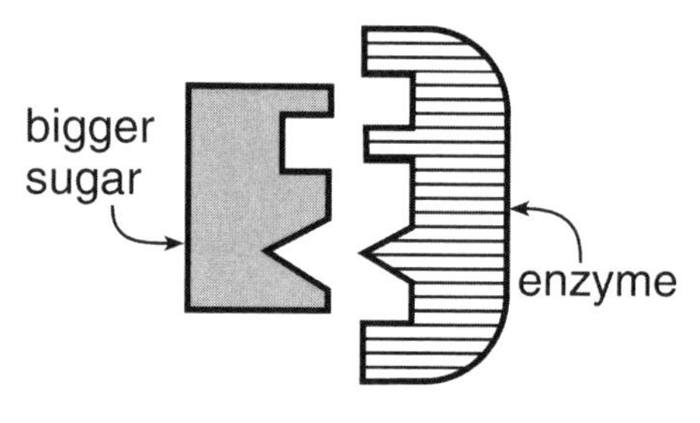

Figure 3

Example 2 Look at Figure 1. Figure 1 shows a specific enzyme and the bigger sugar.

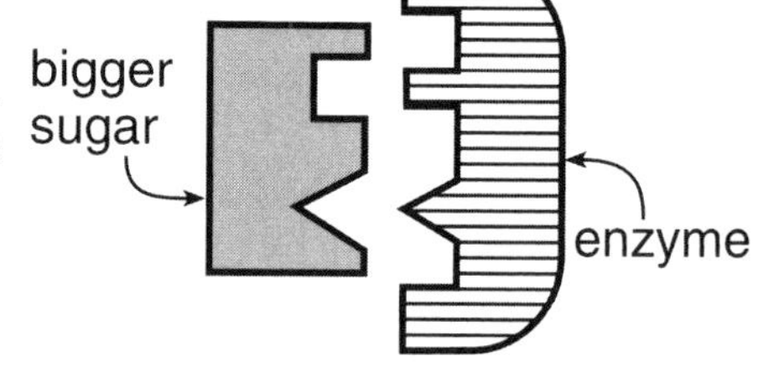

Figure 1

Look at Figure 2. The bigger sugar attaches itself to the enzyme; the enzyme breaks the bigger sugar into two simple sugars.

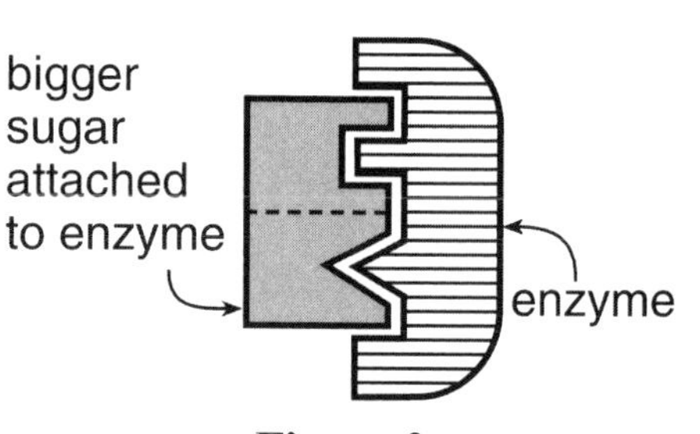

Figure 2

Look at Figure 3. Now the enzyme finished its work in breaking the big sugar into two small simple sugars and the enzyme separates. The enzyme can now break a big sugar into two simple sugars or put two simple sugars together.

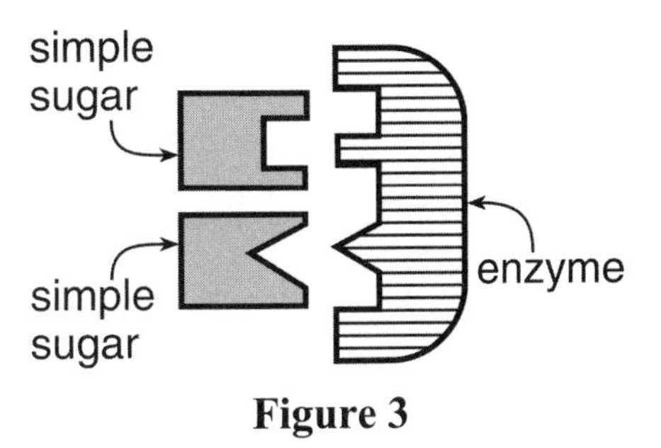

Figure 3

Shape of the enzyme, temperature, and pH affect the rate of the reaction

Shape of the enzyme: The enzyme must have a **specific shape**. A substance (example two simple sugars) must fit exactly into the enzyme, so the enzyme can do its work of speeding up a reaction (putting together two small molecules, example two small sugars, or breaking apart a large molecule).

High temperature and **strong acids and bases** can **change** the **enzyme's shape;** then the enzyme cannot function (work), causing the **rate of reaction** to **decrease**. The **more enzyme** that **doesn't work**, the **slower** is the **rate of reaction.**

Temperature: **Enzymes** have an **optimal temperature** (**best temperature**) at which **they can work** and produce the **fastest rate of reaction** (**highest enzyme activity**).

Look at the graph showing how temperature affects enzyme activity.

1. As the **temperature increases from 0°C to** the **optimal temperature** (best temperature), which is the **highest point on the graph, enzyme activity increases (rate of enzyme action increases or**, you can say, **rate of reaction increases)**.

2. **At** the **optimum (best) temperature,** enzymes and molecules move faster, collide more often, and have the **fastest rate of reaction**.

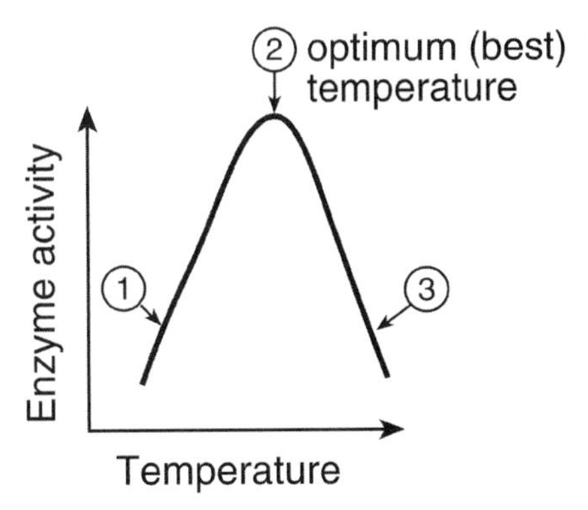

Temperature Affects Enzyme Activity

3. When the **temperature gets higher than** the **optimum temperature**, (the highest point on the graph), the enzyme molecule changes shape or breaks apart, and **enzyme activity decreases quickly (rate of enzyme action decreases quickly** or, you can say, rate of reaction decreases quickly). At a very high temperature (above 50°–60° C), the enzyme becomes deactivated (enzyme does not work).

Look at the graph. The **graph** shows how **temperature affects** the **enzyme activity in a human being;** the higher the enzyme activity, the faster is the rate of enzyme action. The human body has a temperature of about 37°C. Enzymes in the human body work best at that temperature, 37°C,(optimal temperature), which is the highest point on the graph. As the temperature increases from 0°C to 37°C (the optimal temperature), enzyme activity increases and rate of enzyme action (rate of reaction) increases. At 37°C, the highest point on the graph, enzyme activity is highest (rate of enzyme action is fastest). Above 37°C (optimal temperature) enzyme activity (rate of enzyme action) decreases rapidly. The higher the temperature above 37°C, the more the enzyme changes shape (the more poorly it works or the worse it works), therefore enzyme activity (rate of enzyme action) decreases.

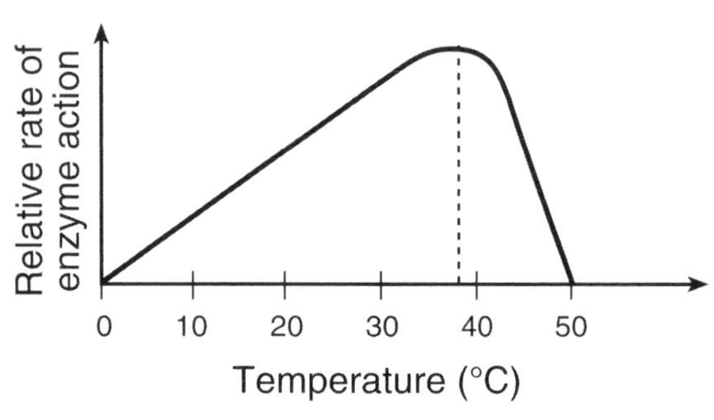

Temperature Affects Enzyme Activity

pH: pH is a scale that measures whether a substance is acidic, basic, or neutral. pH less than 7 (example pH 2, pH 4, or pH 5) is acidic; pH 7 is neutral, pH of more than 7 (example pH 9, pH 11, or pH 13) is basic.

Enzymes have an optimum (best) pH at which they can work best and produce the fastest rate of reaction (highest enzyme activity). When the pH is **higher** or **lower** than the optimal (best) pH, the rate of reaction decreases (enzyme activity is lower).

Look at the graph. The **graph** shows how **pH affects enzyme activity.** As you can see, this enzyme works best at pH 7, which is the highest point on the graph. When pH goes below 7 (example 5) or above 7 (example 10) the rate of enzyme activity decreases.

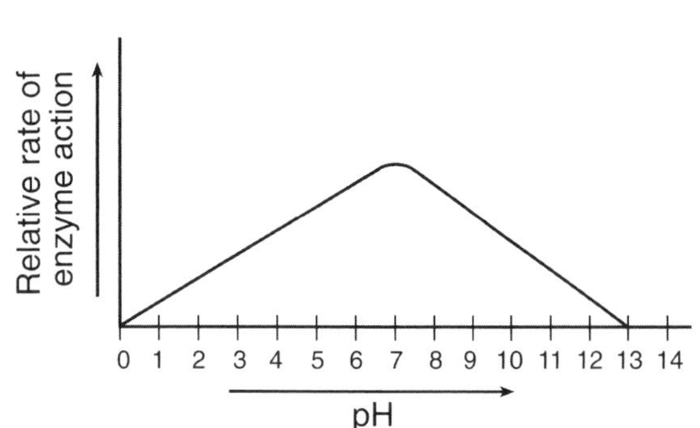

Effect of pH on Rate of Enzyme Activity

When pH is very low (example 2) or very high (example 12), enzymes do not work and there is no reaction.

Note: Enzymes in many cells of the body have an optimum (best) pH of 7, and many cells of the body have a pH of 7.

Enzymes in the stomach have an optimum (best) pH of 3; the stomach has a pH of 3 (see graph). Enzymes in the small intestine work best at pH about 8; the small intestine has a pH of about 8.

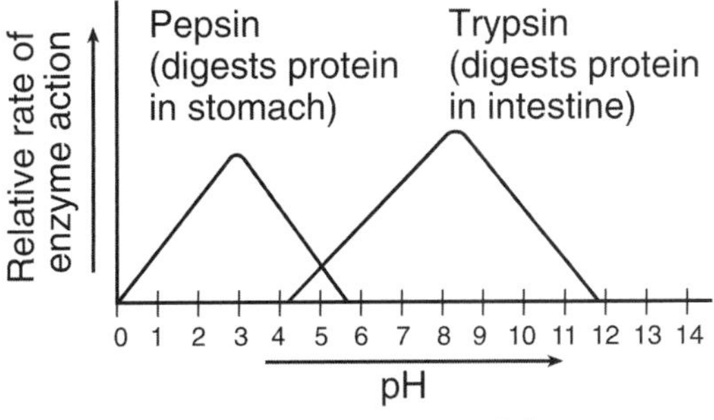

pH Affects Enzyme Activity

PRACTICE QUESTIONS AND SOLUTIONS

Questions 1 and 2: Base your answers to questions 1 and 2 on the diagram below, which represents stages in the digestion of a starch.

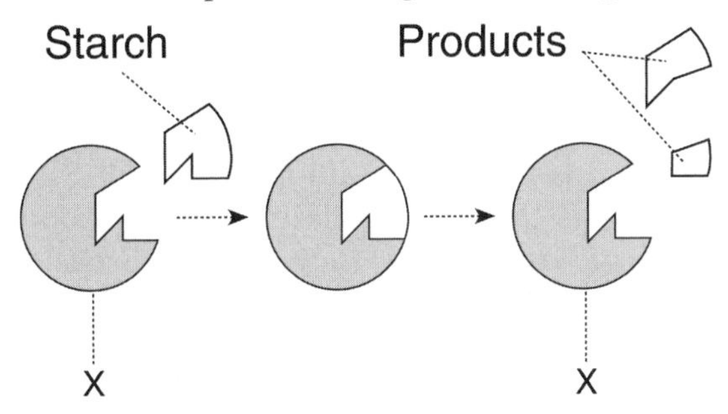

Question 1: The products would most likely contain

(1) simple sugars (2) fats (3) amino acids (4) minerals

Question 2: The structure labeled X most likely represents

(1) an antibody (3) an enzyme
(2) a receptor molecule (4) a hormone

Question 3: Explain why substance X would *not* be likely to digest a different organic compound.

Solution 1: With an enzyme, starch (bigger molecule) is broken down into smaller simple sugars. Answer *1*

Solution 2: In the diagram, X is the enzyme. The shape of the starch fits exactly into the shape of the enzyme (labeled X), see figure above, so the enzyme can do its work of breaking the starch into smaller simple sugars (products). Answer *3*

Solution 3: The shape of a different compound might not fit exactly into the shape of the enzyme; therefore the enzyme cannot do its work of digesting the compound.

Question: In many investigations, both in the laboratory and in natural environments, the pH of substances is measured. Explain why pH is important to living things. In your explanation be sure to identify one example of a life process of an organism that could be affected by a pH change.

Solution: Enzymes in many cells of the body have an optimal (works best at) pH of 7, and many cells of the body have a pH of 7. If the pH is changed to pH 1, 2, or 3, the enzyme would not work that well.

Enzymes in the small intestine work best at pH about 8. The small intestine has a pH of about 8. If the pH of the small intestine was changed to pH 5 or 6, the enzymes would not work that well.

Now Do Homework Questions #52-73, pages 45-50.

Homeostasis

An organism's external (outside the organism) and internal (inside the organism) environment keeps changing. Examples of changing environments are: one day it is 90°F, another day 60°F, and another day 20°F. After eating a meal, there is more sugar in the blood than before the meal. The healthy organism can get sick with a disease when a virus or bacteria enters (gets into) the body.

The **organism** must **notice** these **changes** (example too much sugar or too little sugar in the blood) and **respond to** these **changes**, so the organism's body maintains homeostasis (a constant internal environment).

Homeostasis is also called steady state or **dynamic equilibrium**. The organism corrects the small changes (a little higher temperature, more sugar (glucose) in the blood) by feedback mechanisms (explained below) so the body can maintain homeostasis. Regulation (by using hormones and nerves) helps to maintain homeostasis.

Feedback mechanism maintains homeostasis

Four examples of feedback mechanisms are explained below.

1. Sugar (glucose) in blood is a **feedback mechanism:** If the blood has **too much sugar** (glucose), a message is sent to the **pancreas** to **produce** (make) **insulin** which **takes** (tells the liver to take) the **excess sugar out** of the blood and stores sugar in the liver as glycogen. When there is **too little sugar** (glucose) in the blood, the pancreas stops producing (making) insulin and produces a different hormone called glucagon that releases the stored sugar from the liver and the **sugar goes into** the **blood**. Now there is the right amount of sugar (glucose) in the blood.

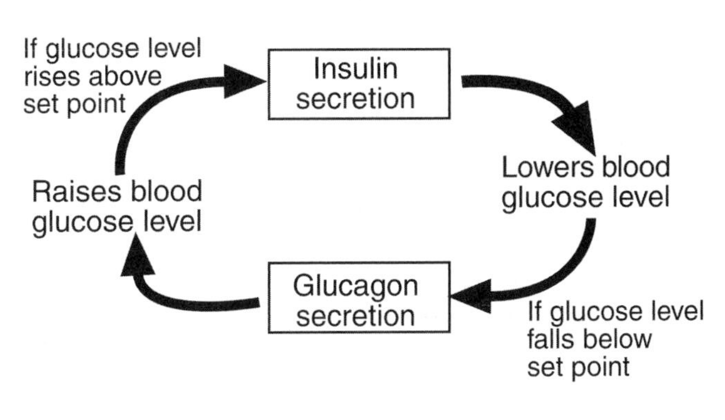

If the feedback mechanism does not work properly, the amount of sugar in the blood will not be regulated (will not be the right amount). To explain: if the pancreas produces the wrong amount of insulin, or the insulin does not work properly, the amount of sugar in the blood is not the right amount and homeostasis is not maintained.

Questions 1 and 2: Base your answers to questions 1 and 2 on the diagram and on your knowledge of biology.

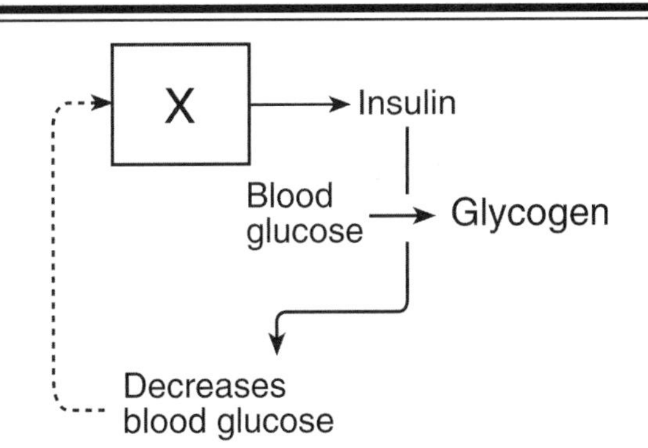

Question 1: Identify the organ labeled X.

Question 2: The dashed line in the diagram represents

1. a digestive process
2. a feedback mechanism
3. cellular differentiation
4. recycling of organic chemicals

Solution 1: X is the pancreas. The pancreas produces insulin. The endocrine system is made of glands (example pancreas, labeled X) that produce hormones (example insulin.)

Solution 2: The feedback mechanism maintains the right amount of glucose (sugar) in the blood. The dashed line shows that when there is too little glucose in the blood, a message is sent to the pancreas to stop producing insulin (the pancreas stops making insulin) Choice *2*

This graph illustrates homeostasis (dynamic equilibrium). The graph shows that the amount of sugar (glucose) in the blood increases (goes up), maybe right after breakfast, then the amount of sugar in the blood decreases (goes down), then increases again and decreases again. The organism corrects these small changes and maintains homeostasis (dynamic equilibrium), a constant internal environment (in this example, a relatively constant level of sugar in the blood).

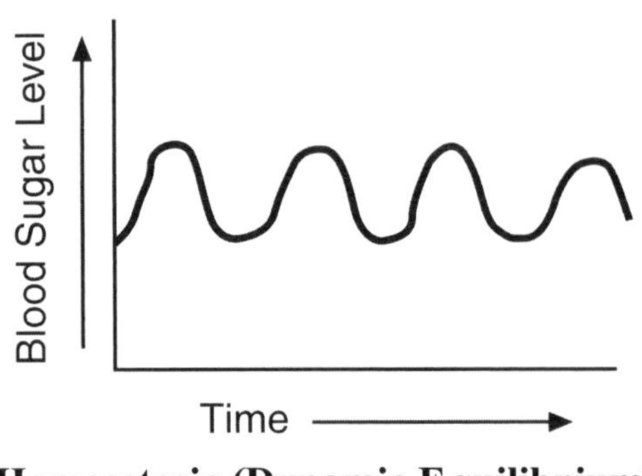

Homeostasis (Dynamic Equilibrium)

2. Temperature is another **feedback mechanism:** If the **temperature** of the organism **goes up** to 98.7°F, the organism perspires **(sweats),** which **lowers** the **temperature** to 98.6°F (37°C); the body temperature is at a constant temperature of 98.6°F(37°C).

If it is very cold outside (0°F), the temperature of the body may go down to 98.5°. The body shivers, causing the temperature to rise to 98.6°, constant body temperature.

3. Carbon dioxide in the **blood** is another **feedback mechanism:** The heart beats at a rate (seen as the pulse) causing the blood to flow, carrying enough oxygen and glucose to the cells and taking away carbon dioxide. When someone is running a marathon or doing **strenuous exercise,** the body **needs more energy.** During strenuous exercise, **more oxygen** is **needed** for cellular respiration (oxygen unites with glucose) to produce more energy (ATP, chemical bond energy), and more carbon dioxide waste is produced (given off), which must be removed. The **increase** in **carbon dioxide** in the blood **sends** a **message** to the brain **to increase** the **breathing rate**, so more oxygen gets into the body. The increase in carbon dioxide causes the **heart rate** and **breathing rate** to **increase** so flow of blood is faster and more oxygen gets into muscle cells and more carbon dioxide is removed. This helps to maintain homeostasis. The **higher heart rate** (heart beating faster) will be seen as a **higher pulse rate**. When there is an increase in pulse rate, it shows that the **heart rate increases,** flow of blood is faster, bringing **more oxygen** into **muscle cells** and **removing more carbon dioxide.** The heart (part of the circulatory system) and breathing (respiratory system) work together to maintain homeostasis. When the carbon dioxide level goes back down, the heart and breathing slow down.

Here is a graph showing the effect of sitting and running on pulse rate for both males and females. The graph is based on a data table.

Data Table

Exercise	Pulse rate	
	Male	Female
Sitting	80	72
Walking	85	90
Running	122	110

Look at the graph below. This is the male bar ▢ ;this is the female bar ■ .

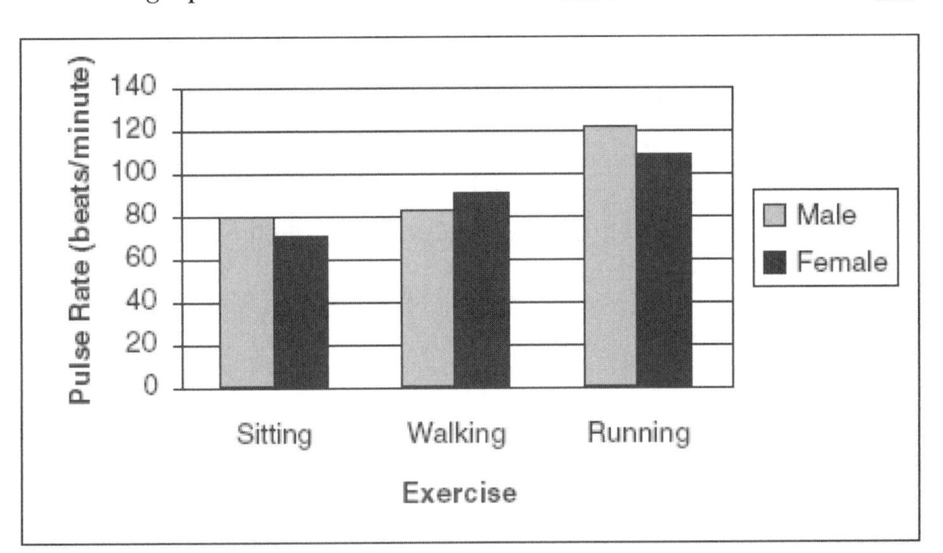

Look at the vertical line (axis) from 0-140 beats per minute. There are 20 beats per minute between each two numbers on the vertical axis (line).

Pulse at sitting: **For males**, look at the top of the male bar ▢ and go across to the vertical axis; the pulse rate for males while sitting is 80 beats per minute.

Pulse at sitting: **For females**, look at the top of the female bar ■ ; draw a dotted line across to the vertical axis (line). You can see the dotted line is a little more than halfway between 60 and 80; the pulse while sitting is 72 beats per minute.

You can draw a conclusion for this experiment, that while sitting males' pulse (80 per minute) is higher than females' pulse (72 per minute).

Pulse at running: **For males,** look at the top of the male bar ▢ ; draw a dotted line across to the vertical axis (line). You see the dotted line is a little more than 120 beats per minute but less than 140 beats per minute; the male pulse rate while running is about 122 beats per minute.

Pulse at running: **For females,** look at the top of the female bar ▇ ; draw a dotted line across to the vertical axis (line). It is half way between 100 and 120; the female pulse rate is about 110 beats per minute.

You can draw a conclusion in this experiment that while running, the pulse for males (122 per minute) is higher than the pulse for females (110 per minute).

By looking at the graph, you can also draw a conclusion that the pulse of people running (both boys, 122 beats per minute, and girls, 110 beats per minute) is higher than the pulse of people sitting (both boys, 80 per minute, and girls, 70 per minute).

When doing exercise like running, breathing is faster and the heart beats faster, which is seen as a higher pulse.

Note: During strenuous exercise, the heart beats faster, causing the blood to flow faster, bringing more oxygen and nutrients (example glucose) to the muscles; blood flows slower to the digestive system (this lets more blood go to the muscles). Faster blood flow brings more glucose and oxygen to the mitochondria in the muscles to produce more ATP.

When there is a lack of oxygen (not enough oxygen) going to the musdes (example leg muscles) of a runner, the runner gets cramps.

If a person's pulse rate falls very low, such as 45 beats per minute (when a person normally has a much higher pulse rate, such as 75), the heart is beating slower, bringing too little oxygen and glucose to the cells (disruption of homeostasis).

4. The **opening** and **closing** of the **guard cells in leaves** is a **feedback mechanism** (explained below). The leaves in a plant help to maintain homeostasis (a constant internal (inside) environment) by keeping the amount of water in the plant constant. If it is a very **dry hot day,** the guard cells that surround the **stomates (openings in** the **leaf) close,** therefore the openings in the leaf are closed and **water cannot go out** of the leaves. This helps to maintain a **constant amount** of **water** in the plant.

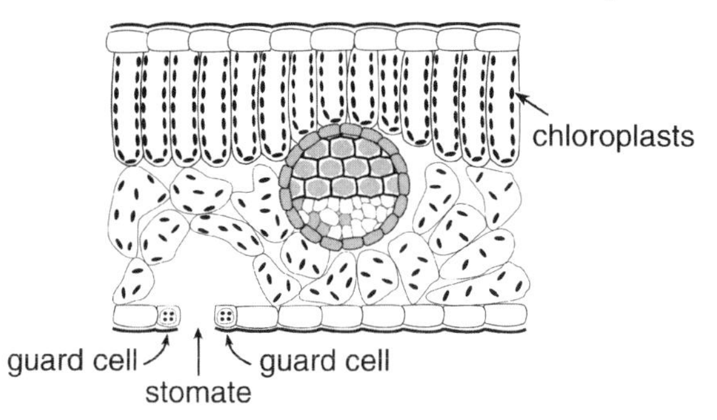

Cross section of leaf

Note: When the openings (stomates) in the leaf are closed, less carbon dioxide can enter the leaf and less photosynthesis takes place.

When the weather is **not** that **hot** and dry, the guard cells open and the **stomates (openings in** the **leaf**) are **open; carbon dioxide enters** the leaves and **water goes out** by evaporation. This maintains homeostasis (a constant internal environment).

As you can see in feedback mechanisms, if there is too much (example **too much sugar** in the blood or **too high temperature** in the body), the body corrects it (by taking away sugar from the blood **(leaving less sugar** in **blood**) or **lowering temperature**) so the body can maintain homeostasis (constant sugar level or constant internal temperature).

Question 1: This diagram illustrates part of

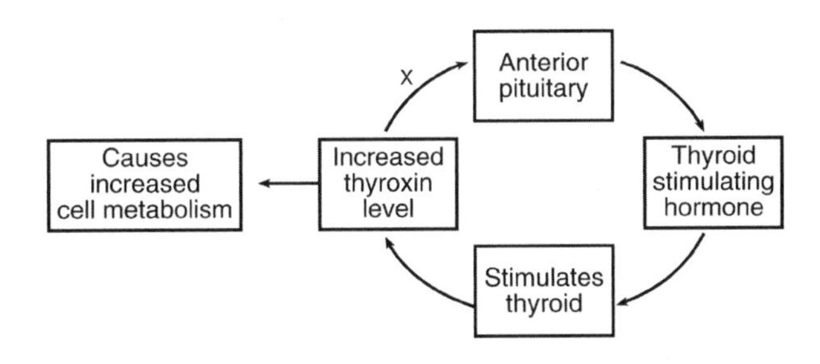

1. a feedback mechanism
2. an enzyme pathway
3. a digestive mechanism
4. a pattern of learned behavior

Question 2: Describe the action represented by the arrow labeled *X* in the diagram and state *one* reason that this action is important.

Solution 1: The diagram illustrates another feedback mechanism. In a feedback reaction, the body notices the changes (examples: increased thyroxin or increased blood sugar) and corrects these changes and maintains homeostasis. Choice *1*

Solution 2: Look at the feedback loop in the question. An **increase in thyroxin** level sends a **message** (represented by ***X***) to the anterior pituitary to **stop (inhibit) producing** (making) **thyroid-secreting hormone** (TSH) and therefore lessen the amount of thyroxin produced. This is important because it maintains homeostasis, the normal (correct) amount of thyroxin in the blood and the correct rate of metabolism.

Now Do Homework Questions #74-109, pages 50-57.

Diseases

There are various causes of diseases. Viruses and bacteria are microbes (microscopic organisms). **Viruses, bacteria, fungi,** and other **parasites** (organisms that harm other organisms) can **infect (invade**, go into) both **plants and animals** and make them **sick.**

Examples of pathogens: (viruses, bacteria, fungi, and parasites which cause disease).

Viruses cause the common cold, AIDS, chicken pox, and influenza (flu).

Bacteria cause streptococcus throat and food poisoning. Treatment is antibiotics.

Fungi cause athlete's foot and ringworm. Treatment is fungicides and antibiotics.

A one-celled parasite causes malaria (a deadly disease). Other parasites are leeches and tapeworms.

Antibiotics, such as penicillin, treat diseases (example bacterial). Antibiotics kill the bacteria that are now in the body, but do not protect the body against future infections.

Diseases interfere with normal **life functions** (example, a person with AIDS has a weakened immune system and cannot protect himself against infectious diseases (caused by pathogens such as viruses and parasites) and cancers (which can cause death).

Diseases interfere with **homeostasis**; **disease** is a **disturbance (disruption**, failure) of homeostasis (maintaining a constant internal environment). For example, when a person gets an infection, the person's temperature goes up (higher than 98.6°F), which is a disturbance (disruption, failure) of homeostasis.

Question: Base your answer on the graph below.

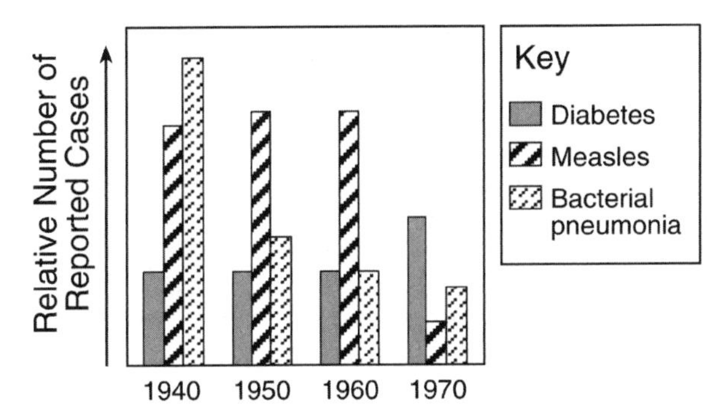

The greatest difference between the incidence of measles and the incidence of bacterial pneumonia occurred in

(1) 1940 (2) 1950 (3) 1960 (4) 1970

Solution: From the key, you see [measles pattern] shows measles and [pneumonia pattern] shows pneumonia.

The taller (higher) the bar, the more people had that disease. The shorter the bar, the fewer people had the disease.

In 1940, there was a small difference between the height of the bars for measles and pneumonia.

In 1960, there was the biggest difference (greatest difference) between the height of the bars for measles and pneumonia.

Answer *3*

How Viruses and Bacteria Attack Organisms

You learned an enzyme must match the shape of the substance (example simple sugar, complex sugar, or starch) that it works on. Similarly, a **protein** on a **virus** or **bacteria must match** the **shape** of the receptor on the cell (example on a human cell) in order that the bacteria or virus attaches itself to the receptor on the human cell and makes the human sick.

Note: Hormones (example insulin), which are proteins, must also match the shape of the receptor on the cell that the hormone works on (example insulin controls the amount of sugar in the blood by working on liver cells, therefore insulin must match the shape of the receptor on liver cells).

In the diagram below, the proteins on the bird influenza virus do not match the shape of the receptors on the human cell, therefore the virus cannot infect the human cell. In order for the virus to infect the human cell (see diagram below), the protein on the virus must change its shape so the protein fits in the receptor on the human cell.

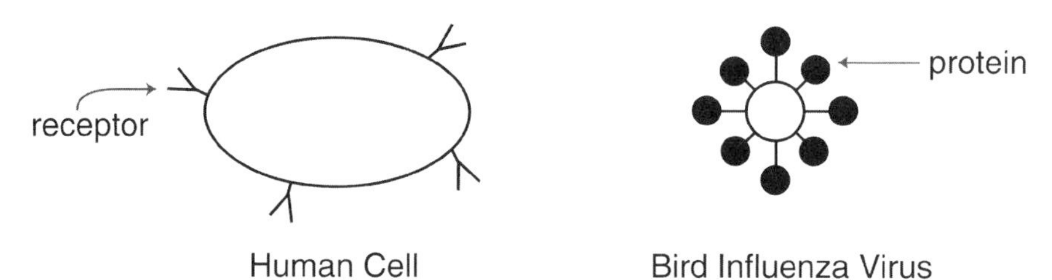

If, however, the protein on the virus changed to ➤, then the new virus protein would fit the receptor on the human cell, and the virus can infect the human cell (see figure at right).

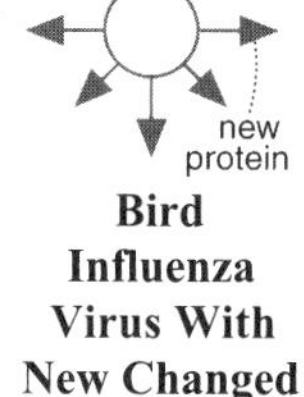

Bird Influenza Virus With New Changed Protein

In chapter 3 you will learn that a change in a living thing (example virus, plant, animal) is caused by a mutation.

The parasite (virus or bacteria) **benefits (+)** while the host (example human) is **harmed (−).** This relationship is called **parasite-host** relationship.

Immune System

The **immune** system has **white blood cells** and **antibodies**, which protect us from pathogens (viruses, bacteria, fungi, and other parasites which cause disease), foreign substances, cancer, or toxins (poisonous wastes from some pathogens, etc).

Let's understand how the immune system works. **White blood cells** and **antibodies** in the blood help **fight disease.** There are **antigens** (see figure) on the surface of pathogens, foreign substances, cancer cells, toxins (poisonous wastes), etc.. The **white blood cells** which are in the blood can notice or **recognize** the **antigen**; then white blood cells attack the pathogen (with the antigen on it). Look at the figure below. The **white blood cells surround** and **engulf** (eat) the **pathogen** or microbe (tiny organism that you can only see under a microscope).

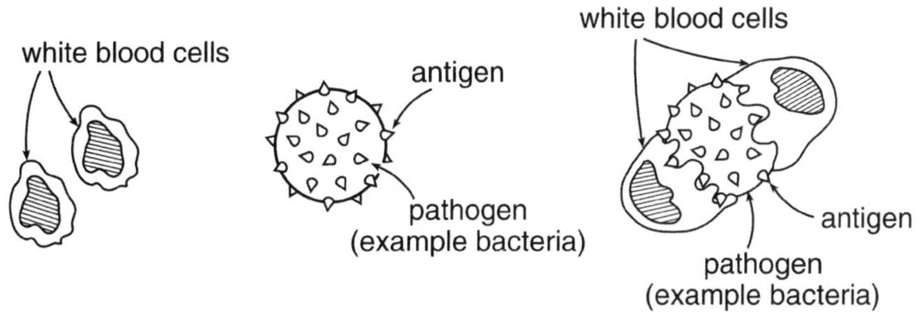

White blood cell recognizes antigen **Then white blood cell engulfs (eats) the pathogen with antigen on it**

The white blood cell engulfing a pathogen can also be shown this way:

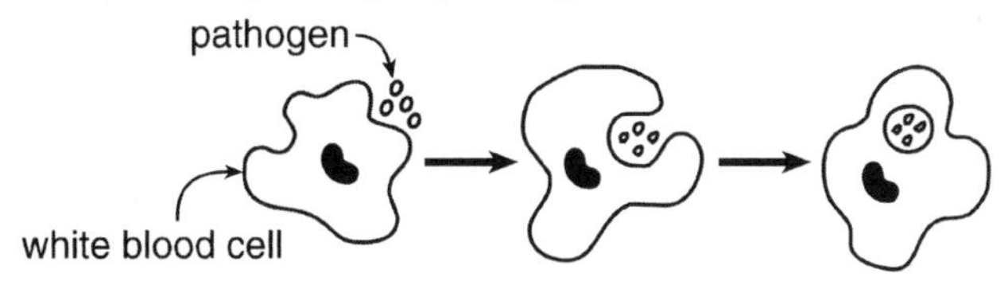

White blood cells surround and engulf the pathogen (organism that causes a disease, example bacteria)

Other **white blood cells (**when they notice the antigen on the pathogen) **produce antibodies** (which are proteins) that attach themselves to the antigen; the antibodies attack the pathogen or mark the pathogen for killing by white blood cells which engulf (eat) the pathogen or microbe (see figure below).

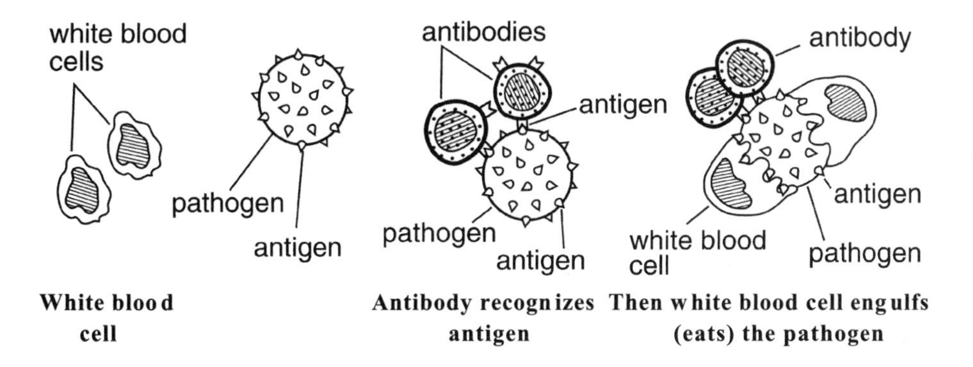

White blood cell **Antibody recognizes antigen** **Then white blood cell engulfs (eats) the pathogen**

Just like **enzymes** must **match** the **shape** of the substance (example starch, sugar) that it works on, or **hormones**, such as insulin, must **match** the **shape** of the **receptor** on the cells they work on**, antibodies** must **match** the **shape** of the **antigen** on the pathogen (viruses, bacteria, fungi, or other parasites which cause diseases) or on cancer cells, etc., in order to work (see figures below).

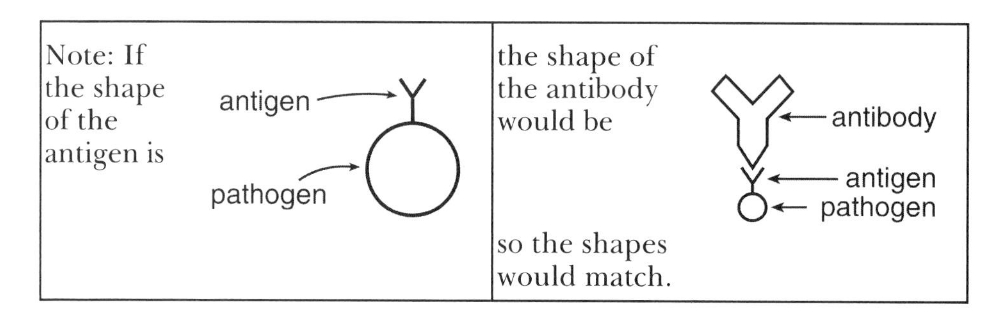

Note: If the shape of the antigen is
antigen
pathogen
the shape of the antibody would be
antibody
antigen
pathogen
so the shapes would match.

Summary: In short, in the immune system, some **white blood cells attack** the **pathogen** and **kill it**; **other white blood cells produce specific antibodies** which attack a specific pathogen (example polio virus) or mark the pathogen for a killing done by other white blood cells. The immune system has white blood cells and antibodies that fight disease. After the person has the disease, most of the white blood cells and antibodies break down. A **few white blood cells** are **left in** the **body** and will **protect against** the **same specific disease in** the **future.**

Note: All of our cells have antigens on them. The antigens only on foreign substances (virus, bacteria) are attacked by white blood cells and antibodies because they are foreign (not belonging to our bodies).

Immunity to a specific disease

After a person gets the disease (example chicken pox), a few white blood cells (which only fight chicken pox) remain in the body and can produce more white blood cells and antibodies to fight off a new attack (invasion) of chicken pox. The antibodies are **specific** for the chicken pox virus and only work on the chicken pox virus. The antibody against the chicken pox virus has a **specific shape** that **fits into the shape** of the **antigen** on the **chicken pox virus** and can **only work against** the **chicken pox virus**, not against measles, polio, etc. This gives the person immunity (protection) against a specific disease (in this case chicken pox), even years later, but not against other diseases.

After a person has had any disease (example measles) a few white blood cells remain in the blood, which can produce more white blood cells and antibodies to fight that specific disease in the future, even ten or fifteen years later.

Vaccination

You got many vaccinations (commonly called shots) to protect you from various diseases (to prevent you from getting various diseases). Examples of some vaccinations are polio vaccine and MMR vaccine.

A **vaccine** is **made of weakened, killed**, or **parts of pathogens** or **microbes** (microscopic (tiny) organisms such as bacteria or viruses) that are injected (injections) into the body or swallowed. The vaccine stimulates the body to produce white blood cells which **produce antibodies** to **fight** a specific **microbe** (such as a specific bacteria or virus) and prevent you from getting a specific disease.

The vaccine can have weakened bacteria or other microbes. The antigen on the **bacteria**, virus, or other microbe has a **specific shape**. The antigen on the bacteria causes the production of white blood cells that make **antibodies** that **match the shape of** the (antigens on the) **bacteria** or other microbes. Since the antibodies have a specific shape, they can only protect against this bacteria, which has a specific shape. A vaccine only protects against one specific disease (or you can say, against one specific microbe, such as one specific virus or one specific bacteria).

As you see, **vaccinations** use **weakened microbes** (or parts of microbes) **to stimulate** the **immune system to react** (to stimulate the body to produce white blood cells which **produce antibodies** to **fight** the **microbe**). These antibodies remain in the blood and will protect you from future attacks (invasions) of the same microbe (example bacteria). Vaccination for a disease (example chicken pox) will only prevent chicken pox, by causing the body to produce (make) specific white blood cells and antibodies to fight against future attacks of chicken pox. The person does not become sick from a vaccine (vaccination) because the vaccine has weakened or dead pathogen (example virus or bacteria).

Now Do Homework Questions #110-130, pages 58-61.

Immune system diseases

The immune system has white blood cells and antibodies. **AIDS** and **some** other **viral diseases** (diseases caused by a virus) **damage** (weaken) the **immune system** (damage the white blood cells), then the body **cannot fight** many (multiple) infectious diseases (which are caused by infectious agents such as viruses or bacteria) and cancer, therefore the person gets very sick and may even die.

AIDS

Cause	Prevention (control)	Diagnosis
HIV (a virus)	do not share needles avoid transfer of body fluids	blood is tested for the AIDS virus; the test detects antibodies to the virus.

Allergy: immune system response to harmless substances

When a person has an **allergic reaction** (allergy), the body's **immune system responds** to usually harmless environmental substances such as **pollen, certain foods, chemicals** from insect bites, etc., just like it would respond to bacteria or viruses; the white blood cells produce antibodies against antigens on pollen, foods, etc. When a person has an allergy, the body produces histamines which can cause runny nose, rashes, swelling, etc.

People with allergies take antihistamines to take away the effect of the histamine, therefore lessening or taking away the runny nose, rashes or swelling.

Immune system attacks on body cells or transplanted organs

Sometimes the **immune system** (white blood cells and antibodies) **attacks** some of the **body's own cells.** In one type of diabetes, the immune system attacks and destroys the pancreas cells which produce insulin.

Similarly, the **immune system** also **attacks transplanted organs**. The transplanted organs (which come from another person) have **foreign** antigens on them, therefore the immune system attacks the transplanted organs.

When a person gets a transplant, he usually gets drugs to reduce the effect of the immune system, so the immune system will not attack the transplant. However, since his immune system is weakened (by the drugs), he is more likely to get sick (example from viruses, bacteria) than a person with a normal immune system.

Causes of disease

There are many causes of disease. As you learned, a **disease** is a **disturbance (disruption, failure)** in **homeostasis.** If the disease is not cured (disruption in homeostasis is not corrected or responded to), in some cases it can even cause death.

CAUSE	EXAMPLE	PREVENTION	TREATMENT
Heredity (inheritance)	Down's Syndrome, PKU, Tay-Sachs	Genetic testing	
Toxic substances	lead poisoning	avoid eating lead paint;	treatment to remove lead from blood
Poor nutrition	scurvy	eat fruits	eat fruits
Organ malfunction	diabetes	limit eating sugar	insulin injections
	heart attack (blood clot in heart)	avoid stress; low-fat diet	drugs to dissolve clots; bypass surgery
Personal behavior	smoking causes lung cancer	stop smoking	
	high-fat diet causes heart disease	limit fats in diet	
Gene mutation	cancer		chemotherapy, radiation

The nucleus contains chromosomes which have genes (you will learn about this in the next chapter). Gene **mutation** (a change in a gene) can **cause cancer**. Cancer is a disease in which the cells keep dividing (uncontrolled cell division), forming more and more abnormal cells. Exposure of cells to certain chemicals or radiation increases mutations (change in genes) and therefore increases the chance of cancer. The immune system can attack and kill the cancer cells. If the immune cells cannot kill the cancer cells, the person can die.

Some diseases (effects of diseases) show up right away, for example birth defects; other diseases (effects of diseases) show up years later, for example lung cancer is caused by years of cigarette smoking.

Now Do Homework Questions #131-137, pages 61-63.

Biological research

Biology research gives us knowledge and helps us find (design) better ways to diagnose, prevent, treat, control, and cure diseases in plants and animals.

Drawing Graphs

Let's see how we can draw graphs based on experimental data.

Problem 1:

The laboratory experiment was to see the effect of time on cellular respiration in yeast. Yeast-glucose solution is in the flask at 35°C. The student counted and recorded the number of gas bubbles in the test tube every five minutes.

Data Table

Time (minutes)	Total Number of Bubbles Released 35°
5	5
10	15
15	30
20	50
25	75

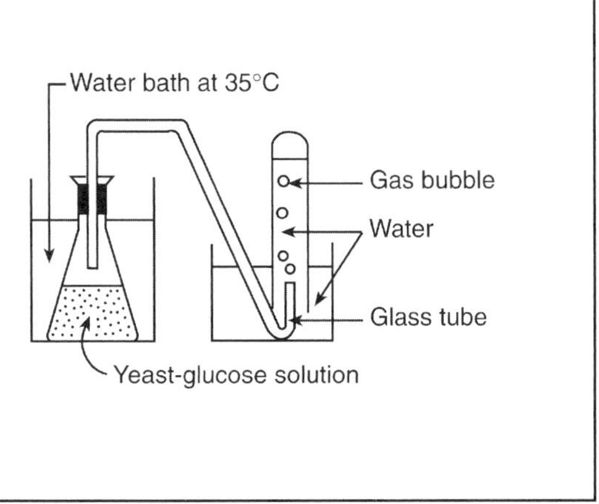

The Effect of Time on Respiration in Yeast

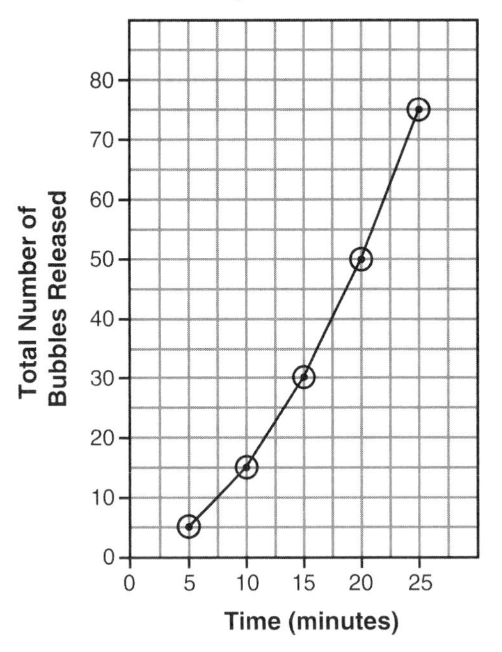

How to draw the line graph:

1. On the x axis, put "Time, minutes". Always include units (in this example minutes) on the axis (see graph). The **thing you change** (in this case time) is always put on the **x axis.** This is the independent variable. Space the lines along the axis equally. "Make an appropriate scale" by spacing the numbers on the graph so that all the data fits on the graph and it is easy to read. There must be an equal number

of minutes between lines (see graph). On the x axis, put five minutes between lines (every two lines) (scale on the x axis), then all the time on the data table fits on the graph and it is easy to read.

2. On the y axis, put "Total Number of Bubbles Released". The **result** you get (total number of bubbles released) is always put on the **y axis**. This is the dependent variable. Space the lines along the axis equally. "Make an appropriate scale" by spacing the numbers on the graph so all the data fits on the graph and it is easy to read. There must be an equal number of bubbles between lines (see graph.) On the y axis, put ten bubbles between lines (every two lines) (scale on the y axis) then all the bubbles on the data table fit on the graph and it is easy to read.

3. Plot the experimental data on the graph. Draw a circle around each point. Draw a line that connects the points. Do not continue the line past the last point.

4. Put a title on the graph which shows what the graph is about. Example: "Effect of time on respiration rate in yeast."

Problem 2:

Let's do a different experiment and compare the number of gas bubbles released at different temperatures. See the effect of temperature on cellular respiration in yeast. (You learned in **cellular respiration** glucose unites with oxygen, producing water, **carbon dioxide gas**, and ATP. Glucose + oxygen → carbon dioxide + water+ ATP).

One flask of yeast-glucose solution is at 20°C; the second flask of yeast-glucose solution is at 35°C. The student counted and recorded the number of gas bubbles at the two different temperatures every five minutes.

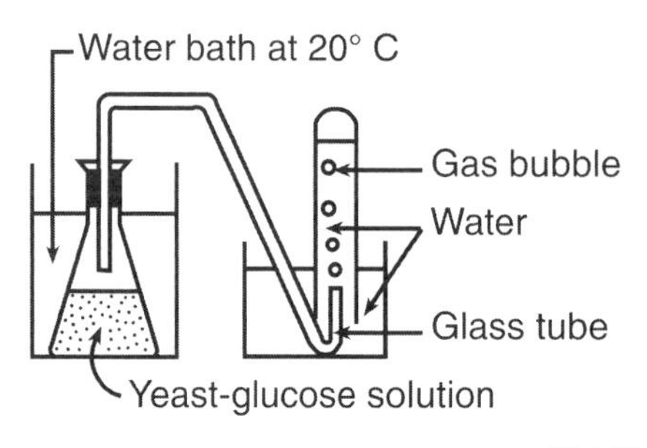

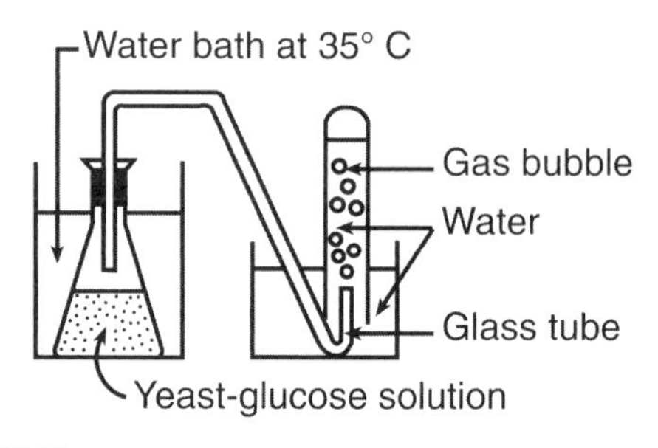

DATA TABLE

Time (minutes)	**Total Number of Bubbles Released**	
	20°C	35°C
5	0	5
10	5	15
15	15	30
20	30	50
25	45	75

Using the information in the data table, construct a line graph on the grid:

1. Look at the graph on the next page. On the x axis, put "Time, minutes". Always include units (in this example minutes) on the axis (see graph). The **thing you change** (in this case time) is always put on the **x axis.** This is the independent variable. Space the lines along the axis equally. "Make an appropriate scale" by spacing the numbers on the graph so that all the data fits on the graph and it is easy to read. There must be an equal number of minutes between lines (see graph). On the x axis, put five minutes between lines (every two lines) (scale on the x axis), then all the time on the data table fits on the graph and it is easy to read.

2. Look at the graph on the next page. On the y axis, put "Total Number of Bubbles Released". The **result** you get (total number of bubbles released) is always put on the **y axis**. This is the dependent variable. Space the lines along the axis equally. "Make an appropriate scale" by spacing the numbers on the graph so all the data fits on the graph and it is easy to read. There must be an equal number of bubbles between lines (see graph.) On the y axis, put ten bubbles between lines (every two lines) (scale on the y axis) then all the bubbles on the data table fit on the graph and it is easy to read.

3. Plot the data for the total number of bubbles released at 20°C on the graph. Surround each point with a small triangle and connect the points with a line. Do not continue the line past the last data point.

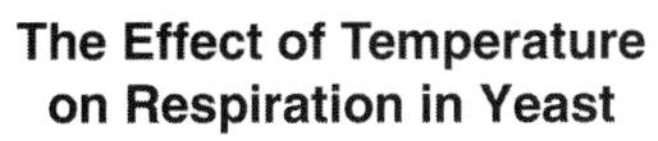

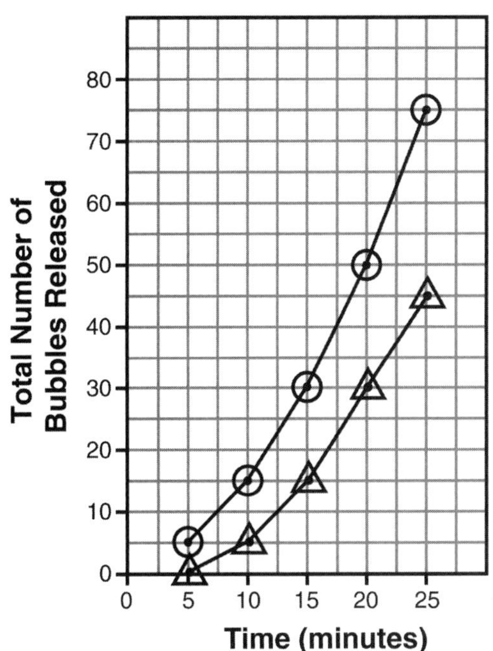

Key
- Yeast respiration at 20°C
- Yeast respiration at 35°C

4. Plot the data for the total number of bubbles released at 35°C on the graph. Surround each point with a small circle and connect the points with a line. Do not continue the line past the last data point.

5. Put a title on the graph which shows what the graph is about. Example: "Effect of temperature on respiration rate in yeast."

As you can see from the graph, as time increases, the number of (gas) bubbles released increases, or, you can say, rate of gas production increases.

You know the gas bubbles produced are carbon dioxide, because, in cellular respiration, glucose unites with oxygen, producing water, carbon dioxide gas, and ATP.

Now Do Homework Questions #138-147, pages 63-66.

NOW LET'S TRY A FEW HOMEWORK QUESTIONS

To assign fewer questions, you may want to skip the questions with underlined numbers.

<u>1</u>. Carbon exists in a simple organic molecule in a leaf and in an inorganic molecule in the air humans exhale. Identify the simple organic molecule formed in the leaf and the process that produces it.

<u>2</u>. The graph below shows the results of an experiment in which a container of oxygen-using bacteria and strands of a green alga were exposed to light of different colors.

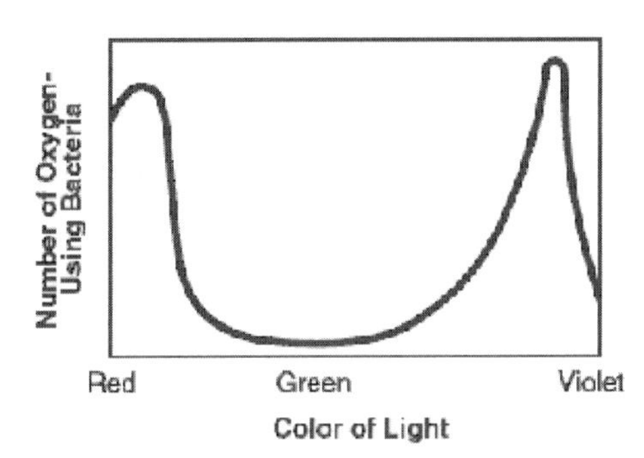

Which statement best explains the results of this experiment?
(1) The rate of photosynthesis is affected by variations in the light.
(2) In all environments light is a vital resource.
(3) The activities of bacteria and algae are not related.
(4) Uneven numbers and types of species can upset ecosystem stability.

<u>3</u>. Which process usually uses carbon dioxide molecules?
(1) cellular respiration
(2) asexual reproduction
(3) active transport
(4) autotrophic nutrition

4. The diagram below represents events associated with a biochemical process that occurs in some organisms.

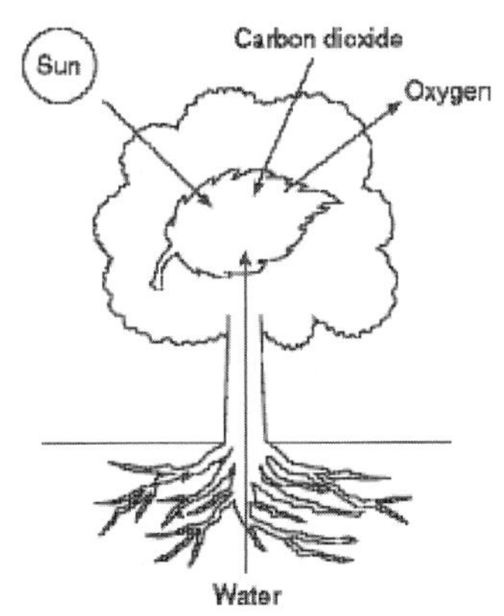

Which statement concerning this process is correct?
(1) The process represented is respiration and the primary source of energy for the process is the Sun.
(2) The process represented is photosynthesis and the primary source of energy for the process is the Sun.

(3) This process converts energy in organic compounds into solar energy which is released into the atmosphere.
(4) This process uses solar energy to convert oxygen into carbon dioxide.

5. The diagram below illustrates the movement of materials involved in a process that is vital for the energy needs of organisms.

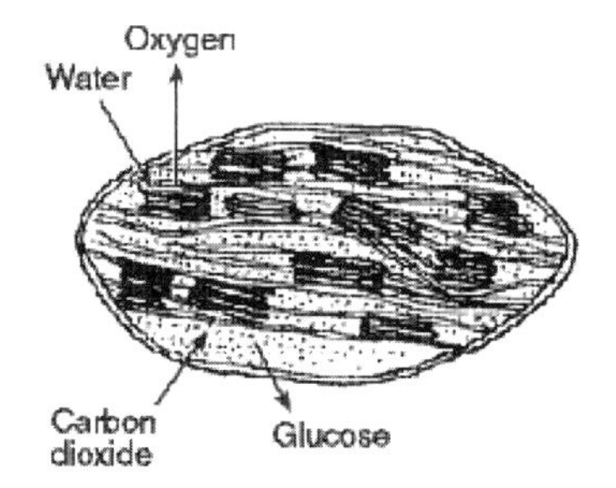

The process illustrated occurs within
(1) chloroplasts
(2) mitochondria
(3) ribosomes
(4) vacuoles

6. The diagrams below represent two different cells and some of their parts. The diagrams are not drawn to scale.

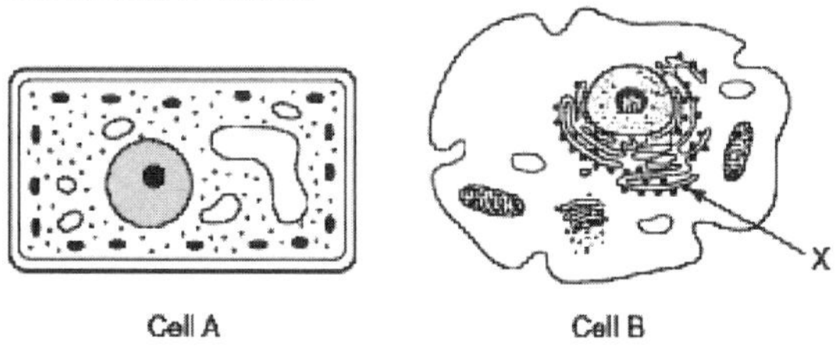

Identify an organelle in cell *A* that is the site of autotrophic nutrition.

7. An experimental setup is shown below.

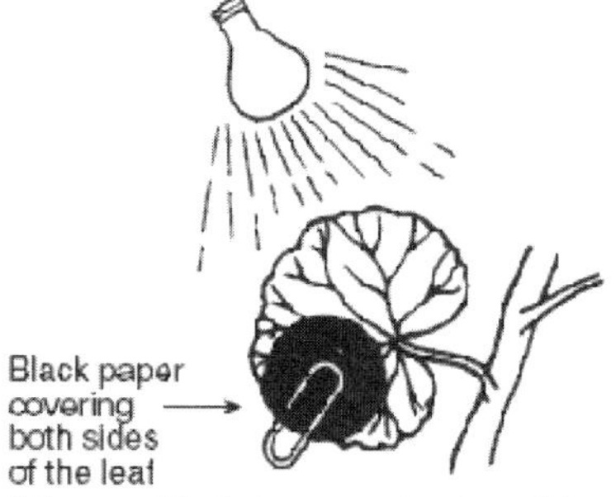

Which hypothesis would most likely be tested using this setup?
(1) Light is needed for the process of reproduction.
(2) Glucose is not synthesized by plants in the dark.
(3) Protein synthesis takes place in leaves.
(4) Plants need fertilizers for proper growth.

8. In the transfer of energy from the Sun to ecosystems, which molecule is one of the first to store this energy?
(1) protein (2) fat (3) DNA (4) glucose

9. The diagram below represents the setup for an experiment. Two black paper discs are opposite each other on both sides of each of two leaves.

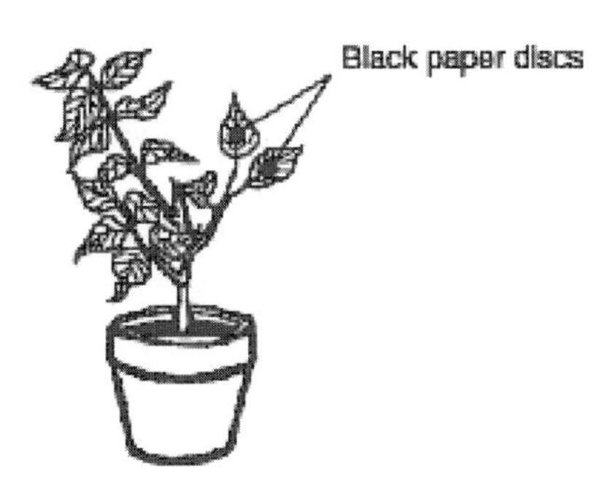

This experimental setup would most likely be used to show that

(1) glucose is necessary for photosynthesis
(2) protein is a product of photosynthesis
(3) light is necessary for photosynthesis
(4) carbon dioxide is a product of photosynthesis

10. The green aquatic plant represented in the diagram below was exposed to light for several hours.

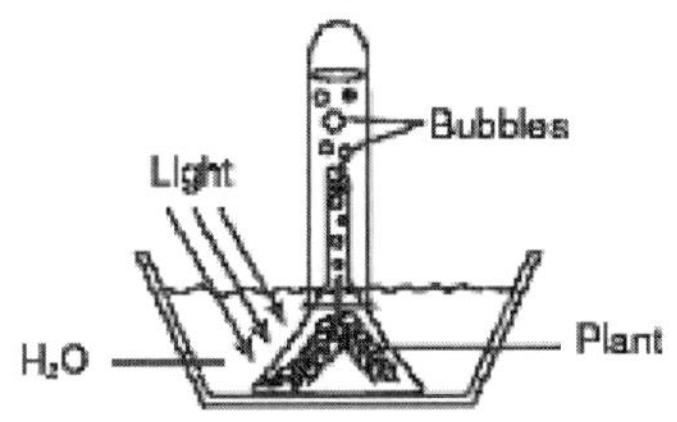

Which gas would most likely be found in the greatest amount in the bubbles?

(1) oxygen
(2) nitrogen
(3) ozone
(4) carbon dioxide

11. Based on the experimental setup below, what could be the hypothesis in this experiment?.

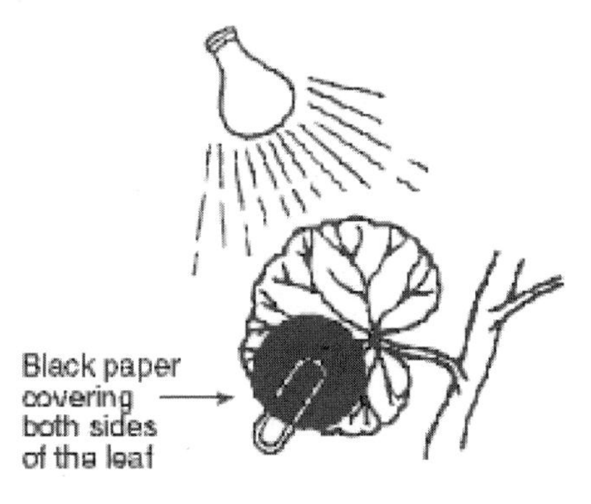

(1) Protein synthesis does not take place in leaves in the dark.
(2) Glucose is not synthesized by plants in the dark.
(3) Photosynthesis only takes place in the dark.
(4) Light is necessary for fat synthesis in plants.

12. Plants in areas with short growing seasons often have more chloroplasts in their cells than plants in areas with longer growing seasons. Compared to plants in areas with longer growing seasons, plants in areas with shorter growing seasons most likely
   (1) make and store food more quickly
   (2) have a higher rate of protein metabolism
   (3) grow taller
   (4) have a different method of respiration

13. Organisms that have the ability to use an atmospheric gas to produce an organic nutrient are known as
   (1) herbivores
   (2) decomposers
   (3) carnivores
   (4) autotrophs

14. To determine which colors of light are best used by plants for photosynthesis, three types of underwater green plants of similar mass were subjected to the same intensity of light of different colors for the same amount of time. All other environmental conditions were kept the same. After 15 minutes, a video camera was used to record the number of bubbles of gas each plant gave off in a 30-second period of time. Each type of plant was tested six times. The average of the data for each plant type is shown in the table below.

Average Number of Bubbles Given Off in 30 Seconds

Plant Type	Red Light	Yellow Light	Green Light	Blue Light
Elodea	35	11	5	47
Potamogeton	48	8	2	63
Utricularia	28	9	6	39

Which statement is a valid inference based on the data?
   (1) Each plant carried on photosynthesis best in a different color of light.
   (2) Red light is better for photosynthesis than blue light.
   (3) These types of plants make food at the fastest rates with red and blue light.
   (4) Water must filter out red and green light.

15. A five-year study was carried out on a population of algae in a lake. The study found that the algae population was steadily decreasing in size. Over the five-year period this decrease most likely led to
   (1) a decrease in the amount of nitrogen released into the atmosphere
   (2) an increase in the amount of oxygen present in the lake
   (3) an increase in the amount of water vapor present in the atmosphere
   (4) a decrease in the amount of oxygen released into the lake

16. A biologist used the Internet to contact scientists around the world to obtain information about declining amphibian populations. He was able to gather data on 936 populations of amphibians, consisting of 157 species from 37 countries. Results showed that the overall numbers of amphibians dropped 15% a year from 1960 to 1966 and continued to decline about 2% a year through 1997.

   What is the importance of collecting an extensive amount of data such as this?
   (1) Researchers will now be certain that the decline in the amphibian populations is due to pesticides.
   (2) The data collected will prove that all animal populations around the world are threatened.

(3) Results from all parts of the world will be found to be identical.
(4) The quantity of data will lead to a better understanding of the extent of the problem.

17. In 1883, Thomas Engelmann, a German botanist, exposed a strand of algae to different wavelengths of light. Engelmann used bacteria that concentrate near an oxygen source to determine which sections of the algae were releasing the most O_2. The results are shown below.

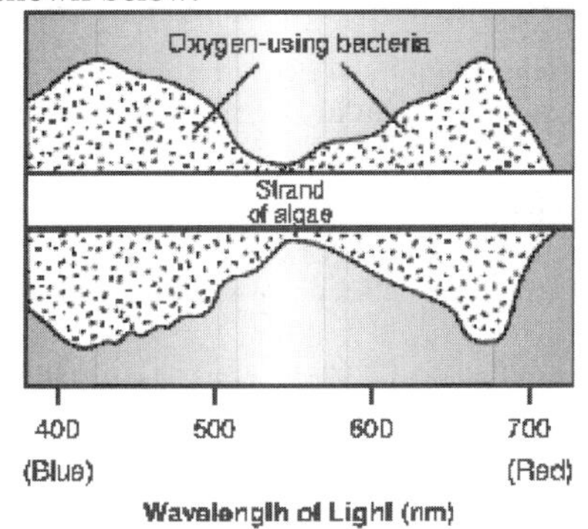

Which statement is a valid inference based on this information?
(1) Oxygen production decreases as the wavelength of light increases from 550 to 650 nm.
(2) Respiration rate in the bacteria is greatest at 550 nm.
(3) Photosynthetic rate in the algae is greatest in blue light.
(4) The algae absorb the greatest amount of oxygen in red light.

Base your answers to the next two questions on the information and diagram below and on your knowledge of biology.

A small water plant (elodea) was placed in bright sunlight for five hours as indicated below. Bubbles of oxygen gas were observed being released from the plant.

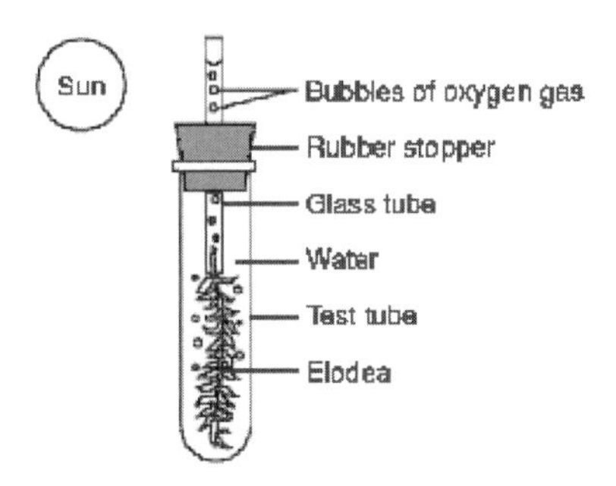

18. Since oxygen gas is being released, it can be inferred that the plant is
(1) producing glucose
(2) making protein
(3) releasing energy from water
(4) carrying on active transport

19. What substance did the plant most likely absorb from the water for the process that produces the oxygen gas?
(1) dissolved nitrogen
(2) carbon dioxide
(3) an enzyme
(4) a hormone

Base your answers to the next two questions on the information below and on your knowledge of biology.

A biology student was given three unlabeled jars of pond water from the same source, each containing a different type of mobile unicellular organism: euglena, ameba, and paramecium. The only information the student has is that the ameba and paramecium are both heterotrophs and the euglena can be either heterotrophic or autotrophic, depending on its environment.

20. State ***one*** way the euglena's two methods of nutrition provide a survival advantage the other unicellular organisms do ***not*** have.

21. Which procedure and resulting observation would help identify the jar that contains the euglena?
    (1) Expose only one side of each jar to light. After 24 hours, only in the jar containing euglena will most of organisms be seen on the darker side of the jar.
    (2) Expose all sides of each jar to light. After 48 hours, the jar with the highest dissolved carbon dioxide content will contain the euglena.
    (3) Over a period of one week, determine the method of reproduction used by each type of organism. If mitotic cell division is observed, the jar will contain euglena.
    (4) Prepare a wet-mount slide of specimens from each jar and observe each slide with a compound light microscope. Only the euglena will have chloroplasts.

---

22. The diagram below represents an autotrophic cell.

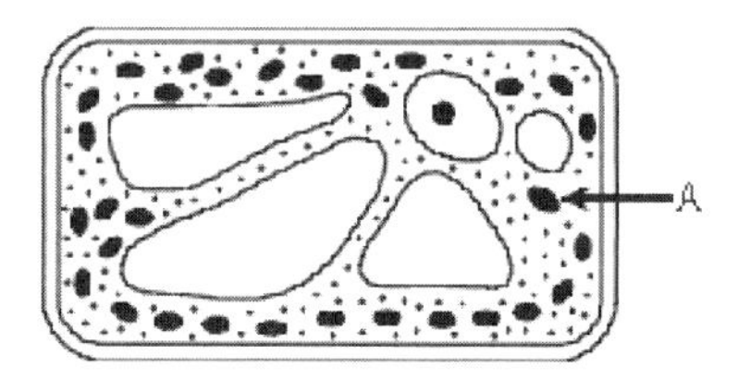

For the process of autotrophic nutrition, the arrow labeled ***A*** would most likely represent the direction of movement of
    (1) carbon dioxide, water, and solar energy
    (2) oxygen, glucose, and solar energy
    (3) carbon dioxide, oxygen, and heat energy
    (4) glucose, water, and heat energy

23. When organisms break the bonds of organic compounds, the organisms can
    (1) use the smaller molecules to plug the gaps in the cell membrane to slow diffusion
    (2) use the energy obtained to digest molecules produced by respiration that uses oxygen
    (3) obtain energy or reassemble the resulting materials to form different compounds
    (4) excrete smaller amounts of solid waste materials during vigorous exercise

24. The diagrams below represent two different cells and some of their parts. The diagrams are not drawn to scale.

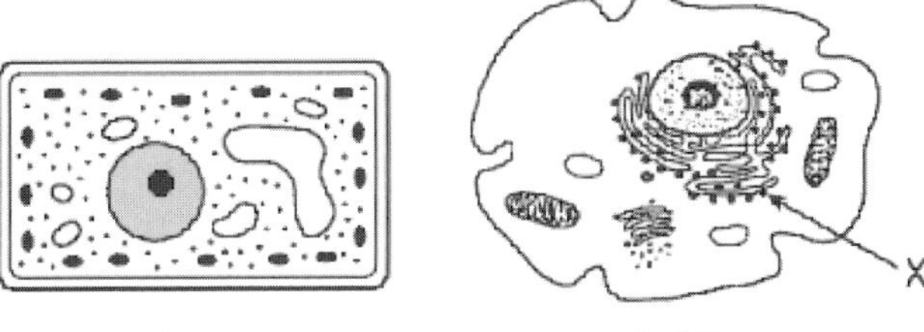

Cell A   Cell B

Which statement best describes these cells?

(1) Cell ***B*** lacks vacuoles while cell ***A*** has them.
(2) DNA would not be found in either cell ***A*** or cell ***B***.
(3) Both cell ***A*** and cell ***B*** use energy released from ATP.
(4) Both cell ***A*** and cell ***B*** produce antibiotics.

25. Which group contains only molecules that are each assembled from smaller organic compounds?

(1) proteins, water, DNA, fats
(2) proteins, starch, carbon dioxide, water
(3) proteins, DNA, fats, starch
(4) proteins, carbon dioxide, DNA, starch

26. In plants, simple sugars are least likely to be

(1) linked together to form proteins
(2) broken down into carbon dioxide and water
(3) used as a source of energy
(4) stored in the form of starch molecules

27. The diagram below represents the synthesis of a portion of a complex molecule in an organism.

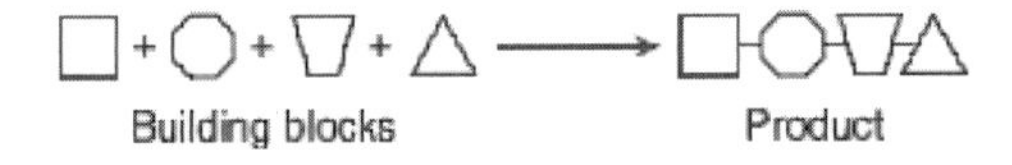

Which row in the chart could be used to identify the building blocks and product in the diagram?

Row	Building Blocks	Product
(1)	starch molecules	glucose
(2)	amino acid molecules	part of protein
(3)	sugar molecules	ATP
(4)	DNA molecules	part of starch

28. Which statement describes starches, fats, proteins, and DNA?

(1) They are used to store genetic information.
(2) They are complex molecules made from smaller molecules.
(3) They are used to assemble larger inorganic materials.
(4) They are simple molecules used as energy sources.

29. Which statement best describes cellular respiration?

(1) It occurs in animal cells but not in plant cells.
(2) It converts energy in food into a more usable form.

(3) It uses carbon dioxide and produces oxygen.
(4) It stores energy in food molecules.

30. All life depends on the availability of usable energy. This energy is released when
(1) organisms convert solar energy into the chemical energy found in food molecules
(2) respiration occurs in the cells of producers and high-energy molecules enter the atmosphere
(3) cells carry out the process of respiration
(4) animal cells synthesize starch and carbon dioxide

31. Which substance is the most direct source of the energy that an animal cell uses for the synthesis of materials?
(1) ATP (2) glucose (3) DNA (4) starch

32. The graphs below show the changes in the relative concentrations of two gases in the air surrounding a group of mice.

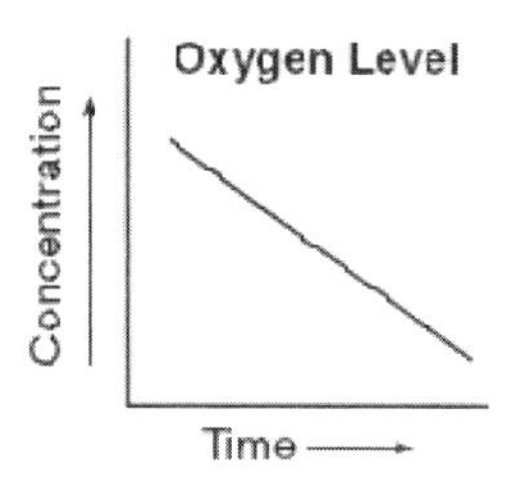

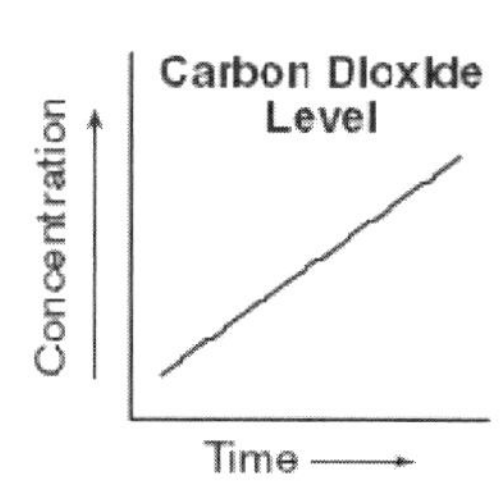

Which process in the mice most likely accounts for the changes shown?
(1) active transport (3) respiration
(2) evaporation (4) photosynthesis

33. Base your answers to the next question on the information and diagram below and on your knowledge of biology.

Two test tubes, ***A*** and ***B***, were set up as shown in the diagram below. Bromthymol blue, which turns from blue to yellow in the presence of carbon dioxide, was added to the water at the bottom of each tube before the tubes were sealed. The tubes were maintained at the temperatures shown for six days. (Average room temperature is 20°C.)

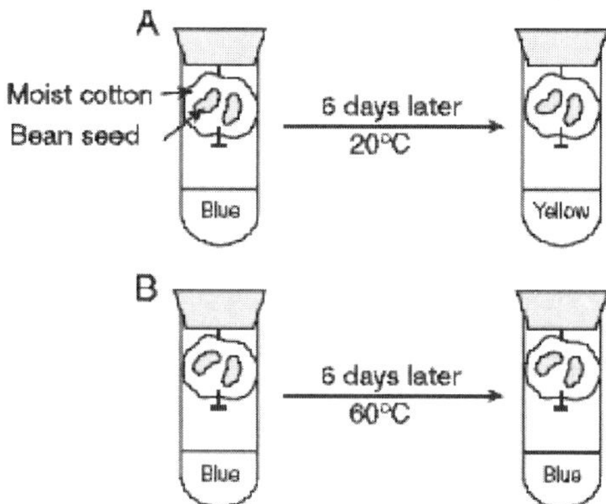

Identify the life process responsible for the change in tube ***A***.

34. The diagram below represents a structure involved in cellular respiration.

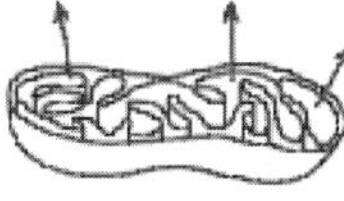

The release of which substance is represented by the arrows?

(1) glucose
(2) oxygen
(3) carbon dioxide
(4) DNA

35. Base your answers to the next question on the diagram below, which illustrates a transport pathway of CO_2 in the human body, and on your knowledge of biology.

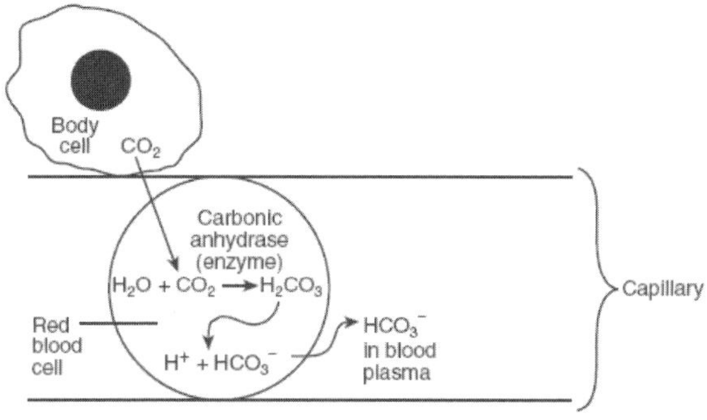

Identify the cellular process that most likely produced the CO_2 in the body cell.

36. The production of energy-rich ATP molecules is the direct result of

(1) recycling light energy to be used in the process of photosynthesis
(2) releasing the stored energy of organic compounds by the process of respiration
(3) breaking down starch by the process of digestion
(4) copying coded information during the process of protein synthesis

37. The arrows in the diagram below indicate the movement of materials into and out of a single-celled organism.

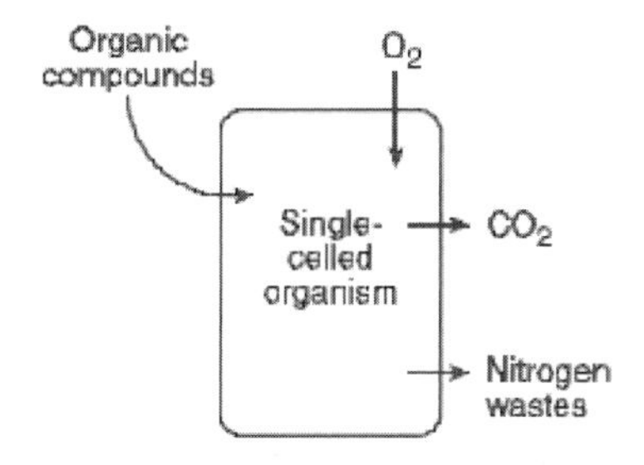

The movements indicated by all the arrows are directly involved in

(1) the maintenance of homeostasis
(2) respiration, only
(3) excretion, only
(4) the digestion of proteins

38. The energy released in this process was originally present in

(1) sunlight and then transferred to sugar
(2) sunlight and then transferred to oxygen
(3) the oxygen and then transferred to sugar
(4) the sugar and then transferred to oxygen

39. Base your answers to the next question on the diagram below, which illustrates a transport pathway of CO_2 in the human body, and on your knowledge of biology.

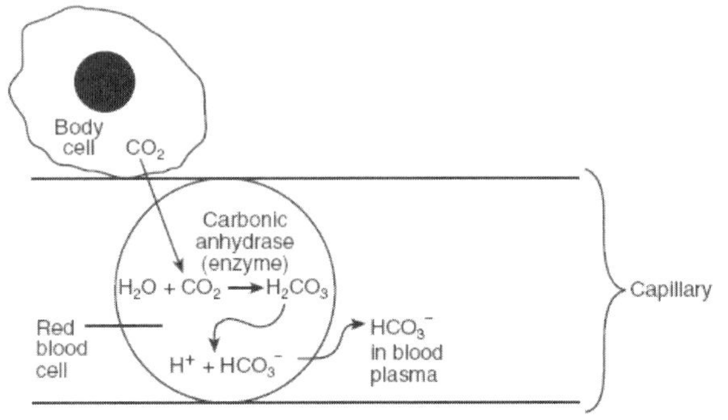

Explain why carbon dioxide moves into red blood cells by diffusion rather than by active transport.
Hint: Review Diffusion in Chapter 1.

40. Which set of terms best identifies the letters in the diagram below?

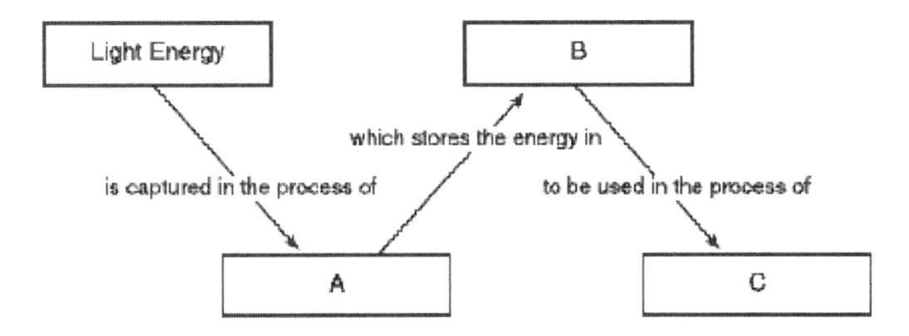

	A	B	C
(1)	photosynthesis	inorganic molecules	decomposition
(2)	respiration	organic molecules	digestion
(3)	photosynthesis	organic molecules	respiration
(4)	respiration	inorganic molecules	photosynthesis

41. The diagrams below represent two different cells and some of their parts. The diagrams are not drawn to scale.

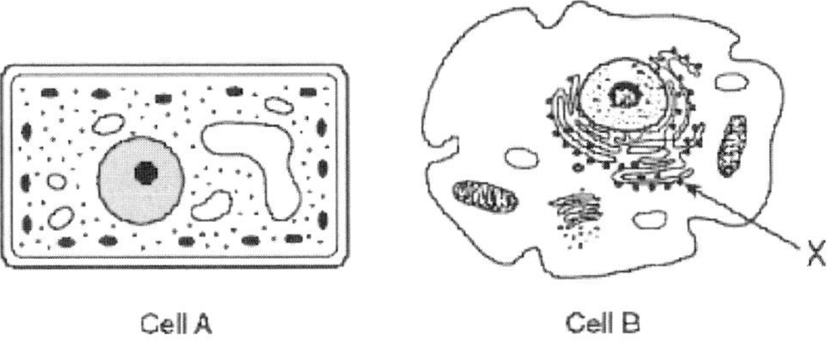

Which statement best describes these cells?

(1) Cell ***B*** lacks vacuoles while cell ***A*** has them.
(2) DNA would not be found in either cell ***A*** or cell ***B***.

(3) Both cell *A* and cell *B* use energy released from ATP.
(4) Both cell *A* and cell *B* produce antibiotics.

42. Information concerning a metabolic activity is shown below.

$$X \xrightarrow{\text{enzyme}} \text{products + energy for metabolism}$$

Substance *X* is most likely
(1) DNA (2) oxygen (3) ATP (4) chlorophyll

Base your answers to the next three questions on the information and diagrams below and on your knowledge of biology.

The laboratory setups represented below were used to investigate the effect of temperature on cellular respiration in yeast (a single-celled organism). Each of two flasks containing equal amounts of a yeast-glucose solution was submerged in a water bath, one kept at 20°C and one kept at 35°C. The number of gas bubbles released from the glass tube in each setup was observed and the results were recorded every 5 minutes for a period of 25 minutes. The data are summarized in the table below.

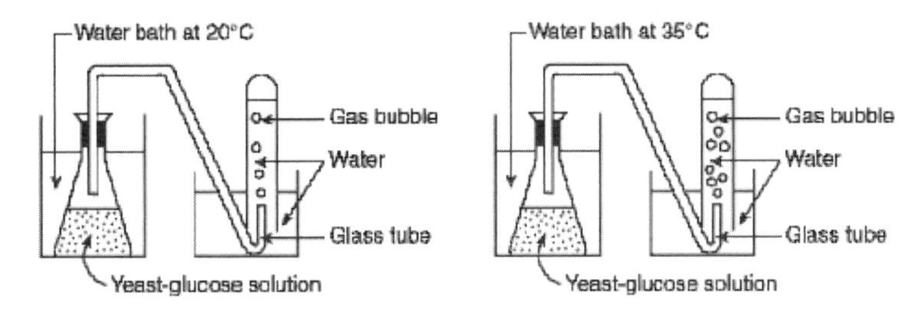

Data Table

Time (minutes)	Total Number of Bubbles Released	
	20°C	35°C
5	0	5
10	5	15
15	15	30
20	30	50
25	45	75

Directions for the next three questions: Using the information in the data table, construct a line graph on the grid provided, following the directions below.

43. Mark an appropriate scale on each axis.

44. Plot the data for the total number of bubbles released at 20°C on the grid provided. Surround each point with a small circle and connect the points.

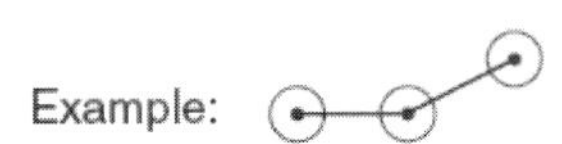

45. Plot the data for the total number of bubbles released at 35°C on the grid. Surround each point with a small triangle and connect the points.

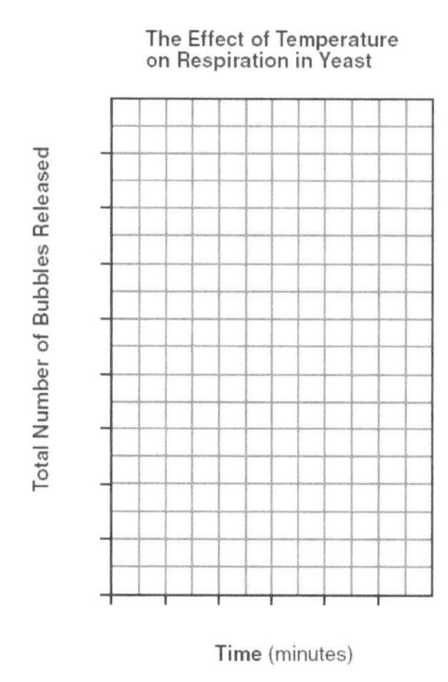

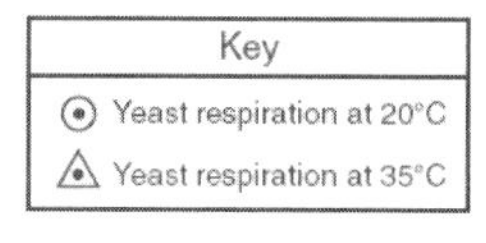

---

46. Arrows ***A***, ***B***, and ***C*** in the diagram below represent the processes necessary to make the energy stored in food available for muscle activity.

Food —A→ Simpler molecules —B→ Mitochondria —C→ ATP in muscle cells

The correct sequence of processes represented by ***A***, ***B***, and ***C*** is
(1) diffusion → synthesis → active transport
(2) digestion → diffusion → cellular respiration
(3) digestion → excretion → cellular respiration
(4) synthesis → active transport → excretion

47. The arrows in the diagram below represent biological processes.

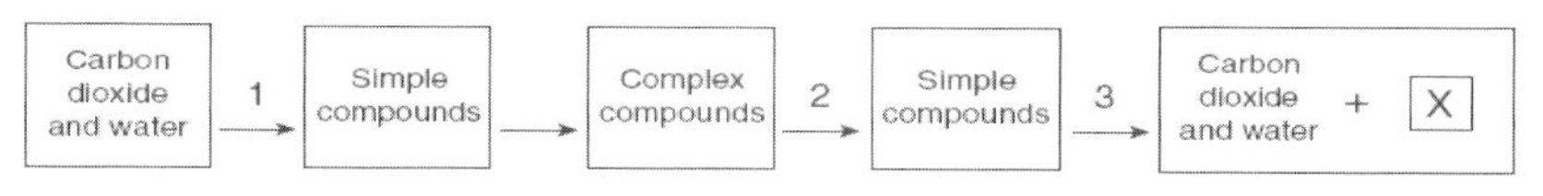

Identify ***one*** type of organism that carries out process ***1***.

48. Certain poisons are toxic to organisms because they interfere with the function of enzymes in mitochondria. This results directly in the inability of the cell to
(1) store information
(2) build proteins
(3) release energy from nutrients
(4) dispose of metabolic wastes

49. In what way are photosynthesis and cellular respiration similar?
(1) They both occur in chloroplasts.
(2) They both require sunlight.

(3) They both involve organic and inorganic molecules.
(4) They both require oxygen and produce carbon dioxide.

50. A marathon runner frequently experiences muscle cramps while running. If he stops running and rests, the cramps eventually go away. The cramping in the muscles most likely results from
(1) lack of adequate oxygen supply to the muscle
(2) the runner running too slowly
(3) the runner warming up before running
(4) increased glucose production in the muscle

51. The direct source of ATP for the development of the muscle of the arm is chemical activities that take place in
(1) mitochondria of the head
(2) mitochondria of the arm
(3) ribosomes of the arm
(4) chloroplasts

52. Which statement describes the ecosystem represented in the diagram below?

(1) This ecosystem would be the first stage in ecological succession.
(2) This ecosystem would most likely lack decomposers.
(3) All of the organisms in this ecosystem are producers.
(4) All of the organisms in this ecosystem depend on the activities of biological catalysts.

53. Enzyme molecules normally interact with substrate molecules. Some medicines work by blocking enzyme activity in pathogens. These medicines are effective because they
(1) are the same size as the enzyme
(2) are the same size as the substrate molecules
(3) have a shape that fits into the enzyme
(4) have a shape that fits into all cell receptors

54. Experiments revealed the following information about a certain molecule:

— It can be broken down into amino acids.
— It can break down proteins into amino acids.
— It is found in high concentrations in the small intestine of humans.

This molecule is most likely
(1) an enzyme
(2) an inorganic compound
(3) a hormone
(4) an antigen

Base your answers to the next three questions on the diagram below, which illustrates a transport pathway of CO_2 in the human body, and on your knowledge of biology.

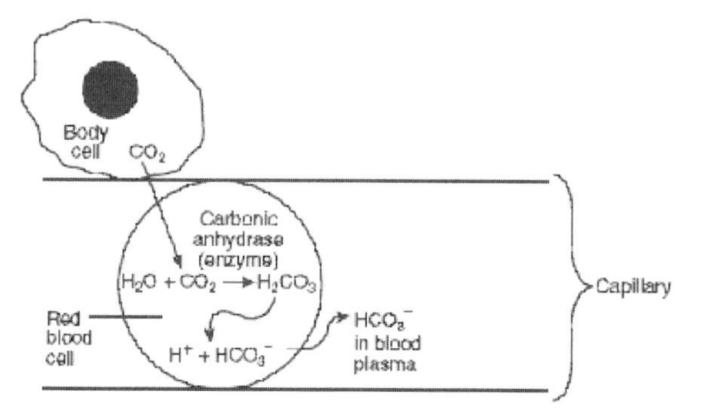

55. Identify the cellular process that most likely produced the CO_2 in the body cell.

56. Explain why carbon dioxide moves into red blood cells by diffusion rather than by active transport.

57. State what would happen to the production of bicarbonate ions (HCO_3^-) if the carbonic anhydrase were not present in red blood cells.

58. Base your answers to the next question on the graph below and on your knowledge of biology.

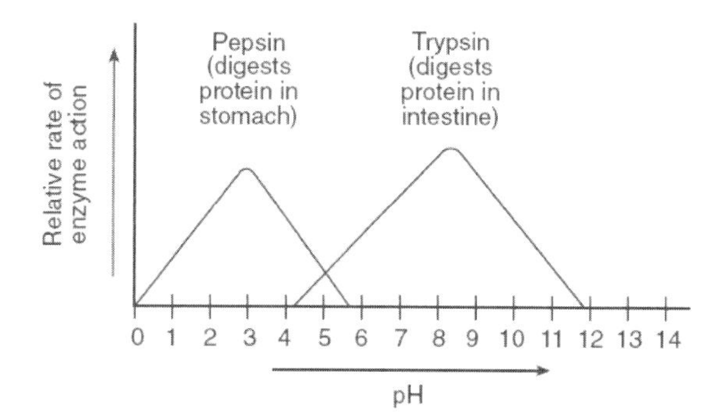

Neither enzyme works at a pH of

(1) 1 (2) 5 (3) 3 (4) 13

59. In many investigations, both in the laboratory and in natural environments, the pH of substances is measured. Explain why pH is important to living things. In your explanation be sure to:

- identify ***one*** example of a life process of an organism that could be affected by a pH change
- state ***one*** environmental problem that is directly related to pH
- identify ***one*** possible cause of this environmental problem

Base your answers to the next two questions on the statement below and on your knowledge of biology.

Some internal environmental factors may interfere with the ability of an enzyme to function efficiently.

60. Identify ***two*** internal environmental factors that directly influence the rate of enzyme action.

61. Explain why changing the shape of an enzyme could affect the ability of the enzyme to function.

62. The graph below shows the effect of temperature on the relative rate of action of enzyme ***X*** on a protein.

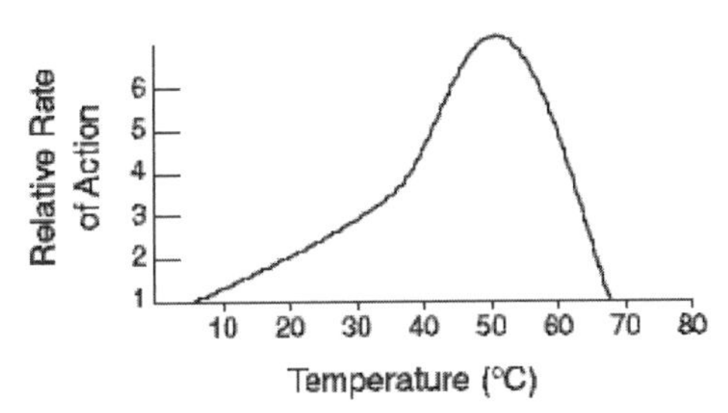

Which change would not affect the relative rate of action of enzyme ***X***?

(1) the addition of cold water when the reaction is at 50°C
(2) an increase in temperature from 70°C to 80°C
(3) the removal of the protein when the reaction is at 30°C
(4) a decrease in temperature from 40°C to 10°C

63. The diagram below represents a beaker containing a solution of various molecules involved in digestion.

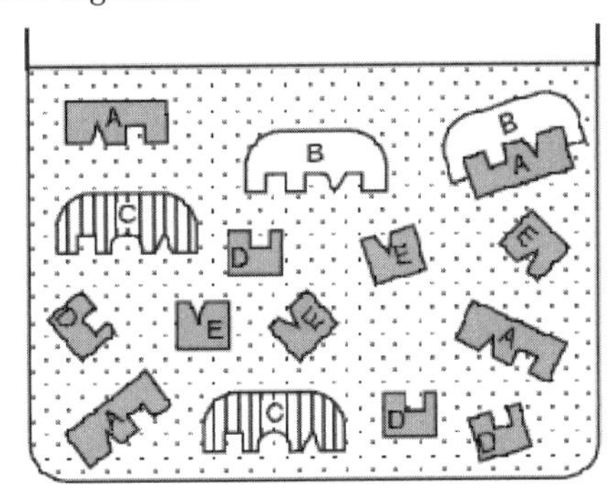

Which structures represent products of digestion?

(1) ***A*** and ***D***
(2) ***B*** and ***C***
(3) ***B*** and ***E***
(4) ***D*** and ***E***

64. The sweet taste of freshly picked corn is due to the high sugar content in the kernels. Enzyme action converts about 50% of the sugar to starch within one day after picking. To preserve its sweetness, the freshly picked corn is immersed in boiling water for a few minutes, and then cooled.

Which statement most likely explains why the boiled corn kernels remain sweet?

(1) Boiling destroys sugar molecules so they cannot be converted to starch.
(2) Boiling kills a fungus on the corn that is needed to convert sugar to starch.
(3) Boiling activates the enzyme that converts amino acids to sugar.
(4) Boiling deactivates the enzyme responsible for converting sugar to starch.

65. Two test tubes, ***A*** and ***B***, were set up as shown in the diagram below. Bromthymol blue, which turns from blue to yellow in the presence of carbon dioxide, was added to the water at the bottom of each tube before the tubes were sealed. The tubes were maintained at the temperatures shown for six days. (Average room temperature is 20°C.)

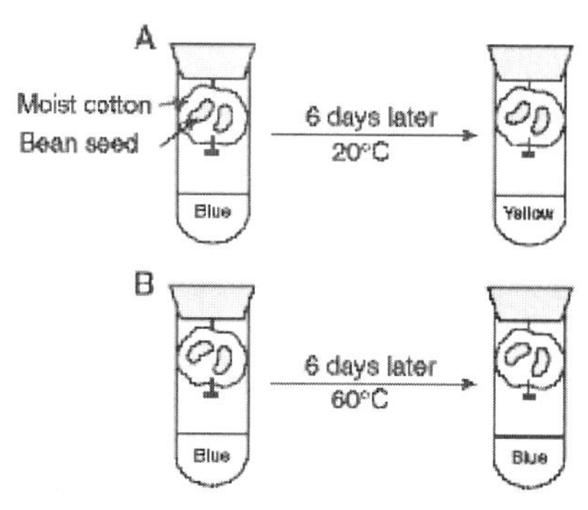

Explain how the temperature difference could lead to the different results in tubes ***A*** and ***B*** after six days.

66. Base your answers to the next question on the graph below and on your knowledge of biology. The graph illustrates a single species of bacteria grown at various pH levels.

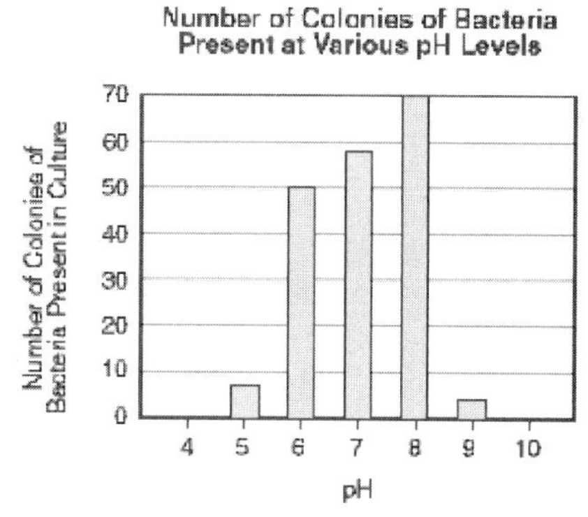

Which statement is supported by data from this graph?

(1) All species of bacteria can grow well at pH 7.
(2) This type of bacterium would grow well at pH 7.5.
(3) This type of bacterium would grow well at pH 2.
(4) Other types of bacteria can grow well at pH 4.

67. Tomato plants in a garden are not growing well. The gardener hypothesizes that the soil is too acidic. To test this hypothesis accurately, the gardener could

(1) plant seeds of a different kind of plant
(2) move the tomato plants to an area with less sunlight
(3) change the pH of the soil
(4) reduce the amount of water available to the plant

68. Base your answers to the next question on the diagram below that represents a human enzyme and four types of molecules present in a solution in a flask.

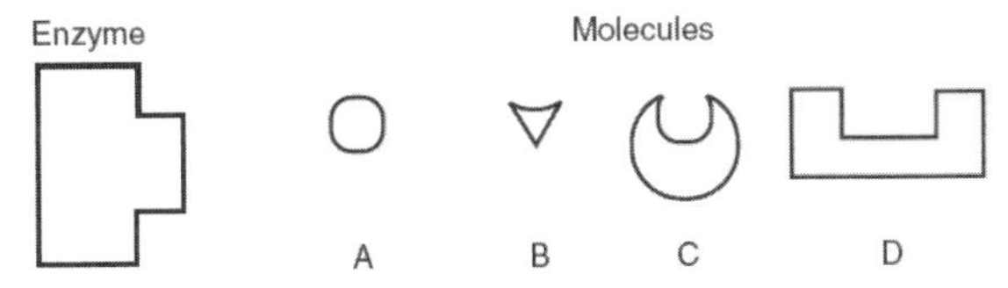

State what would most likely happen to the rate of reaction if the temperature of the solution in the flask were increased gradually from 10°C to 30°C.

69. Base your answers to the next question on the graph below and on your knowledge of biology.

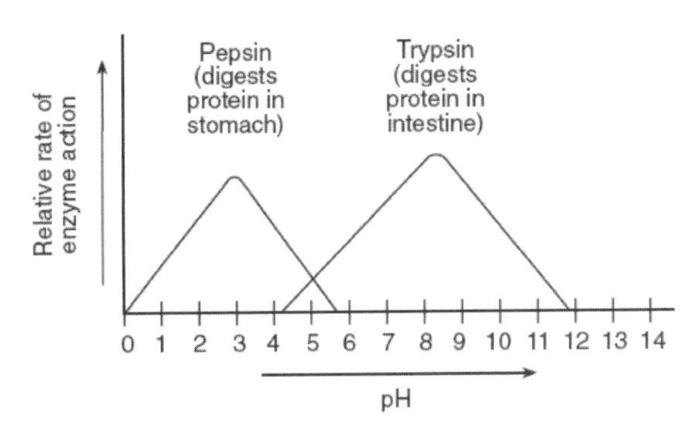

Pepsin works best in which type of environment?

(1) acidic, only
(2) basic, only
(3) neutral
(4) sometimes acidic, sometimes basic

70. A drug is developed that, due to its molecular shape, blocks the action of substance *A*: . Which shape would the drug molecule most likely resemble?

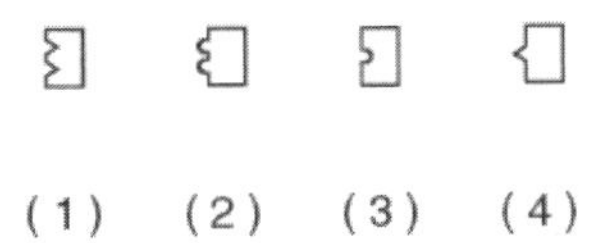

71. In the body of a human, the types of chemical activities occurring within cells are most dependent on the

(1) biological catalysts present
(2) size of the cell
(3) number of chromosomes in the cell
(4) kind of sugar found on each chromosome

72. Identify the gas that would be produced by the process taking place in both laboratory setups.

73. State ***one*** way extremely high temperatures can affect biological catalysts found in these organisms.

74. Base your answers to the next question on the diagram below and on your knowledge of biology.

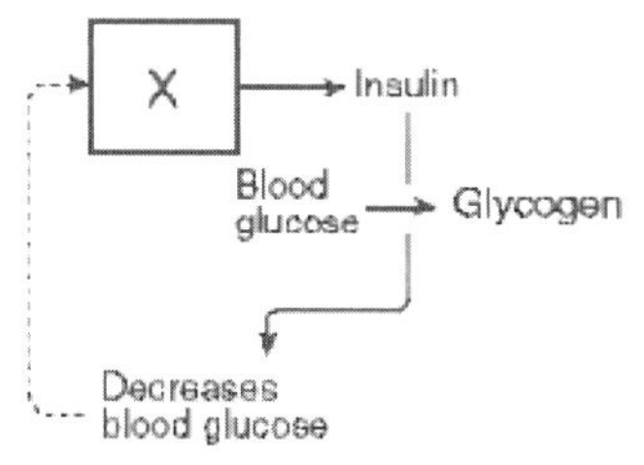

The dashed line in the diagram represents

(1) a digestive process
(2) a feedback mechanism
(3) cellular differentiation
(4) recycling of organic chemicals

Base your answers to the next three questions on the information and data table below and on your knowledge of biology.

The results of blood tests for two individuals are shown in the data table below. The blood glucose level before breakfast is normally 80–90 mg/100 mL of blood. A blood glucose level above 110 mg/100 mL of blood indicates a failure in a feedback mechanism.

Injection of chemical ***X***, a chemical normally produced in the body, may be required to correct this problem.

Data Table

Time	Blood Glucose (mg/100 mL)	
	Individual 1	Individual 2
7:00 a.m.	90	150
7:30 a.m.	120	180
8:00 a.m.	140	220
8:30 a.m.	110	250
9:00 a.m.	90	240
9:30 a.m.	85	230
10:00 a.m.	90	210
10:30 a.m.	85	190
11:00 a.m.	90	170

75. Identify chemical ***X***.

76. State ***one*** reason for the change in blood glucose level between 7:00 a.m. and 8:00 a.m.

77. What term refers to the relatively constant level of blood glucose of individual ***1*** between 9:00 a.m. and 11:00 a.m.?

---

78. Some human white blood cells help destroy pathogenic bacteria by

(1) causing mutations in the bacteria
(2) engulfing and digesting the bacteria
(3) producing toxins that compete with bacterial toxins
(4) inserting part of their DNA into the bacterial cells

79. Base your answers to the next question on the data table below and on your knowledge of biology.

A group of students obtained the following data:

Data Table

Student Tested	Pulse Rate at Rest	Pulse Rate After Exercising
1	70	97
2	75	106
3	84	120
4	60	91
5	78	122

The activity of which body system was measured to obtain these data?

80. Base your answers to the next question on the diagram below of activities in the human body.

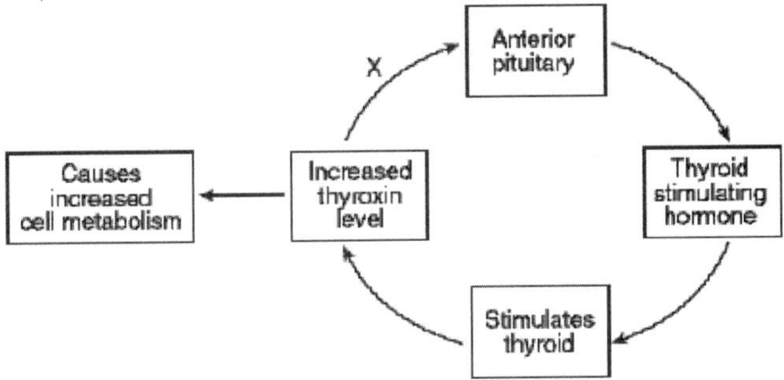

This diagram illustrates part of
(1) a feedback mechanism
(2) an enzyme pathway
(3) a digestive mechanism
(4) a pattern of learned behavior

81. The data table below compares blood flow in various human body structures, both at rest and during strenuous exercise.

Structure	Blood Flow at Rest (mL/min)	Blood Flow During Strenuous Exercise (mL/min)
heart	250	750
skeletal muscle	1200	12,500
digestive organs	1400	600

For each structure, explain ***one*** way that the change in the rate of blood flow in this structure helps maintain homeostasis during exercise.

82. When a person exercises, changes occur in muscle cells as they release more energy. Explain how increased blood flow helps these muscle cells release more energy.

83. Identify ***one*** hormone involved in another biological relationship and an organ that is directly affected by the hormone you identified.

84. Antibody molecules and receptor molecules are similar in that they both
(1) control transport through the cell membrane
(2) have a specific shape related to their specific function
(3) remove wastes from the body
(4) speed up chemical reactions in cells

85. Which graph of blood sugar level over a 12-hour period best illustrates the concept of dynamic equilibrium in the body?

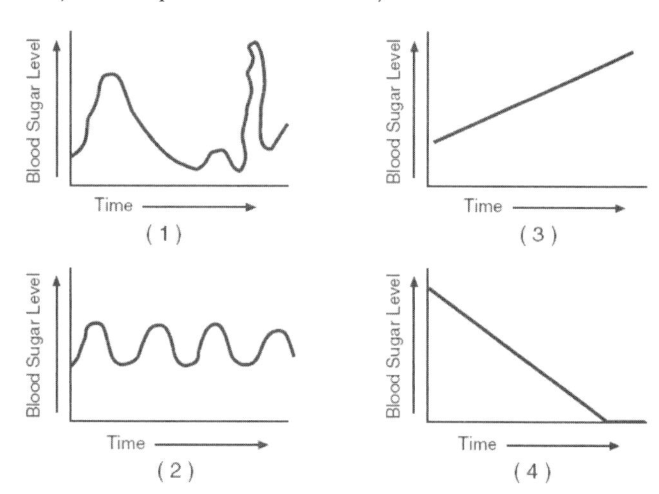

86. Which statement describes all enzymes?
(1) They control the transport of materials.
(2) They provide energy for chemical reactions.
(3) They affect the rate of chemical reactions.
(4) They absorb oxygen from the environment.

87. What effect would exercise have on the system you identified in question 79, above?

88. Which situation is not an example of the maintenance of a dynamic equilibrium in an organism?
(1) Guard cells contribute to the regulation of water content in a geranium plant.
(2) Water passes into an animal cell causing it to swell.
(3) The release of insulin lowers the blood sugar level in a human after eating a big meal.
(4) A runner perspires while running a race on a hot summer day.

89. Which statement does not describe an example of a feedback mechanism that maintains homeostasis?
(1) The guard cells close the openings in leaves, preventing excess water loss from a plant.
(2) White blood cells increase the production of antigens during an allergic reaction.
(3) Increased physical activity increases heart rate in humans.
(4) The pancreas releases insulin, helping humans to keep blood sugar levels stable.

90. Enzymes have an optimum temperature at which they work best. Temperatures above and below this optimum will decrease enzyme activity. Which graph best illustrates the effect of temperature on enzyme activity?

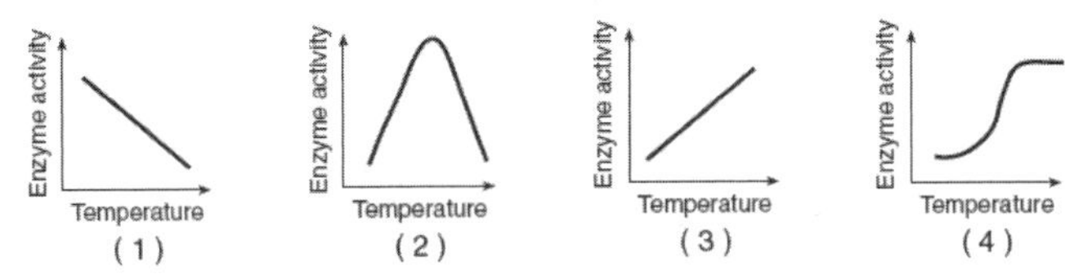

91. For ***each*** of the paired items below and describe how the ***first*** item in the pair regulates the ***second*** item for the maintenance of homeostasis.

insulin—blood sugar level
CO_2 in blood—breathing rate
activity of guard cells—water loss from a leaf

92. Identify a substance in the blood, other than insulin, that could change in concentration and indicate a person is not secreting insulin in normal amounts.

Base your answers to the next two questions on the diagram below and on your knowledge of biology. Each arrow in the diagram represents a different hormone released by the pituitary gland that stimulates the gland indicated in the diagram. All structures are present in the same organism.

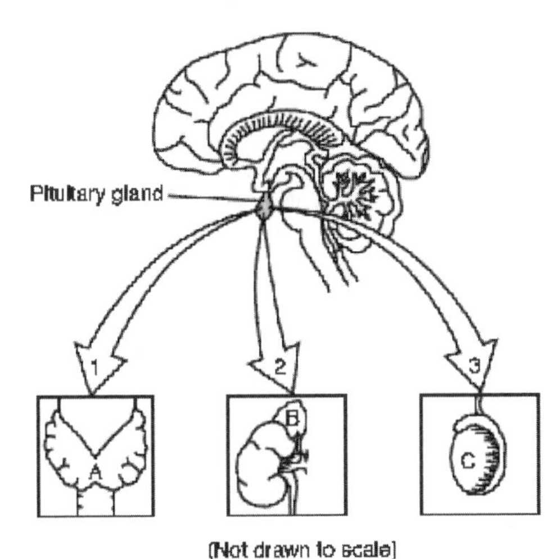

(Not drawn to scale)

93. The pituitary gland may release hormone ***2*** when blood pressure drops. Hormone ***2*** causes gland ***B*** to release a different hormone that raises blood pressure which, in turn, stops the secretion of hormone ***2***. The interaction of these hormones is an example of
    (1) DNA base substitution
    (2) manipulation of genetic instructions
    (3) a feedback mechanism
    (4) an antigen-antibody reaction

94. Why does hormone ***1*** influence the action of gland ***A*** but not gland ***B*** or ***C***?
    (1) Every activity in gland ***A*** is different from the activities in glands ***B*** and ***C***.
    (2) The cells of glands ***B*** and ***C*** contain different receptors than the cells of gland ***A***.

---

(3) Each gland contains cells that have different base sequences in their DNA.
(4) The distance a chemical can travel is influenced by both pH and temperature.

---

95. The diagram below represents a cross section of part of a leaf.

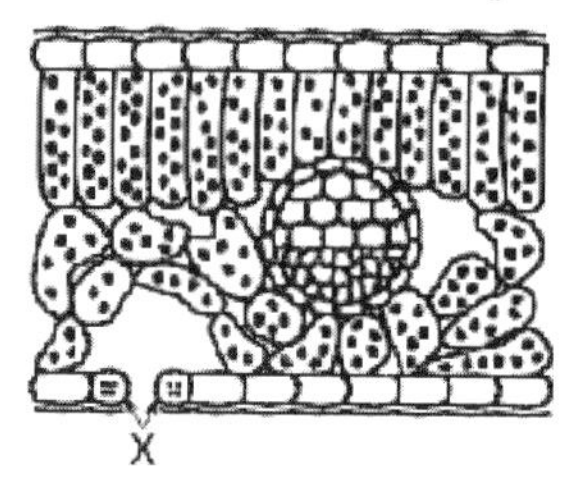

Which life functions are directly regulated through feedback mechanisms associated with the actions of the structures labeled ***X***?
(1) excretion and immunity
(2) digestion and coordination
(3) circulation and reproduction
(4) respiration and photosynthesis

96. Base your answers to the next question on the information below and on your knowledge of biology.

Arsenic and Old Glucocorticoids

Constant exposure to small amounts of arsenic in drinking water has been found to increase the risk of cancer and other diseases. In January of 2001, the EPA (Environmental Protection Agency) lowered the acceptable levels of arsenic in drinking water from 50 ppb (parts per billion) to 10 ppb.

Researchers are now trying to determine how arsenic affects the body. Recent experiments suggest that arsenic may block the activity of hormones. One group of hormones affected by arsenic is glucocorticoids, which are responsible for activating many genes that appear to suppress cancer.

Rat tumor cells were used to determine the effect of arsenic on glucocorticoids. One group of cells was treated with a solution of synthetic glucocorticoid and arsenic, another with a solution of synthetic glucocorticoid and water, and a third group with a solution containing only water. Researchers then measured the activity of one of the genes that is usually activated by glucocorticoids. The genes in the cells treated with the hormone and arsenic mixture and those treated with just water did not become activated. The genes in the cells treated with the hormone and water mixture were activated. Researchers concluded that arsenic blocked the normal activity of the hormone. They are now extending their studies to determine if arsenic acts in a similar manner in other types of cells and in entire organisms.

Identify ***one*** specific hormone in the body, other than glucocorticoid. Explain how disruption of the activity of the hormone you identified might upset a feedback mechanism in the body.

Base your answers to he next four questions on the passage below and on your knowledge of biology.

Decline of the Salmon Population

Salmon are fish that hatch in a river and swim to the ocean where their body mass increases. When mature, they return to the river where they were hatched and swim up stream to reproduce and die. When there are large populations of salmon, the return of nutrients to the river ecosystem can be huge. It is estimated that during salmon runs in the Pacific Northwest in the 1800s, 500 million pounds of salmon returned to reproduce and die each year. Research estimates that in the Columbia River alone, salmon contributed hundreds of thousands of pounds of nitrogen and phosphorus compounds to the local ecosystem each year. Over the past 100 years, commercial ocean fishing has removed up to two-thirds of the salmon before they reach the river each year.

97. Identify the process that releases the nutrients from the bodies of the dead salmon, making the nutrients available for other organisms in the ecosystem.

98. Identify *one* organism, other than the salmon, that would be present in or near the river that would most likely be part of a food web in the river ecosystem.

99. Identify *two* nutrients that are returned to the ecosystem when the salmon die.

100. State *one* impact, other than reducing the salmon population, that commercial ocean fishing has on the river ecosystem.

---

101. Base your answer to the next question on the information and data table below and on your knowledge of biology

Two students collected data on their pulse rates while performing different activities. Their average results are shown in the data table below.

Data Table

Activity	Average Pulse Rate (beats/min)
sitting quietly	70
walking	98
running	120

State the relationship between activity and pulse rate.

Base your answers to the next four questions on the passage below and on your knowledge of biology.

When humans perspire, water, urea, and salts containing sodium are removed from the blood. Drinking water during extended periods of physical exercise replenishes the water but not the sodium. This increase in water dilutes the blood and may result in the concentration of sodium dropping low enough to cause a condition known as hyponatremia.

Symptoms of hyponatremia include headache, nausea, and lack of coordination. Left untreated, it can lead to coma and even death. The body has a variety of feedback mechanisms that assist in regulating water and sodium concentrations in the blood. The kidneys play a major role in these mechanisms, as they filter the blood and produce urine.

102. The best way to reduce the symptoms of hyponatremia would be to
(1) drink more water
(2) eat chocolate
(3) eat salty foods
(4) drink cranberry juice

103. Many runners pour water on their bodies during a race. Explain how this action helps to maintain homeostasis.

104. How would running in a marathon on a warm day most likely affect urine production? Support your answer.

105. Many people today drink sport drinks containing large amounts of sodium. Describe ***one*** possible effect this might have on a person who is not very active.

106. The diagram below represents an interaction between parts of an organism.

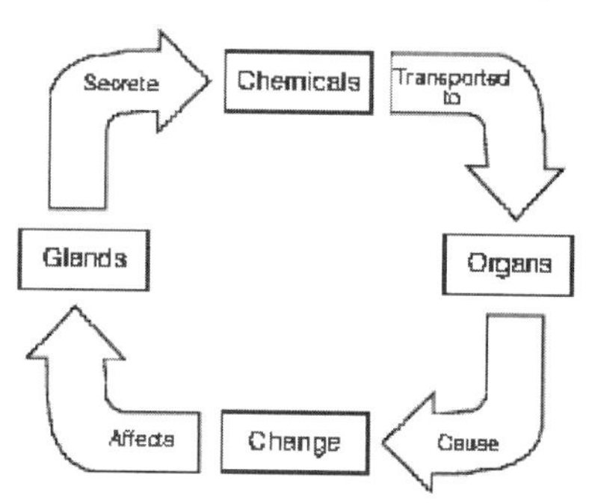

The term ***chemicals*** in this diagram represents
(1) starch molecules
(2) DNA molecules
(3) hormone molecules
(4) receptor molecules

Base your answers to the next three questions on the information below and on your knowledge of biology.

In a test for diabetes, blood samples were taken from an individual every 4 hours for 24 hours. The glucose concentrations were recorded and are shown in the data table below.

Blood Glucose Level Over Time

Time (h)	Blood Glucose Concentration (mg/dL)
0	100
4	110
8	128
12	82
16	92
20	130
24	104

107. State ***one*** likely cause of the change in blood glucose concentration between hour 16 and hour 20.

ıble)

108. How might these results be different if this individual was not able to produce sufficient levels of insulin?
(1) The level of blood glucose would be constant.
(2) The average level of blood glucose would be lower.
(3) The maximum level of blood glucose would be higher.
(4) The minimum level of blood glucose would be lower.

109. The chemical that is responsible for the decrease in blood glucose concentration is released by
(1) muscle cells (2) guard cells (3) the ovaries (4) the pancreas

---

110. Diagram ***A*** below represents a microscopic view of the lower surface of a leaf. Diagram ***B*** represents a portion of the human body.

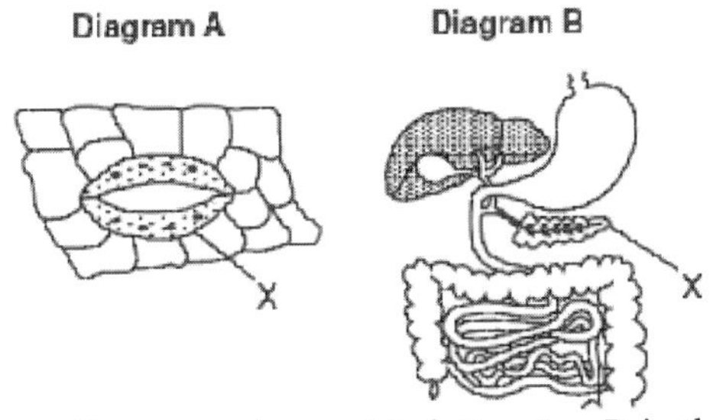

a. Choose one diagram and record its letter, ***A*** or ***B***, in the space provided.
b. Identify the structure labeled ***X*** in the diagram you chose.
c. State ***one*** problem for the organism that would result from a malfunction of the structure you identified.

111. The diagram below represents what can happen when homeostasis in an organism is threatened.

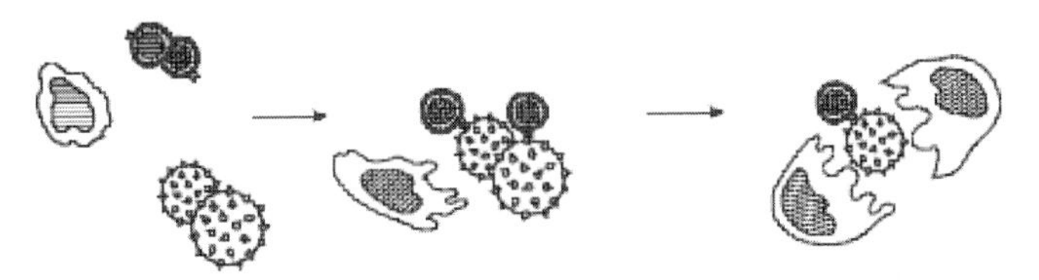

Which statement provides a possible explanation for these events?
(1) Antibiotics break down harmful substances by the process of digestion.
(2) Some specialized cells mark and other cells engulf microbes during immune reactions.
(3) Embryonic development of essential organs occurs during pregnancy.
(4) Cloning removes abnormal cells produced during differentiation.

112. A part of the Hepatitis B virus is synthesized in the laboratory. This viral particle can be identified by the immune system as a foreign material but the viral particle is not capable of causing disease. Immediately after this viral particle is injected into a human it
(1) stimulates the production of enzymes that are able to digest the Hepatitis B virus
(2) triggers the formation of antibodies that protect against the Hepatitis B virus
(3) synthesizes specific hormones that provide immunity against the Hepatitis B virus
(4) breaks down key receptor molecules so that the Hepatitis B virus can enter body cells

113. State ***two*** ways cells of the immune system fight disease.

114. The immune system of humans may respond to chemicals on the surface of an invading organism by
(1) releasing hormones that break down these chemicals
(2) synthesizing antibodies that mark these organisms to be destroyed
(3) secreting antibiotics that attach to these organisms
(4) altering a DNA sequence in these organisms

115. Antibody molecules and receptor molecules are similar in that they both
(1) control transport through the cell membrane
(2) have a specific shape related to their specific function
(3) remove wastes from the body
(4) speed up chemical reactions in cells

116. Which statement best describes what will most likely happen when an individual receives a vaccination containing a weakened pathogen?
(1) The ability to fight disease will increase due to antibodies received from the pathogen.
(2) The ability to fight disease caused by the pathogen will increase due to antibody production.
(3) The ability to produce antibodies will decrease after the vaccination.
(4) The ability to resist most types of diseases will increase.

117. Which activity is not a function of white blood cells in response to an invasion of the body by bacteria?
(1) engulfing these bacteria
(2) producing antibodies to act against this type of bacteria
(3) preparing for future invasions of this type of bacteria
(4) speeding transmissions of nerve impulses to detect these bacteria

118. Vaccinations help prepare the body to fight invasions of a specific pathogen by
(1) inhibiting antigen production
(2) stimulating antibody production
(3) inhibiting white blood cell production
(4) stimulating red blood cell production

119. The purpose of introducing weakened microbes into the body of an organism is to stimulate the
(1) production of living microbes that will protect the organism from future attacks
(2) production of antigens that will prevent infections from occurring
(3) immune system to react and prepare the organism to fight future invasions by these microbes
(4) replication of genes that direct the synthesis of hormones that regulate the number of microbes

120. Many people become infected with the chicken pox virus during childhood. After recovering from chicken pox, these people are usually immune to the disease for the rest of their lives. However, they may still be infected by viruses that cause other diseases, such as measles.

Discuss the immune response to the chicken pox virus. In your answer, be sure to include:

- the role of antigens in the immune response
- the role of white blood cells in the body's response to the virus
- an explanation of why recovery from an infection with the chicken pox virus will not protect a person from getting a different disease, such as measles
- an explanation of why a chicken pox vaccination usually does not cause a person to become ill with chicken pox

121. Smallpox is a disease caused by a specific virus, while the common cold can be caused by over 100 different viruses. Explain why it is possible to develop a vaccine to prevent smallpox, but it is difficult to develop a vaccine to prevent the common cold. In your answer be sure to:

- identify the substance in a vaccine that makes the vaccine effective
- explain the relationship between a vaccine and white blood cell activity
- explain why the response of the immune system to a vaccine is specific
- state ***one*** reason why it would be difficult to develop a vaccine to be used against the common cold

122. Base your answer to the next question on the information below and on your knowledge of biology.

Until the middle of the 20th century, transplanting complex organs, such as kidneys, was rarely successful. The first transplant recipients did not survive. It was not until 1954 that the first successful kidney transplant was performed. Success with transplants increased as research scientists developed techniques such as tissue typing and the use of immunosuppressant drugs. These are drugs that suppress the immune system to prevent the rejection of a transplanted organ. In 2002, there were nearly 15,000 kidney transplants performed in the United States with a greater than 95% success rate.

Describe the relationship of the immune system to organ transplants and the use of immunosuppressant drugs to prevent the rejection of a transplanted organ. In your answer be sure to:

- state ***one*** way the immune system is involved in the rejection of transplanted organs
- explain why the best source for a donated kidney would be the identical twin of the recipient
- explain why immunosuppressant drugs might be needed to prevent rejection of a kidney received from a donor other than an identical twin
- state ***one*** reason a person may get sick more easily when taking an immunosuppressant drug

123. The use of a vaccine to stimulate the immune system to act against a specific pathogen is valuable in maintaining homeostasis because
    (1) once the body produces chemicals to combat one type of virus, it can more easily make antibiotics
    (2) the body can digest the weakened microbes and use them as food
    (3) the body will be able to fight invasions by the same type of microbe in the future
    (4) the more the immune system is challenged, the better it performs

124. The diagram below represents an event that occurs in the blood.

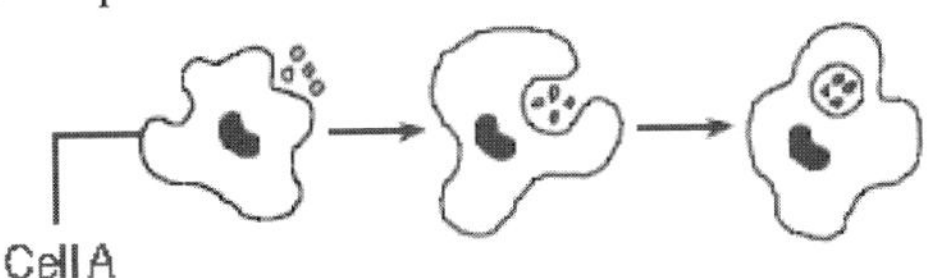

Which statement best describes this event?
    (1) Cell ***A*** is a white blood cell releasing antigens to destroy bacteria.
    (2) Cell ***A*** is a cancer cell produced by the immune system and it is helping to prevent disease.
    (3) Cell ***A*** is a white blood cell engulfing disease-causing organisms.
    (4) Cell ***A*** is protecting bacteria so they can reproduce without being destroyed by predators.

Base your answers to the next three questions on the information below and on your knowledge of biology.

Proteins on the surface of a human cell and on a bird influenza virus are represented in the diagram below.

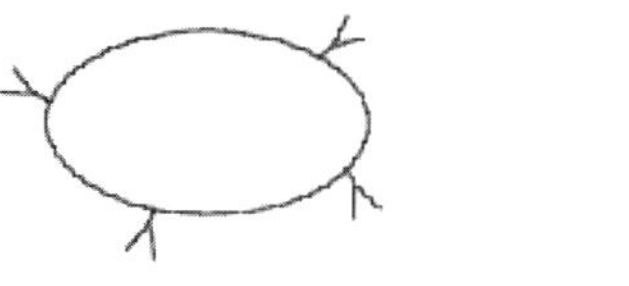

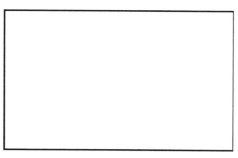

Human Cell   Bird Influenza Virus

125. In the space below, draw a change in the bird influenza virus that would allow it to infect this human cell.

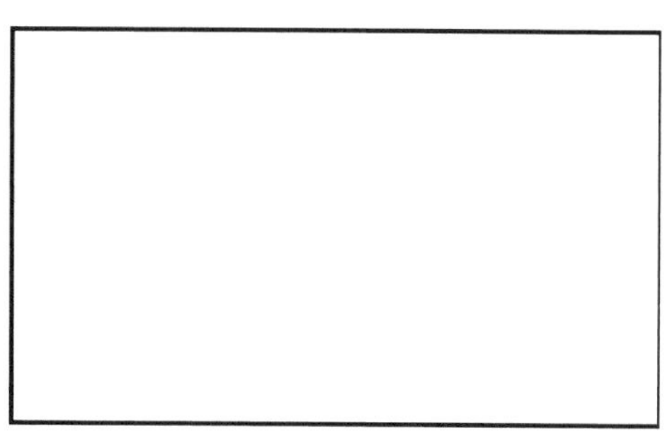

126. Explain how this change in the virus could come about.

127. Identify the relationship that exists between a virus and a human when the virus infects the human.

---

128. Which statement best describes how a vaccination can help protect the body against disease?
   (1) Vaccines directly kill the pathogen that causes the disease.
   (2) Vaccines act as a medicine that cures the disease.
   (3) Vaccines cause the production of specific molecules that will react with and destroy certain microbes.
   (4) Vaccines contain white blood cells that engulf harmful germs and prevent them from spreading throughout the body.

129. Many viruses infect only a certain type of cell because they bind to certain
   (1) other viruses on the surface of the cell
   (2) mitochondria in the cell
   (3) hormones in the cell
   (4) receptor sites on the surface of the cell

130. Food is often treated by freezing (example meat, fish, poultry) or salting (example pickles, sauerkraut) to lower the risk of disease and spoilage. Identify ***one*** type of organism that is controlled by these food preservation methods.

131. AIDS is an infectious disease that has reached epidemic proportions. Describe the nature of this disease and identify two ways to prevent or control the spread of infectious diseases, such as AIDS. In your response be sure to include:

- the type of pathogen that causes AIDS
- the system of the body that is attacked by that pathogen
- the effect on the body when this system is weakened by AIDS
- ***two*** ways to prevent or control the spread of infectious diseases, such as AIDS

132. Base your answers to the next question on the passage below and on your knowledge of biology.

In Search of a Low-Allergy Peanut

Many people are allergic to substances in the environment. Of the many foods that contain allergens (allergy-inducing substances), peanuts cause some of the most severe reactions. Mildly allergic people may only get hives. Highly allergic people can go into a form of shock. Some people die each year from reactions to peanuts.

A group of scientists is attempting to produce peanuts that lack the allergy-inducing proteins by using traditional selective breeding methods. They are searching for varieties of peanuts that are free of the allergens. By crossing those varieties with popular commercial types, they hope to produce peanuts that will be less likely to cause allergic reactions and still taste good. So far, they have found one variety that has 80 percent less of one of three complex proteins linked to allergic reactions. Removing all three of these allergens may be impossible, but even removing one could help.

Other researchers are attempting to alter the genes that code for the three major allergens in peanuts. All of this research is seen as a possible long-term solution to peanut allergies.

Allergic reactions usually occur when the immune system produces

(1) antibiotics against usually harmless antigens
(2) antigens against usually harmless antibodies
(3) antibodies against usually harmless antigens
(4) enzymes against usually harmless antibodies

133. Base your answer to the next question on the information below and on your knowledge of biology.

Until the middle of the 20th century, transplanting complex organs, such as kidneys, was rarely successful. The first transplant recipients did not survive. It was not until 1954 that the first successful kidney transplant was performed. Success with transplants increased as research scientists developed techniques such as tissue typing and the use of immunosuppressant drugs. These are drugs that suppress the immune system to prevent the rejection of a transplanted organ. In 2002, there were nearly 15,000 kidney transplants performed in the United States with a greater than 95% success rate.

Describe the relationship of the immune system to organ transplants and the use of immunosuppressant drugs to prevent the rejection of a transplanted organ. In your answer be sure to:

- state ***one*** way the immune system is involved in the rejection of transplanted organs
- explain why the best source for a donated kidney would be the identical twin of the recipient
- explain why immunosuppressant drugs might be needed to prevent rejection of a kidney received from a donor other than an identical twin
- state ***one*** reason a person may get sick more easily when taking an immunosuppressant drug

134. A single-celled organism is represented in the diagram below. An activity is indicated by the arrow.

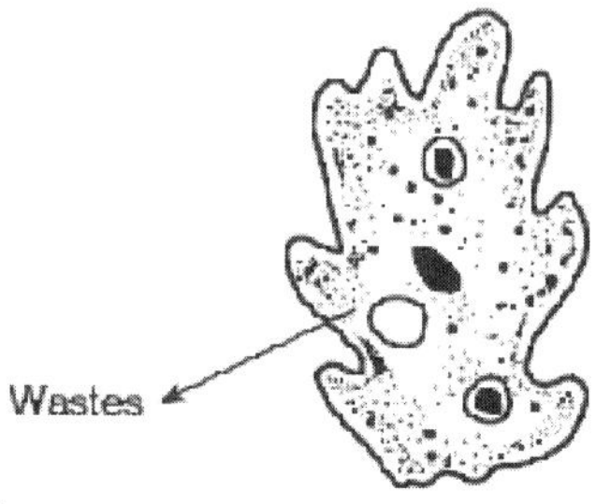

If this activity requires the use of energy, which substance would be the source of this energy?

(1) DNA (2) ATP (3) a hormone (4) an antibody

135. To increase chances for a successful organ transplant, the person receiving the organ should be given special medications. The purpose of these medications is to

(1) increase the immune response in the person receiving the transplant
(2) decrease the immune response in the person receiving the transplant
(3) decrease mutations in the person receiving the transplant
(4) increase mutations in the person receiving the transplant

136. In some people, substances such as peanuts, eggs, and milk cause an immune response. This response to usually harmless substances is most similar to the

(1) action of the heart as the intensity of exercise increases
(2) mechanism that regulates the activity of guard cells
(3) action of white blood cells when certain bacteria enter the body
(4) mechanism that maintains the proper level of antibiotics in the blood

137. Base your answers to the next question on the diagram below and on your knowledge of biology.

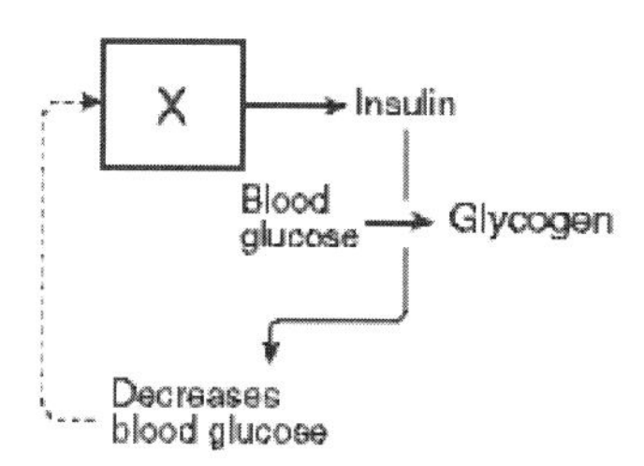

Identify the organ labeled ***X***.

Base your answers to the next two questions on the information and data table below and on your knowledge of biology.

The results of blood tests for two individuals are shown in the data table below. The blood glucose level before breakfast is normally 80–90 mg/100 mL of blood. A blood glucose level above 110 mg/100 mL of blood indicates a failure in a feedback mechanism.

Injection of chemical ***X***, a chemical normally produced in the body, may be required to correct this problem.

Data Table

Time	Blood Glucose (mg/100 mL)	
	Individual 1	Individual 2
7:00 a.m.	90	150
7:30 a.m.	120	180
8:00 a.m.	140	220
8:30 a.m.	110	250
9:00 a.m.	90	240
9:30 a.m.	85	230
10:00 a.m.	90	210
10:30 a.m.	85	190
11:00 a.m.	90	170

Directions for the next two questions: Using the information in the data table, construct a line graph on the grid provided below, following the directions below.

138. Mark an appropriate scale on each labeled axis.

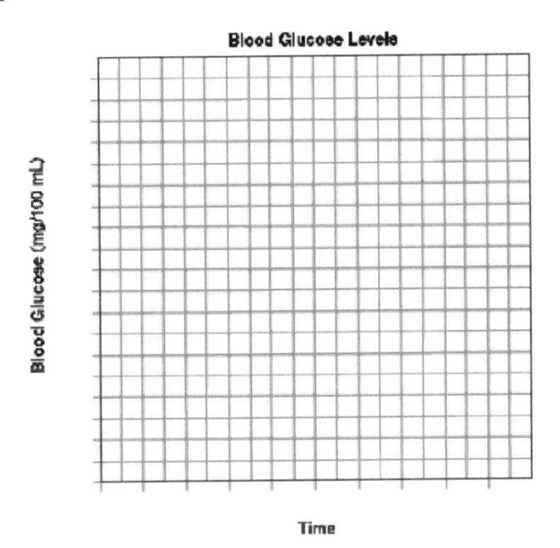

139. Using the information in the data table, construct a line graph on the grid provided below. Plot the blood glucose levels for the individual who will most likely need injections of chemical **X**. Surround each point with a small circle and connect the points.

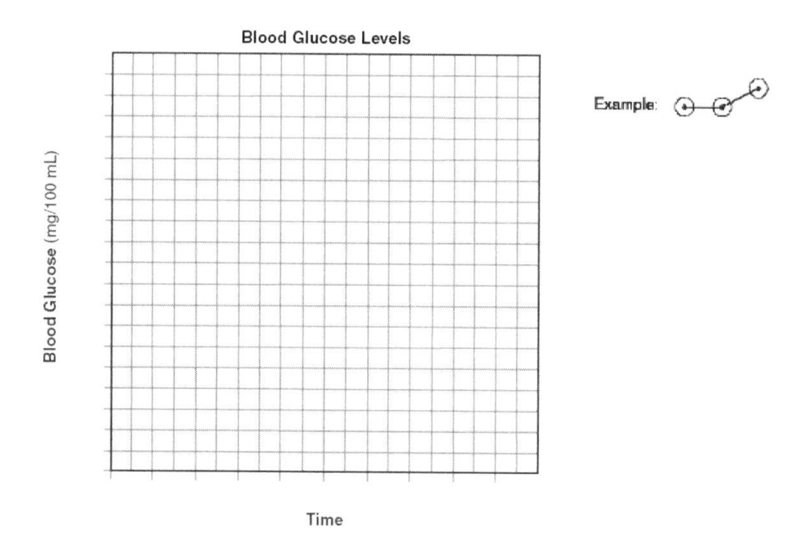

Hint: Review homeostasis and feedback mechanism in this chapter, Chapter 2.

Base your answers to the next two questions on the information and data table below and on your knowledge of biology.

The results of blood tests for two individuals are shown in the data table below. The blood glucose level before breakfast is normally 80–90 mg/100 mL of blood. A blood glucose level above 110 mg/100 mL of blood indicates a failure in a feedback mechanism.

Injection of chemical ***X***, a chemical normally produced in the body, may be required to correct this problem.

Data Table

Time	Blood Glucose (mg/100 mL)	
	Individual 1	Individual 2
7:00 a.m.	90	150
7:30 a.m.	120	180
8:00 a.m.	140	220
8:30 a.m.	110	250
9:00 a.m.	90	240
9:30 a.m.	85	230
10:00 a.m.	90	210
10:30 a.m.	85	190
11:00 a.m.	90	170

140. State ***one*** reason for the change in blood glucose level between 7:00 a.m. and 8:00 a.m.

141. What term refers to the relatively constant level of blood glucose of individual *1* between 9:00 a.m. and 11:00 a.m.?

Base your answers to the next five questions on the information and diagrams below and on your knowledge of biology.

The laboratory setups represented below were used to investigate the effect of temperature on cellular respiration in yeast (a single-celled organism). Each of two flasks containing equal amounts of a yeast-glucose solution was submerged in a water bath, one kept at 20°C and one kept at 35°C. The number of gas bubbles released from the glass tube in each setup was observed and the results were recorded every 5 minutes for a period of 25 minutes. The data are summarized in the table below.

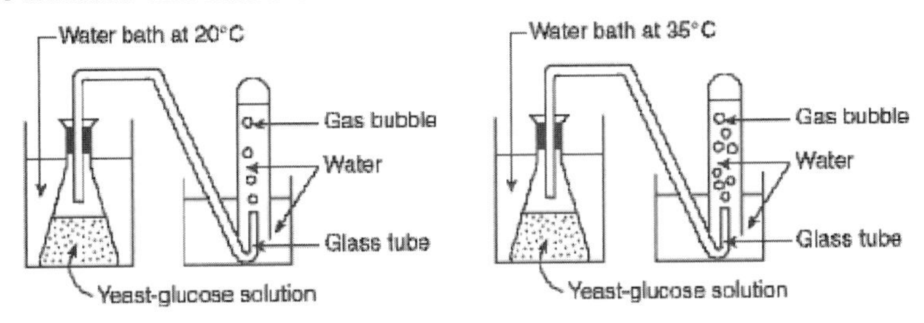

Data Table

Time (minutes)	Total Number of Bubbles Released	
	20°C	35°C
5	0	5
10	5	15
15	15	30
20	30	50
25	45	75

Directions for the next three questions: Using the information in the data table, construct a line graph on the grid provided, following the directions below.

142. Mark an appropriate scale on each axis.

143. Plot the data for the total number of bubbles released at 20°C on the grid provided. Surround each point with a small circle and connect the points.

Example:

144. Plot the data for the total number of bubbles released at 35°C on the grid. Surround each point with a small triangle and connect the points.

The Effect of Temperature on Respiration in Yeast

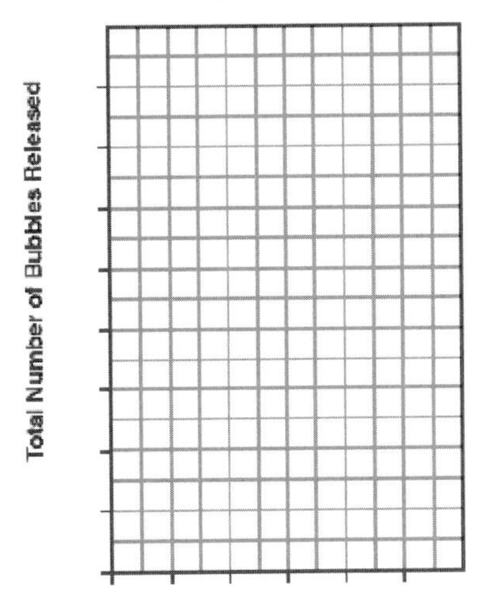

Example:

Key
Yeast respiration at 20°C
Yeast respiration at 35°C

145. State ***one*** relationship between temperature and the rate of gas production in yeast.

146. Identify the gas that would be produced by the process taking place in both laboratory setups.

---

147. The development of an embryo is represented in the diagram below.

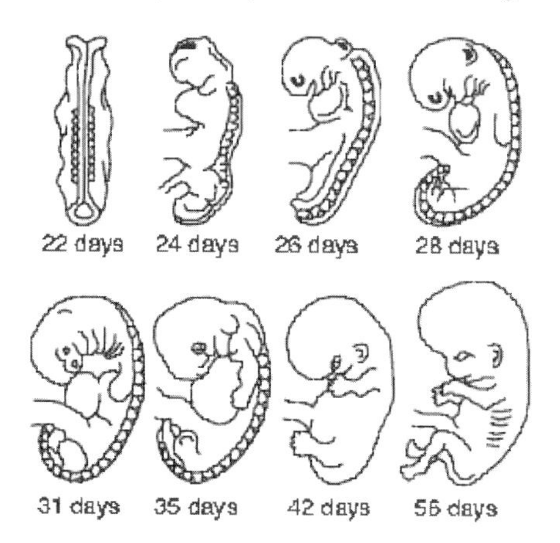

(Not drawn to scale)

These changes in the form of the embryo are a direct result of

(1) uncontrolled cell division and mutations
(2) differentiation and growth
(3) antibodies and antigens inherited from the father
(4) meiosis and fertilization

CHAPTER 3: GENETICS

Some people are tall, some are short, some have brown hair, others blond, some have brown eyes and some blue eyes. Let's explain why people have these traits.

You learned each cell has a nucleus. Threadlike chromosomes are a structure inside the nucleus (see figure). Every chromosome has many genes. (There are many genes on a chromosome). There is a gene for tall, a gene for short, a gene for brown hair, a gene for blond hair. Genes determine how you look (eye color, height, etc.). In the cells of our body (body cells), there are two genes for every trait; for example there are two genes for height. Genes have genetic information (also called coded instruction), instructing (telling) a person to be tall or short, brown eyes or blue eyes etc. In short, genes in plants and animals have genetic information telling (directing, ordering, giving instruction to) a plant or animal how to look (examples: will the plant have red or yellow flowers, big or small leaves, will the plant or animal be tall or short, etc).

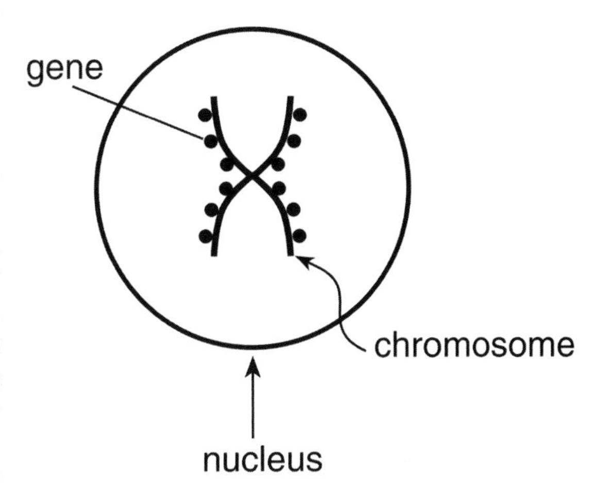

Dots are genes on the chromosome

Genes determine how an organism (plants, animals including humans) look. The genes (genetic information) are inherited (passed down) from your grandparents to your parents and to you, therefore the children resemble the parents and grandparents. You will learn later about gene expression (some genes are used or turned on in different environments).

Heredity is the passage of genetic information by genes (example gene for tall) from one generation to another. An inherited trait (example eye color) can be determined by one or many genes and a single gene can influence more than one trait. An inherited trait that you can easily recognize is height, eye and hair color, etc. Other inheritable traits that are not easily recognized are blood types and color blindness.

A human cell has many thousands of different genes in the nucleus.

Genes passed down from parent to offspring

Genes are passed down from parents to offspring in living things (in plants, animals, and one celled organisms).

Asexual reproduction: In asexual reproduction there in only one parent producing offspring.

For example:

one ameba divides and produces two amebas.

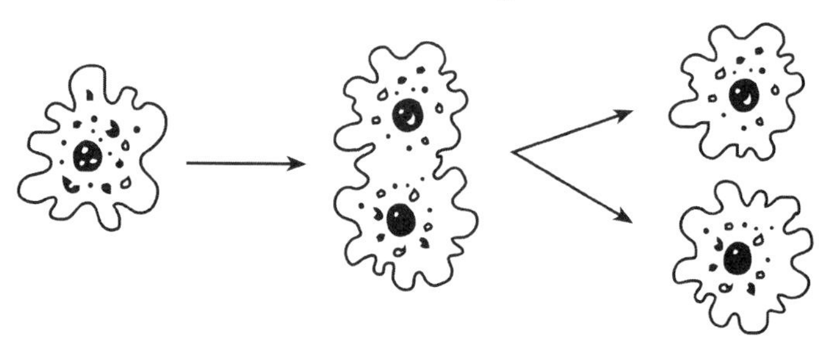

All the genes in the baby amebas come from the parent ameba and therefore the offspring (baby amebas) are identical to the parent ameba. The offspring (baby amebas) has the same genes (genetic code) as the parent ameba.

Sexual reproduction: In sexual reproduction (example humans), there are two parents (male and female). The male produces the **male gamete** (**sperm in** a **human being**) which has the genes-genetic information from the man. The female produces the **female gamete (egg in** a **human being**) which has the genes-genetic information from the lady. The **sperm** (which has the **genes from** the **father**) and the **egg** (which has the **genes from** the **mother) unite** together to **produce a child**. The child has the genes (genetic information) from both the father and mother (half of the genetic information from the mother and half of the genetic information from the father), therefore the child often resembles but is not identical to either the mother or father.

Genetic Recombination: Combining genes from both parents are called **genetic recombination**. Some of the genes from the father and some of the genes from the mother are combined together to form the child's genes therefore the child resembles the parent but is not identical to either parent.

There are also differences (variations) among the children (brothers and sisters) because one child can get a certain combination of genes from the mother and father (example brown hair, blue eyes) and the other child can get a different combination of genes from the mother and father (blond hair, brown eyes). In the next chapter you will learn how this happens.

Let's explain sexual reproduction and genetic recombination in more detail. Every cell in the human being (except the gametes) has two genes for every trait. There are two genes for height, two genes for eye color. For example, the mother has two genes for height, one gene can be for tall and one gene for short. The father has two genes for height, one gene can be for tall and one gene for short. The mother produces the egg (by meiosis which you will learn in the next chapter) which only has one gene for height. The father produces the sperm (by meiosis which you will learn in the next chapter) which also has only one gene for height. The sperm with one gene for height unites with the egg which has only one gene for height to produce a fertilized egg (with two genes for height, one gene from the father and one gene from the mother) that grows into the child. Every cell in the human being (except for the sperm and egg) has two genes for each trait (example height).

Since the sperm only has one gene for height and the egg only has one gene for height, if the sperm with one gene for tall unites with the egg with one gene for tall the child is tall. If from the same parents, a different sperm with one gene for short unites with a different egg with one gene for short, this child is short. Therefore there can be differences (variations) among the children (brothers and sisters), some can be tall and some can be short.

Similarly in sexual reproduction with other animals (dogs, cats, goldfish) or plants. the male gamete (which has the genes from the male) unites with the female gamete (which has the genes from the female) to produce the offspring(child). Similarly, the offspring has the genes from both parents and resemble both parents and not identical to either parent. The offspring (children) look different from each other (variations), some children are tall, some children are short, some flowers are red, others white and others pink, because each offspring gets a different combination of genes from both parents.

PRACTICE QUESTION AND SOLUTION

Question: A child has brown hair and brown eyes. His father has brown hair and blue eyes. His mother has red hair and brown eyes. The best explanation for the child having brown hair and brown eyes is that

(1) a gene mutation occurred that resulted in brown hair and brown eyes

(2) gene expression must change in each generation so evolution can occur

(3) the child received genetic information from each parent

(4) cells from his mother's eyes were present in the fertilized egg

Solution: Since the child receives (gets) genes (genetic information) from the mother and also from the father, therefore the child can have some traits like the mother (brown eyes) and some like the father (brown hair). Answer *3*

Now Do Homework Questions #1-12, pages 27-28.

DNA

The **DNA molecule** (a very large molecule) **has** the **genetic code** (genes) that tells the plant, animal or one celled organism how to look and how to function (work). A gene, which is part (a segment) of DNA, has the genetic code (genetic information) for one trait, such as eye color. Many genes are on one chromosome, which has DNA. DNA controls how proteins are made (DNA controls the production of protein) in a cell.

Look at the figure of the DNA molecule. The DNA molecule is in the shape of a double helix (looks like a twisted ladder or a coil). Look at the figure again. On the sides of the ladder is sugar (S), then phosphate (P), then sugar (S), then phosphate (P), then sugar (S), then phosphate (P).

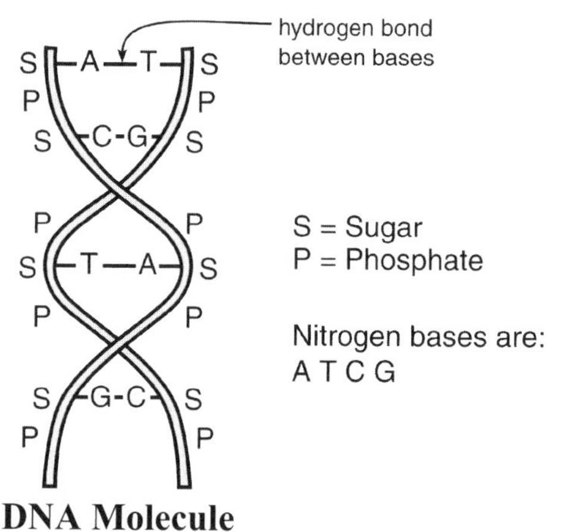

DNA Molecule

On the steps of the ladder are **nitrogen bases: adenine**, represented by the letter **A, thymine** represented by the letter **T, cytosine** represented by the letter **C** and **guanine** represented by the letter **G**. The **nitrogen base A** is always **attached to** the **nitrogen base T** (see figure of DNA), or obviously you can say the **nitrogen base T** is always **attached to** the **nitrogen base A.** Nitrogen bases **A and T** are a **nitrogen base pair** because A always goes with T (A is always attached to T).

The **nitrogen base C** is always **attached to** the **nitrogen base G** (see figure of DNA), or obviously, you can say **nitrogen base G** is always **attached to nitrogen base C.** Nitrogen bases **C and G** are a **nitrogen base pair** because C always goes with G (C is always attached to G).

The two nitrogen bases that are attached (example A attached to T) are connected by weak hydrogen bonds, therefore the bond between the nitrogen bases (example A attached to T) can be broken.

You can also say DNA is made up of groups (subunits) of one sugar and one phosphate (on the sides of the ladder) and one nitrogen base (on the steps of the ladder). Each group (subunit) of DNA has the same sugar and the same phosphate, but the nitrogen bases can be A or T or C or G.

A DNA molecule is made up of thousands of subunits (one sugar, one phosphate, and one nitrogen base).

sub-unit S P — A — T — S P sub-unit
S — C-G — S
P P
S — T — A — S
P P
S — G-C — S
P P

S = Sugar
P = Phosphate
Nitrogen bases are:
A T C G

DNA Molecule

A DNA molecule can also be drawn this way:

Look at the left strand or the right strand of the DNA molecule. In this DNA molecule (see figure at right), the sequence of nitrogen bases in the left strand is ACAG, which is part of a genetic code telling the cell how to make a specific protein. Each three nitrogen bases in sequence (example ACA) order a specific amino acid to be put into a protein. (Amino acids join together to form proteins, which may be enzymes, hormones, part of red blood cells, etc.)

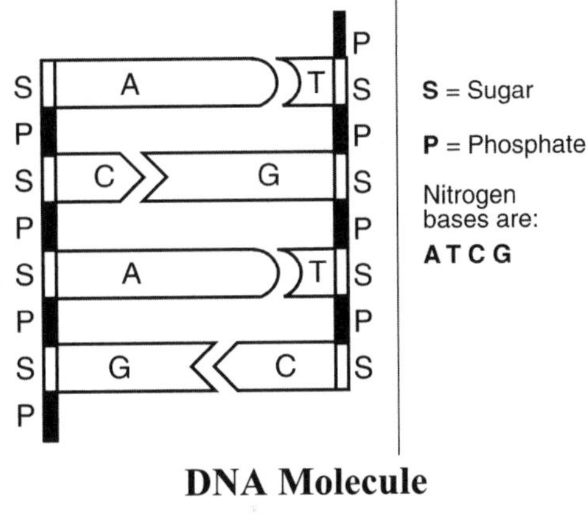

DNA Molecule

In a **DNA molecule,** the **sequence of the nitrogen bases (order of the nitrogen bases) forms the genetic code** telling the cell what to do.

Mutation: When one of the nitrogen bases in DNA, such as base A, is changed, to C or T or G, the **change in DNA** causes a **change in** a **gene** (genetic information), which is called a **mutation**. A **mutation** is a **change** (error) in **DNA** (change in a gene (genetic information), or a change in a chromosome).

A gene, which is part (a segment) of DNA, has the genetic code (genetic information) for one trait, such as eye color. Many genes are on a chromosome (has DNA).

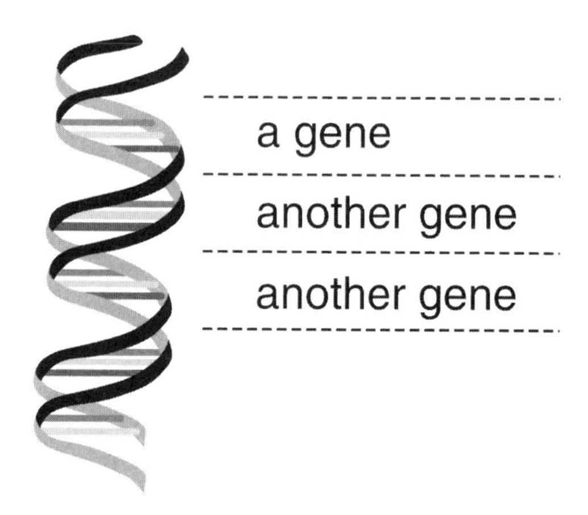

DNA Molecule

Practice Questions and Solutions

Question: Use appropriate letters to write a 9-base DNA sequence that could represent a portion of a gene.

Solution: The nitrogen bases of DNA are A (adenine), T (thymine), C (cytosine), and G (guanine).

Take nine nitrogen bases (make a nitrogen base sequence).

Write any nine nitrogen bases in DNA, making a 9-base DNA sequence, which can make up a part of a gene.

Examples of answers are: AACCTGCTC or CCTACCCAG

Now Do Homework Questions #13-18, pages 29-30.

DNA Replication

A child (or a dog or cat) gets bigger. The cells in your body divide and form more cells. One cell divides and forms two cells, two cells divide and form four cells, four cells divide and form eight cells, so making more cells helps the child (or dog or cat) to grow, to make new cells to replace damaged cells, and to repair damaged cells. When one cell divides to form two cells (you can call it two daughter cells), each cell needs its own DNA molecule (genetic code) to tell that cell what to do. The DNA molecule must replicate (copy) itself and form two DNA molecules, one DNA molecule for one cell and another DNA molecule for the other cell.

Since each cell needs its own DNA molecule, lets see how a DNA molecule replicates (copies) itself and forms more molecules of DNA. Similarly, when a one celled organism (example ameba) reproduces (amebas divide and form two amebas), each ameba needs its own DNA molecule, therefore the DNA molecule must replicate (copy) itself and form more DNA molecules. Figures A-D , at right and below, show the replication (copying) of a DNA molecule to form more DNA molecules. In the DNA molecules at right and below, S = sugar, P = phosphate, and nitrogen bases = ATCG.

DNA Molecule
Figure A

Look at a DNA molecule:

Step 1: Look at Figure A above. The **DNA molecule untwists.** You learned there are weak bonds between the nitrogen bases (base pairs), a weak bond between A and T, C and G, G and C, and T and A. Look at Figure B. The **weak bonds break** and **two strands of DNA separate, forming two single strands**.

weak bonds break
strands separate
Figure B

Step 2: Look at figure C. Each single strand (made of sugar, phosphate, and nitrogen base) is a template (pattern) for a new DNA strand to be produced. The new strand of DNA also has the side made of sugar and phosphate; the new nitrogen base is the base that must attach to the base of the template. If the template base is T, obviously the new base attached to T must be A **(base A pairs with base T**). You learned in a DNA molecule, A is always attached to T. If the template base is G, obviously the new base attached to G must be C **(base C pairs with base G**). If the template base is C, obviously the new base attached to C must be G. If the template base is A, the new base attached to the template base must be T (see Figure C).

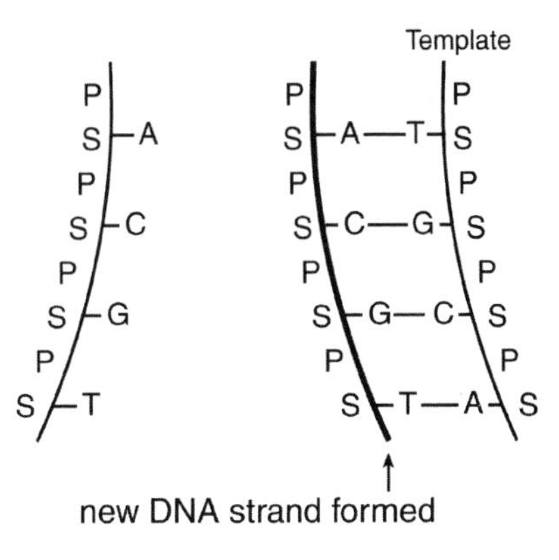

Figure C

Similarly, the other single strand is also a template producing a new strand of DNA (see figure D), just like we discussed in step 2. The new strand of DNA also has the side made of sugar and phosphate; the new nitrogen base is the base that must attach to the base of the template (bases pair up). If the template base is T (see bottom of template), the new base attached to T must be A. You learned in a DNA molecule, A is always attracted to T. If the template base is G, the new base attached to G must be C. If the template base is C, the new base attached to C must be G. If the template base is A, the new base attached to the template base must be T (see Figure D).

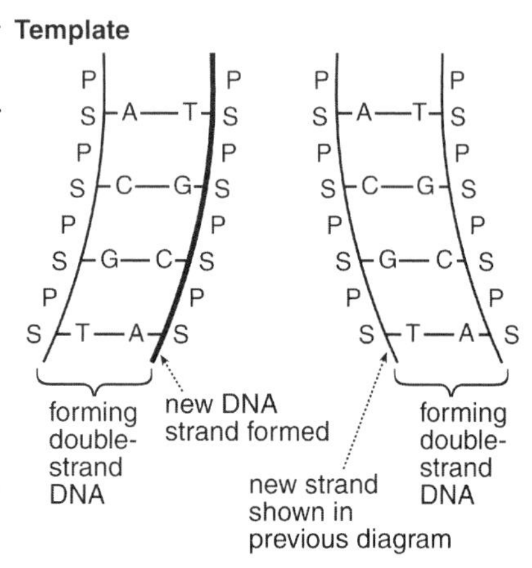

Figure D

Look at figure D. Now you have **two identical double-stranded DNAs; two identical DNA molecules are produced.** When one cell divides and forms two cells, one double-stranded DNA goes to one cell and the identical double-stranded DNA goes to the other cell. (Both cells have the same DNA, which means the same genetic code (same genetic information)). A change in the order of nitrogen bases in DNA (which is a mutation) causes a change in a gene. When an ameba (a one celled organism) divides and forms two amebas, similarly the DNA molecule replicates (copies) itself and one double-stranded DNA molecule goes to one ameba and the other double-stranded DNA molecule goes to the other ameba. Both amebas have the same DNA, which means the same genetic code (genetic information, hereditary information).

DNA controls how proteins are made

Genes are made of DNA, which is in the nucleus. The sequence (order) of bases (example TGC or TAC or TAG) in the gene (or you can say, the sequence of bases in the DNA molecule) will determine which proteins (protein for hormone, cell membrane, enzyme, antibody) will be made (produced). DNA, which is in the nucleus, has the information needed for protein synthesis (joining together of small molecules to form protein (large molecules). **DNA controls how proteins are made** (DNA controls the production of protein) in a cell.

Step 1: The **sequence of bases in** a **gene** (section or piece of **DNA** example CGC, etc.) gives the cell **instructions** to **make** one **messenger RNA (mRNA)** and gives **instructions how** to **make protein** (which amino acids make up a protein).

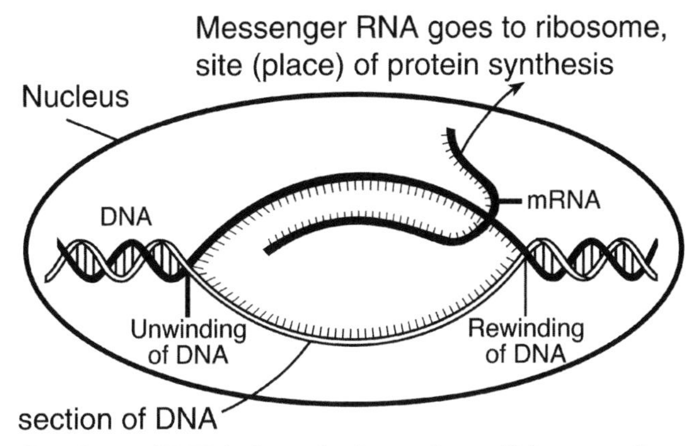

Section of DNA (gene) gives the cell instructions to make messenger RNA

The **messenger RNA** has nitrogen **bases** that **pair up with** the **DNA bases** (in the nucleus), telling how to make one protein (protein synthesis). Look at the chart below. In the DNA base sequence CGC, the cytosine **(C)** base of the DNA molecule **pairs up with** the guanine **(G)** of the messenger RNA or the guanine **(G)** of the DNA molecule **pairs up with** the cytosine **(C)** of the messenger RNA. There is no base thymine (T) in RNA. Instead of thymine (T), RNA has uracil (U). Therefore, **A** in DNA **pairs with U** in messenger RNA. (**T** in DNA **pairs with A** in messenger RNA).

Base Pairs

Base in DNA	C	G	C	A	T	C
Pairs with base in messenger RNA	G	C	G	U	A	G

For example, if DNA has a base sequence ATC,
then messenger RNA (mRNA) has a base sequence UAG.

To summarize Step 1: The **DNA** gives **instructions to make** a **messenger RNA.** The **DNA** also **gives instructions** to the messenger RNA **how to make protein.** The messenger RNA follows the instructions (genetic code) of the DNA; the messenger RNA tells the cell which proteins to make (which amino acids to join together to form a protein).

Step 2: The **messenger RNA leaves** the **nucleus** (see figures above and below) and **goes to** the **ribosomes** where proteins are made. The ribosome is in the cytoplasm (jellylike living material inside the cell). Look at the diagram below. The **messenger RNA has bases such as CGC, AUA,** etc.

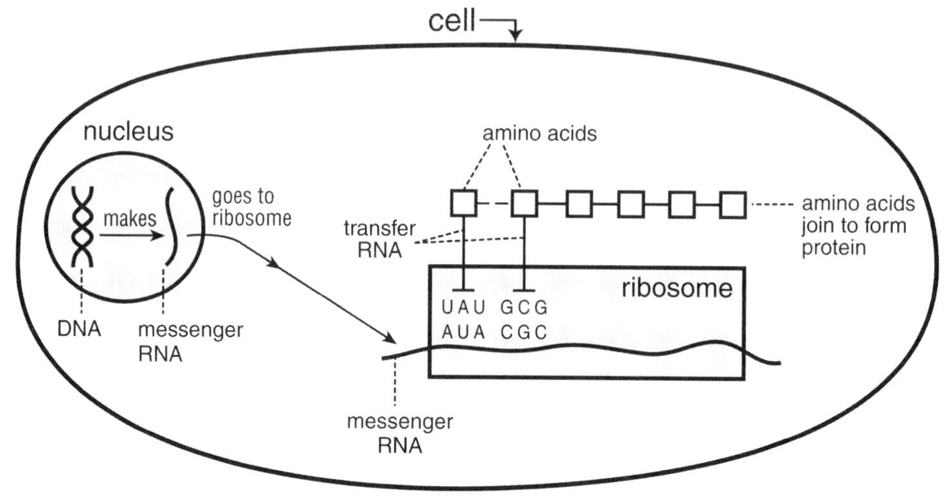

Synthesis of Proteins

Step 3: You saw the messenger RNA, in this example, had CGC and AUA. The **transfer RNA brings** a specific **amino acid to** the **ribosome** (see diagram above). As you can see, the **three nitrogen bases of the transfer RNA,** such as **GCG, pair with** the **three nitrogen bases of the messenger RNA,** such as **CGC.** You learned that G pairs with C and C pairs with G. Similarly, if the messenger RNA has the nitrogen bases AUA, the transfer RNA has the nitrogen bases UAU, because A pairs with U, U pairs with A, and C pairs with G.

Note: The three nitrogen bases, either on the DNA or on the messenger RNA (example CGC), tell which specific amino acid is brought to the ribosome. The transfer RNA brings the amino acid to the ribosome.

Step 4: Look at the diagram again. The **amino acids join together** (synthesis) to **form** a **specific protein** (example enzyme, hormone, cell part, antibody) according to the original instructions from the DNA. To explain: Look at the diagram above.

1. The DNA gives instructions to make a messenger RNA and gives instructions (to the messenger RNA) how to make proteins (which amino acids make up a protein).

2. A messenger RNA carries the message of how proteins are made to the ribosomes in the cell.

3. The amino acids are brought to the ribosome by the transfer RNA.

4. Amino acids join together to form protein.

If some animals and plants have similar proteins, you realize the sequence of the DNA bases that control (direct) the making of these proteins are similar.

You learned, **DNA** (with its genetic code (genetic information)) **controls** (directs) how thousands of **proteins** are **made** (protein synthesis). For example, the **DNA** (DNA code) can have the **sequence of nitrogen bases GTCAAA.** A sequence of **three nitrogen bases**, example GTC, **codes** for **one specific amino acid** (tells which one amino acid should be put into a protein). In the base sequence GTCAAA above, GTC (DNA code) causes the amino acid named glutamine to be put into a protein. In the same sequence GTCAAA, the next three bases (DNA code) AAA cause the next amino acid, phenylalanine, to be put into the protein (protein chain). You do not have to memorize which three nitrogen bases cause which amino acid to be put into protein, just look at the chart below. In the **DNA base sequence** (sequence of nitrogen bases) **GTCAAA,** the three nitrogen bases GTC (DNA code) code for the amino acid glutamine (abbreviated Gln), see chart below. The next three bases AAA (DNA code) code for the next amino acid phenylalanine (Phe), see chart below. If there were three more nitrogen bases (example TTA) after GTCAAA, (example GTCAAA**TTA**) a third amino acid (asparagine, Asn) would be added to the protein. Amino acids join together to form protein (protein chain). Look again at the chart below. As you can see, any three nitrogen bases in DNA cause one specific amino acid to be put into protein (example AAG puts amino acid phenylalanine into a protein, ACC puts amino acid tryptophan into a protein, and AGA puts amino acid serine into a protein). (You do not have to memorize which three bases code for each amino acid, just look at a chart which has the DNA code for each amino acid.)

TABLE

DNA Code	Amino acid	Abbreviation
GTC	Glutamine	Gln
AAA, AAG	Phenylalanine	Phe
TTA	Asparagine	Asn
ACC	Tryptophan	Trp
AGA	Serine	Ser
CCA	Glycine	Gly

Question: In DNA, a sequence of three bases is a code for the placement of a certain amino acid in a protein chain. The table below shows some amino acids with their abbreviations and DNA codes.

Amino Acid	Abbreviation	DNA Code
Phenylalanine	Phe	AAA, AAG
Tryptophan	Try	ACC
Serine	Ser	AGA, AGG, AGT, AGC, TCA, TCG
Valine	Val	CAA, CAG, CAT, CAC
Proline	Pro	GGA, GGG, GGT, GGC
Glutamine	Glu	GTT, GTC
Threonine	Thr	TGA, TGG, TGT, TGC
Asparagine	Asp	TTA, TTG

Which amino acid chain would be produced by the DNA base sequence below?

C-A-A-G-T-T-A-A-A-T-T-A

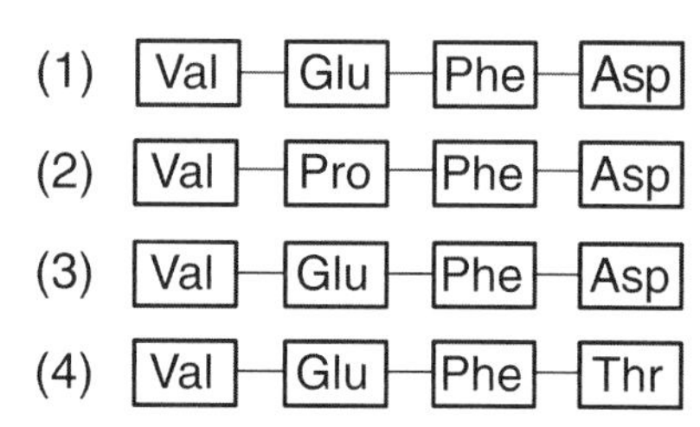

Solution: Look at the **DNA base sequence CAAGTTAAATTA** given in the question; look at the nitrogen bases in groups of three:

CAAGTTAAATTA

Three nitrogen bases code for one amino acid.

Look at the first three bases, CAA, of the DNA base sequence. Now look at the amino acid, abbreviation, and DNA code table above (given in the question) for CAA (DNA code).

Part of Table

Amino Acid	Abbreviation	DNA Code
Valine	Val	CAA, CAG, CAT, CAC

The DNA code CAA is for the amino acid valine, therefore valine is the first amino acid in the protein chain.

Look at the next three nitrogen bases, GTT, of the DNA base sequence. Look again at the table (given in the question) for GTT (DNA code).

The DNA code GTT is for the amino acid glutamine, therefore the next amino acid in the protein chain is glutamine.

The next three nitrogen bases are AAA. Look again at the table (given in the question) for AAA (DNA code). AAA is the code for the amino acid phenylalanine, therefore phenylalanine is the next amino acid in the protein chain.

The last three nitrogen bases are TTA. Look again at the table (given in the question) for TTA (DNA code). TTA is the code for the amino acid asparagine, therefore asparagine is the next amino acid in the protein chain.

In this order, the amino acids valine, glutamine, phenylalanine, and asparagine join together, forming a protein chain.

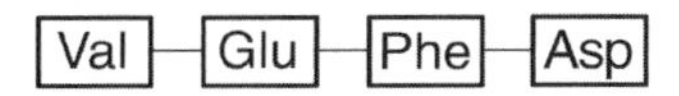

Answer 1

Which amino acids are in the protein chain and the order of the amino acids is determined by the sequence (order) of the bases in the DNA (DNA code).

Question: Base your answers to questions 1-3 on the Universal Genetic Code Chart below and on your knowledge of biology. Some DNA, RNA, and amino acid information from the analysis of a gene present in three different species is shown in the second chart (Species A, Species B, Species C).

Universal Genetic Code Chart

Messenger RNA Codons and Amino Acids for Which They Code

First base	Second base: U	Second base: C	Second base: A	Second base: G	Third base
U	UUU PHE	UCU SER	UAU TYR	UGU CYS	U
U	UUC PHE	UCC SER	UAC TYR	UGC CYS	C
U	UUA LEU	UCA SER	UAA STOP	UGA STOP	A
U	UUG LEU	UCG SER	UAG STOP	UGG TRP	G
C	CUU LEU	CCU PRO	CAU HIS	CGU ARG	U
C	CUC LEU	CCC PRO	CAC HIS	CGC ARG	C
C	CUA LEU	CCA PRO	CAA GLN	CGA ARG	A
C	CUG LEU	CCG PRO	CAG GLN	CGG ARG	G
A	AUU ILE	ACU THR	AAU ASN	AGU SER	U
A	AUC ILE	ACC THR	AAC ASN	AGC SER	C
A	AUA ILE	ACA THR	AAA LYS	AGA ARG	A
A	AUG MET or START	ACG THR	AAG LYS	AGG ARG	G
G	GUU VAL	GCU ALA	GAU ASP	GGU GLY	U
G	GUC VAL	GCC ALA	GAC ASP	GGC GLY	C
G	GUA VAL	GCA ALA	GAA GLU	GGA GLY	A
G	GUG VAL	GCG ALA	GAG GLU	GGG GLY	G

Question 1: Using the Universal Genetic Code Chart, fill in the missing amino acids in the amino acid sequence for species A in the chart below.

Question 2: Using the information given, fill in the missing mRNA bases in the mRNA strand for species B in the chart below.

Question 3: Using the information given, fill in the missing DNA bases in the DNA strand for species C in the chart below.

Species A	DNA strand:	TAC	CGA
	mRNA strand:	AUG	GCU
	Amino acid sequence:	____	____
Species B	DNA strand:	TAC	TTT
	mRNA strand:	____	____
	Amino acid sequence:	MET	LYS
Species C	DNA strand:	____	____
	mRNA strand:	AUG	UUU
	Amino acid sequence:	MET	PHE

The Universal Genetic Code chart shows messenger RNA (mRNA) codons (example UUU, UUC, UUA, UUG) with the amino acids they code. Look at the Genetic Code Chart called Messenger RNA Codons and Amino Acids for Which They Code. Messenger RNA UUU or UUC codes for the amino acid phenylalanine (PHE). The messenger RNA UUA or UUG codes for the amino acid leucine (LEU). The messenger RNAs CUU, CUC, CUA, and CUG also code for the amino acid leucine (LEU) (see genetic code chart above).

Solution 1: In species A you see that the DNA bases TAC pair up with the messenger RNA (mRNA) bases AUG; the messenger RNA codes for amino acids.

Look at the genetic code chart in the question. Messenger RNA AUG codes for the amino acid methionine (MET).

In species A, you see that the next three DNA bases are CGA. They pair up with the messenger RNA bases GCU. Look again at the chart in the question with messenger RNA's and amino acids. Messenger RNA GCU codes for the amino acid alanine (ALA). Fill in the amino acid sequence for species A above, Met Ala.

Solution 2: In species B, you are given the DNA base sequence. If the nitrogen base sequence is TAC, the messenger RNA (mRNA) would be AUG. T pairs with A, A pairs with U, and C pairs with G.

In species B, you see that the next three DNA bases are TTT. The messenger RNA (mRNA) would be AAA. T pairs with A. Fill in the mRNA strand for species B above, AUG AAA.

Solution 3: In species C, you are given the messenger RNA. If the messenger RNA (mRNA) bases are AUG, then the DNA must be TAC. A (of the RNA) pairs with T (of the DNA), U (of the RNA) pairs with A (of the DNA), and G (of the RNA) pairs with C (of the DNA).

In species C, you see that the next three bases in the messenger RNA (mRNA) are UUU. The DNA bases must be AAA. U (of the RNA) pairs with A (of the DNA). Fill in the DNA strand for species C above, TAC AAA.

If species A, B, C, D, and E have these amino acid sequences:

Species	Amino Acid Sequence
A	Met-Ala-Gly-Ser
B	Met-Lys-Arg-Pro
C	Met-Phe-Cys-Pro
D	Met-His-Gln-Arg
E	Met-Lys-Arg-Pro

you realize that species B and E are most closely related because they have the same amino acid sequence (same amino acids in the same order).

The work of the cell (example, making parts of cells such as ribosomes, cell membrane) is carried out by many different types of molecules, mostly proteins. Proteins form cells and parts of cells (example ribosomes, cell membrane) that carry on life functions (examples protein synthesis, transport). Protein also makes enzymes, hormones, and antibodies (enzymes, hormones, and antibodies are made of protein).

Protein molecules are long, usually folded chains made from 20 different kinds of amino acids in a specific sequence (order). The sequence (order) of amino acids influences the shape (folding) of the protein; the shape of the protein determines its function. The **enzyme** (a protein) to digest starch must have a **specific shape** so that the **starch molecule** can fit **exactly into the enzyme** (like a lock and key), then the enzyme (a protein) can do its work and digest starch into sugar.

Similarly, hormones and antibodies also have a specific shape in order for the hormones and antibodies to function (do their work).

Now Do Homework Questions #19-39, pages 30-34.

Mutation

Mutation is a **change** (error) in a **gene** (genetic information) or a change (error) in a chromosome in a living thing (example plants, animals, bacteria). **Mutation** is a **change** in the **DNA sequence**, which means a change in the sequence (order) of nitrogen bases in the DNA (example, TACGCTA changed to TAACGCTA) causes a change in which amino acids are put into the protein produced. Since mutations cause different proteins (examples enzymes, hormones) to be produced, there will be differences (variations) among the organisms (living things).

Living things with mutations get new, changed, or different characteristics. For example: hemophilia is a disease caused by a mutation; when a person with hemophilia gets a cut, it does not stop bleeding,

Example: DNA sequence (order) of nitrogen bases is TACGCTA. Let's make four kind of mutations or changes in the DNA sequence.

Mutation: Change in the Nitrogen Bases of DNA

Base sequence	Type of Mutation
TACGCTA	Original sequence
TA**A**CGCTA	Addition (A was added)
T_CGCTA	Deletion (A was taken away)
TA**G**GCTA	Substitution (G was substituted for C)
CATGCTA	Reversal of order of bases (CAT is reverse of TAC)

A mutation is a change (alteration, error) of the DNA sequence (order) of nitrogen bases. A mutation can be caused by a nitrogen base being added (inserted), deleted (taken away), substituted by another base, or the order of bases can be reversed. If a base is added or deleted, it affects all the bases after that.

A sequence of three nitrogen bases (example TAG) codes for one specific amino acid (tells which one amino acid should be put into a protein).

For example, the original sequence of nitrogen bases on DNA (DNA code) is TACGCTA. TAC (DNA code) causes the amino acid methionine (MET) to be put into the protein. The next three bases GCT (DNA code) cause the amino acid arginine (ARG) to be put into the protein. (You do not have to memorize which three nitrogen bases cause which amino acid to be put into protein.)

If base A is added to the sequence of nitrogen bases (in this example, at the beginning of the sequence), the new sequence of nitrogen bases on DNA (DNA code) is ATACGCTA. As you can see, the first three nitrogen bases are ATA. ATA (DNA code) causes the amino acid named tyrosine (TYR) to be put into a protein (protein chain). The next three nitrogen bases CGC (DNA code) cause the amino acid alanine (ALA) to be put into the protein (protein chain). (You do not have to memorize which three nitrogen bases cause which amino acid to be put into protein.)

You can see in this example, when nitrogen base A is added, the new sequence has two different amino acids (example tyrosine and alanine) made into protein; therefore a different protein is produced. A different protein can be a defective protein (such as an enzyme, hormone, antibody that cannot do its work).

A mutation (adding, deleting, substituting, or reversing the sequence of bases in DNA) causes a change in the protein (example the shape of the protein is changed), therefore the protein (example enzyme) is defective (not working the way it should).

An example of a mutation (substitution of a nitrogen base) is sickle cell disease (sickle cell anemia). A red blood cell is normally round, but in sickle cell disease a red blood cell is sickle shaped and cannot do its function (work).

red blood cell

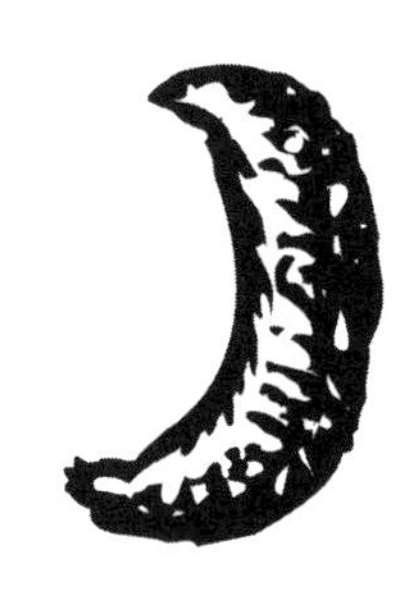

Sickle-shaped red blood cell

A **mutated gene** (altered gene, changed gene) can be **passed on** to **every cell** that **develops from** the **cell** with the **mutated gene**. If one cell with an altered (changed) gene divides and forms two cells, then both cells will have the mutation. Changes (mutation) in DNA are passed on to new cells. Changes in messenger RNA (mRNA) or transfer RNA (tRNA) affect the protein formed, but are not passed on when cells divide.

You learned in sexual reproduction, the male gamete (example sperm) and female gamete (example egg) unite and form a fertilized egg which develops into a child. **Only a mutation** (altered gene) **in** the **sex cells** (male gamete or female gamete) will be **passed on** to the offspring (**child**). A **mutation** in a **stomach** cell or liver cell will **not** be passed on to the child.

Mutations (changes in the sequence (order) of nitrogen bases in DNA) can be **caused** by cells being exposed to too much **radiation** (such as **x rays** or **ultraviolet rays from** the **sun**), and also by **certain chemicals**; mutations increase the chance of getting cancer. Sickle cell disease (sickle cell anemia) is caused by a mutation, a wrong base substituted for the correct base in the

DNA base sequence. Mutations are random and can occur anyplace in the DNA.

Practice Questions and Solutions

Question: A change in the base subunit sequence during DNA replication can result in

(1) variation within an organism

(2) rapid evolution of an organism

(3) synthesis of antigens to protect the cell

(4) recombination of genes within the cell

Solution: A change in the sequence of base subunits in DNA during replication can cause a change in a protein, a defective protein or a different protein.

The new cells that were produced after the change in the base sequence have a different protein than the cells in the body before the mutation. Therefore there is variation within an organism. Answer *1*

Now Do Homework Questions #40-55, pages 34-37.

Different Cells of the Body Use Different Genes, Gene Expression

Each cell in your body and even a fertilized egg cell (made when sperm and egg unite together) **has all the genes** your body needs to carry on digestion, circulation, respiration , excretion which are **life functions**, has **all** the **genes** to **produce** all **cells**, all **tissues**, all **organs** (such as heart, liver, etc.), and all **organ systems** (such as circulatory system, digestive system) , and has genes to **make hormones**, **enzymes**, genes for **height**, **eye color**, etc.

Each cell uses some of the genes (turns the gene on because the cell needs to use it) and does not use other genes (turns them off).

For example, the **gene** to **produce** an **enzyme** to **digest food** in the **stomach** is **used in** the **stomach.** Obviously, that gene is not used in the cells of the heart, lungs, etc. Another example is the gene to produce insulin, which is used (functions) in the cells of the pancreas (when it needs to produce insulin) and is not used in all the other cells of the body. In short, **genes** (which are part of DNA) are **used** (turned on) when the **cell needs** to **use them**, and **genes** are **not used** (turned off) when the **cell doesn't need** to **use** them.

Question: The types of human cells shown below are different from one another, even though they all originated from the same fertilized egg and contain the same genetic information.

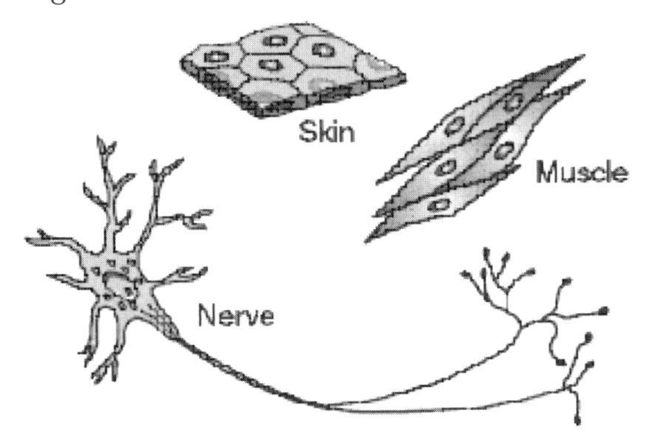

Explain why these genetically identical cells can differ in structure and function.

Solution: All these cells have the same genetic information, which means the same genes. Each **cell** (skin, muscle, or nerve) only **uses** (activates) the **genes** it **needs** for **function** and for **structure** (shape and size of cells). Another possible answer is that the **environment inside** the **body influences** which type of **cells** are **made** (example skin, nerve, or muscle cell).

In short, every body cell has identical DNA (genetic information, genes) in the nucleus of every cell, but each cell only uses (turns on) the genes that are needed to carry on the work (functions) of the cell.

Environment effects on genes and gene expression

The internal (inside the body) and external (outside the body) **environment influence which genes** are **used** (are **activated**) in each cell. The Himalayan **rabbit** has a **gene** for **fur color**. In **cold weather,** the **gene** for **black** fur **color** is turned on, **activated** (used) and the rabbit's fur color is **black** on **ears, feet** and **tail**. Note: This gene gives instructions to make a black protein (black color), which goes into the fur, making the fur black. In **warm weather**, the gene for black fur color is turned off (not used) and the rabbit's fur color is **white**.

Gene expression- In cold weather, the **gene** for **black** fur **color** is **used (activated)** on rabbits' ears, feet, and tail; the gene for black fur is **expressed** (you can see the rabbit has black fur). In **warm weather**, the **gene** for **black** color is **not used** (not activated, **not expressed**) and the fur color is **white.** As you can see, the environment has an effect on gene expression (black or white color) in the Himalayan rabbit.

The **environment** has an **effect** on **gene expression** in **plants**. Plants have genes to produce chlorophyll, a green pigment (green color) that is needed for photosynthesis. When there is **sunlight,** the plant uses (activates) the gene to make chlorophyll (green pigment). The gene for green color is expressed (you can see the **plant** is **green**). When there is always **no sunlight** (always dark), the plant does not use the gene to make chlorophyll (green pigment) and the **plant** is **not green.**

The **environment** also has an **effect** on **human genes (DNA)**. **Identical twins** (meaning they have **identical genes**, identical genetic information) who were brought up in two different environments (example different places such as a big city or a farm) became different. The different environments (example different places) caused different genes to be activated (used) in each of the twins.

Environment influences what types of cells a fertilized egg will develop into (examples skin, nerve, muscle). Environment also can influence which types of cells a "stem cell" can develop into.

Learned activities, subjects and experiences

Learned subjects (example learning to type, history, reading) are not passed on to his or her offspring(children) because the genes were not affected. If a person **practices playing baseball, basketball**, or **tennis**, he improves as a baseball, basketball or tennis player. He is better in **sports**; the genes were not changed, therefore being better in sports is not passed on to his children.

The **genes** (base sequence in DNA) in the sperm and in the egg are **passed on to** the **children** and determine the characteristics of the children (example how they look, exceptional ability or talent in music and art).

Now Do Homework Questions #56-75, pages 37-41.

BIOTECHNOLOGY (SELECTIVE BREEDING AND GENETIC ENGINEERING)

Biotechnology is using technology with biological science (biology). Early examples of biotechnology are using enzymes to make cheese and using yeast to make bread and wine. For thousands of years, humans have used **biotechnological methods** (using technology with biological science) to produce **products** (examples better apples, better peaches) or **organisms** (better cows, horses, dogs) with **desirable traits** (characteristics).

Examples (types) **of biotechnology** are **selective breeding** (explained below) and **genetic engineering** (explained after selective breeding).

Selective Breeding

You can combine an apple plant (tree) that has sweet apples and an apple tree with juicy apples to produce **new** baby apple **trees** with **sweet juicy** apples. From these baby plants(trees) with juicy sweet apples, you can plant more of these new desirable plants. Selective **breeding means select (choose) one plant (**or it can be an **animal) with** a **good characteristic** such as **sweet, and unite it with another plant (**or it can be an **animal) with a different good,** desirable, **characteristic such as juicy and you get a new** plant (or it can be an **animal) with both good characteristics such as sweet and juicy.** Another example of selective breeding is a male dog which can run fast is mated (male gamete and female gamete united together) with a female dog which can do tricks, producing a **puppy** which can **run fast and do tricks.** Another example of using selective breeding is to produce commercial peanuts that people are not allergic to. Note: Some people have very serious allergies to peanuts. Varieties (or types) of peanuts that **don't cause allergies** are **crossed** (united) with regular commercial (grown on farms) peanuts (which can **cause allergies**), producing some new **baby plants** that **don't cause allergies**. Selective breeding can produce new varieties of cultivated (grown by man) plants and domestic (not wild) animals with desired traits.

Selective breeding is also used to produce a particular desired characteristic (example racehorses that are very very fast) in animals or plants. Mate two fast racehorses and you can get a very very fast racehorse. The new racehorses might be even faster than the parents. Similarly, cross two marigold plants with very light yellow flowers. Keep crossing the lightest yellow plants over and over again; finally you can even get a white-flowered plant, which some plant growers want.

A disadvantage of selective breeding is that the offspring (children) can get undesirable traits (traits that are not good) together with the desired trait (trait which the person was trying to get, example very fast racehorses). Racehorses bred for speed can have legs that break easily; dogs that are bred for appearance may have problems with their hips.

Question: Explain the technique illustrated below.

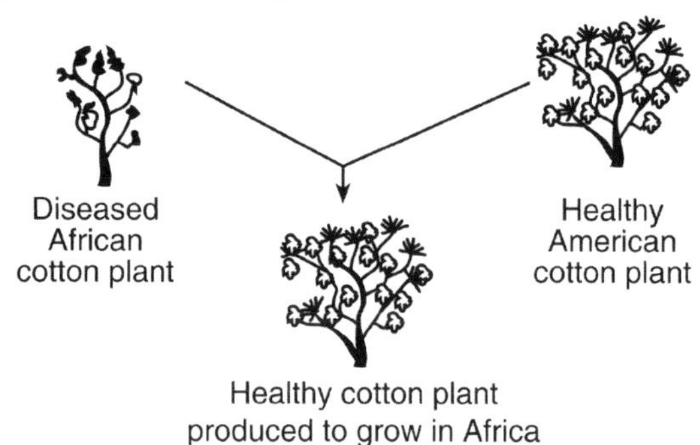

Solution: This technique is called selective breeding. Mate two different parent plants to get offspring which have desirable characteristics from both parents. The diagram shows the offspring that are healthy and grow in Africa.

Now Do Homework Questions #76-84, pages 41-42.

Genetic Engineering

In recent years, **new varieties** of **plants**, bacteria, and **animals** have been produced (**engineered, genetic engineering)** by **manipulating** (**changing**, altering) **their genes** (**genetic information**) to produce **new characteristics**, new combinations of traits or new (better, improved) varieties of plants or animals. Using genetic engineering, organisms (plants or bacteria) can be made to produce new substances (example insulin).

In an organism with a desirable trait (example gene to produce insulin), first identify the gene (find (locate) the piece of DNA that has the gene for the desirable trait). Take out **(remove)** the segment of DNA (the gene) from the organism and put the DNA (which orders the making of insulin) into a new organism, therefore the new organism can also have the desirable trait (able to produce insulin). The insulin can then be extracted from the new organism.

Examples of genetic engineering:

1. The **DNA** from a certain fish **tells** the fish to **produce** a **protein** (like **antifreeze**) which prevents freezing. The gene (segment of **DNA**) to prevent freezing is removed **from** the **fish** and **inserted** (put) **into** the **strawberry plant**, ordering the plant to also produce a protein to **prevent freezing.** Now the new strawberry plant can survive in very cold weather (survive the effect of freezing (frost)), in addition to the characteristics it had before (example producing a lot of strawberries on the plant). The strawberry plant now has the improved characteristic of being able to survive in freezing weather.

2. **Identify the gene** (piece of DNA) that produces insulin in a human cell. An **enzyme** is used to **cut (remove)** a **segment** (piece) of **DNA** (a gene)(which orders **insulin** to be **produced) from** a **human** cell; the **DNA** (from the human cell) is **inserted** (put) **into** a **bacterial cell**. Now the **bacterial cell** will **produce** the **human insulin,** in addition to whatever the bacteria previously produced and did.

When the bacterial cell divides and forms more bacterial cells (by mitosis), all these new bacterial cells can produce insulin in addition to what the bacteria produces and does. The **insulin** can be taken out **(extracted)** from the bacterial cell and used by people. Lots of insulin can be produced at low cost without being contaminated. You learned diabetics use insulin to regulate sugar in the blood.

How is genetic engineering done: Let's explain how this is done in more detail. **First identify the gene** (piece of DNA). Look at the diagram below. Using a specific enzyme, a piece of human DNA that produces insulin is cut **(removed),** and a circular piece of DNA from a bacterial cell is also cut. Different enzymes are used to cut, copy, and move segments of DNA.

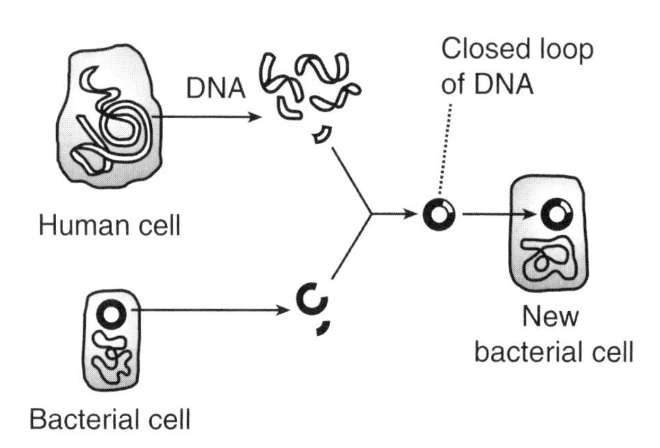

Genetic Engineering

Using enzymes, a piece of human DNA (a gene) joins together with an open loop of bacterial DNA, forming a dosed loop of DNA (made up of both human DNA and bacterial DNA) see diagram above.

Using an enzyme, the closed loop is **inserted** (put) into a new bacterial cell, causing the new bacteria to produce insulin, in addition to whatever the bacteria produces and does.

When **new bacterial cells** (can be shown as 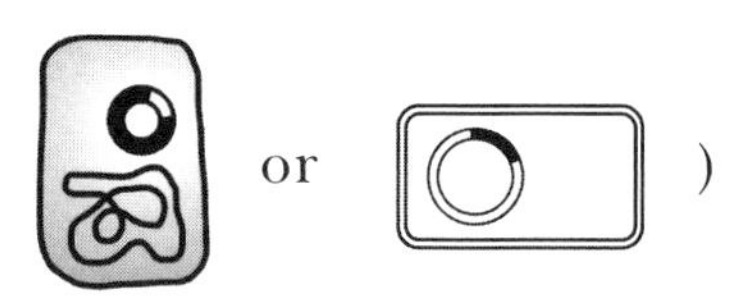)

divide and form **more cells** (by **mitosis**)(see next figure),

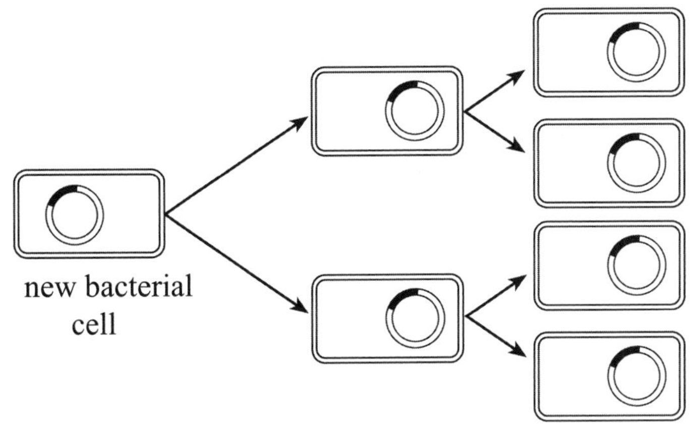

New Bacterial Cell with Inserted Human Insulin Gene Produces Cells with Human Insulin Gene (Produce Insulin)

all these cells will produce insulin in addition to whatever the bacteria produced and did. The **insulin** can be taken out **(extracted)** from the bacterial cells and used by people. Lots of insulin can be produced at low cost without being contaminated. This is asexual reproduction (only one parent), because one bacterial cell alone produces two cells. In genetic engineering, **different enzymes** are used to **cut, insert**, move, copy, etc., a segment of DNA (a gene).

3. Genetic engineering is used to insert (put) genes into crops like corn. Corn crops are damaged by weeds (which compete with the corn for nutrients) and are damaged by insects. One gene put into corn protects the corn from chemicals that are used to kill weeds. Another gene (from bacteria) put into corn makes a protein that kills insects (insect control), therefore, the farmer does not need to spray pesticides (insecticides) to kill the insects. Using genetic engineering (inserting genes into crops) helps the farmer because his crop is not damaged when he uses chemicals to kill weeds and he does not have to spray pesticides (insecticides) to kill the insects.

4. A detective might collect the DNA at a crime scene to determine the criminal. Enzymes can copy and produce more of the DNA that is found at the scene so the detective can figure out who is the criminal.

Inserting, deleting, substituting DNA segments: Inserting, deleting (taking away DNA) or substituting DNA segments (changing nitrogen bases) can alter (change) the genes. For example, by altering (changing) the DNA of a peanut (which causes allergies) the altered (changed) DNA can now produce a protein which causes fewer (less) allergies. An altered gene may be passed on to every cell that develops from it.

Some people are concerned that added genes, example fish genes, put into plants may cause an allergic reaction. Similarly, some people are afraid that any added genes will be harmful if the food with the gene is eaten for a long period of time.

Benefits from knowledge of genetics

Knowledge of genetics or human genes makes possible new fields of health care. For example, finding genes which may have mutations that cause disease will help us in finding ways to prevent the disease and fight it. Substances such as hormones and enzymes from genetically engineered organisms may reduce the cost and side effects of replacing missing body chemicals.

Some people are concerned that people whose genes show that they may get sick (get a disease) may be denied jobs or insurance.

PRACTICE QUESTION AND SOLUTION

Question: Which process is illustrated in the diagram below?

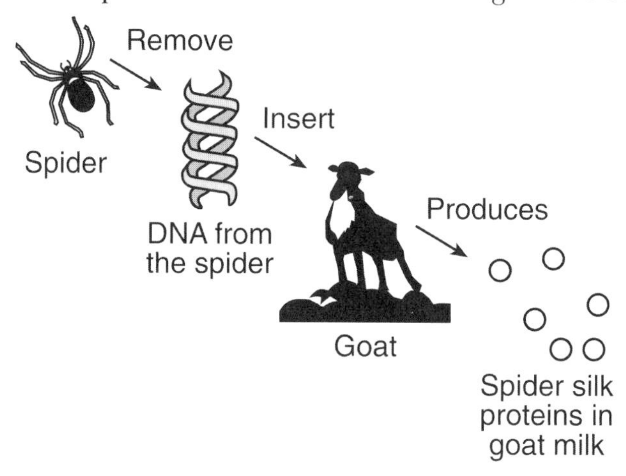

(1) chromatography
(2) direct harvesting
(3) meiosis
(4) genetic engineering

Solution: DNA is removed from a spider and **inserted** (put) **into** a **goat,** telling (ordering) the goat to produce spider silk proteins.

You learned, in genetic engineering, enzymes are used to cut a piece of DNA from one organism (example spider) and to insert the DNA into another organism (example goat). **Genetic engineering** is removing a piece of DNA (a gene) from one organism (example spider) and inserting the DNA from the spider into a new organism (example goat)(see diagram above); the DNA from the spider (the first organism) orders (tells) the goat to make spider silk (a desirable trait). Answer *4*

Now Do Homework Questions #85-101, pages 42-46.

NOW LET'S TRY A FEW HOMEWORK QUESTIONS

1. Which diagram best represents the relative locations of the structures in the list below?

   A–chromosome
   B–nucleus
   C–cell
   D–gene

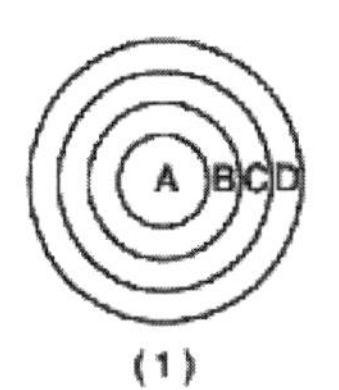

(1)

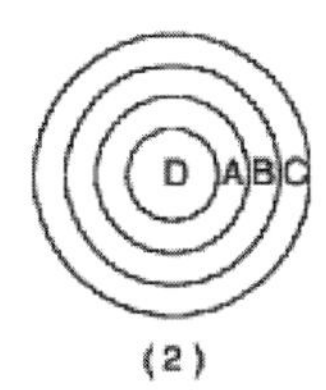

(2)

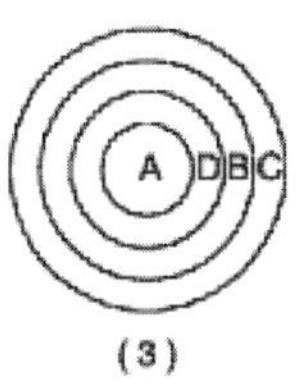

(3)

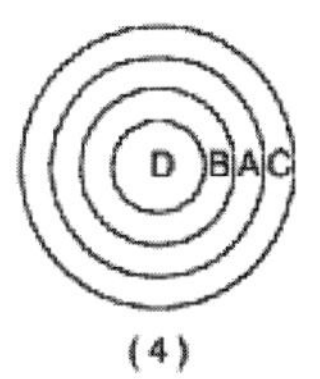

(4)

2. Which sequence of terms represents a ***decrease*** from the greatest number of structures to the least number of structures present in a cell?
   (1) nucleus → gene → chromosome
   (2) gene → nucleus → chromosome
   (3) gene → chromosome → nucleus
   (4) chromosome → gene → nucleus

3. Which statement concerning proteins is not correct?
   (1) Proteins are long, usually folded, chains.
   (2) The shape of a protein molecule determines its function.
   (3) Proteins can be broken down and used for energy.
   (4) Proteins are bonded together, resulting in simple sugars.

4. Which amino acid chain would be produced by the DNA base sequence below?

   C-A-A-G-T-T-A-A-A-T-T-A-T-T-G-T-G-A

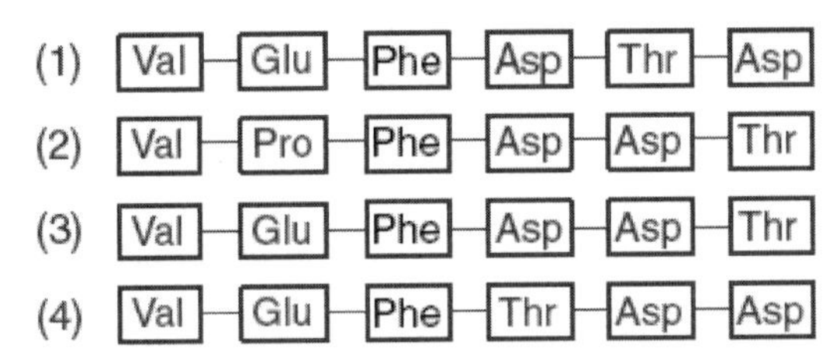

5. The transfer of genes from parents to their offspring is known as
   (1) differentiation (2) heredity (3) immunity (4) evolution

6. The diagram below represents single-celled organism ***A*** dividing by mitosis to form cells ***B*** and ***C***.

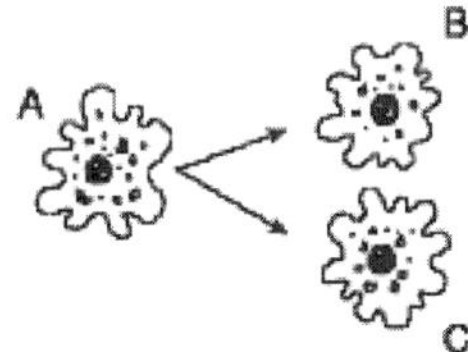

   Cells ***A***, ***B***, and ***C*** all produced protein ***X***. What can best be inferred from this observation?
   (1) Protein ***X*** is found in all organisms.
   (2) The gene for protein ***X*** is found in single-celled organisms, only.
   (3) Cells ***A***, ***B***, and ***C*** ingested food containing the gene to produce protein ***X***.
   (4) The gene to produce protein ***X*** was passed from cell ***A*** to cells ***B*** and ***C***.

7. Strawberries can reproduce by means of runners, which are stems that grow horizontally along the ground. At the region of the runner that touches the ground, a new plant develops. The new plant is genetically identical to the parent because
   (1) it was produced sexually
   (2) nuclei traveled to the new plant through the runner to fertilize it
   (3) it was produced asexually
   (4) there were no other strawberry plants in the area to provide fertilization

8. Chromosomes can be described as
   (1) large molecules that have only one function
   (2) folded chains of bonded glucose molecules
   (3) reproductive cells composed of molecular bases
   (4) coiled strands of genetic material

9. The diagram below represents an incomplete section of a DNA molecule. The boxes represent unidentified bases.

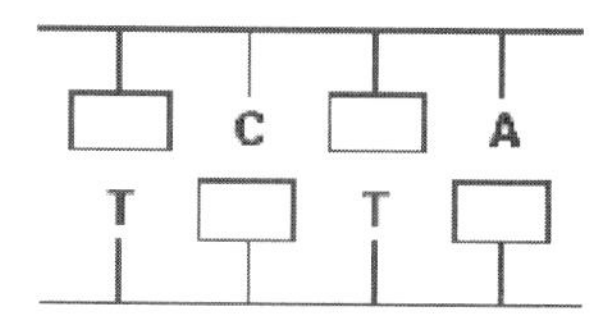

When the boxes are filled in, the total number of bases represented by the letter ***A*** (both inside and outside the boxes) will be
   (1) 1 (2) 2 (3) 3 (4) 4

10. A single pair of goldfish in an aquarium produced a large number of offspring. These offspring showed variations in body shape and coloration. The most likely explanation for these variations is that the
   (1) offspring were adapting to different environments
   (2) offspring were produced from different combinations of genes
   (3) parent fish had not been exposed to mutagenic agents
   (4) parent fish had not reproduced sexually

11. Which statements best describe the relationship between the terms ***chromosomes***, ***genes***, and ***nuclei***?
   (1) Chromosomes are found on genes. Genes are found in nuclei.
   (2) Chromosomes are found in nuclei. Nuclei are found in genes.
   (3) Genes are found on chromosomes. Chromosomes are found in nuclei.
   (4) Genes are found in nuclei. Nuclei are found in chromosomes.

12. Which diagram represents the relative sizes of the structures listed below?

Structures

A	gene
B	cell
C	chromosome
D	nucleus

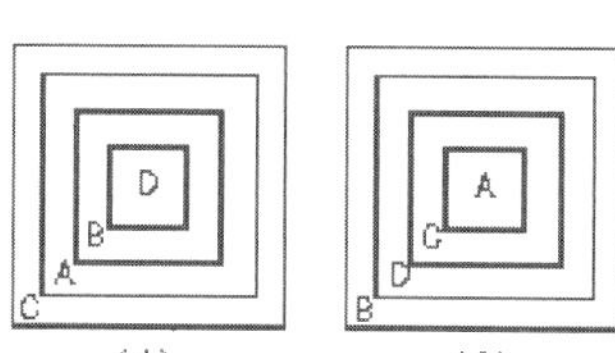

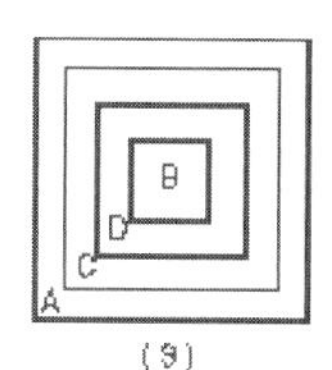

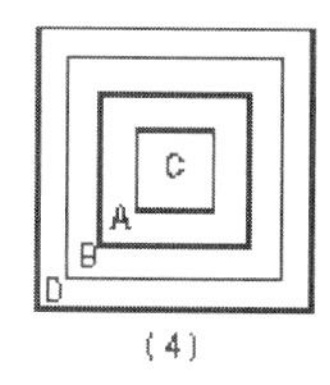

Hint: Smallest box (structure) is inside/part of all the other boxes (structures).

13. Chromosomes can be described as
    (1) large molecules that have only one function
    (2) folded chains of bonded glucose molecules
    (3) reproductive cells composed of molecular bases
    (4) coiled strands of genetic material

14. Three structures are represented in the diagram below.

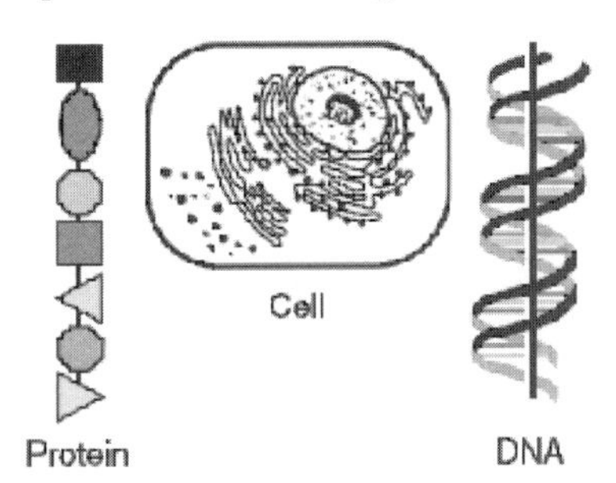

What is the relationship between these three structures?
    (1) DNA is made up of proteins that are synthesized in the cell.
    (2) Protein is composed of DNA that is stored in the cell.
    (3) DNA controls the production of protein in the cell.
    (4) The cell is composed only of DNA and protein.

15. Synthesis of a defective protein may result from an alteration in
    (1) vacuole shape
    (2) the number of mitochondria
    (3) a base sequence code
    (4) cellular fat concentration

16. The Y-chromosome carries the SRY gene that codes for the production of testosterone in humans. Occasionally a mutation occurs resulting in the SRY gene being lost from the Y-chromosome and added to the X-chromosome, as shown in the diagram below.

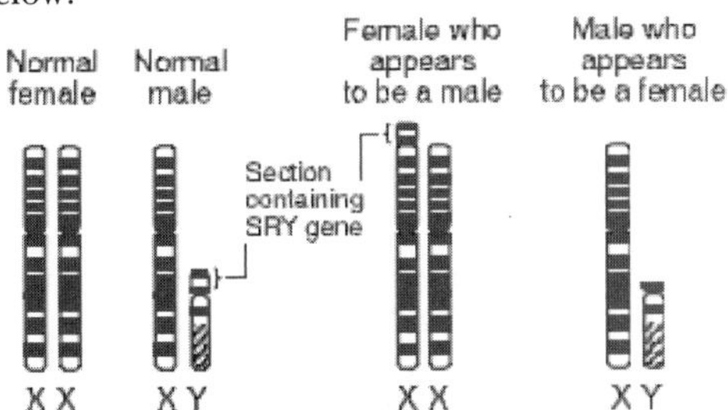

Based on the diagram, which statement is correct?
    (1) The production of testosterone influences the development of male characteristics.
    (2) Reproductive technology has had an important influence on human development.
    (3) Normal female characteristics develop from a single X-chromosome.
    (4) Male characteristics only develop in the absence of X-chromosomes.

17. Asexually reproducing organisms pass on hereditary information as
    (1) sequences of ***A***, ***T***, ***C***, and ***G***
    (2) chains of complex amino acids
    (3) folded protein molecules
    (4) simple inorganic sugars

18. Mustard gas removes guanine (G) from DNA. For developing embryos, exposure to mustard gas can cause serious deformities because guanine
    (1) stores the building blocks of proteins
    (2) supports the structure of ribosomes
    (3) produces energy for genetic transfer
    (4) is part of the genetic code

19. Which statement indicates one difference between the gene that codes for insulin and the gene that codes for testosterone in humans?
    (1) The gene for insulin is replicated in vacuoles, while the gene for testosterone is replicated in mitochondria.
    (2) The gene for insulin has a different sequence of molecular bases than the gene for testosterone.
    (3) The gene for insulin is turned on in liver cells, but the gene for testosterone is not.
    (4) The gene for insulin is a sequence of five different molecular bases while the gene for testosterone is a sequence of only four different molecular bases.

20. Which statement best expresses the relationship between the three structures represented below?

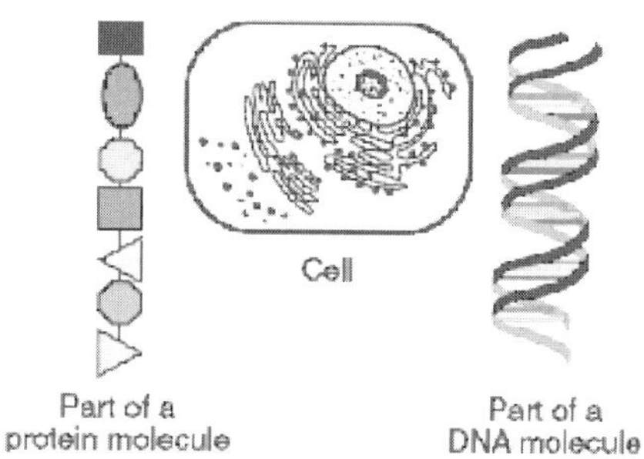

    (1) DNA is produced from protein absorbed by the cell.
    (2) Protein is composed of DNA that is produced in the cell.
    (3) DNA controls the production of protein in the cell.
    (4) Cells make DNA by digesting protein.

21. Base your answer to the following question on the portion of the mRNA codon chart and information below.

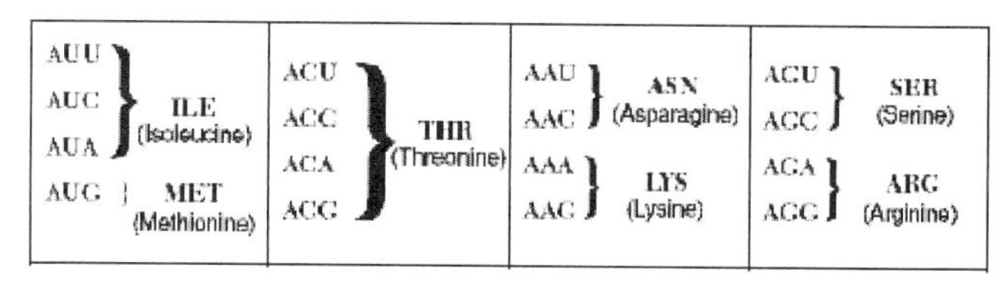

AUU, AUC, AUA } ILE (Isoleucine) AUG } MET (Methionine)	ACU, ACC, ACA, ACG } THR (Threonine)	AAU, AAC } ASN (Asparagine) AAA, AAG } LYS (Lysine)	AGU, AGC } SER (Serine) AGA, AGG } ARG (Arginine)

Series I represents three mRNA codons. Series II includes a mutation of series I:

Series I AGAUCGAGU

Series II ACAUCGAGU

How would the amino acid sequence produced by the mutant strand (series II) compare to the amino acid sequence produced by series I?
    (1) The amino acid sequence would be shorter.
    (2) One amino acid in the sequence would change.
    (3) The amino acid sequence would remain unchanged.
    (4) More than one amino acid in the sequence would change.

22. Some steps involved in DNA replication and protein synthesis are summarized in the table below.

Step A	DNA is copied and each new cell gets a full copy.
Step B	Information copied from DNA moves to the cytoplasm.
Step C	Proteins are assembled at the ribosomes.
Step D	Proteins fold and begin functioning.

In which step would a mutation lead directly to the formation of an altered gene?
(1) *A* (2) *B* (3) *C* (4) *D*

23. Describe how a protein would be changed if a base sequence mutates from GGA to TGA.

24. Identify ***one*** environmental factor that could cause a base sequence in DNA to be changed to a different base sequence.

25. In the human pancreas, acinar cells produce digestive enzymes and beta cells produce insulin. The best explanation for this is that
    (1) a mutation occurs in the beta cells to produce insulin when the sugar level increases in the blood
    (2) different parts of an individual's DNA are used to direct the synthesis of different proteins in different types of cells
    (3) lowered sugar levels cause the production of insulin in acinar cells to help maintain homeostasis
    (4) the genes in acinar cells came from one parent while the genes in beta cells came from the other parent

26. Show ***one*** example of what could happen to the 9-base DNA sequence you wrote in the last question if a mutation occurred in that gene.

27. Genes involved in the production of abnormal red blood cells have an abnormal sequence of
    (1) ATP molecules (3) sugars
    (2) amino acids (4) bases

28. The diagram below shows some of the steps in protein synthesis.

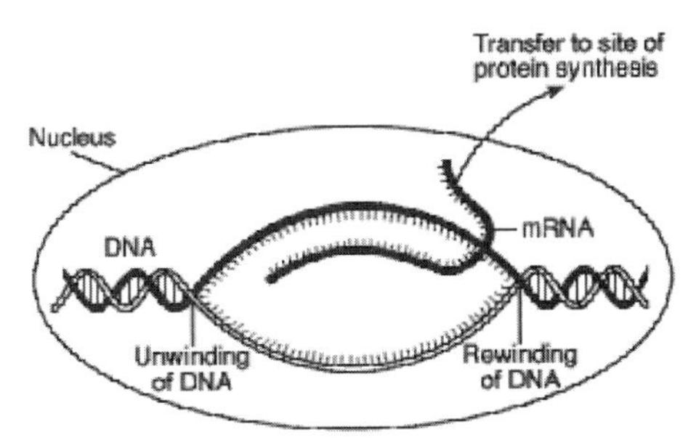

The section of DNA being used to make the strand of mRNA is known as a
    (1) carbohydrate (3) ribosome
    (2) gene (4) chromosome

Base your answers to the next four questions on the Universal Genetic Code Chart below and on your knowledge of biology. Some DNA, RNA, and amino acid information from the analysis of a gene present in five different species is shown in the chart provided.

Universal Genetic Code Chart

Messenger RNA Codons and Amino Acids for Which They Code

First base	Second base: U	Second base: C	Second base: A	Second base: G	Third base
U	UUU, UUC } PHE UUA, UUG } LEU	UCU, UCC, UCA, UCG } SER	UAU, UAC } TYR UAA, UAG } STOP	UGU, UGC } CYS UGA } STOP UGG } TRP	U C A G
C	CUU, CUC, CUA, CUG } LEU	CCU, CCC, CCA, CCG } PRO	CAU, CAC } HIS CAA, CAG } GLN	CGU, CGC, CGA, CGG } ARG	U C A G
A	AUU, AUC, AUA } ILE AUG } MET or START	ACU, ACC, ACA, ACG } THR	AAU, AAC } ASN AAA, AAG } LYS	AGU, AGC } SER AGA, AGG } ARG	U C A G
G	GUU, GUC, GUA, GUG } VAL	GCU, GCC, GCA, GCG } ALA	GAU, GAC } ASP GAA, GAG } GLU	GGU, GGC, GGA, GGG } GLY	U C A G

Hint: *The chart shows mRNA coding (example UUU, UUC, UUA, UUG) with the amino acids they code. Look at the genetic code chart. Messenger RNA UUU or UUC codes for the amino acid phenylalanine. The messenger RNA UUA or UUG codes for the amino acid leucine. The messenger RNAs CUU, CUC, CUA, and CUG code for the amino acid leucine.*

Species A	DNA strand:	TAC	CGA	CCT	TCA
	mRNA strand:	AUG	GCU	GGA	AGU
	Amino acid sequence:	___	___	___	___
Species B	DNA strand:	TAC	TTT	GCA	GGA
	mRNA strand:	___	___	___	___
	Amino acid sequence:	MET	LYS	ARG	PRO
Species C	DNA strand:	___	___	___	___
	mRNA strand:	AUG	UUU	UGU	CCC
	Amino acid sequence:	MET	PHE	CYS	PRO
Species D	DNA strand:	TAC	GTA	GTT	GCA
	mRNA strand:	AUG	CAU	CAA	CGU
	Amino acid sequence:	MET	HIS	GLN	ARG
Species E	DNA strand:	TAC	TTC	GCG	GGT
	mRNA strand:	AUG	AAG	CGC	CCA
	Amino acid sequence	MET	LYS	ARG	PRO

29. Using the Universal Genetic Code Chart, fill in the missing amino acids in the amino acid sequence for species ***A*** in the chart on the next page.
Hint: In species A you see that the DNA bases TAC pair up with the messenger RNA bases AUG; the messenger RNA codes for amino acids.
Look at the chart in the question with messenger RNA's and amino acids. Messenger RNA AUG codes for the amino acid methionine, etc.

30. Using the information given, fill in the missing mRNA bases in the mRNA strand for species ***B*** in the chart provided.
Hint: In species B, you are given the DNA base sequence. If the nitrogen base sequence is TAC, the messenger RNA would be AUG. T pairs with A, A pairs with U, and C pairs with G etc.

31. Using the information given, fill in the missing DNA bases in the DNA strand for species ***C*** in the chart provided.
Hint: Species C gives you the messenger RNA. If the messenger RNA is AUG then the DNA must be TAC, etc.

32. According to the information, which two species are most closely related? Support your answer.
Hint: Species are more closely related if they have similar arrangements of amino acids, similar enzymes, similar proteins and similar DNA sequences. Look at species A-E. Which two species have the same amino acid sequence?

33. Identify ***one*** physical characteristic of plants that can be readily observed and compared to help determine the relationship between two different species of plants.

34. Explain why comparing the DNA of the unknown and known plant species is probably a more accurate method of determining relationships than comparing only the physical characteristic you identified in the previous question.

35. Scientists hypothesize that cabbage, broccoli, cauliflower, and radishes developed along a common evolutionary pathway. Which observation would best support this hypothesis?
    (1) Fossils of these plants were found in the same rock layer.
    (2) Chloroplasts of these plants produce a gas.
    (3) These plants live in the same environment.
    (4) These plants have similar proteins.

36. Two proteins in the same cell perform different functions. This is because the two proteins are composed of
    (1) chains folded the same way and the same sequence of simple sugars
    (2) chains folded the same way and the same sequence of amino acids
    (3) chains folded differently and a different sequence of simple sugars
    (4) chains folded differently and a different sequence of amino acids

37. The function of most proteins depends primarily on the
    (1) type and order of amino acids
    (2) environment of the organism
    (3) availability of starch molecules
    (4) nutritional habits of the organism

38. Which nuclear process is represented below?

A DNA molecule untwists. → The two strands of DNA separate. → Molecular bases pair up. → Two identical DNA molecules are produced.

(1) recombination
(2) fertilization
(3) replication
(4) mutation

39. Complete the chart below by identifying two cell structures involved in protein synthesis and stating how each structure functions in protein synthesis.

Cell Structure	Function in Protein Synthesis

40. Base your answers to the following question on the information below and on your knowledge of biology.

Sickle-cell anemia is an inherited disease that occurs mainly in people from parts of Africa where malaria is common. It is caused by a gene mutation that may be harmful or beneficial.

A person with two mutant genes has sickle-cell disease. The hemoglobin of a person with sickle-cell disease twists red blood cells into a crescent shape. These blood cells cannot circulate normally. Symptoms of the disease include bleeding and pain in bones and muscles. People with sickle-cell disease suffer terribly in childhood and, until modern medicine offered treatment, most of them died before reproducing. An individual who has one mutant gene is protected from malaria because the gene changes the hemoglobin structure in a way that speeds removal of malaria-infected cells from circulation. A person with two normal genes has perfectly good red blood cells, but lacks resistance to malaria.

Define the term ***mutation***.

41. One ***disadvantage*** of a genetic mutation in a human skin cell is that it
    (1) may result in the production of a defective protein
    (2) may alter the sequence of simple sugars in insulin molecules
    (3) can lead to a lower mutation rate in the offspring of the human
    (4) can alter the rate of all the metabolic processes in the human

42. Use appropriate letters to write a 9-base DNA sequence that could represent a portion of a gene.

43. The chart below shows relationships between genes, the environment, and coloration of tomato plants.

Inherited Gene	Environmental Condition	Final Appearance
A	Light	Green
B	Light	White
A	Dark	White
B	Dark	White

Which statement best explains the final appearance of these tomato plants?
    (1) The expression of gene ***A*** is not affected by light.
    (2) The expression of gene ***B*** varies with the presence of light.
    (3) The expression of gene ***A*** varies with the environment.
    (4) Gene ***B*** is expressed only in darkness.

44. A basketball player develops speed and power as a result of practice. This athletic ability will not be passed on to her offspring because
    (1) muscle cells do not carry genetic information
    (2) mutations that occur in body cells are not inherited
    (3) gametes do not carry complete sets of genetic information

(4) base sequences in DNA are not affected by this activity

45. Some steps involved in DNA replication and protein synthesis are summarized in the table below.

Step A	DNA is copied and each new cell gets a full copy.
Step B	Information copied from DNA moves to the cytoplasm.
Step C	Proteins are assembled at the ribosomes.
Step D	Proteins fold and begin functioning.

In which step would a mutation lead directly to the formation of an altered gene?
(1) *A* (2) *B* (3) *C* (4) *D*

46. All cells in an embryo have the same DNA. However, the embryonic cells form organs, such as the brain and the kidneys, which have very different structures and functions. These differences are the result of
(1) having two types of cells, one type from each parent
(2) rapid mitosis causing mutations in embryo cells
(3) new combinations of cells resulting from meiosis
(4) certain genes being expressed in some cells and not in others

47. A mutation changes a gene in a cell in the stomach of an organism. This mutation could cause a change in
(1) both the organism and its offspring
(2) the organism, but not its offspring
(3) its offspring, but not the organism itself
(4) neither the organism nor its offspring

48. Base your answers to the next question on the diagram below, which illustrates some steps in genetic engineering and on your knowledge of biology.

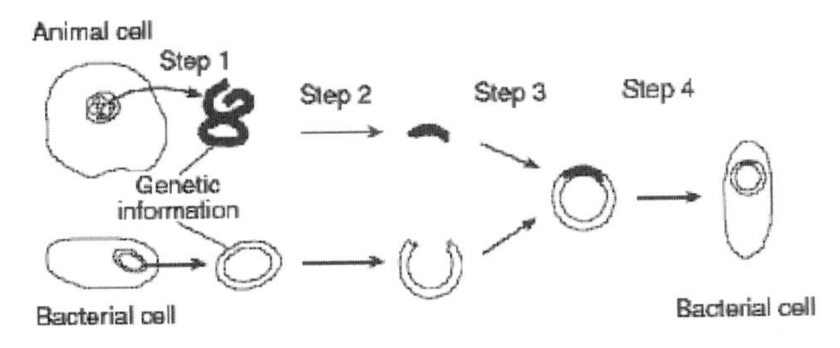

State *one* way that enzymes are used in Step 2.

49. The enzyme pepsin is produced in the cells of the stomach but not in the cells of the small intestine. The small intestine produces a different enzyme, trypsin. The reason that the stomach and small intestine produce different enzymes is that the gene that codes for pepsin is
(1) in the cells of the stomach, but not in the cells of the small intestine
(2) expressed in the stomach but not expressed in the small intestine
(3) mutated in the small intestine
(4) digested by the trypsin in the small intestine

50. One variety of strawberry is resistant to a damaging fungus, but produces small fruit. Another strawberry variety produces large fruit, but is not resistant to the same fungus. The two desirable qualities may be combined in a new variety of strawberry plant by

(1) cloning
(2) asexual reproduction
(3) direct harvesting
(4) selective breeding

51. Which statement best explains the change shown in the diagram below?

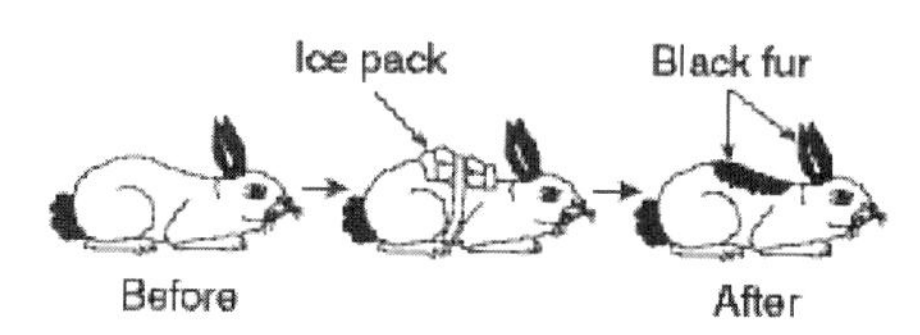

(1) Gene expression in an organism can be modified by interactions with the environment.
(2) Certain rabbits produce mutations that affect genes in specific areas of the body.
(3) Sorting and recombination of genes can be influenced by very cold temperatures.
(4) Molecular arrangement in existing proteins can be altered by environmental factors.

52. Which statement indicates that different parts of the genetic information are used in different kinds of cells, even in the same organism?

(1) The cells produced by a zygote usually have different genes.
(2) As an embryo develops, various tissues and organs are produced.
(3) Replicated chromosomes separate during gamete formation.
(4) Offspring have a combination of genes from both parents.

53. Plants inherit genes that enable them to produce chlorophyll, but this pigment is not produced unless the plants are exposed to light. This is an example of how the environment can

(1) cause mutations to occur
(2) influence the expression of a genetic trait
(3) result in the appearance of a new species
(4) affect one plant species, but not another

54. An error in genetic information present in a body cell of a mammal would most likely produce

(1) rapid evolution of the organism in which the cell is found
(2) a mutation that will affect the synthesis of a certain protein in the cell
(3) an adaptation that will be passed on to other types of cells
(4) increased variation in the type of organelles present in the cell

55. Base your answers to the next question on the information below and on your knowledge of biology.

Scientists are increasingly concerned about the possible effects of damage to the ozone layer.

Damage to the ozone layer has resulted in mutations in skin cells that lead to cancer. Will the mutations that caused the skin cancers be passed on to offspring? Support your answer.

56. Which statement indicates one difference between the gene that codes for insulin and the gene that codes for testosterone in humans?
(1) The gene for insulin is replicated in vacuoles, while the gene for testosterone is replicated in mitochondria.
(2) The gene for insulin has a different sequence of molecular bases than the gene for testosterone.
(3) The gene for insulin is turned on in liver cells, but the gene for testosterone is not.
(4) The gene for insulin is a sequence of five different molecular bases while the gene for testosterone is a sequence of only four different molecular bases.

57. Molecules ***B***, ***C***, and ***D*** are similar in that they are usually
(1) composed of genetic information
(2) involved in the synthesis of antibiotics
(3) composed of amino acids
(4) involved in the diffusion of oxygen into the cell

58. Which statement provides accurate information about the technique illustrated below?

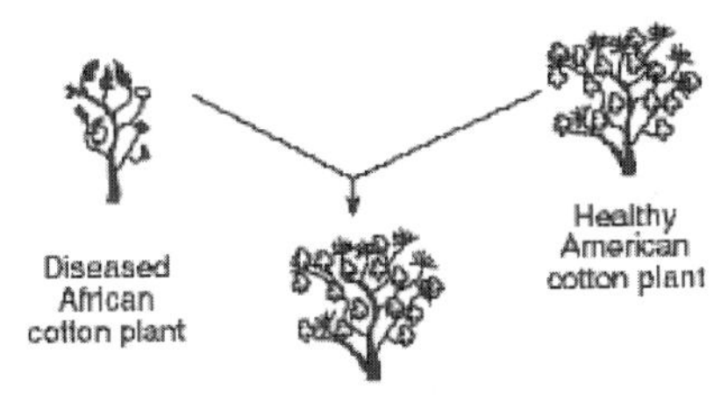

(1) This technique results in offspring that are genetically identical to the parents.
(2) New varieties of organisms can be developed by this technique known as selective breeding.
(3) This technique is used by farmers to eliminate mutations in future members of the species.
(4) Since the development of cloning, this technique is no longer used in agriculture.

59. A change in the base subunit sequence during DNA replication can result in
(1) variation within an organism
(2) rapid evolution of an organism
(3) synthesis of antigens to protect the cell
(4) recombination of genes within the cell

Base your answers to the next two questions on the statement below and on your knowledge of biology.

Selective breeding has been used to improve the racing ability of horses.

60. Define ***selective breeding*** and state how it would be used to improve the racing ability of horses.

61. State ***one*** disadvantage of selective breeding.

---

62. The headline "Improved Soybeans Produce Healthier Vegetable Oils" accompanies an article describing how a biotechnology company controls the types of lipids (fats) present in soybeans. The improved soybeans are most likely being developed by the process of

(1) natural selection
(2) asexual reproduction
(3) genetic engineering
(4) habitat modification

63. Cells that develop from a single zygote all contain identical DNA molecules. However, some of these cells will develop differently because

(1) different groups of cells containing the DNA may be exposed to different environmental conditions.
(2) only the DNA in certain cells will replicate
(3) some of the DNA in some of the cells will be removed by chemical reactions
(4) DNA is functional in only 10% of the cells of the body

64. Which substance from bacteria was most likely inserted into rice plants in the development of the trehalose-producing rice?

(1) sugar
(2) enzymes
(3) DNA
(4) trehalose

65. Base your answers to the next question on the passage below and on your knowledge of biology.

Better Rice

The production of new types of food crops will help raise the quantity of food grown by farmers. Research papers released by the National Academy of Sciences announced the development of two new superior varieties of rice—one produced by selective breeding and the other by biotechnology.

One variety of rice, called ***Nerica*** (New Rice for Africa), is already helping farmers in Africa. Nerica combines the hardiness and weed resistance of rare African rice varieties with the productivity and faster maturity of common Asian varieties.

Another variety, called ***Stress-Tolerant Rice***, was produced by inserting a pair of bacterial genes into rice plants for the production of ***trehalose*** (a sugar). Trehalose helps plants maintain healthy cell membranes, proteins, and enzymes during environmental stress. The resulting plants survive drought, low temperatures, salty soils, and other stresses better than standard rice varieties.

Which strain of rice was produced as a result of genetic engineering? Support your answer.

66. Although all of the cells of a human develop from one fertilized egg, the human is born with many different types of cells. Which statement best explains this observation?

(1) Developing cells may express different parts of their identical genetic instructions.
(2) Mutations occur during development as a result of environmental conditions.
(3) All cells have different genetic material.
(4) Some cells develop before other cells.

67. At warm temperatures, a certain bread mold can often be seen growing on bread as a dark-colored mass. The same bread mold growing on bread in a cooler environment is red in color. Which statement most accurately describes why this change in the color of the bread mold occurs?

(1) Gene expression can be modified by interactions with the environment.
(2) Every organism has a different set of coded instructions.
(3) The DNA was altered in response to an environmental condition.
(4) There is no replication of genetic material in the cooler environment.

68. One variety of wheat is resistant to disease. Another variety contains more nutrients of benefit to humans. Explain how a new variety of wheat with disease resistance and high nutrient value could be developed. In your answer, be sure to:

- identify one technique that could be used to combine disease resistance and high nutrient value in a new variety of wheat
- describe how this technique would be carried out to produce a wheat plant with the desired characteristics
- describe one specific difficulty (other than stating that it does not always work) in developing a new variety using this technique

69. Strawberries can reproduce by means of runners, which are stems that grow horizontally along the ground. At the region of the runner that touches the ground, a new plant develops. The new plant is genetically identical to the parent because
    (1) it was produced sexually
    (2) nuclei traveled to the new plant through the runner to fertilize it
    (3) it was produced asexually
    (4) there were no other strawberry plants in the area to provide fertilization

70. The flounder is a species of fish that can live in very cold water. The fish produces an "antifreeze" protein that prevents ice crystals from forming in its blood. The DNA for this protein has been identified. An enzyme is used to cut and remove this section of flounder DNA that is then spliced into the DNA of a strawberry plant. As a result, the plant can now produce a protein that makes it more resistant to the damaging effects of frost. This process is known as
    (1) sorting of genes
    (2) genetic engineering
    (3) recombination of chromosomes
    (4) mutation by deletion of genetic material

71. Viruses frequently infect bacteria and insert new genes into the genetic material of the bacteria. When these infected bacteria reproduce asexually, which genes would most likely be passed on?
    (1) only the new genes
    (2) only the original genes
    (3) both the original and the new genes
    (4) neither the original nor the new genes

72. Base your answers to the next question on the diagram below, which illustrates some steps in genetic engineering and on your knowledge of biology.

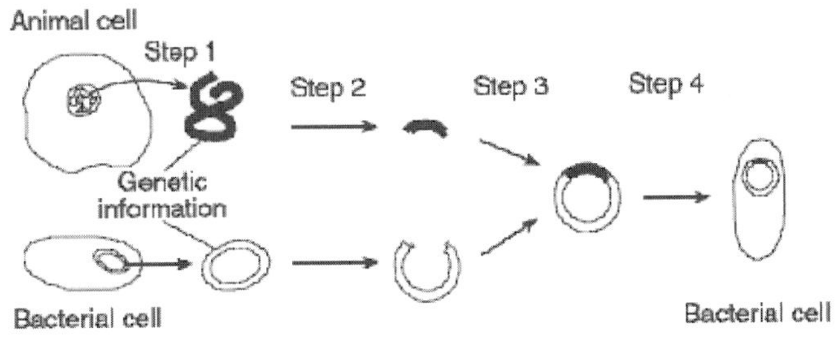

What is the result of Step *3*?
    (1) a new type of molecular base is formed
    (2) different types of minerals are joined together
    (3) DNA from the bacterial cell is cloned
    (4) DNA from different organisms is joined together

73. Base your answers to the next question on the diagram below and on your knowledge of biology.

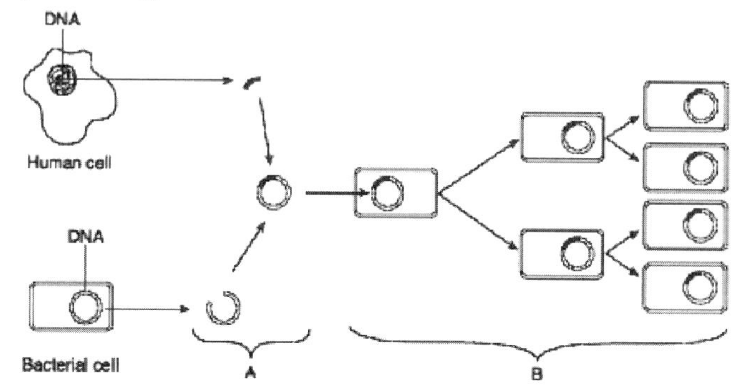

In the procedure indicated by Letter ***A***, DNA segments from humans and bacteria are joined by the action of

(1) starch molecules
(2) simple sugars
(3) enzymes
(4) hormones

74. Which statement best describes the relationship between cells, DNA, and proteins?

(1) Cells contain DNA that controls the production of proteins.
(2) DNA is composed of proteins that carry coded information for how cells function.
(3) Proteins are used to produce cells that link amino acids together into DNA.
(4) Cells are linked together by proteins to make different kinds of DNA molecules.

75. A human liver cell and a human skin cell in the same person have the same genetic sequences. However, these cells are different because the liver cell

(1) has more dominant traits than the skin cell
(2) can reproduce but the skin cell cannot
(3) carries out respiration but the skin cell does not
(4) uses different genes than the skin cell

76. Genes are inherited, but their expressions can be modified by the environment. This statement explains why

(1) some animals have dark fur only when the temperature is within a certain range
(2) offspring produced by means of sexual reproduction look exactly like their parents
(3) identical twins who grow up in different homes have the same characteristics
(4) animals can be cloned, but plants cannot

77. Explain how selective breeding is being used to try to produce commercial peanuts that will not cause allergic reactions in people.

78. Which statement provides accurate information about the technique illustrated below?

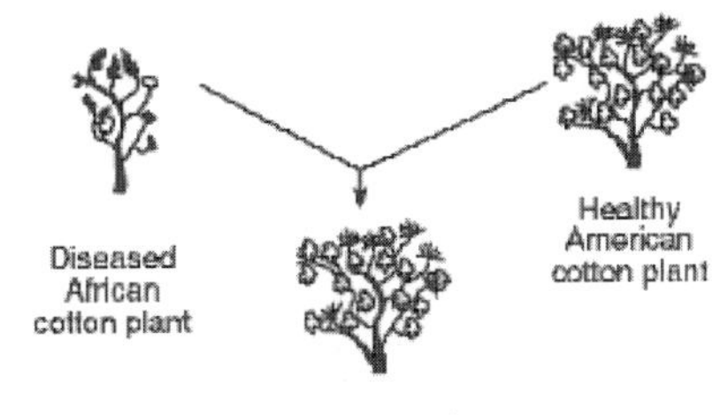

(1) This technique results in offspring that are genetically identical to the parents.
(2) New varieties of organisms can be developed by this technique known as selective breeding.
(3) This technique is used by farmers to eliminate mutations in future members of the species.
(4) Since the development of cloning, this technique is no longer used in agriculture.

Base your answers to the next two questions on the statement below and on your knowledge of biology.

Selective breeding has been used to improve the racing ability of horses.

79. Define ***selective breeding*** and state how it would be used to improve the racing ability of horses.

80. State ***one*** disadvantage of selective breeding.

---

81. In order to produce the first white marigold flower, growers began with the lightest yellow-flowered marigold plants. After crossing them, these plants produced seeds, which were planted, and only the offspring with very light-yellow flowers were used to produce the next generation. Repeating this process over many years, growers finally produced a marigold flower that is considered the first white variety of its species. This procedure is known as
(1) differentiation
(2) cloning
(3) gene insertion
(4) selective breeding

82. For centuries, certain animals have been crossed to produce offspring that have desirable qualities. Dogs have been mated to produce Labradors, beagles, and poodles. All of these dogs look and behave very differently from one another. This technique of producing organisms with specific qualities is known as
(1) gene replication
(2) natural selection
(3) random mutation
(4) selective breeding

Base your answers to the next two questions on the information below and on your knowledge of biology.

A biologist at an agriculture laboratory is asked to develop a better quality blueberry plant. He is given plants that produce unusually large blueberries and plants that produce very sweet blueberries.

83. Describe ***one*** way the biologist could use these blueberry plants to develop a plant with blueberries that are both large and sweet.

84. The headline "Improved Soybeans Produce Healthier Vegetable Oils" accompanies an article describing how a biotechnology company controls the types of lipids (fats) present in soybeans. The improved soybeans are most likely being developed by the process of

(1) natural selection
(2) asexual reproduction
(3) genetic engineering
(4) habitat modification

85. Some farmers currently grow genetically engineered crops. An argument ***against*** the use of this technology is that

(1) it increases crop production
(2) it produces insect-resistant plants
(3) its long-term effects on humans are still being investigated
(4) it always results in crops that do not taste good

86. Knowledge of human genes gained from research on the structure and function of human genetic material has led to improvements in medicine and health care for humans.

- state two ways this knowledge has improved medicine and health care for humans
- identify one specific concern that could result from the application of this knowledge

87. Base your answers to the next question on the passage below and on your knowledge of biology.

Better Rice

The production of new types of food crops will help raise the quantity of food grown by farmers. Research papers released by the National Academy of Sciences announced the development of two new superior varieties of rice—one produced by selective breeding and the other by biotechnology.

One variety of rice, called ***Nerica*** (New Rice for Africa), is already helping farmers in Africa. Nerica combines the hardiness and weed resistance of rare African rice varieties with the productivity and faster maturity of common Asian varieties.

Another variety, called ***Stress-Tolerant Rice***, was produced by inserting a pair of bacterial genes into rice plants for the production of ***trehalose*** (a sugar). Trehalose helps plants maintain healthy cell membranes, proteins, and enzymes during environmental stress. The resulting plants survive drought, low temperatures, salty soils, and other stresses better than standard rice varieties.

Which strain of rice was produced as a result of genetic engineering? Support your answer.

88. One variety of wheat is resistant to disease. Another variety contains more nutrients of benefit to humans. Explain how a new variety of wheat with disease resistance and high nutrient value could be developed. In your answer, be sure to:

- identify one technique that could be used to combine disease resistance and high nutrient value in a new variety of wheat
- describe how this technique would be carried out to produce a wheat plant with the desired characteristics
- describe one specific difficulty (other than stating that it does not always work) in developing a new variety using this technique

89. A product of genetic engineering technology is represented below.

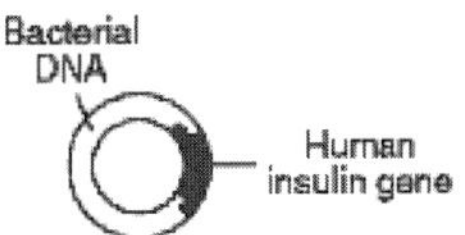

Which substance was needed to join the insulin gene to the bacterial DNA as shown?

(1) a specific carbohydrate
(2) a specific enzyme
(3) hormones
(4) antibodies

90. The diagram below represents a common laboratory technique in molecular genetics.

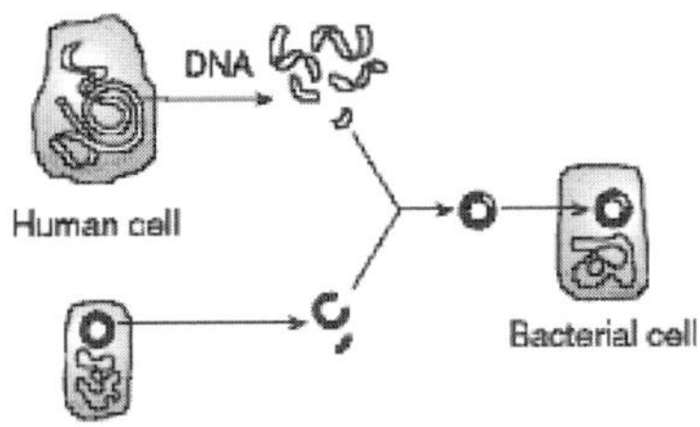

One common use of this technology is the

(1) production of a human embryo to aid women who are unable to have children
(2) change of single-celled organisms to multicellular organisms
(3) introduction of a toxic substance to kill bacterial cells
(4) production of hormones or enzymes to replace missing human body chemicals

91. How does altering the DNA of a peanut affect the proteins in peanuts that cause allergic reactions?

(1) The altered DNA is used to synthesize changed forms of these proteins.
(2) The altered DNA leaves the nucleus and becomes part of the allergy-producing protein.
(3) The altered DNA is the code for the antibodies against the allergens.
(4) The altered DNA is used as an enzyme to break down the allergens in peanuts.

92. Scientists have genetically altered a common virus so that it can destroy the most lethal type of brain tumor without harming the healthy tissue nearby. This technology is used for all of the following except

(1) treating the disease
(2) curing the disease
(3) controlling the disease
(4) diagnosing the disease

93. Base your answers to the next question on the diagram below and on your knowledge of biology.

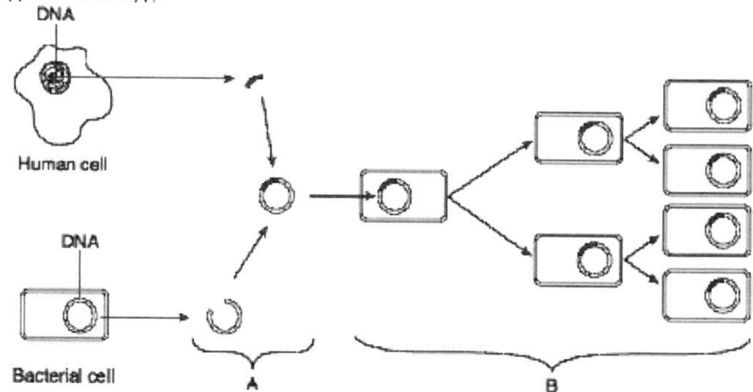

In the procedure indicated by Letter ***A***, DNA segments from humans and bacteria are joined by the action of

(1) starch molecules
(2) simple sugars
(3) enzymes
(4) hormones

94. Base your answers to the next question on the diagram below and on your knowledge of biology.

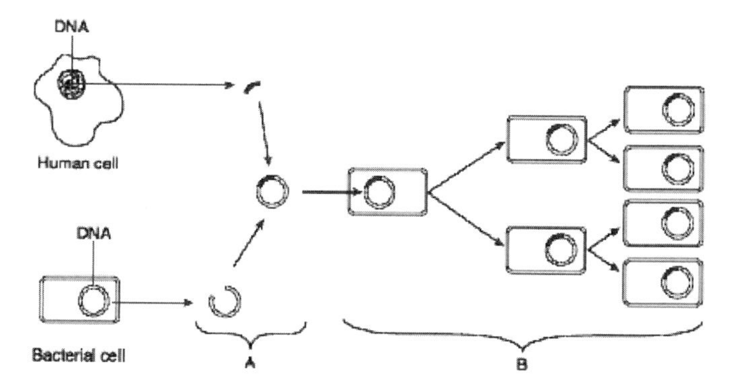

Which process is indicated by Letter ***B***?

(1) natural selection
(2) asexual reproduction
(3) sexual reproduction
(4) gene deletion

Hint: *Look only at Letter* ***B.***

95. Researchers Cohn and Boyer transferred a gene from an African clawed frog into a bacterium. To accomplish this, these scientists had to use

(1) enzymes to cut out and insert the gene
(2) hereditary information located in amino acids
(3) radiation to increase the gene mutation rate of the bacterial cells
(4) cancer cells to promote rapid cell division

96. The DNA of a human cell can be cut and rearranged by using

(1) a scalpel
(2) electrophoresis
(3) hormones
(4) enzymes

97. Base your answer to the next question on the diagram below, which illustrates some steps in genetic engineering and on your knowledge of biology.

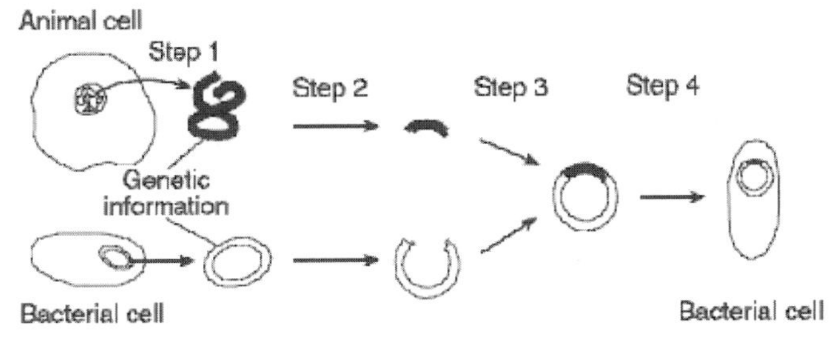

What is the result of Step *3*?
(1) a new type of molecular base is formed
(2) different types of minerals are joined together
(3) DNA from the bacterial cell is cloned
(4) DNA from different organisms is joined together

98. Base your answer to the next question on the diagram below, which illustrates some steps in genetic engineering and on your knowledge of biology.

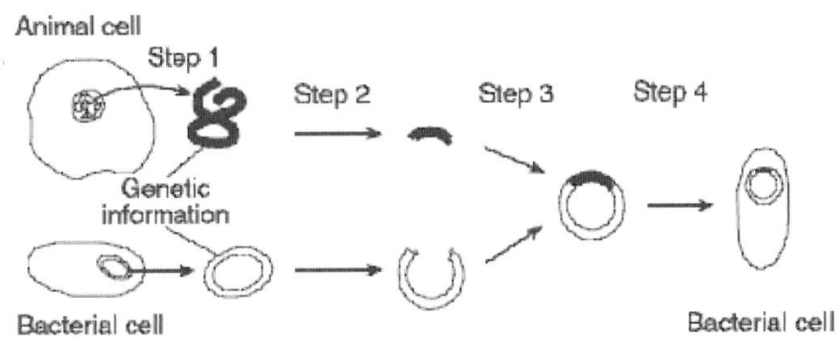

State ***one*** way that enzymes are used in Step ***2***.

99. Which process is illustrated in the diagram below?

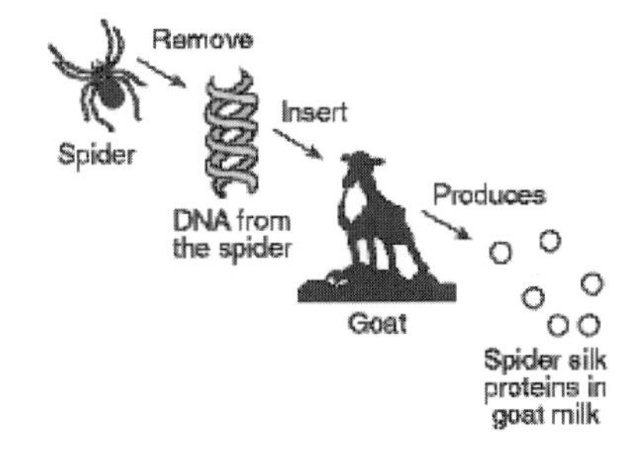

(1) chromatography
(2) direct harvesting
(3) meiosis
(4) genetic engineering

100. The headline "Improved Soybeans Produce Healthier Vegetable Oils" accompanies an article describing how a biotechnology company controls the types of lipids (fats) present in soybeans. The improved soybeans are most likely being developed by the process of
(1) natural selection
(2) asexual reproduction
(3) genetic engineering
(4) habitat modification

CHAPTER 4: REPRODUCTION

A **species** (example human beings) is a group of closely related organisms that **can reproduce** (produce offspring (children)). Members of the same species have some of the same characteristics or you can say some characteristics in common (example most human beings can see, hear, walk, talk). A human being dies, but because members of a species (example humans) reproduce (produce offspring, children), the species (human beings) keeps on living because of reproduction. As you know, producing offspring (children) is not necessary for the individual to survive, but is necessary for the species (example human beings) to keep on living.

REPRODUCTION

There are two types of reproduction, asexual reproduction and sexual reproduction.

Asexual Reproduction

Asexual reproduction: there in only **one parent**, therefore the **offspring** (child) is **identical (genetically identical) to** the **parent.**

For example, one ameba (which has one cell) divides and produces two amebas (two cells) by the process of mitosis.

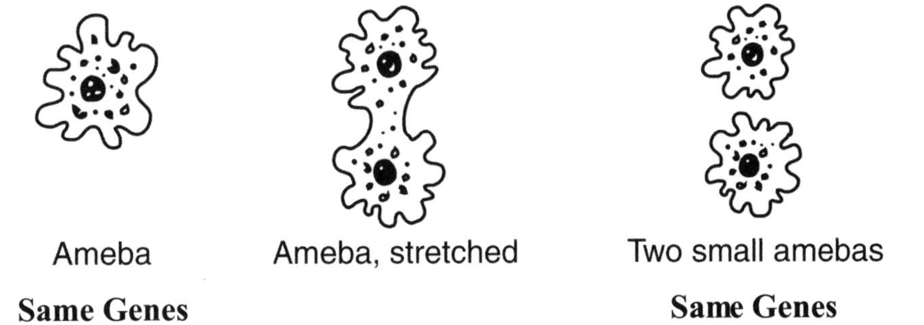

Ameba | Ameba, stretched | Two small amebas

Same Genes | **Same Genes**

All the genes in the baby amebas come from the one parent ameba. The offspring (baby amebas) have the same genes (genetic code, DNA) as the parent ameba (offspring are genetically identical to the parent, having the same genes to make or synthesize the same proteins and same enzymes). Baby's genes are genetically identical, not different (not modified), from the parents' genes.

In **asexual reproduction,** there is only **one parent,** therefore the **offspring is identical (genetically identical) to** the **one parent**.

Examples of asexual reproduction:

1. Organisms that have only one cell (one celled organisms) such as ameba (shown above) divide and form two amebas that are genetically identical (same genes) to each other and to the parent ameba. One celled organisms such as bacteria (or paramecium) also divide and form two bacteria that are genetically identical (same genes).

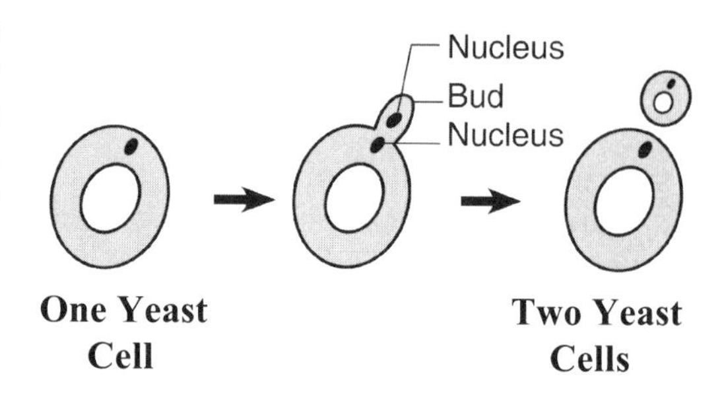

Yeast, which is also one cell, divides and forms two yeast cells. The new cell is smaller in size but has the same identical genes as the parent cell (genetically identical).

One-Celled Organisms That Carry on Asexual Reproduction

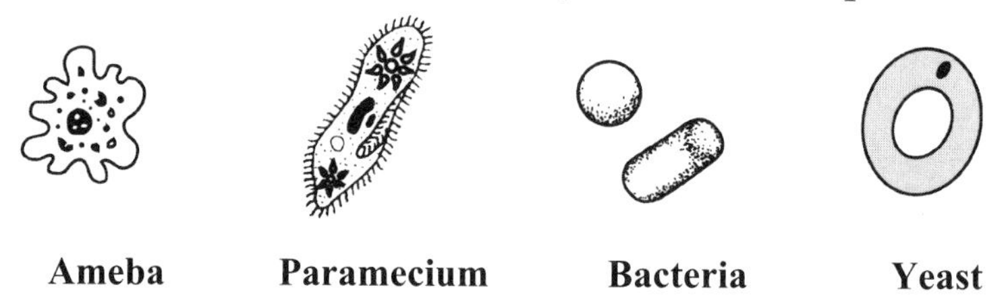

You can see from these examples that one cell (example one ameba) produces two cells (two amebas).

2. Similarly the hydra divides and forms two hydras. The new hydra is smaller than the parent hydra but has the same identical genes as the

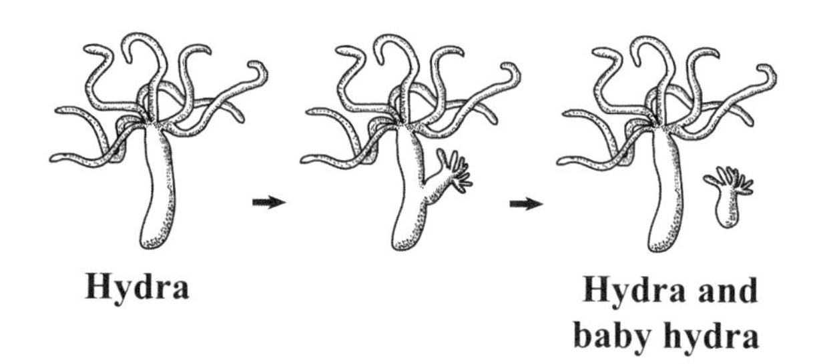

parent organism (genetically identical). To explain, one cell of the parent hydra divides and forms two genetically identical cells, two cells form four cells, four cells form eight cells, eight cells form 16, 16 form 32, 32 form 64, 64 form 128 genetically identical cells, etc. These cells separate from the parent hydra and will become the new hydra (the new organism). Since one cell of the hydra produces the new hydra, it is asexual reproduction.

3. Molds produce spores (one cell). Each spore (one cell) develops into a new mold; a spore is genetically identical to the parent mold. Since one cell produces the new mold, it is asexual reproduction.

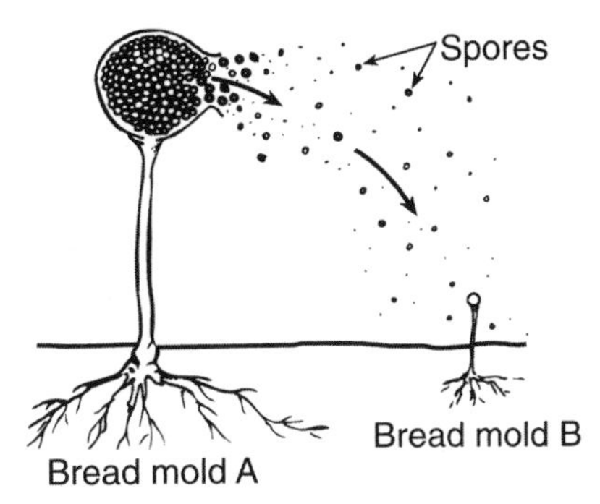

Spore Produces Bread Mold B

4. New plants are produced from parts of the plant, such as leaves, branches, stems and roots of the parent plant. New plants are genetically identical to the parent plant. When these plants are grown naturally, a stem from the original plant (a runner) grows (along the ground) into a new plant, see figure, then the new plant separates from the original plant. To quickly get more trees that produce an improved variety of apples (example sweeter, juicier), a farmer can take stems of the sweet apple tree and plant them. The stems will grow into new apple trees that have the improved variety (sweeter, juicier apples). Since the new apple trees came from stems from one tree, this is asexual reproduction.

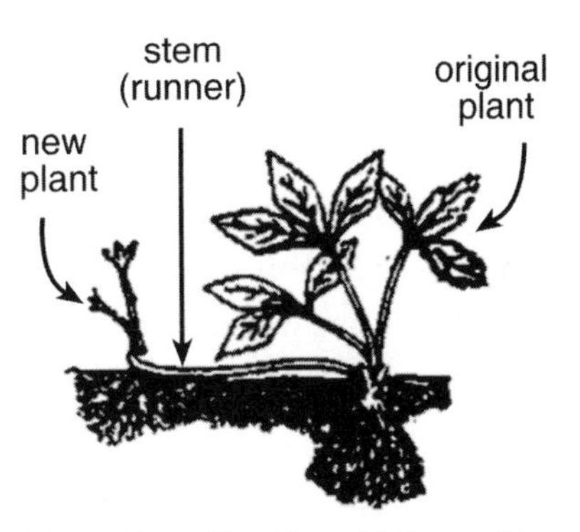

New Plant Produced From Stem

Question: A tree produces only seedless oranges. A small branch cut from this tree produces roots after it is planted in soil. When mature, this new tree will likely produce

(1) oranges with seeds, only
(2) oranges without seeds, only
(3) a majority of oranges with seeds and only a few oranges without seeds
(4) oranges and other kinds of fruit

Solution: As you know, a tree has branches. A branch from the tree (seedless orange tree) is planted and produces a new tree (seedless orange tree) which is the same type, identical to the original tree (seedless orange tree, which obviously produces oranges with no seeds).

Answer *2*

Note: The new baby tree was produced from the branch of only one tree (like one parent), therefore it is asexual reproduction.

---

5. Growth (growing) and repair of damaged cells is done in a living thing (example plants, animals, etc.) by producing more cells. **Animals** (examples humans, dogs, cats) and **plants** (example apple trees) **grow (get bigger),** making more cells. Example one cell divides and forms two genetically identical cells (same genes), two cells divide and form four genetically identical cells, four cells divide and form eight cells, eight cells divide and form 16 genetically identical cells. This process, how one cell divides and forms more genetically identical cells, is called mitosis, which you will learn later in this chapter. Since one cell produces more cells, it is called asexual reproduction.

Cloning is a type of asexual reproduction. Cloning produces new organisms that are identical to the parent (genetically identical, same genes, same number of chromosomes).

a. Cut a part of a plant, such as a stem, from a plant and plant it. It will develop roots and leaves; this new plant will be genetically identical to (same genes as) the parent plant.

b. Take out a cell from a mature plant (see figure below). Grow the cell with a special mixture of growth hormones. The cell will develop into a plant genetically identical to (same genes as) the original plant.

Example of cloning:

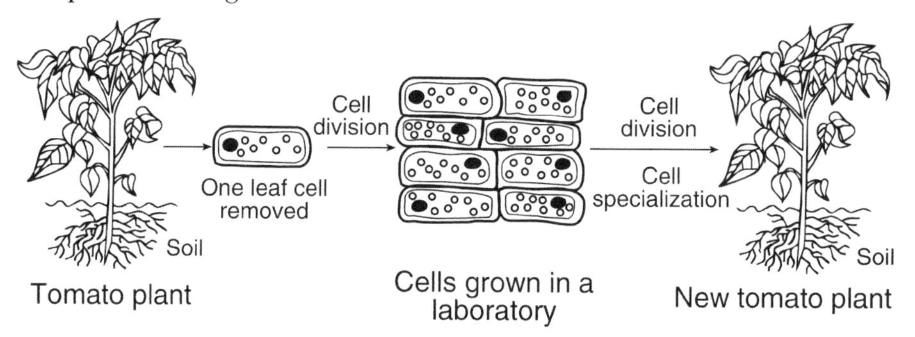

You see one cell produces a clone, a genetically identical copy of the original plant. In this example, one cell produced a tomato plant that is genetically identical to (same genes as) the original tomato plant.

c. You learned a gamete, example egg or sperm, has half the number of genes as a regular cell in the body, or you can say a gamete has half the number of chromosomes as a regular cell in the body. Take out the nucleus of an egg cell (see I in diagram below), which only has half the genes, (example from a sheep) and put in the nucleus of a **body cell,** such as a **skin cell** (see II in diagram below) **which has all the genes** from the same type of organism (also a sheep). The egg cell now has all its genes from one parent (the skin cell from Sheep A) and can grow into a new organism. When the egg cell divides (by mitosis) each cell in the baby sheep (example Dolly) will only have the nucleus (genes, chromosomes, and DNA) from one sheep (sheep A) see diagram below, therefore the baby sheep (example Dolly) will be genetically identical to sheep A. If the body cell is from a female sheep (which has all the genes from a female sheep), all the new cloned sheep will be female, and they cannot be bred (mated) with each other; however, if the skin cell is from a male sheep, all the cloned sheep will be male. In this example, the body cell is from sheep A, a female sheep, so all the cloned sheep are female. Dolly is female.

Cloning A Sheep

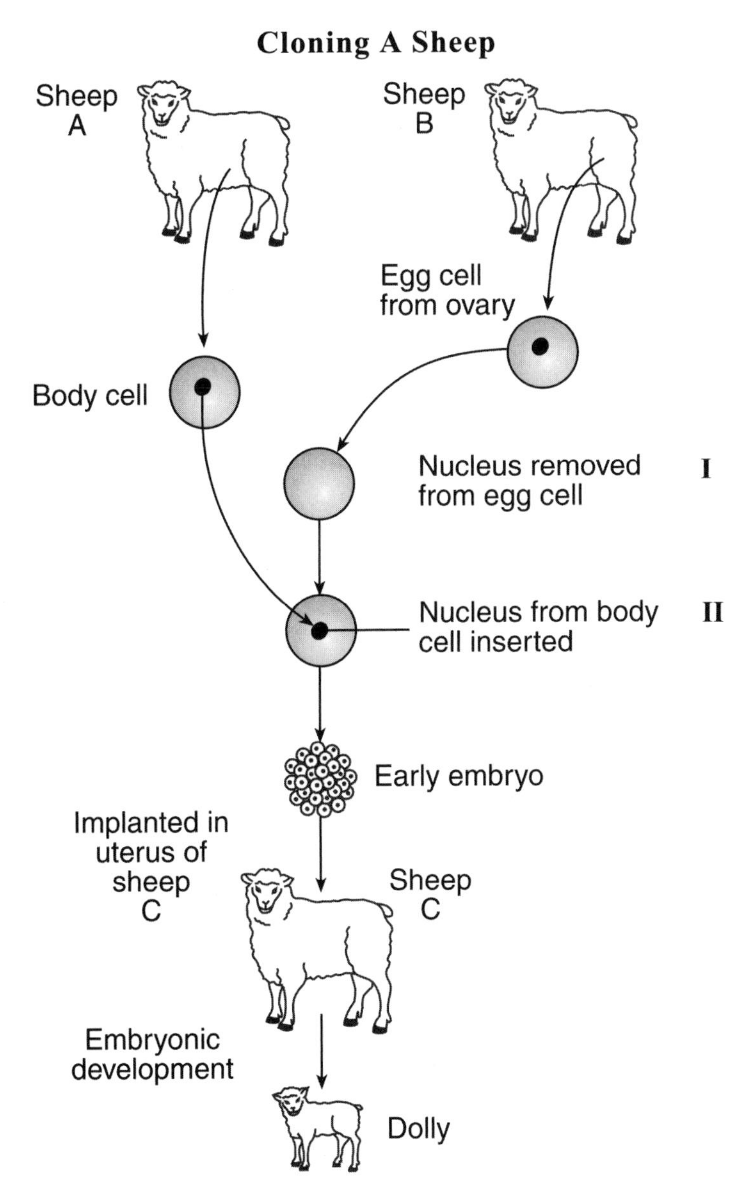

Dolly, a cloned sheep

Note: The group of cells (labeled early embryo in the diagram) is put into sheep C to develop and grow to a baby sheep.

The advantage of cloning is all the new sheep have the same desirable characteristics as the original sheep (the sheep which gave the body cell, in this example sheep A (see figure above)). A disadvantage of cloning is that all the sheep are the same type; therefore, if a disease can infect one sheep and kill it, the same disease can infect all the sheep and kill them.

Sexual reproduction

Sexual reproduction: You learned in the genetics chapter, in sexual reproduction (example humans), there are two parents (male and female). The male produces the male gamete (sperm in a human being) which has the genes-genetic information from the man (you learned genes are located on chromosomes). The female produces the female gamete (egg in a human being) which has the genes-genetic information from the lady. The sperm (which has the genes from the father) and the egg (which has the genes from the mother) unite together to form a fertilized egg (fertilization) which develops into a child. The child has the genes (genetic information, chromosomes, DNA) from both the father and mother (half of the genetic information from the mother and half of the genetic information from the father), therefore the child often resembles but is not identical to either the mother or father. Combining the genes, genetic information, chromosomes, DNA from both the male and female is called recombination; the offspring then have a variety of different characteristics (more variation, more genetic variations, more variations in appearance). When a male sheep is **mated** with a female sheep (**breeding** a male and female sheep), the offspring (baby sheep) gets genes (genetic information, chromosomes, DNA) from both the male sheep and the female sheep, therefore the baby sheep resembles both the male and female sheep but is not identical to either one. Combining the genes from the male and female causes the offspring (example baby sheep) to have a variety of different characteristics (more variation).

Every earthworm and many flowers have both male and female reproductive organs. By mating earthworms with other earthworms or flowers with other flowers, instead of fertilizing themselves, they produce a greater variety of offspring and the chance of survival is higher, even if environmental conditions change, such as flooding, drought, etc.

Now Do Homework Questions #1-20, pages 33-36.

MITOSIS (MITOTIC CELL DIVISION)

When an organism (example, dog, cat, human, plant) grows, they get bigger and have more cells.

Mitosis: One cell divides and forms two cells, two cells divide and form four cells, four cells divide and form eight cells, and eight cells divide and form 16 cells, etc.(see figure below).

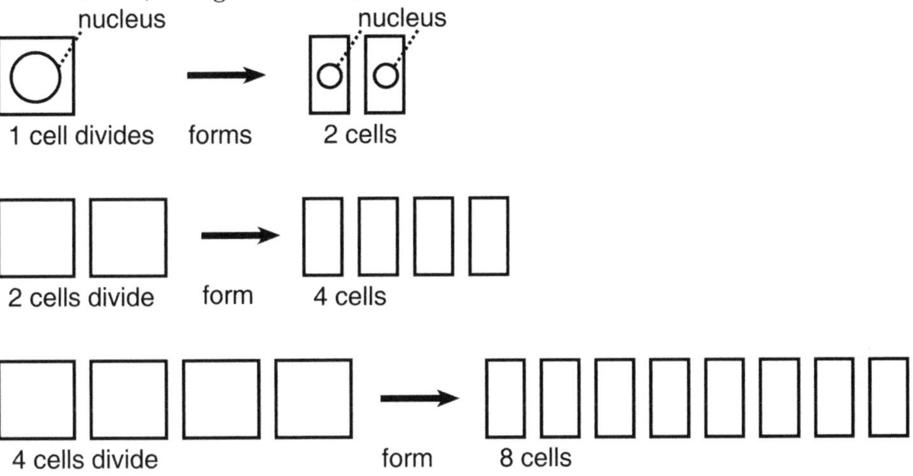

The small circle inside each cell is the nucleus. Chromosomes, genes, and DNA are in the nucleus.

As we said, one cell divides to form two cells.

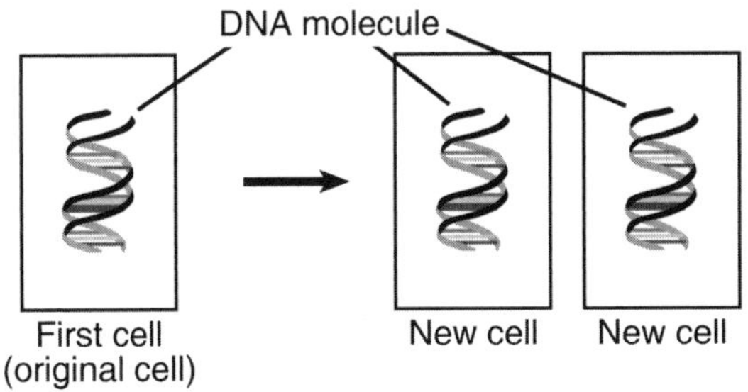

The DNA molecule in the nucleus of the first cell replicates (duplicates) and forms two DNA molecules, one DNA molecule in one new cell, the other DNA molecule in the other new cell. The **DNA** in the **first (original) cell equals** the **DNA** in **one new cell, equals** the **DNA** in the **other new cell.** Chromosomes are made of DNA. The chromosomes duplicate (replicate); the **number** of **chromosomes** in the **original cell equals** the **number** of **chromosomes** in **one new cell equals** the **number** of **chromosomes** in the **other new cell**. If the original cell has eight chromosomes, each new cell has eight chromosomes.

││││││││		││││││││	││││││││
8 chromosomes	⟶	8 chromosomes	8 chromosomes
original cell		new cell	new cell

If the original cell has 15 chromosomes, each new cell has 15 chromosomes.

│││││││││││││││		│││││││││││││││	│││││││││││││││
15 chromosomes	⟶	15 chromosomes	15 chromosomes
original cell		new cell	new cell

Genes are made of DNA and genes are located on the chromosomes. The genes duplicate (replicate); the **number** of **genes** in the **original cell equals** the **number** of **genes** in **one new cell equals** the **number** of **genes** in the **other new cell**. If the original cell has 100 genes, each new cell has 100 genes.

100 genes	⟶	100 genes	100 genes
original cell		new cell	new cell

In mitosis: Each new cell has the **same genetic makeup, same number of genes, same number of chromosomes, same DNA as** the **original cell**. Mitosis (mitotic division) makes new cells; new cells are made for growth (when the organism grows) and repair of damaged cells.

One celled organisms (example ameba) **use mitosis (mitotic division)** for asexual **reproduction** (there is only one parent).
One ameba (one cell) divides and forms two amebas (two cells).
Each ameba (one cell) has the same DNA, same genetic makeup, same number of chromosomes and same number of genes as the original ameba (one cell).

original ameba	⟶ forms	one ameba	one ameba
same number of chromosomes	forms	same number of chromosomes	same number of chromosomes

Let's see what happens in mitosis.

Mitosis

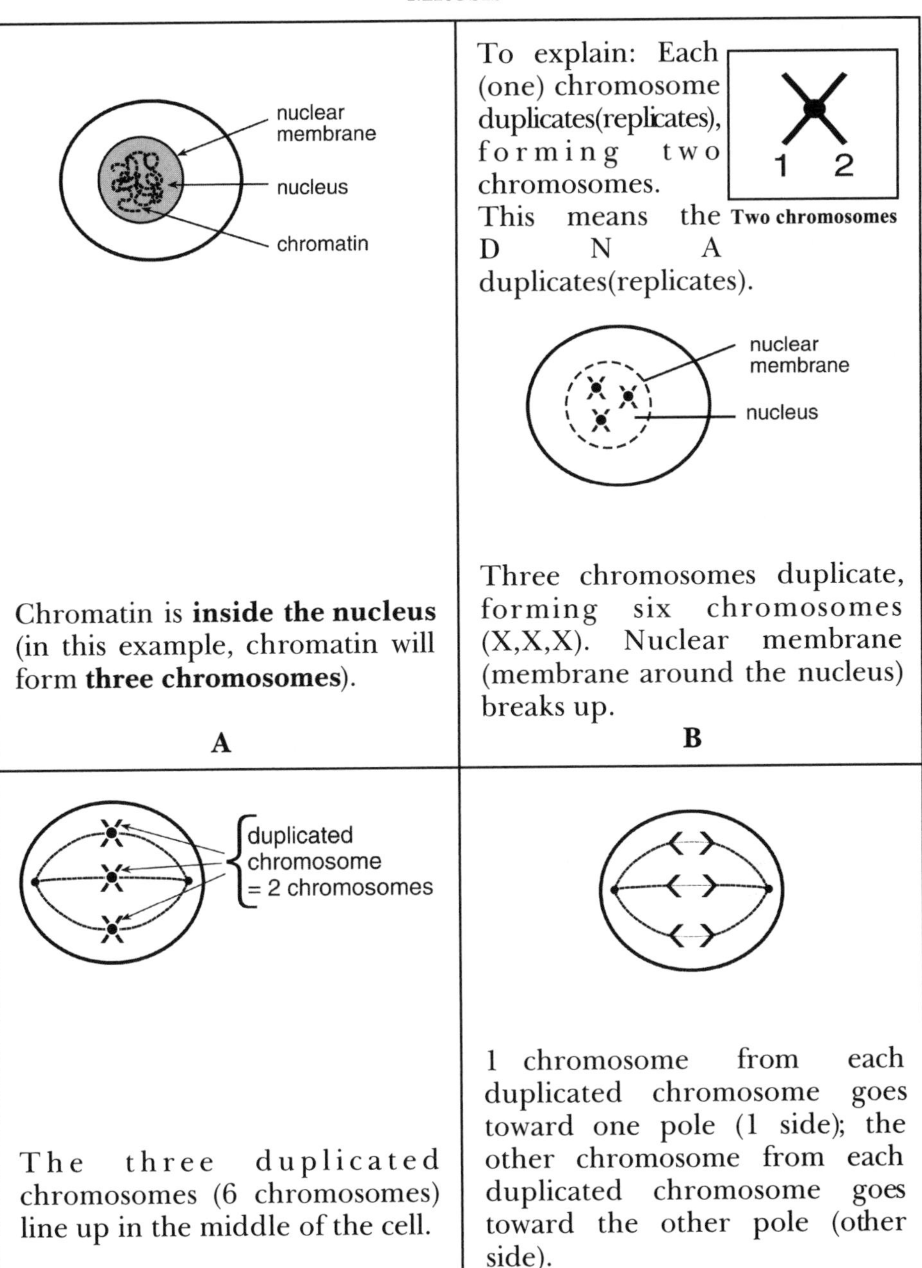

Chromatin is **inside the nucleus** (in this example, chromatin will form **three chromosomes**). **A**	To explain: Each (one) chromosome duplicates(replicates), forming two chromosomes. This means the DNA duplicates(replicates). Three chromosomes duplicate, forming six chromosomes (X,X,X). Nuclear membrane (membrane around the nucleus) breaks up. **B**
The three duplicated chromosomes (6 chromosomes) line up in the middle of the cell. **C**	1 chromosome from each duplicated chromosome goes toward one pole (1 side); the other chromosome from each duplicated chromosome goes toward the other pole (other side). **D**

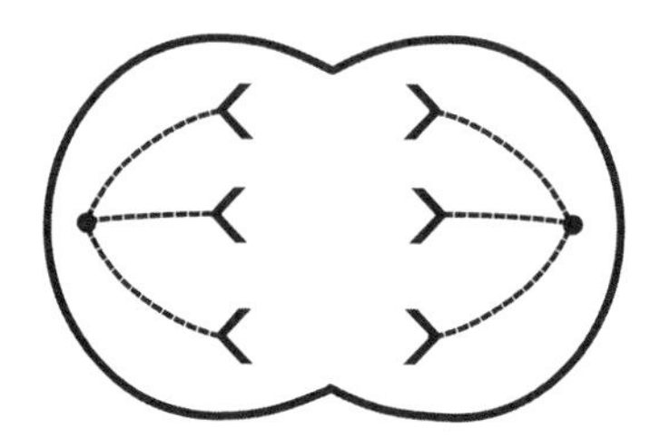

The chromosomes move further apart toward the opposite poles (opposite sides). A nuclear membrane starts to form around each set of chromosomes, forming two identical nuclei. The cytoplasm begins to split. Then, each **new cell will have three chromosomes**.

E

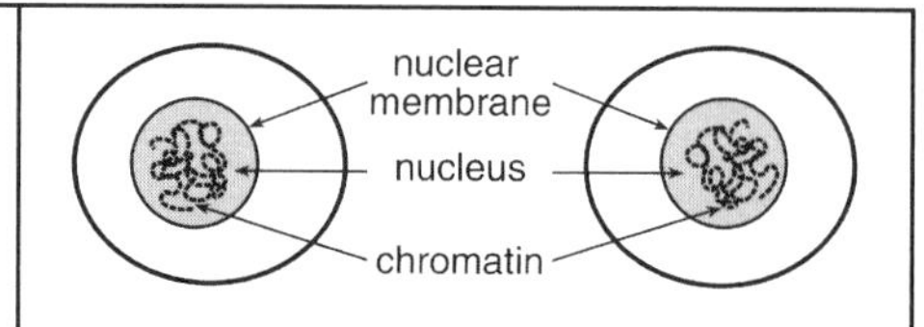

At the end, the chromosomes will look again like chromatin in the nucleus, like in Figure A. The cytoplasm splits, forming two cells. You now have two new cells (daughter cells), identical to the original cell.

F

As you can see in mitosis, the **original cell** and each **new cell** (daughter cell) have the same number of chromosomes (in our example three chromosomes). Note: The original cell and each new cell have the same amount of chromatin, which forms the same number of chromosomes in each cell.
If the original cell would have 10 chromosomes. each new cell would have 10 chromosomes. The new cells (daughter cells) will have the same number of chromosomes as the original cell (also called parent cell).

Question: The chromosome content of a skin cell that is about to form two new skin cells is represented in the diagram.

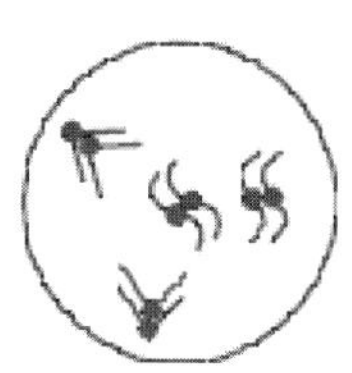

Which diagram best represents the chromosomes that would be found in the two new skin cells produced as a result of this process?

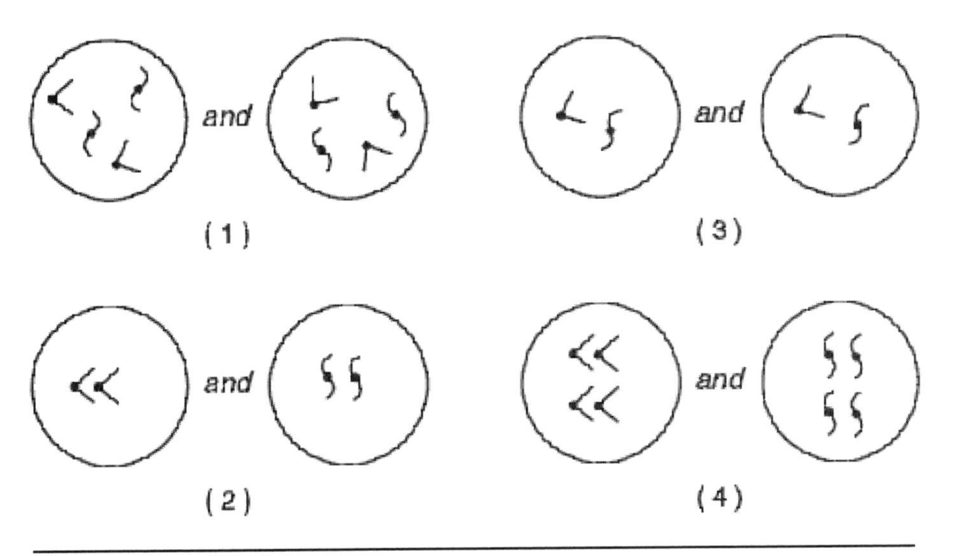

Solution: The one skin cell in the question is in the middle of mitosis and the chromosomes have duplicated (doubled). One chromosome from each duplicated chromosome goes to each cell.

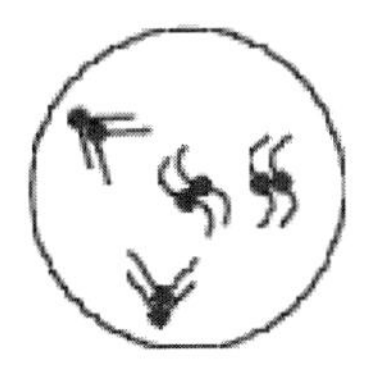

duplicated chromosome, goes to one cell, goes to the other cell

duplicated chromosome, goes to one cell, goes to the other cell

duplicated chromosome, goes to one cell, goes to the other cell

duplicated chromosome, goes to one cell, goes to the other cell

Therefore, one cell has other cell has

Answer *1*

Question: Which activity most directly involves the process represented in the diagram below?

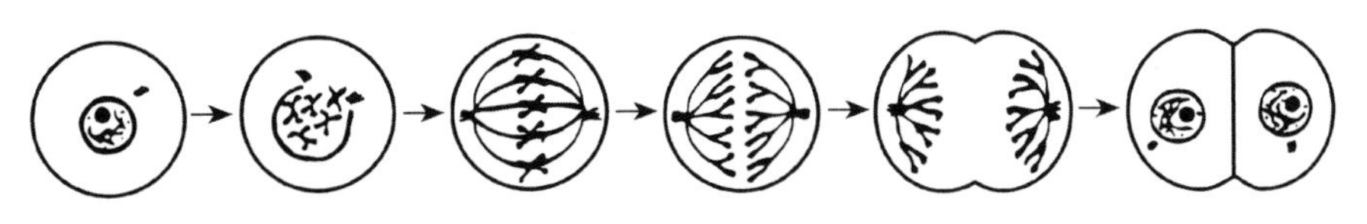

1. a gamete reproducing sexually
2. a white blood cell engulfing bacteria
3. a zygote being produced in an ovary
4. an animal repairing damaged tissue

Solution: The diagram in the question shows mitosis (one cell produces two identical cells). Mitosis is used for growth and repair of damaged tissue. Answer *4*

Practice Questions and Solutions

Question: The *least* genetic variation will probably be found in the offspring of organisms that reproduce using

(1) mitosis to produce a larger population
(2) meiosis to produce gametes
(3) fusion of eggs and sperm to produce zygotes
(4) internal fertilization to produce an embryo

Solution: One cell (example ameba, paramecium) produces two **identical cells by the process of mitosis,** therefore the babies (offspring), example baby amebas, have the least variation (least genetic variation), least difference from the parent amebas. Answer *1*

Choices 2, 3, and 4 are wrong because they are parts of sexual reproduction (choice 2 produces gametes, choice 3 is sperm and egg unite together, choice 4 is fertilization.) You learned in sexual reproduction there is variation.

Now Do Homework Questions #21-31, pages 37-38.

Meiosis (Meiotic Division)

In each cell, each trait (example eye color) has two genes, on two chromosomes (one pair of chromosomes).
For example, there is one gene for blue eye color on one chromosome and one gene for brown eye color on the other chromosome. The two genes are at the same place on the two chromosomes (see diagram).

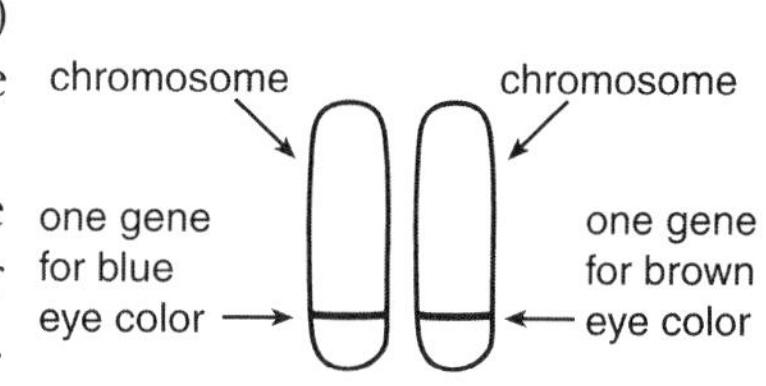

As you know, in every cell, each trait, such as height, has two genes (example gene for tall, gene for short) on two chromosomes (one pair of chromosomes). For example, there can be one gene for tall on one chromosome and one gene for short on the other chromosome.

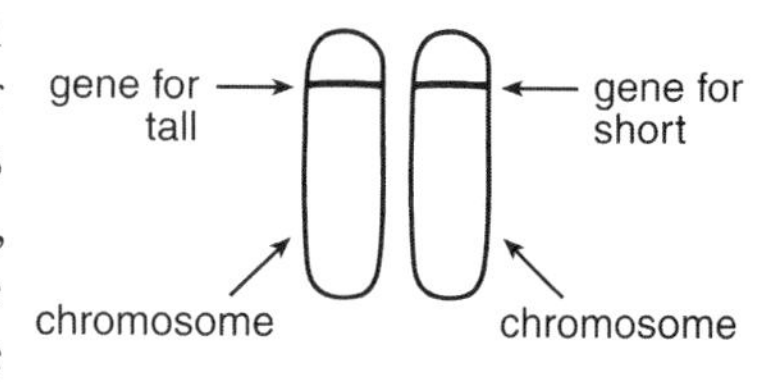

Meiosis produces sperm and eggs.

Meiosis produces sperm in a male: This is a body cell, a cell from the testes (from a male). **This cell** (called the **parent cell** because it will **produce new cells)** has four chromosomes, or two pairs of chromosomes (see Figure A). You know 2 chromosomes = 1 pair of chromosomes, therefore 4 chromosomes = 2 pairs of chromosomes. (Note: Chromatin in the nucleus will form chromosomes.)

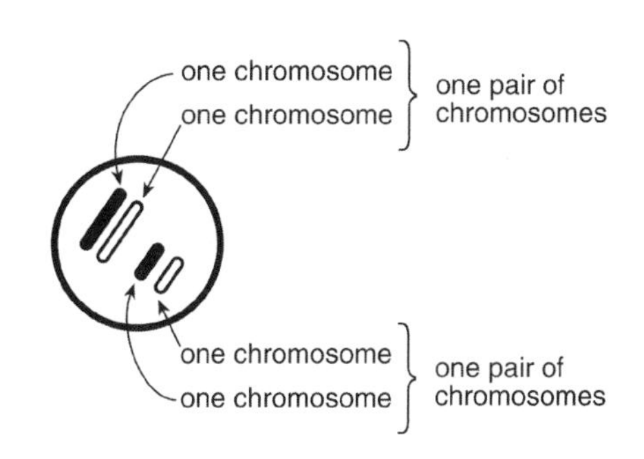

Figure A - Stage 1

Each chromosome in Figure A duplicates (replicates) or doubles, which means one chromosome in Figure A doubles and becomes two chromosomes in Figure B (one doubled chromosome). This cell now has eight chromosomes or four pairs of chromosomes (see Figure B).

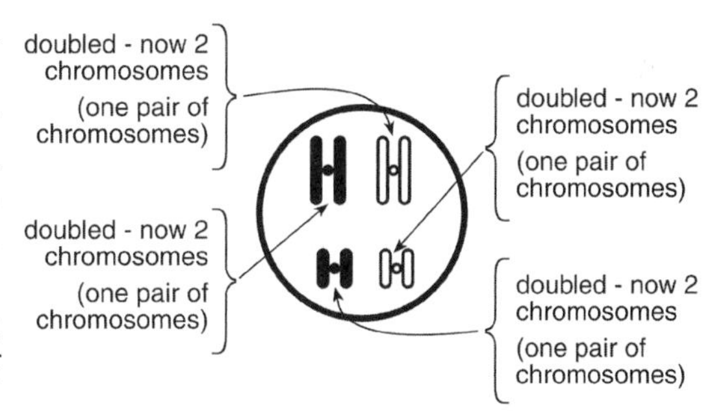

Figure B - Stage 2

By mitosis, the cell (Figure B) divides forming two cells (Figures C and D). One duplicated (doubled) (example big) chromosome from Figure B goes to one cell (see Figure C); the other duplicated (doubled) (big) chromosome from Figure B goes to the other cell (see Figure D). Similarly, one doubled (example small) chromosome goes to one cell (see Figure C); the other small doubled chromosome goes to the other cell (see Figure D). Each cell (cell C or cell D) has four chromosomes (two pairs) like the parent cell which had four chromosomes (two pairs) (see Figure A).

First Division

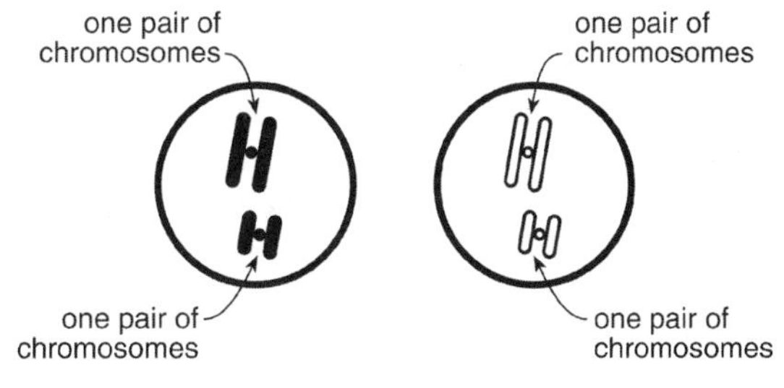

Figure C_{male} **Figure D_{male}**
Stage 3

Now, each **cell** (Figure C and Figure D) will divide again; both the chromosomes and the cytoplasm will divide again.

One chromosome from each pair in Figure C_{male} and Figure D_{male} goes to one gamete, Figure E_{male} (sperm in this example or egg in another example). The other chromosome from that pair goes to another gamete (sperm in this example). Each sperm must only have two chromosomes.

Second Division

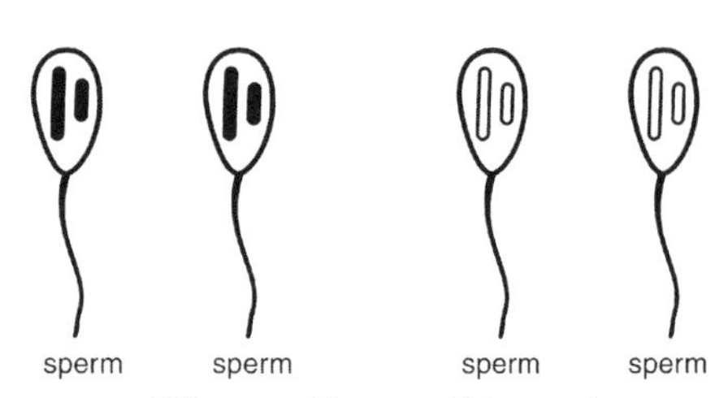

Figure E_{male} - Stage 4

Continued description of Figure E: Each gamete, also called sex cell (sperm in this example), now has two chromosomes, **half the number of chromosomes** of the parent cell (in this case, the parent cell had four chromosomes). The sex cell (example sperm) always has the haploid number of chromosomes, meaning half the number of chromosomes of the parent cell.

The second division of meiosis (going from Figures C and D to E) causes the gametes to have half the number of chromosomes as the parent cell (in this example, parent cell has four chromosomes, each sperm cell has two chromosomes).

Meiosis (reduction division) causes the **gametes** (example sperm or egg) to have **half the number of chromosomes, half the DNA**, as the parent cell. If the parent cell has 50 chromosomes, the gamete (example sperm or egg cell) will have 25 chromosomes; if the parent cell has 100 genes, the gamete will have 50 genes.

Meiosis Produces Eggs In A Female: Meiosis in a female is almost the same as meiosis in a male. The main difference is, in the female, meiosis produces one egg and three nonfunctioning cells; in the male, meiosis produces four sperms. Figures A, B, C, and D in the female are almost the same as Figures A, B, C, and D in the male (see previous figures).

First Division

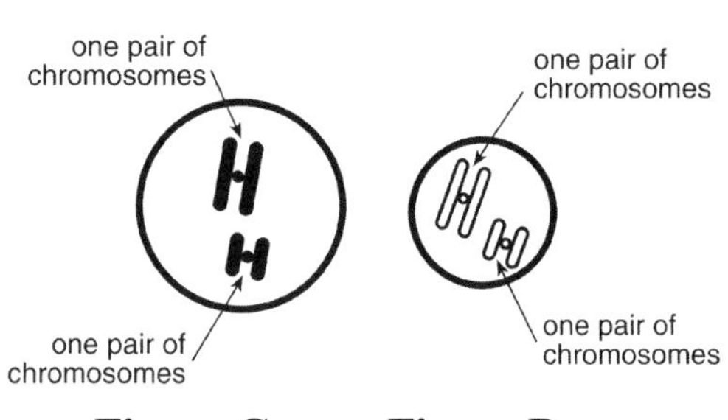

Figure C_{female} Figure D_{female}
Stage 3

Now each cell (Figures C_{female} and D_{female}) will divide again; both the chromosomes and cytoplasm will divide again.

One chromosome from each pair in Figures C_{female} and D_{female} goes to one new cell (Figure E_{female}); the other chromosome from that pair goes to another new cell (one egg cell and three nonfunctional cells are formed). Each cell must only have two chromosomes.

Second Division

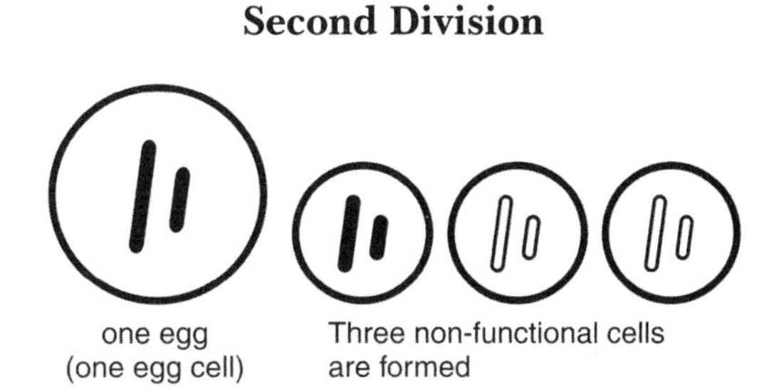

Figure E_{female} - Stage 4

Difference: **But**, in the female reproductive system, after the second division, one egg (large cell) and three nonfunctioning cells (small cells that don't function) are produced. The cytoplasm in the female reproductive system divides unequally, producing one large egg (egg cell) and three small, nonfunctional cells.

Each cell (egg and nonfunctional cells) has only two chromosomes, half the number of chromosomes as the parent cell (see Figure A on previous page). Parent cell has four chromosomes. Egg and three nonfunctional cells always have the haploid number of chromosomes (half the number of chromosomes of the parent cell).

In the second division, going from C_{female} and D_{female} to E_{female}, meiosis causes the sex cell (example egg cell) and three nonfunctional cells to have half the number of chromosomes as the parent cell. You saw in this example, the parent cell has four chromosomes; the egg cell and three nonfunctional cells each have two chromosomes.

Error in meiosis: Down's syndrome is a condition caused by an error in meiosis, which causes a gamete to have an extra chromosome number 21. The zygote and body cells then have the extra chromosome number 21. A person with Down's syndrome has mental retardation and an unusual facial appearance.

Question: The diagram below illustrates some of the changes that occur during gamete formation.

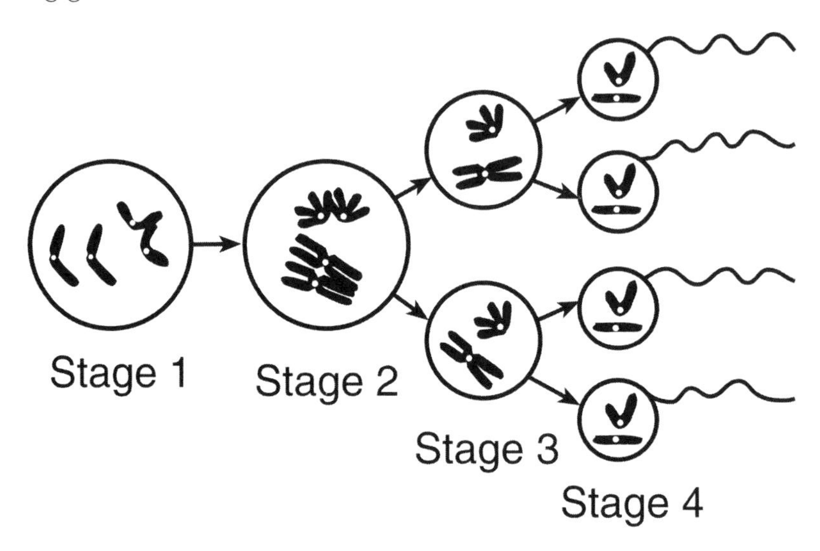

Which graph best represents the changes in the amount of DNA in one of the cells at each stage?

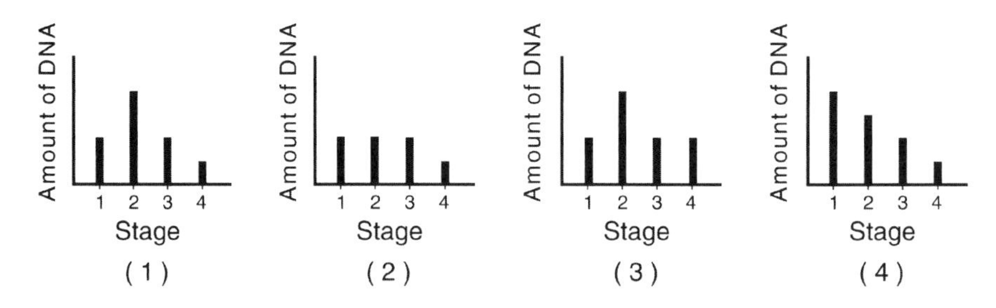

Solution: Look at the top diagram showing stages 1-4 of gamete formation. or = one chromosome.

Stage 1 Cell forms **four chromosomes**.

Stage 2 Cell has double the number of chromosomes as stage 1, which is **eight chromosomes.**

Stage 3 Cell in stage 2 divides, forming two cells. Each cell in stage 3 has the same number of chromosomes as stage 1 (**four chromosomes).**

Stage 4 Each cell in stage 3 divides again. Each cell in stage 4 has half the number of chromosomes as stage 1, which is **two chromosomes.** Look at the graphs. Choose the bar graph that shows the changes in amount of DNA (chromosomes) in one cell at each stage. On the vertical axis is amount of DNA. Chromosomes are made of DNA.

Look at choice (1) in the question. Choice (1) is the correct answer.

Stage 1 on the bar graph shows the amount of DNA in stage 1 of gamete formation; stage 1 has four chromosomes (DNA that will form four chromosomes).

Stage 2 on the bar graph shows the amount of DNA in stage 2 of gamete formation; the bar in stage 2 is twice as long as in stage 1, because the bar represents eight chromosomes.

Stage 3 on the bar graph shows the amount of DNA in stage 3 of gamete formation; the bar in stage 3 is the same height as stage 1, because the bar represents four chromosomes (DNA that will form four chromosomes).

Stage 4 is half the height of stage 1, because the bar represents two chromosomes (DNA that will form two chromosomes).

The bar graph in choice (1) shows the changes in the amount of DNA in stages 1-4 of gamete formation (meiosis). Answer *1*

Wrong choices:
Choice (2) shows the first three stages have the same amount of DNA (same number of chromosomes), which is not correct.
Choice (3) shows that stages 1, 3, and 4 all have the same amount of DNA (same number of chromosomes), which is not correct.
Choice (4) shows that stage 1 has the most DNA (most chromosomes), which is not correct, and stage 3 has half the DNA (half the chromosomes) of stage 1, which is not correct.

Now Do Homework Questions #32-39, pages 39-40.

Meiosis and Crossing Over Cause Variation

You learned meiosis produces gametes (sex cells). In a human being, the male gamete is the sperm; the female gamete is the egg.

Meiosis and crossing over cause many differences, many variations among organisms as explained below. Let's look at Figure A, etc., again, and see how meiosis causes variation.

Figure A shows a body cell from the testes or ovaries.

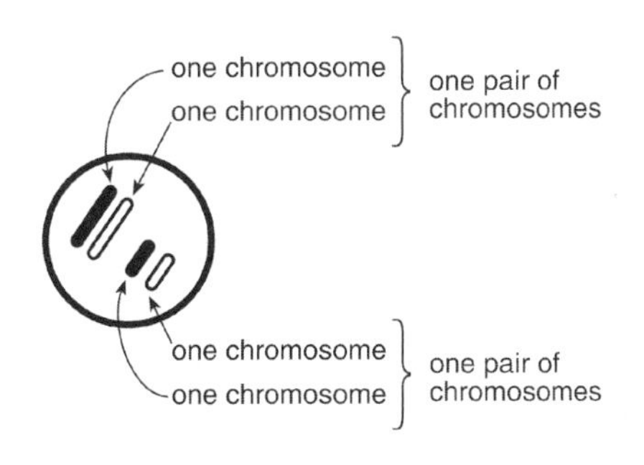

Figure A - Stage 1

Each chromosome in the cell in Figure A duplicates (doubles) and you get Figure B.

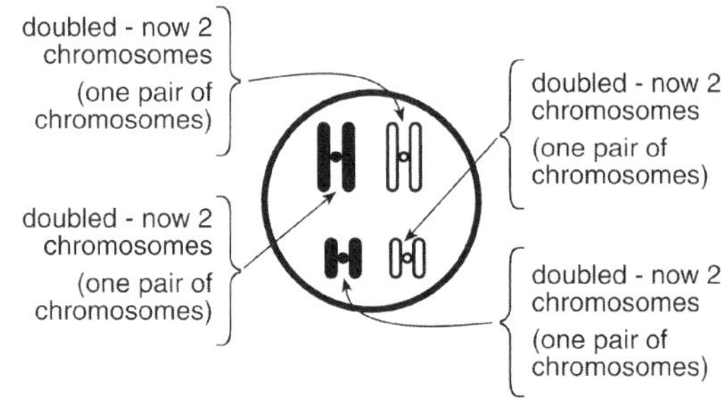

Figure B - Stage 2

Now the cell (Figure B) divides (chromosomes and cytoplasm divide) and forms two cells (C and D). There are two ways Figure B can divide. For example, one doubled chromosome (example big black chromosome) from Figure B goes to one cell and one doubled chromosome (example big white chromosome) from Figure B goes to the other cell (see Figures C_1 and D_1). The small duplicated black chromosome goes to the cell with the big black chromosome (Figure C_1) and the small duplicated white chromosome goes to the cell with the large white chromosome (Figure D_1). C_1 and D_1 then divide again to give four cells (E_1). **Each cell** in **E_1** has either **two black chromosomes or two white chromosomes** (see Figure E_1).

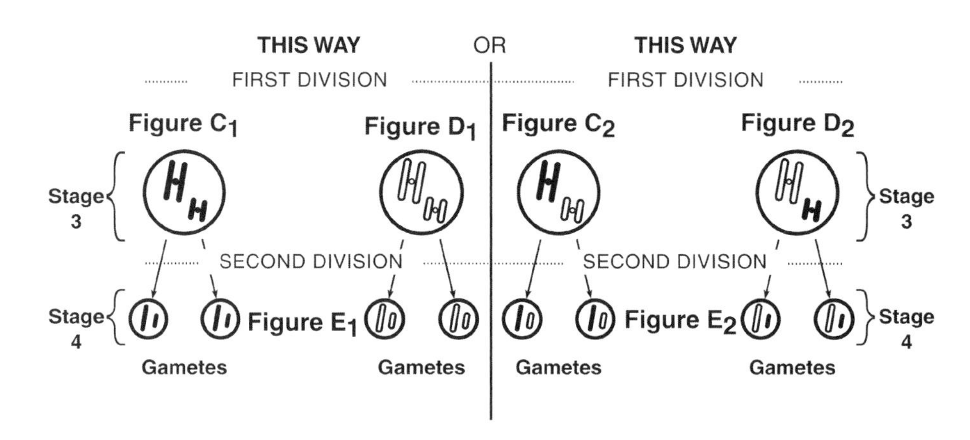

Or the chromosomes in the cell in Figure B can divide differently (see diagram at right), and you get a different combination of chromosomes (figures C_2, D_2, and E_2 instead of C_1, D_1, and E_1). Let's explain. Look at Figure B. One large duplicated chromosome in black and one small duplicated chromosome in white go to one cell (Figure C_2); one large duplicated chromosome in white and one small duplicated chromosome in black go to the other cell (D_2). C_2 and D_2 then divide again to give four cells (E_2). **Each cell** in **E_2** has two chromosomes, **one black and one white chromosome** (see Figure E_2).

You see, in **meiosis**, from the one parent cell (Figure A - Stage 1) you get **different chromosome variations** (different combinations of chromosomes) in gametes (sperm or eggs) or in nonfunctional cells in female meiosis (see stage 4 above and the diagram below) :

The more chromosomes there are, the more different ways the chromosomes can be arranged in gametes (sperm or egg) and recombined to form the zygote (fertilized egg), and the more variations the organism will have.

A human body cell has more chromosomes than the previous example, which had four chromosomes. Body cells of a human being have 46 chromosomes. Each gamete (sperm or egg) has half the number of chromosomes, 23 chromosomes. Because there are a large number of chromosomes, the number of variations (different combinations of chromosomes or genes) in humans is very large.

Crossing Over

During **meiosis,** pairs of chromosomes line up, and sometimes **parts of chromosomes intertwine and cross over.** Part of a chromosome gets exchanged for part of another chromosome (see figure below).

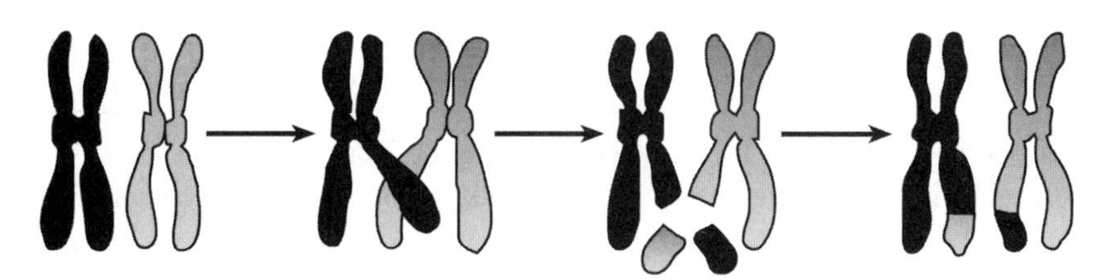

Crossing Over

The second chromosome (chromosome #2) crossed over the third chromosome (chromosome #3) and exchanged part of the chromosome. As you can see, chromosome #2 has a part of chromosome #3 and chromosome #3 got a part of chromosome #2.

Crossing over is when **parts of chromosomes are exchanged, producing** many different varieties of chromosomes; crossing over causes organisms (such as human beings) to have a greater variety of **different characteristics.**

In short, meiosis and crossing over cause variations in the offspring (example tall person with brown eyes and blond hair or short person with brown eyes and brown hair).

Now Do Homework Questions #40-43, pages 40-41.

Comparison of Mitosis and Meiosis

Mitosis	Meiosis
Produces new **cells** for **growth** and **repair**	**Produces** gametes (sex cells), **sperm** and **egg** for **reproduction**
New cells have the **same number** of **chromosomes** (diploid number of chromosomes), same DNA, same genetic information, same genes as original cell. Example: In humans, original cell has 46 chromosomes, new cells have 46 chromosomes.	Sex cells, sperm and egg, have **half** the **number** of **chromosomes** (haploid number of chromosomes), half the DNA, half the genetic information of the original cell. Example: In humans, original cell in testes or ovaries has 46 chromosomes; sperm or egg have half the number of chromosomes: sperm has 23 chromosomes, egg has 23 chromosomes.
One cell **produces two cells;** there is only **one cell division**.	One cell **produces four cells**; there are **two cell divisions** (example: one cell in testes produces four sperm).
One (each) cell produces two **genetically identical cells.**	One cell in testes produces four sperm (cells); one cell in ovary produces one egg and three nonfunctional cells (total four cells). **Meiosis causes variation**- each sperm or egg is different (**different genes**) from other sperm or eggs.

Fertilization

Meiosis produces gametes and causes the **gametes**, also called **sex cells** (example **sperm** or **egg**) to have **half the number** of **chromosomes** as the cells in the body. Body cells in a human being have 46 chromosomes; gametes (sperm or egg) have 23 chromosomes.

Fertilization: Sperm and egg unite together to form a fertilized egg (zygote).

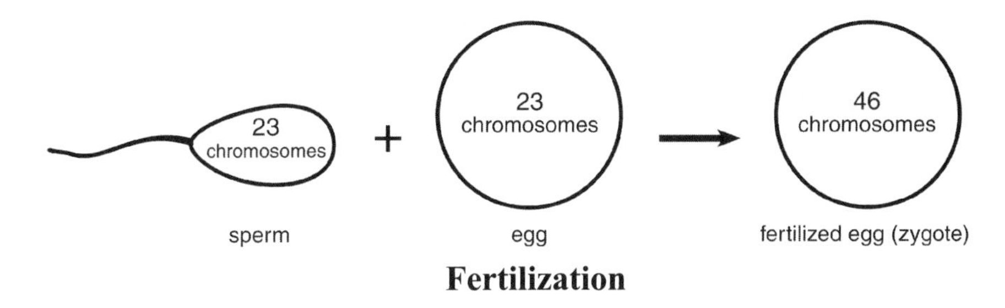

Fertilization

Let n = 23, number of chromosomes in the sex cell (example sperm or egg), haploid number of chromosomes (half the number of chromosomes as the cells in the body).
The sperm (which has 23 chromosomes) unites with the egg (which has 23 chromosomes) to form a fertilized egg (also called a zygote) which has 46 chromosomes. The fertilized egg has double the number of chromosomes as the sperm or egg, 46 chromosomes (diploid number of chromosomes), which we can call 2n.

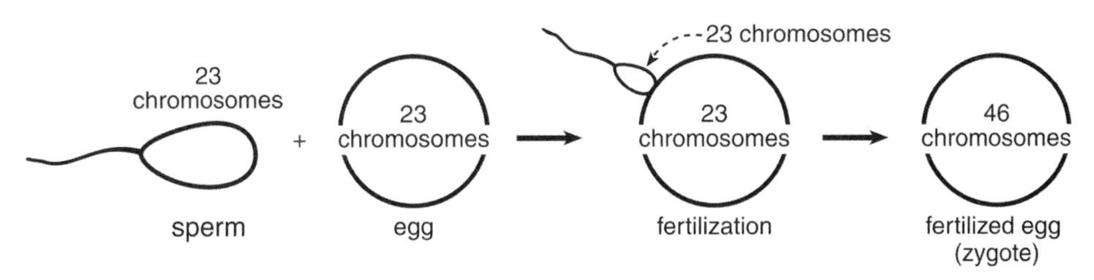

Fertilization

This fertilization diagram above shows sperm and egg uniting. Genes are on the chromosomes. Combining genes from sperm and egg is called recombination. **Fertilization** (sperm from the male parent unites with the egg from the female parent) and **recombination** (combining genes from the sperm with genes from the egg) **cause variation in** the **offspring.** Each sperm and each egg has a different combination of genes.

Question: In sexually reproducing species, the number of chromosomes in each body cell remains the same from one generation to the next as a direct result of

1. meiosis and fertilization
2. mitosis and mutation
3. differentiation and aging
4. homeostasis and dynamic equilibrium

Solution: Human body cells have 23 pairs of chromosomes (46 chromosomes). Meiosis causes sperm or egg to have 23 chromosomes. Fertilization combines sperm (23 chromosomes) with egg (23 chromosomes) to form a fertilized egg (zygote) with 46 chromosomes. Every cell in the body except sperm and eggs (examples, skin cell, nerve cell) has 46 chromosomes. Answer *1*

Example: In this organism, all cells except sperm and egg have four chromosomes (two pairs of chromosomes), including cells of the testes and ovaries.

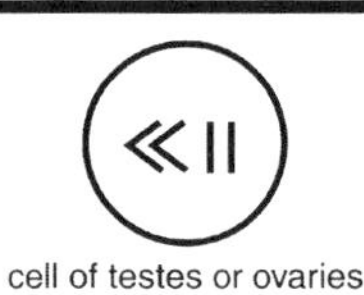
cell of testes or ovaries

The sperm gets one chromosome from each pair (see figure of cell above).

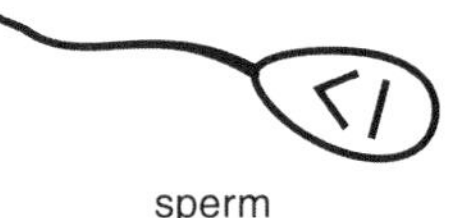
sperm

The egg gets one chromosome from each pair (see figure of cell above).

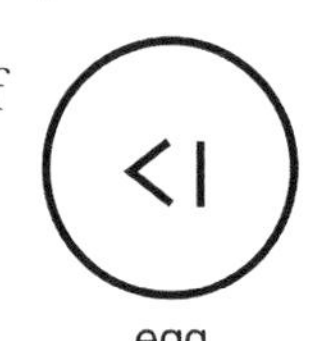
egg

Fertilization means sperm and egg unite together to form afertilized egg (zygote).

Fertilization

The sperm, with half the number of chromosomes (half the DNA, half the genetic material) and the egg, with half the number of chromosomes (half the DNA, half the genetic material) unite together to form a fertilized egg (zygote), with double the number of chromosomes, double the DNA, double the genetic material.

All the cells in the organism (except sperm and egg) will have the same number of chromosomes, the same chromosomes, as the fertilized egg,

Zygote Formation, Growth, and Differentiation

Sperm, with half the number of chromosomes (half the DNA, half the genetic information) as cells of the body, unites with the egg, with half the number of chromosomes (half the DNA, half the genetic information) (see fertilization, number 1 in early development figure below) to form a fertilized egg (zygote) (see number 2 in early development figure below). The zygote has double the number of chromosomes etc. as the sperm or egg.

Look again at the figure below. The one-celled fertilized egg (zygote) keeps dividing by mitosis to form more and more cells, to grow and develop. **Development** means all the changes that happen after the egg becomes fertilized (examples: growth (making more cells), and forming tissues, organs, and organ systems). By mitosis, one cell forms two cells, two cells form four cells, four cells form eight cells, and eight cells form 16 cells, etc. The zygote and all cells in the body (except sperm and egg) have the same genetic makeup (same genes, same DNA, and same number of chromosomes).

The different cells start to become different. Some specialized cells will form tissues (example skin, muscle) and then organs (example heart, kidneys), and organ systems (example circulatory system). This process is called differentiation.

Every cell in the body, except sperms and eggs, has all the genes. You learned in the genetics chapter each cell uses some of the genes (turns on the genes because the cell needs to use them) and does not use other genes (turns the genes off).

Different genes were activated (used) in different cells, causing different proteins to be produced and causing some cells to become skin, muscles, and nerves.

Also, different genes were used (activated) in different cells causing cells to become pancreas (which will produce hormones) and liver and circulatory system (including heart and blood vessels).

Early Development

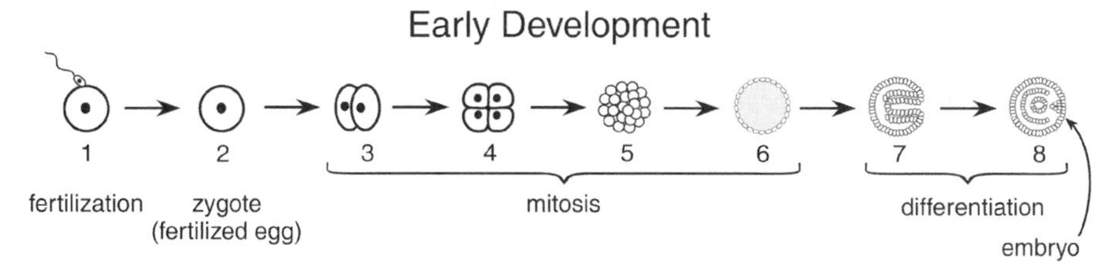

#7 in the early development figure above shows three layers, outer layer, middle layer, and inner layer. The three layers of #7 are enlarged and labeled at right. The three layers develop (differentiate, differentiation) into different systems. The outer layer of cells forms the nervous system, etc. The middle layer forms the circulatory system, testes and ovaries. The inner layer forms the lining of the digestive tract. The three-layered stage develops into the embryo, #8.

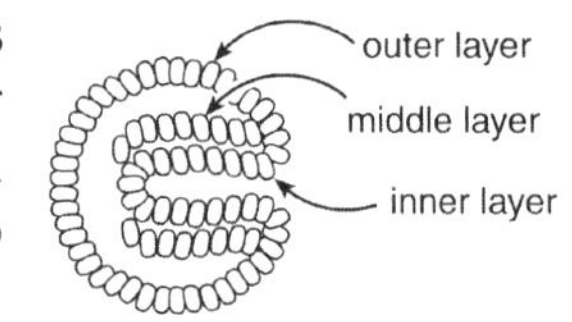

Part of Differentiation

Practice Questions and Solutions

Question: The diagram below represents a series of events in the development of a bird.

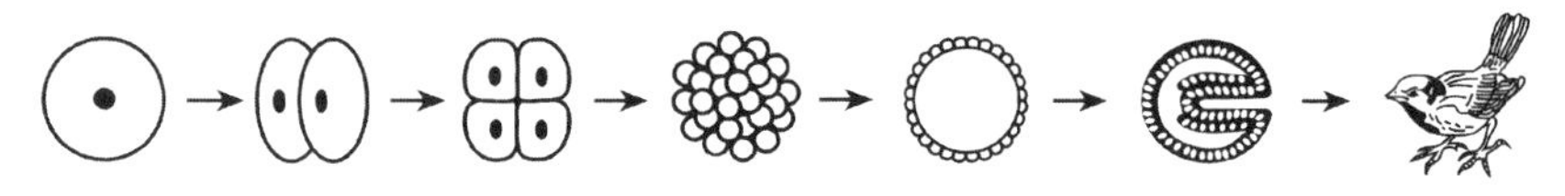

Which series of terms best represents the sequence of processes shown?

(1) meiosis → growth → differentiation
(2) meiosis → differentiation → growth
(3) mitosis → meiosis → differentiation
(4) mitosis → differentiation → growth

Solution: We labeled the diagram in the question.

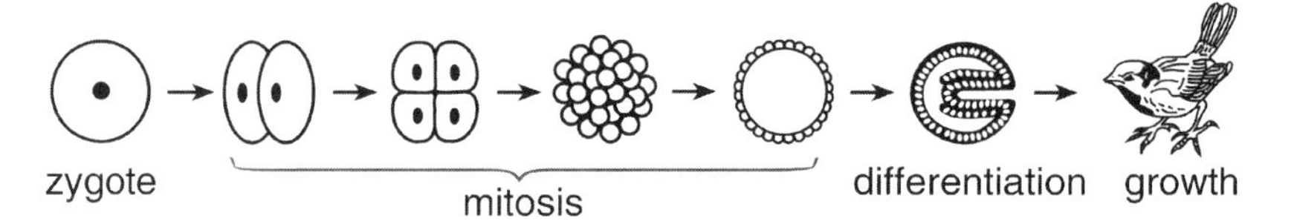

The one-celled fertilized egg (zygote) keeps dividing by **mitosis** to form more and more cells. The sixth figure in the diagram in the question (or above) shows three layers of cells, outer, middle, and inner. The outer layer forms the nervous system, etc. The middle layer forms the circulatory system, testes and ovaries. The inner layer forms the lining of the digestive tract. This is called **differentiation.** The last figure in the diagram (the bird) shows **growth.** The three-layered embryo grew into a bird and the bird grew.

Choice *4*, mitosis, then differentiation, then growth. Answer *4*

Now Do Homework Questions #44-61, pages 41-45.

HUMAN REPRODUCTION AND DEVELOPMENT

The function of the reproductive system is to produce more organisms (plants, animals, including dogs, cats, humans) of the same type.

Female Reproductive System: The ovaries produce egg cells (female gametes). Look at the figures at right and study them. The egg goes from the ovary to the oviduct and then to the uterus. If sperm are present, egg unites with a sperm (fertilization, egg fertilized) in the oviduct and forms a fertilized egg (zygote). The fertilized egg will develop into an embryo (see early development figure above, #2 fertilized egg (zygote) to #8 embryo). The fertilized egg goes down the oviduct to the uterus.

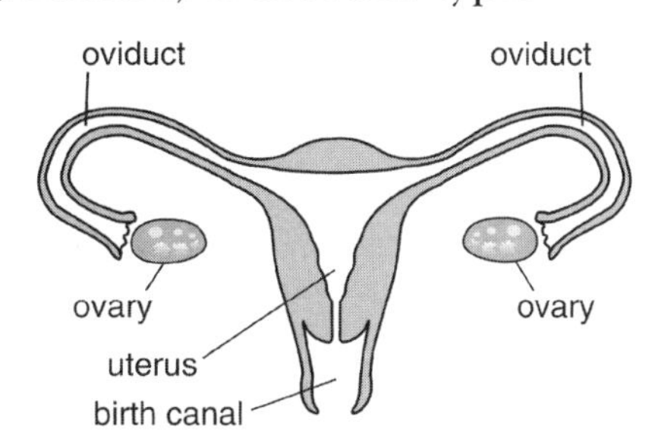

Female Reproductive System

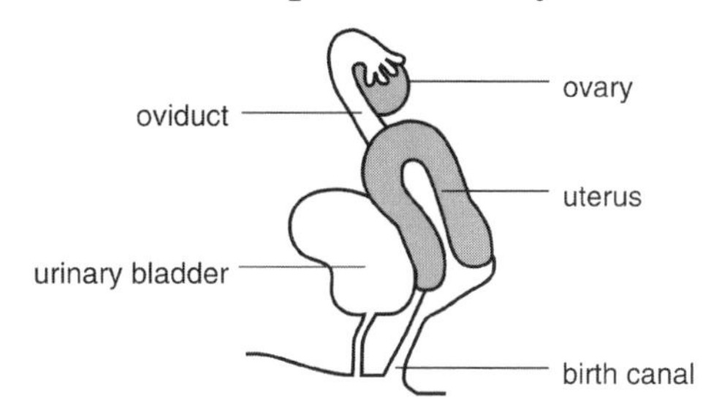

Female Reproductive System Side View

The **embryo attaches** itself (implants itself) **to** the wall of the **uterus** and develops (development) in the uterus. After the embryo is attached to the uterus, a **placenta** (which is an organ made of mother's and baby's tissues) **is formed** (see figure at right). In the placenta, nutrients and oxygen from the mother's blood diffuse (go) into the embryo's blood. Wastes from the embryo's blood diffuse (go) into the mother's blood in the placenta. Now the embryo has the nutrients and oxygen it needs and gets rid of its wastes. Note: When the embryo is more developed, after about two months, it is a fetus (see figure at right).

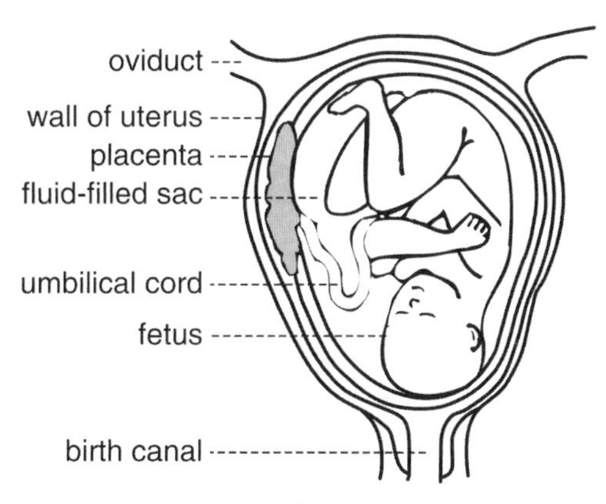

Fetus with Placenta

At the beginning of pregnancy, a **fluid-filled sac** in the uterus is formed (see figure), which cushions and protects the embryo.

As you can see in the figure, the umbilical cord attaches the fetus to the mother (placenta). When a baby is born, the uterus (muscles of the uterus) contracts and pushes out the baby.
Note: From the time the egg is fertilized in the oviduct, mitosis (cells dividing) continues to take place, producing more and more cells , different organs, and different systems.

Hormones: When a girl reaches sexual maturity (puberty), the **ovaries produce** two hormones, estrogen and progesterone:

a. estrogen (a female sex hormone) which influences the **development** of the **secondary sex characteristics** (such as developing mammary glands (breasts) that provide milk for the newborn, and widening of the hips) and also **regulates** the **reproductive cycle (thickens** the **lining of the uterus).**

b. progesterone (a female sex hormone), which regulates the menstrual **cycle** and prepares and **maintains** the **lining of** the **uterus** for a pregnancy.

The structures and functions of the **human female reproductive system**, as in almost all other mammals (animals that give milk for their babies, examples dogs, cows, etc.), are designed to

a. **produce gametes (eggs) in ovaries**
b. allow for **internal fertilization** (**fertilization inside** the mother),
c. support (help) the **internal development of** the embryo and fetus (the embryo and fetus **develop (development) inside** the **mother** in the uterus)
d. **provide essential materials** (examples oxygen, glucose) from the mother **through** the **placenta** to the embryo and fetus
e. provide nutrition (giving mother's **milk to** the **newborn).**

Male Reproductive System:
The testes produce the sperm (male gamete). Look at the figure of the male reproductive system and study it. The sperm go from the testes to the sperm duct and then to the urethra, which is in the penis. There are glands which produce (secrete) fluids that help sperm swim from the testes through the sperm ducts to the urethra, which is inside the penis; then the penis transfers the sperm to the female reproductive system.

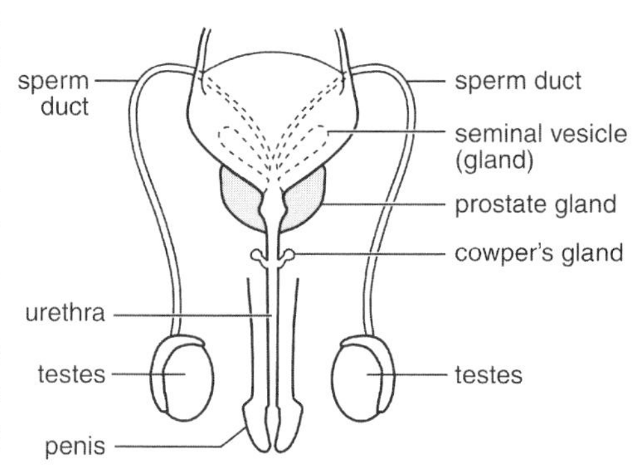

Male Reproductive System

Note: The urethra is an organ (structure) in both the excretory system and the reproductive system.

Hormone: When a boy reaches sexual maturity (puberty), the **testes produce testosterone. Testosterone,** male sex hormone, influences (regulates) the development of secondary sexual characteristics (examples, facial hair (a beard) and a deep voice); testosterone is also important in reproduction (it causes sperm to mature and be able to unite with the egg to form a fertilized egg (zygote)).

The structure and functions of the **human male reproductive system,** and also the reproductive system of other mammals (example dogs and cats), are designed to **produce gametes (sperm)** in testes and deliver sperm (bring the sperm) to the egg so sperm can fertilize the egg (fertilization).

Question: The diagram represents a human reproductive system.

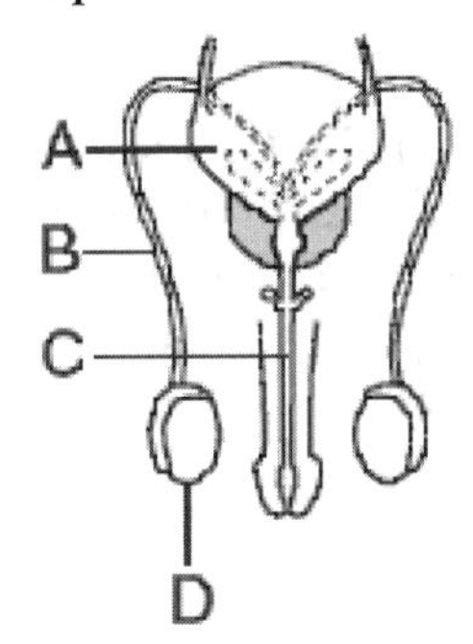

Meiosis occurs within structure

(1) A (2) B (3) C (4) D

Solution: Structure D is the testes. The testes produce sperm by the process of meiosis. (Meiosis causes the sperm to have 23 chromosomes, haploid number of chromosomes, half the number of chromosomes as a body cell (such as a skin cell), which has 46 chromosomes).Answer *4*

Fertilization: The sperm enters the female reproductive system from the birth canal (vagina), goes up to the uterus, and to the **oviduct**, where **sperm** and **egg unite** together to **form** a **fertilized egg (zygote). Fertilization** (sperm and egg unite together) takes place in the **oviduct.**

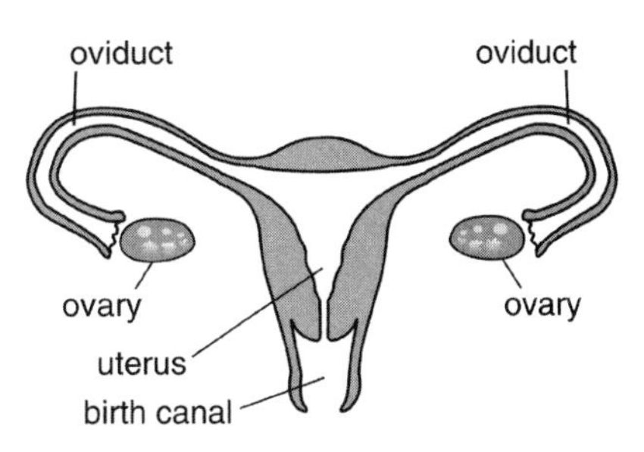

Female Reproductive System

Hormones regulate the female reproductive cycle

Let's discuss the female hormones in a little more detail. When a girl (female) reaches maturity (sexual maturity, puberty), the ovaries produce the hormones estrogen and progesterone. Hormones regulate the human reproductive cycle (explained below).

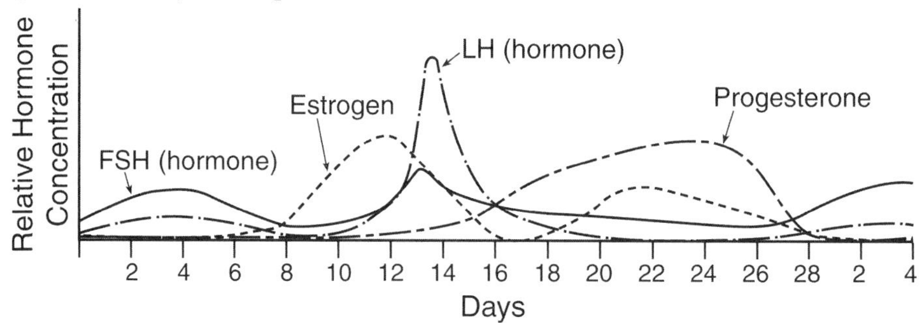

Female Reproductive Cycle

The female reproductive cycle begins when the ovary produces the egg. On about day 14 of the reproductive cycle the egg leaves (is released from) the ovary (see diagram above). Hormones control the timing of the reproductive cycle, when the egg leaves (is released from) the ovary and when the lining of the uterus will be made and maintained for the embryo to attach to. Hormones are needed for eggs to be released from the ovary and for the uterus to keep the embryo alive.

Hormones estrogen and progesterone (both from the ovary) with a few other hormones from the pituitary gland (bottom of brain) regulate the reproductive cycle. Estrogen thickens the lining of the uterus and causes the lining to have more blood vessels. Progesterone maintains the lining once it forms. The thick lining of the uterus is needed for the embryo to attach to it. Let's explain the graph of the female reproductive cycle (see graph above). Looking at the graph, you see in the reproductive cycle there are changes in the amount of the hormone estrogen, the hormone progesterone, LH (hormone) and FSH (hormone).

Look at the graph again. There is very little estrogen on Day 4 and more estrogen on Day 12; there is a lot of progesterone on Day 25 and very little progesterone on Day 9. When there is more estrogen, the lining of the uterus gets thicker and has more blood vessels. When there is more progesterone, the lining of the uterus is maintained more.

At the end of the human reproductive cycle, if the egg is not fertilized, the amounts of estrogen and progesterone decrease (see diagram above) and the lining of the uterus breaks down and is discharged from the body (menstruation). The reproductive cycle from start to finish takes about 28 days. Sometimes the reproductive cycle can be a little longer, such as 32 days, or a little shorter, such as 26 days.

If the egg is fertilized, the hormone progesterone level would remain high. Progesterone is called the pregnancy hormone. Progesterone maintains the lining of the uterus where the embryo or fetus stays until it is born.

When a person ages, the hormones decrease (become less) and the reproductive cycle does not take place.

Now Do Homework Questions #84-86, pages 49-50.

Reproduction: How fertilized egg, embryo, fetus, and baby develop

Now that we understand the male and female reproductive systems and the hormones, let's understand how a fertilized egg, embryo, fetus, and baby develop (understand human reproduction and development).

The **ovaries produce egg** cells (female gametes). The egg goes from the ovary to the oviduct.

The **testes produce** the **sperm** (male gamete). The sperm go from the testes to the sperm duct and then to the urethra, which is in the penis. There are glands which produce fluids that help sperm swim from the testes through the sperm ducts to the urethra in the penis. The penis transfers sperm to the female.

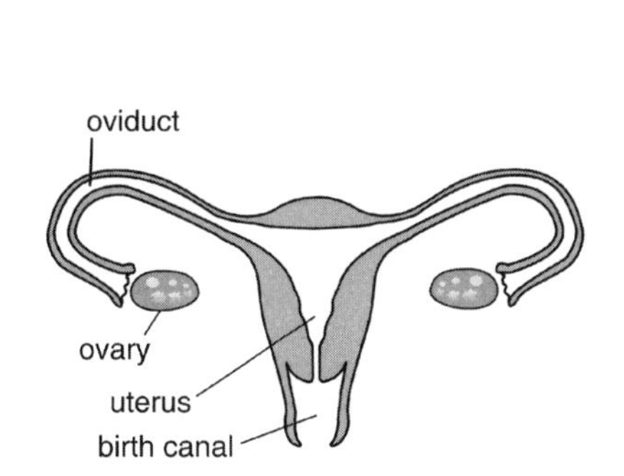

Female Reproductive System

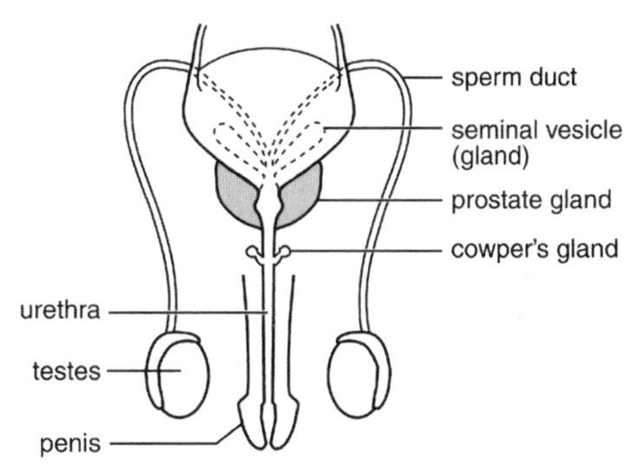

Male Reproductive System

The sperm enters the female reproductive system from the birth canal (vagina), goes up to the uterus, and to the **oviduct**, where **sperm** and **egg unite** together to **form** a **fertilized egg (zygote)**. **Fertilization** (sperm and egg uniting together) takes place in the **oviduct.** The fertilized egg goes from the oviduct to the uterus.

Note: Look at the early development diagram below. From the time the egg is fertilized in the oviduct, mitosis (cells dividing) continues to take place, producing more and more cells, then different organs and different systems (differentiation); it develops into an embryo.

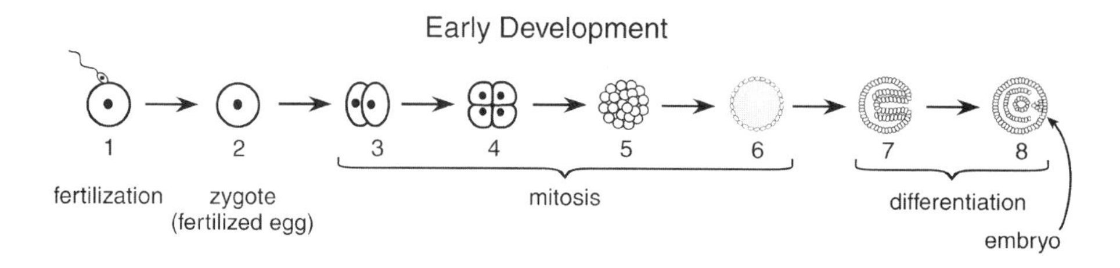

The embryo attaches itself (implants itself) to the wall of the uterus and develops (development). Then a placenta (which is an organ made of mother's and baby's tissues) is formed. **In** the **placenta, nutrients and oxygen from** the **mother's blood diffuse into** the **embryo's blood** (or into the fetus' blood). Wastes from the embryo's blood diffuse into the mother's blood in the placenta. Now the embryo has the nutrients and oxygen it needs and gets rid of its wastes. Note: You learned when the embryo is more developed, after about two months, it is a fetus (see figure at right). In short, in the placenta there is an exchange of material between the maternal (mother's) blood (example nutrients) and the fetal blood (blood of fetus)(example wastes). At the beginning of pregnancy, a **fluid-filled sac** in the uterus is formed (see figure above), which cushions and protects the embryo.

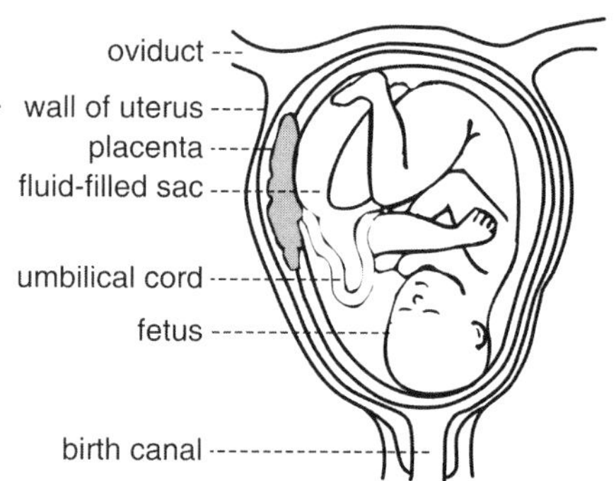

Fetus with Placenta

The figure below shows the **development of** an **embryo** (the embryo differentiates (**differentiation)** into muscles, skeleton, and systems (nervous system, circulatory system, etc.)), and the embryo grows larger (**growth**). Look at the figure below.

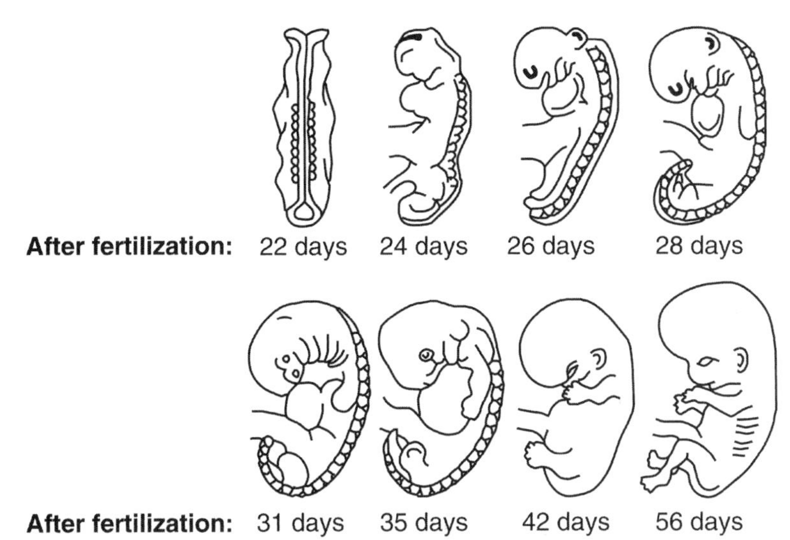

Development of an Embryo

Problems in the embryo

In humans, the embryonic (embryo's) development of essential organs occurs in early stages of pregnancy. The **embryo** may **have** risks **(problems) from faults in** its **genes.** The **embryo** can also have **problems from** its **mother's** exposure to **environmental factors** such as mother eating an **inadequate** (poor) **diet**, **using alcohol/drugs/tobacco**, and having **toxins** (example lead poisoning), or **infections** (such as German measles or AIDS) in her pregnancy. The alcohol, drugs, tobacco, and infections enter the embryo or fetus through (across) the placenta. The alcohol, drugs, tobacco etc. are more dangerous for the embryo than the mother because the embryo's organs (examples heart, brain) are forming and these drugs, etc. can hurt these organs while they are forming; and also the embryo is very small and therefore the amount of alcohol has a bigger effect. If the mother uses alcohol, drugs, or tobacco, it can cause a baby to be brain damaged, drug addicted, or have low birth weight. If the mother has an inadequate diet, has toxins (example lead poisoning) or infections (example rubella or AIDS) it can hurt (harm) the baby.

When a baby is born, the uterus (muscles of the uterus) contracts and pushes out the baby.

After the **baby** is **born,** there is cell **differentiation, growth** (body grows) and **development** and the child becomes sexually mature (example testes produce sperm and the hormone testosterone, or ovaries produce eggs and the hormones estrogen and progesterone) and again fertilization can take place and a baby is born.

In short, after the baby is born, there is cell differentiation, growth (body grows) and development until the person is an adult. **Human reproduction** and **development** are **influenced by** these factors: **gene expression** (genes to produce eggs are turned on in the ovaries, genes to produce sperms are turned on in the testes etc.), **hormones** (examples estrogen, progesterone and testosterone regulate the reproductive cycle), **environment,** etc. Late in years the person ages, gets weaker, and dies.
Human development, birth, growth, aging, and death are a predictable pattern of events.

Now Do Homework Questions #87-97, pages 50-52.

Reproductive Technology

Reproductive technology is using scientific methods to produce organisms with desired characteristics. It is used for medical purposes, in agriculture, and for ecological uses (ecology).

In medicine, in vitro fertilization (combining sperm and egg in the laboratory and implanting (putting) the fertilized egg into the mother) is used for people to be able to have children when the oviduct is blocked.

In agriculture, useful genes have been inserted into plants (genetic engineering); the inserted genes are also present in the seeds of the plant (seeds are fertilized eggs (zygotes)) and passed along to the offspring (the new baby plant). One example is a gene for an antifreeze protein from fish that is inserted into tomatoes; it protects the tomatoes from being damaged by freezing. Another example is a gene that protects corn plants from a weedkiller; then the weedkiller does not damage the corn plants but only kills the weeds.

In ecology (ecological applications), a new way has been found to help increase the number of animals of some endangered species. An animal may normally only produce one baby at a time. The **endangered animal** is **treated with hormones** to make it **produce** a **large number of eggs.** The **many eggs** are **fertilized** in the laboratory **or cloned** (the nucleus of the egg cell of the endangered animal is removed and replaced by the nucleus of a skin cell of that endangered animal; these cloned eggs can develop into a baby without using sperm). The fertilized eggs or cloned eggs are then implanted (put) into other non-endangered animals (cloned eggs from an endangered ox-type animal have been implanted into similar animals, such as cows), **producing** many **babies of** the **endangered animal**.

NOW LET'S TRY A FEW HOMEWORK QUESTIONS

1. The diagram below represents a yeast cell that is in the process of budding, a form of asexual reproduction.

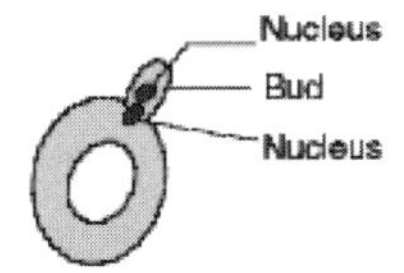

Which statement describes the outcome of this process?
   (1) The bud will develop into a zygote.
   (2) The two cells that result will each contain half the species number of chromosomes.
   (3) The two cells that result will have identical DNA.
   (4) The bud will start to divide

2. A tree produces only seedless oranges. A small branch cut from this tree produces roots after it is planted in soil. When mature, this new tree will most likely produce
   (1) oranges with seeds, only
   (2) oranges without seeds, only
   (3) a majority of oranges with seeds and only a few oranges without seeds
   (4) oranges and other kinds of fruit

3. Strawberries can reproduce by means of runners, which are stems that grow horizontally along the ground. At the region of the runner that touches the ground, a new plant develops. The new plant is genetically identical to the parent because
   (1) it was produced sexually
   (2) nuclei traveled to the new plant through the runner to fertilize it
   (3) it was produced asexually
   (4) there were no other strawberry plants in the area to provide fertilization

4. Which process is indicated by letter ***B***?

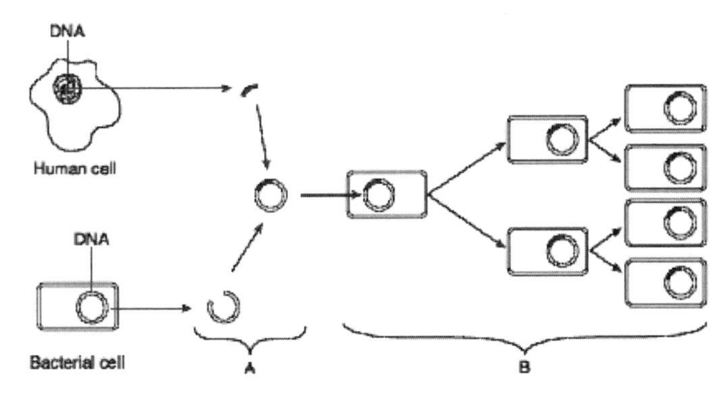

   (1) natural selection
   (2) asexual reproduction
   (3) sexual reproduction
   (4) gene deletion

5. Which statement describes asexual reproduction?
   (1) Adaptive traits are usually passed from parent to offspring without genetic modification.
   (2) Mutations are not passed from generation to generation.
   (3) It always enables organisms to survive in changing environmental conditions.
   (4) It is responsible for many new variations in offspring.

6. A variation causes the production of an improved variety of apple. What is the best method to use to obtain additional apple trees of this variety in the shortest period of time?
   (1) selective breeding
   (2) natural selection
   (3) asexual reproduction
   (4) hormone therapy

7. Cloning an individual usually produces organisms that
   (1) contain dangerous mutations
   (2) contain identical genes
   (3) are identical in appearance and behavior
   (4) produce enzymes different from the parent

8. A certain bacterial colony originated from the division of a single bacterial cell. Each cell in this colony will most likely
   (1) express adaptations unlike those of the other cells
   (2) replicate different numbers of genes
   (3) have a resistance to different antibiotics
   (4) synthesize the same proteins and enzymes

9. The chromosome content of a skin cell that is about to form two new skin cells is represented in the diagram below.

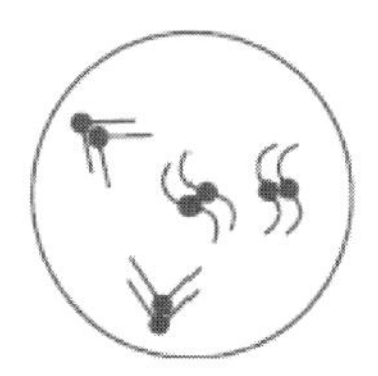

Which diagram best represents the chromosomes that would be found in the two new skin cells produced as a result of this process?

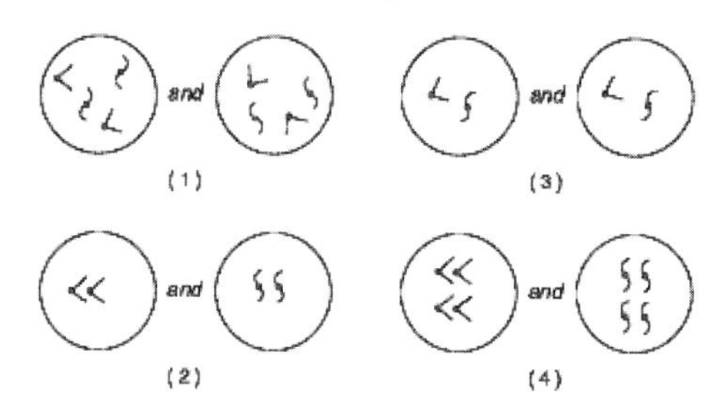

10. The sequence of events occurring in the life cycle of a bacterium is listed below.
   (A) The bacterium copies its single chromosome.
   (B) The copies of the chromosome attach to the cell membrane of the bacterium.
   (C) As the cell grows, the two copies of the chromosome separate.
   (D) The cell is separated by a wall into equal halves.
   (E) Each new cell has one copy of the chromosome.

   This sequence most closely resembles the process of
   (1) recombination
   (2) zygote formation
   (3) mitotic cell division
   (4) meiotic cell division

11. The diagram below represents the cloning of a carrot plant.

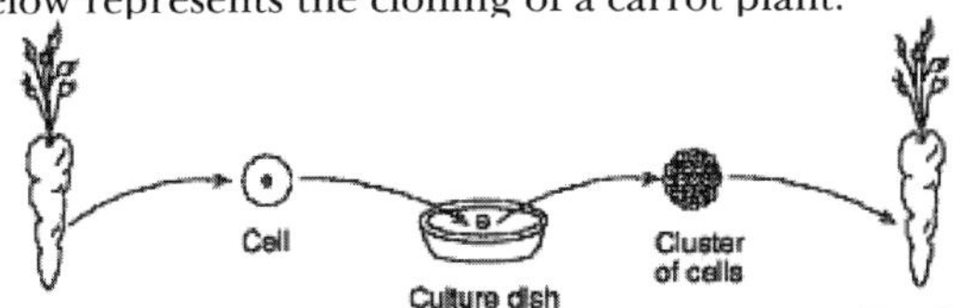

Compared to each cell of the original carrot plant, each cell of the new plant will have

(1) the same number of chromosomes and the same types of genes
(2) the same number of chromosomes, but different types of genes
(3) half the number of chromosomes and the same types of genes
(4) half the number of chromosomes, but different types of genes

---

Base your answers to the next two questions on the diagram below and on your knowledge of biology. The diagram represents a single-celled organism, such as an ameba, undergoing the changes shown.

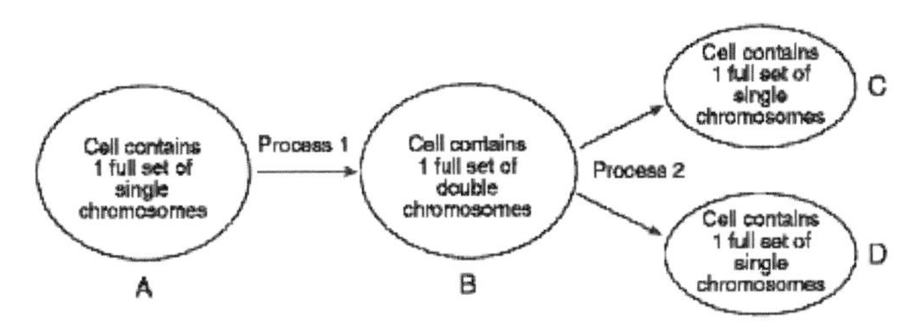

12. Process *1* is known as
(1) replication (2) meiosis (3) differentiation (4) digestion

13. The genetic content of ***C*** is usually identical to the genetic content of
(1) ***B*** but not ***D***
(2) both ***B*** and ***D***
(3) ***D*** but not ***A***
(4) both ***A*** and ***D***

---

14. The equation below represents a chemical reaction that occurs in humans.

$$\text{Substance X} + \text{Substance Y} \xrightarrow{\text{enzyme C}} \text{Substance W}$$

What data should be collected to support the hypothesis that enzyme ***C*** works best in an environment that is slightly basic?

(1) the amino acid sequence of enzyme ***C***
(2) the amount of substance ***W*** produced in five minutes at various pH levels
(3) the shapes of substances ***X*** and ***Y*** after the reaction occurs
(4) the temperature before the reaction occurs

15. The diagrams below represent cells that transport chromosomes.

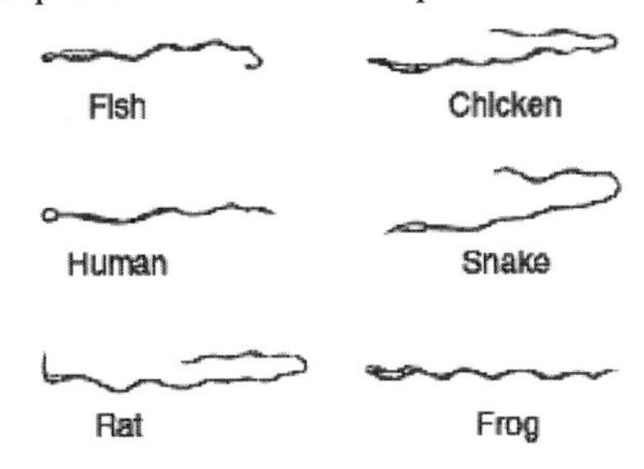

These cells are specialized for
(1) oxygen transport
(2) transmitting chemical signals over long distances
(3) sexual reproduction
(4) injecting antibodies into harmful bacteria

16. Individuals of some species, such as earthworms, have both male and female sex organs. In many cases, however, these individuals do not fertilize their own eggs.

State ***one*** genetic advantage of an earthworm mating with another earthworm for the production of offspring.

17. Which process will increase variations that could be inherited?
(1) mitotic cell division
(2) active transport
(3) recombination of genes
(4) synthesis of proteins

18. Which process usually results in offspring that exhibit new genetic variations?

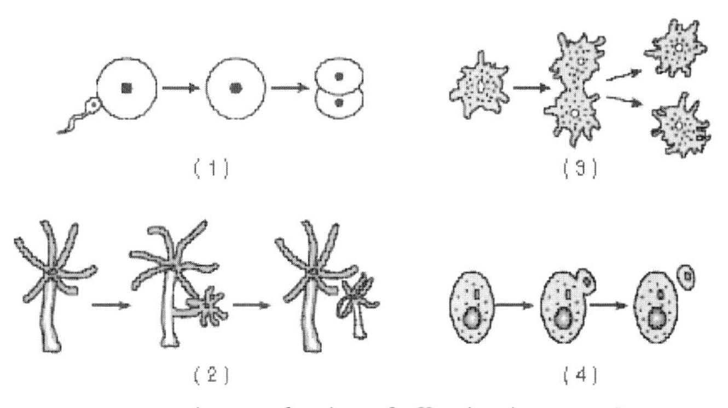

19. Which statement concerning production of offspring is correct?
(1) Production of offspring is necessary for a species to survive, but it is not necessary for an individual to survive.
(2) An organism can reproduce without performing any of the other life processes.
(3) Production of offspring is necessary for an individual organism to survive, while the other life processes are important for a species to survive.
(4) Reproduction is a process that requires gametes in all species.

20. Base your answers to the next question on the information below and on your knowledge of biology.

A biologist at an agriculture laboratory is asked to develop a better quality blueberry plant. He is given plants that produce unusually large blueberries and plants that produce very sweet blueberries.

The biologist is successful in producing the new plant. State ***one*** method that can be used to produce many identical blueberry plants of this new type.

21. The diagram below represents some stages of early embryonic development.

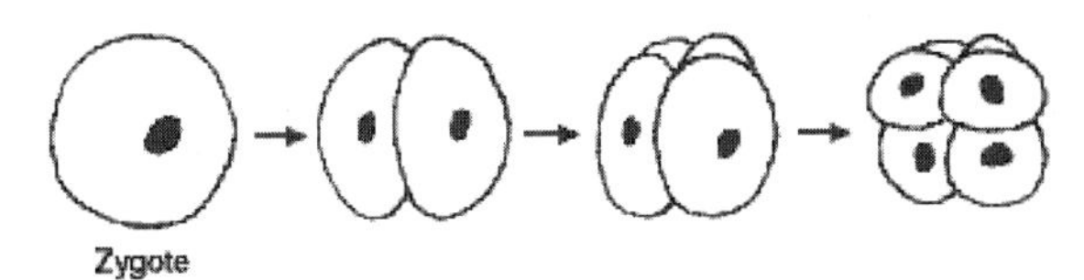

Which process is represented by the arrows in the diagram?

(1) meiosis
(2) fertilization
(3) mitosis
(4) evolution

22. Which cell is normally produced as a direct result of meiosis?

(1) a uterine cell having half the normal species number of chromosomes
(2) an egg having the full species number of chromosomes
(3) a zygote having the full species number of chromosomes
(4) a sperm having half the normal species number of chromosomes

23. The diagram below illustrates the process of cell division.

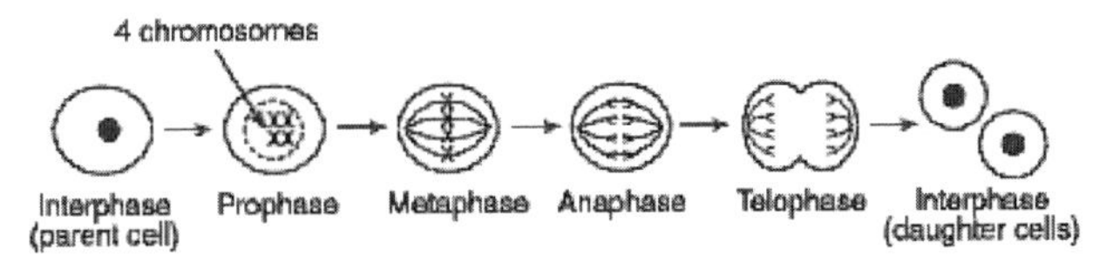

What is the significance of anaphase in this process?

(1) Anaphase usually ensures that each daughter cell has the same number of chromosomes as the parent cell.
(2) Anaphase usually ensures that each daughter cell has twice as many chromosomes as the parent cell.
(3) In anaphase, the cell splits in half.
(4) In anaphase, the DNA is being replicated.

24. Which reproductive structure is correctly paired with its function?

(1) uterus—usual site of fertilization
(2) testis—usual location for egg development
(3) ovary—delivers nutrients to the embryo
(4) sperm—transports genetic material

25. Compared to human cells resulting from mitotic cell division, human cells resulting from meiotic cell division would have

(1) twice as many chromosomes
(2) the same number of chromosomes
(3) one-half the number of chromosomes
(4) one-quarter as many chromosomes

26. Which statement correctly describes the genetic makeup of the sperm cells produced by a human male?

(1) Each cell has pairs of chromosomes and the cells are usually genetically identical.
(2) Each cell has pairs of chromosomes and the cells are usually genetically different.
(3) Each cell has half the normal number of chromosomes and the cells are usually genetically identical.
(4) Each cell has half the normal number of chromosomes and the cells are usually genetically different.

27. Which statement is true of both mitosis and meiosis?
    (1) Both are involved in asexual reproduction.
    (2) Both occur only in reproductive cells.
    (3) The number of chromosomes is reduced by half.
    (4) DNA replication occurs before the division of the nucleus.

28. The diagram below represents a nucleus containing the normal chromosome number for a species.

Which diagram best illustrates the normal formation of a cell that contains all of the genetic information needed for growth, development, and future reproduction of this species?

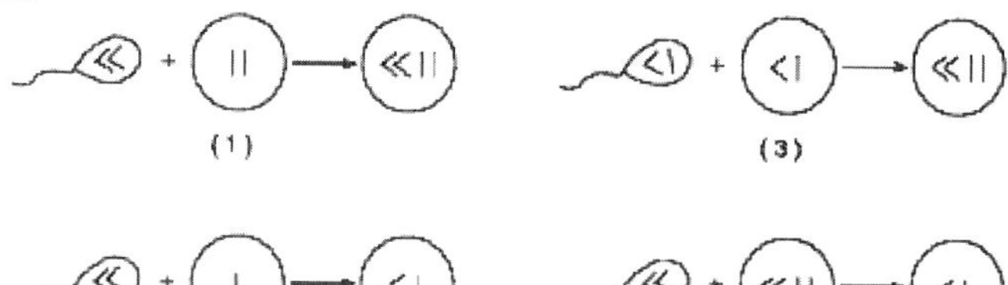

29. The sequence of events occurring in the life cycle of a bacterium is listed below.
    (A) The bacterium copies its single chromosome.
    (B) The copies of the chromosome attach to the cell membrane of the bacterium.
    (C) As the cell grows, the two copies of the chromosome separate.
    (D) The cell is separated by a wall into equal halves.
    (E) Each new cell has one copy of the chromosome.

    This sequence most closely resembles the process of
    (1) recombination
    (2) zygote formation
    (3) mitotic cell division
    (4) meiotic cell division

30. As a human red blood cell matures, it loses its nucleus. As a result of this loss, a mature red blood cell lacks the ability to
    (1) take in material from the blood
    (2) release hormones to the blood
    (3) pass through artery walls
    (4) carry out cell division

31. Marine sponges contain a biological catalyst that blocks a certain step in the separation of chromosomes. Which cellular process would be directly affected by this catalyst?
    (1) mitosis
    (2) diffusion
    (3) respiration
    (4) photosynthesis

32. The diagram below shows a process that can occur during meiosis.

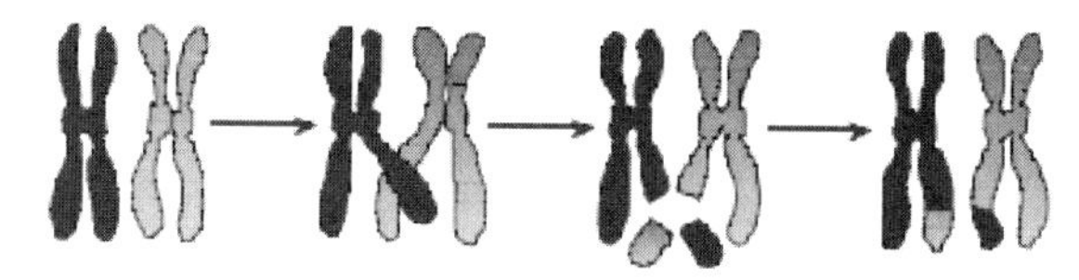

The most likely result of this process is
    (1) a new combination of inheritable traits that can appear in the offspring
    (2) an inability to pass either of these chromosomes on to offspring

(3) a loss of genetic information that will produce a genetic disorder in the offspring
(4) an increase in the chromosome number of the organism in which this process occurs

33. Which two structures of a frog would most likely have the same chromosome number?
(1) skin cell and fertilized egg cell
(2) zygote and sperm cell
(3) kidney cell and egg cell
(4) liver cell and sperm cell

34. Some body structures of a human male are represented in the diagram below.

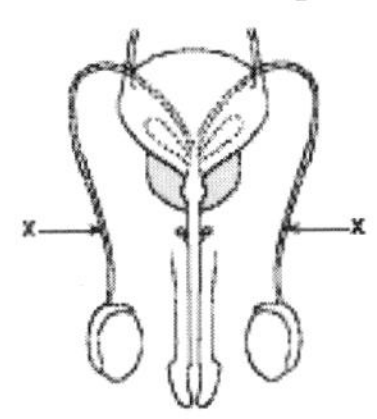

An obstruction in the structures labeled ***X*** would directly interfere with the
(1) transfer of sperm to a female
(2) production of sperm
(3) production of urine
(4) transfer of urine to the external environment

35. Compared to human cells resulting from mitotic cell division, human cells resulting from meiotic cell division would have
(1) twice as many chromosomes
(2) the same number of chromosomes
(3) one-half the number of chromosomes
(4) one-quarter as many chromosomes

36. Which statement describes the reproductive system of a human male?
(1) It releases sperm that can be used only in external fertilization.
(2) It synthesizes progesterone that regulates sperm formation.
(3) It produces gametes that transport food for embryo formation.
(4) It shares some structures with the excretory system.

37. Variation in the offspring of sexually reproducing organisms is the direct result of
(1) sorting and recombining of genes
(2) replication and cloning
(3) the need to adapt and maintain homeostasis
(4) overproduction of offspring and competition

38. A sperm cell from an organism is represented in the diagram below.

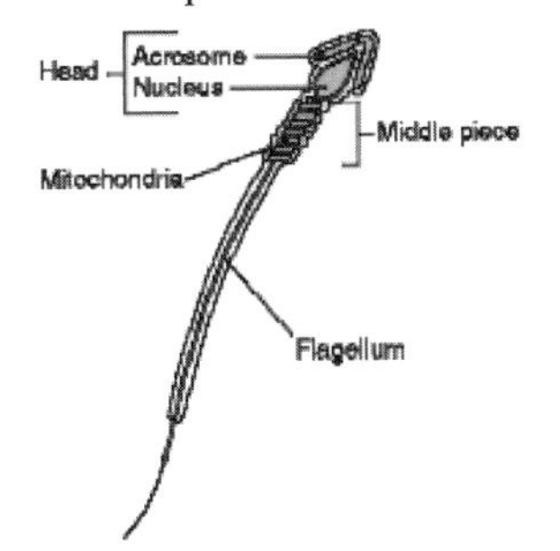

Which statement regarding this sperm cell is not correct?

(1) The acrosome contains half the normal number of chromosomes.
(2) Energy to move the flagellum originates in the middle piece.
(3) The head may contain a mutation.
(4) This cell can unite with another cell resulting in the production of a new organism.

39. A technique used to produce new plants is represented in the diagram below.

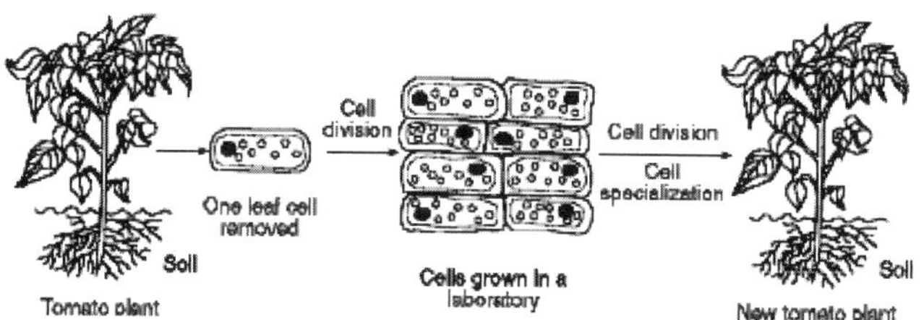

Which statement is best supported by the information in the diagram?

(1) The one leaf cell removed formed a zygote that developed into a new plant by mitotic cell division.
(2) This procedure is used to produce new tomato plants that are clones of the original tomato plant.
(3) The cell taken from the leaf produced eight cells, each having one-half of the genetic information of the original leaf cell.
(4) The new tomato plant will not be able to reproduce sexually because it was produced by mitotic cell division.

40. A human is a complex organism that develops from a zygote. Briefly explain some of the steps in this developmental process. In your answer be sure to:

- explain how a zygote is formed
- compare the genetic content of the zygote to that of a body cell of the parents
- identify one developmental process involved in the change from a zygote into an embryo
- identify the structure in which fetal development usually occurs
- identify two factors that can affect fetal development and explain how each factor affects fetal development

41. Offspring that result from meiosis and fertilization each have

(1) twice as many chromosomes as their parents
(2) one-half as many chromosomes as their parents
(3) gene combinations different from those of either parent
(4) gene combinations identical to those of each parent

42. The sequence of diagrams below represents some events in a reproductive process.

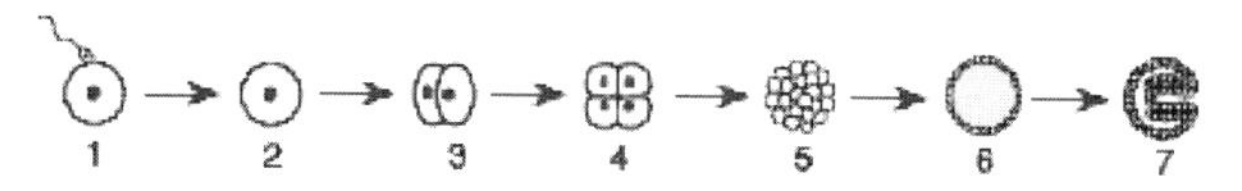

To regulate similar events in human reproduction, what adaptations are required?

(1) the presence of genes and chemicals in each cell in stages *1* to *7*
(2) an increase in the number of genes in each cell in stages *3* to *5*
(3) the removal of all enzymes from the cells in stage *7*
(4) the elimination of mutations from cells after stage *5*

43. In sexually reproducing species, the number of chromosomes in each body cell remains the same from one generation to the next as a direct result of
   (1) meiosis and fertilization
   (2) mitosis and mutation
   (3) differentiation and aging
   (4) homeostasis and dynamic equilibrium

44. Some cells involved in the process of reproduction are represented in the diagram below.

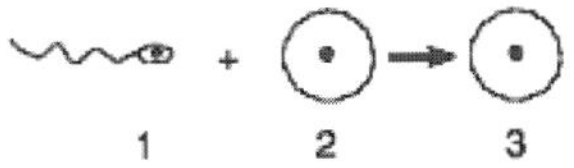

The process of meiosis formed
   (1) cell *1*, only
   (2) cells *1* and *2*
   (3) cell *3*, only
   (4) cells *2* and *3*

45. Meiosis and fertilization are important for the survival of many species because these two processes result in
   (1) large numbers of gametes
   (2) increasingly complex multicellular organisms
   (3) cloning of superior offspring
   (4) genetic variability of offspring

46. Certain insects are kept under control by sterilizing the males with x rays so that sperm production stops. Explain how this technique reduces the survival of this insect species.

47. A cell resulting from the fertilization of an egg begins to divide. Two cells are formed that normally remain attached and could develop into a new individual. If the two cells become separated, which statement describes what would most likely occur?
   (1) The cells would each have all of the needed genetic information, and both could survive.
   (2) The cells would each have only one-half of the needed genetic information, so both would die.
   (3) One cell would have all of the needed genetic information and would survive, but the other would have none of the needed genetic information and would die.
   (4) Each cell would have some of the needed genetic information, but would be unable to share it, so both would die.

48. Scientists have successfully cloned sheep and cattle for several years. A farmer is considering the advantages and disadvantages of having a flock of sheep cloned from a single individual. Discuss the issues the farmer should take into account before making a decision. Your response should include:

- how a cloned flock would be different from a noncloned flock
- one advantage of having a cloned flock
- one disadvantage of having a cloned flock
- one reason that the farmer could not mate these cloned sheep with each other to increase the size of his flock
- one reason that the offspring resulting from breeding these sheep with an unrelated sheep would not all be the same

49. Reproduction in humans usually requires
   (1) the process of cloning
   (2) mitotic cell division of gametes
   (3) gametes with chromosomes that are not paired
   (4) the external fertilization of sex cells

50. The diagram below represents a series of events in the development of a bird.

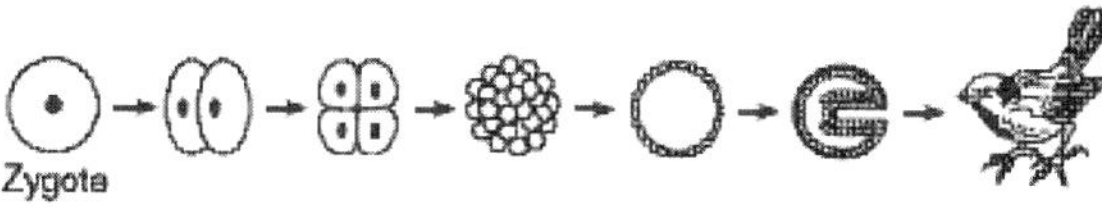

Which series of terms best represents the sequence of processes shown?

(1) meiosis → growth → differentiation
(2) meiosis → differentiation → growth
(3) mitosis → meiosis → differentiation
(4) mitosis → differentiation → growth

51. Which developmental process is represented by the diagram below?

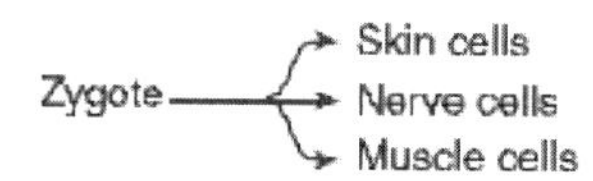

(1) fertilization
(2) differentiation
(3) evolution
(4) mutation

52. Part of embryonic development in a species is illustrated in the diagram below.

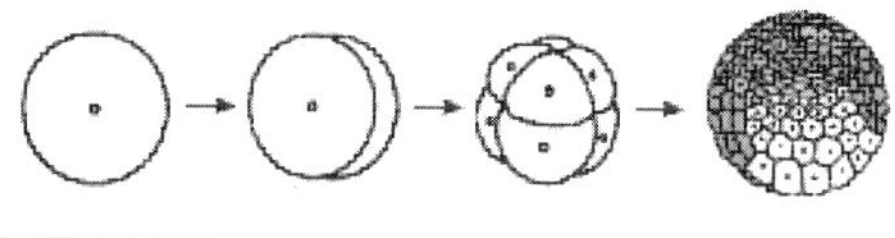

Which set of factors plays the most direct role in controlling the events shown in the diagram?

(1) genes, hormones, and cell location
(2) antibodies, insulin, and starch
(3) ATP, amino acids, and inorganic compounds
(4) abiotic resources, homeostasis, and selective breeding

53. The diagram below represents a nucleus containing the normal chromosome number for a species.

Which diagram best illustrates the normal formation of a cell that contains all of the genetic information needed for growth, development, and future reproduction of this species?

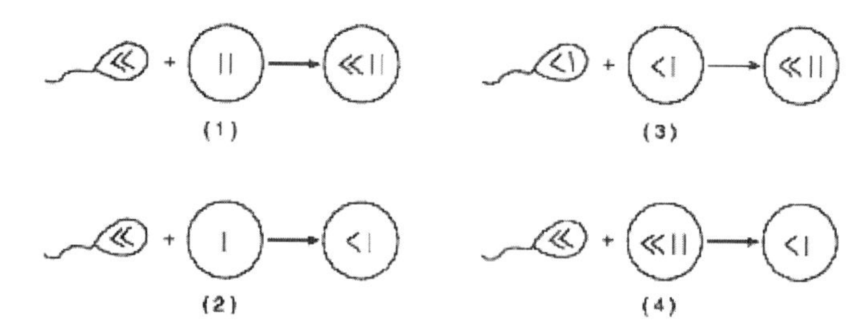

54. A human is a complex organism that develops from a zygote. Briefly explain some of the steps in this developmental process. In your answer be sure to:

- explain how a zygote is formed
- compare the genetic content of the zygote to that of a body cell of the parents
- identify one developmental process involved in the change from a zygote into an embryo
- identify the structure in which fetal development usually occurs
- identify two factors that can affect fetal development and explain how each factor affects fetal development

55. Which sequence represents the correct order of processes that result in the formation and development of an embryo?
    (1) meiosis → fertilization → mitosis
    (2) mitosis → fertilization → meiosis
    (3) fertilization → meiosis → mitosis
    (4) fertilization → mitosis → meiosis

56. The diagram below represents stages in the processes of reproduction and development in an animal.

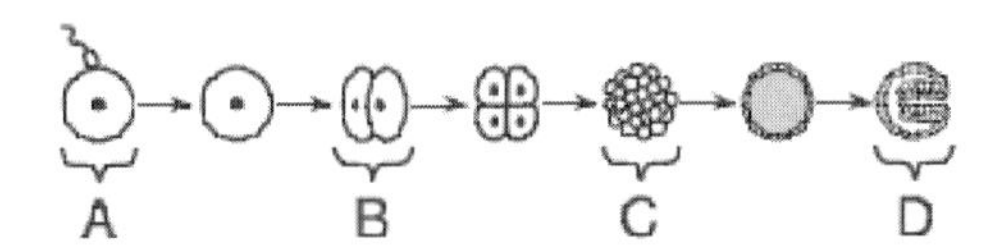

Cells containing only half of the genetic information characteristic of this species are found at
    (1) *A* (2) *B* (3) *C* (4) *D*

Hint: *Gametes are haploid.*

57. The diagrams below represent cells that transport chromosomes.

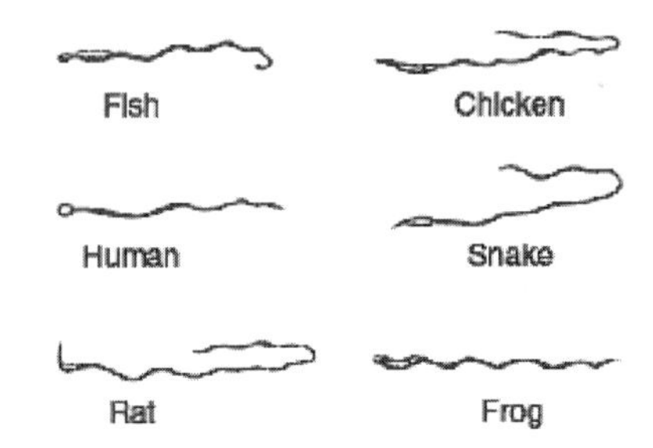

These cells are specialized for
    (1) oxygen transport
    (2) transmitting chemical signals over long distances
    (3) sexual reproduction
    (4) injecting antibodies into harmful bacteria

58. A technique used to produce new plants is represented in the diagram below.

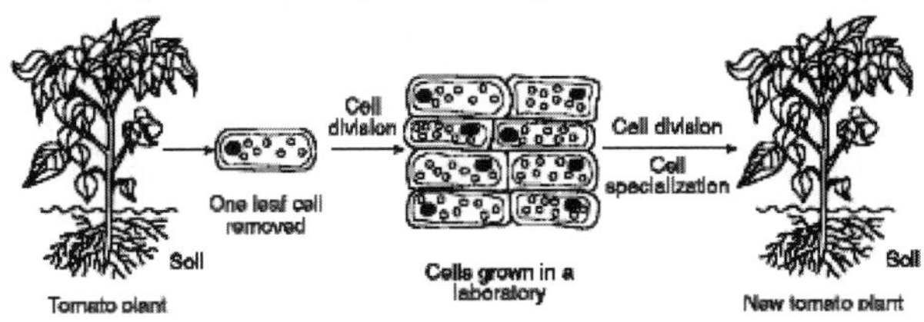

Which statement is best supported by the information in the diagram?

(1) The one leaf cell removed formed a zygote that developed into a new plant by mitotic cell division.
(2) This procedure is used to produce new tomato plants that are clones of the original tomato plant.
(3) The cell taken from the leaf produced eight cells, each having one-half of the genetic information of the original leaf cell.
(4) The new tomato plant will not be able to reproduce sexually because it was produced by mitotic cell division.

59. The diagram below represents early stages of embryo development.

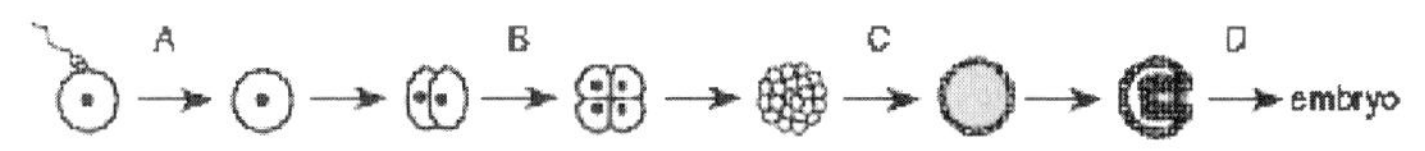

The greatest amount of differentiation for organ formation most likely occurs at arrow

(1) ***A*** (2) ***B*** (3) ***C*** (4) ***D***

60. The diagram and chart below represent some of the changes a zygote undergoes during its development.

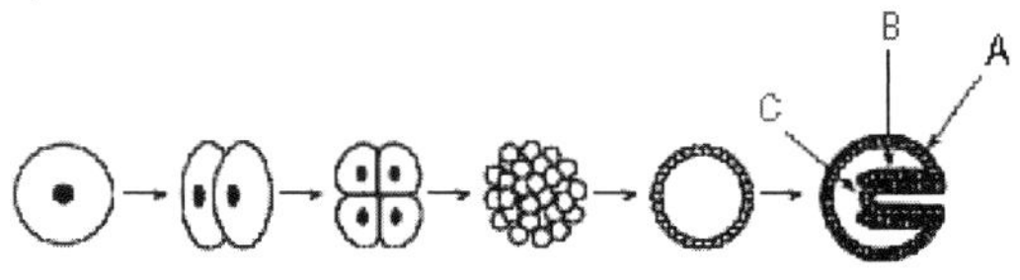

Layer	Develops Into
A	skin and nervous system
B	muscles and blood vessels
C	digestive and respiratory systems

The processes that are most directly responsible for these changes are

(1) sorting and recombination of genetic information
(2) mitosis and differentiation
(3) meiosis and adaptation
(4) fertilization and cycling of materials

61. Tissues develop from a zygote as a direct result of the processes of

(1) fertilization and meiosis
(2) fertilization and differentiation
(3) mitosis and meiosis
(4) mitosis and differentiation

62. A large number of sperm cells are produced by males every day. This large number of sperm cells increases the chance that

(1) at least one sperm cell will be reached when the eggs swim toward the sperm cells in the ovary
(2) several sperm cells will unite with an egg so the fertilized egg will develop properly
(3) some of the sperm cells will survive to reach the egg
(4) enough sperm cells will be present to transport the egg from where it is produced to where it develops into a fetus

63. The structure that makes nutrients most directly available to a human embryo is the

(1) gamete (2) ovary (3) stomach (4) placenta

Directions for the next three questions: The diagrams below represent organs of two individuals. The diagrams are followed by a list of sentences. For each phrase in the next three questions, select the sentence from the list below that best applies to that phrase. Then record its number in the space provided.

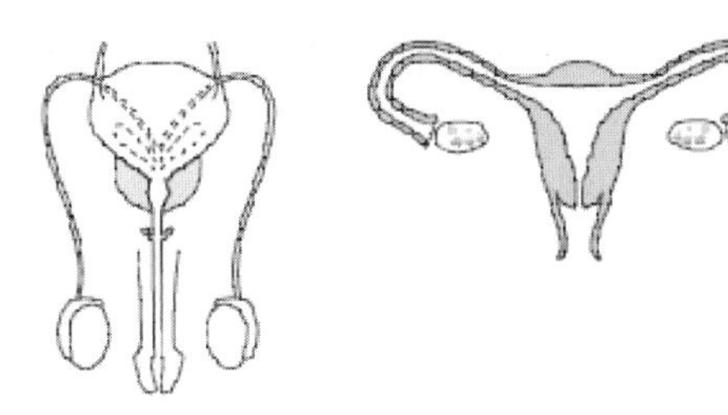

SENTENCES

1. The phrase is correct for both Individual ***A*** and Individual ***B***.
2. The phrase is not correct for either Individual ***A*** or Individual ***B***.
3. The phrase is correct for Individual ***A***, only.
4. The phrase is correct for Individual ***B***, only

64. Contains organs involved in internal fertilization

65. Contains a structure in which a zygote divides by mitosis

66. Which phrase best describes a process represented in the diagram below?

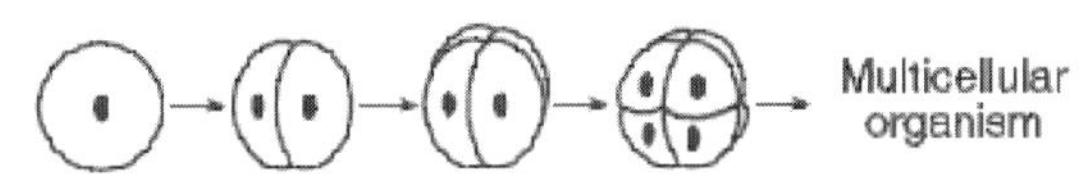

(1) a zygote dividing by mitosis
(2) a zygote dividing by meiosis
(3) a gamete dividing by mitosis
(4) a gamete dividing by meiosis

67. Estrogen has a direct effect on the
(1) formation of a zygote
(2) changes within the uterus
(3) movement of an egg toward the sperm
(4) development of a placenta within the ovary

68. Most mammals have adaptations for
(1) internal fertilization and internal development of the fetus
(2) internal fertilization and external development of the fetus
(3) external fertilization and external development of the fetus
(4) external fertilization and internal development of the fetus

---

Base your answers to the next three questions on the diagram below, which represents systems in a human male and on your knowledge of biology.

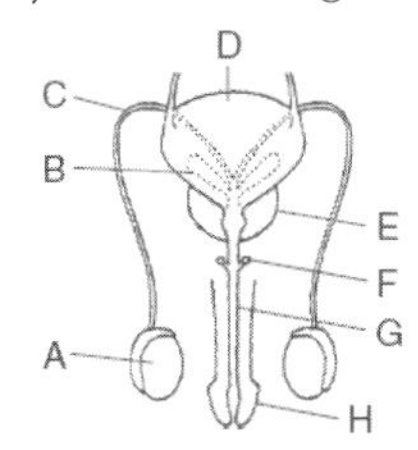

69. Which sequence represents the path of sperm leaving the body?
(1) $A \rightarrow C \rightarrow G$ (2) $A \rightarrow C \rightarrow B$ (3) $E \rightarrow F \rightarrow H$ (4) $D \rightarrow F \rightarrow G$

70. Which structures aid in the transport of sperm by secreting fluid?
(1) *A* and *H* (2) *B* and *E* (3) *C* and *D* (4) *D* and *H*

---

71. The diagram below represents the reproductive system of a mammal.

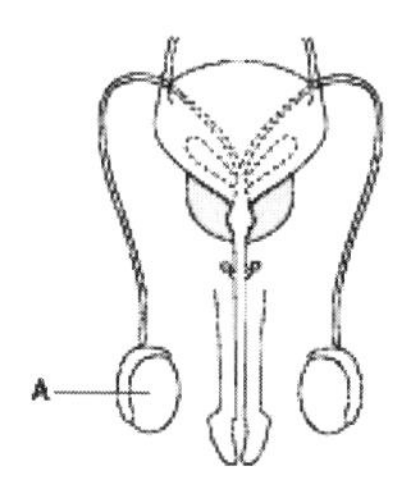

The hormone produced in structure *A* most directly brings about a change in
(1) blood sugar concentration
(2) physical characteristics
(3) the rate of digestion
(4) the ability to carry out respiration

72. The diagram below represents a human reproductive system.

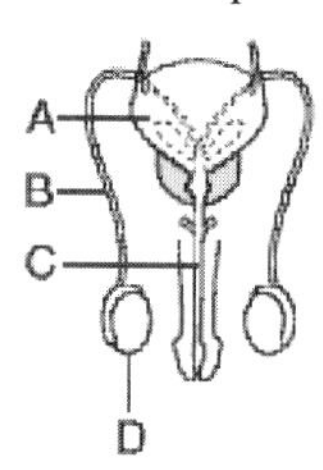

Meiosis occurs within structure
(1) *A* (2) *B* (3) *C* (4) *D*

73. The data in the table below indicate the presence of specific reproductive hormones in blood samples taken from three individuals. An *X* in the hormone column indicates a positive lab test for the appropriate levels necessary for normal reproductive functioning in that individual.

Data Table

Individuals	Hormones Present		
	Testosterone	Progesterone	Estrogen
1		X	X
2			X
3	X		

Which processes could occur in individual *3*?
(1) production of sperm, only
(2) production of sperm and production of eggs
(3) production of eggs and embryonic development
(4) production of eggs, only

74. The diagram below represents some stages of early embryonic development.

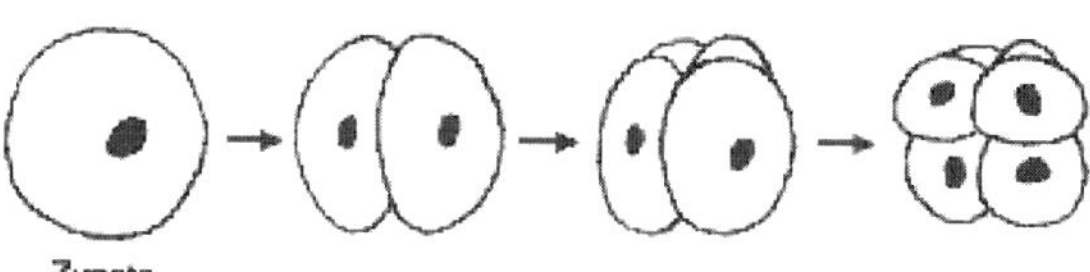

Which process is represented by the arrows in the diagram?

(1) meiosis (3) mitosis
(2) fertilization (4) evolution

75. The ***least*** genetic variation will probably be found in the offspring of organisms that reproduce using
    (1) mitosis to produce a larger population
    (2) meiosis to produce gametes
    (3) fusion of eggs and sperm to produce zygotes
    (4) internal fertilization to produce an embryo

76. The diagram below illustrates the process of cell division.

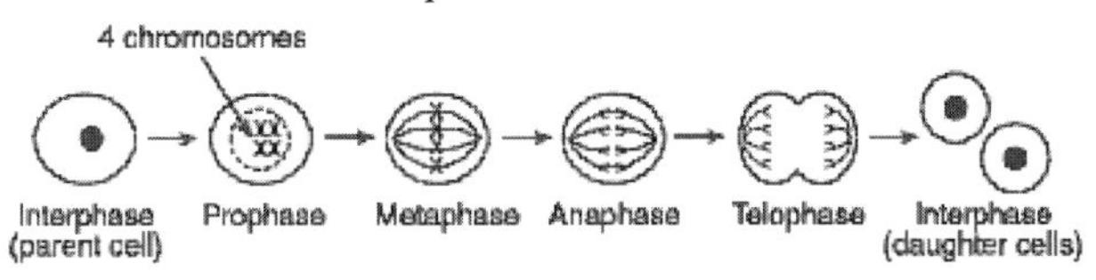

What is the significance of anaphase in this process?
    (1) Anaphase usually ensures that each daughter cell has the same number of chromosomes as the parent cell.
    (2) Anaphase usually ensures that each daughter cell has twice as many chromosomes as the parent cell.
    (3) In anaphase, the cell splits in half.
    (4) In anaphase, the DNA is being replicated.

77. The diagram below represents the human female reproductive system.

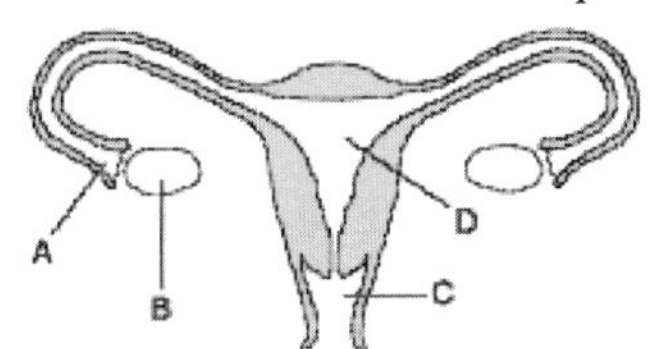

Exposure to radiation or certain chemicals could alter the genetic information in the gametes that form in structure

(1) ***A*** (2) ***B*** (3) ***C*** (4) ***D***

78. Which hormones most directly influence the uterus during pregnancy?
    (1) testosterone and insulin
    (2) progesterone and testosterone
    (3) estrogen and insulin
    (4) progesterone and estrogen

79. The human female reproductive system is adapted for
    (1) production of zygotes in ovaries
    (2) external fertilization of gametes
    (3) production of milk for a developing embryo
    (4) transport of oxygen through a placenta to a fetus

80. The letters in the diagram below represent structures in a human female.

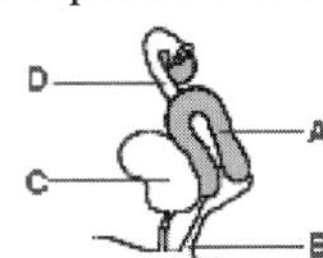

Estrogen and progesterone increase the chance for successful fetal development by regulating activities within structure

(1) *A* (2) *B* (3) *C* (4) *D*

81. Kangaroos are mammals that lack a placenta. Therefore, they must have an alternate way of supplying the developing embryo with
   (1) nutrients
   (2) carbon dioxide
   (3) enzymes
   (4) genetic information

82. The diagram below represents a system in the human body.

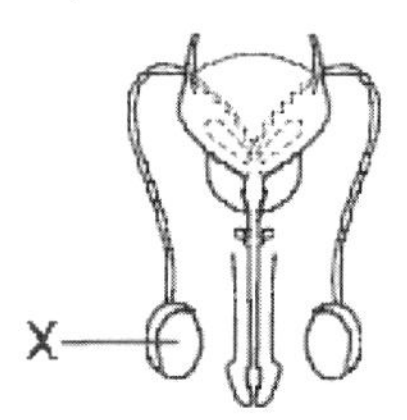

The primary function of structure ***X*** is to
   (1) produce energy needed for sperm to move
   (2) provide food for the sperm to carry to the egg
   (3) produce and store urine
   (4) form gametes that may be involved in fertilization

83. The direct source of ATP for the development of a fetus is
   (1) a series of chemical activities that take place in the mitochondria of fetal cells
   (2) a series of chemical activities that take place in the mitochondria of the uterine cells
   (3) the transport of nutrients by the cytoplasm of the stomach cells of the mother
   (4) the transport of nutrients by the cytoplasm of the stomach cells of the fetus

84. Base your answers to the question on the diagram below, which represents systems in a human male and on your knowledge of biology.

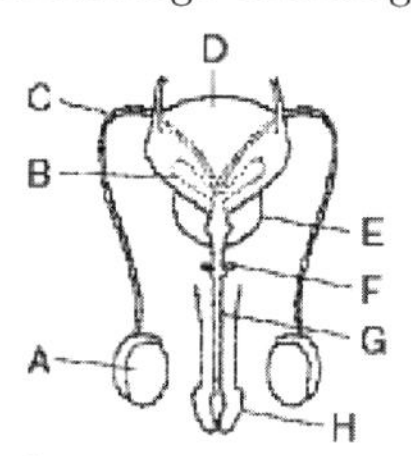

Which structure has both reproductive and excretory functions?

(1) *A* (2) *G* (3) *C* (4) *D*

85. The diagram below represents the human female reproductive system.

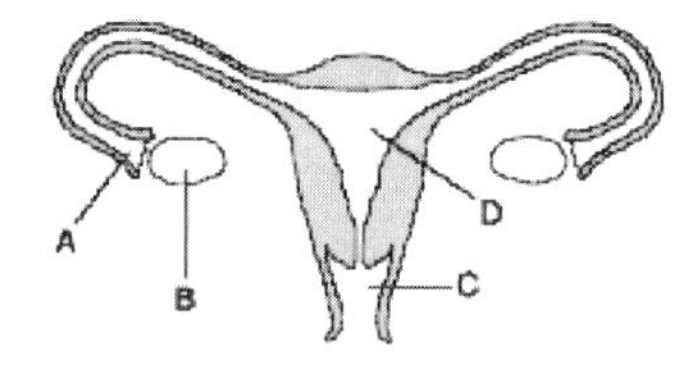

Exposure to radiation or certain chemicals could alter the genetic information in the gametes that form in structure

(1) *A* (2) *B* (3) *C* (4) *D*

86. Some chemical interactions in a human are shown in the graph below.

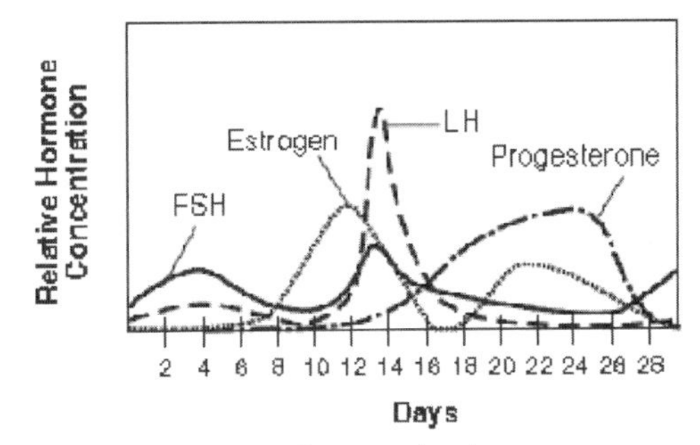

This graph represents hormones and events in the
(1) process of fetal growth and development
(2) process of meiotic cell division during sperm development
(3) reproductive cycle of males
(4) reproductive cycle of females

87. The diagram below represents part of the human female reproductive system.

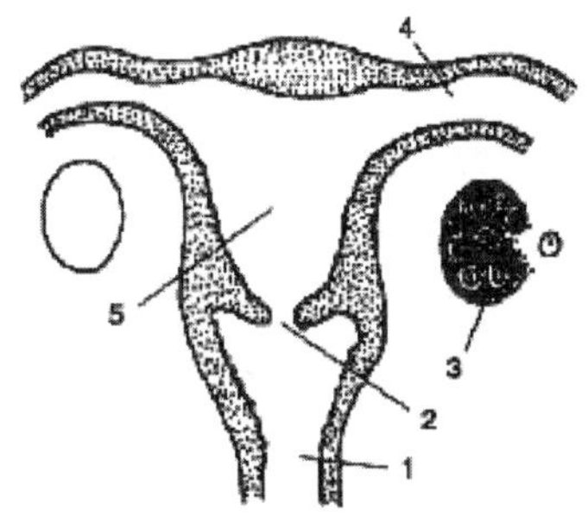

Fertilization and development normally occur in structures
(1) *1* and *5* (2) *2* and *4* (3) *3* and *1* (4) *4* and *5*

88. Which process normally occurs at the placenta?
(1) Oxygen diffuses from fetal blood to maternal blood.
(2) Materials are exchanged between fetal and maternal blood.
(3) Maternal blood is converted into fetal blood.
(4) Digestive enzymes pass from maternal blood to fetal blood.

89. Base your answer to this question on the list below and on your knowledge of biology. The list includes two processes involved in the development of a human fetus.

PROCESSES
mitosis
differentiation

Select ***one*** process from the list and describe its role in the development of a human fetus. In your answer be sure to:
- identify the process you selected
- state the role of this process in fetal development
- identify the organ in the mother where this process occurs

Process: ______________________________

90. To prevent harm to the fetus, women should avoid tobacco, alcohol, and certain medications during pregnancy. State ***one*** specific way that ***one*** of these substances could harm the fetus.

91. Which statement about embryonic organ development in humans is accurate?

(1) It is affected primarily by the eating habits and general health of the father.
(2) It may be affected by the diet and general health of the mother.
(3) It will not be affected by any medication taken by the mother in the second month of pregnancy.
(4) It is not affected by conditions outside the embryo.

92. The diagram below represents stages in the processes of reproduction and development in an animal.

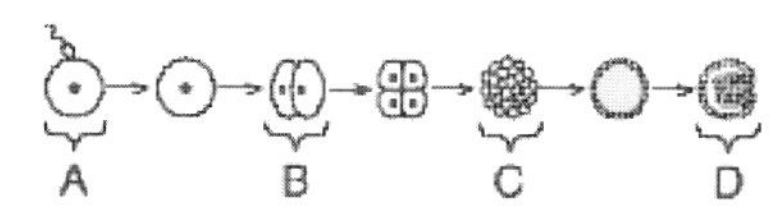

Cells containing only half of the genetic information characteristic of this species are found at

(1) ***A*** (2) ***B*** (3) ***C*** (4) ***D***

Hint: *Gametes are haploid.*

93. Explain how harmful substances in the blood of a pregnant female can enter a fetus even though the blood vessels of the mother and fetus are not directly connected.

94. German measles is a disease that can harm an embryo if the mother is infected in the early stages of pregnancy because the virus that causes German measles is able to

(1) be absorbed by the embryo from the mother's milk
(2) be transported to the embryo in red blood cells
(3) pass across the placenta
(4) infect the eggs

Base your answers to the next three questions on the diagram below, which represents some stages in the development of an embryo, and on your knowledge of biology.

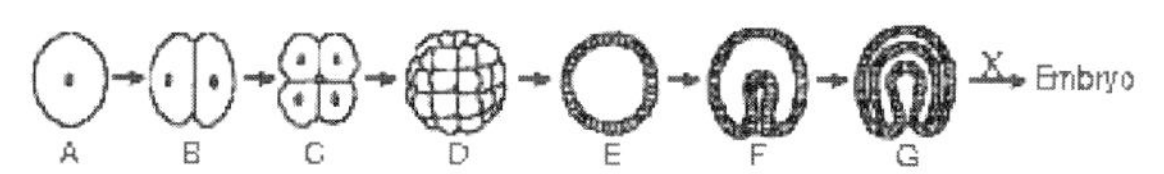

95. This entire sequence (A through embryo) started with

(1) the periodic shedding of a thickened uterine lining
(2) mitotic cell division in a testis
(3) meiotic cell division in the placenta
(4) the process of fertilization

96. If cell ***A*** has 46 chromosomes, how many chromosomes will most likely be found in each cell of stage ***G***?

(1) 23 (2) 46 (3) 69 (4) 92

97. The arrow labeled ***X*** represents the process of

(1) meiosis
(2) recombination
(3) differentiation
(4) cloning

CHAPTER 5: EVOLUTION

In the dictionary, evolve or evolution means change (slowly over time). New life forms appeared over time. Three billion years ago, there were the first simple one-celled (single-celled, unicellular) organisms, then later more complex single-celled organisms (living things). One billion years ago, there were simple multicellular (many-celled) organisms. After that, there were complex multi-cellular organisms, such as shellfish, then other fish, then amphibians (example frogs), then reptiles (example dinosaurs), then birds, then mammals, and then humans. As time went on, there was an increase in diversity (more different types or more species) of complex multicellular organisms.

Evolution is **change over time** (example change of species over time, how a species, such as a horse, changes over time). A species is a group of similar organisms that can interbreed (produce offspring together) . Geologic evolution means how the Earth (geology) changed over time. The Earth has existed for 4½ billion years, which is called geologic time.

Evolutionary Trees

Look at the evolutionary tree (evolutionary pathways), showing ancestors and the species evolving (changing or becoming different) from the ancestors. Letters ADFGEHIXY represent different species.

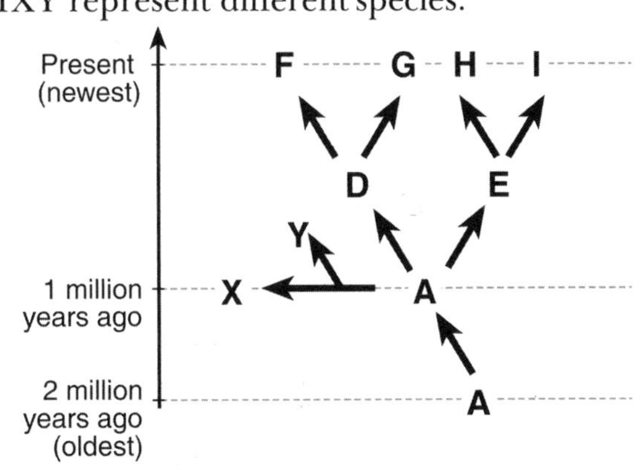

1. The **bottom** of the **tree** (A) is the **oldest ancestor**, which is the oldest species. The **top** of the **tree** (letters **F, G, H, I**) is the **newest species** (example, species that exist today). Look at the diagram. A is the original ancestor to F and also to G. D is a newer ancestor to F and G. F and G might have similar DNA (genes) to D and also similar DNA and genes to A (oldest original ancestor). A is the original ancestor to H and I. E is a newer ancestor to H and I. H and I might have similar DNA (genes) to E and also to A (the original ancestor).

2. Look at the arrow from A to A. In this example, A lasted from two million years ago to one million years ago, meaning **A lasted one million years.** Species A lasted the **longest period of time** in this evolutionary tree

(one million years). There were probably changes in the environment over the one million years, but species A was more able to adapt (adjust) to the changes in the environment and survive.

3. By looking at the evolutionary tree, you can see which **species evolved into other species** and which **species** are **closely related** (similar DNA, genes, and proteins). Species A lasted one million years. One million years ago (maybe environment caused it, or changes in the genes (mutation), or sorting, or recombination), species A evolved (changed) into two different species, species D and species E (see evolutionary tree diagrams below). Since **species D and species E directly branch from species A (common** (same) **ancestor)**, **D and E** are **closely related**. *This can be compared to children born from the same parents (common ancestor); the children are closely related.*

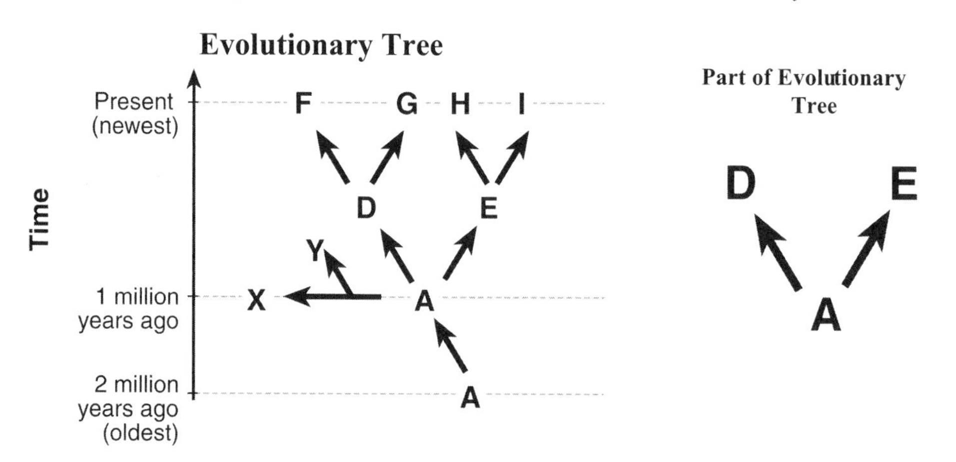

Look at the evolutionary tree diagram above. Since **species F and species G directly branch from species D (common** (same) **ancestor)**, **F and G** are **closely related**. *This can be compared to children born from the same parents (common ancestor);the children are closely related*. By looking at the diagram, you reach the conclusion (valid inference) that F and G are closely related. Since **species H and species I directly branch from species E (common** (same) **ancestor)**, **H and I** are **closely related**. *This can be compared to children born from the same parents (common ancestor);the children are closely related.* By looking at the diagram, you reach the conclusion (valid inference) that H and I are closely related.

Look at the evolutionary tree again. **F and I** are **NOT closely related** (less closely related) because **F branches from D** and **I branches from E** (different ancestors D and E). But also follow the line of arrows F D A and I E A and you see F and I both come from the same original ancestor A. In short, **F and I** are related but **not** as **closely related** because F and I **have different ancestors D and E,** but the **same common original ancestor A.** This can be compared to children born from different parents (ancestors) but the same grandparents (older ancestor). The children are related, but not as closely related as if they were born from the same parents (ancestors).

4. **Species X** and **species Y ended** at a **certain time** (became **extinct**) (see evolutionary tree diagram above). **Species X** and **species Y** do **not** exist **today** (at the present time); species X and Y end before the present.

Look at the two evolutionary trees in the diagram below. Letters M, D, E, F, G, H, I, B, C, J, K, and L represent different species.

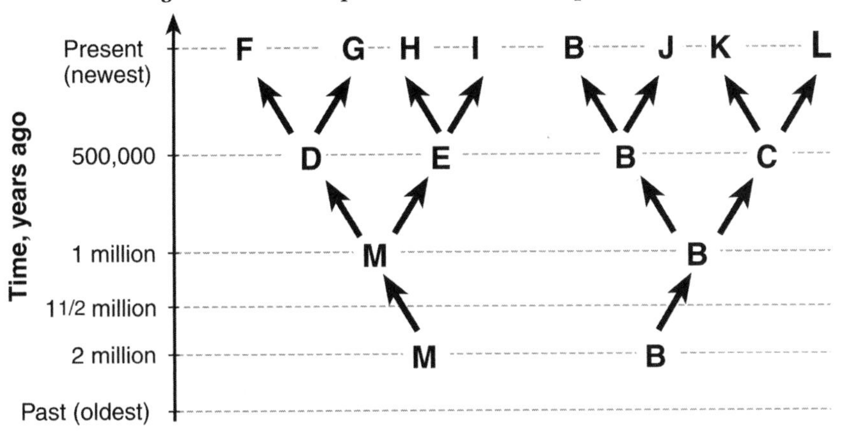

1. Looking at both trees, species B lasted the longest period of time, from the oldest time shown to the present, because species B was able to adapt (adjust) to the changes in its environment.

2. The left edge of the diagram shows "Time, years ago". 1 million means 1 million years ago; 500,000 means 500,000 years ago.
Look at the left evolutionary tree in the diagram above. F and I branched out from M a long time ago (one million years ago), therefore they had time to become more different (evolve). F and G branched out from D not as long ago, closer to the present time (500,000 years ago); therefore they had less time to become different (evolve), and are more closely related.

Evolutionary trees show evolutionary relationships (which species evolved into other species and which species are closely related, etc.).

Question: The evolutionary pathways of ten different species are represented in the diagram.

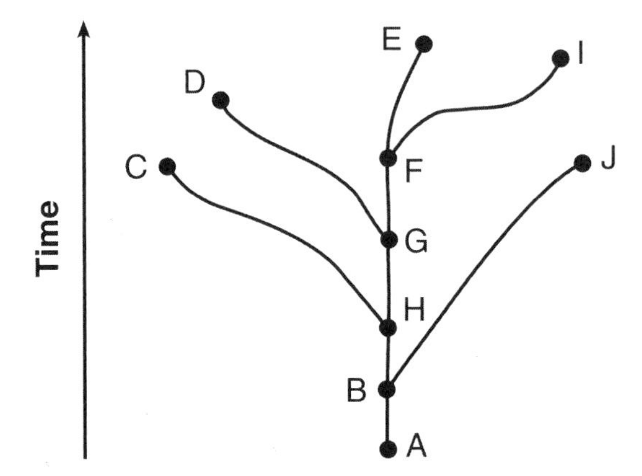

Which two species are most closely related?

1. C and D 2. E and I 3. G and J 4. A and F

Solution: E and I are most **closely related** because E and I **directly branch** from F (common or same ancestor). Answer *2*

Note: Since all the species (B, J, H, C, G, D, F, I, E) branch directly or indirectly from A, they are all related to the oldest ancestor A.

REALIZE that the same evolutionary tree can also be drawn this way:

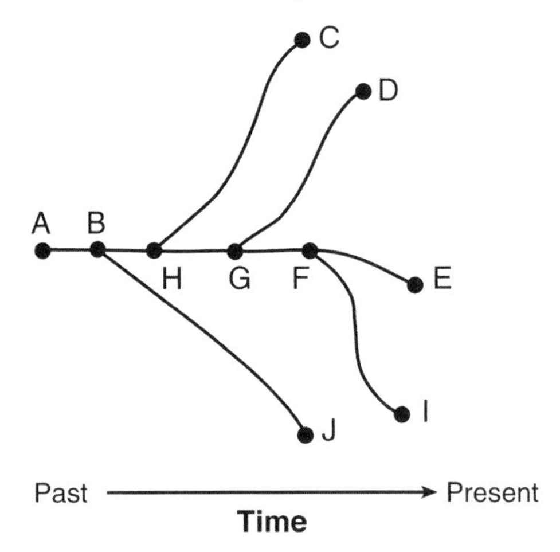

E and I are most **closely related** because E and I **directly branch** from F (common or same ancestor). Answer *2*

Question: Base your answers to questions 1, 2, and 3 on the diagram below. The diagram shows an interpretation of relationships based on evolutionary theory. The letters represent different species.

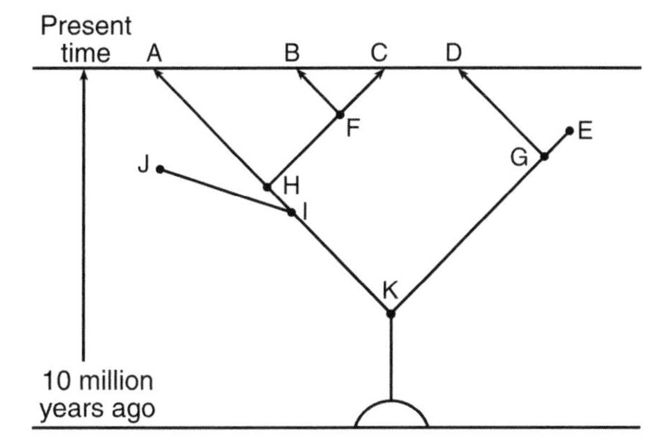

Question 1: Explain why species B and C are more closely related than species A and C are.

Question 2: The diagram indicates that a common ancestor for species C and E is species

1. F 2. G 3. H 4. K

Question 3: Which species are least likely to be vital parts of a present-day ecosystem?

1. A and E 2. C and D 3. E and J 4. B and F

Solution 1: Look at the diagram. Species B and Species C are more closely related than Species A and Species C because B and C branched out (diverged) from ancestor F more recently (closer to the present time), therefore B and C had less time to become different. Species A and Species C branched out from H a longer time ago, therefore A and C had more time to become different (by mutation, gene sorting, and gene shuffling).

Solution 2: Follow the line back from point C and point E to where the lines meet. The lines meet at point K, which is the common ancestor. Answer *4*

Solution 3: E and J died out (became extinct). E and J do not exist at the present time. therefore E and J cannot be part of a present-day ecosystem. Answer *3*

Determining Which Species are Closely Related

We can tell **evolutionary relationships** (which species evolved into other species and which species are closely related) by comparing (looking at) their bones, embryos, cell organelles (such as ribosomes or mitochondria), DNA, genes, and proteins (such as hormones or enzymes). The species that have similar bones, or similar methods of reproduction, or similar embryos, or similar cell organelles, or similar DNA (similar base sequence of DNA) or similar sequence of amino acids in proteins (example Arg-Leu-Glu-Gly, etc.) are more closely related. Species that have less similar bone structures, or less similar embryos, or less similar DNA, etc. are less closely related.

If you get a certain medicine from one type of plant (one species), and that plant dies, you might be able to get a very similar medicine from a different plant (different species) that is closely related.

Living things with similar bone structures (discussed later in the chapter), or similar embryos, or similar cell organelles, similar biochemistry, similar DNA (similar base sequence of DNA), or similar proteins (hormones and enzymes) have common ancestors and are more closely related.

Gel Electrophoresis: Gel electrophoresis is a **very reliable way,** very good way of **determining evolutionary relationships**. Gel electrophoresis, which you will learn more about in chapter 9, is a method used to analyze DNA; enzymes break up the DNA into pieces of DNA, and an electric current separates the smaller pieces of DNA from the bigger (larger) pieces of DNA. **Animals** that have **similar bands of DNA** (lines are in similar patterns) on the gel electrophoresis are **more closely related.** Animals that have bands of DNA that are not similar are not closely related. Gel electrophoresis shows which species evolved into other species and which species are closely related.

Fossils

Fossils are **remains** (example bones, shells, etc.) or **traces** (example footprints) of an **organism** that was **once living** (lived a long time ago and died). When an animal dies, the soft parts of the animal decay (break down, disappear) and the **hard parts** (example **bones or shells**) are **left**. People found fossils (bones) from dinosaurs that lived 100 million years ago.

Sediments (sand) accumulate on the bottom of the sea, forming rock (sedimentary rock). The bottom layers of the rock are the oldest; in the bottom layers of rock are found fossils (example bones) of animals that lived a very long time ago.

Higher up in the rock are found fossils of animals that existed later on. Near the top of the rock are found fossils of animals that existed most recently.

Now Do Homework Questions #1-27, pages 25-34.

Theory of Evolution

Charles Darwin is famous for the **theory of evolution.** The **modern theory** (explained below) takes Darwin's theory of evolution together with the ideas of later scientists. Darwin knew there were variations (different characteristics and different traits), such as tall or short, strong or weak organisms in a species, but he did not know that mutations and genes (or combinations of genes) caused the different traits.

1. Overproduction: In each generation, there are more organisms produced than can survive. There is a limited supply of resources, such as food and living space, which means there is not enough food to support all the offspring produced.

2. Limited Resources: There is a **limited** (finite) **supply** of resources (examples food, water, oxygen, shelter) needed for life. There are not enough resources (example food, etc.)for all the organisms to live.
There is **competition** for food. There is a **struggle to survive** (struggle to get food).
Result of limited resources: The organism with the better traits, such as better able to get food in that environment, or better able to run fast (better genes to get food or better genes to run fast) will more easily survive and reproduce, producing many more organisms with the better traits (genes). The organisms pass these traits (with advantageous genes) to the offspring. There will be a **higher percentage** of organisms with the **better trait**, better genes (**more adaptive value**). The organisms that do **not** have the **good trait** (**less adaptive value**) might **die out** and might not even have time to reproduce.
Evolution takes place because there is a much higher percentage with the better trait (better genes) and a very small percentage with the poorer traits (poorer genes), and a new species can evolve with the better traits (better genes).

3. Genetic variability (genetic variation) of offspring: Different individuals in a species and different offspring (children) have different characteristics (example, stronger, weaker, run faster or slower, fight their enemies better or worse), which are caused by **mutations** and **genetic variation (**different varieties of genes, or you can say, different combinations of genes, different genetic combinations). In sexual reproduction, the offspring gets genes from the father and genes from the mother, which gives the offspring different combinations of genes, different varieties (variations) of genes. **Genetic variation** (different genes or different combinations of genes) is **caused by sexual reproduction, meiosis** (producing sperm and egg), **fertilization** and **recombination** (combining genes of sperm and egg), genetic shuffling (sorting and recombination), crossing over, mutation (change in genes), etc.

Result also of genetic variation: Organisms with a better trait (such as better able to fight, better able to get food, or better able to run away from enemies) more easily survive. The **genes** for these advantageous **traits** help them survive. The organisms pass these traits with the advantageous genes (genetic variation) to the offspring. Some of the offspring which do not have the better traits (do not have the better genes) die out, and therefore more of the future generation will evolve (evolution) with the better traits (better genes). A better trait has **adaptive value**, which means the **trait helps** the **organism** to **survive.** Evolution takes place because a much higher percentage of organisms have better traits (better genes) and very few organisms have the poorer traits.

4. Natural selection: The organisms that are **best adapted (having adaptations** such as stronger muscles, better able to get food, having better genes, better traits) **survive** better in their environment (survival of the fittest), producing more organisms that are also better adapted (have better traits).
There are many different environments, and therefore there are many different best adapted organisms (organisms that have adaptations that help them survive in their environment).

Some examples of adaptations that help organisms survive in their environment are:

1. Ducks have an adaptation (webbed feet) that helps them swim.

2. Woodpeckers live in forests. Woodpeckers have an adaptation (chisel-like beaks) that they use to drill holes in trees both to get food (insects) and to make a home (nest). Woodpeckers also have the adaptation (two toes facing front, two toes facing back, and sharp claws) which helps the woodpecker hold onto the tree even when drilling holes in the tree to find food.

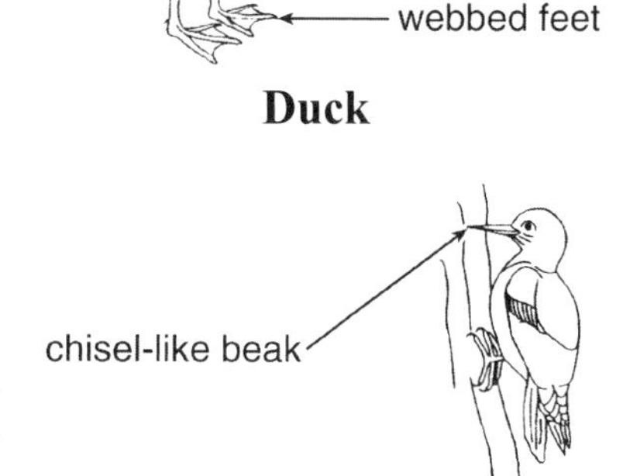

Duck

Woodpecker

3. Polar bear. Polar bears have an adaptation (white color) which camouflages (blends them into the white color of the snow and ice around them). Polar bears also have an adaptation (thick fur) that keep them warm; another adaptation (fur on bottom of their feet) keeps the feet warm even when walking on snow or ice.

Polar Bear

Result of Natural Selection: The organisms that are **best adapted (having adaptations** such as stronger muscles, better able to get food, having better genes, better traits) **survive** better in their environment (survival of the fittest), producing more organisms that are also better adapted (have better traits). The **population** of this **species changes (evolves)** with a **higher percentage** of organisms (higher frequency of organisms) **having** adaptations (more adaptive value), better traits. If the organism has a short reproductive cycle (example bacteria reproduce about every 20 minutes) a better trait spreads more quickly through the population (will be explained later in the chapter). **Organisms** that do **not** have adaptations **(adaptive value) might die off**, and might not reproduce (produce children). **Evolution** takes place because there is a much higher percentage of organisms (more organisms) with better traits (better genes), having adaptations (more adaptive value), and a very small percentage of organisms with the poorer traits (poorer genes), and even a **new species** can **evolve** with the **better traits** (better genes). A **species** (example wooly mammoth) with very **poor traits**, with no adaptive value, can become **extinct** (die out).

Evolution in a changing environment

Environment: Evolution takes place more often **in** a **changing environment**; there is less evolution when the environment is stable (not changing). When the environment changes (unstable), the organism must handle environmental conditions or environmental changes, such as temperature, disease, or predators. If the land becomes flooded with water (too much water), animals must adapt (adjust) to the changed environment to survive. Those animals that cannot adapt (adjust) to the flooding die out (become extinct) and only those animals that can adapt (adjust) to the flood survive. In a species, the animals with the adaptive traits (adaptive genes, having adaptations) to protect against flooding will survive and the animals of that species that do not have the adaptive traits (adaptive genes) to protect against flooding will die out.

Species with organisms with **few variations** might not survive (or would become extinct) because none of the organisms might be able to protect themselves against the changing environment (such as flooding, or drought-very little rainfall, or extremely hot temperature). Even more so, if there is a very small population with very few varieties, none of the offspring might survive.

Species with organisms with **many different variations** (many different traits and many different combinations of traits, many different combinations of genes (genetic variability)) will **more likely survive in a changing environment** (example flooding, or drought-very little rainfall, or freezing temperature). Since there are so many different organisms with different variations (different traits): black or white organisms, organisms with different types of feet, different types of beaks for eating, some can run faster or slower, some are better fighters, some stronger or weaker, some of these organisms will be able to survive and some of the organisms will die out. The organisms that are able to survive in the changed environment (example flooding) produce offspring also able to survive in the changed environment and the species or population (all members of a species living in one area) survives (does not become extinct).

Let's see which population is most likely to survive when there is a big environmental change. A population is all members of a species living in one area, such as all gray squirrels living in a park. The graph below shows the percent of variation for a given trait (examples color of moths) in four different populations of the same species. The populations inhabit similar environments.

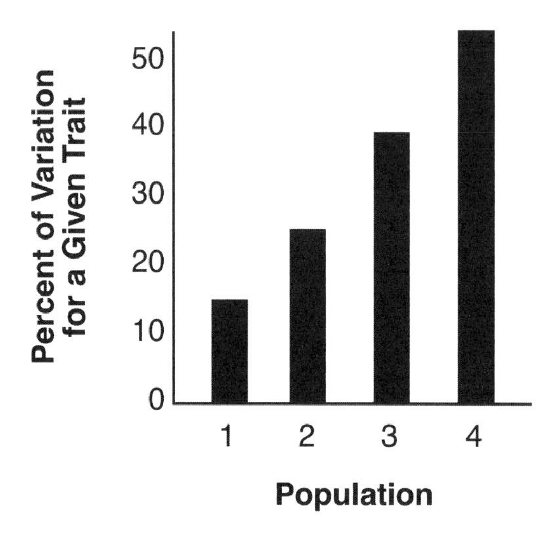

On the vertical axis is percent of variation (how much variation there is for a given trait).
Population 1 has the shortest bar, which means population 1 has the smallest amount of variation.
Population 2 has a taller (longer) bar which means population 2 has more variation.
Population 3 has an even taller (longer) bar, which means popul;ation 3 has even more variation.
Population 4 has the tallest (longest) bar, which means it has the most variation.
When there is a big environmental change, the population (example population 4) with the most variation is most likely to survive. Population 4 is most likely to survive.

Let's read the percentage of variation for a given trait on the bar graph. There is 10% between each two numbers on the vertical scale (see bar graph). Drawing a dotted line from the top of each bar to the vertical axis may make it easier to see the percent of variation.
For population 1, the bar is more than 10% but less than 20%; the variation in population 1 is about 13%.
For population 4, the bar is more than 50% but less than 60%; the variation in population 4 is about 53%.

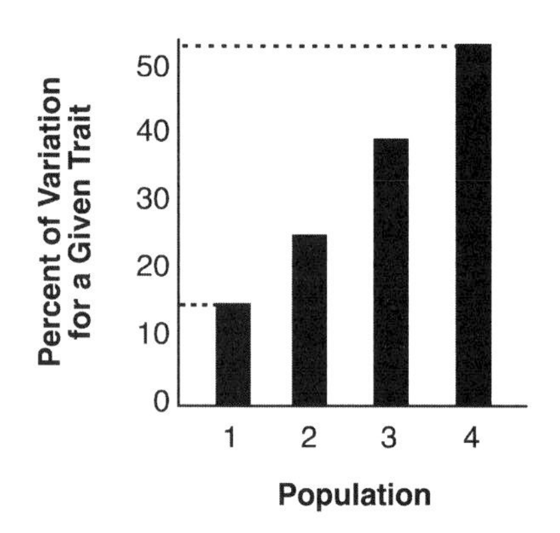

Environment influences which moths will survive: Environment (or environmental change) influences which color of moths will survive more in that environment. In England, there is a type of moth which can be either black or white in color. Before there were factories, when there was no black soot and there was white lichen on the trees, the white moths had the adaptive value (white color) of being hidden (camouflaged) from the birds (birds eat moths) because the white colored moths had the same color as their environment (white colored lichens and no soot). More of the white moths survived (natural selection) and there were more white moths than black moths.

Later, factories were built. They produced large amounts of soot (black colored), which covered the trees, making the trees black. At that time, black moths had the adaptive value (black color) of being hidden (camouflaged) from birds (birds eat moths) because the black colored moths had the same color as the environment (black soot). More of the black moths survived (natural selection) and there were more black moths than white moths.

Many years later, when there were antipollution laws and there was less black soot, more of the white moths survived.

Use the graph to answer the four questions below.

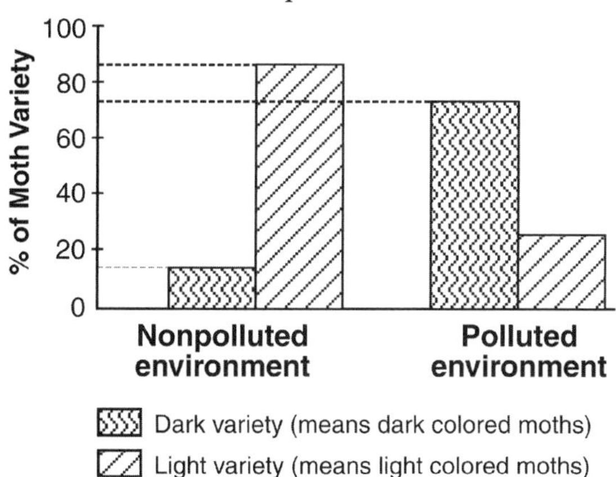

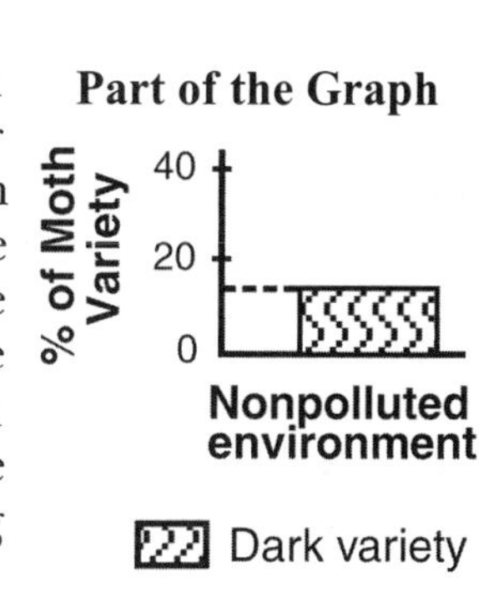

1. What is the percentage of dark-colored moths in the nonpolluted environment? Look at the bar graph above and at the right. There is 20% between each two numbers on the vertical axis (line). In the non polluted environment, look at the top of the bar for the dark variety ▩, and go across to the vertical line. You see the dotted line is more than halfway between 0% and 20%, or about ¾ of the way up between 0% and 20%; with equal spacing between the numbers, it is about 16%.

2. What is the percentage of light-colored moths in the nonpolluted environment? Look at the bar graph above. There is 20% between each two numbers on the vertical axis (line). In the nonpolluted environment, look at the top of the bar for the light variety ▨ and go across to the vertical line. You see the dotted line is at a little more than 80% and less than 100%; with equal spacing between the numbers, it is about 84%.

3. What is the percentage of light-colored moths in the the polluted environment? Look at the bar graph above. There is 20% between each two numbers on the vertical axis (line). In the polluted environment, look at the top of the bar for the light variety ▨ and go across to the vertical line. You see it is at a little more than 20% and less than 40%; with equal spacing between 20% and 40%, it is about 24%.

4. What is the percentage of dark-colored moths in the polluted environment? Look at the bar graph above. There is 20% between each two numbers on the vertical axis (line). In the polluted environment, look at the top of the bar for the dark variety ▩, and go across to the vertical line. You see the dotted line is more than halfway between 60% and 80%, or about ¾ of the way up between 60% and 80%; with equal spacing between the numbers, it is about 76%.

Question: When a particular white moth lands on a white birch tree, its color has a high adaptive value. If the birch trees become covered with black soot, the white color of this particular moth in this environment would most likely

1. retain its adaptive value
2. increase in adaptive value
3. change to a more adaptive black color
4. decrease in adaptive value

Solution: The **white color** of the moth has high **adaptive value on** a **white tree** (white moth hidden from enemies). When the tree became covered with black soot, the **white color** of the moth does not hide the moth on a **black** sooted tree (white moth is not hidden from enemies) and therefore it decreases in adaptive value. Answer *4*

Now Do Homework Questions #28-57, pages 34-39.

Geographic isolation brings about evolution

Environment: You learned, **evolution** takes place more often **in** a **changing environment**; there is less evolution when the environment is stable (not changing).

Geographic isolation: **Geographic isolation** is when different members of a **species** become **separated** (isolated) by **oceans** or **mountains,** etc.; they are in **different environments**, having **different foods**, and may **adapt** and **become new species (evolution** takes place).

Examples of **geographic isolation** are the Galapagos **finches** and **African cichlid fish** (see below):

Finches: A few finches (a type of bird) originally left the South American mainland, flew over the Pacific Ocean, and came to the isolated Galapagos Islands, which are 600 miles away from the mainland. The islands also are **separated by water** (Pacific **Ocean**). There were **different types of food** for the finches on the **different islands**; some foods were plants, other food were animals. Charles Darwin saw that the finches on all the islands were very similar but the finches have different types of bills and eat different types of food.

Finch	Type of Bill	Type of Food
sharp-billed ground finch	crushing	plants
warbler finch	probing	animals

The finches (example large ground finch, woodpecker finch, and vegetarian finch) had **evolved** (**evolution**) into **different species** (types) of finch which have different types of bills and eat different types of food; these finches cannot interbreed with each other. On each island, those finches with adaptations that matched the food on that island survived and had offspring (babies) (natural selection); those finches not adapted to the food on that island died out.

Finch Diversity

The diagram below shows different species of finches (examples large ground finch, vegetarian finch, large tree finch), their bills, and the food they eat. The title of the diagram is Finch Diversity, which means different varieties (species) of finches.

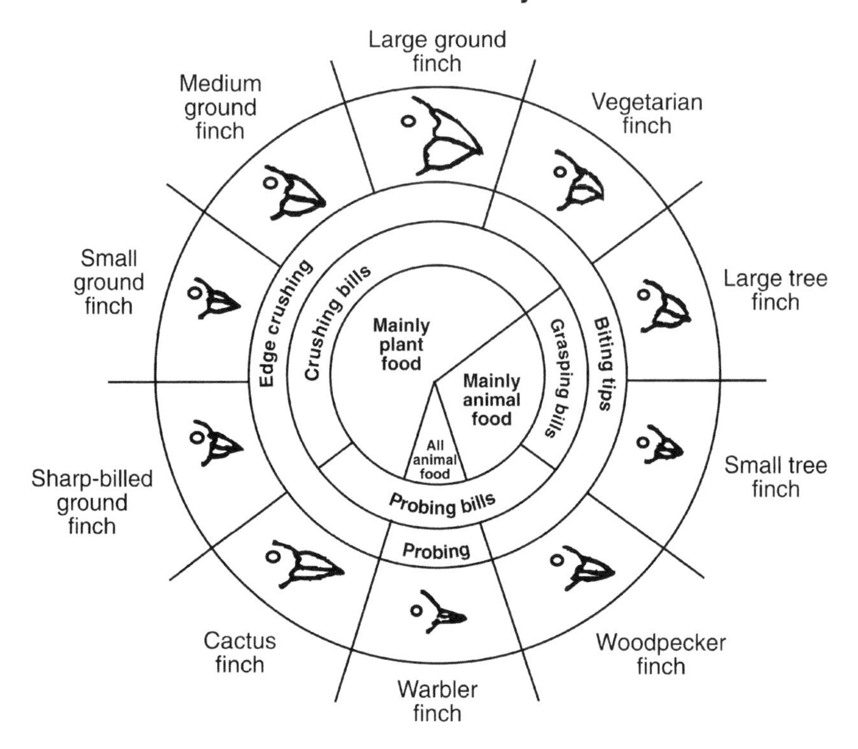

Look at the finch diversity diagram above. The **outer circle** shows **drawings of** the **bills** of each species of finch. The **next two circles**, inside the drawings of the bills, describe the **type of bill** each species has. The **innermost circle** shows the **type of food** each species eats. For example, the large tree finch and the small tree finch have grasping bills and biting tips and eat mainly animal food. Both species compete for the same food with the same type of bill.

Similarly, the sharp-billed ground finch, the small ground finch, the medium ground finch, and the large ground finch all have edge crushing and crushing bills and mainly eat plant food. These species compete for the same food with the same type of bill.

Question: The cactus finch, warbler finch, and woodpecker finch all live on one island. Based on the information in the finch diversity diagram above, which one of these finches is least likely to compete with the other two for food? Support your answer with an explanation.

Solution: The cactus finch is least likely (unlikely) to compete with the others for food. Look at the finch diversity diagram. The cactus finch eats mainly plant food, while the other two (warbler finch and woodpecker finch) eat mainly or all animal food.

Question: In the *Beaks of Finches* laboratory activity, students were each assigned a tool to use to pick up seeds. In round one, students acting as birds used their assigned tools to pick up small seeds from their own large dishes (the environment) and place them in smaller dishes (their stomachs). The seeds collected by each student were counted. Some students were able to collect many seeds, while others collected just a few.
In round two, students again used their assigned tools to collect seeds. This time several students were picking up seeds from the same dish of seeds.

Question 1: Explain how this laboratory activity illustrates the process of natural selection.

Question 2: One factor that influences the evolution of a species that was *not* part of this laboratory activity is

(1) struggle for survival (3) competition
(2) variation (4) overproduction

Question 3: Identify *one* trait, other than beak characteristics, that could contribute to the ability of a finch to feed successfully.

Solution 1: The tools the students used represented (modeled) the beaks of the finches. The students with the better tools (meaning birds with better beaks) are able to get more food and more survived. This is natural selection.

Solution 2: The lab discussed variations in beaks, competition for seeds, and struggle for survival. The lab did not discuss overproduction. Overproduction means too many offspring are produced and therefore there is not enough food to support all of them.

Solution 3: The stronger birds can compete better for food. The birds with better vision (seeing better) can find the food more easily. Faster birds can get to the food more quickly.

African cichlid fish (example of geographic isolation): The three great (large) lakes in Africa contain very many different varieties (a great diversity) of species of cichlid fish. One suggested explanation for the large number of species in each lake is geographic isolation, as shown in the next figure.

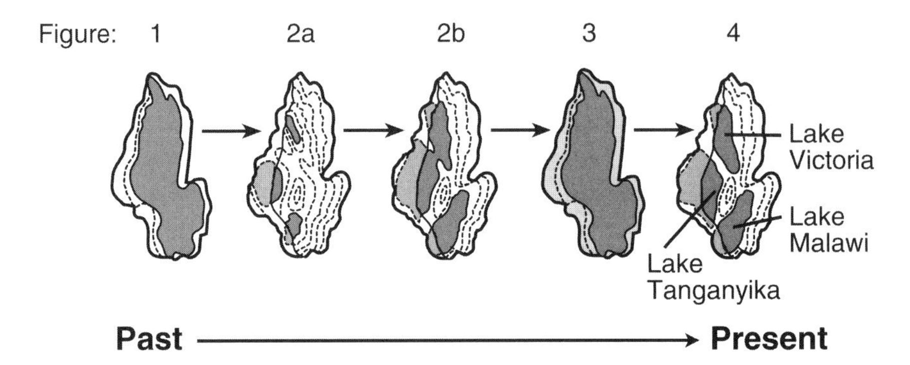

1. All three lakes that we have today were once one lake (Figure 1).
2. The water level dropped, separating the lakes. Each lake had different conditions and evolved different species of cichlid fish adapted to its environment (Figures 2a and 2b).
3. Water rose, allowing the fish in the three lakes to mix (Figure 3).
4. The water dropped again, separating the lakes; each lake now still has fish from all the lakes mixed together, a great diversity (great variety) of cichlid fish (Figure 4).

Now Do Homework Questions #58-81, pages 40-46.

Causes of Variation, Causes of Evolution

1. Mutation. You learned the order (sequence) of nitrogen bases in DNA determines the traits (characteristics) of the organism (example tall, short, brown eyes, blue eyes). You learned a mutation is a change in the nitrogen bases of the DNA molecule or a change in the sequence (order) of bases in DNA or a change in a gene (segment or piece of DNA).
Mutations rarely happen (not often). Mutations are caused by errors (changes) in DNA and also by radiation and some chemicals.
If a mutation happens in a body cell (example stomach), the offspring do not get the gene for the mutation. If the mutation is in the sex cells (example sperm or egg), the gene for the mutation (mutated gene) is passed on to the offspring (children) and future generations. This helps to bring about evolution (change over time, such as a change of species over time, because the offspring get the mutated gene).
Most mutations are harmful and the offspring (children) can die. A few mutations are good (beneficial) and help the offspring survive and then reproduce.

Note: It is interesting to know that if an offspring (child) gets two genes for sickle cell disease, the offspring gets the disease. But, if the offspring (child) only gets one gene (not two genes) for sickle cell disease, the offspring does not get sickle cell disease and is protected from a different disease called malaria.

2. **Genetic shuffling.** You learned in **sexual reproduction genetic shuffling,** which includes meiosis (producing eggs and sperm), fertilization (sperm and egg uniting (unite together) to form a fertilized egg), recombination (combining genes of the sperm and egg), and crossing over, **causes** the **offspring** to get a **different combination** of **genes** and therefore to have **different traits** (characteristics) from the parents and different traits from each other.

Genetic shuffling causes the offspring to be different (having different genes), which is a change from the parent and therefore evolution takes place. Most of the variations found in the offspring are due to genetic shuffling.

Note: In **asexual reproduction,** organisms have only one parent , and the offspring is identical (genetically identical) to the parent. In asexual reproduction, evolution is slow, because it can only happen by mutation, not by genetic shuffling. Genetic shuffling, which causes most variations and most of evolution, only happens in sexual reproduction.

Evolutionary Changes

In sexually reproducing organisms, **mutation** (change in DNA) and **genetic shuffling** in sex cells can **cause evolution** (changes in species over time, such as changes in structure, function, or behavior).

Structural Change: Look at the next figure. A shows the shoulder (joint). The bones in the forelimbs (front limbs) in five different species (example human, whale, cat, etc.) are similar, but they have different functions. Look at the human limb, whale limb, cat limb, bat limb, and bird limb in the figure. These **limbs have similar bones,** one big bone close to the shoulder, two thinner bones further away from the shoulder, and a "hand", usually with five "fingers". Since these animals (example human, cat etc.) have similar arrangements of bones, they had (evolved from, evolution) a **common ancestor**.

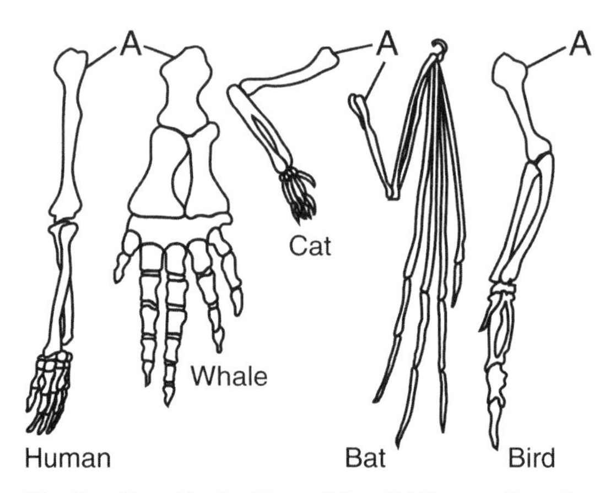

Similar Forelimbs From Five Different Species

Functional change: A change in function is a change in how part of the body works. Functional change is caused by a mutation (change in DNA). A rattlesnake is poisonous because it produces a poison to kill the animals it eats. This is an evolutionary change because most snakes do not produce poison.

Behavioral change: In many animal species (including antelopes, wild goats, deer, walrus) males fight each other and the strongest male, the winner, is then able to mate with many females. This behavior of fighting evolved so only the strongest males give their genes to the offspring, causing the offspring to be stronger.

Now Do Homework Questions #82-94, pages 46-48.

Variations (different varieties of organisms)

You learned variations (different varieties of organisms) are caused by mutations or/and genetic shuffling (meiosis, fertilization, genetic recombination, or crossing over).

Bacteria: Some bacteria are pathogens (cause diseases).

First: When people become sick with a bacterial infection, they are given an antibiotic. A few bacteria are **resistant to** the **antibiotic,** meaning the **antibiotic cannot kill** these **few resistant bacteria.** Most of the other bacteria are not resistant (nonresistant) and the antibiotic can kill them.

The resistant bacteria probably have either a mutation (mutated gene, change in a gene) or variation to be resistant to this antibiotic (example penicillin). These few resistant bacteria had the gene to be resistant to the antibiotic even before the person was given the antibiotic. The resistant bacteria survived (survival of the fittest (the resistant bacteria)). Most of the other bacteria were not resistant (nonresistant) and died out (a small number escaped the antibiotic and survived).

Next: The bacteria then produced many new bacteria (offspring) in a short period of time. A bigger percentage of the new bacteria are resistant to the antibiotic because most of the nonresistant bacteria died out.

After that: The antibiotic is given again to the new population of bacteria, consisting mostly of resistant bacteria and now only a few nonresistant bacteria. Most of the nonresistant bacteria are killed. More (a larger percentage) of the bacteria survive because most of them were resistant. The antibiotic is an **agent of selection**; the antibiotic does not cause the mutation (to resist antibiotics) but it selects which bacteria will live and which will be killed. Those bacteria which are resistant (resistant bacteria) will not be killed by the antibiotic.

Insects: Some insects carry (transmit) diseases (such as malaria); other insects may damage crops. The **figure below** shows the **effect of spraying pesticides on leaves.**

A. Look at the leaf drawings. Figure I shows a leaf with insects on it (10 nonresistant and 2 resistant insects). Pesticides (insecticides) are poisons which are used to kill insects. Pesticides (insecticides) were sprayed on the leaf (first spraying of insecticides)(see A in diagram).

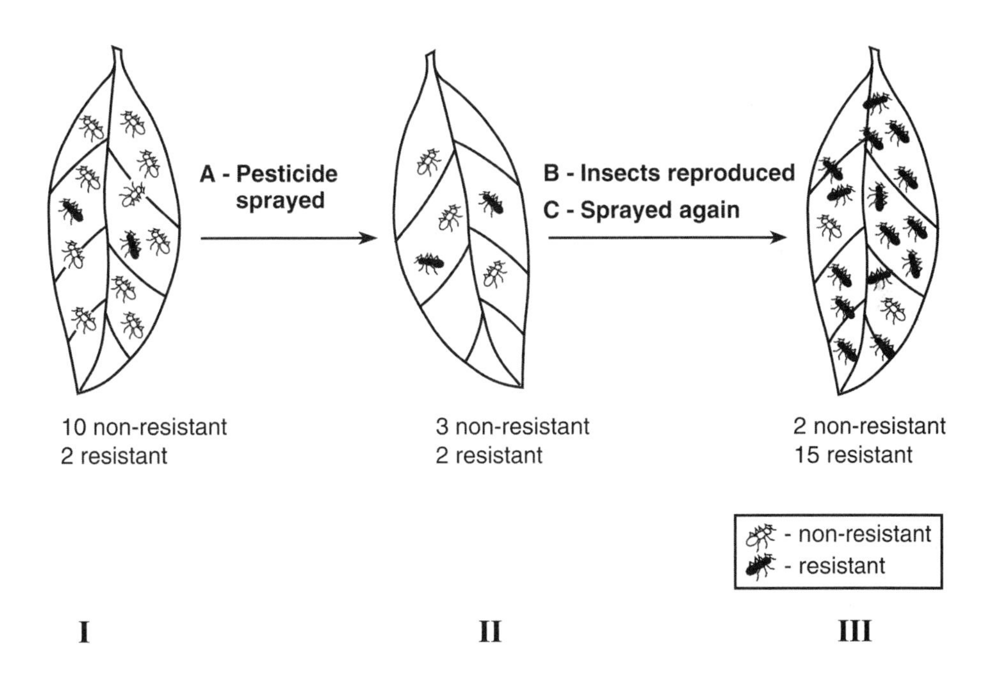

B. Next, look again at Figure II and see B in the diagram. The surviving insects reproduced (insects have short reproduction cycles); many new insects were produced in a short period of time. The resistant insects had many offspring; a large part of the insect population is resistant to the insecticide because most of the nonresistant insects died out.

C. After the insects reproduced, the insects were sprayed again with insecticide (see C in diagram). The resistant insects survived the insecticide and reproduced, producing more resistant insects. You see in Figure III that there are a lot of resistant insects on the leaf; most of the nonresistant insects were killed. More of the insects (a bigger percentage) survive because most of them were resistant (see Figure III); therefore the insecticide was not that useful in killing the insects.
In short, the same organism (living thing, example insects), has different varieties; some of them are resistant and some are not resistant. Most of the nonresistant organisms (such as nonresistant bacteria or nonresistant insects) die out, and the resistant organisms survive, reproduce and produce more resistant organisms.

Other ways of controlling pests (example insects): Besides using insecticides (pesticides) to kill insects, fruit growers can also bring in organisms that kill the insects, or genetically engineered plants (put a gene in the plants so the insects will not eat the plants), therefore the plants are not damaged by the insects, or can bring in sterile male insects that mate with female insects but cannot reproduce (produce offspring).

Question: The diagram below shows the effect of spraying a pesticide on a population of insects over three generations.

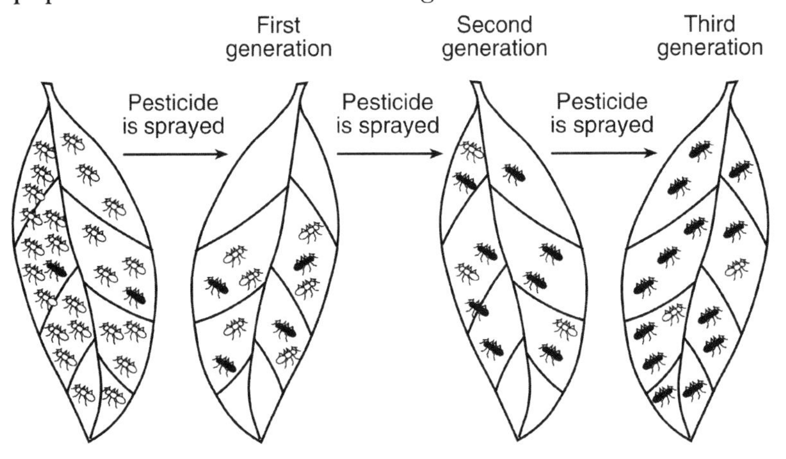

Which concept is represented in the diagram?

1. survival of the fittest
2. dynamic equilibrium
3. succession
4. extinction

Solution: In the first leaf, there are only two dark colored insects on the leaf and many light colored insects. The two dark colored insects must represent the resistant insects, which have the resistant gene caused by a mutation or genetic shuffling. Look at the diagram. After spraying the pesticide, the second leaf has more resistant insects (four) and fewer nonresistant insects (because they died out from the pesticide). After spraying again, the third leaf now has more resistant insects (eight) and less nonresistant insects (three). After spraying again the third time, the fourth leaf has more resistant insects (13) and less nonresistant insects (two). As you can see, the resistant insects (dark colored) survived and reproduced (survival of the fittest (the resistant insects)); most of the nonresistant insects (light colored) died out. Answer *1*

Now Do Homework Questions #95-101, pages 48-50.

Faster or Slower Evolution

The **rate of evolution (faster or slower evolution)** depends on the type of organism (examples bacteria, humans) and how much the environment changes (big or small environmental changes). **Evolution** takes place **faster** when the organism **reproduces** in **less time (short reproductive cycle)**, therefore producing a lot (large number) of organisms in a short period of time. **Bacteria** have a short reproductive cycle; they reproduce about every twenty minutes. For example, bacteria which are resistant to an antibiotic would quickly spread the resistance trait through the population because they produce offspring resistant to antibiotics every twenty minutes. Reproducing every twenty minutes means **there are** a **lot** of **generations** in a small amount of time, there is a greater possibility of mutation through all these generations, therefore evolution takes place quickly (fast). **Insects** also have a **short reproductive cycle** (many generations in a small amount of time), there is a greater possibility of mutation and genetic shuffling through all these generations, therefore evolution takes place quickly (fast).

Evolution also **happens quickly when** there is a **big change in** the **environment**. **Evolution** happens **slowly** when there is a **small change** in the environment; the horse went through many changes over the past 60 million years ago before evolving into the horse that we have today.

Extinction

Extinction is when a species is **no longer here**. A **species** becomes **extinct** (dies out) when it has **very poor traits** (very poor genes) and no adaptive value (cannot adapt to a changing environment). When the environment changes (unstable), some organisms are able to adapt and survive, and some organisms are **NOT** able to adapt and die out (become extinct). Examples of environmental changes are high temperature, freezing temperature, a lot of rainfall, drought (very little rainfall), pollution, etc. If a **species** has **organisms** with **very little variation** (very few varieties), the species with very few variations might **not** be able to adapt to a changing environment, might not survive and will die out (become **extinct**). An example of an extinct species is the wooly mammoth.

In short, the causes of extinction are:

1. species has very few variations
2. species cannot adapt to a changing environment (no adaptive value, lack adaptations)
3. in a changing environment, if the species has very few varieties and no adaptive value, the species will become extinct (die)

Organisms in a **species** with a **lot of variation** (different varieties of organisms in the species with different traits) are more **able to survive** in a changing environment because some of the organisms will **adapt** and be able to handle the changed environment. Large populations also help a species to survive because they might have more variations.

Organisms in a **species** with **very little variation** (very few varieties) might **not** be able to adapt to a changing environment, might not survive and will die out (become **extinct**). An example of an extinct species is the wooly mammoth. Millions of species became extinct. From the fossils (example bones, shells) in rocks, we see many species existed in the past which do not exist today (species became extinct).

Look at the evolutionary tree of the horse (see below). **Species A and species B** became **extinct** (do not exist today, ended at a certain time) because species A and species B did not have the adaptive value (adaptive characteristics) to survive.

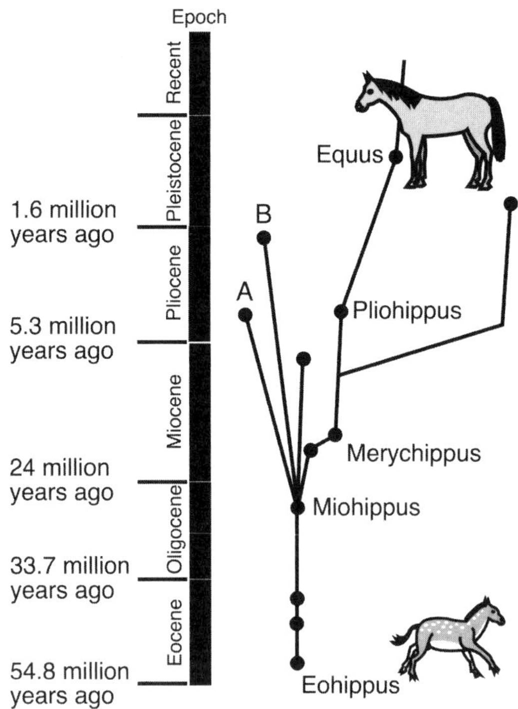

Evolutionary Tree of the Horse

Endangered species: An endangered species is a species that is in danger of dying out (becoming extinct). Some endangered species have so few individuals left (example the Florida panther has less than 100 animals) that they have no variation and they can not adapt to a change in their environment. If very few organisms of a species are left (endangered species) or if it dies out (becomes extinct), we will **lose** all the species' **genetic material (genes**); these lost genes could have been used in medicine, knowledge, and helping organisms adapt to the environment and survive.

Now Do Homework Questions #102-111, pages 50-51.

NOW LET'S TRY A FEW HOMEWORK QUESTIONS

To assign fewer questions, you may want to skip the questions with underlined numbers.

1. The relationship of some mammals is indicated in the diagram below.

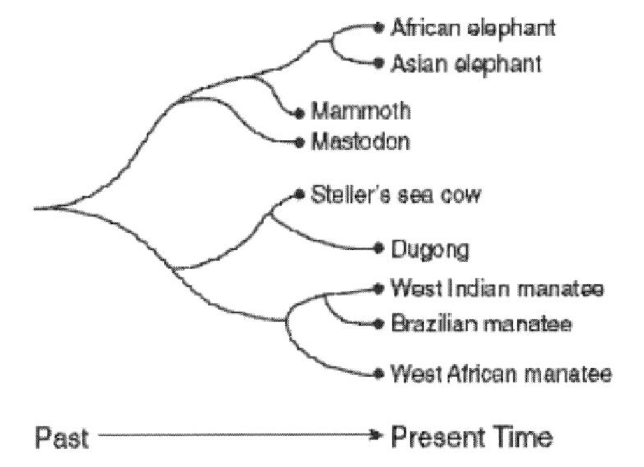

Which statement about the African elephant is correct?

(1) It is more closely related to the mammoth than it is to the West African manatee.
(2) It is more closely related to the West Indian manatee than it is to the mastodon.
(3) It is not related to the Brazilian manatee or the mammoth.
(4) It is the ancestor of Steller's sea cow.

2. The evolutionary pathways of five species are represented in the diagram below.

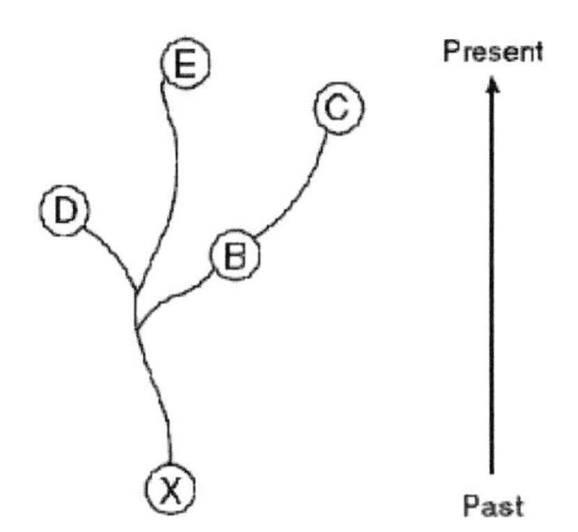

Which statement is supported by the diagram?

(1) Species ***C*** is the ancestor of species ***B***.
(2) Species ***D*** and ***E*** evolved from species ***B***.
(3) Species ***X*** evolved later than species ***D*** but before species ***B***.
(4) Both species ***C*** and species ***D*** are related to species ***X***.

3. Base your answers to this question on the information below and on your knowledge of biology.

Based on their analysis of the differences in amino acid sequences of one kind of protein, scientists prepared the evolutionary tree shown below.

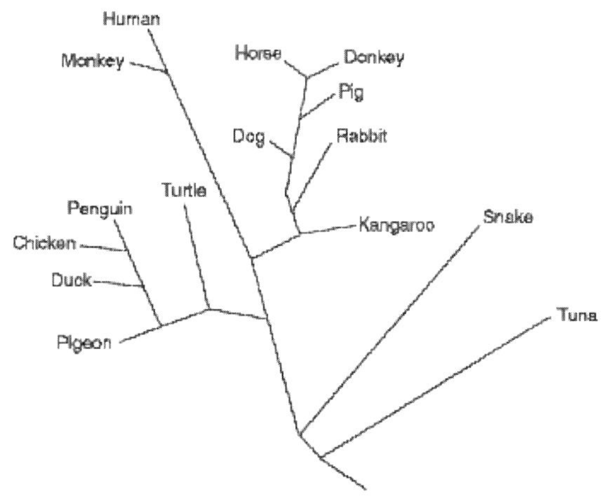

According to this diagram, the DNA of which pair of organisms would show the greatest similarity?

(1) penguin and turtle
(2) horse and donkey
(3) snake and tuna
(4) turtle and rabbit

Hint*: See which two organisms have more things in common (the same).*

4. The diagrams below represent organs of two individuals. The diagrams are followed by a list of sentences. For the phrase given at the bottom, select the sentence from the list below that best applies to that phrase. Then record its number in the space provided.

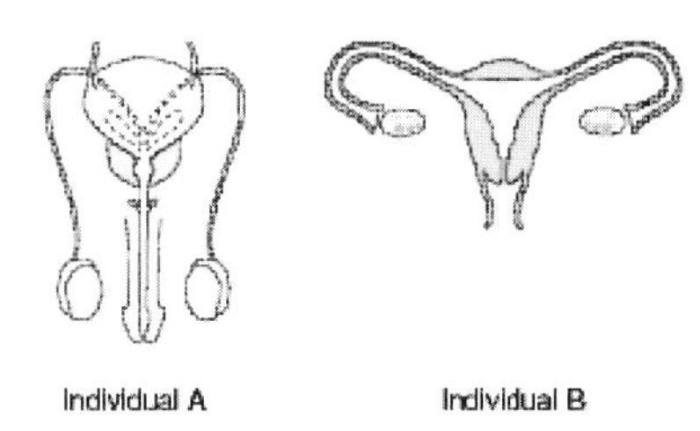

Individual A Individual B

SENTENCES

1. The phrase is correct for both Individual ***A*** and Individual ***B***.
2. The phrase is not correct for either Individual ***A*** or Individual ***B***.
3. The phrase is correct for Individual ***A***, only.
4. The phrase is correct for Individual ***B***, only

Contains a structure in which a zygote divides by mitosis

5. Base your answer to the next question on the chart below and on your knowledge of biology.

A	B	C
The diversity of multicellular organisms increases.	Simple, single-celled organisms appear.	Multicellular organisms begin to evolve.

According to most scientists, which sequence best represents the order of biological evolution on Earth?

(1) $A \rightarrow B \rightarrow C$
(2) $B \rightarrow C \rightarrow A$
(3) $B \rightarrow A \rightarrow C$
(4) $C \rightarrow A \rightarrow B$

6. The chart below contains a number of characteristics for three different organisms. The characteristics can be used in classifying these organisms.

Characteristics	Organism A	Organism B	Organism C
Number of cells	unicellular	multicellular	unicellular
Type of nutrition	autotrophic	autotrophic	heterotrophic
Nuclear membrane	absent	present	absent
DNA	present	present	present

Which *two* organisms would be expected to have the most similar genetic material? Support your answer using information from the chart.

7. The diagram below illustrates possible evolutionary pathways of some species.

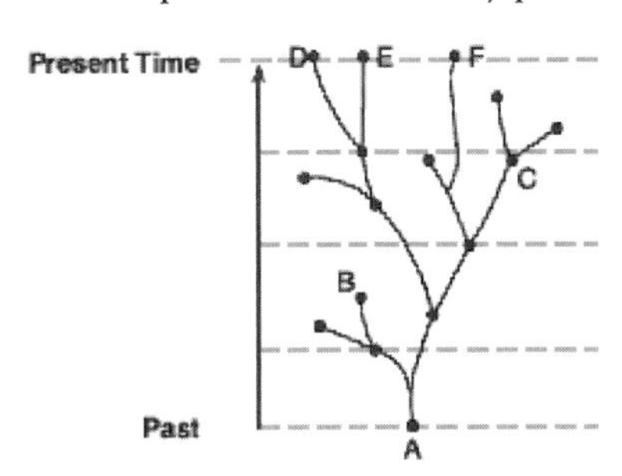

Which statement is a valid inference-conclusion based on fact or observation (what you see), based on the information in the diagram?

(1) Species ***A*** is the common ancestor of all life on Earth.
(2) Species ***D*** is more closely related to species ***E*** than to species ***F***.
(3) Species ***B*** is the ancestor of species ***F***.
(4) Species ***C*** is the ancestor of species that exist at the present time.

8. Base your answers to the next question on the diagram below and on your knowledge of biology. Letters ***A*** through ***L*** represent different species of organisms. The arrows represent long periods of geologic time.

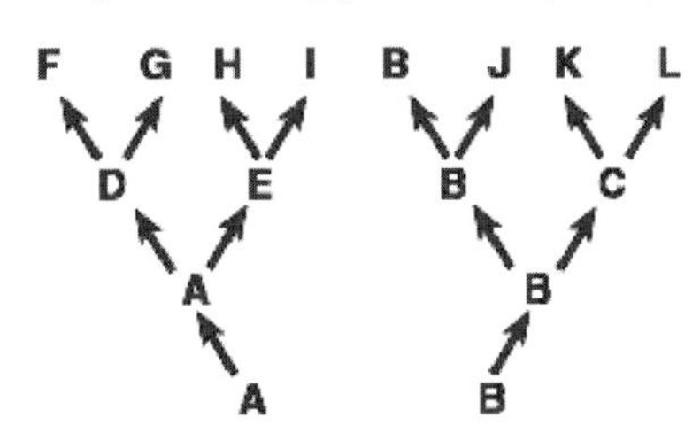

Which two species are the most closely related?

(1) ***J*** and ***L*** (2) ***G*** and ***L*** (3) ***F*** and ***H*** (4) ***F*** and ***G***

9. ***R***, ***S***, and ***T*** are three species of birds. Species ***S*** and ***T*** show similar coloration. The enzymes found in species ***R*** and ***T*** show similarities. Species ***R*** and ***T*** also exhibit many of the same behavioral patterns.

Show the relationship between species ***R***, ***S***, and ***T*** by placing the letter representing each species at the top of the appropriate branch on the diagram below.

10. In the diagram below, Letters ***A*** through ***J*** represent different species of organisms. The vertical distances between the dotted lines represent long periods of time in which major environmental changes occurred.

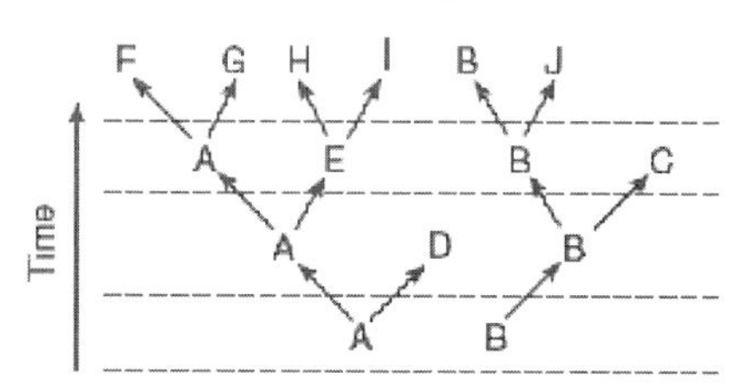

Which species appears to have been most successful in surviving changes in the environment over time?

(1) ***A*** (2) ***B*** (3) ***C*** (4) ***H***

11. According to some scientists, patterns of evolution can be illustrated by the diagrams below.

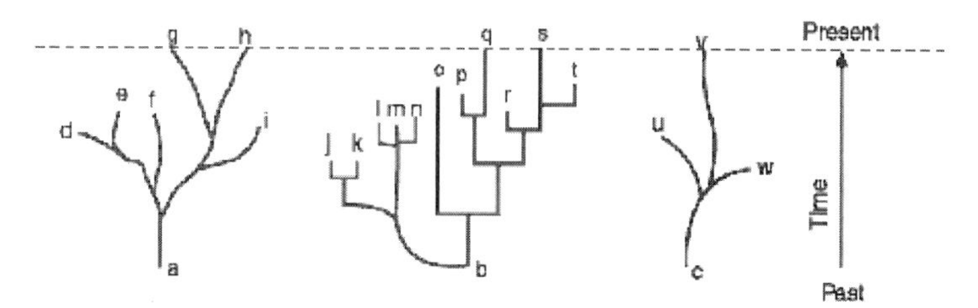

Which statement best explains the patterns seen in these diagrams?

(1) The organisms at the end of each branch can be found in the environment today.
(2) The organisms that are living today have all evolved at the same rate and have undergone the same kinds of changes.
(3) Evolution involves changes that give rise to a variety of organisms, some of which continue to change through time while others die out.
(4) These patterns cannot be used to illustrate the evolution of extinct organisms.

12. The diagram below shows the evolution of some different species of flowers.

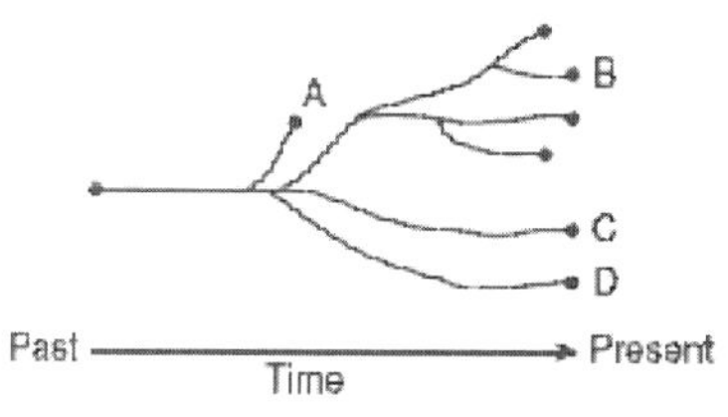

Which statement about the species is correct?

(1) Species ***A***, ***B***, ***C***, and ***D*** came from different ancestors.
(2) Species ***C*** evolved from species ***B***.
(3) Species ***A***, ***B***, and ***C*** can interbreed successfully.
(4) Species ***A*** became extinct.

Base your answers to the next two questions on the information below and on your knowledge of biology.

Based on their analysis of the differences in amino acid sequences of one kind of protein, scientists prepared the evolutionary tree shown below.

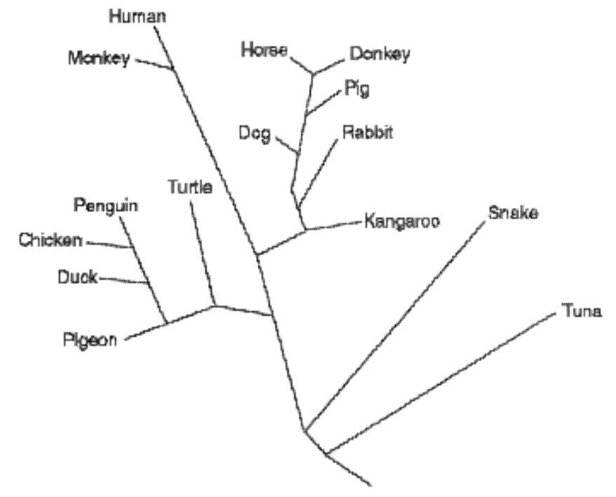

13. Older systems of classification always placed penguins, chickens, ducks, and pigeons in the bird group and turtles and snakes in the reptile group. Does this diagram support the older system of classification? Explain your answer.

14. According to this diagram, is the pig more closely related to the dog or the kangaroo? Justify your answer.

---

15. According to the diagram below, which three species lived on Earth during the same time period?

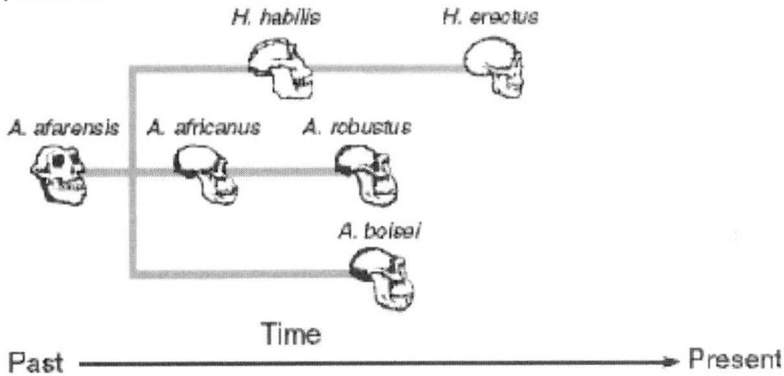

(1) ***robustus, africanus, afarensis***
(2) ***habilis, erectus, afarensis***
(3) ***habilis, robustus, boisei***
(4) ***africanus, boisei, erectus***

16. A current proposal in the field of classification divides life into three broad categories called domains. This idea is illustrated below.

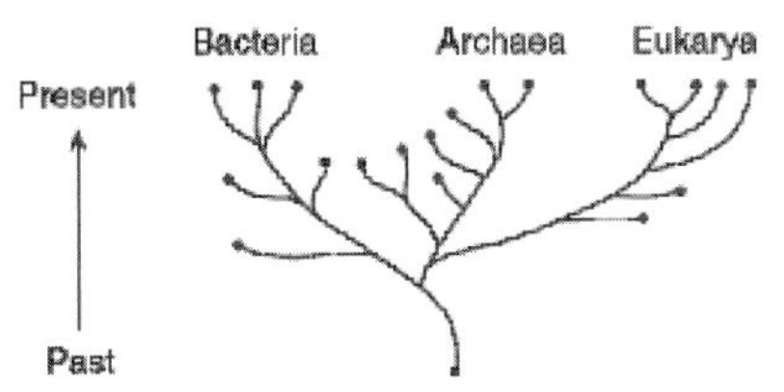

Which concept is best supported by this diagram?

(1) Evolutionary pathways proceed only in one set direction over a short period of time.
(2) All evolutionary pathways will eventually lead to present-day organisms.
(3) All evolutionary pathways are the same length and they all lead to present-day organisms.
(4) Evolutionary pathways can proceed in several directions with only some pathways leading to present-day organisms.

17. Base your answer to the following question on the diagram below, which represents the relationships between animals in a possible canine family tree, and on your knowledge of biology.

Canine Family Tree

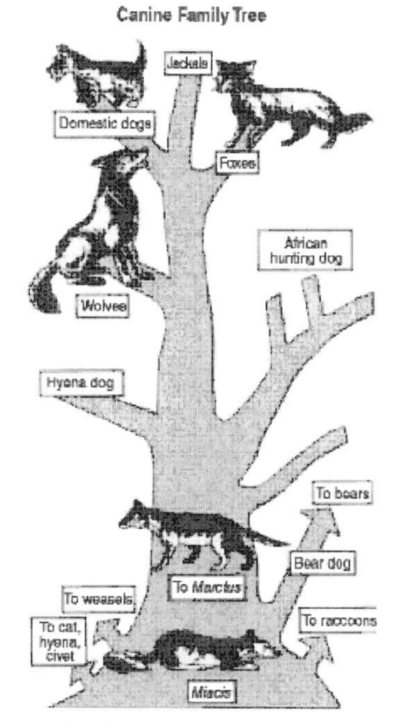

According to the diagram, which group of organisms has the most closely related members?

(1) cats, weasels, and wolves
(2) bears, raccoons, and hyena dogs
(3) jackals, foxes, and domestic dogs
(4) African hunting dogs, hyena dogs, and domestic dogs

18. Base your answer to the following question on the information and data table below and on your knowledge of biology.

Body Structures and Reproductive Characteristics of Four Organisms

Organism	Body Structures	Reproductive Characteristics
pigeon	feathers, scales 2 wings, 2 legs	lays eggs
A	scales 4 legs	lays eggs
B	fur 2 leathery wings, 2 legs	gives birth to live young provides milk for offspring
C	fur 4 legs	lays eggs provides milk for offspring

Explain why it would be difficult to determine which one of the other three organisms from the table should be placed in box *1*.

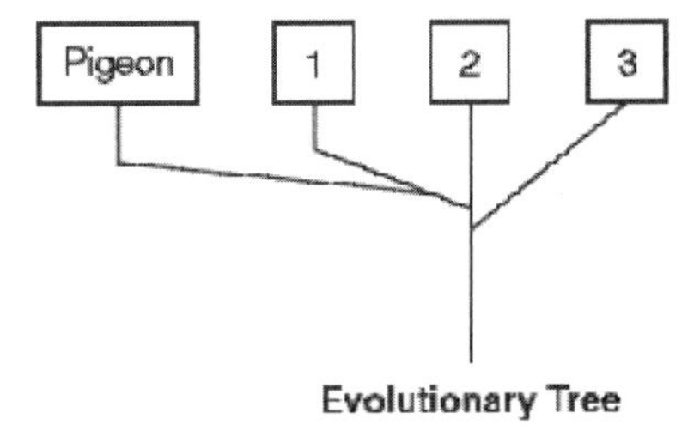

19. Base your answer to the next question on the diagram below, which represents possible relationships between animals in the family tree of the modern horse, and on your knowledge of biology.

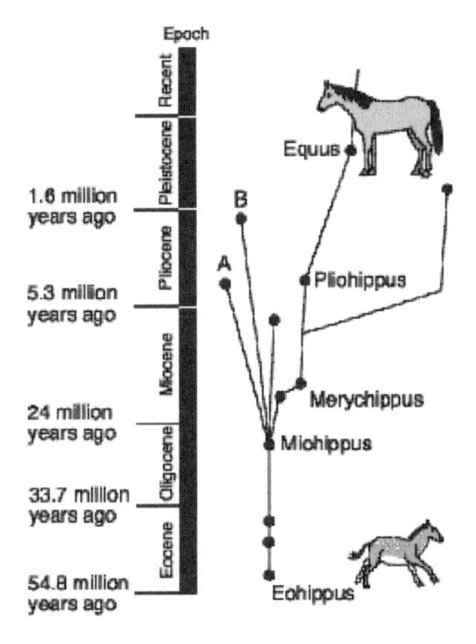

Miohippus has been classified as a browser (an animal that feeds on shrubs and trees) while ***Merychippus*** has been classified as a grazer (an animal that feeds on grasses). One valid inference that can be made regarding the evolution of modern horses based on this information is that

(1) ***Eohippus*** inhabited grassland areas throughout the world
(2) ***Pliohippus*** had teeth adapted for grazing
(3) ***Equus*** evolved as a result of the migration of ***Pliohippus*** into forested areas due to increased competition
(4) Ecological succession led to changes in tooth structure during the Eocene Epoch

20. Relationships between plant species may most accurately be determined by comparing the

(1) habitats in which they live
(2) structure of guard cells
(3) base sequences of DNA
(4) shape of their leaves

21. The diagram below represents possible evolutionary relationships between groups of organisms.

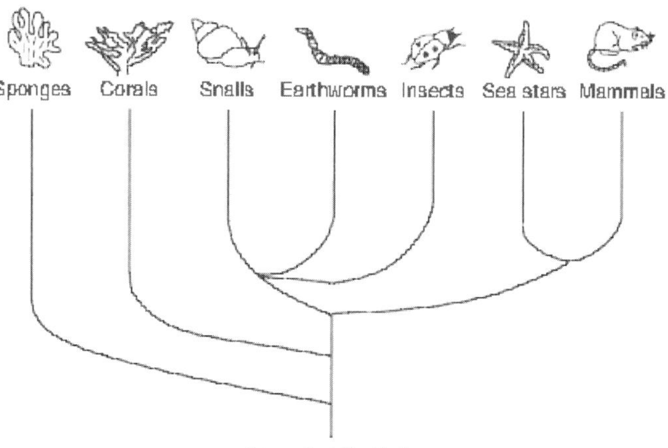

Which statement is a valid conclusion that can be drawn from the diagram?
(1) Snails appeared on Earth before corals.
(2) Sponges were the last new species to appear on Earth.
(3) Earthworms and sea stars have a common ancestor.
(4) Insects are more complex than mammals.

22. To determine evolutionary relationships between organisms, a comparison would most likely be made between all of the characteristics below except
(1) methods of reproduction
(2) number of their ATP molecules
(3) sequences in their DNA molecules
(4) structure of protein molecules present

Base your answers to the next three questions on the information below and on your knowledge of biology.

A series of investigations was performed on four different plant species. The results of these investigations are recorded in the data table below.

Characteristics of Four Plant Species

Plant Species	Seeds	Leaves	Pattern of Vascular Bundles (structures in stem)	Type of Chlorophyll Present
A	round/small	needle-like	scattered bundles	chlorophyll a and b
B	long/pointed	needle-like	circular bundles	chlorophyll a and c
C	round/small	needle-like	scattered bundles	chlorophyll a and b
D	round/small	needle-like	scattered bundles	chlorophyll b

23. Based on these data, which two plant species appear to be most closely related? Support your answer.

Plant species ____________ and __________

24. What additional information could be gathered to support your answer to the last question?

25. State ***one*** reason why scientists might want to know if two plant species are closely related.

26. Which observation could best be used to indicate an evolutionary relationship between two species?
   (1) They have similar base sequences.
   (2) They have similar fur color.
   (3) They inhabit the same geographic regions.
   (4) They occupy the same niche.

27. Base your answer to the following question on the chart below and on your knowledge of biology.

Species	Sequence of Four Amino Acids Found in the Same Part of the Hemoglobin Molecule of Species
human	Lys–Glu–His–Phe
horse	Arg–Lys–His–Lys
gorilla	Lys–Glu–His–Lys
chimpanzee	Lys–Glu–His–Phe
zebra	Arg–Lys–His–Arg

Which evolutionary tree best represents the information in the chart?

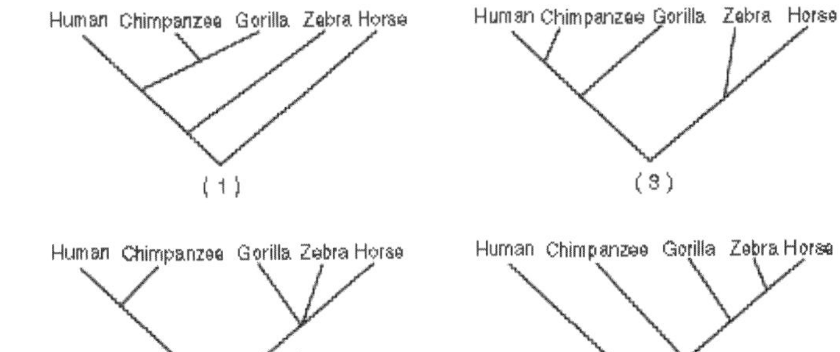

28. Which species is most likely to survive changing environmental conditions?
   (1) a species that has few variations
   (2) a species that reproduces sexually
   (3) a species that competes with similar species
   (4) a species that has a limited life span

29. Meiosis and fertilization are important for the survival of many species because these two processes result in
   (1) large numbers of gametes
   (2) increasingly complex multicellular organisms
   (3) cloning of superior offspring
   (4) genetic variability of offspring

30. The theory of biological evolution includes the concept that
   (1) species of organisms found on Earth today have adaptations not always found in earlier species
   (2) fossils are the remains of present-day species and were all formed at the same time
   (3) individuals may acquire physical characteristics after birth and pass these acquired characteristics on to their offspring
   (4) the smallest organisms are always eliminated by the larger organisms within the ecosystem

31. Which statement is most closely related to the modern theory of evolution?
(1) Characteristics that are acquired during life are passed to offspring by sexual reproduction.
(2) Evolution is the result of mutations and recombination, only.
(3) Organisms best adapted to a changed environment are more likely to reproduce and pass their genes to offspring.
(4) Asexual reproduction increases the survival of species.

32. The diagram below represents four different species of wild birds. Each species has feet with different structural adaptations.

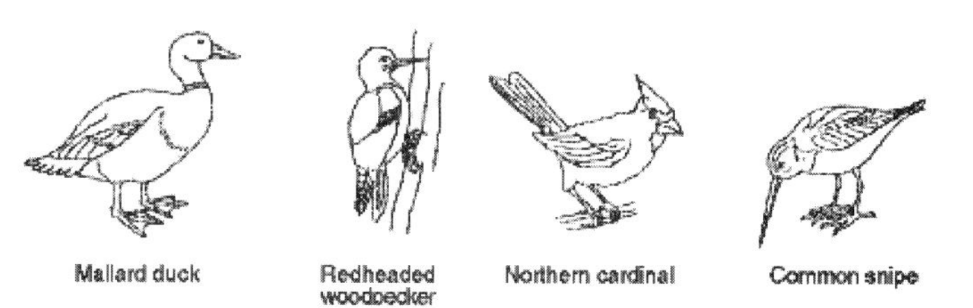

The development of these adaptations can best be explained by the concept of
(1) inheritance of resistance to diseases that affect all these species
(2) inheritance of characteristics acquired after the birds hatched from the egg
(3) natural selection
(4) selective breeding

33. Which statement best describes a current understanding of natural selection?
(1) Natural selection influences the frequency of an adaptation in a population.
(2) Natural selection has been discarded as an important concept in evolution.
(3) Changes in gene frequencies due to natural selection have little effect on the evolution of species.
(4) New mutations of genetic material are due to natural selection.

34. The teeth of carnivores are pointed and are good for puncturing and ripping flesh. The teeth of herbivores are flat and are good for grinding and chewing. Which statement best explains these observations?
(1) Herbivores have evolved from carnivores.
(2) Carnivores have evolved from herbivores.
(3) The two types of teeth most likely evolved as a result of natural selection.
(4) The two types of teeth most likely evolved as a result of the needs of an organism.

35. A hawk has a genetic trait that gives it much better eyesight than other hawks of the same species in the same area. Explain how this could lead to evolutionary change within this species of hawk over a long period of time. In your answer, be sure to include an explanation of:

- competition within the hawk population
- survival of various individuals in the population
- how the frequency of the better-eyesight trait would be expected to change over time within the population
- what would most likely happen to the hawks having the better-eyesight trait if they also had unusually weak wing muscles

36. Scientists compared fossil remains of a species that lived 5,000 years ago with members of the same species living today. Scientists concluded that this species had changed very little over the entire time period. Which statement best accounts for this lack of change?

(1) The environment changed significantly and those offspring without favorable characteristics died.
(2) The environment changed significantly, but the species had no natural enemies for a long period of time.
(3) The environment did not change significantly and those offspring expressing new characteristics survived their natural enemies.
(4) The environment did not change significantly and those offspring expressing new characteristics did not survive.

37. Which factor is ***least*** likely to contribute to an increase in the rate of evolution?
(1) presence of genetic variations in a population
(2) environmental selection of organisms best adapted to survive
(3) chromosomal recombinations
(4) a long period of environmental stability

38. Which population of organisms would be in greatest danger of becoming extinct?
(1) A population of organisms having few variations living in a stable environment.
(2) A population of organisms having few variations living in an unstable environment.
(3) A population of organisms having many variations living in a stable environment.
(4) A population of organisms having many variations living in an unstable environment.

39. The Florida panther, a member of the cat family, has a population of fewer than 100 individuals and has limited genetic variation. Which inference based on this information is valid?
(1) These animals will begin to evolve rapidly.
(2) Over time, these animals will become less likely to survive in a changing environment.
(3) These animals are easily able to adapt to the environment.
(4) Over time, these animals will become more likely to be resistant to disease.

Base your answers to the next four questions on the information below and on your knowledge of biology.

Color in peppered moths is controlled by genes. A light-colored variety and a dark-colored variety of a peppered moth species exist in nature. The moths often rest on tree trunks, and several different species of birds are predators of this moth. Before industrialization in England, the light-colored variety was much more abundant than the dark-colored variety and evidence indicates that many tree trunks at that time were covered with light-colored lichens. Later, industrialization developed and brought pollution which killed the lichens leaving the tree trunks covered with dark-colored soot. The results of a study made in England are shown below.

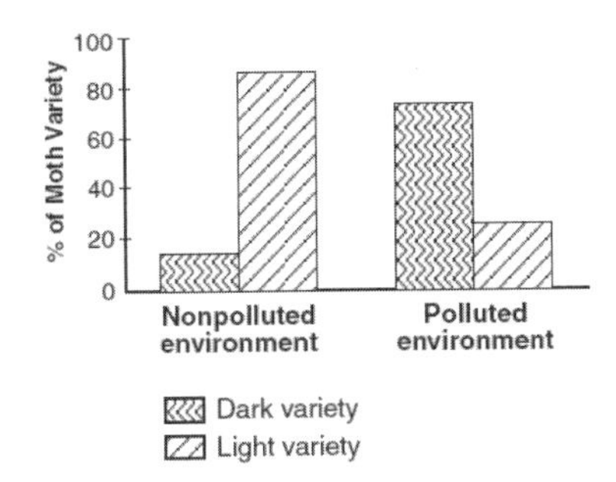

40. When a particular white moth lands on a white birch tree, its color has a high adaptive value. If the birch trees become covered with black soot, the white color of this particular moth in this environment would most likely
   (1) retain its adaptive value
   (2) increase in adaptive value
   (3) change to a more adaptive black color
   (4) decrease in adaptive value

41. Which conclusion can best be drawn from the information given?
   (1) The trait for dark coloration better suits the peppered moth for survival in non-polluted environments.
   (2) The trait for light coloration better suits the peppered moth for survival in polluted environments.
   (3) The variation of color in the peppered moth has no influence on survival of the moth.
   (4) A given trait may be a favorable adaptation in one environment, but not in another environment.

42. State ***one*** possible reason that a larger number of the dark-colored variety were present in the polluted environment.

43. The percentage of light-colored moths in the polluted environment was closest to
   (1) 16   (2) 24   (3) 42   (4) 76

44. Which concept is best illustrated in the flowchart below?

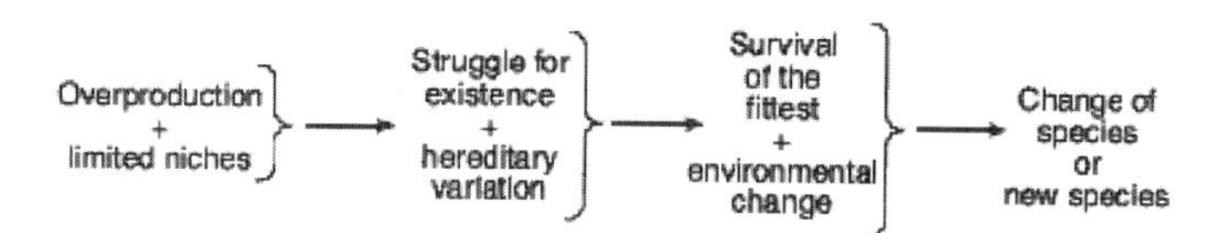

   (1) natural selection
   (2) genetic manipulation
   (3) dynamic equilibrium
   (4) material cycles

45. Certain insects resemble the bark of the trees on which they live. Which statement provides a possible biological explanation for this resemblance?
   (1) The insects needed camouflage so they developed protective coloration.
   (2) Natural selection played a role in the development of this protective coloration.
   (3) The lack of mutations resulted in the protective coloration.
   (4) The trees caused mutations in the insects that resulted in protective coloration.

46. The diagram below represents four different species of bacteria.

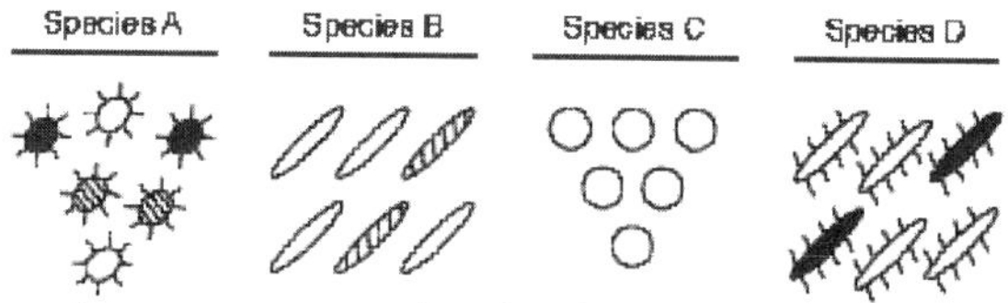

Which statement is correct concerning the chances of survival for these species if there is a change in the
environment?
   (1) Species ***A*** has the best chance of survival because it has the most genetic diversity.
   (2) Species ***C*** has the best chance of survival because it has no gene mutations.

(3) Neither species ***B*** nor species ***D*** will survive because they compete for the same resources.
(4) None of the species will survive because bacteria reproduce asexually.

Base your answers to the next three questions on the information below and on your knowledge of biology.

Honeybees have a very cooperative way of living. Scout bees find food, return to the hive, and do the "waggle dance" to communicate the location of the food source to other bees in the hive. The waggle, represented by the wavy line in the diagram below, indicates the direction of the food source, while the speed of the dance indicates the distance to the food. Different species of honeybees use the same basic dance pattern in slightly different ways as shown in the table below.

Number of Waggle Runs in 15 Seconds		Distance to Food (feet)
Giant Honeybee	Indian Honeybee	
10.6	10.5	50
9.6	8.3	200
6.7	4.4	1000
4.8	2.8	2000

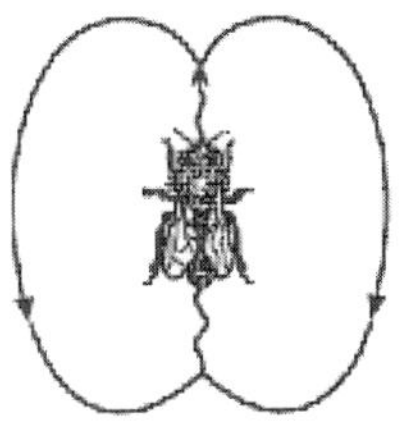

47. State the relationship between the distance to the food source and the number of waggle runs in 15 seconds.

48. Explain how waggle-dance behavior increases the reproductive success of the bees.

49. The number of waggle runs in 15 seconds for each of these species is most likely due to
(1) behavioral adaptation as a result of natural selection
(2) replacement of one species by another as a result of succession
(3) alterations in gene structure as a result of diet
(4) learned behaviors inherited as a result of asexual reproduction

50. The illustration below shows an insect resting on some green leaves.

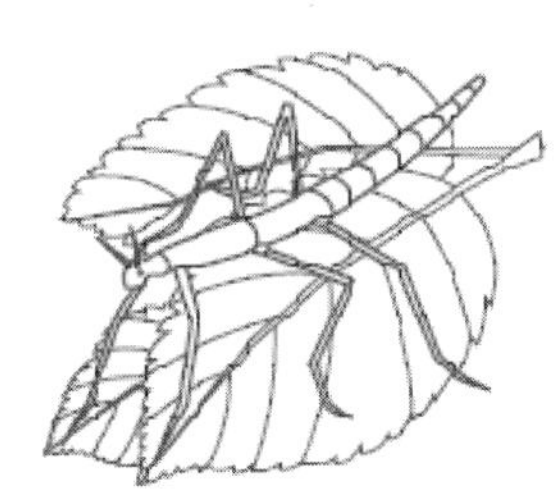

The size, shape, and green color of this insect are adaptations that would most likely help the insect to
(1) compete successfully with all birds
(2) make its own food
(3) hide from predators
(4) avoid toxic waste materials

Hint: *Predators are animals that kill and eat other animals.*

51. In 1993, there were only 30 panthers in Florida. They were all closely related and many had reproductive problems. To avoid extinction and restore health to the population, biologists introduced 8 female panthers from Texas. Today, there are more than 80 panthers in Florida and most individuals have healthy reproductive systems. The success of this program was most likely due to the fact that the introduced females

(1) produced more reproductive cells than the male panthers in Texas
(2) solved the reproductive problems of the species by asexual methods
(3) increased the genetic variability of the panther population in Florida
(4) mated only with panthers from Texas

52. Woolly mammoths became extinct thousands of years ago, while other species of mammals that existed at that time still exist today. These other species of mammals most likely exist today because, unlike the mammoths, they

(1) produced offspring that all had identical inheritable characteristics
(2) did not face a struggle for survival
(3) learned to migrate to new environments
(4) had certain inheritable traits that enabled them to survive

53. In an area of Indonesia where the ocean floor is littered with empty coconut shells, a species of octopus has been filmed "walking" on two of its eight tentacles. The remaining six tentacles are wrapped around its body. Scientists suspect that, with its tentacles arranged this way, the octopus resembles a rolling coconut. Local predators, including sharks, seem not to notice the octopus as often when it behaves in this manner. This unique method of locomotion has lasted over many generations due to

(1) competition between octopuses and their predators
(2) ecological succession in marine habitats
(3) the process of natural selection
(4) selective breeding of this octopus species

54. Limited resources contribute to evolutionary change in animals by increasing

(1) genetic variation within the population
(2) competition between members of the species
(3) the carrying capacity for the species
(4) the rate of photosynthesis in the population

55. The diagram below represents a process involved in reproduction in some organisms.

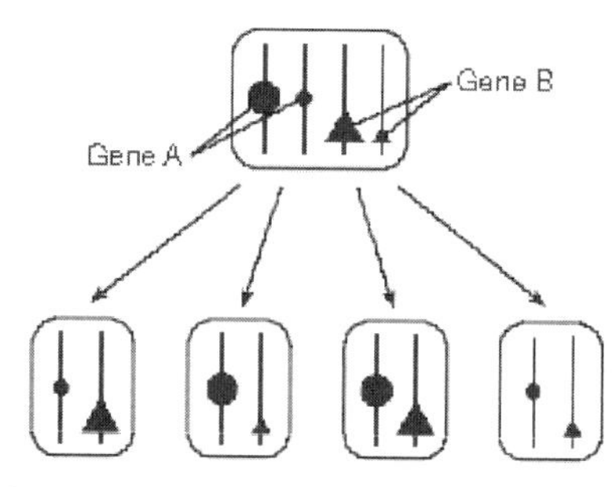

This process is considered a mechanism of evolution because

(1) mitosis produces new combinations of inheritable traits
(2) it increases the chances of DNA alterations in the parent
(3) it is a source of variation in the offspring produced
(4) meiosis prevents recombination of lethal mutations

Base your answers to the next two questions on the information below and on your knowledge of biology.

Two adaptations of the monarch butterfly that aid in its survival are the production of a certain chemical and a distinctive coloration that other animals can easily recognize. When a monarch butterfly is eaten, the presence of the chemical results in a bad taste to the predator. Although the viceroy butterfly does not contain the chemical that tastes bad to a predator, it does resemble the monarch in size, shape, and coloration.

56. Explain how the combination of this chemical and the distinctive coloration aid in the survival of the monarch butterfly.

57. How do the characteristics of the viceroy butterfly aid in its survival?

58. The diagram below shows variations in beak sizes and shapes for several birds on the Galapagos Islands.

Finch Diversity

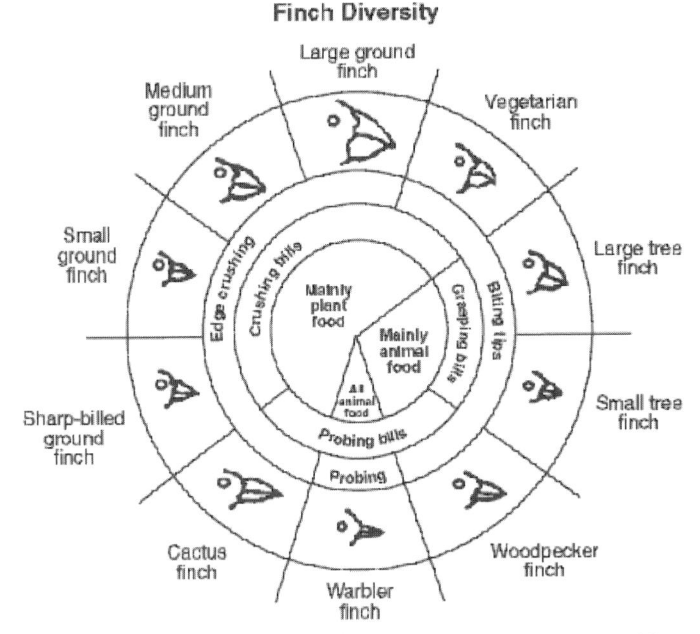

Using information provided in the chart, identify *two* birds that would most likely compete for food in times of food shortage and explain why they would compete.

59. Even though the finches on the various Galapagos Islands require different biotic and abiotic factors for their survival, these finches would most likely be grouped in the same
(1) species, but found in different habitats
(2) kingdom, but found in different ecological niches
(3) species and found in the same biosphere
(4) population, but found in different ecosystems

60. Galapagos finches evolved partly due to
(1) cloning and recombination
(2) migration and selective breeding
(3) mutation and asexual reproduction
(4) variation and competition

Base your answers to the next two questions on the information below and on your knowledge of biology. The diagram below represents the relationship between beak structure and food in several species of finches in the Galapagos Islands.

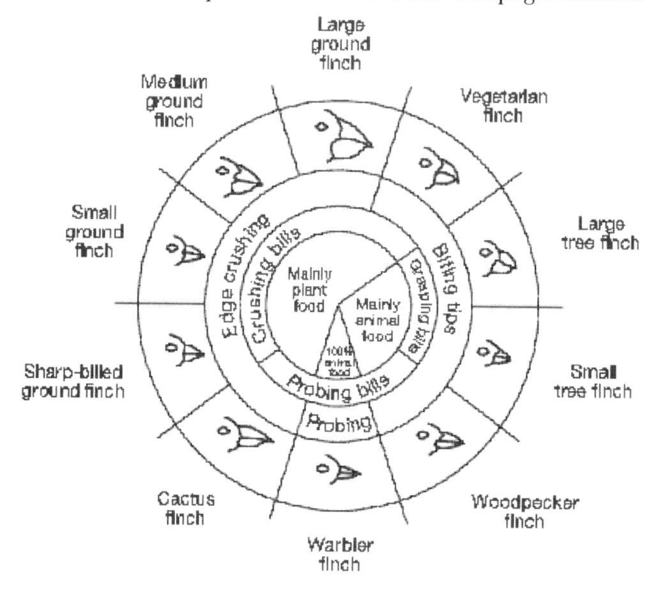

From: Galapagos: A Natural History Guide

Variations in Beaks of Galapagos Islands Finches

61. Which factor most directly influenced the evolution of the diverse types of beaks of these finches?
   (1) predation by humans
   (2) available food sources
   (3) oceanic storms
   (4) lack of available niches

62. State *one* reason why the large tree finch and the large ground finch are able to coexist on the same island.

63. Which process is correctly matched with its explanation?

	Process	Explanation
(1)	extinction	adaptive characteristics of a species are not adequate
(2)	natural selection	the most complex organisms survive
(3)	gene recombination	genes are copied as a part of mitosis
(4)	mutation	overproduction of offspring takes place within a certain population

Base your answers to the next two questions on the finch diversity chart below, which contains information concerning the finches found on the Galapagos Islands.

Finch Diversity

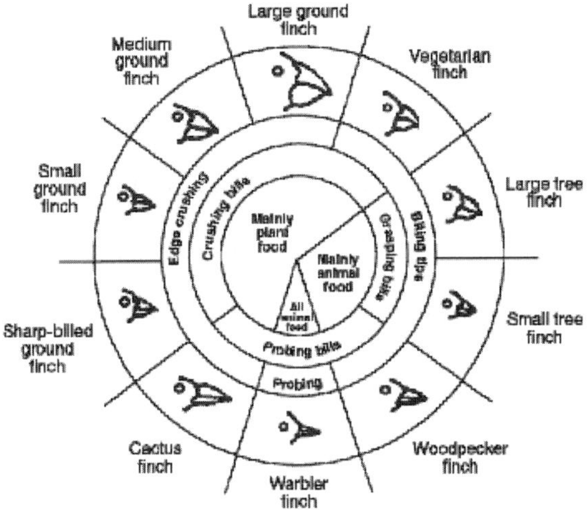

Large ground finch
Vegetarian finch
Large tree finch
Small tree finch
Woodpecker finch
Warbler finch
Cactus finch
Sharp-billed ground finch
Small ground finch
Medium ground finch
Edge crushing
Crushing bills
Grasping bills
Biting tips
Probing bills
Probing
Mainly plant food
Mainly animal food
All animal food

64. Identify ***one*** bird that would most likely compete for food with the large tree finch. Support your answer.

65. Identify ***one*** trait, other than beak characteristics, that would contribute to the survival of a finch species and state ***one*** way this trait contributes to the success of this species.

Base your answers to the next two questions on the information below and on your knowledge of biology.

In birds, the ability to crush and eat seeds is related to the size, shape, and thickness of the beak. Birds with larger, thicker beaks are better adapted to crush and open seeds that are larger.

One species of bird found in the Galapagos Islands is the medium ground finch. It is easier for most of the medium ground finches to pick up and crack open smaller seeds rather than larger seeds. When food is scarce, some of the birds have been observed eating larger seeds.

66. Describe ***one*** change in beak characteristics that would most likely occur in the medium ground finch population after many generations when an environmental change results in a permanent shortage of small seeds.

67. Explain this long-term change in beak characteristics using the concepts of:

- competition
- survival of the fittest
- inheritance

Base your answers to the next three questions on the information below and on your knowledge of biology.

In the ***Beaks of Finches*** laboratory activity, students were each assigned a tool to use to pick up seeds. In round one, students acting as birds used their assigned tools to pick up small seeds from their own large dishes (the environment) and place them in smaller dishes (their stomachs). The seeds collected by each student were counted. Some students were able to collect many seeds, while others collected just a few.

In round two, students again used their assigned tools to collect seeds. This time several students were picking up seeds from the same dish of seeds.

68. Explain how this laboratory activity illustrates the process of natural selection.

69. One factor that influences the evolution of a species that was not part of this laboratory activity is
    (1) struggle for survival
    (2) variation
    (3) competition
    (4) overproduction

70. Identify ***one*** trait, other than beak characteristics, that could contribute to the ability of a finch to feed successfully.

71. The cactus finch, warbler finch, and woodpecker finch all live on one island. Based on the information in the diagram below, which one of these finches is least likely to compete with the other two for food? Support your answer with an explanation.

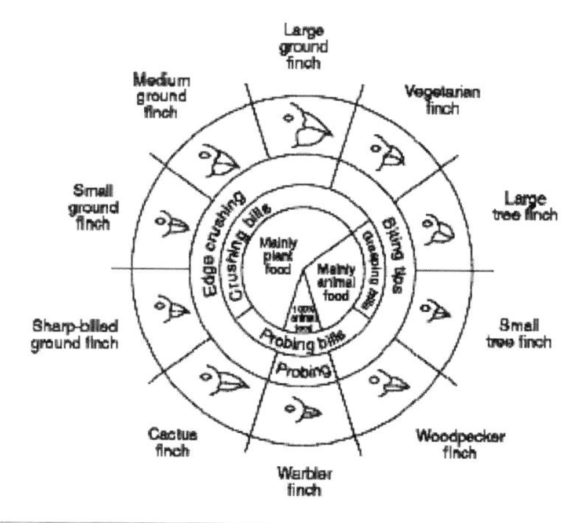

Base your answers to the next two questions on the information below and on your knowledge of biology.

Evolutionary changes have been observed in beak size in a population of medium ground finches in the Galapagos Islands. Given a choice of small and large seeds, the medium ground finch eats mostly small seeds, which are easier to crush. However, during dry years, all seeds are in short supply. Small seeds are quickly consumed, so the birds are left with a diet of large seeds. Studies have shown that this change in diet may be related to an increase in the average size of the beak of the medium ground finch.

72. The most likely explanation for the increase in average beak size of the medium ground finch is that the
    (1) trait is inherited and birds with larger beaks have greater reproductive success
    (2) birds acquired larger beaks due to the added exercise of feeding on large seeds
    (3) birds interbred with a larger-beaked species and passed on the trait
    (4) lack of small seeds caused a mutation which resulted in a larger beak

73. In exceptionally dry years, what most likely happens in a population of medium ground finches?
    (1) There is increased cooperation between the birds.
    (2) Birds with large beaks prey on birds with small beaks.

(3) The finches develop parasitic relationships with mammals.
(4) There is increased competition for a limited number of small seeds.

74. Beak structures differ between individuals of one species of bird. These differences most likely indicate
(1) the presence of a variety of food sources
(2) a reduced rate of reproduction
(3) a large supply of one kind of food
(4) an abundance of predators

75. Researchers discovered four different species of finches on one of the Galapagos Islands. DNA analysis showed that these four species, shown in the illustration below, are closely related even though they vary in beak shape and size. It is thought that they share a common ancestor.

Which factor most likely influenced these differences in beak size and shape?
(1) Birds with poorly adapted beaks changed their beaks to get food.
(2) Birds with yellow beaks were able to hide from predators.
(3) Birds with successful beak adaptations obtained food and survived to have offspring.
(4) Birds with large, sharp beaks become dominant.

76. Species of finches are represented in the diagram below.

Finch Diversity

Large ground finch
Vegetarian finch
Large tree finch
Small tree finch
Woodpecker finch
Warbler finch
Cactus finch
Sharp-billed ground finch
Small ground finch
Medium ground finch
Edge crushing
Crushing bills
Grasping bills
Biting tips
Probing bills
Probing
Mainly plant food
Mainly animal food
All animal food

State the name of ***one*** species of finch from the diagram that is most likely to compete with the small tree finch if they lived on the same island. Support your answer with an explanation.

Base your answers to the next two questions on the diagram below and on your knowledge of biology.

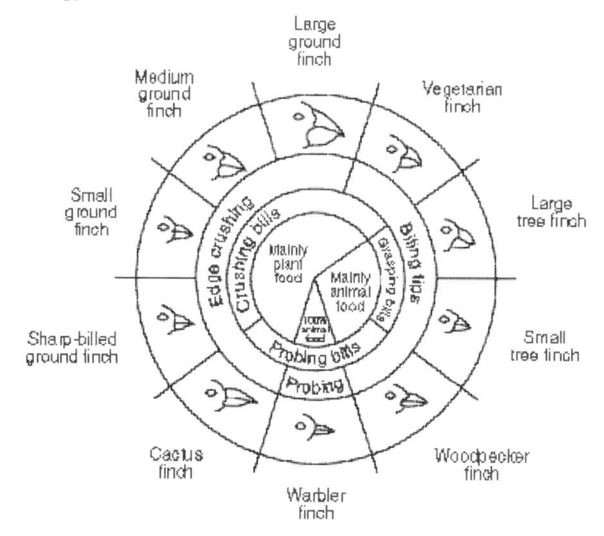

Variations in Beaks of Galapagos Islands Finches

77. The only finch that is completely carnivorous has a beak adapted for
    (1) probing, only
    (2) probing and edge crushing
    (3) probing and biting
    (4) biting and edge crushing

78. Which two finches would compete the least for food?
    (1) small ground finch and large ground finch
    (2) large ground finch and sharp-billed ground finch
    (3) small tree finch and medium ground finch
    (4) vegetarian finch and small ground finch

Base your answers to the next three questions on the information below and on your knowledge of biology.

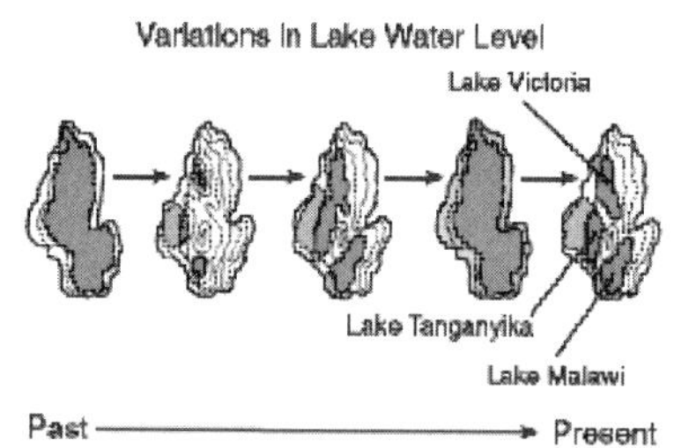

The three great lakes in Africa (Victoria, Tanganyika, and Malawi) contain a greater number of fish species than any other lakes in the world. Lake Malawi alone has 200 species of cichlid fish. The diversity of cichlid species in these African lakes could have been caused by changes in water level over thousands of years.

According to one hypothesis, at one time the three lakes were connected as one large lake and all the cichlids could interbreed. When the water level fell, groups of cichlids were isolated in smaller lakes as shown in the diagram. Over time, the groups of cichlids developed genetic differences. When the water levels rose again, the isolated populations were brought back into contact. Due to significant genetic differences, these populations were unable to interbreed. Variations in water level over thousands of years resulted in today's diversity of cichlid species.

79. Which discovery would support this explanation of cichlid diversity?
    (1) The water level changed little over time.
    (2) The local conditions in each of the small lakes were very different.
    (3) Differences between cichlid species are small and interbreeding is possible.
    (4) Once formed, the lakes remained isolated from each other.

80. As the water level of the lakes changed, many species of cichlids survived while others became extinct. State why some species survived while others became extinct.

81. Each cichlid population is genetically different from the other cichlid populations. State ***one*** reason for these genetic differences.

82. Mutations are often referred to as the "raw materials" of evolution. State ***one*** reason that mutations are often referred to as the "raw materials" of evolution.

83. In a group of mushrooms exposed to a poisonous chemical, only a few of the mushrooms survived. The best explanation for the resistance of the surviving mushrooms is that the resistance
    (1) was transmitted to the mushrooms from the poisonous chemical
    (2) resulted from the presence of mutations in the mushrooms
    (3) was transferred through the food web to the mushrooms
    (4) developed in response to the poisonous chemical

84. Thousands of years ago, giraffes with short necks were common within giraffe populations. Nearly all giraffe populations today have long necks. This difference could be due to
    (1) giraffes stretching their necks to keep their heads out of reach of predators
    (2) giraffes stretching their necks so they could reach food higher in the trees
    (3) a mutation in genetic material controlling neck size occurring in some skin cells of a giraffe
    (4) a mutation in genetic material controlling neck size occurring in the reproductive cells of a giraffe

85. Base your answer to the following question on the information below and on your knowledge of biology.

    Sickle-cell anemia is an inherited disease that occurs mainly in people from parts of Africa where malaria is common. It is caused by a gene mutation that may be harmful or beneficial.

    A person with two mutant genes has sickle-cell disease. The hemoglobin of a person with sickle-cell disease twists red blood cells into a crescent shape. These blood cells cannot circulate normally. Symptoms of the disease include bleeding and pain in bones and muscles. People with sickle-cell disease suffer terribly in childhood and, until modern medicine offered treatment, most of them died before reproducing. An individual who has one mutant gene is protected from malaria because the gene changes the hemoglobin structure in a way that speeds removal of malaria-infected cells from circulation. A person with two normal genes has perfectly good red blood cells, but lacks resistance to malaria.

    Define the term mutation.

86. Which statement about having one sickle-cell gene is correct?
    (1) It is fatal to anyone who inherits the gene.
    (2) It is beneficial to anyone who inherits the gene.

(3) It is beneficial in certain environments.
(4) It is beneficial or harmful depending on whether it is common or rare.

87. Explain why the percentage of the population with one mutant sickle-cell gene is higher in areas where malaria is common.

88. Which ***two*** processes result in variations that commonly influence the evolution of sexually reproducing species?
(1) mutation and genetic recombination
(2) mitosis and natural selection
(3) extinction and gene replacement
(4) environmental selection and selective breeding

89. A mutation occurs in the liver cells of a certain field mouse. Which statement concerning the spread of this mutation through the mouse population is correct?
(1) It will spread because it is beneficial.
(2) It will spread because it is a dominant gene.
(3) It will not spread because it is not in a gamete.
(4) It will not spread because it is a recessive gene.

90. Base your answer to the following question on the diagram below and on your knowledge of biology.

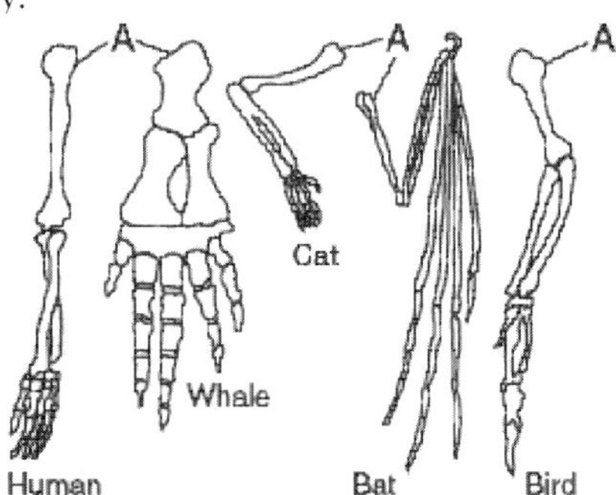

The similarities of the bones labeled ***A*** provide evidence that
(1) the organisms may have evolved from a common ancestor
(2) all species have one kind of bone structure
(3) the cells of the bones contain the same type of mutations
(4) all structural characteristics are the same in animals

91. The bones in the forelimbs of three mammals are shown below.

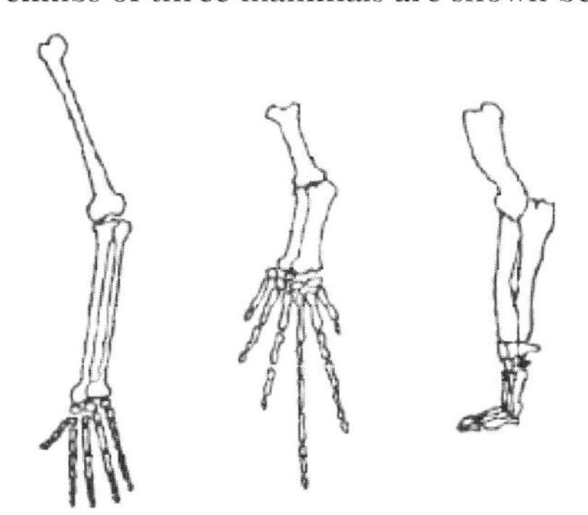

For these mammals, the number, position, and shape of the bones most likely indicates that they may have
(1) developed in a common environment
(2) developed from the same earlier species
(3) identical genetic makeup
(4) identical methods of obtaining food

92. The diagrams below show the bones in the forelimbs of three different organisms.

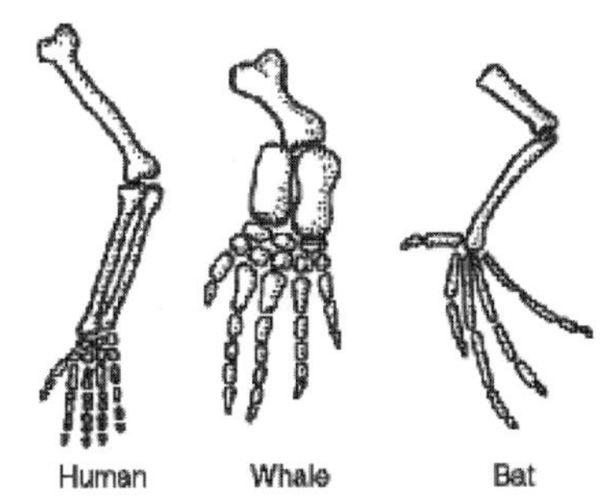

Differences in the bone arrangements support the hypothesis that these organisms
(1) are members of the same species
(2) may have descended from the same ancestor
(3) have adaptations to survive in different environments
(4) all contain the same genetic information

93. The presence of some similar structures in all vertebrates suggests that these vertebrates
(1) all develop at the same rate
(2) evolved from different animals that appeared on Earth at the same time
(3) all develop internally and rely on nutrients supplied by the mother
(4) may have an evolutionary relationship

94. A species in a changing environment would have the best chance of survival as a result of a mutation that has a
(1) high adaptive value and occurs in its skin cells
(2) low adaptive value and occurs in its skin cells
(3) high adaptive value and occurs in its gametes
(4) low adaptive value and occurs in its gametes

95. Two cultures, each containing a different species of bacteria, were exposed to the same antibiotic. Explain how, after exposure to this antibiotic, the population of one species of bacteria could increase while the population of the other species of bacteria decreased or was eliminated.

96. A new chemical was discovered and introduced into a culture containing one species of bacteria. Within a day, most of the bacteria were dead, but a few remained alive. Which statement best explains why some of the bacteria survived?
(1) They had a genetic variation that gave them resistance to the chemical.
(2) They were exposed to the chemical long enough to develop a resistance to it.
(3) They mutated and became a different species after exposure to the chemical.
(4) They absorbed the chemical and broke it down in their digestive systems.

Base your answers to the next two questions on the information below and on your knowledge of biology.

A small village was heavily infested with mosquitoes. The village was sprayed weekly with an insecticide for a period of several months. The results of daily counts of the mosquito population are shown in the graph below.

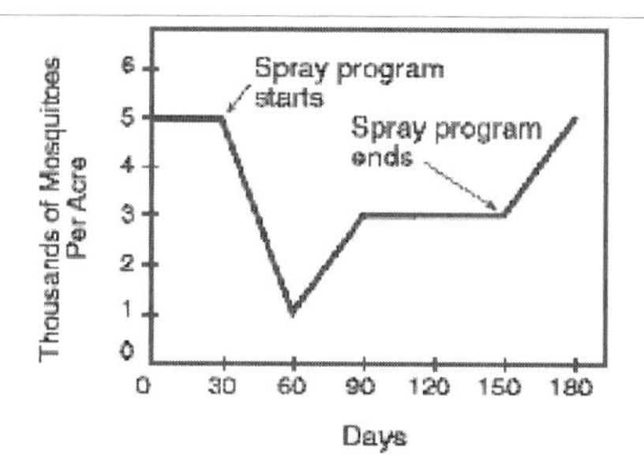

97. Which statement best explains why some mosquitoes survived after the first spraying?
   (1) Some mosquitoes were adapted to the climatic change that occurred over the several-month period of spraying.
   (2) All of the mosquitoes contained DNA unique to the species.
   (3) The spraying of the insecticide represented a change in the environment to which all adult mosquitoes were adapted.
   (4) A natural variation existed within the mosquito population.

98. Which statement best explains the decreased effectiveness of the insecticide?
   (1) The insecticide caused mutations that resulted in immunity in the mosquito.
   (2) Mosquitoes resistant to the insecticide lived and produced offspring.
   (3) The insecticide reacted chemically with the DNA of the mosquitoes and was destroyed.
   (4) All of the mosquitoes produced antibodies that activated the insecticide.

99. The diagram below shows the effect of spraying a pesticide on a population of insects over three generations.

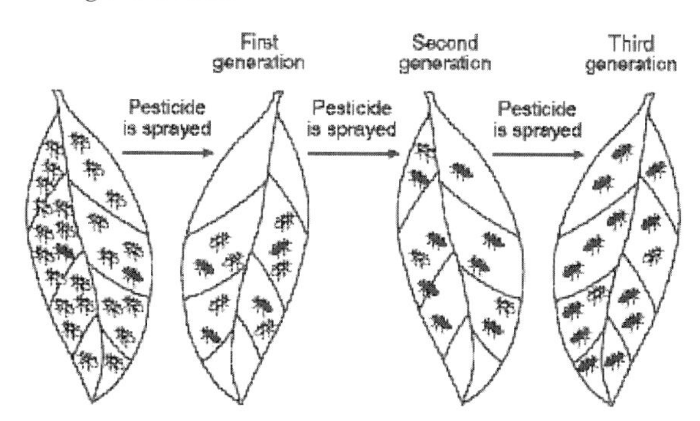

Which concept is represented in the diagram?
   (1) survival of the fittest
   (2) dynamic equilibrium
   (3) succession
   (4) extinction

100. Growers of fruit trees have always had problems with insects. Insects can cause visible damage to fruits, making them less appealing to consumers. As a result of this damage, much of the fruit cannot be sold. Insecticides have been useful for controlling these insects, but, in recent years, some insecticides have been much less effective. In some cases, insecticides do nothing to stop the insect attacks.

Provide a biological explanation for this loss of effectiveness of the insecticides. In your answer, be sure to:

- identify the original event that resulted in the evolution of insecticide resistance in some insects
- explain why the percentage of resistant insects in the population has

increased

- describe one alternative form of insect control, other than using a different insecticide, that fruit growers could use to protect their crops from insect attack

101. An insect pest known as the medfly significantly reduced the orange crop in California. Pesticides were used to control the medfly. Using the concept of natural selection, explain how the continued use of a certain pesticide may become ineffective in controlling this fly. Your answer must include the concepts of:

- variation
- adaptive value of a variation (adaptation)
- survival
- reproduction

102. Which population of organisms would be in greatest danger of becoming extinct?
   (1) A population of organisms having few variations living in a stable environment.
   (2) A population of organisms having few variations living in an unstable environment.
   (3) A population of organisms having many variations living in a stable environment.
   (4) A population of organisms having many variations living in an unstable environment.

103. Extinction of a species could result from
   (1) evolution of a type of behavior that produces greater reproductive success
   (2) synthesis of a hormone that controls cellular communication
   (3) limited genetic variability in the species
   (4) fewer unfavorable mutations in the species

104. A certain species has little genetic variation. The rapid extinction of this species would most likely result from the effect of
   (1) successful cloning
   (2) gene manipulation
   (3) environmental change
   (4) genetic recombination

105. In 1993, there were only 30 panthers in Florida. They were all closely related and many had reproductive problems. To avoid extinction and restore health to the population, biologists introduced 8 female panthers from Texas. Today, there are more than 80 panthers in Florida and most individuals have healthy reproductive systems. The success of this program was most likely due to the fact that the introduced females
   (1) produced more reproductive cells than the male panthers in Texas
   (2) solved the reproductive problems of the species by asexual methods
   (3) increased the genetic variability of the panther population in Florida
   (4) mated only with panthers from Texas

106. Woolly mammoths became extinct thousands of years ago, while other species of mammals that existed at that time still exist today. These other species of mammals most likely exist today because, unlike the mammoths, they
   (1) produced offspring that all had identical inheritable characteristics
   (2) did not face a struggle for survival
   (3) learned to migrate to new environments
   (4) had certain inheritable traits that enabled them to survive

107. Which statement describing a cause of extinction includes the other three?
(1) Members of the extinct species were unable to compete for food.
(2) Members of the extinct species were unable to conceal their presence by camouflage.
(3) Members of the extinct species lacked adaptations essential for survival.
(4) Members of the extinct species were too slow to escape from predators.

108. A certain plant species, found only in one particular stream valley in the world, has a very shallow root system. An earthquake causes the stream to change its course so that the valley in which the plant species lives becomes very dry. As a result, the species dies out completely. The effect of this change on this plant species is known as
(1) evolution
(2) extinction
(3) mutation
(4) succession

109. Base your answers to the next question on the diagram below, which represents possible relationships between animals in the family tree of the modern horse, and on your knowledge of biology.

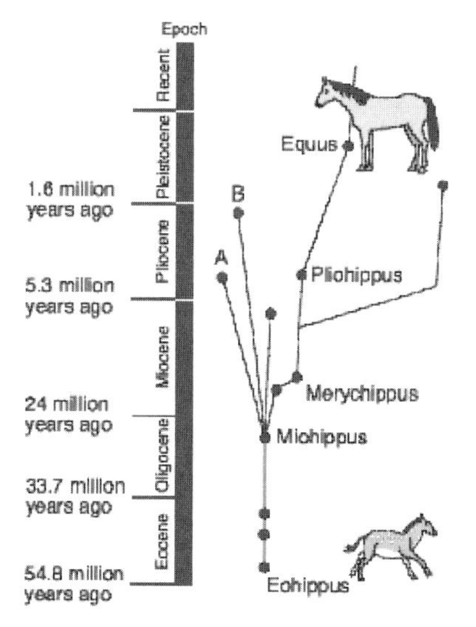

One possible conclusion that can be drawn regarding ancestral horses ***A*** and ***B*** is that
(1) ***A*** was better adapted to changes that occurred during the Pliocene Epoch than was ***B***
(2) the areas that ***B*** migrated to contained fewer varieties of producers than did the areas that ***A*** migrated to
(3) competition between ***A*** and ***B*** led to the extinction of ***Pliohippus***
(4) the adaptive characteristics present in both ***A*** and ***B*** were insufficient for survival

110. When is extinction of a species most likely to occur?
(1) when environmental conditions remain the same and the proportion of individuals within the species that lack adaptive traits increases
(2) when environmental conditions remain the same and the proportion of individuals within the species that possess adaptive traits increases
(3) when environmental conditions change and the adaptive traits of the species favor the survival and reproduction of some of its members
(4) when environmental conditions change and the members of the species lack adaptive traits to survive and reproduce

111. Species of bacteria can evolve more quickly than species of mammals because

bacteria have

(1) less competition
(2) more chromosomes
(3) lower mutation rates
(4) higher rates of reproduction

CHAPTER 6: ECOLOGY

Ecology is the study of how **organisms** (living things) **interact** with the **living organisms** and **nonliving things** that surround them in their environment. Example: **fish interact** with smaller **fish** and **worms** (fish eat smaller fish and worms) and fish take in **oxygen** in their environment.

Environment

Environment is everything that **surrounds an organism** (see figure below). Example: everything around the deer is its environment; everything around people is the people's environment, and everything around you (examples plants, animals, air, water) is your environment.

Environment

Ecosystem

Ecosystem is part of the environment. Examples of an ecosystem: forest, lake, pond, or desert. An **ecosystem** is an area where the **living things interact** with the **living** and **nonliving things**.

a. In a forest, humans eat deer, and deer eat grass (living things); they all drink water and take in oxygen (nonliving). A forest is an ecosystem where living things interact with living and nonliving things.

b. Look at the lake ecosystem below. Ducks, frogs, and fish eat the plants (living); they all drink water and take in oxygen (nonliving). The lake is an ecosystem.

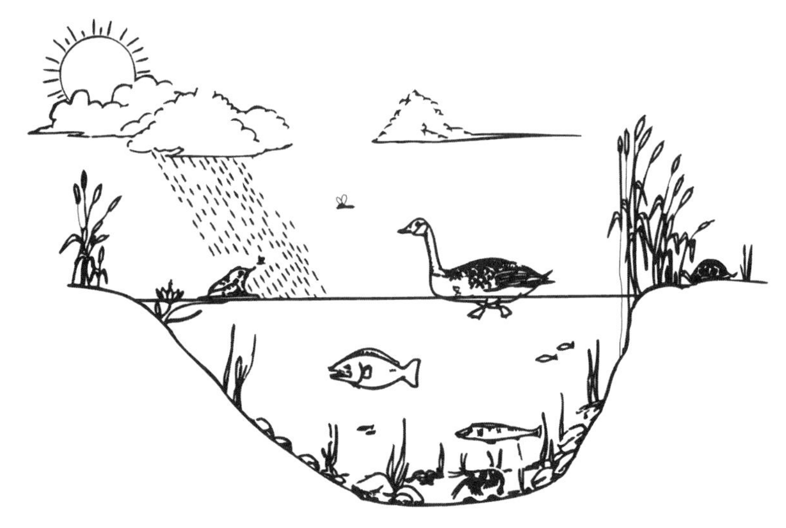

Lake Ecosystem

c. tropical rain forest (explained below).
d. desert (explained below)

Climate has the biggest effect on the type of ecosystem. If it is a hot, rainy climate, there will be a tropical rain forest with big trees with broad leaves. If the climate is very dry, like a desert, plants such as cactus or mesquite (which require very little water) will survive. A tropical rain forest is an ecosystem, where living and nonliving things interact; a desert is also an ecosystem, where living and nonliving things interact.

The type of ecosystem (examples forest, lake) and the number of living things in the ecosystem depend also on:

Abiotic factors: **nonliving** things in an ecosystem (examples: water, oxygen, sunlight, temperature, minerals).

Biotic factors: **living** things in an ecosystem (examples: geese, frogs, fish). Living things (biotic factors) can be divided into groups: producers (plants which make their own food), herbivores (animals that eat plants, example cows), carnivores (animals that eat animals, example lions, wolves), and decomposers (organisms that eat wastes and dead organisms).

Population is all members of **one species** that live **in one area.** Examples of a population are: all deer (a species) in a forest, or all pigeons (a species) in New York City, or all salmon (a species of fish) in a lake.

Community is **all populations (all species)** in a given **area.**
Examples:

1. **deer** (a species), **squirrels** (a species), and **oak trees** (a species) **in** a **forest** make up a community;
2. salmon (a species of fish), frogs (a species), and geese (a species) in a lake make up a community.

Habitat is where the **animal lives.** Example: Fish live in a pond.
Biosphere is the part of the **Earth** where **life exists.**

In the lake ecosystem figure above there is:

POPULATION	COMMUNITY	ECOSYSTEM
one population of geese one population of salmon one population of frogs	geese, salmon, and frogs in the lake together make up one community	geese, salmon, and frogs in the lake interact with each other and with water and oxygen

The diagram at right shows the relationship between humans (H) and an ecosystem (E). The large circle (E) represents an ecosystem (examples forest or desert). H (humans) are part of an ecosystem (humans can live in a forest or desert), therefore H (humans) is the small circle H inside the large circle E.

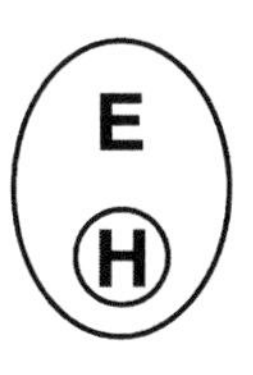

For an **ecosystem** (examples: lake, ocean, forest, or pond) to **remain stable** (self-sustaining or constant) it must have:

a. there must be a **constant supply of energy**; energy comes from the **sun**. The sun's energy is **not recycled** (the energy that the plant took in cannot be used over and over again). The sun's energy must be continuously added to the ecosystem (example lake or forest or pond). You learned in photosynthesis **plants use the sun's energy to make glucose** (a simple sugar, has stored energy, a food). Plants are producers (autotrophs) because they make their own food.

b. **animals eat plants or animals** and **get energy from plants or animals.** To explain: the **animals** (example deer) **eat** the **plants** and **get** the **energy from** the **plants**, which originally got the sun's energy (energy from the sun). Other **animals** (example wolves) **eat** the **animals** (example deer) and **get** the **energy from** the **animals** (example deer) which got the energy from the plants. Some other animals, including humans, eat both plants and animals, and get energy from both.

In short, the sun must keep giving off energy (adding energy to the ecosystem), therefore plants can keep taking in the sun's energy to make glucose, a simple sugar (has stored energy). The animals eat the plants (which have energy); energy flows (goes) from the sun to the plants to the animals. The animals only get a small percentage of the plants' energy because a lot of the plants' energy is given off as heat.

c. **recycling** of **materials** (molecules) between organisms (living things) and their environment.

When organisms (example plants and animals) die, their bodies decompose (break down) and materials such as carbon dioxide and nitrogen compounds go back into the environment (recycled) to be used again by living things. **Decomposers** (example **some bacteria** and **fungi**) eat wastes and dead organisms and **decompose (break down)** the **wastes** and **dead organisms** (dead material) to simple materials like carbon dioxide and nitrogen compounds (ammonia which has nitrogen in it), returning **(recycling) materials** (carbon dioxide and nitrogen compounds) **to** the **air** and **soil**, to the environment.

Balanced aquarium

A balanced aquarium is a stable (self-sustaining) ecosystem.

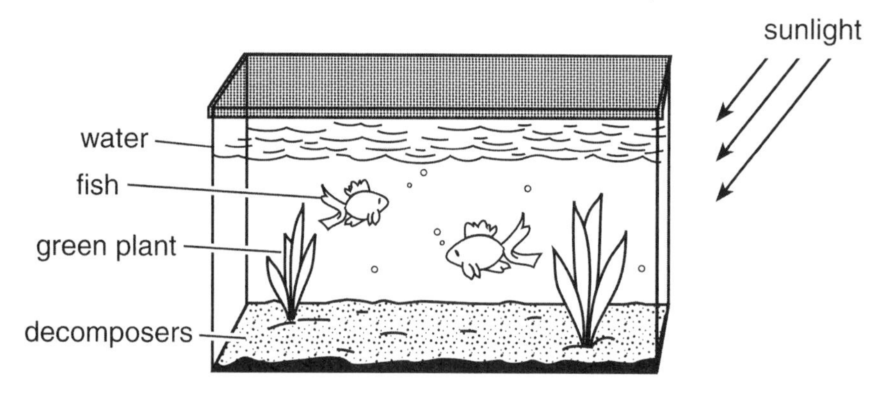

Balanced Aquarium

1. The sun provides energy for photosynthesis. Plants take in the sun's energy to make glucose (simple sugar, has stored energy, food).
2. The fish eat the plants, which have glucose (stores energy).
3. The fish carry on respiration, taking in oxygen and giving off carbon dioxide.
4. The carbon dioxide is taken in by the plants. The plants in photosynthesis take in carbon dioxide, give off oxygen and produce glucose (simple sugar, has stored energy). The animals take in oxygen and give off carbon dioxide, which the plants take in.
5. The animals again eat the plants and get energy.
6. The decomposers break down dead organisms or wastes, producing carbon dioxide and nitrogen compounds, which are used by the plants. Carbon dioxide is used for photosynthesis (you learned this in #4).

The aquarium is a stable ecosystem that can survive by itself.

Practice Questions and Solutions

Question: In an ocean, the growth and survival of seaweed, small fish, and sharks depends on abiotic factors such as
(1) sunlight, temperature, and minerals
(2) sunlight, pH, and type of seaweed
(3) number of decomposers, carbon dioxide, and nitrogen
(4) number of herbivores, carbon and food

Solution: Abiotic factors are nonliving things in an environment, such as sunlight, temperature, minerals, water, and oxygen. Answer *1*

Wrong choices:

Choice *2:* Seaweed (which is a plant) is living (biotic).
Choice *3:* Decomposers are living (biotic).
Choice *4:* Herbivores are living (biotic).

Question: Untreated organic wastes were accidentally discharged into a river from a sewage treatment plant. State why this accident would be expected to benefit the decomposers in the river below the sewage plant.

Solution: Decomposers eat **wastes (food** for decomposers). The more wastes put into the river, the more food there is for the decomposers.

Now Do Homework Questions #1-11, pages 33-34.

Limiting Factors

1. **Resources-** There is a **limited (finite) amount** of resources, such as **food** (example grass, nuts, berries, seeds, etc.)**, water, minerals**, **oxygen** (in water) and **energy** in any ecosystem (example lake, forest, field, desert). The amount of energy from the sun is limited when large trees block sunlight from reaching smaller plants (examples grass, shrubs) and some cannot survive because of lack of sunlight; the amount of sunlight decreases (becomes less) when the sunlight goes deeper in the ocean, and some plants cannot survive because of lack of sunlight. When a population gets large, there might not be enough resources (examples food, water, or oxygen) for the entire population, and some living things might die. The amount of food, water, or oxygen becomes the **limiting factor** that limits the number of living things that can survive.

2. **Predators-** Predators (such as wolves or coyotes) eat prey (such as deer, moose, beaver). Predators (example wolves or coyotes) are a limiting factor because they eat deer and therefore limit the number of deer that can survive. The more wolves there are, the fewer deer survive.

3. **Disease-** Disease is a limiting factor because it limits the number of living things that can survive. Some living things die from disease.

In short, **limiting factors limit** the **size** of the **population**.

Limiting factors can also **control** the **type of organisms** that live there.

Example 1: **Salt** in the ocean limits the type of fish that can live there. Fresh water fish would die in salt water.

Example 2: **lack of water** in the desert limits the type of plants that can grow there. Cactus plants need very little water and can live in the desert. Apple trees need more water and cannot live there.

Competition

Competition is when **organisms fight** for the **things** they **need to live**. To explain: Competition is when organisms of the same or different species live in the same place (area, habitat, ecosystem) and use or fight for the same limited resources (food, water, or oxygen).

Examples of competition:

1. Wolves eat deer. When the number of wolves increases, there might not be enough deer to eat. The wolves compete for the deer, and some wolves starve and die.
2. Competition for sunlight: Large trees and tiny shrubs or grass compete for sunlight. Large trees take in sunlight and block or prevent the small shrubs or grass from getting sunlight, then the shrubs or grass can die from lack of sunlight.

Carrying Capacity

Carrying capacity of an area is the **number** of **organisms** (living things) of a species that **can live there** or **survive.** How many organisms of a species can live there (the carrying capacity) depends on the amount of food, oxygen, water, minerals, and energy, and the work of decomposers. Decomposers eat wastes and dead organisms and decompose (break down) the wastes and dead material to simple substances like carbon dioxide and ammonia, which has nitrogen, returning (recycling) materials to the air and soil. These recycled materials help more plants to grow and provide food for more animals. Carrying capacity (number of organisms of a species that can live in an area) is also limited by the number of predators (animals that kill and eat other animals) and by disease. The **more predators** and the **more disease** causes the number of living organisms (living things) to decrease (become less), which means the **carrying capacity** (**the number of living things that can survive in the area**) **becomes less**. Example: In an ecosystem, the carrying capacity for deer depends on how much grass there is (food), the number of wolves that eat them (predators) and disease.

Connection between limited resources and carrying capacity. The limited amount of resources (example food, water, minerals, decomposers) limits the number of organisms of a species (examples deer or squirrels or pigeons) that can survive. The carrying capacity is the most organisms of a species that can survive, or the most organisms that the habitat (place where it lives) can support, with a limited number of resources.

The graph below represents the growth of a population of flies in a jar.

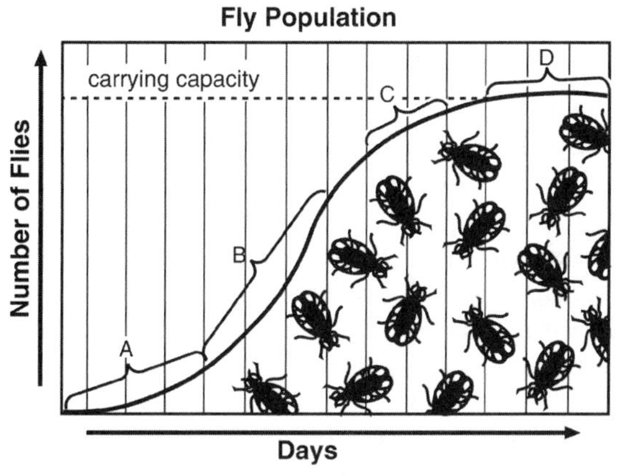

Growth of a Population (Reached carrying capacity at D)

The dashed line or you can say the letter D on the graph, represents the carrying capacity (the most fruit flies that can survive in that area, the jar).

The population increased from the beginning until it reached its carrying capacity (from A until D). If after point D (see point D at the end of the graph) many more organisms are produced above the carrying capacity (dashed line), population increases, they will not have enough food, etc. to live and therefore some will die (population decreases). Now the cycle of increasing population and decreasing population will begin again. Since there is the same amount of food and some organisms died out, now more can survive and the population now increases. Now there is not enough food for the new population and some will die (the population decreases), but generally the population stays about the same. This will be shown in the graph below.

Look at the graph below. D (the dashed line) in the previous graph and D in the graph below represents the carrying capacity. After it reaches the carrying capacity at D, the population increases (1), then decreases (2), but the overall population (from #1 to #6 on the graph) stays about the same (dynamic equilibrium), explained below.

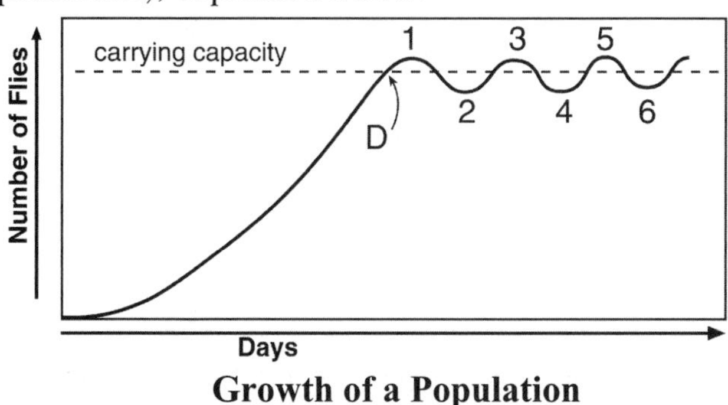

Growth of a Population

The dashed line (or letter D) represents the carrying capacity (the most organisms in the population that can survive in this ecosystem with the limited number of resources).

#1 shows the population increases slightly above the dashed line (the carrying capacity). Note: some of the organisms will die because it is above the dashed line (carrying capacity) and there is not enough of limited resources (example not enough food) for all to survive.

#2 shows the population decreased slightly. Note: Since some of the organisms died, then there will be enough food for more organisms to live (increase in number of organisms).

The cycle repeats itself. #3 increases in number, #4 decreases in number, #5 increases in number, and #6 decreases in number, but as you can see from the graph, generally the population stays about the same.

When a population keeps increasing slightly above the carrying capacity (see #1), drops slightly below the carrying capacity (see #2), increases above the carrying capacity (see #3), decreases below the carrying capacity (see #4), increases again (see #5), then decreases again (see #6) because of limiting factors, such as amount of available food, but the population stays about the same, this is called dynamic equilibrium (going up a little and down a little, but population stays about the same).

Question: A population of gray squirrels lived in the trees surrounding four houses in a city. The houses and trees were removed, and a tall office building was constructed in their place. Some of the squirrels were able to survive by relocating to the trees in a park nearby.

Question 1: State *one* specific way the relocated squirrels would most likely interact with a gray squirrel population that has lived in the park for many years.

Question 2: State *one* specific way the relocated squirrels will change an abiotic factor in the park ecosystem.

Question 3: State *one* specific natural factor in the park ecosystem that will limit the growth of the squirrel population and support your answer.

Solution 1: Relocated squirrels compete with park squirrels for food or space or mates. Or, relocated squirrels can mate with park squirrels.

Solution 2: Relocated squirrels will use water. Or, relocated squirrels will take up space. Or, relocated squirrels will add wastes to the soil, which makes the soil better for plants.

Solution 3: Predators will eat the squirrels. Or, competition with other gray squirrels will keep the population from increasing because of lack of food, water, etc. Or, diseases will spread because of the denser squirrel population.

Now Do Homework Questions #12-32, pages 34-40.

Niches

Niche is the animal's **habitat** (where it lives) **and** its **feeding style** (what it eats and where it eats). Two species cannot occupy (have) the same niche (same habitat (where it lives) and same feeding style (what it eats and where it eats)) because both species would compete for the same food in the same place and one species would die off.

Niche is the **role** the species plays: the kind of food it eats, where it eats, the place it lives and reproduces, and how this organism affects other organisms. Birds and bees affect flowers. Birds' and bees' role or job is to bring pollen (which has the male sex cells of a plant) from one flower to another flower, which helps to produce new plants. The birds and bees also eat the nectar of the flower.

Example 1: The figure below shows three species of birds, Cape May warblers, Bay-breasted warblers, and Yellow-rumped warblers. **Each species feeds** (eats) on a **different place** in the **tree**, upper part of the tree, middle of the tree and lower part of the tree (see figure below), and therefore there is no competition for the food on the tree and all three species can survive. Each species feeds (eats) on a different part of the tree, which means each species has its own niche. The niche of the Cape May is on the top part of the tree, the niche of the Bay-breasted is in the middle of the tree, the niche of the Yellow-rumped is on the lower part of the tree.

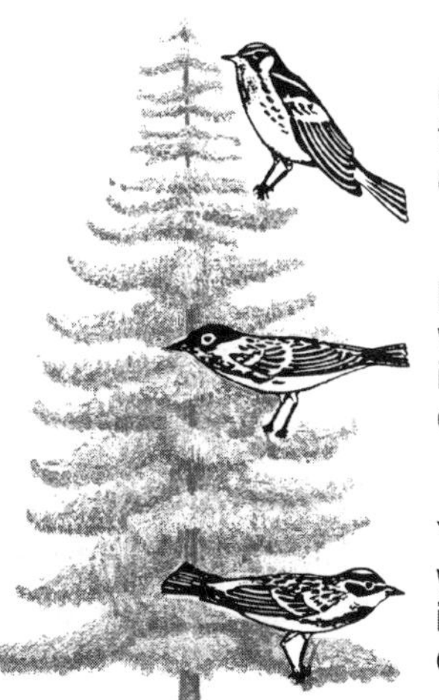

Cape May warblers feed in the upper area of the tree.

Bay-breasted warblers feed in the middle of the tree.

Yellow-rumped warblers feed in the lower part of the tree.

Three Different Feeding Niches in a Tree

Example 2: Species A and Species B live in the same tree. Species A feeds on (eats) ants and termites, while Species B feeds on (eats) caterpillars. Species A eats different organisms (food) than Species B, and therefore there is no competition for food and both species can survive. Each species feeds on different organisms, which means each species has its own niche.

The niche of Species A is Species A feeds on (eats) ants and termites; the niche of Species B is Species B feeds on (eats) caterpillars.

Now Do Homework Questions #33-40, page 41.

How organisms get food

1. **Producers,** also called **autotrophs,** make their own food. Plants and algae are producers (autotrophs); they make their own food. As you learned. plants take in carbon dioxide and water and take in sunlight (sun's energy) to produce glucose (a simple sugar, has stored energy, food). Energy flows from the sun's energy to the producers (plants), to the herbivores (animals that eat plants), and then to the carnivores (animals that eat animals)(see below).

2. **Consumers,** also called **heterotrophs, eat** (feed on) **plants** or **animals** (or both plants and animals) for food. They cannot produce their own food. There are two types of consumers, herbivores and carnivores:

a. **Herbivores** are **animals** that **eat** (feed on) **plants**.
Example: cow (herbivore) eats grass.

b. **Carnivores** are **animals** that **eat** (feed on) **animals.** There are two types of carnivores:

Predators kill and **eat** other **animals** (prey).
Examples:
1. Cat is a predator that kills and eats the mouse. (Mouse is the prey.)
2. Hawk is a predator that kills and eats the owl. (Owl is the prey.)
3. Frog is a predator that kills and eats the insect. (Insect is the prey.)
4. Big fish is a predator that eats little fish. (Little fish is the prey.)

Scavengers eat dead animals.
Example: Vulture eats dead animals.

3. **Decomposers** can be considered **part of consumers or** a **category by itself. Decomposers** (examples some **bacteria** and **fungi**) eat wastes and dead organisms and **decompose** (**break down** chemically) the **wastes** and **dead material** into simple materials like carbon dioxide and simple nitrogen compounds, etc., returning **(recycling) materials to** the **air** and **soil,** to the environment. The more wastes and dead organisms that would be broken down (decomposed or decayed), the more material (carbon dioxide and nitrogen compounds) would be released (recycled) to the air and soil. Decomposers break down wastes and dead material by a **process called decomposition,** making carbon dioxide and nitrogen compounds that are used by plants.
Note: Decomposers do not provide (give) energy to other organisms.

In an ecosystem, there is a connection between carnivores, herbivores, producers, and decomposers. When the **predators** (animals that kill and eat other animals) such as **wolves** (a type of carnivore), **die off** or are **removed**, more **deer** (a type of herbivore) will **survive** (because there are fewer predators (wolves) to kill the deer).

Food chain

A **food chain** shows which **organisms** are **eaten** by other **organisms,**

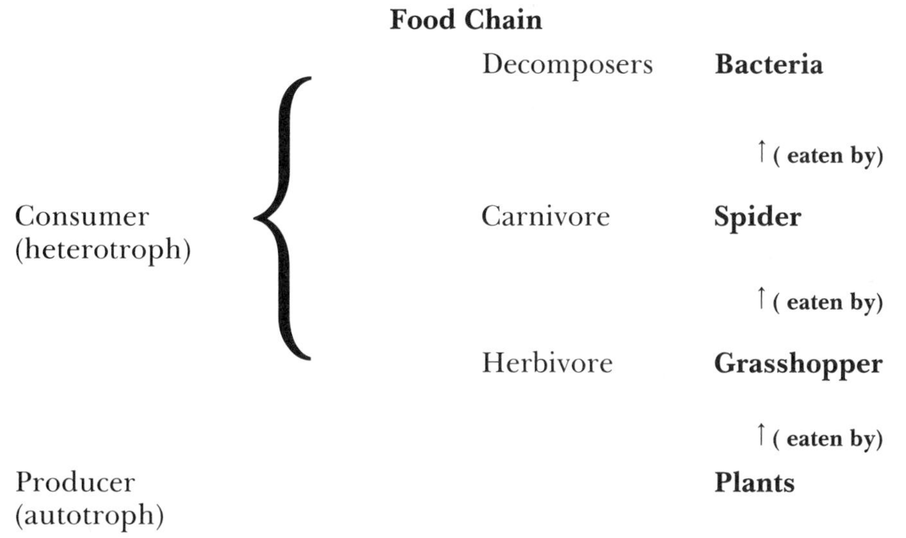

In short, in the food chain, the plants are eaten by the grasshoppers, which are eaten by the spiders, which are eaten (broken down, decomposed) by the bacteria. On food chain diagrams, generally the arrow is written without the words "(eaten by)"; you should know the arrow means "eaten by".

The food chain can also be written as shown below, going across (generally food chains are written without the words "(eaten by)"):

Plants $\xrightarrow{\text{(eaten by)}}$ Grasshopper $\xrightarrow{\text{(eaten by)}}$ Spider $\xrightarrow{\text{(eaten by)}}$ Bacteria

Plants $\longrightarrow$ Grasshopper $\longrightarrow$ Spider $\longrightarrow$ Bacteria

Plants are eaten by the grasshopper (or you can say the grasshopper eats plants). The grasshopper is eaten by the spider (or you can say the spider eats the grasshopper).

Another example of a food chain is:

seaweed $\longrightarrow$ fish $\longrightarrow$ seal $\longrightarrow$ polar bear

Seaweed (a plant), is eaten by fish, fish is eaten by seal, seal is eaten by polar bear.

In a food chain, energy goes (is transferred) **from producers** to **consumers** and to **decomposers** (which break down wastes and dead organisms). Note: Decomposers break down wastes and dead organisms at all levels of the food chain (examples plants (producers) and animals (consumers)).

Question: Which relationship best describes the interaction between lettuce and a rabbit?

1. predator–prey
2. producer–consumer
3. parasite–host
4. decomposer–scavenger

Solution: Lettuce is a plant, a producer, which is eaten by the rabbit, a consumer. Answer *2*

Question: Carbon dioxide containing carbon-14 is introduced into a balanced aquarium ecosystem. After several weeks, carbon-14 will most likely be present in

1. the plants, only
2. the animals, only
3. both the plants and animals
4. neither the plants nor animals

Solution: Plants take in carbon dioxide. In photosynthesis, plants take in the sun's energy, carbon dioxide and water to make glucose (a simple sugar).

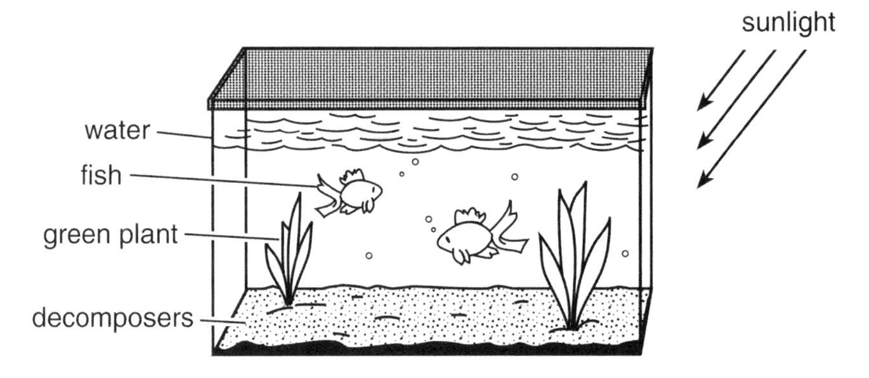

Balanced Aquarium

The carbon dioxide in this balanced aquarium ecosystem contains carbon-14. Therefore, when the plants take in carbon dioxide that contains carbon-14, the plant has carbon-14. (The plant uses the carbon dioxide with carbon-14 to make glucose; the glucose in the plant has carbon-14.) When the animal eats the plant, the animal also gets carbon-14. As you can see, both the plants and the animals will have carbon-14. Answer *3*

Now Do Homework Questions #41-61, pages 42-46.

Food Web

A food web is made of many food chains interconnected together (in an ecosystem). In a **food web** or in any **food chain**, **energy flows** from **producers** (example plants) **to consumers** (herbivores or carnivores). Producers (example plants) make their own food. In photosynthesis, plants take in carbon dioxide and water and take in sunlight (sun's energy) to produce glucose (a simple sugar, has stored energy, food). Carbon dioxide, water, and sunlight are called abiotic (nonliving) factors. In a food chain or a food web, energy flows from the sun's energy to the producers (plants), to the herbivores and then to the carnivores. Producers have the most energy; herbivores have less energy, and carnivores have the least energy.

Food Web

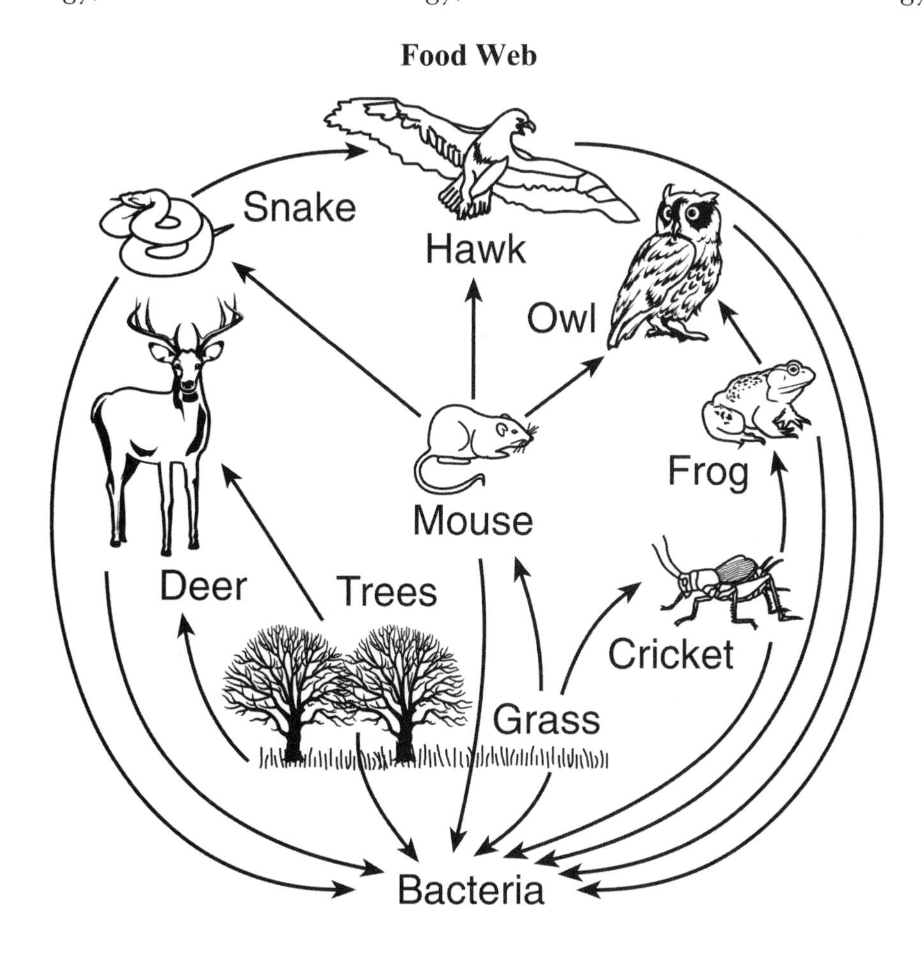

Look at the food web; the food web is made up of many food chains, as explained below. Below are listed four food chains in the food web. There are also other food chains in this food web.

Food Chains in the Food Web

			decomposer (bacteria)
			↑
decomposer (bacteria)	decomposer (bacteria)	decomposer (bacteria)	owl
↑	↑	↑	↑
snake	hawk	owl	frog
↑	↑	↑	↑
mouse	mouse	mouse	cricket
↑ (eaten by)	↑	↑	↑
grass	grass	grass	grass

Let's look at the first food chain listed above and in the food web.

Grass (eaten by) ⟶ Mouse (eaten by) ⟶ Snake (eaten by) ⟶ Decomposer (Bacteria)

Grass is eaten by mouse, mouse is eaten by snake, snake is "eaten by" (broken down, used for energy) by decomposers.

Look again at the first food chain above. Grass is eaten by the mouse, or you can say the mouse eats the grass. The **eater** (the organism that eats) example **mouse,** is at the **head of** the **arrow** (see figure 1 below).

The mouse is eaten by the snake or you can say, the snake eats the mouse. The **eater** (the organism that eats) example **snake**, is at (next to) the **head of** the **arrow** (see figure 2 below).

The snake is eaten by bacteria (a decomposer), or you can say, the bacteria "eat" (decompose, break down) the snake. The eater (the organism that eats) example bacteria, is at the head of the arrow (see figure 3 below).

Note: The decomposers break down and use wastes and dead plants and animals at all levels of the food chain and food web.

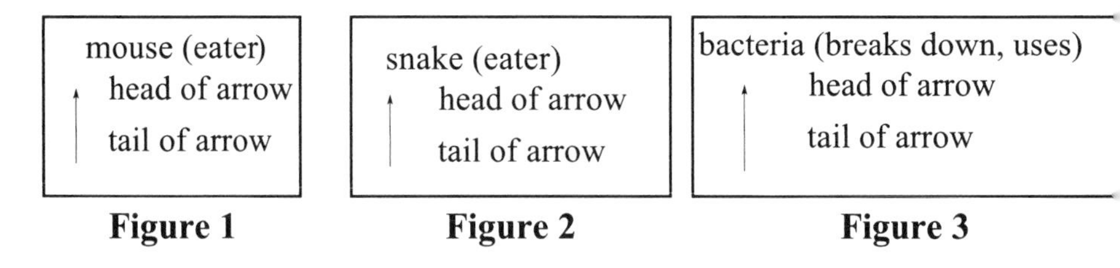

Figure 1 **Figure 2** **Figure 3**

You can see from the food web that grass is a producer and makes its own food. Look at the food web. There is no head of an arrow (no arrowhead) next to grass, which shows that grass is not an "eater" (grass does not eat food). Grass is a producer (autotroph) and makes its own food (glucose, a simple sugar).

We already looked at the first food chain. Now let's look above at the other three food chains in the chart called "Food Chains in the Food Web", and at the food web. See where each food chain is in the web.
Food chain 2: grass is eaten by mouse, mouse is eaten by hawk, hawk is broken down by decomposers.
Food chain 3: grass is eaten by mouse, mouse is eaten by owl, owl is broken down by decomposers.
Food chain 4: grass is eaten by cricket, cricket is eaten by frog, frog is eaten by owl, owl is broken down by decomposers.
Learn and **know** how to **read food chains in** a **food web.**

The **food web shows:**

1. which organisms are eaten by which other organisms. For example, in the food web above, you see grass is eaten by cricket, cricket is eaten by frog, frog is eaten by owl, owl is eaten by bacteria (decomposers).

2. how changes in the population of one kind of organism affect other organisms. Look at the food web diagram above:

a. The owl eats frogs and mice. If the frogs increase in number, there would be more food for the owls and the owl population would increase (more would survive and reproduce), then the mouse population would decrease.

b. grass is eaten by both the cricket and the mouse. If the mouse population increases, the mice would eat more grass and there would be less grass for the crickets. Some crickets would die (cricket population decreases), because there is not enough grass for the crickets to survive; the crickets are negatively affected.

3. some organisms can eat different foods. Therefore, even if one food is hardly available, the organism can eat the other food and stay alive. The ecosystem remains stable. For example: Owls can eat mice and frogs. If there is a decline in the frog population (such as because of disease), the owls can still eat mice and the ecosystem remains stable.

4. some organisms can eat only one kind of food. Crickets only eat grasses (see food web diagram above). Therefore, if the grasses were removed, the crickets would die.

5. look at the food web again. Grass is eaten by crickets, crickets are eaten by frogs, and frogs are eaten by owls.

grass → cricket → frog → owl

If the crickets are killed off, then the frogs have less or no food to eat and the frogs might die. This disrupts the food chain.

In short, if one organism (example cricket) is removed or killed, the food chain is disrupted. The animal (example frog) that eats only that organism (example cricket) would not have food and might die.
Note: You learned, if the crickets are killed off, then the frogs have less or no food to eat and the frogs might die. If the frogs have died out (see food web diagram above) the owls might not have enough food and might also

die out. OR, if there are enough mice (see food web diagram above), the owls might eat more mice and survive.

Question: A partial food web is represented in the diagram below.

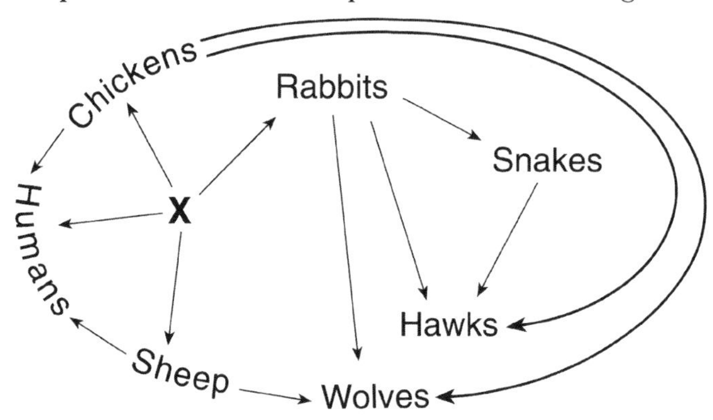

Letter X most likely represents
(1) autotrophs (2) carnivores (3) decomposers (4) parasites

Solution: Use either Method 1 (start with eaten by) or Method 2 (look at the head of the arrow).

Method 1: Look at the food web. **X** is **eaten by rabbits**, chickens, humans, and sheep. X is eaten by rabbits, or you can say, rabbits eat X (**X can be grass**; grass is an **autotroph**, because grass makes its own food). Rabbits eat grass, which is an autotroph (X). Answer *1*

Method 2: Look at the food web. Rabbit is **next to** the **head** of the **arrow** therefore rabbit is the **eater** and rabbits eat X. (**X can be grass**; grass is an **autotroph**, because grass makes its own food). Rabbits eat grass, which is an autotroph (X). Answer *1*

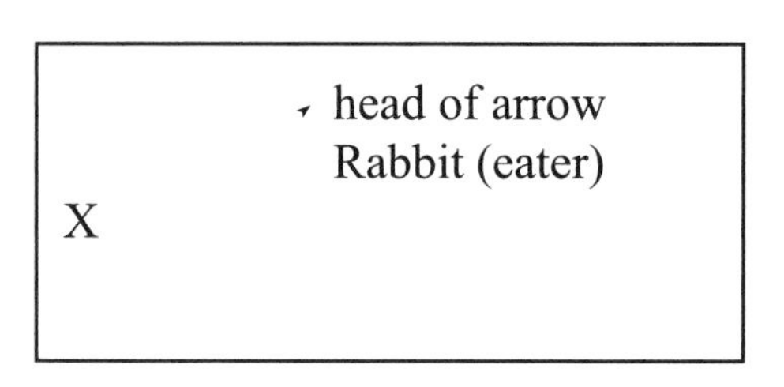

Look at the food web above. Similarly, sheep is **next to** the **head** of the **arrow** therefore sheep is the **eater** and sheep eat X. (**X can be grass**; grass is an **autotroph**, because grass makes its own food). Sheep eat grass, which is an autotroph (X). Answer *1*

Question: A food web is shown below.

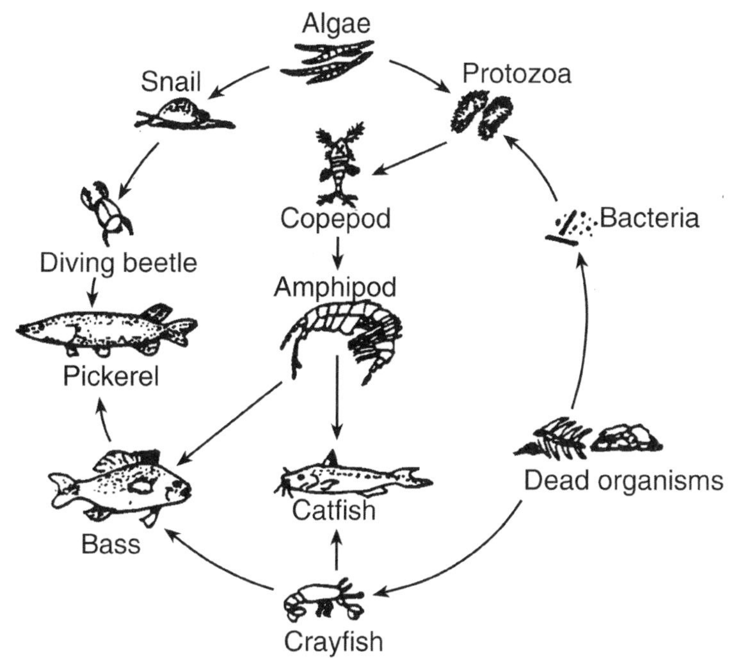

Which organisms feed on *both* producers and decomposers?
(1) amphipods (2) catfish (3) crayfish (4) protozoa

Solution: Method 1: In this food web, algae are the only producers; bacteria are the only decomposers. Look at the food web. Bacteria are eaten by protozoa, or you can say protozoa eat bacteria (a decomposer). **Protozoa eat (feed on)** bacteria **(decomposers)**.
Look at the food web again. Algae are eaten by the protozoa, or you can say **protozoa** eat (**feed on**) the **algae**. There is no arrowhead next to the algae because the algae is not an eater (does not eat food). Algae are producers; they make their own food. **Protozoa** eat (**feed on)** algae **(producers)**.
Protozoa feed on producers and decomposers. Answer *4*

Method 2: The **head** of the **arrow** is at (next to) the **eater.** Look at the food web again. The head of the arrow from algae is at (next to) protozoa. Protozoa are eaters (eat algae). Algae are producers; they make their own food. **Protozoa** eat (**feed on**) algae (**producers**).
Look at the food web again. The head of the arrow from bacteria is at (next to) protozoa. **Protozoa** are eaters (eat, **feed on**) bacteria (**decomposers**).
Protozoa feed on producers and decomposers. Answer *4*

Now Do Homework Questions #62-87, pages 46-55.

Pyramid of Energy

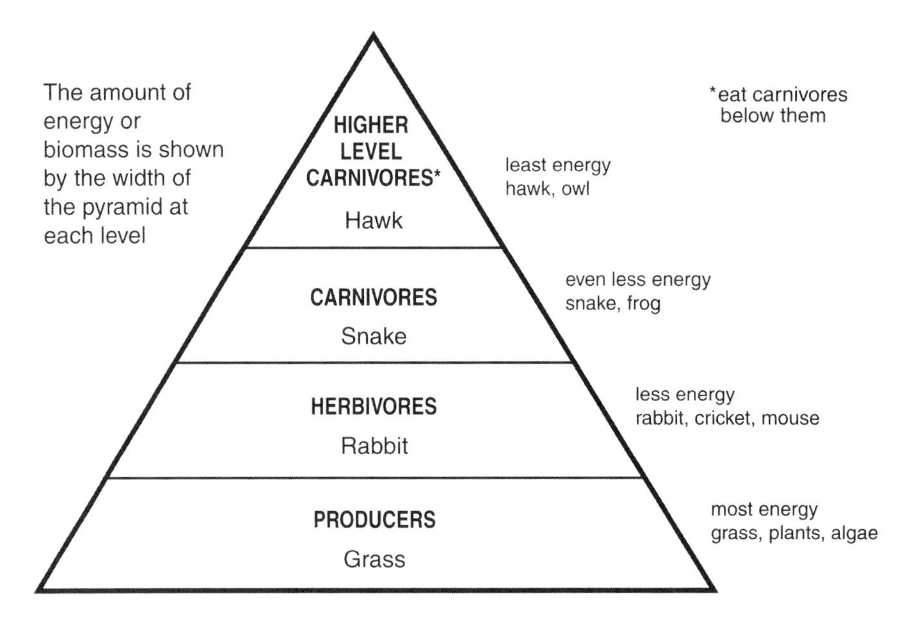

Producers (example plants and algae) **make their own food.** In photosynthesis, plants and algae take in carbon dioxide, water, and sunlight (sun's energy) to make food (which has stored energy in nutrients (example glucose)). Herbivores and carnivores are consumers; **consumers do not make their own food.**

Look at the energy pyramid (pyramid of energy, food pyramid) above. In the energy pyramid, the bottom level, the producers, is widest. This shows that the bottom level (producers) has the **most biomass** (amount of living material), **most energy** (from nutrients, such as glucose) and therefore the **most energy available for transfer** to the next higher level, the herbivores, and then to the carnivores (next level above herbivores). The **top** of a food chain or **food pyramid** (energy pyramid) is always a **carnivore (**see pyramid above); a carnivore is a **predator** (animal (example cat or hawk) that eats other animals) or a **scavenger** (animal (example vulture) that eats dead animals).

In all organisms, in all levels of the food pyramid, stored energy (example glucose) is changed to ATP (usable form of energy). Look at the food pyramid (pyramid of energy) above. The producer level is widest (showing it has the most energy), herbivores is less wide (less energy), carnivores is even less wide (even less energy), and carnivores that eat carnivores is narrowest (least energy). As you go up the food pyramid (food chain or food web) each level has less energy, less ATP, less biomass, and a smaller number of organisms. Organisms at each level use energy for the organism's life activities and give off energy as heat to the environment. The energy given off as heat can be shown in a food chain (see below). The wavy arrows ⟿ represent the energy given off (released) as heat.

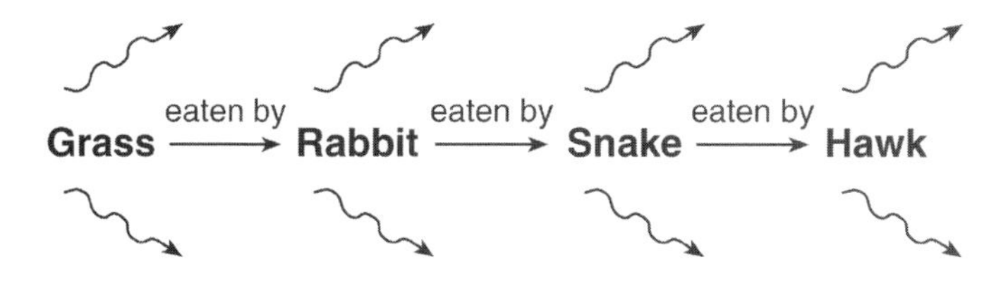

The food chain can also be drawn without the words "eaten by":

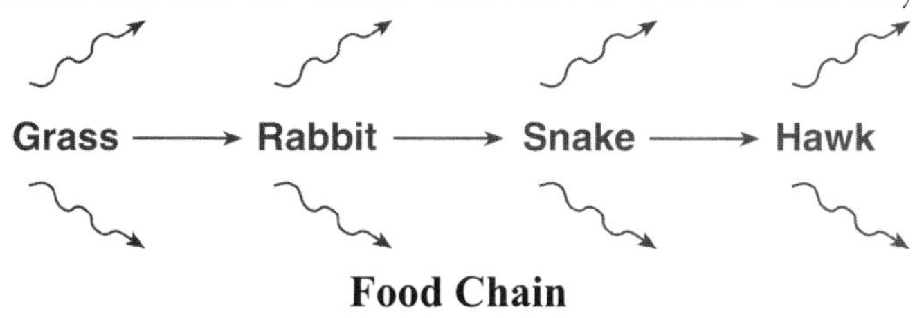

Food Chain

You learned organisms at each level use energy for the organism's life activities and give off energy as heat to the environment. Therefore, only about 10% of the energy from each level is left and is passed on to the next higher level. For example, (see energy pyramid above) herbivores only have about 10% of the energy of producers, carnivores only have about 10% of the energy of the herbivores and carnivores that eat carnivores only have about 10% of the energy of the carnivores. As you go up the energy pyramid (food pyramid), each level has less energy (less food) and therefore can support a smaller number of organisms (living things). The top of the pyramid has the least energy (least food), smallest amount of biomass, and smallest number of organisms (example, seven hawks top of pyramid, 50 snakes below it, 100 rabbits below and 1000 blades of grass on the bottom of the pyramid).

Energy flows (goes) from the sun to photosynthetic organisms (example plants and algae), then to herbivores, and then to carnivores. Energy flows from organisms (plants and animals) at all levels of the pyramid to decomposers.

Similarly in a food chain, just like in the food pyramid above:

Grass $\xrightarrow{\text{eaten by}}$ rabbit $\xrightarrow{\text{eaten by}}$ snake $\xrightarrow{\text{eaten by}}$ hawk

Grass $\longrightarrow$ rabbit $\longrightarrow$ snake $\longrightarrow$ hawk

Grass is eaten by rabbits, rabbits are eaten by snakes, and snakes are eaten by hawks.

Grass has the most energy, most biomass (mass of all the grass in the area) and biggest number of organisms (blades of grass), rabbit has less energy, less biomass and smaller number of organisms, snake has even less energy, less biomass and even a smaller number of organisms and the hawk has the least energy, least biomass and the smallest number of organisms.

Look at the energy pyramids below. (An energy pyramid can be drawn either way).

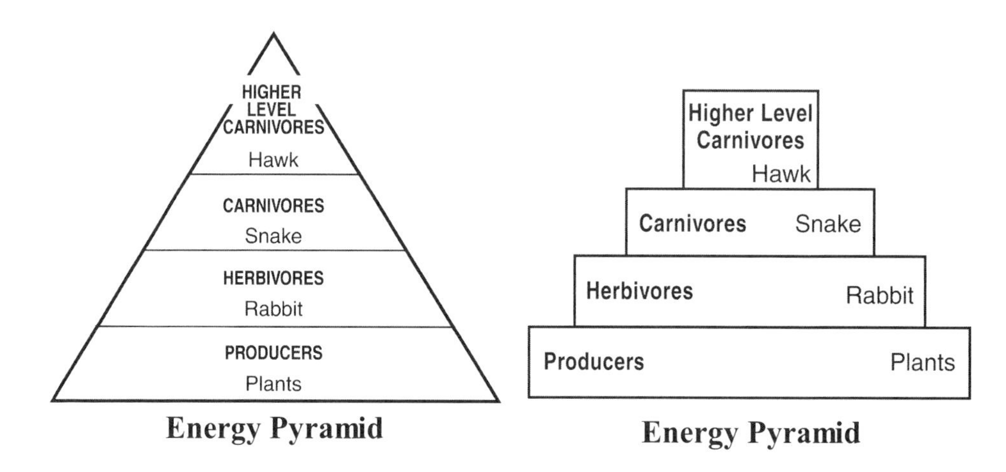

Energy Pyramid

Energy Pyramid

Producers (autotrophs), which are on the bottom of the food pyramid, are the widest (largest number of living things). Herbivores, above the producers, are less wide (smaller number of living things), and carnivores, above the herbivores, are even less wide (even smaller number of living things). In a **stable ecosystem**, there must be more producers (example plants) than herbivores, more herbivores than carnivores, therefore the bottom of the pyramid (the producers) must be widest, the next level up (herbivores) less wide, and the level above that (carnivores) even less wide. The diagrams of the energy pyramid shows a stable ecosystem. But, if there are a small number of producers, some of the herbivores would die out because they would not have enough food, causing the ecosystem to be unstable.

Base your answers to questions 1-4 on the energy pyramid below and on your knowledge of biology.

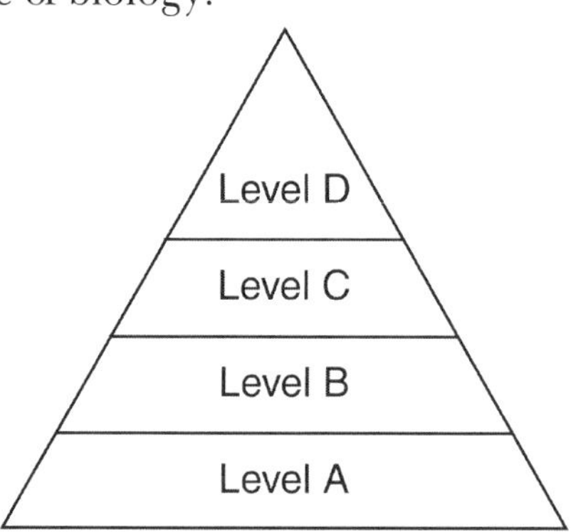

Question 1: Energy from nutrients is transferred to ATP in
(1) Level A only (3) Levels B, C, and D only
(2) Levels B and C only (4) Levels A, B, C, and D

Question 2: The greatest amount of available energy is transferred from level
(1) A to level B (2) A to level C (3) B to level A (4) D to level A

Question 3: Which energy levels could contain carnivores?
(1) A and B (2) B and C (3) C and D (4) D and A

Question 4: In a community where grass, cats, insects, and mice are found, which of these organisms would fill level A?

Solution 1: In cellular respiration, stored energy in nutrients (example glucose) is changed into ATP (usable form of energy) in all organisms at all levels of the food pyramid. Answer *4*

Solution 2: Level A is widest in the pyramid. This shows that level A has the most energy available for transfer to the next level up, level B.
Answer *1*

Solution 3: The first level (level A) is producers. The second level (level B) is herbivores and the third and fourth levels (levels C and D) are carnivores. Answer *3*

Solution 4: Level A is the producers. Plants, example grass, are producers.

Now Do Homework Questions #88-106, pages 56-61.

Recycling of materials

As you learned, for an **ecosystem** to be **stable** (self-sustaining, can survive by itself), there must be a **constant source** of **energy** (sun) and **materials** (examples oxygen and water) **must be recycled** (returned to the ecosystem or environment so they are not used up).

1. **Carbon Dioxide-Oxygen Cycle:** Plants take in carbon dioxide for photosynthesis and give off oxygen, recycling (returning) the oxygen to the environment. Animals (example humans, birds) take in oxygen in respiration (cellular respiration) and give off carbon dioxide, recycling (returning) the carbon dioxide to the environment. Then the plants take in the carbon dioxide for photosynthesis.

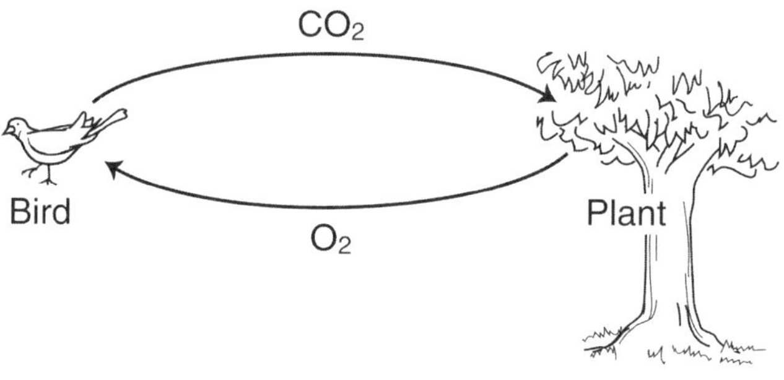

Carbon Dioxide - Oxygen Cycle

2. **Nitrogen Cycle:** Decomposers break down wastes or dead materials to simpler substances, such as nitrogen compounds that can be used by plants (autotrophs). This process is called decomposition (decay).

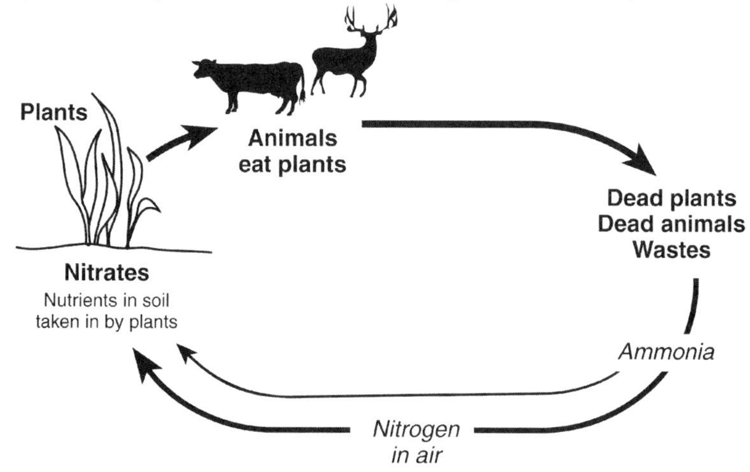

Dead plants and animals and **wastes** are **changed by bacteria into** ammonia, nitrogen, and finally **nitrates**, or ammonia directly to nitrates.

Nitrogen Cycle

Look at the nitrogen cycle above. Some producers (plants) eventually die. Some producers (plants) are eaten by consumers (animals) and the animals eventually die. The **bacteria of decay (decomposers)** convert **(change) dead material into ammonia**, and the other bacteria change ammonia **into nitrate** (a mineral in the soil which plants take in) (see nitrogen cycle above). Or you can say, different types of bacteria change dead material into nitrates. The producers (example **plants) take in** the **nitrate,** which is used to make plant proteins.

The cycle starts again. When **plants or animals die,** the **nitrate** in the dead plants or animals is **recycled (returned to** the **environment** (soil)) and **taken in by** other **plants** to make plant protein. When the **new plants or animals die,** the **nitrate** again is recycled (returned to the soil) and taken in by other plants.

3. **Water Cycle:** Water is evaporated from lakes, oceans, etc. Water comes back to the earth as rain, snow, sleet or hail.
Let's explain the water cycle. Water is evaporated from lakes forming water vapor. The water vapor condenses (changes to water (small drops of water)) forming clouds. When the clouds can no longer hold the water, the water comes down as rain, snow, hail, etc.

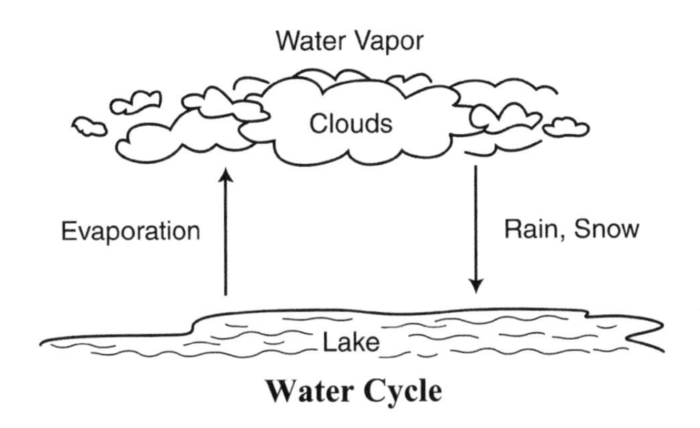

Water Cycle

Now Do Homework Questions #107-115, pages 61-62.

Relationships among species

Two species living together have an effect on one another. Generally one of the species benefits (shown by (+)). The species that is harmed is shown by (−) and a species that is not affected is shown by (0).

1. **Parasitism(+,−):** One organism benefits (+) and one organism is harmed (−). One organism (**parasite) benefits (+)** and one organism **(host) is harmed (−).**

Examples:

a. Bacteria or virus and humans (or other plants and animals). The bacteria or virus (parasite) benefits (+); the human, etc. (host) is harmed (−).

b. Athlete's foot fungus and humans. The **fungus** (a parasite) **benefits** (+) because it gets nutrients from the human being. The **human** (the host) is **harmed** (−); the human's foot gets itchy and swollen.

c. Tapeworm and some animals. The tapeworm lives in the digestive tract of some animals and absorbs some of the nutrients (food) from the animal, therefore the **tapeworm** (parasite) **benefits** (+). The **animal** that the tapeworm lives in **loses** some of its nutrients because the tapeworm absorbs them and the **animal** (host) is **harmed** (−).

2. **Commensalism(+, 0):** One organism benefits (+) and the other organism is not affected (not harmed and not benefitted)(0).

Example:

a. Barnacles on whales. The barnacle benefits (+) because the barnacle attaches itself to a whale and gets carried through the water; the water has a lot of food for the barnacles. The whale is not affected (0); the whale does **NOT** benefit and is not harmed.

b. Orchids on trees. Orchids benefit (+) by growing on tree trunks and branches of trees; the orchid has a place to grow, has a high place with sunlight, and the roots of the orchid absorb moisture from the air. The tree is not affected (0) by the orchid; the tree does **NOT** benefit and is not harmed.

3. **Mutualism(+,+):** Both organisms benefit.
Examples:

a. Nitrogen-fixing bacteria on the roots of leguminous plants (pea plants, bean plants): Bacteria benefit (+) from the plants because bacteria live and reproduce on the roots of the plant. The plants benefit (+) because the plants get nitrates which they need because nitrogen-fixing bacteria change nitrogen in the air into nitrates.

b. Protozoa (unicellular (one-celled) organisms) in the digestive tract (intestine) of termites: Protozoa benefit (+) from the termites because protozoa are inside the termites, where they are protected from predators (enemies). The termites benefit (+) from the protozoa because the termites ingest (take into the termite, eat) wood, but the **protozoa must digest the wood** so the termites can use the wood as food.

c. Lichen made up of algae and fungus: The fungus benefits (+) from the algae because the algae produce food by photosynthesis for both the algae and fungus. The algae benefit (+) because the fungus provides moisture and minerals for the algae, and the fungus attaches the algae (together with the fungus) to a rock.

Now Do Homework Questions #116-119, page 62.

Biodiversity

Biodiversity means how many different types of organisms (genetic variations in one species or different species) are in an area (example different types of trees such as maple, oak, or pine or different types of animals, such as dogs, cows, etc.).

You learned in sexual reproduction, the offspring have genetic recombination (different combinations of genes from both parents), genetic variation (different varieties of genes) and therefore more biodiversity.

In an **ecosystem** that has a **large number of different species** (example different species of trees or different species of predators (animals that eat other animals)), if one type of tree is attacked by insects or disease, the other types of trees still survive. Therefore, the **ecosystem is stable**. A forest (or a rainforest) has a variety of different kinds of trees; therefore, it is usually a stable ecosystem. An **increase in biodiversity** (increase in variety of living things) causes an **increase in stability** of an ecosystem. Example: People brought back wolves to Yellowstone National Park (where they had been killed off many years before); bringing in wolves increased biodiversity. If a **farm** or field has only **one type of living thing** (example: apples or cotton or corn), insects or disease that attack only apples will destroy the entire apple farm; insects or disease that attack only cotton will destroy the entire cotton farm. The **ecosystem is unstable.**

Greater biodiversity (greater variety of species of plants and animals in one place, example forest) makes an ecosystem **more stable. Loss of biodiversity** (losing varieties of species in one place) makes an ecosystem **less stable.**

In **sexual reproduction,** the offspring have genetic recombination (different combinations of genes from both parents) and genetic variation (different varieties of genes), therefore **more biodiversity,** which makes an ecosystem **more stable.**

sexual reproduction —*causes*→ genetic variation —*causes*→ biodiversity —*causes*→ ecosystem stability (stable)

Humans can cause loss of biodiversity (losing species)

Examples of **how humans cause loss of biodiversity:**

1. Planting one crop (apples or cotton) instead of the many species that grew there naturally, therefore losing many species.

2. Overhunting (too much hunting) of one species (examples: coyotes, wolves, or mountain lions), therefore losing that species from the ecosystem and making the ecosystem less stable.

Results of overhunting: Shooting mountain lions or giving bounties (money) for killing mountain lions caused more deer to survive (increase in deer population). Too many deer ate too much of the grass, there was not enough grass left for the deer to eat, and many deer starved and died. Overgrazing (eating too much of the grass) by the deer also caused soil erosion (wearing away of the topsoil).

3. Cutting down forests causes many plant and animal species to be lost, losing their genes (genetic material) that could be useful for medicines, etc.

Solutions to **prevent loss of biodiversity** (losing species):

1. Solution to example 1: to keep the entire crop from being destroyed by one type of insect or disease, plant a variety of crops in the same place.

2. Solution to example 2: do not overhunt and do not shoot mountain lions.

3. Solution to example 3: stop cutting down forests.

Benefits of biodiversity:

1. Agriculture: Useful genes can be put into plants and animals, which are used by people.

Example: Tomatoes can be protected from freezing by giving the tomato plant a gene from fish that live in ice cold water.

2. Medicine: Different medicines come from different species of plants.

Loss of biodiversity (reducing the number of species) can cause a loss of genes (genetic material) and a loss of plants and animals. Loss of biodiversity would destroy these genes and these organisms (plants or animals) and keep them from being used to benefit people. A **loss of biodiversity** (example some plants dying out) can result in a **shortage** of **food,** a **lack** of **material for building** (example wood) **and** for **research,** and a **loss of medicines**. For example, some medicines come from plants. If the plants die out, we cannot get the medicines.

Now Do Homework Questions #120-126, pages 62-63.

ECOLOGICAL SUCCESSION

Land **communities** and water communities **change over time by** a process called **ecological succession.** Each community (all species in a given area) changes (modifies) the environment, making it more suitable for a new, different community. Ecological succession (or succession) is when one community (example grass) is replaced by another community (example pine trees) until a **stable** community, a **climax** community (means not changing, not being replaced by other plants or animals) is reached. Examples of a climax community (stable community) are a beech-maple forest or oak-hickory forest, etc. One type of climate produces a beech-maple forest (climax community); a different climate produces a different forest, an oak-hickory forest (a different climax community).

Ecological succession: bare rock changes to beech-maple forest.

An example of ecological succession is when bare rock changes over time to a beech-maple forest. Look at the ecological succession figure below. The order of the succession is shown from left to right. It means that the next species (species at right) replaces the previous species (species at left); as time goes on, grass replaces lichen, shrubs replace grass, pine trees replace shrubs, oak trees replace pine trees, and beech and maple trees replace oak trees. Only the newest species survive.

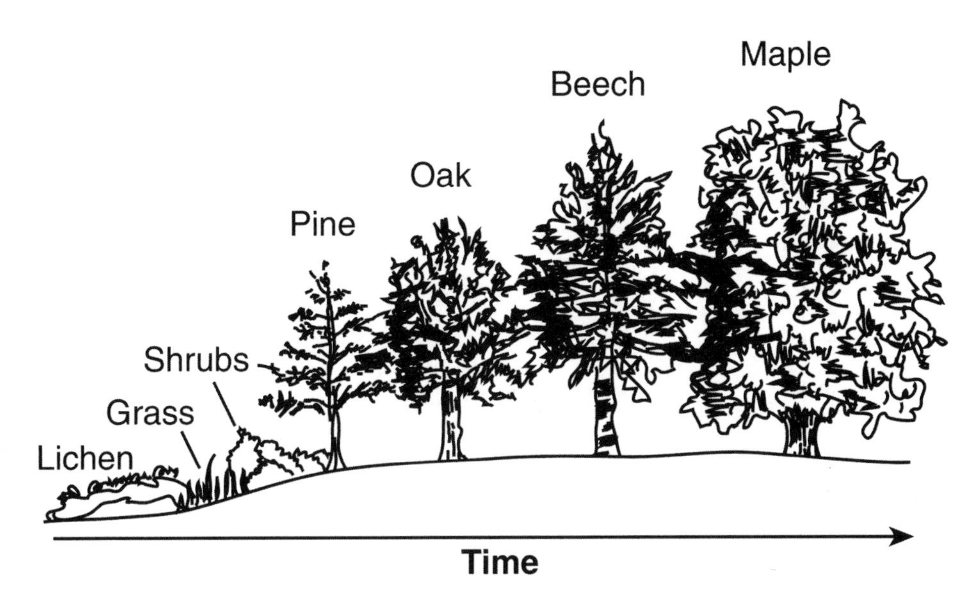

Succession: Bare Rock Changes to Beech-Maple Forest

1. **Lichens** can grow **on rock.** Lichens change (modify) their environment by breaking up rocks into soil. (When the soil becomes too moist and too rich lichens will die out).

2. Because lichens break up rocks, now there is more soil and grasses and weeds grow. Grasses and weeds enrich (make better) the soil therefore shrubs can now grow there. Shrubs shade the grass and the grass dies.

3. Pine trees grow in rich shaded soil. Pine trees shade the shrubs under them and the shrubs die.

4. Oak trees shade the pine trees and the pine trees die out.

5. Beeches and maples replace oak trees because beech and maple can grow in heavily shaded areas, but baby oak trees cannot grow there and die. Beech and maple trees live there for a long time; beech-maple forest is the **climax community**.

Each community (example pine trees) changes the environment and makes the new environment better suited for a different community (example oak trees) and less suitable for itself (pine trees). For example, white pine trees produce so much shade that shrubs and even the baby pine trees cannot grow there. The pine trees are replaced by oak trees. The oak trees are replaced by beech and maple trees (climax community).

A **climax community** (example beech-maple forest) **remains** until a catastrophe (such as fire, flood, or volcanic eruption) takes place and hardly any plants are left. Succession begins again until a new climax community is formed. If the climate or soil is the same (after and before the fire, etc.), it will be the same climax community. If the climate or soil is different, it will be a different climax community.

The graph below shows an example of ecological succession.

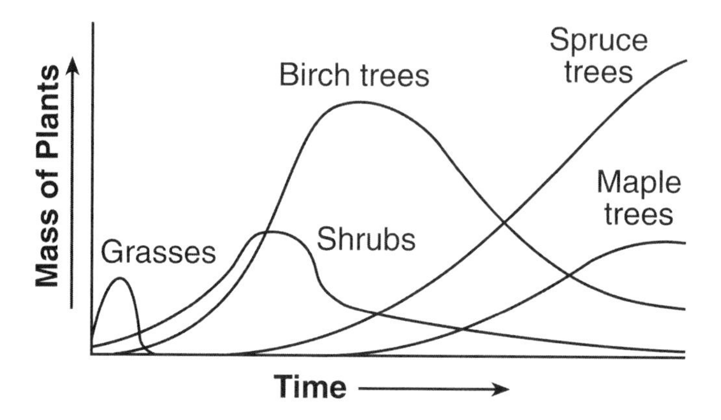

Question: A fire burns an oak forest down to bare ground. Over the next 150 years, if the climate remains constant, this area will most likely

(1) remain bare ground (3) become a rain forest
(2) return to an oak forest (4) become a wetland

Solution: After the fire, succession begins again (starting with lichens or grasses) until a new climax community is formed. Since the **climate** after the fire **is the same** as before the fire, there will be the **same climax community**, an **oak forest.** Answer *2*

As you saw, in the process of ecological succession, the plants change (example: lichens to grass to shrubs to pines to oak to beech and maple) and therefore, the animals living there also change.

Ecological succession: a lake community changes to a forest

An example of ecological succession is when a lake community changes over time to a forest. Look at the ecological succession figure below.

You are starting with a lake community (a lake with plants, fish, etc.)(see Figure A below). Sediments from broken rock start to fill the lake and form a shallow lake (see Figure B).

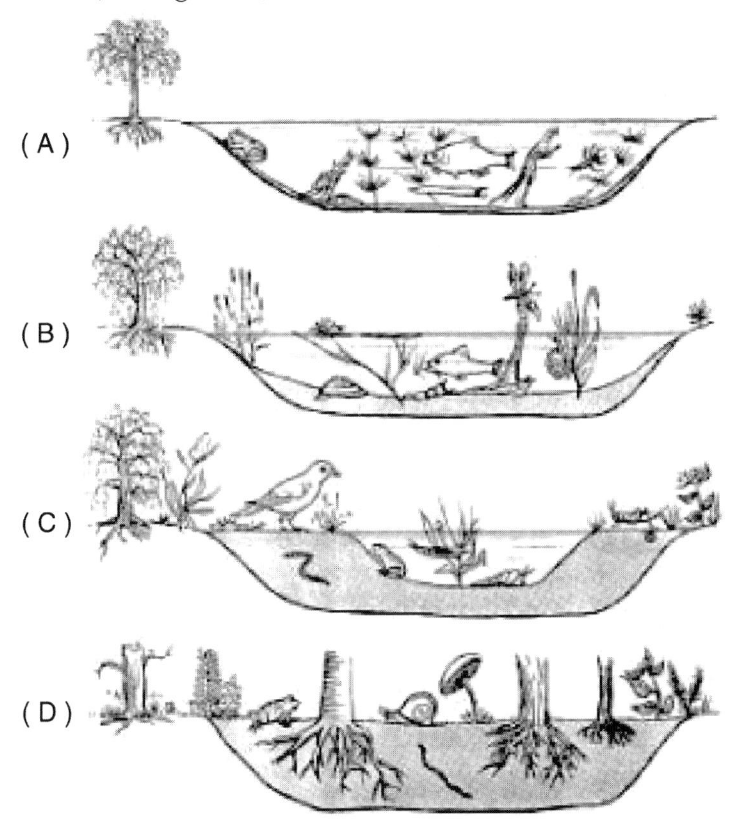

Succession: Lake Changes to a Forest

More and more sediments fill up the lake until the lake becomes a swamp (see Figure C).

More and more sediments fill the lake completely and it becomes a forest (see Figure D). The forest remains there for a long time. The forest is the climax community.

In succession, one community replaces another community. In the lake, the community of fish in water (see Figure A) is finally replaced by trees and mushrooms, etc . (see Figure D). The lake community was changed into a forest community (climax community).

In short, in the process of ecological succession, one community (example lake) changes into another community until a climax community (stable community, stable ecosystem, example forest) forms that can last for hundreds or thousands of years.

Now Do Homework Questions #127-138, pages 63-65.

Drawing Graphs

Let's see how we can draw graphs based on experimental data. Data (from an experiment) is written on a data table (see below). Draw the graph based on the data table.

A student added two species of single-celled organisms, *Paramecium caudatum* and *Didinium nasutum,* to the same culture medium. Each day, the number of individuals of each species was determined and recorded. The results are shown in the data table below.

Culture Population

Day	Number of *Paramecium*	Number of *Didinium*
0	25	2
1	60	5
2	150	10
3	50	30
4	25	20
5	0	2
6	0	0

How to draw the line graph:

1. On the x axis, put "Days". **The thing you change** (in this case number of days) is always put on the **x axis**. This is the independent variable. Space the lines along the axis equally. "Make an appropriate scale" by spacing the numbers on the graph so that all the data fits on the graph and it is easy to read. There must be an equal number of days between lines (see graph.) On the x axis, put one day between lines (every two lines) (scale on the x axis), then all the days in the data table fit on the graph and it is easy to read. Start counting the lines (every two lines) **after** the axis).

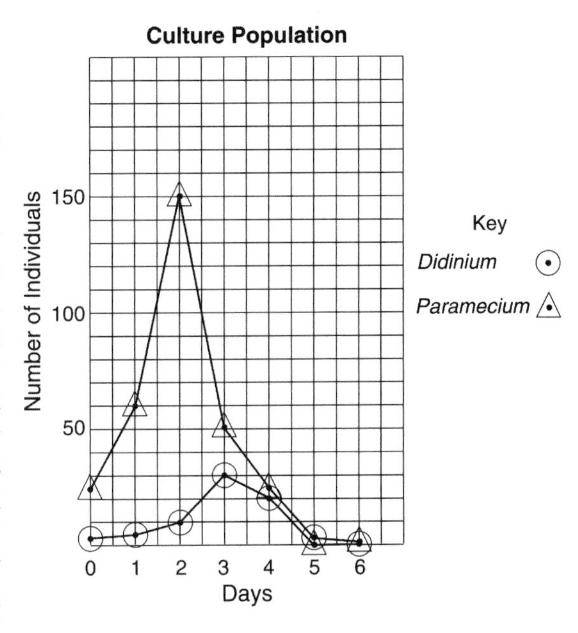

2. On the y axis, put "Number of Individuals". The **result** you get (number of individuals) is always put on the **y axis**. This is the dependent variable. Space the lines along the axis equally. There must be an equal number of individuals between lines (see graph)."Make an appropriate scale" by spacing the numbers on the graph so that all the data fits on the graph and it is easy to read.

   This method can help you to figure out how to make the scale: From one column in the data table, you see you have a number of paramecium (0 to 150). Look at the y axis of the graph; it has 21 lines. To find the **number** of paramecium (or number of seconds in another example) that each line should show, divide

   $$\frac{150\ (paramecium)}{21\ lines} = \text{about } \frac{7\frac{1}{2}\ paramecium}{line}$$

   . One line should show at least 7½ paramecium; therefore, **make each line** a **little bigger, easier number**, such as 10 paramecium.

   On the y axis, put 50 individuals between lines (every five lines) (scale on the y axis), then all the individuals on the data table fit on the graph and it is easy to read. Start counting the lines (every five lines) **after** the axis (see graph).

3. Plot the data for the number of *Didimium* on the graph. Surround each point with a small circle; draw a line that connects the points.

Do not continue the line past the last data point. Make a key showing that ⊙ means *Didimium*.

4. Plot the data for the number of *Paramecium* on the graph. Surround each point with a small triangle; draw a line that connects the points. Do not continue the line past the last point. Make a key showing that ⚠ means *Paramecium*.

5. Put a title on the graph which shows what the graph is about. Example: "Culture Population".

Now Do Homework Questions #139-142, page 66.

Now let's try a few Homework Questions

To assign fewer questions, you may want to skip the questions with underlined numbers.

1. Identify *one* abiotic factor that would directly affect the survival of organism *A* shown in the diagram below.

2. In an ocean, the growth and survival of seaweed, small fish, and sharks depends on abiotic factors such as
   (1) sunlight, temperature, and minerals
   (2) sunlight, pH, and type of seaweed
   (3) number of decomposers, carbon dioxide, and nitrogen
   (4) number of herbivores, carbon, and food

3. Which statement best describes what happens to energy and molecules in a stable ecosystem?
   (1) Both energy and molecules are recycled in an ecosystem.
   (2) Neither energy nor molecules are recycled in an ecosystem.
   (3) Energy is recycled and molecules are continuously added to the ecosystem.
   (4) Energy is continuously added to the ecosystem and molecules are recycled.

4. Which component of a stable ecosystem can *not* be recycled?
   (1) oxygen (2) water (3) energy (4) nitrogen

5. Which statement describes the ecosystem represented in the diagram below?

   (1) This ecosystem would be the first stage in ecological succession.
   (2) This ecosystem would most likely lack decomposers.
   (3) All of the organisms in this ecosystem are producers.
   (4) All of the organisms in this ecosystem depend on the activities of biological catalysts.

Base your answers to the next three questions on the information below and on your knowledge of biology.

Untreated organic wastes were accidentally discharged into a river from a sewage treatment plant. The graph below shows the dissolved oxygen content of water samples taken from the river at specific distances downstream from the plant, both before, and then three days after the discharge occurred.

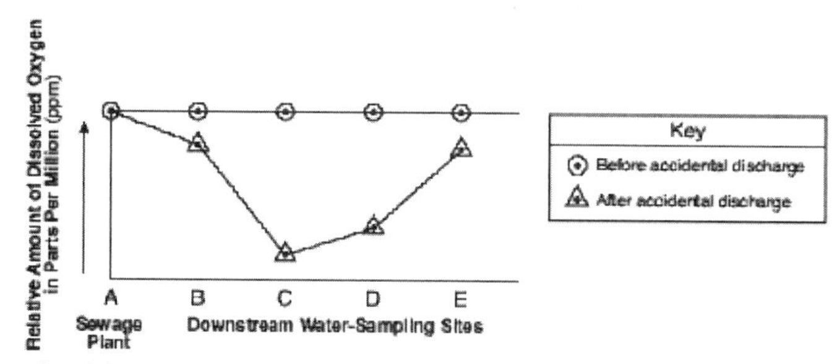

6. State why this accident would be expected to benefit the decomposers in the river below the sewage plant.

7. Explain why an energy-releasing process occurring in the mitochondria of the decomposer organisms is most likely responsible for the change indicated by the data shown at sampling site *C* in the graph.

8. State *one* reason why the statement below is correct.

b/c the Relative Amount of Dissolved Oxygen

"The effects of the accidental discharge are not expected to last for a long time."

---

9. Which statement describes a role of fungi in an ecosystem?
   (1) They transfer energy to decaying matter.
   (2) They release oxygen into the ecosystem.
   (3) They recycle chemicals from dead organisms.
   (4) They synthesize organic nutrients from inorganic substances.

10. One biotic factor that affects consumers (animals) in an ocean ecosystem is
    (1) number of autotrophs
    (2) temperature variation
    (3) salt content
    (4) pH of water

11. Abiotic factors that could affect the stability of an ecosystem could include
    (1) hurricanes, packs of wolves, and temperature
    (2) blizzards, heat waves, and swarms of grasshoppers
    (3) droughts, floods, and heat waves
    (4) species of fish, number of decomposers, and supply of algae

12. Four environmental factors are listed below.

    A. energy
    B. water
    C. oxygen
    D. minerals

    Which factors limit environmental carrying capacity in a land ecosystem?
    (1) *A*, only
    (2) *B*, *C*, and *D*, only
    (3) *A*, *C*, and *D*, only
    (4) *A*, *B*, *C*, and *D*

13. What impact do the amounts of available energy, water, and oxygen have on an ecosystem?
    (1) They act as limiting factors.
    (2) They are used as nutrients.
    (3) They recycle the residue of dead organisms.
    (4) They control environmental temperature.

14. The reason that organisms can *not* produce populations of unlimited size is that
    (1) the resources of Earth are finite
    (2) there is no carrying capacity on Earth
    (3) species rarely compete with one another
    (4) interactions between organisms are unchanging

15. Cattail plants in freshwater swamps in New York State are being replaced by purple loosestrife plants. The two species have very similar environmental requirements. This observation best illustrates
    (1) variations within a species
    (2) dynamic equilibrium
    (3) random recombination
    (4) competition between species

16. One biotic factor that limits the carrying capacity of any habitat is the
    (1) availability of water
    (2) level of atmospheric oxygen
    (3) activity of decomposers
    (4) amount of soil erosion

17. The size of a mouse population in a natural ecosystem tends to remain relatively constant due to
    (1) the carrying capacity of the environment
    (2) the lack of natural predators
    (3) cycling of energy
    (4) increased numbers of decomposers

18. The graph below represents the growth of a population of flies in a jar.

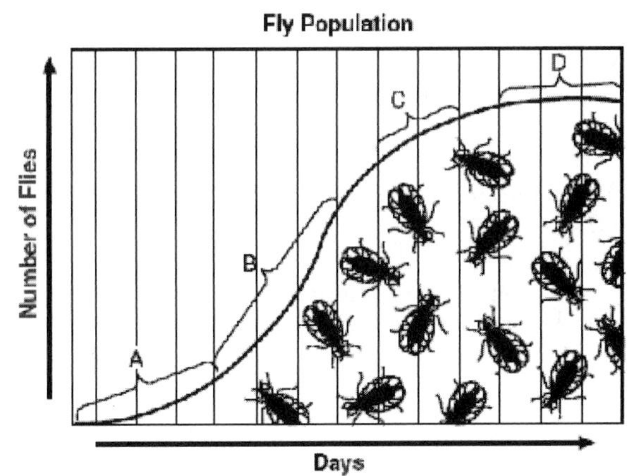

Which letter indicates the part of the graph that represents the carrying capacity of the environment in the jar?

(1) *A* (2) *B* (3) *C* (4) *D*

19. Which statement best describes the fruit fly population in the part of the curve labeled *X* in the graph shown below?

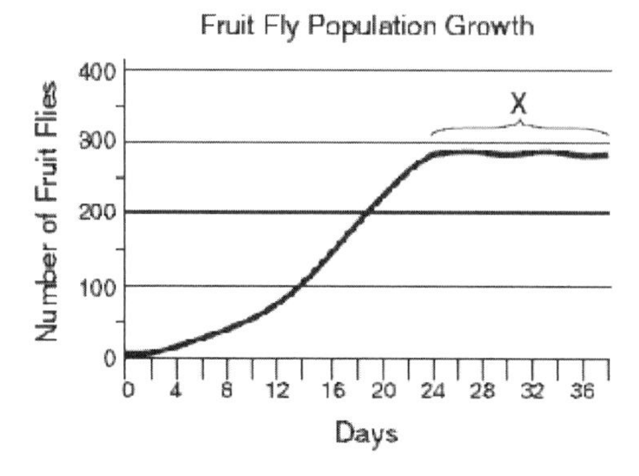

(1) The fruit fly population has reached the number of organisms the habitat can support.
(2) The fruit fly population can no longer mate and produce fertile offspring.
(3) The fruit fly population has an average life span of 36 days.
(4) The fruit fly population is no longer able to adapt to the changing environmental conditions.

20. The growth of a population is shown in the graph below.

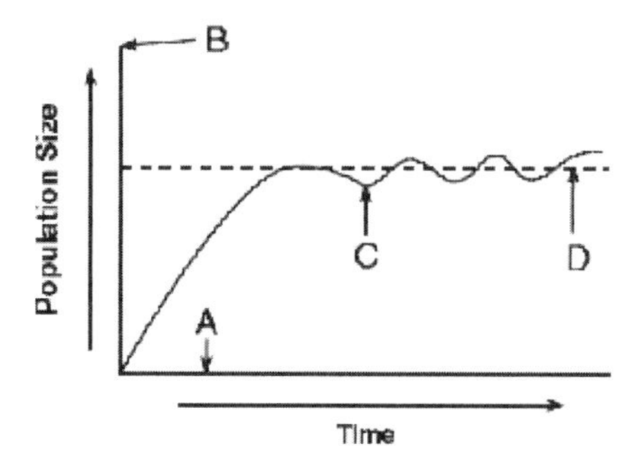

Which letter indicates the carrying capacity of the environment for this population?

(1) *A* (2) *B* (3) *C* (4) *D*

21. A population of chipmunks migrated to an environment where they had little competition. Their population quickly increased but eventually stabilized, as shown in the graph.

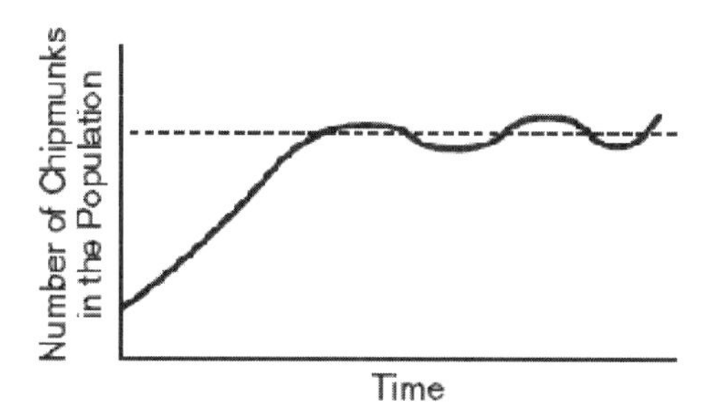

Which statement best explains why the population stabilized?

(1) Interbreeding between members of the population increased the mutation rate.

(2) The population size became limited due to factors such as availability of food.

(3) An increase in the chipmunk population caused an increase in the producer population.

(4) A predator species came to the area and occupied the same niche as the chipmunks.

Base your answers to the next two questions on the information and graph below and on your knowledge of biology.

A population of paramecia (single-celled aquatic organisms) was grown in a 200-mL beaker of water containing some smaller single-celled organisms. Population growth of the organisms for 28 hours is shown in the graph below.

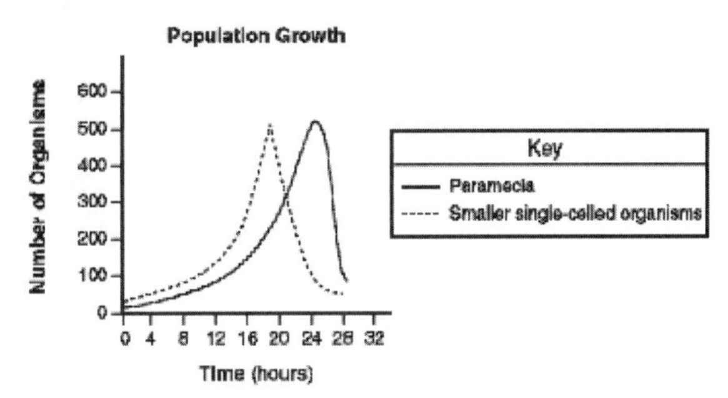

22. Which factor most likely accounts for the change in the paramecium population from 8 to 20 hours?

(1) an increase in the nitrogen content of water

(2) an increase in wastes produced

(3) an increase in available food

(4) an increase in water pH

23. One likely explanation for the change in the paramecium population from 26 hours to 28 hours is that the

(1) carrying capacity of the beaker was exceeded

(2) rate of reproduction increased

(3) time allowed for growth was not sufficient

(4) oxygen level was too high

---

24. The graph below shows the number of birds in a population.

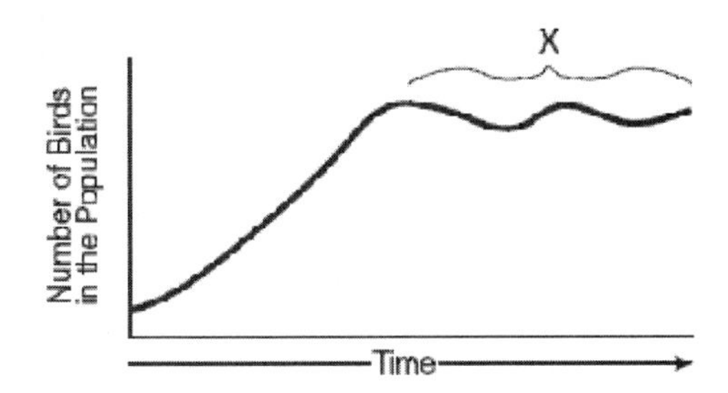

Which statement best explains section *X* of the graph?

(1) Interbreeding between members of this population increased the mutation rate.

(2) An increase in the bird population caused an increase in the producer population.

(3) The population reached a state of dynamic equilibrium due to limiting factors.

(4) Another species came to the area and provided food for the birds.

25. The graph below represents the growth of a population of flies in a jar.

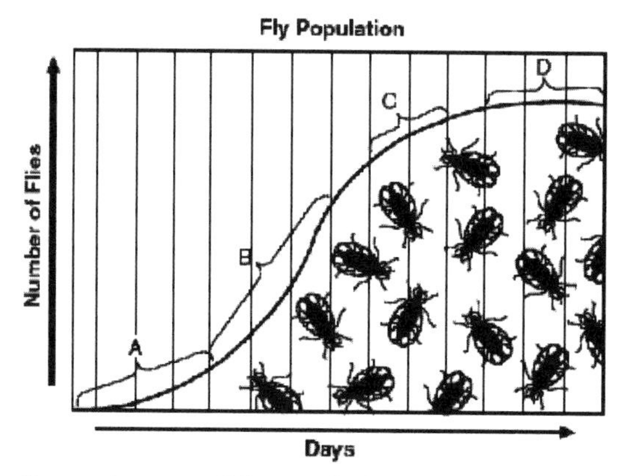

Which letter indicates the part of the graph that represents the carrying capacity of the environment in the jar?

(1) *A* (2) *B* (3) *C* (4) *D*

26. A graph of the population growth of two different species is shown below.

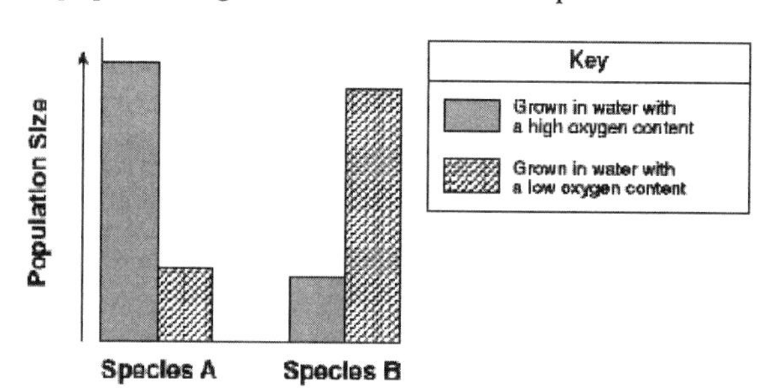

Which conclusion can be drawn from information in the graph?

(1) Oxygen concentration affects population sizes of different species in the same manner.

(2) Species *A* requires a high oxygen concentration for maximum population growth.

(3) Species *B* requires a high oxygen concentration to stimulate population growth.

(4) Low oxygen concentration does not limit the population size of either species observed.

27. On which day did the population represented in the graph below reach the carrying capacity of the ecosystem?

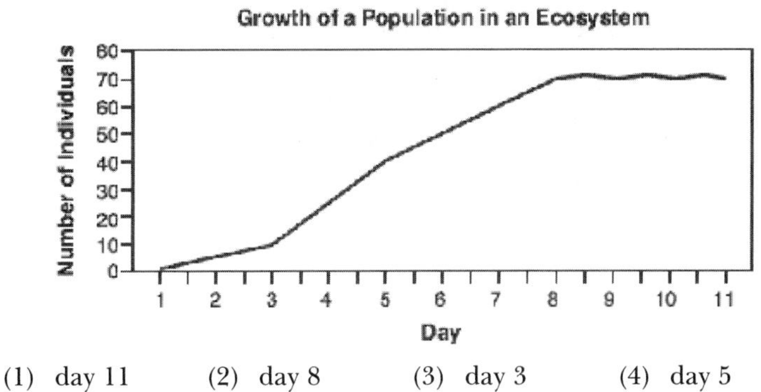
Growth of a Population in an Ecosystem
Number of Individuals
80
70
60
50
40
30
20
10
0
1
2
3
4
5
6
7
8
9
10
11
Day

(1) day 11 (2) day 8 (3) day 3 (4) day 5

28. The ecological niches of three bird species are shown in the diagram below.

Cape May warblers feed in the upper area of the tree.
Bay-breasted warblers feed in the middle of the tree.
Yellow-rumped warblers feed in the lower part of the tree.

What is the advantage of each bird species having a different niche?
(1) As the birds feed higher in the tree, available energy increases.
(2) More abiotic resources are available for each bird.
(3) Predators are less likely to feed on birds in a variety of locations.
(4) There is less competition for food.

29. Two closely related species of birds live in the same tree. Species *A* feeds on ants and termites, while species *B* feeds on caterpillars. The two species coexist successfully because
(1) each occupies a different niche
(2) they interbreed
(3) they use different methods of reproduction
(4) birds compete for food

30. The feeding niches of three bird species are shown in the diagram below.

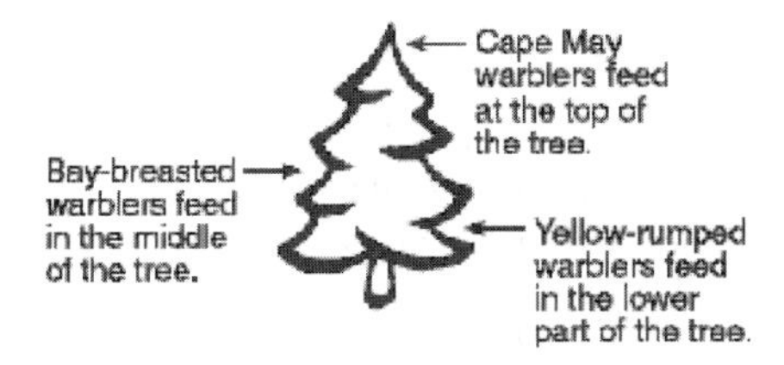
Cape May warblers feed at the top of the tree.
Bay-breasted warblers feed in the middle of the tree.
Yellow-rumped warblers feed in the lower part of the tree.

What is the advantage of these different feeding niches for the birds?
(1) less competition for food
(2) fewer abiotic resources for each bird species

(3) fewer biotic resources for each bird species
(4) less energy available as the birds feed higher in the tree

31. Information concerning nests built in the same tree by two different bird species over a ten-year period is shown in the table below.

Distance of Nest Above Ground (m)	Total Number of Nests Built by Two Different Species	
	A	B
less than 1	5	0
1–5	10	0
6–10	5	0
over 10	0	20

What inference best describes these two bird species?
(1) They most likely do not compete for nesting sites because they occupy different niches.
(2) They do not compete for nesting sites because they have the same reproductive behavior.
(3) They compete for nesting sites because they build the same type of nest.
(4) They compete for nesting sites because they nest in the same tree at the same time.

32. A scientist studied iguanas inhabiting a chain of small ocean islands. He discovered two species that live in different habitats and display different behaviors. His observations are listed in the table below.

Observations of Two Species of Iguanas

Species A	Species B
spends most of its time in the ocean	spends most of its time on land
is rarely found more than 10 meters from shore	is found many meters inland from shore
eats algae	eats cactus and other land plants

Which statement best describes these two species of iguanas?
(1) Both species evolved through the process of ecological succession.
(2) Each species occupies a different niche.
(3) The two species can interbreed.
(4) Species *A* is a scavenger and species *B* is a carnivore.

33. In a stable, long-existing community, the establishment of a single species per niche is most directly the result of
(1) parasitism
(2) interbreeding
(3) competition
(4) overproduction

34. Carbon dioxide containing carbon-14 is introduced into a balanced aquarium ecosystem. After several weeks, carbon-14 will most likely be present in
(1) the plants, only
(2) the animals, only
(3) both the plants and animals
(4) neither the plants nor animals

35. In the transfer of energy from the Sun to ecosystems, which molecule is one of the first to store this energy?
(1) protein (2) fat (3) DNA (4) glucose

36. If humans remove carnivorous predators such as wolves and coyotes from an ecosystem, what will probably be the first observable result?
   (1) The natural prey will die off.
   (2) Certain plant populations will increase.
   (3) Certain herbivores will exceed carrying capacity.
   (4) The decomposers will fill the predator niche.

Base your answers to the next three questions on the lake ecosystem represented below and on your knowledge of biology.

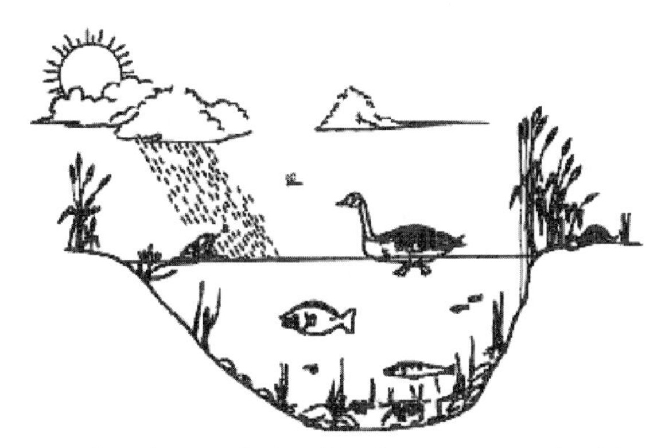

37. Identify *one* predator/prey relationship that may occur in this ecosystem.
38. State *one* piece of evidence from the diagram that indicates that light penetrates to the bottom of the lake.
39. Identify the type of organism that is not visible in the diagram but must be present in this ecosystem to recycle the remains of dead organisms.

40. Base your answers to the next question on the data table below and on your knowledge of biology.

Dietary Preferences of Finches

Species of Finch	Preferred Foods
A	nuts and seeds
B	worms and insects
C	fruits and seeds
D	insects and seeds
E	nuts and seeds

Which *two* species would most likely be able to live in the same habitat without competing with each other for food?
   (1) *A* and *C*
   (2) *B* and *C*
   (3) *B* and *D*
   (4) *C* and *E*

41. The dissolved carbon dioxide in a lake is used directly by
   (1) autotrophs (2) parasites (3) fungi (4) decomposers

42. Base your answers to the next question on the information below and on your knowledge of biology.

   Analysis of a sample taken from a pond showed variety in both number and type of organisms present. The data collected are shown in the table below.

Type of Organisms	Number Present
bass	two
frogs	forty
phytoplankton	thousands
insect larvae	hundreds

If the frogs feed on insect larvae, what is the role of the frogs in this pond ecosystem?

(1) herbivore (2) parasite (3) consumer (4) host

43. Organisms from a particular ecosystem are shown below.

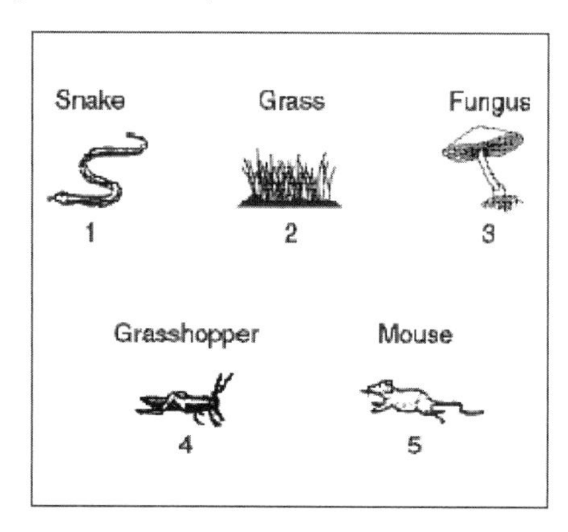

Which statement concerning an organism in this ecosystem is correct?

(1) Organism *2* is heterotrophic.
(2) Organism *3* helps recycle materials.
(3) Organism *4* obtains all of its nutrients from an abiotic source.
(4) Organism *5* must obtain its energy from organism *1*.

44. What is the role of bacteria and fungi in an ecosystem?

45. Two food chains are represented below.

Food chain *A*: aquatic plant → insect → frog → hawk
Food chain *B*: grass → rabbit → hawk

Decomposers are important for supplying energy for

(1) food chain *A*, only
(2) food chain *B*, only
(3) both food chain *A* and food chain *B*
(4) neither food chain *A* nor food chain *B*

Base your answers to the next three questions on the passage below which describes an ecosystem in New York State and on your knowledge of biology.

The Pine Bush ecosystem near Albany, New York, is one of the last known habitats of the nearly extinct Karner Blue butterfly. The butterfly's larvae feed on the wild green plant, lupine. The larvae are in turn consumed by predatory wasps. The four groups below represent other organisms living in this ecosystem.

Group A	Group B	Group C	Group D
algae mosses ferns pine trees oak trees	rabbits tent caterpillars moths	hawks moles hognosed snakes toads	soil bacteria molds mushrooms

46. The Karner Blue larvae belong in which group?
    (1) *A* (2) *B* (3) *C* (4) *D*

47. Which food chain best represents information in the passage?
    (1) lupine → Karner Blue larvae → wasps
    (2) wasps → Karner Blue larvae → lupine
    (3) Karner Blue larvae → lupine → wasps
    (4) lupine → wasps → Karner Blue larvae

48. Which group contains decomposers?
    (1) *A* (2) *B* (3) *C* (4) *D*

49. A graph of the population growth of two different species is shown below.

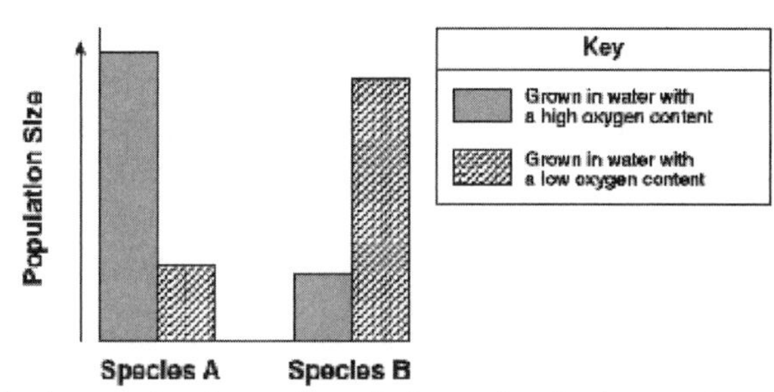

Which conclusion can be drawn from information in the graph?
    (1) Oxygen concentration affects population sizes of different species in the same manner.
    (2) Species *A* requires a high oxygen concentration for maximum population growth.
    (3) Species *B* requires a high oxygen concentration to stimulate population growth.
    (4) Low oxygen concentration does not limit the population size of either species observed.

Base your answers to the next two questions on the information below and on your knowledge of biology.

A student uses a covered aquarium to study the interactions of biotic and abiotic factors in an ecosystem. The aquarium contains sand, various water plants, algae, small fish, snails, and decomposers. The water contains dissolved oxygen and carbon dioxide, as well as tiny amounts of minerals and salts.

50. Identify *one* source of food for the decomposers in this ecosystem.

51. Describe *one* specific way the use of this food by the decomposers benefits the other organisms in the aquarium.

52. What would most likely happen if most of the bacteria and fungi were removed from an ecosystem?
    (1) Nutrients resulting from decomposition would be reduced.
    (2) Energy provided for autotrophic nutrition would be reduced.
    (3) The rate of mutations in plants would increase.
    (4) Soil fertility would increase.

53. The teeth of carnivores are pointed and are good for puncturing and ripping flesh. The teeth of herbivores are flat and are good for grinding and chewing. Which statement best explains these observations?
(1) Herbivores have evolved from carnivores.
(2) Carnivores have evolved from herbivores.
(3) The two types of teeth most likely evolved as a result of natural selection.
(4) The two types of teeth most likely evolved as a result of the needs of an organism.

Base your answers to the next two questions on the information below and on your knowledge of biology.

A student uses a covered aquarium to study the interactions of biotic and abiotic factors in an ecosystem. The aquarium contains sand, various water plants, algae, small fish, snails, and decomposers. The water contains dissolved oxygen and carbon dioxide, as well as tiny amounts of minerals and salts.

54. Identify *one* source of food for the decomposers in this ecosystem.

55. Describe *one* specific way the use of this food by the decomposers benefits the other organisms in the aquarium.

---

56. Two food chains are represented below.

Food chain *A*: aquatic plant → insect → frog → hawk
Food chain *B*: grass → rabbit → hawk

Decomposers are important for supplying energy for
(1) food chain *A*, only
(2) food chain *B*, only
(3) both food chain *A* and food chain *B*
(4) neither food chain *A* nor food chain *B*

57. Even before a flower bud opens, certain plant chemicals have colored the flower in patterns particularly attractive to specific insects. At the same time, these chemicals protect the plant's reproductive structures by killing or inhibiting pathogens and insects that may feed on the plant.

Which statement about the plant and the other organisms mentioned is correct?
(1) Chemicals affect plants but not animals.
(2) Organisms of every niche may be preyed on by herbivores.
(3) Any chemical produced in a plant can protect against insects.
(4) Organisms may interact with other organisms in both positive and negative ways.

58. Base your answers to the next question on the information below and on your knowledge of biology. Analysis of a sample taken from a pond showed variety in both number and type of organisms present. The data collected are shown in the table below.

Type of Organisms	Number Present
bass	two
frogs	forty
phytoplankton	thousands
insect larvae	hundreds

If the frogs feed on insect larvae, what is the role of the frogs in this pond ecosystem?
(1) herbivore (2) parasite (3) consumer (4) host

59. Base your answers to the next question on the information below and on your knowledge of biology.

> Thirty grams of hay (dried grasses) were boiled in 500 milliliters of water, placed in a culture dish, and allowed to stand. The next day, a small sample of pond water was added to the mixture of boiled hay and water. The dish was then covered and its contents observed regularly. Bacteria fed on the nutrients from the boiled hay. As the populations of bacteria increased rapidly, the clear mixture soon became cloudy. One week later, microscopic examination of samples from the culture showed various types of protozoa (single-celled organisms) eating the bacteria.

The protozoa that fed on the bacteria can best be described as
(1) producers (2) herbivores (3) parasites (4) consumers

60. One arctic food chain consists of polar bears, fish, weed, and seals. Which sequence demonstrates the correct flow of energy between these organisms?
(1) seals → weed → fish → polar bears
(2) fish → weed → polar bears → seals
(3) weed → fish → seals → polar bears
(4) polar bears → fish → seals → weed

61. Base your answer to the next question on the diagram below that shows some interactions between several organisms located in a meadow environment and on your knowledge of biology.

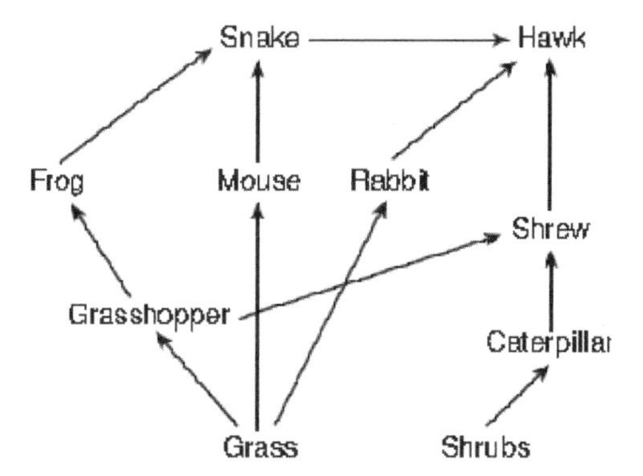

Identify *one* cell structure found in a producer in this meadow ecosystem that is not found in the carnivores.

Hint: *Review organelles (structures inside a cell) in Chapter 1.*

62. The diagram below represents a food web.

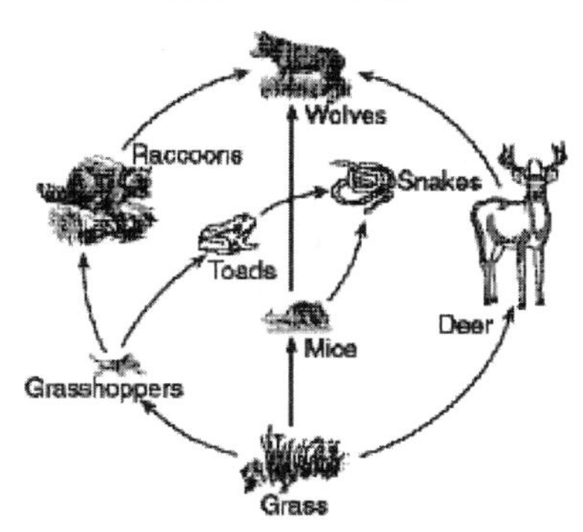

Two of the herbivores represented in this food web are
(1) toads and snakes
(2) deer and mice
(3) wolves and raccoons
(4) grasshoppers and toads

63. Base your answer to the next question on the information below and on your knowledge of biology

Gardeners sometimes use slug traps to capture and kill slugs. These traps were tested in a garden with a large slug population. Organisms found in the trap after one week are shown in the table below.

Organisms in Trap

Organism	Number in Trap
slugs	8
snails	1
aphids	13
centipedes	1
ground beetles	98

How many organisms in the trap were herbivores?
(1) 5 (2) 9 (3) 22 (4) 99

Base your answers to the next three questions on the diagram below and on your knowledge of biology.

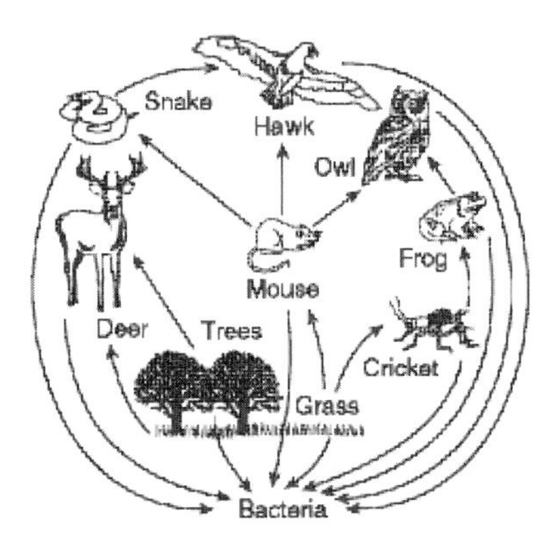

64. What is an appropriate title for this diagram?
(1) Energy Flow in a Community
(2) Ecological Succession
(3) Biological Evolution
(4) A Food Chain

65. Which organism carries out autotrophic nutrition?
(1) hawk (2) cricket (3) grass (4) deer

66. State what would most likely happen to the cricket population if all of the grasses were removed.

---

67. The diagram below illustrates the relationships between organisms in an ecosystem.

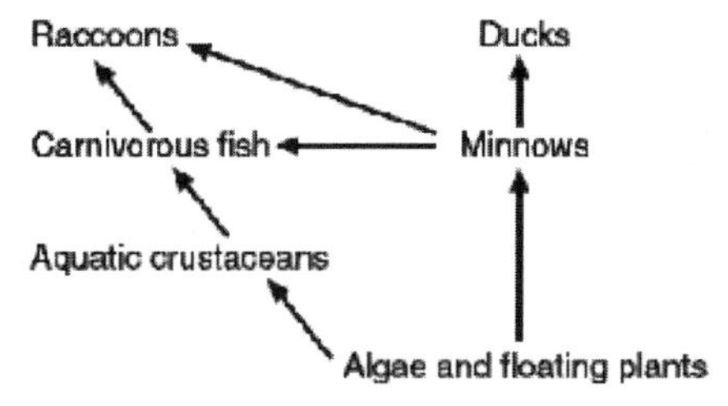

Which change would most likely reduce the population size of the carnivorous fish?

(1) an increase in the autotroph populations
(2) a decrease in the duck population
(3) an increase in the raccoon population
(4) a decrease in pathogens of carnivorous fish

68. Base your answer to the next question on the food web shown below and on your knowledge of biology.

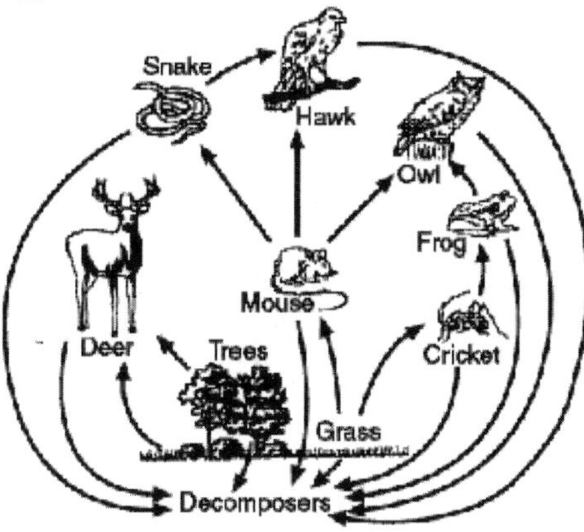

A pesticide is sprayed to kill the crickets. State *one* effect this spraying might have on the food web.

69. A food web is represented below.

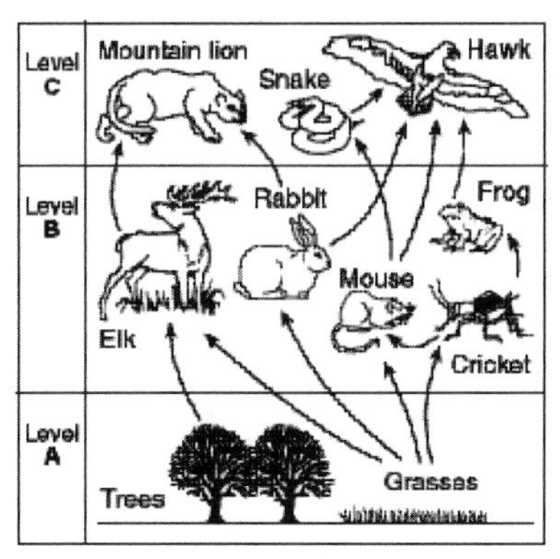

Which statement best describes energy in this food web?

(1) The energy content of level B depends on the energy content of level C.
(2) The energy content of level A depends on energy provided from an abiotic source.
(3) The energy content of level C is greater than the energy content of level A.
(4) The energy content of level B is transferred to level A.

70. What would most likely happen if most of the bacteria and fungi were removed from an ecosystem?

(1) Nutrients resulting from decomposition would be reduced.
(2) Energy provided for autotrophic nutrition would be reduced.
(3) The rate of mutations in plants would increase.
(4) Soil fertility would increase.

71. Nutritional relationships between organisms are shown in the diagram below.

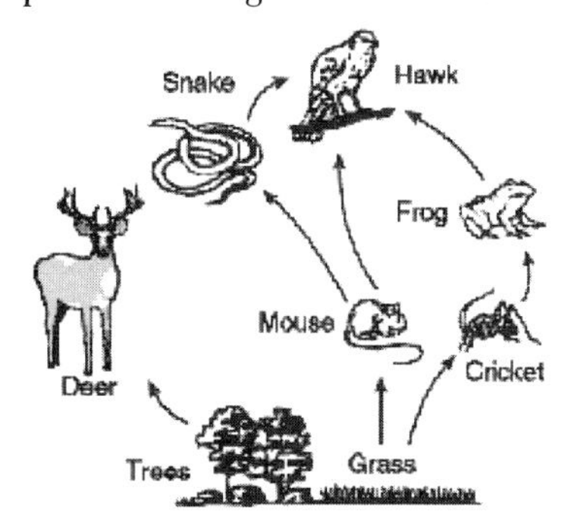

The mouse population would most likely decrease if there were

(1) an increase in the frog and tree populations
(2) a decrease in the snake and hawk populations
(3) an increase in the number of decomposers in the area
(4) a decrease in the amount of available sunlight

72. A food web is shown below.

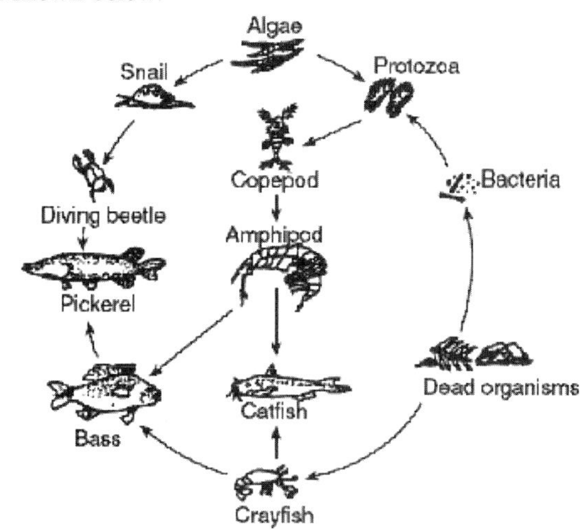

Which organisms feed on both producers and decomposers?

(1) amphipods (2) catfish (3) crayfish (4) protozoa

Base your answers to the next three questions on the information and table below and on your knowledge of biology.

The variety of organisms known as plankton contributes to the unique nutritional relationships in an ocean ecosystem. Phytoplankton include algae and other floating organisms that perform photosynthesis. Plankton that cannot produce food are known as zooplankton. Some nutritional relationships involving these organisms and several others are shown in the table below.

Nutritional Relationships in a North Atlantic Ocean Community

	Food Eaten by Animals in Community				
Animals in Community	**Codfish**	**Phytoplankton**	**Small Fish**	**Squid**	**Zooplankton**
codfish			X		
sharks	X			X	
small fish		X			X
squid	X		X		
zooplankton		X			

73. Humans are currently overfishing codfish in the North Atlantic. Explain why this could endanger both the shark population and the squid population in this community.

74. According to the table, which organism can be classified as both an herbivore and a carnivore?

75. Complete the food web below by placing the names of the organisms in the correct locations.

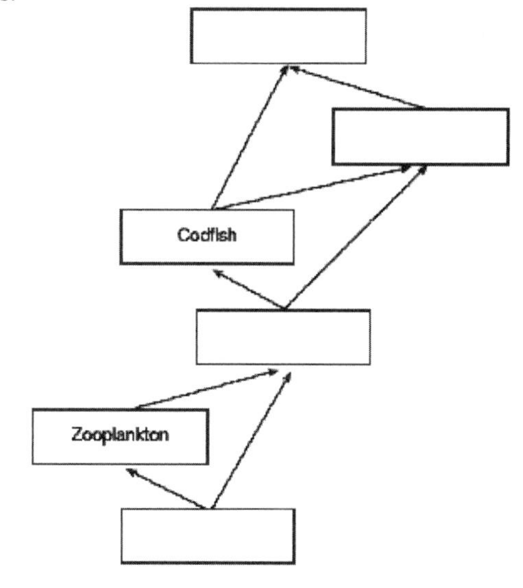

76. Which statement describes the ecosystem represented in the diagram below?

(1) This ecosystem would be the first stage in ecological succession.
(2) This ecosystem would most likely lack decomposers.
(3) All of the organisms in this ecosystem are producers.
(4) All of the organisms in this ecosystem depend on the activities of biological catalysts.

77. A partial food web is represented in the diagram below.

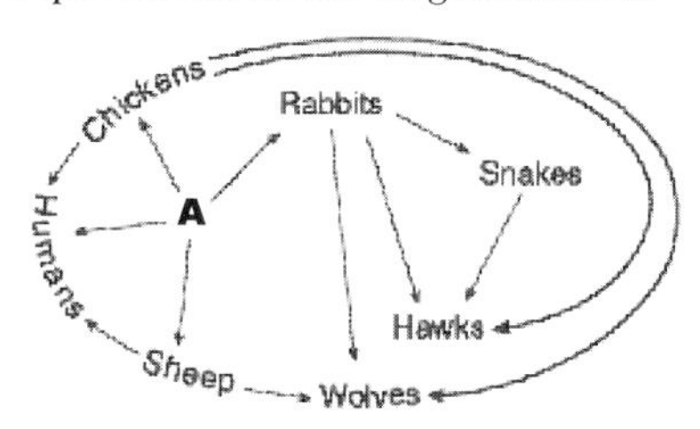

Letter *A* most likely represents
(1) autotrophs
(2) carnivores
(3) decomposers
(4) parasites

Base your answers to the next two questions on the food web and graph below and on your knowledge of biology. The graph represents the interaction of two different populations, *A* and *B*, in the food web.

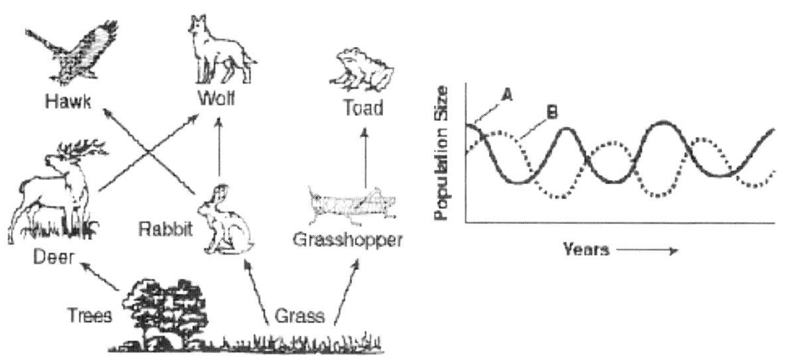

78. Population *A* is made up of living animals. The members of population *B* feed on these living animals. The members of population *B* are most likely
(1) scavengers (2) autotrophs (3) predators (4) parasites

79. Identify *one* heterotroph from the food web that could be a member of population *A*.

---

80. What would most likely happen if most of the bacteria and fungi were removed from an ecosystem?
(1) Nutrients resulting from decomposition would be reduced.
(2) Energy provided for autotrophic nutrition would be reduced.
(3) The rate of mutations in plants would increase.
(4) Soil fertility would increase.

81. A food web is represented in the diagram below.

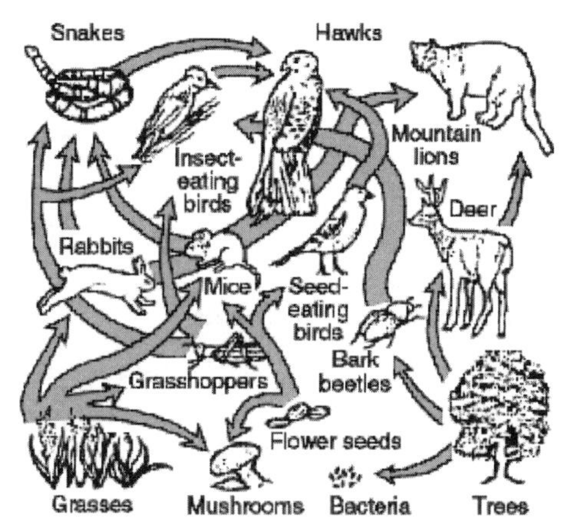

Which organisms are correctly paired with their roles in this food web?
(1) mountain lions, bark beetles – producers
hawks, mice – heterotrophs
(2) snakes, grasshoppers – consumers
mushrooms, rabbits – autotrophs
(3) all birds, deer – consumers
grasses, trees – producers
(4) seeds, bacteria – decomposers
mice, grasses – heterotrophs

82. In an ecosystem, the herring population was reduced by fishermen. As a result, the tuna, which feed on the herring, disappeared. The sand eels, which are eaten by herring, increased in number. The fishermen then over-harvested the sand eel population. Cod and seabirds then decreased. Which food web best represents the feeding relationships in this ecosystem?

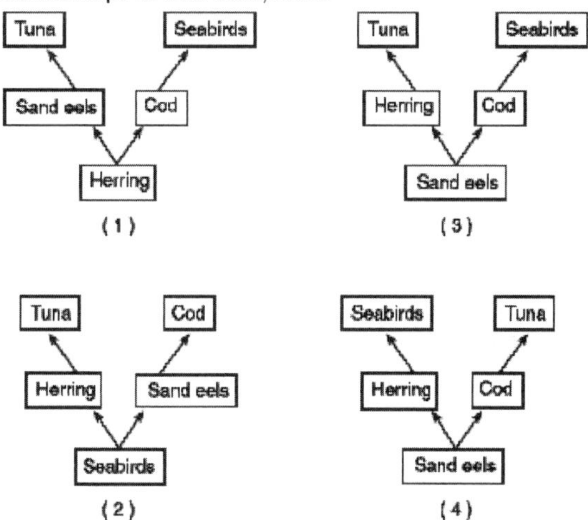

83. The removal of nearly all the predators from an ecosystem would most likely result in
   (1) an increase in the number of carnivore species
   (2) a decrease in new predators migrating into the ecosystem
   (3) a decrease in the size of decomposers
   (4) an increase in the number of herbivores

84. Nutritional relationships between organisms are shown in the diagram below.

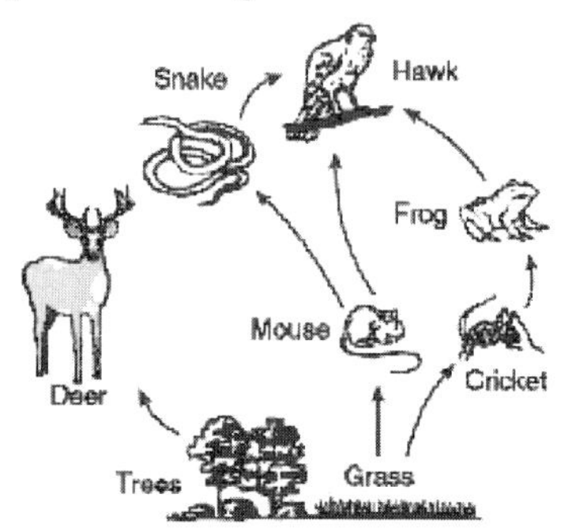

The mouse population would most likely decrease if there were
   (1) an increase in the frog and tree populations
   (2) a decrease in the snake and hawk populations
   (3) an increase in the number of decomposers in the area
   (4) a decrease in the amount of available sunlight

85. A food web is represented in the diagram below.

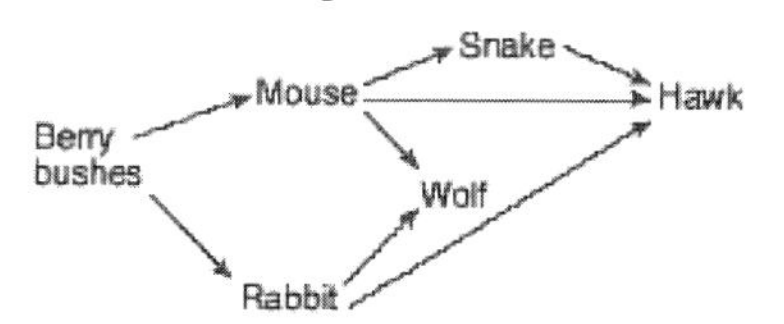

Which population in this food web would most likely be *negatively* affected by an increase in the mouse population?
   (1) snake   (2) rabbit   (3) wolf   (4) hawk

86. Some interactions in a desert community are shown in the diagram below.

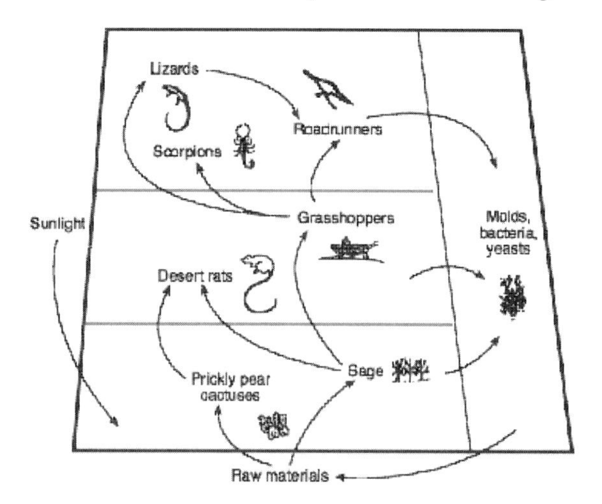

Which statement is a valid inference based on the diagram?

(1) Certain organisms may compete for vital resources.
(2) All these organisms rely on energy from decomposers.
(3) Organisms synthesize energy.
(4) All organisms occupy the same niche.

87. Base your answer to the next question on the diagram below that shows some interactions between several organisms located in a meadow environment and on your knowledge of biology.

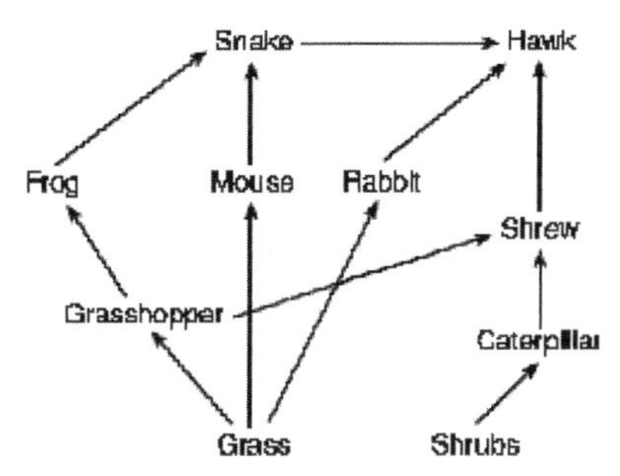

A rapid decrease in the frog population results in a change in the hawk population. State how the hawk population may change. Support your answer.

88. The graph below represents the amount of available energy at successive nutrition levels in a particular food web.

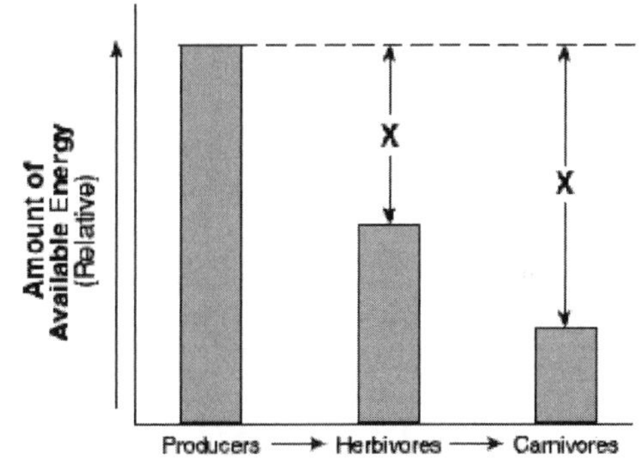

The *X*s in the diagram represent the amount of energy that was most likely

(1) changed into inorganic compounds
(2) retained indefinitely by the herbivores
(3) recycled back to the producers
(4) lost as heat to the environment

89. An energy pyramid containing autotrophs and other organisms from a food chain is represented below.

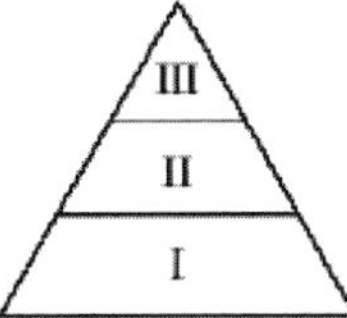

Carnivores would most likely be located in

(1) level *I*, only
(2) level *I* and level *II*
(3) level *III*, only
(4) level *II* and level *III*

90. The diagram below represents a food pyramid.

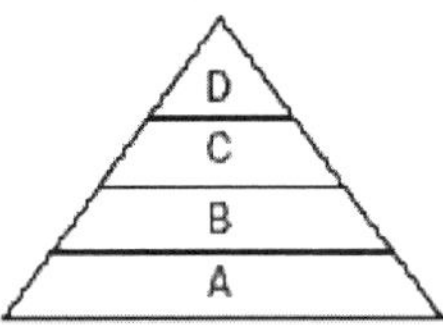

The concentration of the pesticide DDT in ndividual organisms at level *D* is higher than the oncentration in individuals at level *A* because DT is

(1) synthesized by organisms at level *D*
(2) excreted by organisms at level *A* as a toxic waste
(3) produced by organisms at level *C* which are eaten by organisms at level *D*
(4) passed through levels *A*, *B*, and *C* to organisms at level *D*

91. The diagram below represents a pyramid of energy that includes both producers and consumers

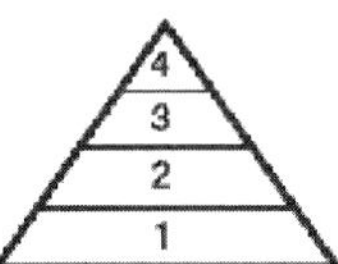

The greatest amount of available energy is found at level

(1) *1* (2) *2* (3) *3* (4) *4*

92. Which level of the energy pyramid below would contain the plant species of this salt marsh?

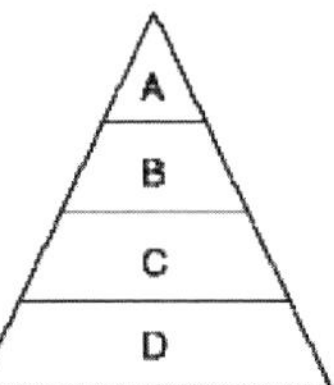

(1) *A* (2) *B* (3) *C* (4) *D*

93. The diagram below represents some energy transfers in an ecosystem.

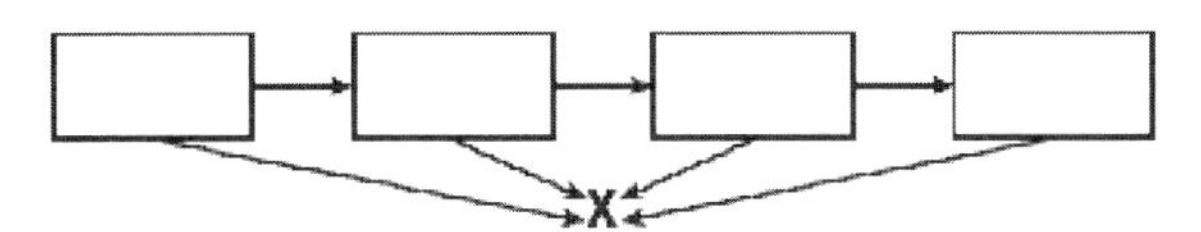

Which type of organism is most likely represented by letter *X*?

(1) decomposer
(2) autotroph
(3) producer
(4) herbivore

94. Carbon dioxide containing carbon-14 is introduced into a balanced aquarium ecosystem. After several weeks, carbon-14 will most likely be present in

(1) the plants, only`
(2) the animals, only
(3) both the plants and animals
(4) neither the plants nor animals

95. The diagram below represents an energy pyramid constructed from data collected from an aquatic ecosystem.

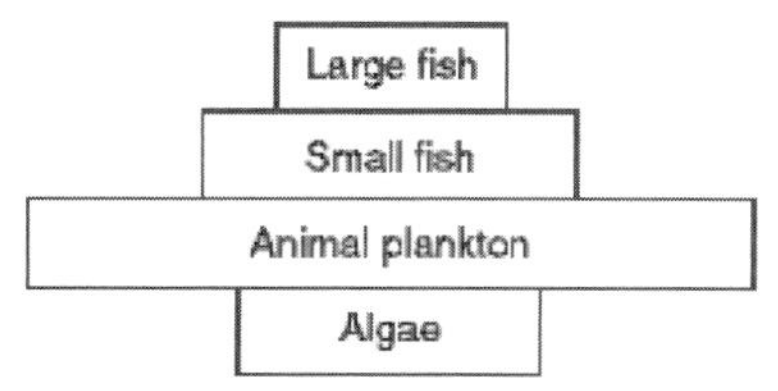

Which statement best describes this ecosystem?
 (1) The ecosystem is most likely unstable.
 (2) Long-term stability of this ecosystem will continue.
 (3) The herbivore populations will continue to increase in size for many years.
 (4) The producer organisms outnumber the consumer organisms.

96. An energy pyramid is represented below.

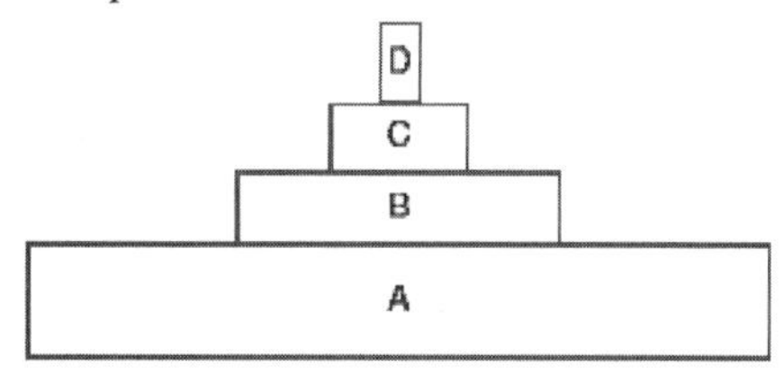

How much energy would be available to the organisms in level *C*?
 (1) all of the energy in level *A*, plus the energy in level *B*
 (2) all of the energy in level *A*, minus the energy in level *B*
 (3) a percentage of the energy contained in level *B*
 (4) a percentage of the energy synthesized in level *B* and level *D*

97. A food web is represented below.

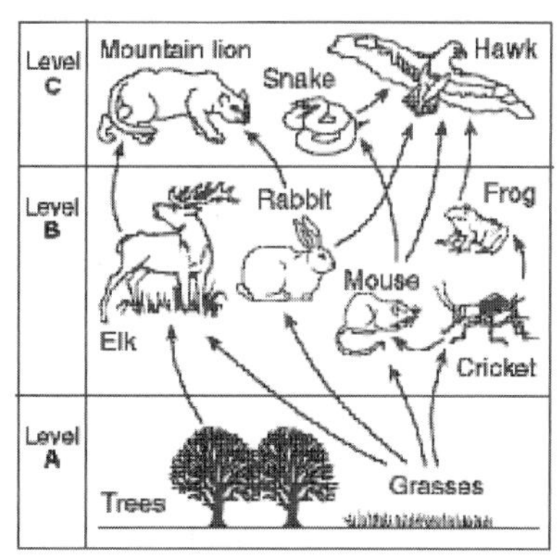

Which statement best describes energy in this food web?
 (1) The energy content of level *B* depends on the energy content of level *C*.
 (2) The energy content of level *A* depends on energy provided from an abiotic source.
 (3) The energy content of level *C* is greater than the energy content of level *A*.
 (4) The energy content of level *B* is transferred to level *A*.

98. Which diagram best represents the organisms arranged as an energy pyramid?

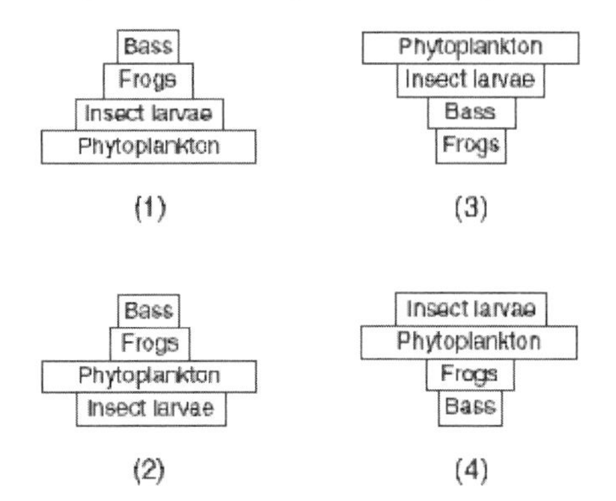

99. Mice store only a small amount of the energy they obtain from plants they eat. State what might happen to some of the remaining energy they obtain from the plants.

100. The diagram below represents an energy pyramid.`

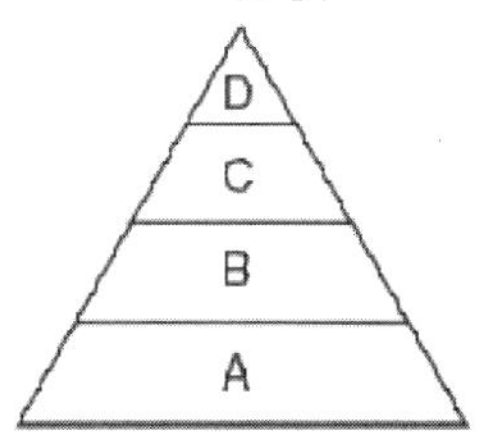

At each successive level from *A* to *D*, the amount of available energy

(1) increases, only
(2) decreases, only
(3) increases, then decreases
(4) remains the same

101. Species *A*, *B*, *C*, and *D* are all different heterotrophs involved in the same food chain in an ecosystem. The chart below shows the population of each species at the same time on a summer day.

Species	Population
A	847
B	116
C	85
D	6

Which statement best describes one of these species of heterotrophs?

(1) Species *A* is the most numerous because it can make its own food.
(2) Species *B* probably feeds on species *D*.
(3) Species *C* and *B* interbred to produce species *A*.
(4) Species *D* is most likely the top predator in the food chain.

102. Which process provides the initial energy to support all the levels in the energy pyramid shown below?

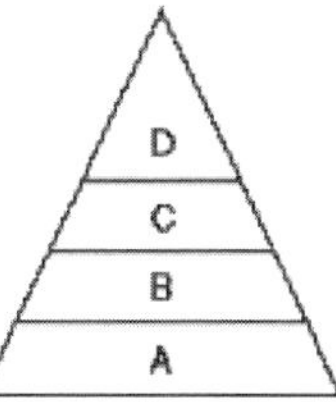

(1) circulation
(2) photosynthesis
(3) active transport
(4) digestion

103. Which statement about the pyramid of energy shown below is correct?

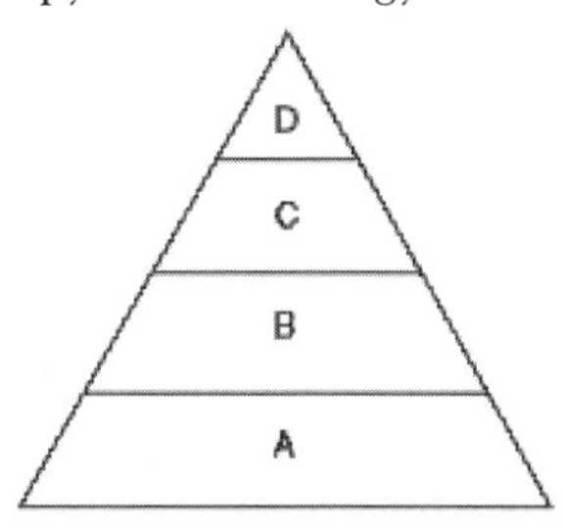

(1) The amount of energy needed to sustain the pyramid enters at level *D*.
(2) The total amount of energy at level *D* is less than the total amount of energy at level *B*.
(3) The total amount of energy decreases with each successive feeding level from *D* to *A*.
(4) The amount of energy is identical in each level of the pyramid.

104. Base your answers to the next question on the information below and on your knowledge of biology.

> Thirty grams of hay (dried grasses) were boiled in 500 milliliters of water, placed in a culture dish, and allowed to stand. The next day, a small sample of pond water was added to the mixture of boiled hay and water. The dish was then covered and its contents observed regularly. Bacteria fed on the nutrients from the boiled hay. As the populations of bacteria increased rapidly, the clear mixture soon became cloudy. One week later, microscopic examination of samples from the culture showed various types of protozoa (single-celled organisms) eating the bacteria.

Label each level of the energy pyramid below with an organism mentioned in the paragraph that belongs at that level.

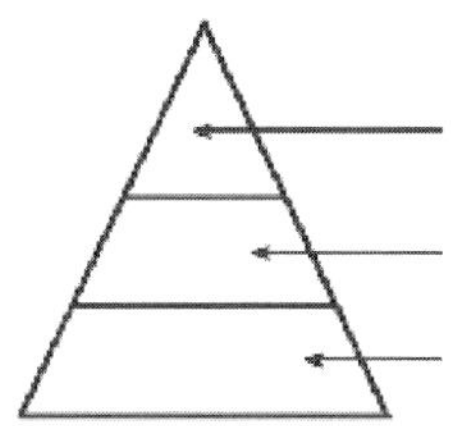

105. An energy pyramid is shown below.

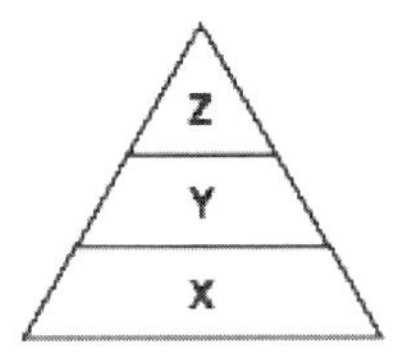

Identify *one* organism shown in the food web that would be found at level *X*.

106. Which condition would most likely upset the stability of an ecosystem?
(1) a cycling of elements between organisms and the environment
(2) energy constantly entering the environment
(3) green plants incorporating sunlight into organic compounds
(4) a greater mass of animals than plants

107. In the diagram below, what does *X* most likely represent?

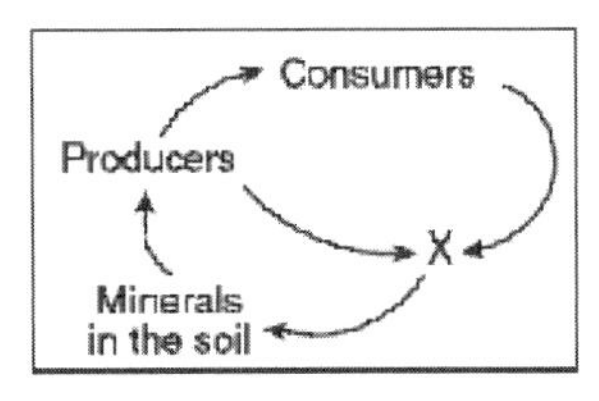

(1) autotrophs (3) decomposers
(2) herbivores (4) carnivores

Base your answers to he next four questions on the passage below and on your knowledge of biology.

Decline of the Salmon Population

Salmon are fish that hatch in a river and swim to the ocean where their body mass increases. When mature, they return to the river where they were hatched and swim up stream to reproduce and die. When there are large populations of salmon, the return of nutrients to the river ecosystem can be huge. It is estimated that during salmon runs in the Pacific Northwest in the 1800s, 500 million pounds of salmon returned to reproduce and die each year. Research estimates that in the Columbia River alone, salmon contributed hundreds of thousands of pounds of nitrogen and phosphorus compounds to the local ecosystem each year. Over the past 100 years, commercial ocean fishing has removed up to two-thirds of the salmon before they reach the river each year.

108. Identify the process that releases the nutrients from the bodies of the dead salmon, making the nutrients available for other organisms in the ecosystem.

109. Identify *one* organism, other than the salmon, that would be present in or near the river that would most likely be part of a food web in the river ecosystem.

110. Identify *two* nutrients that are returned to the ecosystem when the salmon die.

111. State *one* impact, other than reducing the salmon population, that commercial ocean fishing has on the river ecosystem.

Base your answers to the next four questions on the information below and on your knowledge of biology.

A student uses a covered aquarium to study the interactions of biotic and abiotic factors in an ecosystem. The aquarium contains sand, various water plants, algae, small fish, snails, and decomposers. The water contains dissolved oxygen and carbon dioxide, as well as tiny amounts of minerals and salts.

112. Explain how oxygen is cycled between organisms in this ecosystem.

113. Describe *one* specific way the fish population changes the amount of *one* specific abiotic factor (other than oxygen) in this ecosystem.

114. Identify *one* source of food for the decomposers in this ecosystem.

115. Describe *one* specific way the use of this food by the decomposers benefits the other organisms in the aquarium.

---

116. Many species of plants interact with harmless underground fungi. The fungi enable the plants to absorb certain essential minerals and the plants provide the fungi with carbohydrates and other nutrients. This describes an interaction between a
(1) parasite and its host
(2) predator and its prey
(3) scavenger and a decomposer
(4) producer and a consumer

117. What is the role of the algae component of a lichen in an ecosystem?
(1) decomposer
(2) parasite
(3) herbivore
(4) producer

118. A particular species of unicellular organism inhabits the intestines of termites, where the unicellular organisms are protected from predators. Wood that is ingested by the termites is digested by the unicellular organisms, forming food for the termites. The relationship between these two species can be described as
(1) harmful to both species
(2) parasite/host
(3) beneficial to both species
(4) predator/prey

119. Even before a flower bud opens, certain plant chemicals have colored the flower in patterns particularly attractive to specific insects. At the same time, these chemicals protect the plant's reproductive structures by killing or inhibiting pathogens and insects that may feed on the plant.

Which statement about the plant and the other organisms mentioned is correct?
(1) Chemicals affect plants but not animals.
(2) Organisms of every niche may be preyed on by herbivores.
(3) Any chemical produced in a plant can protect against insects.
(4) Organisms may interact with other organisms in both positive and negative ways.

120. Human activities have had a major impact on biodiversity. Scientists cannot solve this problem alone. Concerned individuals need to be involved in restoring and maintaining biodiversity.

Explain how a loss of biodiversity today can affect the survival of humans in the future.

121. When habitats are destroyed, there are usually fewer niches for animals and plants. This action would most likely *not* lead to a change in the amount of
(1) biodiversity
(2) competition
(3) interaction between species
(4) solar radiation reaching the area

122. The rapid destruction of tropical rain forests may be harmful because
(1) removing trees will prevent scientists from studying ecological succession
(2) genetic material that may be useful for future medical discoveries will be lost
(3) energy cycling in the environment will stop
(4) the removal of trees will limit the construction of factories that will pollute the environment

123. Cutting down a rain forest and planting agricultural crops, such as coffee plants, would most likely result in
(1) a decrease in biodiversity
(2) an increase in the amount of energy recycled
(3) a decrease in erosion
(4) an increase in the amount of photosynthesis

124. Some data concerning bird species are shown in the chart below.

Number of Bird Species	Location
26	northern Alaska
153	southwest Texas
600	Costa Rica

Which statement is a valid inference based on information in the chart?
(1) The different species in northern Alaska can interbreed.
(2) There are conditions in Costa Rica that account for greater biodiversity there.
(3) The different species in southwest Texas evolved from those in northern Alaska.
(4) The greater number of species in Costa Rica is due to a greater number of predators there.

125. A greater stability of the biosphere would most likely result from
(1) decreased finite resources
(2) increased deforestation
(3) increased biodiversity
(4) decreased consumer populations

126. Explain why most ecologists would agree with the statement "A forest ecosystem is more stable than a cornfield."

127. Stage *D* in the diagram below is located on land that was once a bare field.

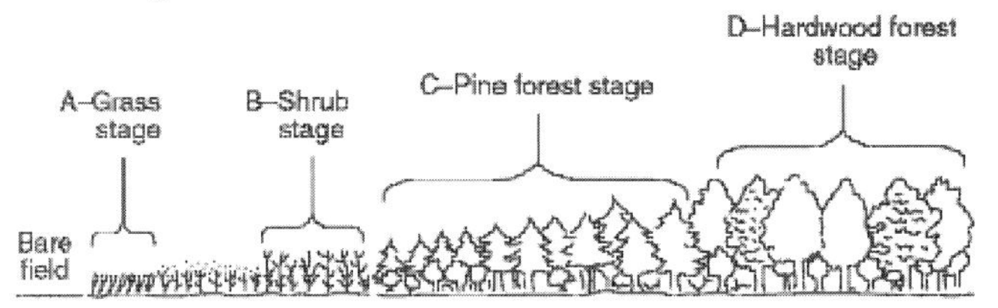

The sequence of stages leading from bare field to stage *D* best illustrates the process known as
(1) replication (2) recycling (3) feedback (4) succession

128. A fire burns an oak forest down to bare ground. Over the next 150 years, if the climate remains constant, this area will most likely
(1) remain bare ground
(2) return to an oak forest
(3) become a rain forest
(4) become a wetland

129. Base your answers to the next question on the information below and on your knowledge of biology.

> Lichens are composed of two organisms, a fungus that cannot make its own food and algae that contain chlorophyll. Lichens may live on the bark of trees or even on bare rock. They secrete acids that tend to break up the rock they live on, helping to produce soil. As soil accumulates from the broken rock and dead lichens, other organisms, such as plants, may begin to grow.

The ability of lichens to alter their environment, enabling other organisms to grow and take their places in that environment, is one step in the process of
(1) biological evolution
(2) ecological succession

(3) maintenance of cellular communication
(4) differentiation in complex organisms

130. Which concept is represented in the graph below?

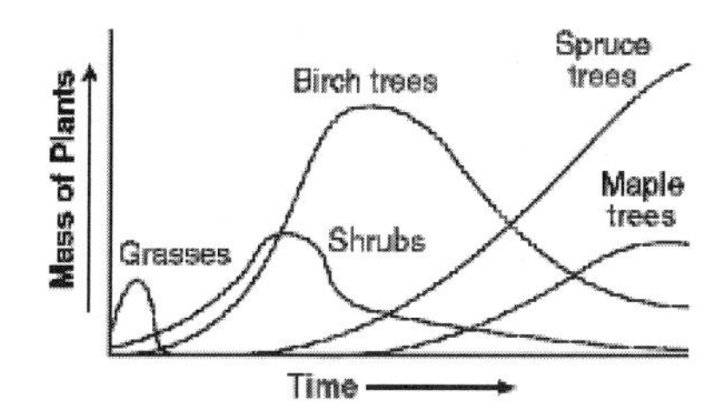

(1) ecological succession in a community
(2) cycling of carbon and nitrogen in a forest
(3) energy flow in a food chain over time
(4) negative human impact on the environment

Base your answers to the next two questions on the information below and on your knowledge of biology.

A pond in the Adirondack Mountains of New York State was once a fishing spot visited by many people. It was several acres in size, and fishermen in boats were a common sight. Over time, the pond has become smaller in area and depth. Places where there was once open water are now covered by grasses and shrubs. Around the edges of the pond there are cattails and other wetland plants.

131. Identify the ecological process responsible for the changes to this pond.

132. Predict what will most likely happen to this pond area over the next hundred years if this process continues.

---

133. Many years ago, a volcanic eruption killed many plants and animals on an island. Today the island looks much as it did before the eruption. Which statement is the best possible explanation for this?
(1) Altered ecosystems regain stability through the evolution of new plant species.
(2) Destroyed environments can recover as a result of the process of ecological succession.
(3) Geographic barriers prevent the migration of animals to island habitats.
(4) Destroyed ecosystems always return to their original state.

134. Lichens and mosses are the first organisms to grow in an area. Over time, grasses and shrubs will grow where these organisms have been. The grasses and shrubs are able to grow in the area because the lichens and mosses
(1) synthesize food needed by producers in the area
(2) are at the beginning of every food chain in a community
(3) make the environment suitable for complex plants
(4) provide the enzymes needed for plant growth

Base your answers to the next two questions on the diagram below, which represents the changes in an ecosystem over a period of 100 years, and on your knowledge of biology.

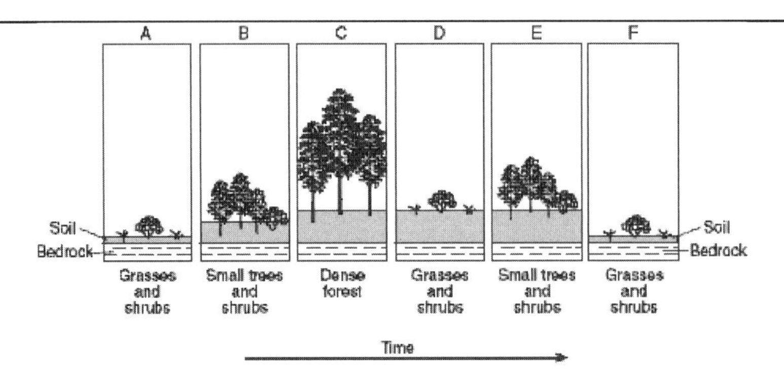

135. State *one* biological explanation for the changes in types of vegetation observed from *A* through *C*.

136. Predict what would happen to the soil and vegetation of this ecosystem after stage *F*, assuming no natural disaster or human interference.

137. Which concept is represented in the graph below?

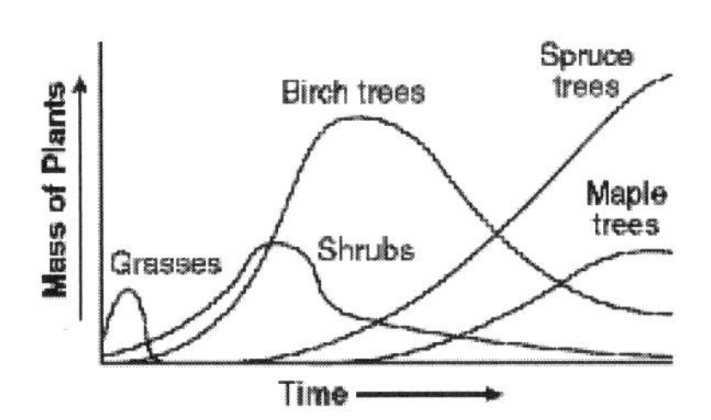

(1) ecological succession in a community
(2) cycling of carbon and nitrogen in a forest
(3) energy flow in a food chain over time
(4) negative human impact on the environment

138. Years after the lava from an erupting volcano destroyed an area, grasses started to grow in that area. The grasses were gradually replaced by shrubs, evergreen trees, and finally, by a forest that remained for several hundred years. This entire process is an example of

(1) feedback
(2) ecological succession
(3) plant preservation
(4) deforestation

Base your answers to the next four questions on the information and data table below and on your knowledge of biology.

A student added two species of single-celled organisms, *Paramecium caudatum* and *Didinium nasutum*, to the same culture medium. Each day, the number of individuals of each species was determined and recorded. The results are shown in the data table below.

Culture Population

Day	Number of *Paramecium*	Number of *Didinium*
0	25	2
1	80	5
2	150	10
3	50	30
4	25	20
5	0	2
6	0	0

Directions (for the next two questions): Using the information in the data table, construct a line graph on the grid provided, following the directions below.

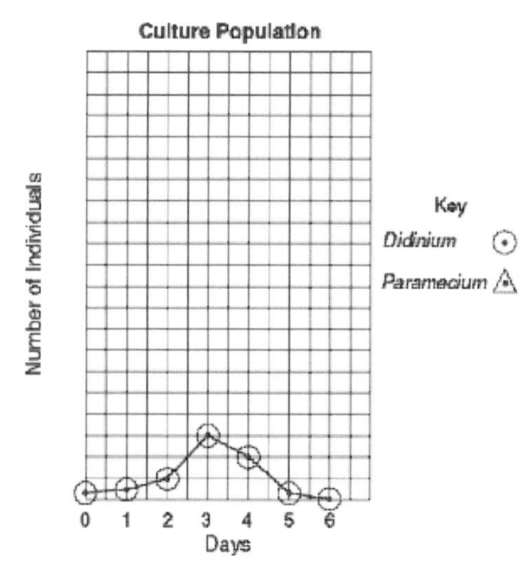

139. Mark a scale on the axis labeled "Number of Individuals" that is appropriate for the plotted *Didinium* population and for plotting the *Paramecium* population.

140. Plot the data for *Paramecium* on the grid. Surround each data point with a small triangle and connect the points.

141. What evidence in the data indicates that *Didinium* could be a predator of the *Paramecium*?

142. State *two* possible reasons that the two populations died off between days **4** and **6**.

CHAPTER 7: HOW HUMANS AFFECT THE ENVIRONMENT

In this environment (see figure below), there are human beings, fish, birds, a moose, frogs, trees, sun, water, and air. As you learned, the living organisms (example humans) have an effect on both the biotic (living organisms), such as birds, fish, and deer, and on the abiotic (nonliving things), such as water and air.

Humans Affect Living Things

In the picture above, you see the fisherman is trying to catch fish, the man is trying to shoot a bird, and another man hunts a deer. You see that human beings have an effect on living organisms.

Limited Resources

On the Earth, you realize we have limited (a finite amount of) natural resources such as oil, coal, iron ore, aluminum, silver, sunlight, water, food, and soil.

Two types of resources are renewable resources and nonrenewable resources (see below).

Renewable resources can be **replaced** in a reasonable amount of time:

a. **sunlight** (sun's energy, also called solar energy) shines on the Earth every day.

b. **water** - water in lakes and oceans evaporates (goes up as water vapor) and water vapor comes down as rain.

c. **animals** (example fish) and **plants** (example crops, such as corn, wheat, sugar) are renewable because they reproduce (producing baby animals and plants). **Wood** (from cutting down trees) and **foods** such as corn, wheat, eggs, meat, chicken, and fish are renewable (replaceable) because plants and animals reproduce. However, if we catch too many fish in a lake and there are hardly any fish left to reproduce, then there would be almost no baby fish to grow up and reproduce; the fish are then nonrenewable.

Nonrenewable Resources cannot be replaced in a reasonable amount of time:

a. **fossil fuels** (coal, petroleum, oil, natural gas) - When oil, coal, petroleum, or natural gas is depleted (used up) , we cannot produce more oil, coal, petroleum, or natural gas in a reasonable amount of time. Note: Petroleum is made into oil, gasoline, and natural gas (oil, gasoline, and natural gas are made from petroleum).
Note: We drill for oil, getting oil out from the ground or offshore (from the bottom of the ocean). The advantage (positive effect) of getting oil out is that it is used as a fuel to provide energy. More oil drilling in the U.S. will produce more oil (increase the oil supply), create jobs, and lower oil prices. The disadvantage of drilling for oil is there can be oil spills, either during drilling, or in the pipeline or tanker carrying the oil. Oil spills on land or water, pollutes the environment and kills many plants and animals.

b. **minerals - iron ore, aluminum, silver**- When they are used up, there is no more left.

The Earth has limited (finite) resources (examples water, plants, animals, oil, coal, iron ore). When we increase human consumption (we keep on using more resources), some resources can be depleted (used up) and cannot be renewed (replaced); other resources which are renewable can still be in danger if they are being used up too quickly and cannot be renewed (replenished) that quickly.

Question: In order to reduce consumption of nonrenewable resources, humans could

(1) burn coal to heat houses instead of using oil
(2) heat household water with solar radiation
(3) increase industrialization
(4) use a natural-gas grill to barbecue instead of using charcoal

Solution: In order not to use up the nonrenewable resources like fossil fuels (coal, oil, natural gas) humans could use renewable resources like sunlight (sun's energy, **solar radiation**, **solar energy**). Answer *2*

Preserve Resources

Let's see what we can do to preserve our resources (to make sure that our resources such as wood, oil, and coal do not get used up). Remember the three R's: **Reduce, Reuse,** and **Recycle**.

Reduce: Try to use less of the resource (example oil, coal, or gasoline, which provide energy) or not to use the resource:

a. On an air conditioner, when you set the temperature to 75°F (24°C) instead of 68°F (20°C) in the room, you save electricity (electrical energy).

b. When you set the thermostat for your heating system for 68°F (20°C) when you are home but only 60°F (15°C) when you are not home, you save energy when you are not home.

c. Use energy efficient (Energy Star) air conditioners, refrigerators, washing machines, etc. to save energy.

d. You can walk instead of using a car; you save fuel (such as gasoline).

e. Use renewable resources instead of using up (consuming) nonrenewable resources:

1. use solar energy (solar radiation), a renewable resource, instead of oil, a nonrenewable resource, for heating water in homes. Note: A disadvantage of solar energy is that it cannot be used all over the world, but only in places where there is a lot of sunlight.

2. use electricity (renewable if it is made using falling water or wind) instead of gasoline (nonrenewable) to run cars.

Reuse: Reuse the same thing over and over again instead of throwing it away. Use dishes, silverware, pots and towels instead of paper or plastic plates and cups, towel paper, and disposable foil pans, which are thrown away. Reusing helps to save resources such as trees (used to make paper) and oil (used to make plastic).

Recycle: Take products such as plastic and glass bottles, cans, etc. and put them into recycling bins so the plastic, glass, or metal **can be used over again** instead of just throwing them away in the garbage. Recycling helps to save our natural resources such as trees, oil, glass, metal, etc.

In short, you can help preserve (save) our resources: **reduce** (use less of our resources), **reuse** (use our resources over and over again), and **recycle** (do not throw away plastic and glass containers or metal cans, but recycle them so the plastic, glass, or metal can be used over again).

Natural Processes in an Ecosystem Affect Humans

These natural processes are:

1. maintaining good quality of the atmosphere (air). The air must have the right amounts of oxygen and carbon dioxide for living things. Plants take in carbon dioxide and give off oxygen (photosynthesis) and animals take in oxygen and give off carbon dioxide (respiration), maintaining a good amount of oxygen and carbon dioxide in the air. Rain also clears (washes out pollutants from) the air and moistens the air.

2. forming soils. Let's understand how soil is formed. Soil is formed (made) when water freezes in cracks in rocks (breaking up the rocks), tree roots grow into rocks, and acids (which are made by both lichens and other plants) all **break up rocks** into **small pieces;** the small pieces of rock together with decomposed (broken down chemically) dead plants and dead animals form soil. Plants (producers) can grow in soil and it is a home (habitat) for decomposers.

3. controlling the water cycle, providing a fresh supply of water for all plants and animals. Water **evaporates** from lakes, oceans, streams, land, and plants, forming water vapor. Water vapor rises and **forms clouds**; clouds are made of tiny liquid drops of water. Water in the clouds comes down as **rain**, snow, ice, etc. The rainwater goes back into the soil, lakes, and streams. Plants take in water. The water evaporates and starts the cycle over again (forms clouds, then comes down as rain, etc., which goes into the soil, lakes, streams, and rivers). Now plants and animals (examples humans, dogs, deer) have a fresh supply of water.

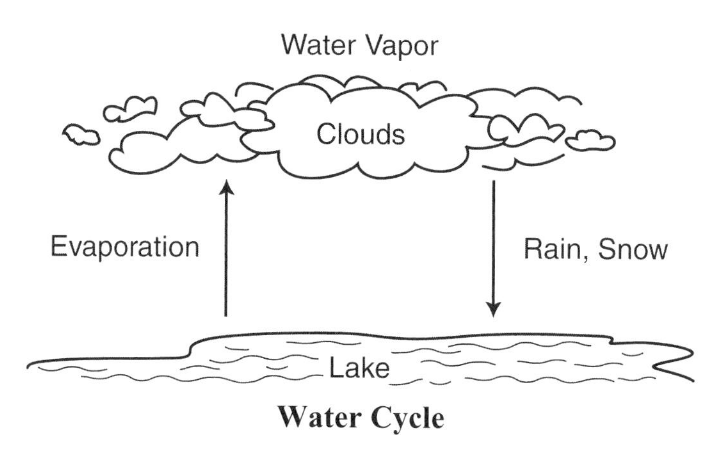

Water Cycle

4. removal of wastes and recycling of materials. Decomposers (example **some bacteria** and **fungi**) **remove wastes** by **breaking down** chemically **dead organisms** and **wastes** into simple materials like carbon dioxide and nitrogen compounds, returning **(recycling) materials to** the **air** and **soil**. The carbon dioxide goes into the air to be used by plants in photosynthesis; the nitrogen compounds go back into the soil to be used by plants (recycling of materials).

Note: When the decomposers break down wastes, the materials (nitrogen compounds) go back into the soil, enriching the soil (making the soil better).

5. energy flow. In the process of photosynthesis, **plants** use the **sun's energy** to make food (sun's energy is transferred to the plants). As you go up the food chain, food web, or food pyramid, **energy** is **transferred** (**energy flow**) from organisms at one level (plants, which are producers) to organisms at the next level higher up (herbivores, which eat plants), and then to the level above that (carnivores, which eat animals), but most of the

energy is lost as heat (see energy pyramid, chapter 6). Energy cannot be recycled (the same energy cannot be used over again), therefore you need a constant source (supply) of energy (the sun). In short, energy flows from sun to producers, to herbivores, to carnivores. Energy flows from all levels of the pyramid to decomposers.

You will learn later how humans change these basic processes (maintaining air quality, forming soils, controlling the water cycle, removing wastes and recycling nutrients, and energy flow, listed above); many times these changes (example, too much pollution) that humans make **harm the world** (earth, air, water resources, etc.). We (humans) should try to protect the Earth, living organisms, plants, flowers, grasses, animals, especially native plants and animals, and our ecosystems and environment.

PRACTICE QUESTIONS AND SOLUTIONS

Question: Currently, Americans rely heavily on the burning of fossil fuels as sources of energy. As a result of increased demand for energy sources, there is a continuing effort to find alternatives to burning fossil fuels.

Discuss fossil fuels and alternative energy sources. In your answer, be sure to:

1. state *one disadvantage* of burning fossil fuels for energy
2. identify *one* energy source that is an alternative to using fossil fuels.
3. state *one* advantage of using this alternative energy source
4. state *one disadvantage* of using this alternative energy source

Solution: 1. Fossil fuels (petroleum, oil. coal, natural gas) are nonrenewable resources (they get used up and we cannot produce more). Other disadvantages of fossil fuels are: Fossil fuels cause pollution (burning fossil fuels such as coal causes pollution). Fossil fuels cause global warming and acid rain (which you will learn about later in the chapter).

2. Solar energy. Other energy sources are wind and water.
3. Solar energy is a renewable resource. It does not get used up.
4. Solar energy (sun's energy) can only be used where there is a lot of sunlight, only in daytime.

Now Do Homework Questions #1-15, pages 35-37.

Population Growth

Human beings are part of the Earth's ecosystem, which is made up of the living and nonliving things in an area (example, pond, lake, field). Human beings can intentionally or unintentionally change the equilibrium of the ecosystem (how the ecosystem works). Humans modify (can change, damage, upset, disrupt) ecosystems and environments by an increase in population, consumption of (using too many) natural resources (such as coal and oil), technology (tools and machinery) and industrialization (industries). When human population increases, it can cause pollution (example more industries burning more fuels and giving off more pollution), more sewage (toilet wastes), scarcity of (too little) habitat (homes) for animals (because humans destroy forests to build houses, schools, etc.), and depletion (using up) of resources. Increasing human population has a negative effect (harmful effect) on the stability of the environment and on ecosystems. Humans destroy habitats by destroying forests, by direct harvesting (destroying or removing a species from its habitat), polluting the atmosphere, importing species, and land use (which will be explained later in the chapter.) Humans are modifying (changing) the ecosystem so much, that it might not be possible to go back to the way the ecosystem was originally.

Human population increased slowly for thousands of years. Look at the graph. A few hundred years ago, humans began to have more food and better health care and people lived longer. Over the last few hundred years, the human population increased tremendously. Look at the graph. The number of people increased from one billion people to almost six billion people in a few hundred years.

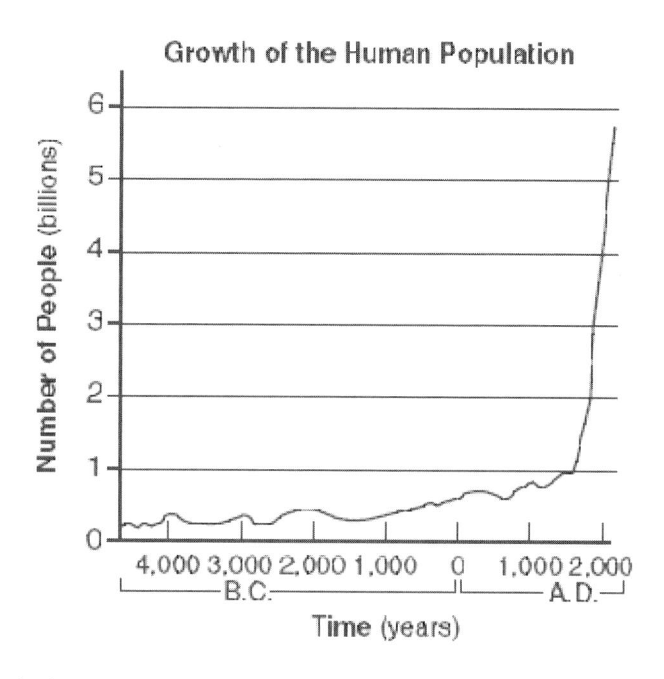

When population increases, people need more place to live. They cut down (destroy) forests, grasslands, and wild areas where animals live (the animals' home (habitat)), to build apartment houses, schools, stores, parking lots, etc.

Solution to **Population Growth:** Instead, people should build buildings with more floors, to provide housing for more people without destroying habitats.

The population is increasing so rapidly that some scientists are concerned that it could soon reach (or go beyond) the Earth's carrying capacity and then there might not be enough food, water, space and oxygen for all the people, and some people will starve. Note: Human population carrying capacity is the number of people that can live or survive in an area with enough food, water, oxygen, space, etc. We had, in Chapter 6, two graphs showing carrying capacity and what happens to organisms (living things) after that.

Some people want to slow down the population growth so it does not reach or does not exceed the carrying capacity of the Earth; this will prevent people from starving.

PRACTICE QUESTIONS AND SOLUTIONS

Question: Which situation has had the most *negative* effect on the stability of the environment and on the ecosystems of Earth?

1. use of air pollution controls
2. use of natural predators to control insect pests
3. recycling glass, plastic, and metal
4. increasing human population

Solution: Increasing human population causes pollution, destroys habitats (example forests and meadows) to build houses, schools etc., to make farms, and uses up too many natural resources (example coal and oil). This destroys the stability of the ecosystem. Answer *4*

Now Do Homework Questions #16-24, pages 37-38.

HUMAN ACTIVITIES AFFECT THE ENVIRONMENT

Human Activities that are Helpful to (Have a Positive Effect on) Earth's Ecosystem, Environment

1. We should try to make sure the Earth, living organisms, plants, flowers, grasses, animals, especially native plants and animals, and our ecosystems and environment do not get harmed (damaged, destroyed).
2. In a vacant lot, plant many different plants that are native (were not imported) to the area. This helps biodiversity (many different species of plants in the area).

Human Activities that are Harmful to (Have a Negative Effect on) Earth's Ecosystem, Environment

Human beings destroy habitats (homes, such as forests, where organisms, live); this causes a loss of biodiversity. Humans destroy habitats (habitat destruction) **through deforestation, direct harvesting, imported species, land use, disrupting food chains, erosion, pollution, atmospheric changes,** etc. (explained below).

1. Deforestation (destroying forests), or **plowing through meadows:**

a. **Deforestation (destroying forests)** : Living organisms (plants and animals) live in the forests. The forest is the home (habitat) of these living organisms. **Human beings destroy forests** by logging (**cutting down trees**) **to build apartment buildings, stores, parking lots, one-crop farms, etc**.

Deforestation

The living organisms (animals) now have no home and need to find another place to live. Many of these living organisms die. The human beings **harmed** the **ecosystem** by **causing** many living **organisms** to **die.** (These organisms were food for other organisms.)

When people build roads into or through a forest, people may damage the forest. By tearing up the forest and using the forest land as a commercial source of food, medicine, or other products, they ruin (damage) the forest ecosystem, destroy habitats, and reduce biodiversity.

Solution: Do not destroy the habitat. **Cut down fewer trees** so you will not destroy the habitat (home for living organisms).

b. **Plowing through** a **meadow**: Many organisms live in a meadow (land that has grasses). Human beings cut down a meadow and turn it into a field which only grows one crop (example cotton). Many of the **organisms** that live in the meadow will **die.** Humans harm the ecosystem by plowing

through the meadow, causing many organisms to die, and then planting on the entire field only one type of crop. (One type of insect can destroy the entire crop).
Note: In a food chain or food web, other organisms needed these organisms (that once lived in the meadow) for food. When we plow the meadow, we kill off organisms, disrupting the food chain and food web.

Solution: Do not plow through a meadow. Try not to destroy the habitat (the meadow).

Effects of Deforestation (destroying forests) or **plowing through a meadow.** Two effects of destroying forests (deforestation) or destroying meadows are **loss of biodiversity** and an **increase in carbon dioxide** (explained below):

a. Loss of Biodiversity: The forests (or meadows) have many different species of plants and animals, which means the forests have biodiversity. When humans cut down the forests or plow through the meadow, biodiversity (having many different species of plants and animals) is lost, which badly affects ecosystems, food chains and food webs. When plant and animal species are lost, genes (genetic material) are lost which could be used for medicine and research. Biodiversity is needed to find new medicines and to find new genes, which can be used (genetically engineered) to produce better and pest resistant crops.

Prevention and Solution of Loss of Biodiversity: Humans can help to lessen (reduce) or solve the loss of biodiversity; **do not destroy habitats** and **fix (rehabilitate) habitats** that were damaged.

b. Increase in Carbon Dioxide: In photosynthesis, plants such as trees and grasses, etc. take in carbon dioxide and give off oxygen. Cutting down trees or destroying grasses means less carbon dioxide will be taken in by the trees and the amount of carbon dioxide in the air will increase. You will learn later that an increase in carbon dioxide causes global warming (which means the temperature of the globe (Earth) increases and some living organisms cannot live (will die) at the higher temperatures.)

2. Direct Harvesting: Direct harvesting is **destroying or removing a species from its habitat.** Humans exploit (use selfishly) wild plants and animals for their products (examples elephant tusks, plywood) and as pets; this disrupts ecosystems.

a. Humans harmed elephants and walruses by killing them for their ivory tusks. Humans killed baby harbor seals for their pelts.

Humans Killed Elephants for Their Ivory Tusks

b. Humans removed Colombian parrots from their own country and brought them to the United States to be used as pets. Many parrots died when they were moved (carried) from Colombia to the United States.

c. Humans harm the ecosystem by cutting down trees of the tropical rain forest and manufacturing plywood.

d. Human beings (hunters) shot millions of passenger pigeons, causing the passenger pigeon to become extinct (all the passenger pigeons were killed off). When living things (example passenger pigeon) become extinct, there is a loss of biodiversity and a loss of genes (genetic material). Too much hunting, trapping, and fishing can cause the animals to become extinct (examples dodo bird and passenger pigeon).

e. Humans are hunting too many blue whales and the blue whale is an endangered species (species that is in danger of becoming extinct).

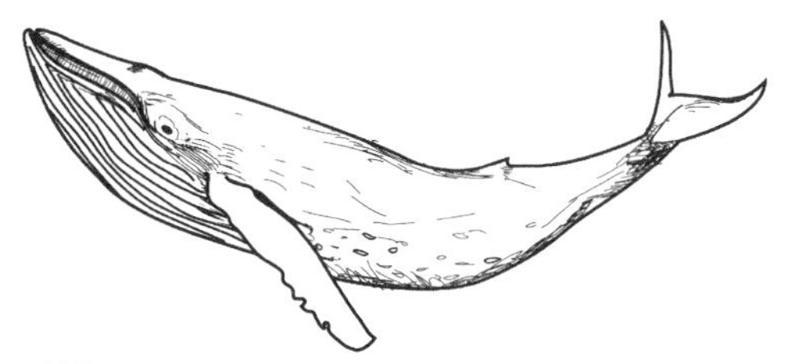

Humans Hunted Blue Whales;
Blue Whale Is An Endangered Species

Other endangered species are polar bears and giant pandas, and many more.

Polar Bear
Endangered Species

Giant Panda
Endangered Species

We want to save endangered species (to preserve biodiversity and to save all the genes (genetic material)). There are laws protecting endangered species. Today we sometimes have problems, because, even though humans are not supposed to hunt these animals, humans still capture and kill the organisms (poaching) because the humans want to get certain products.

Prevention and Solution to **Direct Harvesting:** There are laws protecting endangered species. **Do not overhunt (hunt too much), overfish and overtrap.** Do not hunt too many animals of the same type, too many fish of the same type, or trap too many animals of the same type. Do not remove animals (example parrots) from their own country to another country because many animals die in their new environment (new place).

3. Imported Species: Humans sometimes **imported species** (such as the rabbit or gypsy moth) from one environment and released the imported species (examples rabbit, gypsy moth) in the new environment. These imported species **harm the environment**, ecosystems.

Imported Species

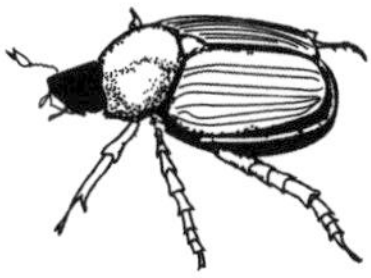

Japanese beetle

Gypsy moth

Zebra mussel

Asian long-horned beetle

a. Before the 1850's, Australia had no **rabbits. Human beings imported and released 24 rabbits into Australia.** The rabbits multiplied. The **rabbit population increased tremendously** because the **rabbits** that had been imported **had no natural enemies.** The imported rabbits successfully competed for food, eating the vegetation that other animals living there would have eaten; some of the organisms (animals) that lived there did not have enough food and died.

b. **Japanese beetle** and **gypsy moth** (see figure above) were accidentally imported and released into the United States. They became pests. They have no natural enemies; they keep multiplying, eat the organisms living here in the United States and sometimes even kill off the organisms living here.

c. **Zebra mussel** (a small shellfish) (see figure above) and the **goby** (a small fish) were both accidentally imported from the Black Sea into the Great Lakes in the United States. Cargo ships that are not carrying a large load need to make the ship heavier; therefore they pump water (example from the Black Sea) into ballast tanks in the ship. People did not realize that the water in the tanks contained living organisms, such as the zebra mussels or gobies.

When the ships reached the Great Lakes, where they were going to take on a large cargo (a big load), they pumped out the tanks (which contained water from the Black Sea that had zebra mussels or gobies). Therefore, the Great Lakes now has the imported zebra mussel and gobies, which came from the Black Sea.

The imported zebra mussels attach themselves to the water intake pipes of power plants and block the opening of the pipes (clog the pipes). **Zebra mussels** are filter feeders and **eat** (take in) **plankton and other tiny organisms;** they **disrupt the food chain** by **removing** some of the **plankton**. **Zebra mussels take in PCB's** and other **pollutants** with their food and the pollutants concentrate in their bodies. Organisms that eat the zebra mussels will have more concentrated pollutants (an increase in pollutants).

Gobies **take away food** and **places to lay eggs** from native fish in the Great Lakes. Gobies also eat the eggs of other fish (sport fish); fishermen have a problem because they like to catch these sport fish. Gobies eat zebra mussels, therefore gobies concentrate even more the PCB's and other pollutants that the zebra mussels collect. When larger fish eat the gobies, the larger fish increase their concentration of PCB's and other pollutants more than the gobies and zebra mussels.

d. **Asian long-horned beetle** (see figure above) was imported to the United States with wooden material. The beetle attacked trees in New York; it makes holes in trees and kills the trees.

Solution to **Imported Species:**

a. **Use a disease organism** that **only affects** (example kills) the **imported, released species**. The Australian rabbits were controlled by a disease organism. This method can be used to control pests.

b. Another **pest control method** that is safe is setting **traps** that have a **chemical scent** that **attracts** the **pests.** No other species is harmed by these traps.

c. Breed and release native species that will kill the pest but will not kill other species.

Solutions that are not as good:

d. Import another species to control (kill off) the original imported species. The problem is that the new imported species may kill native organisms and also eat a lot of vegetation needed by the native organisms.

e. Use pesticides or poisons to kill the imported species; the poisons may also kill native species.

Prevention of **Imported Species:** Laws restrict importing of fruits and vegetables from other countries to prevent diseases and import of organisms (example insects) that damage plants and animals.

Question: When brown tree snakes were accidentally introduced onto the island of Guam, they had no natural predators. These snakes sought out and ate many of the eggs of insect-eating birds. What probably occurred following the introduction of the brown tree snakes?

(1) the bird population increased
(2) the insect population increased
(3) the bird population began to seek a new food source
(4) the insect population began to seek a new food source

Hint: Insect-eating birds means birds that eat insects.

Solution: Since the snakes had no natural predators (no natural enemies), the snake population increased. Snakes ate the eggs of the birds (these birds ate insects), therefore the bird population decreased and more insects survived. The insect population increased. Answer 2

Now Do Homework Questions #25-47, pages 38-42.

Let's continue with human activities that are harmful to (have a negative effect on) Earth's ecosystems, environment.

4. Land Use: When the population increases, humans need more land to live on, to build schools, to build factories, and even for recreation. Humans chop down (destroy) the forests, wilderness and grass areas to build houses and schools. etc. The animals can no longer live there and must find a new habitat (home) to live. Many of these animals will die.

Solution: Build apartment buildings with more stories, therefore more people can live in the same space. Then humans might not have to destroy forests and grasslands to provide more homes for human beings.

5. Disrupting Food Chains: If people kill off any species (example species of fish), the animals that eat the killed off species of fish might have very little food and might die.
Look at the food chain:

algae → (eaten by) small fish → (eaten by) big fish → (eaten by) seagulls and pelicans (birds)

In the food chain, you see fish are eaten by seagulls. If humans kill too many fish, the sea gulls therefore will not have enough food and will die. Killing the fish disrupts (breaks up) the food chain.

If humans kill off all the mosquitoes in an area (see food chain below), the organisms (example bats) that eat the mosquitoes might die because they do not have mosquitoes to eat (food chain is disrupted).

mosquito → bat

If people overhunt, overfish, or remove a species, the animals that eat that overhunted species might die from lack of food. Killing off or removing an animal from a food chain or food web disrupts (breaks up) the food chain.

6. Erosion (soil removed from land): Human activities such as deforestation or poor farming methods cause more soil from the land to be removed (erosion). Soil removed from the land (erosion) causes the remaining soil to be poorer for growing crops (plants that are grown, usually for food, or to sell something, examples wheat, cotton).

In the next section, you will learn how **humans** also cause air pollution, water pollution, soil pollution, atmospheric changes (global warming and ozone depletion), and toxic wastes, which **harm** the Earth, environment, ecosystems, and living organisms. Also, when humans collect leaves and grass and dump them into landfills, it is harmful to the environment because the materials in the leaves are not recycled (used again).

In short, **human activities** that are **harmful to** Earth's **environment** and **ecosystem** are habitat destruction, direct harvesting, imported species, land use, disrupting food chains, and erosion (explained above).

Question: Base your answers to these two questions on the information below and on your knowledge of biology:

Our national parks are areas of spectacular beauty. Current laws usually prohibit activities such as hunting, fishing, logging, mining, and drilling for oil and natural gas in these areas. Congress is being asked to change these laws to permit such activities.

Question 1: For *each* activity listed above, state *one* way that activity could harm the ecosystem.

Question 2: For *each* activity, state *one* way allowing the activity could benefit society.

Solution 1: Possible Correct Answers:

Activity	How it Harms the Ecosystem (Negative Effects)
1. hunting, fishing	Hunting and fishing caused animals (and fish) to die, therefore other animals might not have enough food to eat and might die; killing fish and animals disrupts food chains
2. logging (chopping down trees)	could destroy habitats (places which might be the home of some living organisms)
3. mining (getting minerals from the ground, such as copper, lead, iron)	could destroy habitats
4. oil drilling	danger of oil spills (could pollute the ecosystem); damages habitats

Solution 2: Possible Correct Answers:

Activity	How it Benefits (Helps) Society (Positive Effects)
1. hunting, fishing	people have food to eat; killing off deer prevents deer-car accidents; helps keep ecosystem in balance - deer do not have enough food and would starve - there are no large predators (example wolves) to control the deer population
2. logging (chopping down trees)	prevents forest fires (too many trees help fires to spread)
3. mining (getting minerals from the ground, such as copper, lead, iron)	getting minerals
4. oil drilling	oil used by power plants to provide energy. gasoline used in cars, buses, etc. comes from oil using domestic oil (from the U.S.) instead of foreign oil. creates jobs. getting more oil causes prices to go down (lower prices)

Question: Lawn wastes, such as grass clippings and leaves, were once collected with household trash and dumped into landfills. Identify *one* way that this practice was harmful to the environment.

Solution: When humans collect leaves and grass and dump them into landfills (places where garbage is dumped and buried), it is harmful to the environment because the materials in the leaves are not recycled (used again by plants). If the leaves and grass are composted (made into fertilizer by humans) or allowed to decompose where they fell, the decomposers (bacteria, fungi) eat the dead leaves and grass clippings and decompose (break them down) into carbon dioxide and simple nitrogen compounds, returning (recycling) materials to the air and soil, to the environment; therefore, do not put leaves and grass into landfills.

Controlling Pests

As you know, we use pesticides (insecticides) to kill insects. Some insects cause diseases or damage crops. The problem with pesticides is that they might also damage crops and harm plants and animals, including humans. Pesticides can also harm ecosystems and the environment.

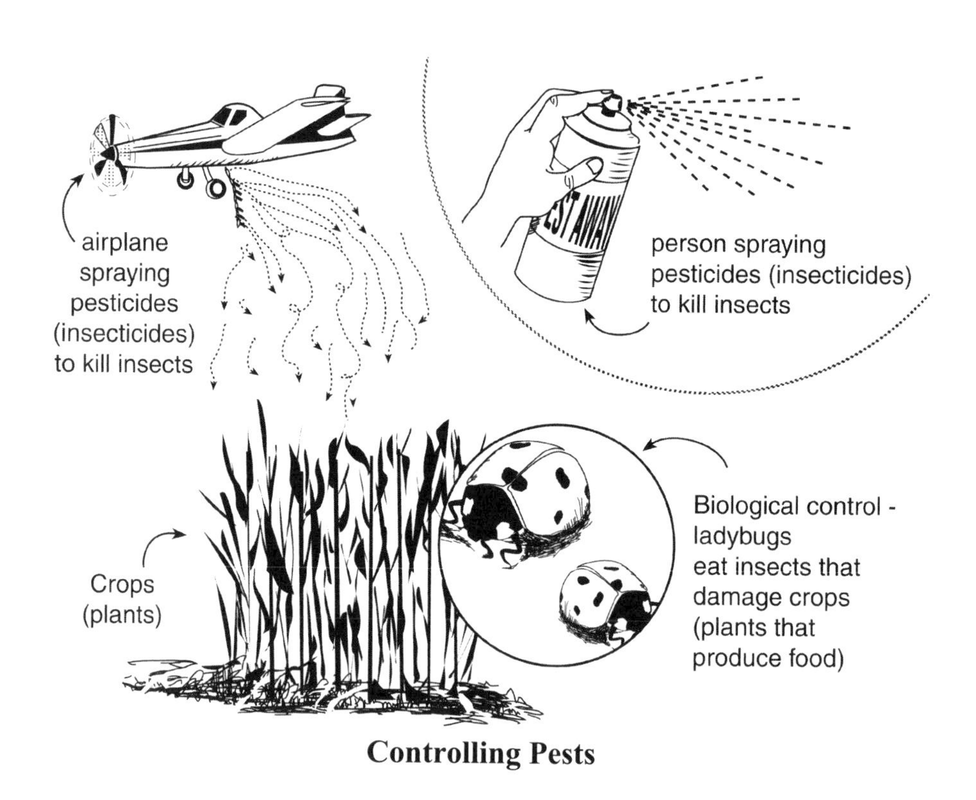
airplane spraying pesticides (insecticides) to kill insects
person spraying pesticides (insecticides) to kill insects
Crops (plants)
Biological control - ladybugs eat insects that damage crops (plants that produce food)

Controlling Pests

We want to control the pest (example insects, such as mosquitos) problem without hurting the environment or living organisms. **Biological control uses living organisms** or something (example hormones) from a living organism **to control pests** (example **kill pests** or **prevent** the **pests from reproducing**). **Biological control does not use chemicals,** therefore biological control **does not harm** the **environment** or living things.

Examples of **biological control:**

1. **increasing** the population of **native organisms** which kill the pests:

   a. in a swamp where mosquitoes live, increase the population of **native fish** that feed on (eat) mosquito larvae (larvae develop into adult mosquitoes).

   b. providing nesting places (similar to building birdhouses) for bats, which causes more **bats** to come to the area; the bats eat thousands of insects.

2. **using parasites** (example some bacteria) **or predators:**

   a. **using bacteria** to **kill mosquito larvae** which **carry** (have) the West Nile **virus**. When mosquitos with the West Nile virus bite a human being, the human being gets very sick and can die. New York City has used bacteria which only kill mosquito larvae (larvae develop into adult mosquitoes), therefore there were less mosquitos, and fewer people got infected with the virus.

   b. using predators (animals that kill other animals). **Use** native **fish** (increase the population of native fish) which feed on **(eat)** (kill) **mosquito larvae** which develop into mosquitos. Some mosquitos carry (have) malaria, encephalitis, and West Nile virus. The mosquitos spread the diseases when they bite people.

3. **trapping insects:** sex hormones from the pests are put into traps to attract the pests; insects come into the trap, cannot leave, and die. Trapping controlled the Mediterranean fruit fly.

4. **releasing sterilized male insects** (male insects that cannot produce baby insects): they mate with females, but produce no offspring (baby insects). This method has been used to control the screwworm fly, a parasite that attacks cattle.

5. **using selective breeding or genetic engineering** (see genetics chapter): by using selective breeding or genetic engineering, breed (make) plants (example, a type of corn) that kill pests (insects).

Now Do Homework Questions #48-55, pages 42-43.

HUMAN ACTIVITIES: TECHNOLOGY AND INDUSTRIALIZATION AFFECT OUR ENVIRONMENT

Technology uses tools and machinery to **make things** (produce goods) that people need faster, cheaper, and in larger quantities. **Industries** (example manufacturing cars, air conditioners, etc.) use the technology (tools and machinery) to make the products (cars, air conditioners) that people need. Industry uses energy to make steam (heat energy) and to make electricity. Humans can have a big (significant) effect on ecosystems because humans can modify (change) their environment by using technology (example, humans burn fuel to produce electricity, which pollutes the air).

Producing Electricity

Industries need electricity (electrical energy) to run factories. Power plants use **fossil fuels** (examples coal, oil, natural gas), **nuclear energy, water, wind,** and **solar energy** to **make electricity** (explained below). The electricity goes from the power plants through wires to the industries (factories).

Fossil Fuels: Fossil fuels (oil, coal, natural gas) burn and produce energy.

Nuclear Energy: Nuclear reactions produce a tremendous amount of energy (much more than burning fossil fuels). In nuclear reactions, mass (material, a thing) is changed into energy.

In a nuclear reaction (called fission), one atom (example, one atom of uranium or one atom of plutonium) splits into two or more pieces and gives off a lot of energy, which is used for electric power (electricity).

Nuclear reactions (such as uranium or plutonium splits into pieces) are done in a nuclear reactor.

The chart below shows how nuclear energy benefits or harms the environment.

Nuclear Energy

How it Benefits (Helps) the Ecosystem or Environment (Positive Effects)	How it Harms the Ecosystem or Environment (Negative Effects)
1. does not use up fossil fuel (examples coal, oil, etc.)	1. thermal pollution: Water from nearby lakes and rivers is used to cool nuclear reactors (extremely hot); when the warmed water goes back into the lake, the temperature of the lake increases. Warmer water holds less oxygen, therefore many organisms will die from lack of oxygen or will need to find a new home (habitat). Also, with the increase in temperature of the lake, some organisms cannot live at the higher temperature and will die (decrease in biodiversity) or will be forced to find a new home (habitat).
2. little air pollution	2. Wastes (radioactive wastes) from nuclear reactors are very radioactive. They must be stored for more than 100,000 years.
3. does not produce carbon dioxide (carbon dioxide causes global warming)	3. Accidents and fires in reactors can give off dangerous levels of radioactivity (gives off rays and particles).
4. does not cause acid rain (does not produce sulfur dioxide and nitrogen oxides which unite with water in the air to form acids)	

The chart below shows the advantages and disadvantages of using fossil fuels, nuclear energy, water, wind and solar (sun's energy) to produce electricity.

Energy Sources to Make Electricity

Power Plants Use (Source of Energy)	Advantages	Disadvantages
Fossil fuel (oil, coal, natural gas)	can be used anywhere, any time	nonrenewable, causes pollution
Nuclear energy	no carbon dioxide given off	accidents give off dangerous amounts of radiation, thermal pollution, storing radioactive wastes
Water	no pollution	only available in certain areas (example near waterfalls or dams)
Wind	no pollution	only when wind blows
Solar (sun's energy)	renewable (does not get used up; sun's energy is replaced).	can only be used where there is a lot of sunlight; only in daytime

Power plants use fossil, nuclear, water, wind, and solar energy to produce electricity. Note: Fossil fuel can also produce heat energy. Solar energy is also used to heat homes, etc. Nuclear energy is also used in bombs (atomic bomb and hydrogen bomb).

Now Do Homework Questions #56-62, pages 43-44.

Positive and Negative Effects of Industrialization

Humans built and run industries. Industries have both positive and negative effects.

Positive Effects: Industries are important (useful, beneficial, have a positive effect) because they help to provide us with the materials (examples clothes, televisions, radios, VCRs, air conditioners, furniture) we want.

Negative Effects: The problem (negative effect, harmful effect) with industrialization (industries) is

1. **uses up** our **natural resources** (examples coal, oil, water, etc., or in some cases iron, aluminum, etc.).
2. **destroys habitat** (example forests and meadows) to build factories.
3. **pollution** (pollutes our environment)
   a. air (**air pollution**) which includes acid rain, smog, global warming, and ozone depletion, which will be discussed further below
   b. water (**water pollution**) } Industry dumps toxic wastes
   c. soil (**soil pollution**) } into water and soil

Power plants burn coal and oil to provide energy (electricity) for machines; burning coal and oil uses up our natural resources and gives off pollution to our environment.

AIR POLLUTION

Air pollution includes acid rain, smog, global warming, and **ozone depletion** (see below). Industries burn fossil fuel for energy, giving off (emitting) pollutants through smokestacks (chimneys) into the air. **Fossil fuels** (examples coal and oil) **burn**, **giving off pollutants,** such as **carbon dioxide** and **gases** that contain **sulfur** or **nitrogen**. When it rains, the rain water carries the pollutants from the air into rivers, lakes, soil, etc. Air pollution harms living organisms and damages the habitat (example trees).

Air Pollution

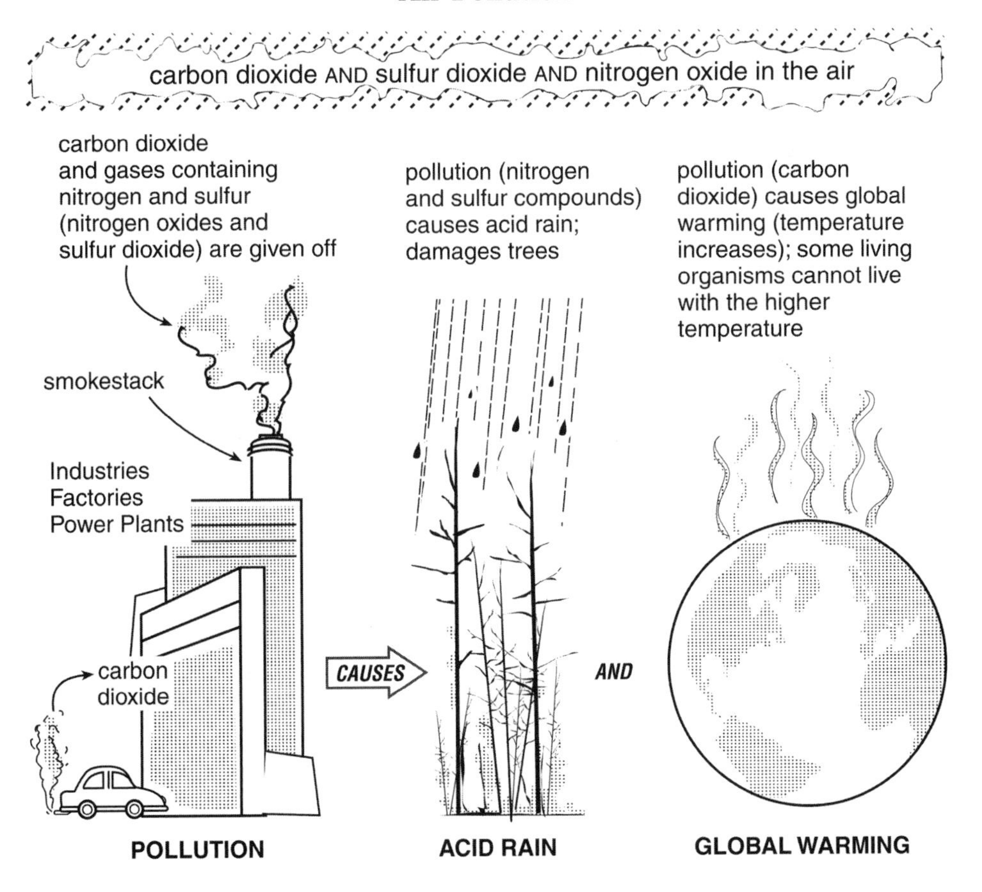
carbon dioxide AND sulfur dioxide AND nitrogen oxide in the air
carbon dioxide and gases containing nitrogen and sulfur (nitrogen oxides and sulfur dioxide) are given off
smokestack
Industries
Factories
Power Plants
carbon dioxide
CAUSES
pollution (nitrogen and sulfur compounds) causes acid rain; damages trees
AND
pollution (carbon dioxide) causes global warming (temperature increases); some living organisms cannot live with the higher temperature
POLLUTION
ACID RAIN
GLOBAL WARMING

Industries burn fossil fuels (example coal and oil) giving off pollutants (carbon dioxide in large amounts, carbon monoxide, and gases containing sulfur and nitrogen) see figure above. Motor vehicles (cars, buses, and trucks) also give off carbon dioxide and sometimes give off carbon monoxide (when the exhaust system is not working properly). The amount of carbon dioxide in the air is increasing over the years (see table below).

Carbon Dioxide in the Air

Year	Parts per Million CO_2
1960	320
1970	332
1980	350
1990	361
2000	370

Global warming: Too much carbon dioxide in the atmosphere **causes global warming,** which means the **temperature** of the globe (Earth) **increases** and some living organisms cannot live (will die) at the higher temperature (see figure above).

Acid Rain: Industries burn fossil fuels, giving off sulfur and nitrogen compounds (**sulfur dioxide and nitrogen oxides)**, which combine **with water** vapor in the air, **producing acids. Rain** and snow, etc. combine with acids, forming **acid rain** (see figure above). Acid rain is very **acidic** (has a low pH); acid rain damages plants. Because acid rain is so acidic, when it goes into streams and lakes, it **lowers** the **pH** of the streams and lakes, causing some algae, fish (and other organisms) to die (loss of biodiversity). (Note: The dead organisms decompose (meaning dead organisms break down chemically into new, different, simpler substances, like carbon dioxide, ammonia, etc.); decomposing uses up oxygen. This leaves less oxygen for the other organisms and some organisms might also die (suffocate)).
Acid rain also damages buildings (especially limestone and marble) and machinery.

Prevention and Solution for Acid Rain: Factories give off sulfur and nitrogen compounds (sulfur dioxide and nitrogen oxides) through the smoke stack (chimney) to the atmosphere. **Factories** should **remove sulfur and nitrogen compounds** (sulfur dioxide and nitrogen oxides) from the exhaust gases (before the exhaust gases go out of the smokestack (chimney)) to **prevent** these gases from going into the atmosphere (air) and causing **acid rain.** Humans can take this action (make sure that factories remove sulfur and nitrogen compounds) to help lessen (reduce) acid rain.

Smog forms when pollutants in the air from cars and industries (factories) react with sunlight. Smog is a brown haze that is harmful to living things, including people.

Global Atmospheric Changes

Let's explain two **global** (worldwide) **atmospheric changes** (**global warming** and **ozone depletion**) a little more fully. People all over the world (globe) must try to solve the problems of global warming and ozone depletion.

Global Warming: You learned **fossil fuels burn, giving off carbon dioxide** (also called a greenhouse gas). Too much carbon dioxide in the atmosphere (air) causes global warming, which means the **temperature** of the atmosphere **(air) increases** and the temperature of the globe (both land and water) increases and some living organisms cannot live (will die) at the higher temperature.
Too much **carbon dioxide** in the atmosphere has a **negative effect (harmful effect).** Too much **carbon dioxide in** the **atmosphere** (air) **causes global warming,** which means the **temperature of the atmosphere** (air) **increases** and the temperature of the globe (both land and water) increases (higher temperature). The higher temperature can cause climate changes such as **melting of the ice caps** at the North and South Poles. When the ice caps melt, more water goes into the ocean and the sea level rises (water gets higher in the ocean), causing **floods** on the land (on habitats, which are homes of living organisms and also on farms that grows crops such as wheat, corn, etc.). **Flooding destroys homes, living organisms, and crops** (example food).

Prevention and Solution for Global Warming: To prevent global warming, **plant more trees** because, in photosynthesis, plants take in carbon dioxide, removing carbon dioxide from the air. Also, to prevent global warming, **use less fossil fuels,** because fossil fuels burn, giving off carbon dioxide. **Use** alternative **(different) energy sources,** such as nuclear, water, wind, or solar, because they do not give off carbon dioxide.

Ozone Depletion: Industry gives off pollutants, and some of the pollutants (example **CFC** gases) **destroy** some of the **ozone shield (ozone layer).** The ozone shield is a layer of ozone gas in the upper atmosphere that protects us from (absorbs) the ultraviolet radiation of the sun.

CFC (chlorofluorocarbon) gases **destroy** some of the **ozone shield (ozone layer)** causing **ozone depletion** (thinning of the ozone layer) letting **more ultraviolet radiation** come down to earth. Ultraviolet radiation can **cause mutations** (changes) in the DNA (genetic makeup, genes) of organisms. Ultraviolet radiation can cause **skin cancer** and kills some plants.

CFC (chlorofluorocarbon) gases used as coolants in refrigerators and in air conditioners, and at one time used as propellants in aerosol cans, cause ozone depletion.

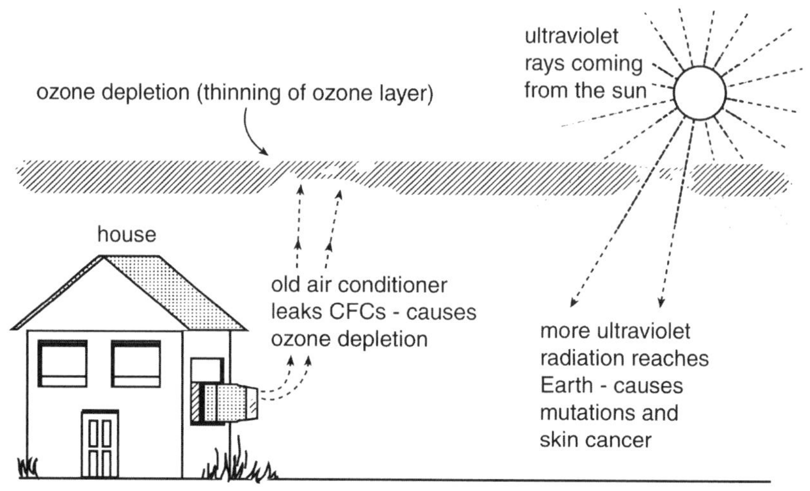

CFC's Cause Ozone Depletion

Prevention and Solution for Ozone Depletion: Do not use or limit the use of **CFC** gases. Use safe products instead of CFCs for coolants in refrigerators and in air conditioners and as propellants in aerosol cans.

Practice Questions and Solutions

Question: State *two* specific ways in which an ocean ecosystem will change (other than fewer photosynthetic organisms) if populations of photosynthetic organisms die off as a result of damage to the ozone layer.

Solution: Plants take in carbon dioxide and give off oxygen in photosynthesis. **When plants die**, plants stop taking in carbon dioxide and giving off oxygen, therefore there will be **more carbon dioxide and less oxygen in the air.**

Other possible answers: **when plants die**, there is a **decrease in biodiversity** and now some animals (consumers) will not have enough food (plants) to eat and some consumers (animals) will die (**decrease in consumers**). When plants die, there will also be a **decrease** in **available energy**, because the plants that died cannot carry on photosynthesis and cannot produce glucose (stored energy).

Now Do Homework Questions #63-87, pages 44-48.

WATER POLLUTION

Water pollution includes nitrogen and phosphorus pollution, sediment pollution, toxic wastes, and **thermal pollution** (see below). **Wastes** from homes, factories, power plants, and mines (example coal mines, coal is taken from the ground), as well as sewage (toilet wastes) from cities, are dumped into the water. When it rains, the rainwater carries fertilizer, animal wastes, and pesticides from farms into lakes and streams, polluting the lakes and streams **(water pollution)**. Note: The water running off the land is called runoff.

Nitrogen and Phosphorus Pollution: Animal wastes, fertilizer, and sewage in the water contain nitrogen and phosphorus compounds; animal wastes and sewage are like fertilizers and cause bigger and more plants, algae, and bacteria to grow in the water.

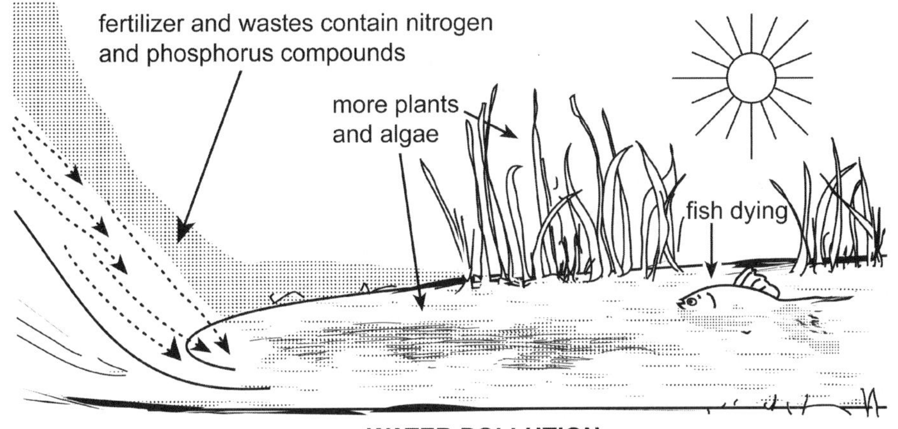

WATER POLLUTION

a. Some plants die. Decomposers (bacteria and fungi) decompose (break down) the plants. The decomposers use the dead plants for food. The decomposers carry on cellular respiration, uniting the dead plants with the oxygen in the water to produce energy for the decomposers. Now the water has less oxygen and some organisms, such as fish and plants, will die from lack of oxygen.

b. There are lots of algae in the water; the algae block some of the sunlight. Plants (example algae) carry on photosynthesis in the presence of sunlight: carbon dioxide + water → glucose + oxygen. When there is less sunlight, less photosynthesis takes place and less oxygen is produced. There is now not enough oxygen for all the fish to survive, and some fish die. Killing off organisms disrupts food chains and might cause a loss of biodiversity.

Prevention and Solution for Nitrogen and Phosphorus Pollution: Humans can take action to **lessen** the **nitrogen** and **phosphorus** problem in lakes and streams. The farmer should use the **least amount** of **fertilizer** (contains nitrogen and phosphorus) **and least amount** of **pesticides**

necessary so the rain water carries less pollutants (fertilizers and pesticides) into the lakes and streams.
Also **treat sewage** (human waste from toilets) before it goes into the water (example lakes) to **remove nitrogen and phosphorus** compounds. This helps to control (to limit or lessen) pollution (pollution control).

Sediment Pollution: Deforestation (cutting down trees) and poor farming practices cause **erosion (soil is removed).** The soil **goes into lakes** and **streams** as sediments. Sediment pollution (sediments going into the water) is similar to nitrogen and phosphorus pollution . When sediments are deposited in the water, some sunlight is blocked, less photosynthesis takes place, **less** (not enough) **oxygen** is produced, and some fish die (fish kill). Similarly, organisms and decomposers use oxygen in cellular respiration (glucose and oxygen → water and carbon dioxide), therefore there is less oxygen available for fish; some fish die (fish kill).

Prevention and Solution for Sediment Pollution:

1. **Use good farming practices** that **hold the soil** and prevent the sediments from going into the water.
2. **Do not cut down trees.** Trees hold the soil.

Toxic Wastes: There are also **toxic** (poisonous) **wastes** such as **weed killers**, **pesticides** (insect killers, DDT), **heavy metals** (mercury, lead, copper) and **PCBs dumped into** the water (streams. lakes, and oceans).

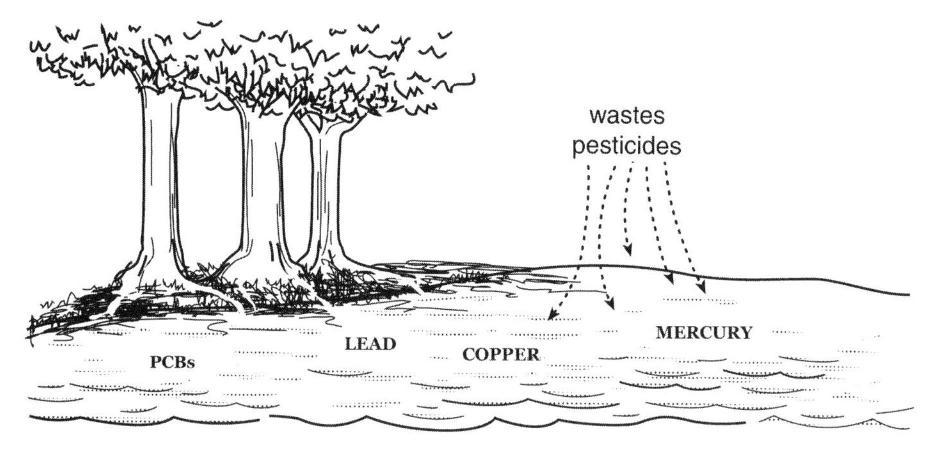

WATER POLLUTION - TOXIC WASTES

The **small plants and algae eat** the **toxic wastes.** When the plants and algae are eaten by herbivores (animals that eat plants), the toxic materials are stored and concentrated in the fat tisues of the **herbivores;** the **carnivores** that eat the herbivores accumulate more of the **toxic material.** Animals at the top of a food chain may even accumulate large (dangerous) amounts of the toxins. Examples: Pregnant women are told to limit their consumption (not to eat too much) of certain types of fish (tuna, swordfish) because of the amount of mercury in the fish.

Thermal Pollution: Some industries and power plants **use water to cool** their **machinery**. The **warmed water goes** back **into** the **lake** or river, causing the **temperature of** the **lake** or river to **increase. Many organisms cannot live** in the **warmer water (higher temperature)** and will die (decrease in biodiversity) or will be forced to find a new home (habitat). Also, **warmer water** holds **less oxygen,** therefore many **organisms** will **die** from lack of oxygen in the lake or will need to find a new home (habitat).

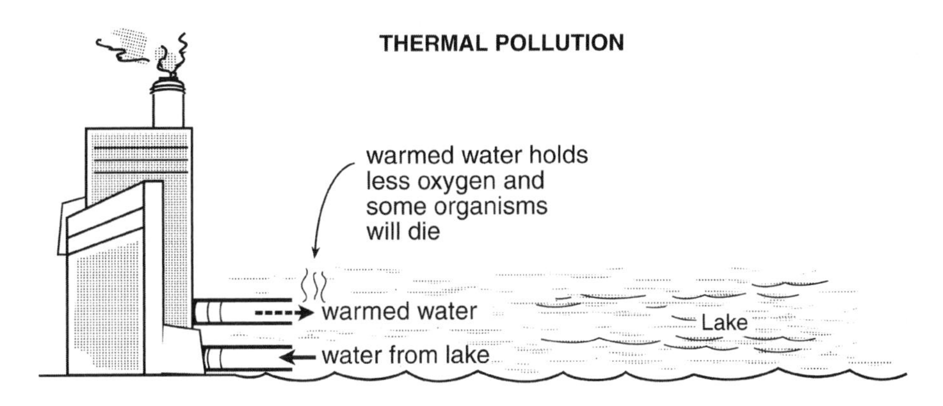

This graph shows that as the temperature of water increases (gets higher), the amount of oxygen in the water decreases (gets less), therefore many organisms will die from lack of oxygen.

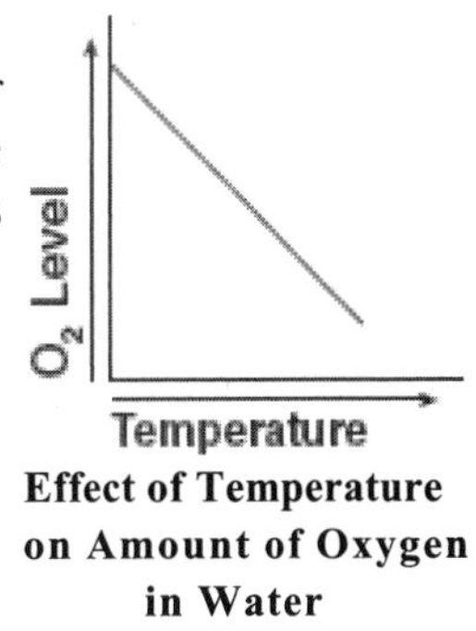

Effect of Temperature on Amount of Oxygen in Water

Now Do Homework Questions #88-102, pages 49-52.

Decisions

We the people will decide what will happen to our ecosystems, environment, plants and animals, earth and earth's atmosphere now and in the future. What we do today to our environment affects the future.

1. **Individuals have choices** of what to do and individuals must **make the right decisions** on how to protect and preserve our environment. A wrong

decision can be irreversible. If people kill a species (example passenger pigeons) until it is extinct (no more left), it cannot be brought back.
We must **preserve our resources:**

A. **reduce, reuse, and recycle.**

B. **use energy carefully** (use energy efficient equipment, such as energy efficient air conditioners). Shut off air conditioners when you are not using them.

C. encourage people and make laws for people **not to harm the environment.** Avoid or lessen destruction of habitats, direct harvesting and importing species. Lessen air, water, and soil pollution.
Laws were made to stop CFC pollution (pollution from chlorofluorocarbon gases), which depleted some areas of the ozone layer, which affects today and the future.

2. We must be **aware** of what is happening in the environment, **protect the environment,** and **make** the **right decisions.**

A. The passenger pigeon and the dodo bird became extinct because people were killing them. If we were aware of the problem right away and there were laws protecting the **passenger pigeon** and the **dodo bird,** they would not be extinct today. Humans are concerned with animals becoming extinct and food chains being disrupted. Therefore, laws are made to control (limit) the number of animals of each kind (example how many deer or how many bears) can be hunted and killed each year. What happened in the past (killing the passenger pigeon and the dodo bird) affects today and the future. Humans should be aware of what is happening, make the right decisions, and should make laws to keep animals from becoming extinct.

B. A **new chemical** can be **sprayed** on **fruits, vegetables**, or **fields** where food animals (example cows, chickens) eat. It prevents other organisms from eating or damaging the fruit or vegetables or kills off the organisms that eat the fruits and vegetables; or, the chemical helps to produce more and better fruits, vegetables, or meat. **Humans** have to **make** a **decision whether** the **chemical** should be **sold** and **used** by the public. There are environmental concerns that we have.

Environmental Concerns:

1. The chemical may be toxic (poisonous) to humans and wildlife (wild animals). A chemical that is supposed to kill a specific organism may also kill off other plants and animals that we don't want to be killed.

2. The chemical may pollute the environment.

3. The chemical might kill off animals, which interferes with (disrupts) food chains and food webs.

4. The chemical may not break down (biodegrade) so the chemical may stay in the environment and harm the environment.

Note: Some chemicals are biodegradable (break down), forming harmless materials; some chemicals are not biodegradable (do not break down), and might harm the environment.

3. When individuals and society (people and governments, such as Congress) **make decisions** about the environment, they must assess (analyze) **risks, costs and benefits**, and **trade offs**.

A. Examples of **trade offs**:

1. use less air conditioning - less comfort.
2. use better equipment to avoid pollution - more expensive.
3. remove nitrogen and phosphorus compounds from sewage before it goes into lakes and streams to prevent too much algae and plants from growing in lakes and streams - expensive to remove nitrogen and phosphorus compounds from sewage.

B. **Making laws:** We have to decide whether we should make laws (laws should be passed) to reduce (lessen) pollution and lessen toxic chemicals that go into the environment (only allow a smaller amount of toxic chemicals to go into the environment). You have to weigh the benefit against the risk.

Benefits: less pollution and less toxic chemicals in the environment.

Risks: 1. Products will cost more money.

2. Businesses will move their factories to another state where there are fewer laws. Then people will lose jobs (less employment), and this state will lose tax money from the businesses.

Note: There were laws passed which helped to reduce (lessen) pollution, to protect endangered wildlife (species), and to make sure landfills (where chemical wastes and garbage are dumped) are safe and do not pollute our environment.

C. Individuals encourage the government to make laws to prevent pollution by cars, buses, etc. There is a law in many states that cars must be inspected (checked) every year to make sure the exhaust system works properly and does not give off too much carbon monoxide (pollution). This helps to maintain good quality of the air (benefit). If there is something wrong with the exhaust system, it costs money to fix it (cost).

D. Another **decision** to make is should farmers be permitted to **grow genetically engineered crops** (crops that the genes were changed (altered) to produce new characteristics, such as juicier or tastier fruit, or new substances).

Positive effects (benefits): The benefit is the genetically engineered crop might have tastier and juicier fruit, and might have insect-resistant plants (plants that resist insects, that insects cannot eat or damage).

Negative effects (harmful effects): The possible negative effect is that we do not know what long-term effect eating genetically engineered crops might have on humans. If humans eat the genetically engineered crop today, will this crop harm humans now, 10 years from now, or 30 years from now (long term effect). Genetic engineering is a recent (new) method (technology), and we do not know what will happen in the future to organisms that eat the genetically engineered crop today.

Another negative effect of genetic engineering is some genetically engineered bacteria might compete with regular bacteria for food, etc., causing some bacteria to die, which disrupts food chains.

A genetically engineered insect-resistant plant produces insecticide inside the plant (example inside the apple or potato). The insecticide cannot be washed off. The negative effect is that the plant with insecticide inside may be harmful to people who eat it. The positive effect is that the plant is protected from insects, and the farmer does not have to spray the crop (plants that are grown, usually for food, or to sell something, examples wheat, cotton). with insecticides.

Making Decisions

4. Some people feel that it is **not ethical** to insert genes from other species into humans. Do you agree or disagree? Do you prefer more experiments inserting genes of other species into humans?

5. Society must decide on **proposals** (suggested ideas) **for new technology**. When a company or person introduces a new technology (using machines, etc.) people or their representatives discuss and decide whether to allow the technology to be used. We must be aware of what is happening to the environment and make the right decisions. A wrong decision can be irreversible (cannot be changed back to the way it was).

6. **Today's decisions affect the future:** Today's decisions affect the future (the next generation and generations after that). Today's decisions make some things possible and other things impossible in the future.

A. **loss of species** such as the passenger pigeon, dodo bird, and others **disrupts** our **ecosystems** and food chains. Individuals should try to protect species so they will not become extinct. Maybe **make laws** to **protect species from becoming extinct.**

B. When individuals today **reduce, reuse, and recycle,** it preserves our resources for now and in the future. If individuals are not careful, we can use up our nonrenewable natural resources.

C. Individuals should try **not** to **destroy habitats** or destroy less habitats. This helps to preserve biodiversity now and in the future. (**Biodiversity** is **useful** to produce **new medicines** and to **provide genes** to make better or more disease-resistant plants or animals.)

D. Individuals should be aware of our ecological (**ecology**) problems and should work together collectively to **solve** these **problems** in the best way to protect our ecosystems.

Now Do Homework Questions #103-112, pages 52-53.

Drawing Graphs

Let's see how we can draw graphs based on experimental data. Data (from an experiment) is written on a data table (see below); draw the graph based on the data table.

Answer questions 1 through 5 based on the information below and on your knowledge of biology. The average level of carbon dioxide in the atmosphere has been measured for the past several decades. The data collected are shown in the table below:

Average CO_2 Levels in the Atmosphere

Year	CO_2 (In parts per million)
1960	320
1970	332
1980	350
1990	361
2000	370

Questions 1 and 2: Using the information in the data table, construct a line graph on the grid, following the directions below.

Question 1: Mark an appropriate scale on each labeled axis.

Question 2: Plot the data on the grid. Surround each point with a small circle and connect the points.

Example:

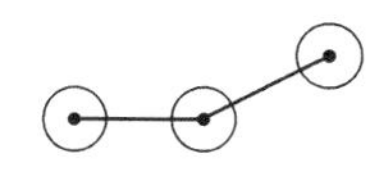

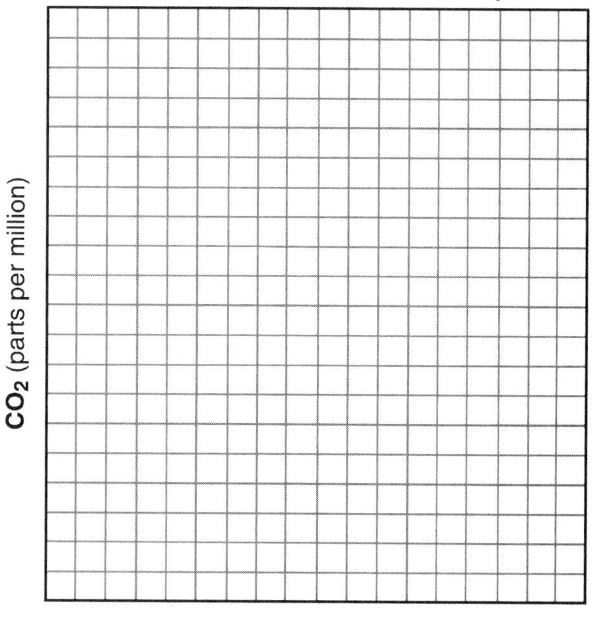

Question 3: Calculate the net change in CO_2 levels in parts per million (ppm) during the years 1960 through 2000.

Question 4: Identify *one* specific human activity that could be responsible for the change in carbon dioxide levels from 1960 to 2000.

Question 5: State *one* possible negative effect this change in CO_2 level has had on the environment of Earth.

Solution 1: Look at the line graph in the question. You see, on the x axis is the thing you change (the independent variable), which is the year, such as 1960, 1970, etc. There must be an equal number of years between lines. "Make an appropriate scale" by spacing the numbers on the graph so that all of the data fits on the graph and it is easy to read. On the x axis, put 10 years between lines (every four lines) (scale on the x axis), then all the years on the data table fit on the graph and it is easy to read. Start counting the lines **after** the axis (see graph).

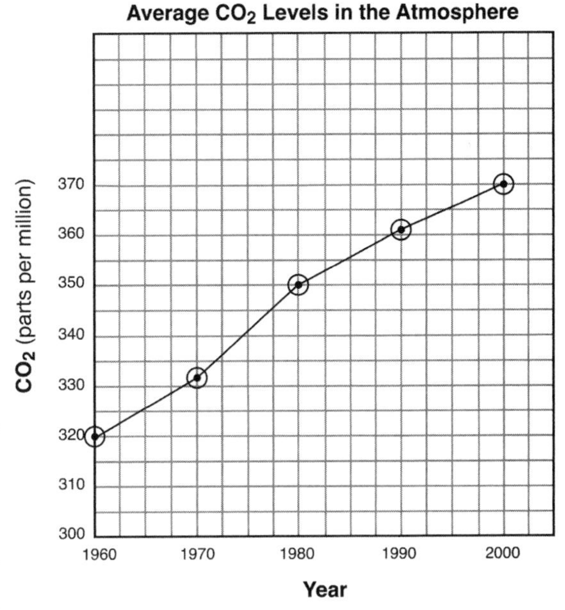

Look at the line graph again. On the y axis was put CO_2 (parts per million). The result you get , the amount of CO_2 (carbon dioxide) in parts per million is always put on the y axis. This is the dependent variable. There must be an equal number of parts per million of carbon dioxide between lines (every two lines). Space the lines along the axis equally. "Make an appropriate scale" by spacing the numbers on the graph so that all of the data fits on the graph and it is easy to read. On the y axis, put 10 parts per million of CO_2 between lines (every two lines) (scale on the y axis), then all the parts per million on the data table fit on the graph and it is easy to read. Start counting the lines **after** the axis (see graph).

Solution 2: Plot the experimental data. Draw a circle around each point. Draw a line that connects the points. Do not continue the line past the last point (see graph).

Solution 3: Look at the data table or the graph.

In 2000, there were 370 parts per million (ppm) CO_2.
In 1960, there were 320 parts per million (ppm) CO_2.
370 ppm − 320 ppm = 50 ppm (parts per million). The change in CO_2 from 1960 to 2000 is 50 parts per million (ppm).

Solution 4: When there are more industries, more fossil fuel (coal, oil) is burned and more carbon dioxide is produced.

Or

Trees and plants take in carbon dioxide and give off oxygen. Cutting down trees means less carbon dioxide will be taken in by the trees and the amount of carbon dioxide in the air will increase.

Solution 5: An increase in carbon dioxide causes global warming, which means the temperature of the globe (Earth) increases and some living organisms cannot live (will die) at the higher temperatures.

Comparing the graph of carbon dioxide levels in the atmosphere (see previous graph in solutions 1 and 2 above) and the graph of atmospheric temperature change below, you can see the overall relationship that as carbon dioxide concentration increases (in the graph above), the atmospheric temperature also increases (in the graph below).

Average Atmospheric Temperature Change per Year

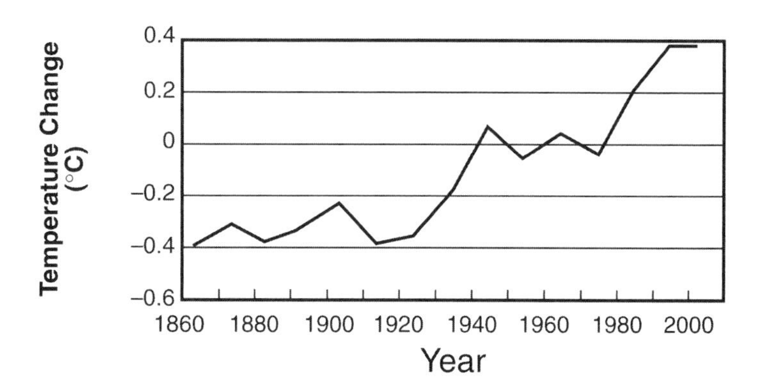

Now Do Homework Questions #113-117, page 54.

NOW LET'S TRY A FEW HOMEWORK QUESTIONS

To assign fewer questions, you may want to skip the questions with underlined numbers.

1. Humans are responsible for some of the negative changes that occur in nature because they
   (1) have encouraged the development of wildlife refuges and parks
   (2) have passed laws to preserve the environment
   (3) are able to preserve scarce resources
   (4) are able to modify habitats more than any other species

Base your answers to the next four questions on the information below and on your knowledge of biology.

A decade after the Exxon Valdez oil tanker spilled millions of gallons of crude [oil] off Prince William Sound in Alaska, most of the fish and wildlife species that were injured have not fully recovered.

Only two out of the 28 species, the river otter and the bald eagle, listed as being injured from the 1989 spill are considered to be recovered said a new report, which was released by a coalition of federal and Alaska agencies working to help restore the oil spill region.

Eight species are considered to have made little or no progress toward recovery since the spill, including killer whales, harbor seals, and common loons [a type of bird].

Several other species, including sea otters and Pacific herring, have made significant progress toward recovery, but are still not at levels seen before the accident the report said.

More than 10.8 million gallons of crude oil spilled into the water when the tanker Exxon Valdez ran aground 25 miles south of Valdez on March 24, 1989.

The spill killed an estimated 250,000 seabirds, 2,800 sea otters, 300 harbor seals, 250 bald eagles, and up to 22 killer whales.

Billions of salmon and herring eggs, as well as tidal plants and animals, were also smothered in oil.

– Reuters

2. Identify *two* species that appear to have been least affected by the oil spill.

3. The oil spilled by the Exxon Valdez tanker is an example of a
   (1) nonrenewable resource and is a source of energy
   (2) renewable resource and is a source of ATP
   (3) nonrenewable resource and synthesizes ATP
   (4) renewable resource and is a fossil fuel

4. The impact that the oil spill made on the environment is still being experienced. State information from the reading passage that supports this statement.

5. Which autotrophic organisms were *negatively* affected by the oil spill?

6. Over the past few decades, many oil companies have discovered oil below the sea floor near the coasts of many states. Some states, however, refuse to permit offshore oil drilling, fearing it might damage the environment.

   Discuss both sides of this issue. In your answer, be sure to:
   - state *one* way in which offshore oil drilling might have a long-term negative effect on the environment
   - state *one* way in which offshore oil drilling could benefit society

7. Which human activity will most likely have a negative effect on global stability?
   (1) decreasing water pollution levels
   (2) increasing recycling programs
   (3) decreasing habitat destruction
   (4) increasing world population growth

8. The negative effect humans have on the stability of the environment is most directly linked to an increase in
   (1) recycling activities by humans
   (2) supply of finite resources
   (3) predation and disease
   (4) human population size

9. Which situation has had the most *negative* effect on the ecosystems of Earth?
   (1) use of air pollution controls
   (2) use of natural predators to control insect pests
   (3) recycling glass, plastic, and meta
   (4) increasing human population

10. Which practice would most likely deplete a nonrenewable natural resource?
   (1) harvesting trees on a tree farm
   (2) burning coal to generate electricity in a power plant
   (3) restricting water usage during a period of water shortage
   (4) building a dam and a power plant to use water to generate electricity

11. The graph below shows the percentage of solid wastes recycled in New York State between 1987 and 1997.

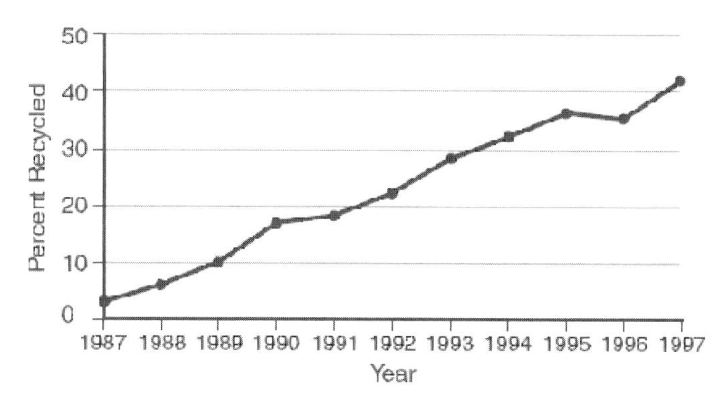

Discuss the impacts of recycling. In your answer be sure to:
- Explain what recycling is and provide *one* example of a material that is often recycled.
- State *one* specific positive effect recycling has on the environment.

12. Which human activity would be *least* likely to disrupt the stability of an ecosystem?
   (1) disposing of wastes in the ocean
   (2) using fossil fuels
   (3) increasing the human population
   (4) recycling bottles and cans

13. An individual has placed an editorial in the community newspaper stating that the local recycling program should be discontinued. Respond to this editorial by explaining the importance of the local recycling program for the environment. In your explanation be sure to:
- State *one* effect the increasing human population will have on the availability of natural resources
- State *one* reason why recycling is important
- Identify *two* natural resources or products made from natural resources that can be recycled

14. Many farmers plant corn, and then harvest the entire plant at the end of the growing season. One negative effect of this action is that
    (1) soil minerals used by corn plants are not recycled
    (2) corn plants remove acidic compounds from the air all season long
    (3) corn plants may replace renewable sources of energy
    (4) large quantities of water are produced by corn plants

15. Which human activity is correctly paired with its likely future consequence?
    (1) overfishing in the Atlantic – increase in supply of flounder and salmon as food for people
    (2) development of electric cars or hybrid vehicles – increased rate of global warming
    (3) use of fossil fuels – depletion of underground coal, oil, and natural gas supplies
    (4) genetically engineering animals – less food available to feed the world's population

16. When habitats are destroyed, there are usually fewer niches for animals and plants. This action would most likely *not* lead to a change in the amount of
    (1) biodiversity
    (2) competition
    (3) interaction between species
    (4) solar radiation reaching the area

17. Which long-term change could directly cause the other three?
    (1) pollution of air and water
    (2) increasing human population
    (3) scarcity of suitable animal habitats
    (4) depletion of resources

18. The graph below shows how the human population has grown over the last several thousand years.

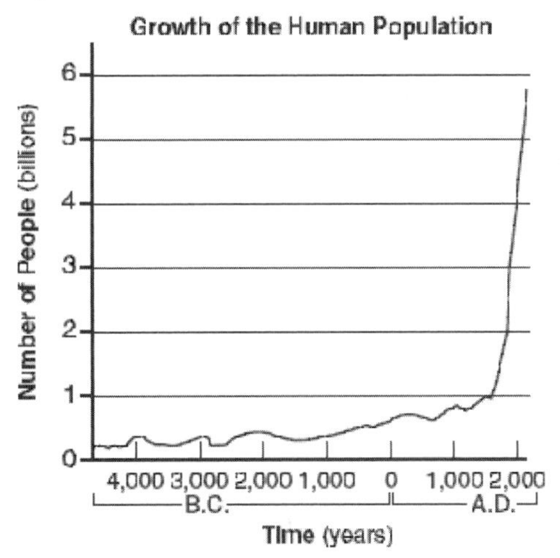

Which statement is a valid inference that can be made if the human population continues to grow at a rate similar to the rate shown between 1000 A.D. and 2000 A.D.?
    (1) Future ecosystems will be stressed and many animal habitats may be destroyed.
    (2) Global warming will decrease as a result of a lower demand for fossil fuels.

(3) One hundred years after all resources are used up, the human population will level off.
(4) All environmental problems can be solved without a reduction in the growth rate of the human population.

19. Which factor is primarily responsible for the destruction of the greatest number of habitats?
(1) human population growth
(2) decreased use of renewable resources
(3) spread of predatory insects
(4) epidemic diseases

20. Deforestation will most directly result in an immediate increase in
(1) atmospheric carbon dioxide
(2) atmospheric ozone
(3) wildlife populations
(4) renewable resources

21. The negative effect humans have on the stability of the environment is most directly linked to an increase in
(1) recycling activities by humans
(2) supply of finite resources
(3) predation and disease
(4) human population size

22. Which situation has had the most *negative* effect on the ecosystems of Earth?
(1) use of air pollution controls
(2) use of natural predators to control insect pests
(3) recycling glass, plastic, and meta
(4) increasing human population

23. Which human activity will most likely have a negative effect on global stability?
(1) decreasing water pollution levels
(2) increasing recycling programs
(3) decreasing habitat destruction
(4) increasing world population growth

24. The Susquehanna River, which runs through the states of New York, Pennsylvania, and Maryland, received the designation "America's Most Endangered River" in 2005. One of the river's problems results from the large number of sewage overflow sites that are found along the course of the river. These sewage overflow sites are a direct result of an increase in
(1) global warming
(2) human population
(3) recycling programs
(4) atmospheric changes

25. Humans are responsible for some of the negative changes that occur in nature because they
(1) have encouraged the development of wildlife refuges and parks
(2) have passed laws to preserve the environment
(3) are able to preserve scarce resources
(4) are able to modify habitats more than any other species

26. The rapid destruction of tropical rain forests may be harmful because
(1) removing trees will prevent scientists from studying ecological succession
(2) genetic material that may be useful for future medical discoveries will be lost
(3) energy cycling in the environment will stop
(4) the removal of trees will limit the construction of factories that will pollute the environment

27. Cutting down a rain forest and planting agricultural crops, such as coffee plants, would most likely result in
   (1) a decrease in biodiversity
   (2) an increase in the amount of energy recycled
   (3) a decrease in erosion
   (4) an increase in the amount of photosynthesis

28. Deforestation will most directly result in an immediate increase in
   (1) atmospheric carbon dioxide
   (2) atmospheric ozone
   (3) wildlife populations
   (4) renewable resources

29. Base your answer to the following question on the information below and on your knowledge of biology.

> The dodo bird inhabited the island of Mauritius in the Indian Ocean, where it lived undisturbed for years. It lost its ability to fly and it lived and nested on the ground where it ate fruits that had fallen from trees. There were no mammals living on the island.
>
> In 1505, the first humans set foot on Mauritius. The island quickly became a stopover for ships engaged in the spice trade. The dodo was a welcome source of fresh meat for the sailors and large numbers of dodos were killed for food. In time, pigs, monkeys, and rats brought to the island ate the dodo eggs in the ground nests.

Which statement describes what most likely happened to the dodo bird within 100 years of the arrival of humans on Mauritius?
   (1) Dodo birds developed the ability to fly in order to escape predation and their population increased.
   (2) The dodo bird population increased after the birds learned to build their nests in trees.
   (3) Human exploitation and introduced species significantly reduced dodo bird populations.
   (4) The dodo bird population became smaller because they preyed upon the introduced species.

30. State *one* biological benefit of preserving endangered species.

31. State *one* way, other than cloning, that gaurs, large ox-like animals from South Asia, currently endangered, might be saved from extinction.

32. Select *one* of the following ecological problems.

ECOLOGICAL PROBLEMS

Acid rain
Increased amounts of nitrogen and phosphorous in a lake
Loss of biodiversity

For the ecological problem that you selected, briefly describe the problem and state *one* way to reduce it. In your answer be sure to:
- State the ecological problem you selected
- State how humans have caused the problem you selected
- Describe *one* specific effect that the problem you selected will have on the ecosystem
- State *one* specific action humans could take to reduce the problem you selected

33. In 1859, a small colony of 24 rabbits was brought to Australia. By 1928 it was estimated that there were 500 million rabbits in a 1-million square mile section of Australia. Which statement describes a condition that probably contributed to the increase in the rabbit population?
   (1) The rabbits were affected by many limiting factors.
   (2) The rabbits reproduced by asexual reproduction.
   (3) The rabbits were unable to adapt to the environment.
   (4) The rabbits had no natural predators in Australia.

34. Oak trees in the northeastern United States have survived for hundreds of years, in spite of attacks by native insects. Recently, the gypsy moth, which has a caterpillar stage that eats leaves, was imported from Europe. The gypsy moth now has become quite common in New England ecosystems. As a result, many oak trees are being damaged more seriously than ever before.

   State *one* biological reason that this imported insect is a more serious problem for the trees than other insects that have been present in the area for hundreds of years.

35. Base your answers to the next question on the information below and on your knowledge of biology.

   Throughout the world, in nearly every ecosystem, there are animal and plant species present that were introduced into the ecosystem by humans or transported to the ecosystem as a result of human activities. Some examples are listed in the chart below.

Examples of Introduced Species

Organism	New Location
purple loosestrife (plant)	wetlands in New York State
zebra mussel	Great Lakes
brown tree snake	Guam

   Identify *one* introduced organism and write its name in the space provided. Describe *one* way in which this organism has altered an ecosystem in the new location.

36. Which long-term change could directly cause the other three?
   (1) pollution of air and water
   (2) increasing human population
   (3) scarcity of suitable animal habitats
   (4) depletion of resources

37. Explain how the introduction of foreign species can often cause environmental problems. In your answer be sure to:
   - State how zebra mussels and gobies were introduced into the United States
   - State *one* way either the zebra mussels or gobies have become a problem in their new environment
   - Describe how both zebra mussels and gobies contribute to increasing the concentration of PCB's in sport fish

38. Which factor is primarily responsible for the destruction of the greatest number of habitats?
   (1) human population growth
   (2) decreased use of renewable resources
   (3) spread of predatory insects
   (4) epidemic diseases

39. Which human activity would have the *least* negative impact on the quality of the environment?
    (1) adding animal wastes to rivers
    (2) cutting down tropical rain forests for plywood
    (3) using species-specific sex attractants to trap and kill insect pests
    (4) releasing chemicals into the groundwater

40. State *one* specific way the removal of trees from an area has had a negative impact on the environment.

41. Rabbits are herbivores that are not native to Australia. Their numbers have increased steadily since being introduced into Australia by European settlers. One likely reason the rabbit population was able to grow so large is that the rabbits
    (1) were able to prey on native herbivores
    (2) reproduced more slowly than the native animals
    (3) successfully competed with native herbivores for food
    (4) could interbreed with the native animals

42. Base your answers to the next question on the information below and on your knowledge of biology.

    The hedgehog, a small mammal native to Africa and Europe, has been introduced to the United States as an exotic pet species. Scientists have found that hedgehogs can transfer pathogens to humans and domestic animals. Foot-and-mouth viruses, Salmonella, and certain fungi are known pathogens carried by hedgehogs. As more and more of these exotic animals are brought into this country, the risk of infection increases in the human population.

    State *one* negative effect of importing exotic species to the United States.

43. The ivory-billed woodpecker, long thought to be extinct, was recently reported to be living in a southern swamp area. The most ecologically appropriate way to ensure the natural survival of this population of birds is to
    (1) feed them daily with corn and other types of grain
    (2) destroy their natural enemies and predators
    (3) move the population of birds to a zoo
    (4) limit human activities in the habitat of the bird

44. Millions of acres of tropical rain forest are being destroyed each year. Which change would most likely occur over time if the burning and clearing of these forests were stopped?
    (1) an increase in the amount of atmospheric pollution produced
    (2) a decrease in the source of new medicines
    (3) an increase in the amount of oxygen released into the atmosphere
    (4) a decrease in the number of species

Base your answers to the next three questions on the passage below and on your knowledge of biology.

Overstaying Their Welcome: Cane Toads in Australia

Everyone in Australia is in agreement that the cane toads have got to go. The problem is getting rid of them. Cane toads, properly known as Bufo marinus, are the most notorious of what are called invasive species in Australia and beyond. But unlike other species of the same classification, cane toads were intentionally introduced into Australia. The country simply got much more and much worse than it bargained for.

Before 1935, Australia did not have any toad species of its own. What the country did have, however, was a major beetle problem. Two species of beetles in particular, French's Cane Beetle and the Greyback Cane Beetle, were in the process of decimating [destroying] the northeastern state of Queenland's sugar cane crops. The beetle's larvae were eating the roots of the sugar cane and stunting, if not killing, the plants.

The anticipated solution to this quickly escalating problem came in the form of the cane toad. After first hearing about the amphibians in 1933 at a conference in the Caribbean, growers successfully lobbied to have the cane toads imported to battle and hopefully destroy the beetles and save the crops....

The plan backfired completely and absolutely. As it turns out, cane toads do not jump very high, only about two feet actually, so they did not eat the beetles that for the most part lived in the upper stalks of cane plants. Instead of going after the beetles, as the growers had planned, the cane toads began going after everything else in sight — insects, bird's eggs and even native frogs. And because the toads are poisonous, they began to kill would-be predators. The toll on native species has been immense....

Source: Tina Butler, mongabay.com, April 17, 2005

45. State *one* reason why the cane toads were imported to Australia.

46. Identify *one* adaptation of cane toads that made them successful in their new environment.

47. State *one* specific example of how the introduction of the cane toads threatened biodiversity in Australia.

48. Which animal has modified ecosystems more than any other animal and has had the greatest negative impact on world ecosystems?
    (1) gypsy moth
    (2) zebra mussel
    (3) human
    (4) shark

49. Lawn wastes, such as grass clippings and leaves, were once collected with household trash and dumped into landfills. Identify *one* way that this practice was harmful to the environment.

Base your answers to the next two questions on the information below and on your knowledge of biology.

Our national parks are areas of spectacular beauty. Current laws usually prohibit activities such as hunting, fishing, logging, mining, and drilling for oil and natural gas in these areas. Congress is being asked to change these laws to permit such activities.

50. Choose *one* of the activities listed above. State *one* way that activity could harm the ecosystem.

51. State *one* way allowing the activity you chose could benefit society.

52. Communities have attempted to control the size of mosquito populations to prevent the spread of certain diseases such as malaria and encephalitis. Which control method is most likely to cause the *least* ecological damage?
    (1) draining the swamps where mosquitoes breed
    (2) spraying swamps with chemical pesticides to kill mosquitoes
    (3) spraying oil over swamps to suffocate mosquito larvae
    (4) increasing populations of native fish that feed on mosquito larvae in the swamps

Base your answers to the next two questions on the information below and on your knowledge of biology

Gardeners sometimes use slug traps to capture and kill slugs. These traps were tested in a garden with a large slug population. Organisms found in the trap after one week are shown in the table below.

Organisms in Trap

Organism	Number in Trap
slugs	8
snails	1
aphids	13
centipedes	1
ground beetles	98

53. State *one* reason that slug traps are not the best method to control slugs.

54. In a process known as biological control, natural predators that prey on plant or animal pests are used to control the populations of the pests. Identify *one* organism shown in this food web that could be used as a biological control to replace the slug traps.

    Hint: *Review Food Web.*

55. State *two* advantages of relying on chemicals released by plants rather than using man-made chemicals for insect control.

56. State *one* positive effect on an ecosystem of using nuclear fuel to generate electricity.

57. State *one* negative effect on an ecosystem of using nuclear fuel to generate electricity.

58. Which phrase would be appropriate for area *A* in the chart below?

Technological Device	Positive Impact	Negative Impact
Nuclear power plant	Provides efficient, inexpensive energy	A

    (1) produces radioactive waste
    (2) results in greater biodiversity
    (3) provides light from radioactive substances
    (4) reduces dependence on fossil fuels

59. Water from nearby rivers or lakes is usually used to cool down the reactors in nuclear power plants. The release of this heated water back into the river or lake would most likely result in
    (1) an increase in the sewage content in the water
    (2) a change in the biodiversity in the water
    (3) a change in the number of mutations in plants growing near the water
    (4) a decrease in the amount of sunlight necessary for photosynthesis in the water

60. A major reason that humans can have such a significant impact on an ecological community is that humans
    (1) can modify their environment through technology
    (2) reproduce faster than most other species
    (3) are able to increase the amount of finite resources available
    (4) remove large amounts of carbon dioxide from the air

Base your answers to the next two questions on the information below and on your knowledge of biology.

Each year, a New York State power agency provides its customers with information about some of the fuel sources used in generating electricity. The table below applies to the period of 2002–2003.

Fuel Sources Used

Fuel Source	Percentage of Electricity Generated
hydro (water)	86
coal	5
nuclear	4
oil	1
solar	0

61. Identify *one* fuel source in the table that is considered a fossil fuel.

62. Identify *one* fuel source in the table that is classified as a renewable resource.

63. Increased industrialization will most likely
    (1) decrease available habitats
    (2) increase environmental carrying capacity for native species
    (3) increase the stability of ecosystems
    (4) decrease global warming

64. Changes in the chemical composition of the atmosphere that may produce acid rain are most closely associated with
    (1) insects that excrete acids
    (2) runoff from acidic soils
    (3) industrial smoke stack emissions
    (4) flocks of migrating birds

65. Increased production of goods makes our lives more comfortable, but causes an increase in the demand for energy and other resources. One negative impact of this situation on ecosystems is an increase in
    (1) living space for wildlife
    (2) renewable resources
    (3) the diversity of plant species
    (4) pollution levels in the atmosphere

66. Base your answers to the question on the information below and on your knowledge of biology.

    The average level of carbon dioxide in the atmosphere has been measured for the past several decades. The data collected are shown in the table below.

Average CO_2 Levels in the Atmosphere

Year	CO_2 (in parts per million)
1960	320
1970	332
1980	350
1990	361
2000	370

Identify *one* specific human activity that could be responsible for the change in carbon dioxide levels from 1960 to 2000.

67. Methods used to reduce sulfur dioxide emissions from smokestacks are an attempt by humans to
   (1) lessen the amount of insecticides in the environment
   (2) eliminate diversity in wildlife
   (3) lessen the environmental impact of acid rain
   (4) use nonchemical controls on pest species

68. Which situation is a result of human activities?
   (1) decay of leaves in a forest adds to soil fertility
   (2) acid rain in an area kills fish in a lake
   (3) ecological succession following volcanic activity reestablishes an ecosystem
   (4) natural selection on an island changes gene frequencies

69. In many investigations, both in the laboratory and in natural environments, the pH of substances is measured. Explain why pH is important to living things. In your explanation be sure to:

   - identify *one* example of a life process of an organism that could be affected by a pH change
   - state *one* environmental problem that is directly related to pH
   - identify *one* possible cause of this environmental problem

   ***Hint**: Review pH in Chapter 2.*

70. Base your answers to the next question on the diagram below and on your knowledge of biology. The diagram shows some of the gases that, along with their sources, contribute to four major problems associated with air pollution.

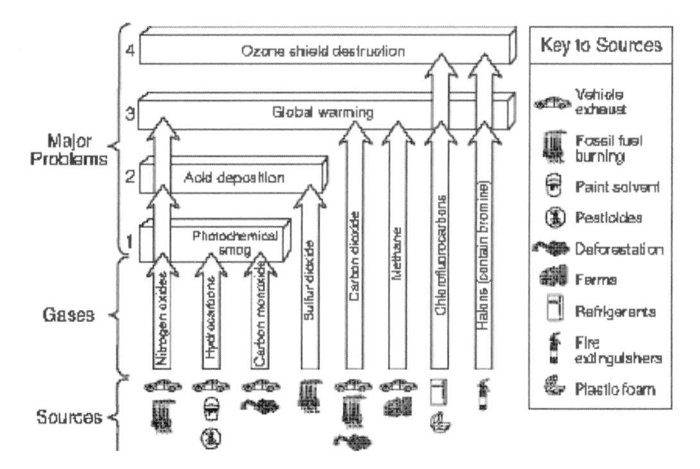

   Select *one* of the four major problems from the diagram and record the number of the problem on the line provided. Identify a gas that contributes to the problem you selected and state *one* way in which the amount of this gas can be reduced.

71. The ozone layer of Earth's atmosphere helps to filter ultraviolet radiation. As the ozone layer is depleted, more ultraviolet radiation reaches Earth's surface. This increase in ultraviolet radiation may be harmful because it can directly cause
   (1) photosynthesis to stop in all marine organisms
   (2) abnormal migration patterns in waterfowl
   (3) mutations in the DNA of organisms
   (4) sterility in most species of mammals and birds

72. Continued depletion of the ozone layer will most likely result in
   (1) an increase in skin cancer among humans
   (2) a decrease in atmospheric pollutants
   (3) an increase in marine ecosystem stability
   (4) a decrease in climatic changes

73. The diagram below shows the growth pattern of some skin cells in the human body after they have been exposed to ultraviolet radiation.

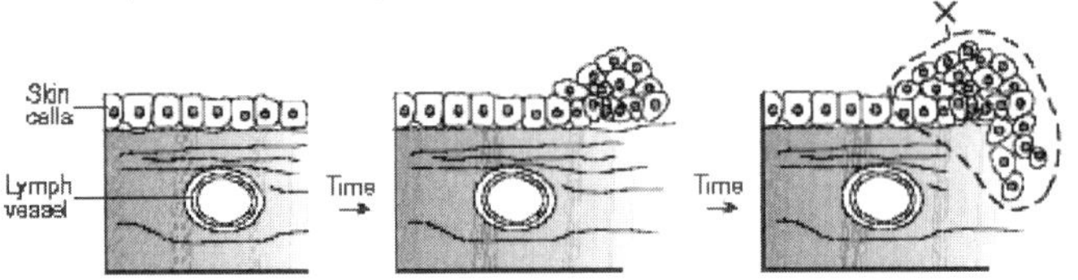

The cells in area *X* are most likely
   (1) red blood cells
   (2) cancer cells
   (3) white blood cells
   (4) sex cells

Base your answers to the next three questions on the passage below and on your knowledge of biology.

Great Effects on the Great Lakes Due to Global Warming

Trees such as the jack pine, yellow birch, red pine, and white pine may no longer be able to grow in the Great Lakes region because summers are becoming warmer. However, other trees such as black walnut and black cherry may grow in the area, given enough time. The change in weather would favor these new tree species.

The Great Lakes region is the only place in the world where the endangered Kirtland's Warbler breeds. This bird species nests in young jack pine trees (5 to 23 years old). The vegetation must have specific characteristics or the birds will not nest. A specific area of Michigan is one of the few preferred areas. If the jack pines can no longer grow in this area, the consequences for the Kirtland's Warbler could be devastating.

Recent research findings also suggest that algae production in Lake Ontario and several other Great Lakes will be affected as warmer weather leads to warmer lake water. An increase in water temperature reduces the ability of water to hold dissolved oxygen. These changes have implications for the entire Great Lakes food web. Changes in deep-water oxygen levels and other habitat changes may prevent the more sensitive cold-water fish from occupying their preferred niches in a warmer climate.

All other factors being equal, climatic changes may not have a negative effect on every species in the Great Lakes region. This is because the length of the growing season would be increased. Some temperature-sensitive fish could move to cooler, deeper water when the surface water temperatures become too high. The total impact of global warming is difficult to predict.

74. Explain how the habitat of the Kirtland's Warbler may be changed as a result of global warming.

75. Identify *one* producer found in the water of Lake Ontario.

76. Which graph best shows the relationship between changes in temperature in the Great Lakes waters and the amount of dissolved oxygen those waters can hold?

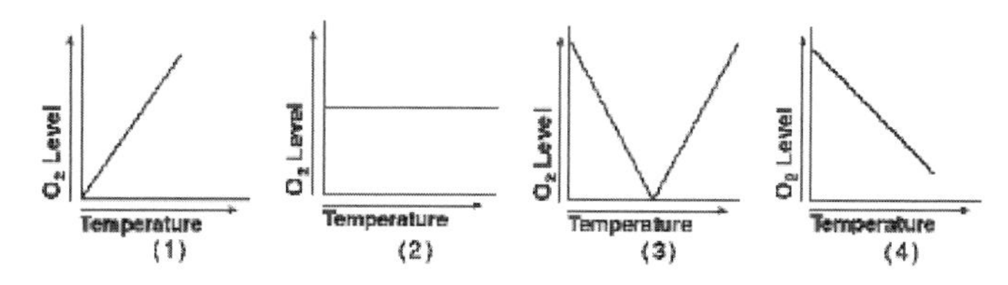

77. For over 100 years scientists have monitored the carbon dioxide concentrations in the atmosphere in relation to changes in the atmospheric temperature. The graphs below show the data collected for these two factors.

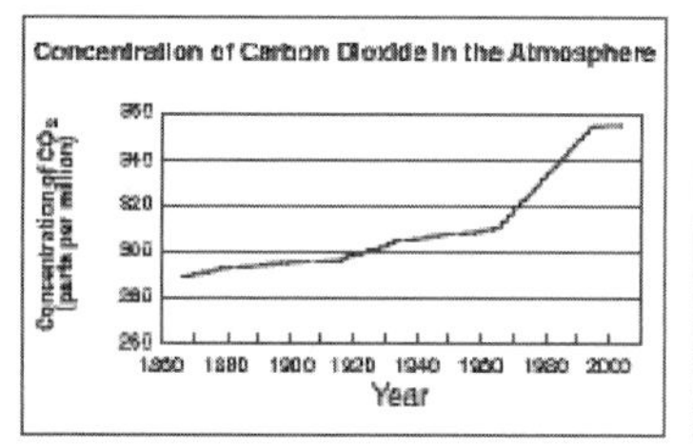

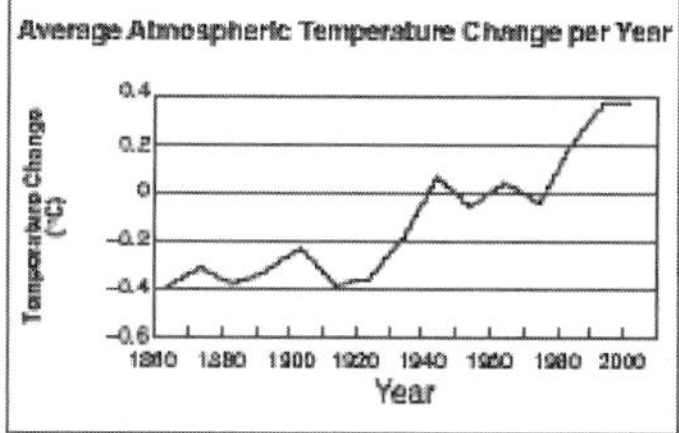

Discuss the overall relationship between carbon dioxide concentration and changes in atmospheric temperature and the effect of these factors on ecosystems. Your answer must include:

- A statement identifying the overall relationship between the concentration of carbon dioxide and changes in atmospheric temperature
- *One* way in which humans have contributed to the increase in atmospheric carbon dioxide
- *One* specific *negative* effect the continued rise in temperature would be likely to have on an ecosystem
- *One* example of how humans are trying to reduce the problem of global warming

Base your answers to the next two questions on the information below and on your knowledge of biology.

The ice fields off Canada's Hudson Bay are melting an average of three weeks earlier than 25 years ago. The polar bears are therefore unable to feed on the seals on these ice fields during the last three weeks in spring. Polar bears have lost an average of 10% of their weight and have 10% fewer cubs when compared to a similar population studied just 20 years ago. Scientists have associated the early melting of the ice fields with the fact that the average world temperature is about 0.6°C higher than it was a century ago and this trend is expected to continue.

78. What ecological problem most likely caused the earlier melting of the ice fields in the Hudson Bay area of Canada?

79. State *one* specific long-term action that humans could take that might slow down or reduce the melting of the ice fields.

Base your answers to the next two questions on the information below and on your knowledge of biology.

The average level of carbon dioxide in the atmosphere has been measured for the past several decades. The data collected are shown in the table below.

Average CO_2 Levels in the Atmosphere

Year	CO_2 (in parts per million)
1960	320
1970	332
1980	350
1990	361
2000	370

80. State *one* possible negative effect this change in CO_2 level has had on the environment of Earth.

81. Identify *one* specific human activity that could be responsible for the change in carbon dioxide levels from 1960 to 2000.

82. Which process helps reduce global warming?
    (1) decay (3) photosynthesis
    (2) industrialization (4) burning

83. In lakes in New York State that are exposed to acid rain, fish populations are declining. This is primarily due to changes in which lake condition?
    (1) size (3) pH
    (2) temperature (4) location

Base your answers to the next three questions on the information below and on your knowledge of biology.

Carbon, like many other elements, is maintained in ecosystems through a natural cycle. Human activities have been disrupting the carbon cycle.

84. State *one* reason why the amount of carbon dioxide in the atmosphere has increased in the last 100 years.

85. Identify *one* effect this increase in carbon dioxide could have on the environment.

86. Describe *one* way individuals can help slow down or reverse the increase in carbon dioxide.

87. Base your answers to the next question on the information below and on your knowledge of biology.

Each year, a New York State power agency provides its customers with information about some of the fuel sources used in generating electricity. The table below applies to the period of 2002–2003.

Fuel Sources Used

Fuel Source	Percentage of Electricity Generated
hydro (water)	86
coal	6
nuclear	4
oil	1
solar	0

State *one* specific environmental problem that can result from burning coal to generate electricity.

88. Identify *one* farming practice that could be a source of environmental pollution.

89. Select *one* of the following ecological problems.

Ecological Problems

Acid rain
Increased amounts of nitrogen and phosphorous in a lake
Loss of biodiversity

For the ecological problem that you selected, briefly describe the problem and state *one* way to reduce it. In your answer be sure to:

- State the ecological problem you selected
- State how humans have caused the problem you selected
- Describe *one* specific effect that the problem you selected will have on the ecosystem
- State *one* specific action humans could take to reduce the problem you selected

---

Base your answers to the next three questions on the information below and on your knowledge of biology.

Untreated organic wastes were accidentally discharged into a river from a sewage treatment plant. The graph below shows the dissolved oxygen content of water samples taken from the river at specific distances downstream from the plant, both before, and then three days after the discharge occurred.

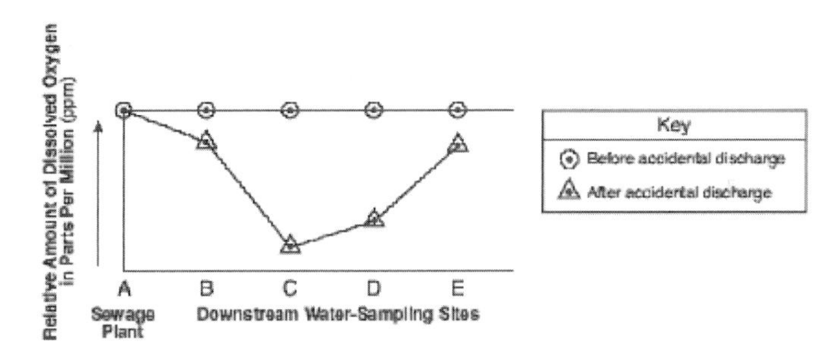

90. State why this accident would be expected to benefit the decomposers in the river below the sewage plant.

91. Explain why an energy-releasing process occurring in the mitochondria of the decomposer organisms is most likely responsible for the change indicated by the data shown at sampling site C in the graph.

92. State *one* reason why the statement below is correct.

"The effects of the accidental discharge are not expected to last for a long time."

Base your answers to the next four questions on the information below and on your knowledge of biology.

In recent years, the striped bass population in Chesapeake Bay has been decreasing. This is due, in part, to events known as "fish kills," a large die-off of fish. Fish kills occur when oxygen-consuming processes in the aquatic ecosystem require more oxygen than the plants in the ecosystem produce, thereby reducing the amount of dissolved oxygen available to the fish.

One proposed explanation for the increased fish kills in recent years is that human activities have increased the amount of sediment suspended in the water of Chesapeake Bay, largely due to increased erosion into its tributary streams. The sediment acts as a filter for sunlight, which causes a decrease in the intensity of the sunlight that reaches the aquatic plants in the Chesapeake Bay ecosystem.

93. Identify *one* abiotic factor in the Chesapeake Bay ecosystem involved in the fish kills.

94. Identify the process carried out by organisms that uses oxygen and contributes to the fish kills.

95. State *one* way humans have contributed to the decrease of the striped bass population in Chesapeake Bay.

96. State how a *decrease* in the amount of light may be responsible for fish kills in the Chesapeake Bay area.

Base your answers to the next two questions on the information below and on your knowledge of biology.

A factory in Florida had dumped toxic waste into the soil for 40 years. Since the company is no longer in business, government officials removed the toxic soil and piled it up into large mounds until they can finish evaluating how to treat the waste.

97. State *one* way these toxins could move from the soil into local ecosystems, such as nearby lakes and ponds.

98. State *one* way these toxins might affect local ecosystems.

---

99. When living organisms obtain water and food from their environment, they may also take in toxic pesticides. Low concentrations of some pesticides may not kill animals, but they may damage reproductive organs and cause sterility. The data table below shows concentrations of a pesticide in tissues of organisms at different levels of a food chain.

Concentration of Pesticide in Tissues

Organisms	Pesticide Concentration (parts per million)
producers	0.01–0.03
herbivores	0.25–1.50
carnivores	4.10–313.80

What does this information suggest to a person who is concerned about health and is deciding on whether to have a plant-rich or an animal-rich diet? Support your answer using the information provided.

100. Compounds containing phosphorus that are dumped into the environment can upset ecosystems because phosphorus acts as a fertilizer. The graph below shows measurements of phosphorus concentrations taken during the month of June at two sites from 1991 to 1997.

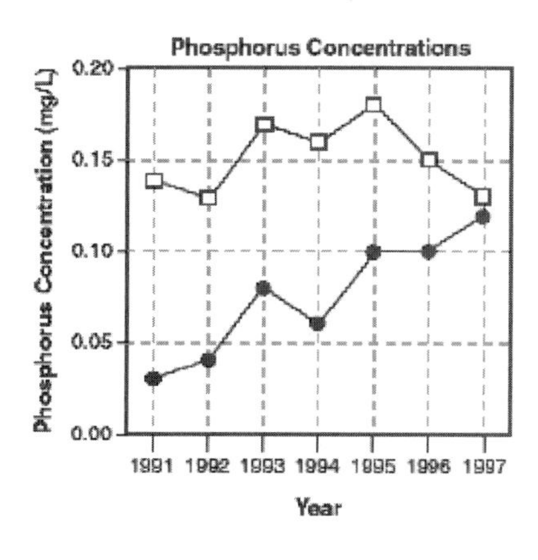

Which statement represents a valid inference based on information in the graph?

(1) There was no decrease in the amount of compounds containing phosphorus dumped at site *2* during the period from 1991 to 1997.
(2) Pollution controls may have been put into operation at site *1* in 1995.
(3) There was most likely no vegetation present near site *2* from 1993 to 1994.
(4) There was a greater variation in phosphorous concentration at site *1* than there was at site *2*.

101. Which result of technological advancement has a positive effect on the environment?

(1) development of new models of computers each year, with disposal of the old computers in landfills
(2) development of new models of cars that travel fewer miles per gallon of gasoline
(3) development of equipment that uses solar energy to charge batteries
(4) development of equipment to speed up the process of cutting down trees

102. An industry releases small amounts of a chemical pollutant into a nearby river each day. The chemical is absorbed by the microscopic water plants in the river. It causes the plants no apparent harm. Explain how this small amount of the chemical in the microscopic plants could enter the food chain and endanger the lives of birds that live nearby and feed on the fish from the river each day.

103. Which action illustrates an increased understanding and concern by humans for ecological interrelationships?

(1) importing organisms in order to stabilize existing ecosystems
(2) eliminating pollution standards for industries that promote technology
(3) removing natural resources at a rate equal to the needs of the population
(4) implementing laws to regulate the number of animals hunted and killed each year

104. Base your answers to the next question on the information below and on your knowledge of biology.

> You are the owner of a chemical company. Many people in your community have been complaining that rabbits are getting into their gardens and eating the flowering plants and vegetables they have planted. Your company is developing a new chemical product called Bunny Hop-Away that repels rabbits. This product would be sprayed on the plants to prevent the rabbits from eating them. Certain concerns need to be considered before you make the product available for public use.

State *two* environmental concerns that should be considered before the product is sold and used by the public.

Base your answers to the next three questions on the information and graph below.

> Reducing toxic chemicals released into the environment often requires laws. When making decisions about whether or not to support the passing of such laws, individuals must weigh the benefits against the potential risks if the law is not passed.
>
> The amounts of toxic chemicals released into the environment of New York State over a ten-year period are shown in the graph below.

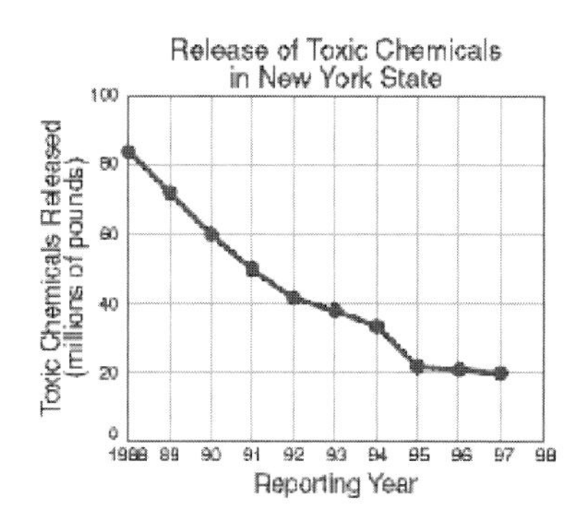

105. State *one* possible *negative* effect of passing a law to reduce the release of toxic chemicals.

106. State *one* possible explanation for why the amount of toxic chemicals released remained relatively constant between 1995 and 1997.

107. State *one* other type of environmental problem that has been reduced by passing laws.

---

108. Farming reduces the natural biodiversity of an area, yet farms are necessary to feed the world's human population. This situation is an example of

(1) poor land use
(2) a trade-off
(3) conservation
(4) a technological fix

109. In most states, automobiles must be inspected every year to make sure that the exhaust fumes they emit do not contain high levels of pollutants such as carbon monoxide. This process is a way humans attempt to

(1) control the water cycle
(2) recycle nutrients from one ecosystem to another
(3) control energy flow in natural ecosystems
(4) maintain the quality of the atmosphere

110. Some farmers currently grow genetically engineered crops. An argument *against* the use of this technology is that
   (1) it increases crop production
   (2) it produces insect-resistant plants
   (3) its long-term effects on humans are still being investigated
   (4) it always results in crops that do not taste good

111. Bacteria that are removed from the human intestine are genetically engineered to feed on organic pollutants in the environment and convert them into harmless inorganic compounds. Which row in the table below best represents the most likely negative and positive effects of this technology on the ecosystem?

Row	Negative Effect	Positive Effect
(1)	Inorganic compounds interfere with cycles in the environment.	Human bacteria are added to the environment.
(2)	Engineered bacteria may out-compete native bacteria.	The organic pollutants are removed.
(3)	Only some of the pollutants are removed.	Bacteria will make more organic pollutants.
(4)	The bacteria will cause diseases in humans.	The inorganic compounds are buried in the soil.

112. In the United States, there has been relatively little experimentation involving the insertion of genes from other species into human DNA. One reason for the lack of these experiments is that
   (1) the subunits of human DNA are different from the DNA subunits of other species
   (2) there are many ethical questions to be answered before inserting foreign genes into human DNA
   (3) inserting foreign DNA into human DNA would require using techniques completely different from those used to insert foreign DNA into the DNA of other mammals
   (4) human DNA always promotes human survival, so there is no need to alter it

Base your answers to the next four questions on the information below and on your knowledge of biology.

Insecticides are used by farmers to destroy crop-eating insects. Recently, scientists tested several insecticides to see if they caused damage to chromosomes. Six groups of about 200 cells each were examined to determine the extent of chromosome damage after each group was exposed to a different concentration of one of two insecticides. The results are shown in the data table below.

Cell Damage After Exposure to Insecticide

Insecticide	Insecticide Concentration (ppm)	Number of Cells with Damaged Chromosomes
Methyl parathion	0.01	7
	0.10	15
	0.20	30
Malathion	0.01	3
	0.10	4
	0.20	11

Directions for the next three questions: *Using the information in the data table, construct a line graph on the grid provided, following the directions below.*

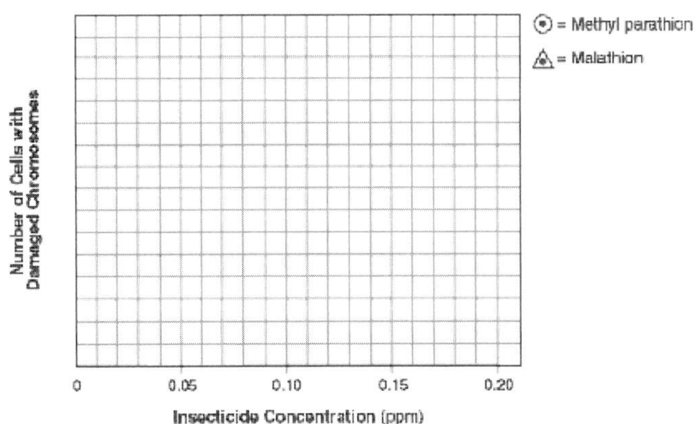

113. Mark an appropriate scale on the axis labeled, "Number of Cells with Damaged Chromosomes."

114. Plot the data for methyl parathion on the grid. Surround each point with a small circle and connect the points.

Example:

115. Plot the data for malathion on the grid. Surround each point with a small triangle and connect the points.

Example:

116. Which insecticide has a more damaging effect on chromosomes? Support your answer.

117. For over 100 years scientists have monitored the carbon dioxide concentrations in the atmosphere in relation to changes in the atmospheric temperature. The graphs below show the data collected for these two factors.

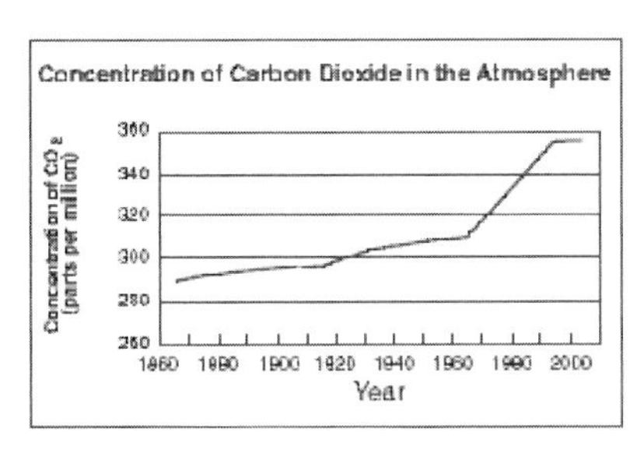

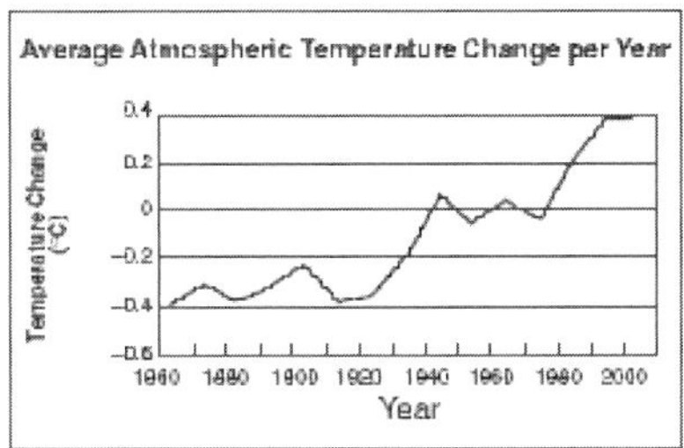

CHAPTER 8: SCIENCE INQUIRY

Science Inquiry: Science inquiry is the way scientists get their ideas to explain how the world works. **Scientific inquiry** (the **way scientists work**) involves **questioning, observing, inferring (conclusions based on observations** (what they see)) **experimenting, finding evidence, collecting** and **organizing data** (information), drawing **valid conclusions (conclusions** that are supported by **(based on) observations (evidence**), and **peer review** (other scientists review the work to make sure it is correct).
Inquiry involves finding information (examples from internet, libraries, discussions with other scientists), processing (understanding) information from a variety of sources (examples scientific journals (which are very reliable), books, magazines). Inquiry involves making sure the **information** you get is **correct** and reliable (**valid**).

Science inquiry explains what happens in the world (natural phenomena), based on scientific knowledge (what we already know in science) and what new things the scientist observes today. Science explanation is based on past and present (new knowledge). Today we have new technology (example electron microscope) to observe things we previously could not.
For a scientific explanation to be accepted, it has to be in agreement with the results of experiments and observations. The **explanation** (explaining something) **must** be able to **predict** what will happen in other situations.
A **group of ideas** which **explain many observations** on **one topic** is a **theory**. A theory (scientific theory) is a statement that is supported by (based on) many observations (a variety of data from many experiments). If in the future new observations or ideas or explanations are discovered which are different than the previous theory or explanation, the theory or explanation must be corrected (modified) or replaced. For example, earlier scientific studies showed the hippo is a close relative (closely related) to the pig, but new studies showed that the hippo is a close relative to the whale. Therefore, the previous (earlier) explanation was changed (replaced) with the newer explanation that the hippo is a close relative (closely related) to the whale.
A theory becomes generally accepted if different scientists from different fields of study come to the same conclusion.

Values

Scientific knowledge by itself is not enough to tell us what to do. We may be able to study genes and determine who is likely to get certain diseases. Should those people (who are likely to get a disease) not be **hired for jobs** or not be allowed to **get health insurance**? Science cannot give us an answer to these questions. We must **use** our **values** and **ethics** to **decide** what the right answer is.

Scientific Literacy

To decide whether to buy a product (example a drug to lose weight quickly), a person might look for other sources of information, such as scientific journals (which are very reliable), books, magazines, and websites to see if the product actually works and if the product is safe.

The question is can you rely on what you read in the book or magazine or see on the website. Was the book, magazine, or website written by a doctor or an expert in the field, or was it written by a person who knows nothing or very little about the drug, its side effects, and its usefulness.

Find out, or have a scientist find out, whether the company carried out controlled scientific experiments showing that the product actually works. Was the product tested on five people or 500,000 people? In what percentage of them did it work? What are the side effects (dangers)?

Rules for Experiments

1. The same experiment should be done on a lot of people (**large sample**), to make sure the **results** are correct and reliable.

2. The **experiment** should be **repeated many times** (repeated trials) and give the **same results**.

3. The experiment should have a **control** (controlled experiment). The **control group** should be identical to the other group, called the experimental group (example both groups eat the same food), but only the experimental group (for example) takes a diet pill; the **control group** does not take the diet pill but takes a **pill** that looks identical to the diet pill but has nothing in it to cause loss of weight. The experiment on the next page shows how a controlled experiment is done.

4. Use statistics to make sure the results show a big enough loss of weight that it is caused by the diet pill.

Research Plan

You want to do an experiment. Make a **research plan**. This involves **getting background information** (example from scientific journals, internet, books, etc., and **seeing which experiments were already done**), then **making** a **hypothesis** (educated guess about what will happen in the experiment), and **designing an experiment** to see if the hypothesis is correct.

The **experiment** should have a **large sample size**, be **repeated many times** and **get the same results**, and have a **control**. The person who does the experiment may sometimes have to develop new methods or build new types of scientific equipment.

Now Do Homework Questions #1-11, pages 41-42.

Designing Experiments

Design an experiment to see if gibberellic acid causes plants to grow taller (does gibberellic acid makes plants grow taller).

Our **HYPOTHESIS** (educated guess) is that plants grow taller with gibberellic acid.

What we should do **(PROCEDURE):** Take a lot of plants (**large sample**) of the **same type**. Keep the plants under the same conditions **(constant conditions)** such as same size pot, with the same amount of soil, same amount of water, vitamins, and light.

Divide the plants into two groups. Now one group of plants gets the gibberellic acid; every plant in this group gets the same amount of the acid. This is the **experimental group**. The other group of plants does not get the acid. The group that does not get the acid is the **control group;** it does not get the thing you are testing (in this example the acid).

A **valid experiment** is an experiment that has only one variable (one thing that is different between the experimental and control groups). In this experiment, the acid is the only one variable, the only thing different between the experimental and control groups.

Then look at the plants **(OBSERVATION)**(what we see). Measure the height of each plant and write down the average height of the plants in each group every day. Record your findings. Make a data table (see below). By using a data table and graphs, you can more easily notice if the variable (acid) helps the plants grow taller.

Each day, measure the height of the plants and record the average height of the plants. On Day 1, the plants are a certain height. On day 2 (we changed the day to the next day) the height is different. On day 3, we again changed the day and again the height is different. The height depends on which day it is.

Data Table

Average Height of Plants (cm)

	Day 1	Day 2	Day 3	Day 4	Day 5
No acid	*height*	*height*			
With acid	*height*	*height*			

The **independent variable** (the thing that we change) is the days. The **dependent variable** is the height, which depends on which day it is.

Draw a line graph. On a line graph, the x axis is the days, the independent variable; the y axis is height, the dependent variable (see graph below). In a valid experiment, there is only one independent variable (the thing that we change).

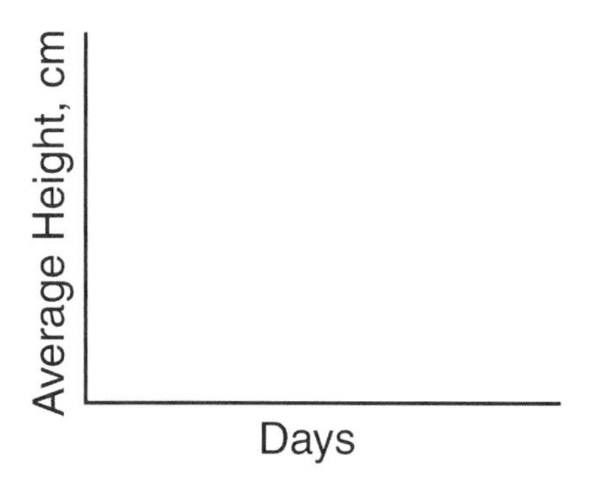

Here is a filled in data table (with the heights of the plants) and line graph.

Data Table

Average Height of Plants (cm)

	Day 1	Day 2	Day 3	Day 4	Day 5
No acid (Group B)	5	6	6.5	7	7.5
With acid (Group A)	5	7	10	13	15

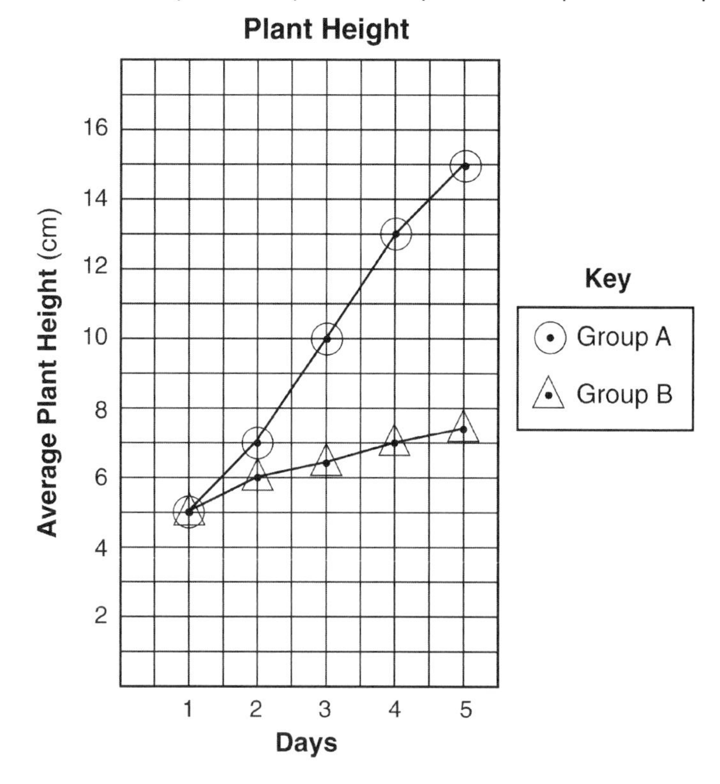

Draw a CONCLUSION. By looking at the data table and the graph showing the height of the plants with and without the gibberellic acid, you can reach a conclusion. The **conclusion** is the **plants with** gibberellic **acid** are **taller,** which **supports** the **hypothesis** (educated guess) that **plants** grow **taller with** the **acid**.

The **conclusion** is **valid because** the **conclusion** (plants with acid are taller) is supported by (**based on**) the **observations** (what we see, evidence) and **data** (example data in data table, in this example, plants with the acid are taller).
Note: An invalid conclusion is when the conclusion is NOT supported by (based on) the observations (evidence).

In this experiment, you saw the **conclusion** (example, plants with acid are taller) is **valid** for the **type of plant** you **used in this experiment**; you cannot apply the condusion to other types of plants. For **other plants** like large trees or microscopic plants, the **conclusion** would be **invalid** (not valid) because you did not do experiments on the other types of plants. To make the conclusion possibly valid for all types of plants, you would have to test (do experiments on, observe) many different types of plants. It would also not be valid to say that this acid works on animals, because this experiment was not done on animals.

Similarly, if a medicine works on one kind of animal (example mice), you cannot generalize and make a conclusion that it will work on all animals; that conclusion (that it will work on all animals) would be invalid (not valid); it is not supported by the evidence (observations).

If, in a different experiment, you see at the end of the experiment that the hypothesis (educated guess) was wrong (the data do not agree with or support the hypothesis), you know you have to investigate (look into and study it) more and find a new hypothesis; then you will carry on an experiment to see if the new hypothesis is correct.

Now Do Homework Questions #12-20, pages 42-44.

Scientific Method

It might be **easier** for you to **use** the **scientific method** (explained below) to **design experiments** and **do** all types of **experiments**, including experiments involving science, math, and technology.

Problem: Do plants grow taller with gibberellic acid?

Materials: (what materials do you use?): pots (number of pots, such as 20 pots, type of pots, etc.), soil, identical plants, gibberellic acid.

Hypothesis (educated guess): Plants grow taller with gibberellic acid.

Procedure (what we should do or what we do): Take a lot of plants of the **same type** (large sample). Divide them into two groups, one group gets the gibberellic acid (experimental group) and the other group does not get the acid (control group). Keep them under the same conditions (constant conditions): same size pot, with the same amount of soil, same amount of water, vitamins, and light.
We will observe and measure the height of each plant and get the average height of the plants in each group every day and record the heights.

Observations (what you see): Make a **data table** and a **line graph** showing what you observed (saw), which is the height of the plants with and without the gibberellic acid.

Data Table

Average Height of Plants (cm)

	Day 1	Day 2	Day 3	Day 4	Day 5
No acid	*height*	*height*			
With acid	*height*	*height*			

The **independent variable** (the thing that we change) is the days. The **dependent variable** is the height, which depends on which day it is. Draw a line graph. On a line graph, the x axis is the days, the independent variable; the y axis is height (average height), the dependent variable (see graph below). In a valid experiment, there is only one independent variable (the thing that we change).

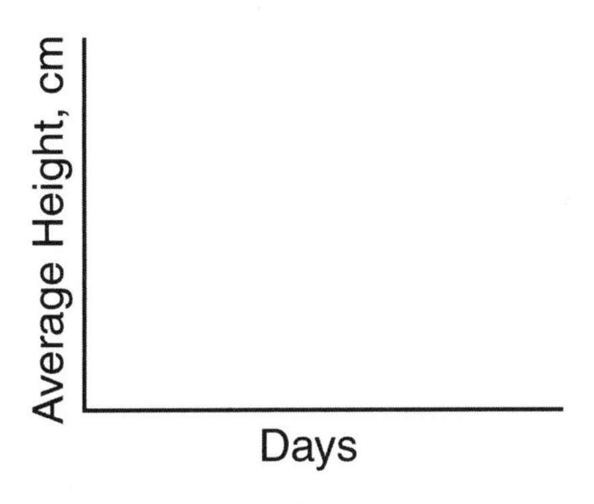

Here is a filled in data table (with the heights of the plants) and line graph.

Data Table

Average Height of Plants (cm)

	Day 1	Day 2	Day 3	Day 4	Day 5
No acid (Group B)	5	6	6.5	7	7.5
With acid (Group A)	5	7	10	13	15

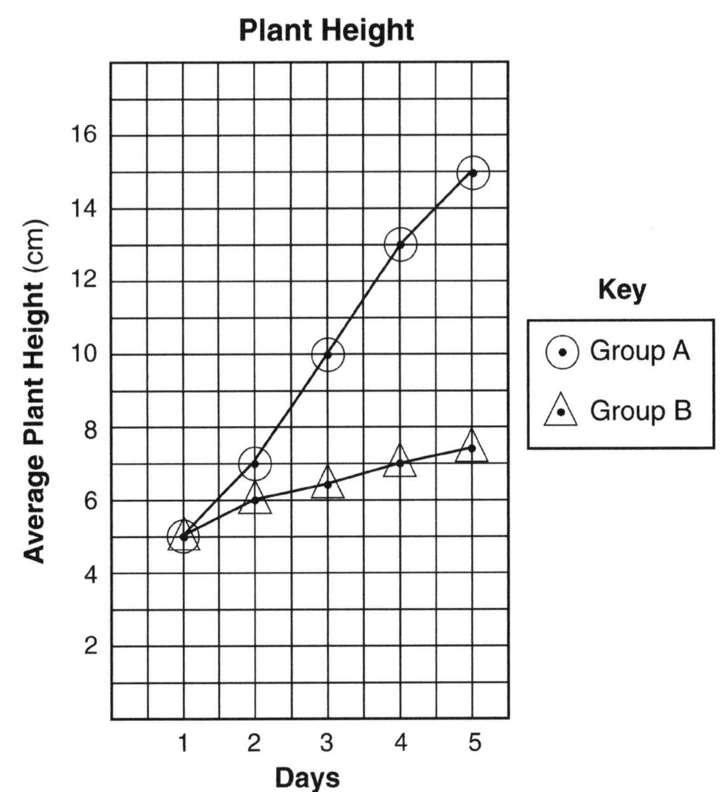

Conclusion: By looking at the data table and the graph showing the height of the plants with and without the gibberellic acid, you can reach a conclusion. The **conclusion** is the **plants with** gibberellic **acid** are **taller,** which **supports** the **hypothesis** (educated guess) that **plants** grow **taller with** the **acid**.

The **conclusion** is **valid because** the **conclusion** (plants with acid are taller) is supported by (**based on**) the **observations** (what we see, evidence) and **data** (example data in data table, in this example, plants with the acid are taller).

If the plants with the gibberellic acid are not taller, then the conclusion would be that plants with the acid are not taller (the acid did not help plants grow); then we would do the experiment again with a new hypothesis. The experiment would still be a valid experiment because the conclusion (plants are not taller with acid) is based on what you see.

PRACTICE QUESTIONS AND SOLUTIONS

Question: The photograph below shows a pill bug. Pill bugs are small animals frequently found in wooded areas near decomposing organic material.

Describe some parts of an experiment to determine the preference of pill bugs for light or darkness. In your answer be sure to:

A. state a hypothesis
B. identify the independent variable in the experiment
C. identify *two* conditions that should be kept the same in all experimental setups.
D. state *one* example of experimental data that would support your hypothesis.

Solution:

Answer for **A and D.**

If **A** your hypothesis is **pill bugs prefer light**, then
D experimental data to support your hypothesis would be **more pill bugs went to the lighted area.**

OR

Answer for **A and D**.

If **A** your hypothesis is **pill bugs prefer dark**, then
D experimental data to support your hypothesis would be **more pill bugs went to the dark area.**

Answer for **B.** The independent variable is light, if there is light or no light.

Answer for **C.** Conditions that should be kept the same include temperature, type of container, or same species of pill bug, or same number of pill bugs in each container.

Question: You have been assigned to design an experiment to determine the effects of light on the growth of tomato plants. In your experimental design be sure to:

- state *one* hypothesis to be tested
- identify the independent variable in the experiment
- describe the type of data to be collected

Solution: Hypothesis: More light (example more hours of light or brighter light) makes plants grow more (example, taller or have more leaves or have bigger leaves).
Independent variable (the thing you change): light. A valid experiment can only have one independent variable (example light).
Data (what you measure): height of plants or number of leaves or size of leaves.

Now Do Homework Questions #21-40, pages 44-49.

Valid Experiments

A **valid** experiment is based on what you see (evidence, observations).
A **conclusion** is **valid** if it is supported by **(based on) observations** (what you see, evidence) and **data**. This means that the conclusions agree with the observations. In a valid experiment, there is only one variable (one thing different between the experimental and control groups, example gibberellic acid), and everything else is the same in the experimental and control groups. The result (what happened in the experiment, example taller plants with gibberellic acid) was caused by the variable (gibberellic aicd) and not by something else.

Example of a valid experiment: A procedure we can use to see if drug A would be effective in treating cancer in white mice is injecting 1 mL of drug A into 100 white mice with cancer and 1 mL of distilled water into another group of 100 white mice with cancer. A valid experiment has both an experimental group and a control group. In this valid experiment, the experimental group takes 1 mL of drug A in water and the control group just takes 1 mL of water. There is only one variable (taking or not taking drug A), but all the other things are the same (100 white mice with cancer in both the experimental and control groups).

Valid Experiments: You learned in a valid experiment:

1. **conclusion** is **based on** the **observations** (**evidence,** what you see)
2. the **experiment** should **have** a **control**
3. **only one variable** (example giving gibberellic acid to plants or drug A to mice) is being tested; **everything else** in the experimental and control groups **is** the **same.**
4. the same experiment should be done on a lot of people (**large sample**) to make sure the results are correct and reliable
5. the **experiment** should be **repeated many times** and give the **same results**.
6. the **result** should **predict** for **new situations**.

Question: A science researcher is reviewing another scientist's experiment and conclusion. The reviewer would most likely consider the experiment *invalid* if

(1) the sample size produced a great deal of data
(2) other individuals are able to duplicate the results
(3) it contains conclusions not explained by the evidence given
(4) the hypothesis was not supported by the data obtained

Solution: A valid experiment has a conclusion based on evidence (what you see, observations). An **invalid (not valid)** experiment has the conclusion **NOT based on** (not explained by) **evidence** (what you see, observations). Answer *3*

Question: The table contains information about glucose production in a species of plant that lives in the water of a salt marsh.

Temperature (°C)	**Glucose Production** (mg/hr)
10	5
20	10
30	15
40	5

What evidence from the data table shows that a salt-marsh plant is sensitive to its environment?

Solution: We see from the data table that the amount of glucose produced changed with the temperature (at 10°C, only 5 mg/hr is produced, while at 30°C 15 mg/hr is produced). This shows that the plant is sensitive to (affected by) its environment (temperature).

A **valid inference** is a **conclusion** made by **interpreting** the **observations** (evidence, what you see) or interpreting what is in the data (example data table, information, facts).

Question: A student studied the location of single-celled photosynthetic organisms in a lake for a period of several weeks. The depth at which these organisms were found at different times of the day varied greatly. Some of the data collected are shown in the table below.

Data Table

Light Conditions at Different Times of the Day	**Average Depth of Photosynthetic Organisms** (cm)
full light	150
moderate light	15
no light	10

A valid inference based on these data is that

1. most photosynthetic organisms live below a depth of 150 cm
2. oxygen production increases as photosynthetic organisms move deeper into the lake
3. photosynthetic organisms respond to changing light levels
4. photosynthetic organisms move up and down to increase their rate of carbon dioxide production

Solution: When there is full light (a lot of light), the organisms are far down in the water (150 cm down). When there is moderate light (some light) the organisms move higher in the water (15 cm down instead of 150 cm down). When the organisms have no light, they even move further up (10 cm down instead of 15 cm down). A valid inference (interpretation) is that the organisms responds to changes in light levels. When there is full light, the organisms are further down, when there is moderate light the organisms are higher up, and when there is no light, the organisms are even higher up. Answer *3*

Question: Recently, scientists noted that stained chromosomes from rapidly dividing cells, such as human cancer cells. contain numerous dark, dotlike structures. Chromosomes from older human cells that have stopped dividing have very few, if any, dotlike structures. The best generalization regarding these dotlike structures is that they

1. will always be present in cells that are dividing
2. may increase the rate of mitosis in human cells
3. definitely affect the rate of division in all cells
4. can cure all genetic disorders

Solution: Choice 2. Dotlike structures are found in rapidly dividing cells but not in cells that stopped dividing. It is possible (maybe) that these dotlike structures caused cells to divide rapidly (increased rate of mitosis (cell division)).
Wrong choices: Choices 1, 3, and 4 are wrong. These three choices say that there are definite (certain) answers, (Choice 1. ...always..., Choice 3. ...definitely..., Choice 4. ...all...), but we are not certain.

Now Do Homework Questions #41-49, pages 49-52.

REVIEW OF GRAPHS

There are a few different types of graphs (line graphs, bar graphs, circle graphs, etc.).

Drawing Line Graphs

Let's see how we can draw graphs based on experimental data. Data (from an experiment) is written on a data table. Draw the graphs based on the data table.

Problem 1 (at the end of chapter 1): The experiment used five tubes to study the effect of temperature on protein digestion (amount or how much protein is digested). The results of the experiment are shown in the data table below.

Protein Digestion at Different Temperatures

Tube #	Temperature (°C)	Amount of Protein Digested (grams)
1	5	0.5
2	10	1.0
3	20	4.0
4	37	9.5
5	85	0.0

How to draw the line graph:

1. On the x axis, put "Temperature, °C". **The thing you change** (in this case temperature) is always put on the **x axis.** This is the independent variable. Space the lines along the axis equally. "Make an appropriate scale" by spacing the numbers on the graph so that all the data fits on the graph and it is easy to read. There must be an equal number of degrees between lines (see graph). On the x axis, put 10°C between lines (every two lines) (scale on the x axis), then all the temperatures on the data table fit on the graph and it is easy to read.

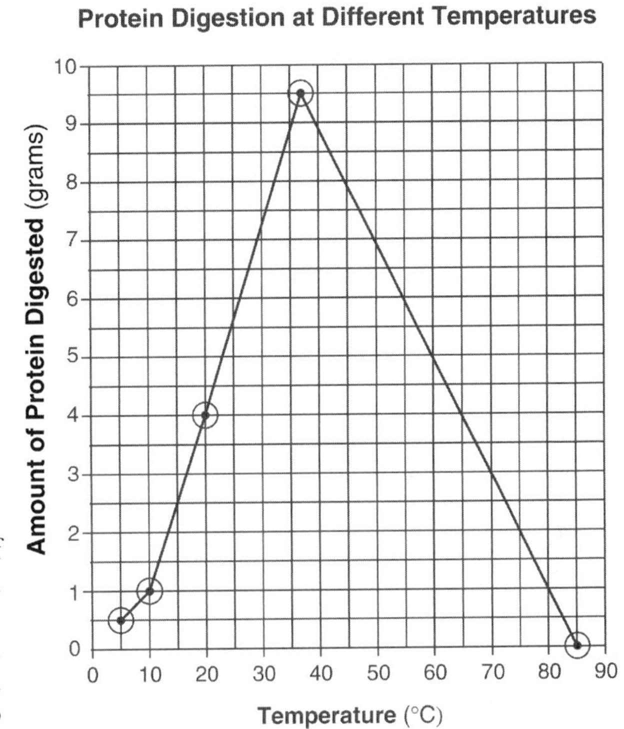

2. On the y axis, put "Amount of Protein Digested (grams)". The **result** you get (amount of protein digested) is always put on the **y axis**. This is the dependent variable. Space the lines along the axis equally. "Make an appropriate scale" by spacing the numbers on the graph so all the data fits on the graph and it is easy to read. There must be an equal number of grams of protein digested between lines (see graph.) On the y axis, put one gram of protein between lines (every two lines) (scale on the y axis), then all the grams of protein digested in the data table fits on the graph and is easy to read.
3. Plot the experimental data on the graph. Draw a circle around each point. Draw a line that connects the points. Do not continue the line past the last point.
4. Put a title on the graph which shows what the graph is about. Examples: "Effect of temperature on protein digestion" or "Protein digestion at different temperatures."

Problem 2 (at the end of chapter 2):

The laboratory experiment was to see the effect of time on cellular respiration in yeast. Yeast-glucose solution is in the flask at 35°C. The student counted and recorded the number of gas bubbles in the test tube every five minutes.

Data Table

Time (minutes)	Total Number of Bubbles Released 35°
5	5
10	15
15	30
20	50
25	75

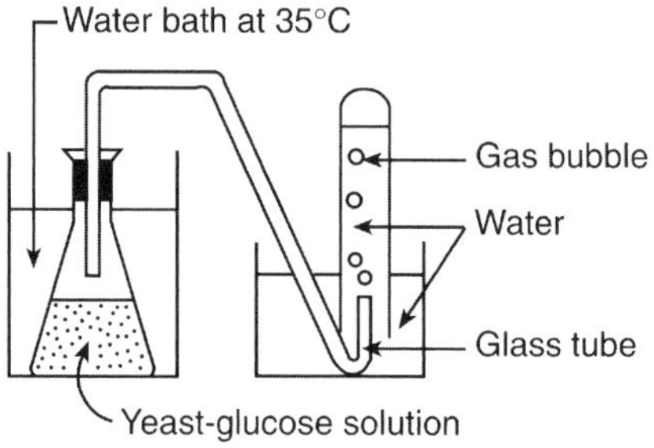

Water bath at 35°C
Gas bubble
Water
Glass tube
Yeast-glucose solution

How to draw the line graph:

1. On the x axis, put "Time, minutes". Always include units (in this example minutes) on the axis (see graph). The **thing you change** (in this case time) is always put on the **x axis.** This is the independent variable. Space the lines along the axis equally. "Make an appropriate scale" by spacing the numbers on the graph so that all the data fits on the graph and it is easy to read. There must be an equal number of minutes between lines (see graph). On the x axis, put five minutes between lines (every two lines)(scale on the x axis), then all the time on the data table fits on the graph and it is easy to read.

2. On the y axis, put "Total Number of Bubbles Released". The **result** you get (total number of bubbles released) is always put on the **y axis**. This is the dependent variable. Space the lines along the axis equally. "Make an appropriate scale" by spacing the numbers on the graph so all the data fits on the graph and it is easy to read. There must be an equal number of bubbles between lines (see graph.) On the y axis, put ten bubbles between lines (every two lines) (scale on the y axis) then all the bubbles on the data table fit on the graph and it is easy to read.

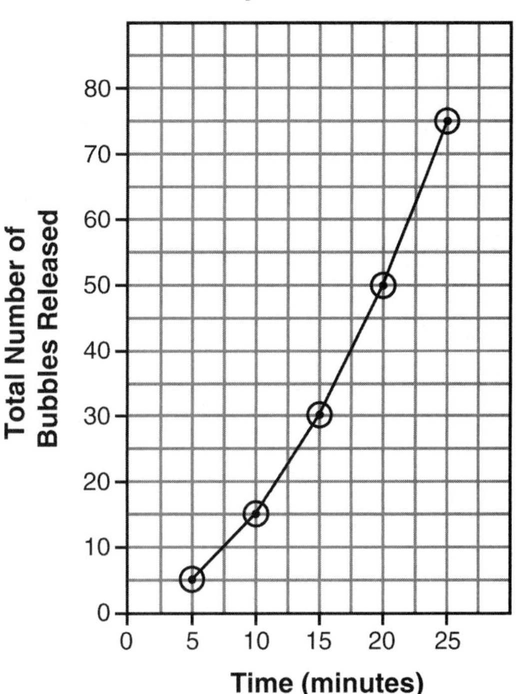

3. Plot the experimental data on the graph. Draw a circle around each point. Draw a line that connects the points. Do not continue the line past the last point.
4. Put a title on the graph which shows what the graph is about. Example: "Effect of time on respiration rate in yeast."

Note: By looking at the data table, we could predict that the number of bubbles at 22½ minutes would be between the number of bubbles at 20 minutes (50 bubbles) and 25 minutes (75 bubbles), therefore the number of bubbles at 22½ minutes would be about 63 bubbles. On the graph you see that the number of bubbles at 22½ minutes equals about 63 bubbles.

Problem 3 (at the end of chapter 2):
Let's do a different experiment and compare the number of gas bubbles released at different temperatures. See the effect of temperature on cellular respiration in yeast. (You learned in **cellular respiration** glucose unites with oxygen, producing water, **carbon dioxide gas**, and ATP. Glucose + oxygen ⟶ carbon dioxide + water+ ATP).

One flask of yeast-glucose solution is at 20°C; the second flask of yeast-glucose solution is at 35°C. The student counted and recorded the number of gas bubbles at the two different temperatures every five minutes.

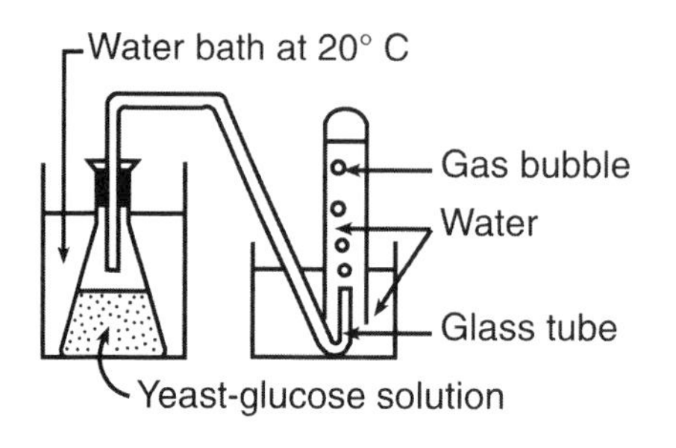

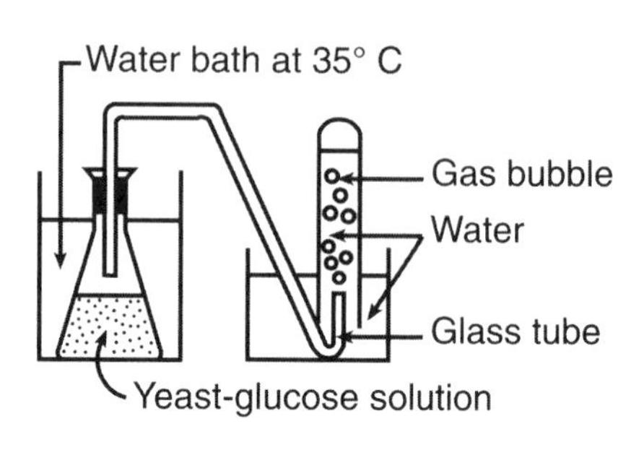

Time (minutes)	Total Number of Bubbles Released	
	20°C	35°C
5	0	5
10	5	15
15	15	30
20	30	50
25	45	75

Using the information in the data table, construct a line graph on the grid:

1. On the x axis, put "Time, minutes". **The thing you change** (in this case time) is always put on the **x axis**. This is the independent variable. Space the lines along the axis equally. There must be an equal number of minutes between lines. (See graph.)
   Note: "Make an appropriate scale" by spacing the numbers on the graph so that all the data fits on the graph and it is easy to read.
2. On the y axis, put "Total Number of Bubbles Released". The **result** you get (total number of bubbles released) is always put on the **y axis**. This is the dependent variable. Space the lines along the axis equally. There must be an equal number of degrees between lines (see graph).
3. Plot the data for the total number of bubbles released at 20°C on the graph. Surround each point with a small triangle and connect the points with a line. Do not continue the line past the last point.
4. Plot the data for the total number of bubbles released at 35°C on the graph. Surround each point with a small circle and connect the points with a line. Do not continue the line past the last point.
5. Put a title on the graph which shows what the graph is about. Example: "Effect of temperature on respiration rate in yeast."

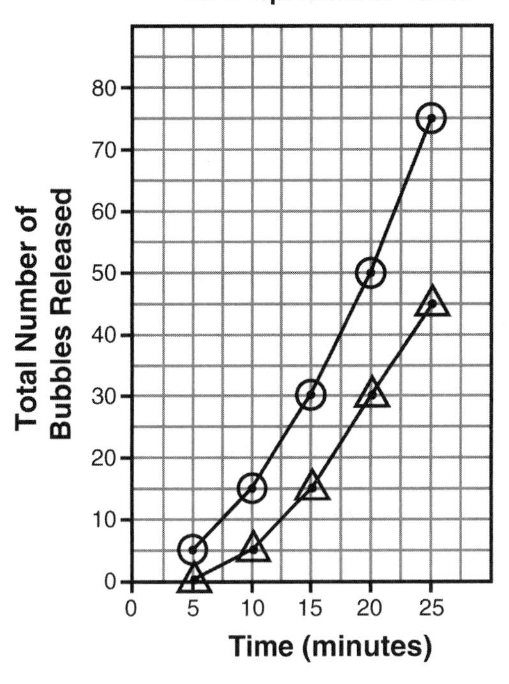

As you can see from the graph, as time increases, the number of (gas) bubbles released increases, or, you can say, rate of gas production increases. You know the gas bubbles produced are carbon dioxide, because, in cellular respiration, glucose unites with oxygen, producing water, **carbon dioxide gas**, and ATP.

Try sample line graphs problem 4 and problem 5:

Problem 4 (at the end of Chapter 6):
A student added two species of single-celled organisms, *Paramecium caudatum* and *Didinium nasutum,* to the same culture medium. Each day, the number of individuals of each species was determined and recorded. The results are shown in the data table below.

Culture Population

Day	Number of *Paramecium*	Number of *Didinium*
0	25	2
1	60	5
2	150	10
3	50	30
4	25	20
5	0	2
6	0	0

How to draw the line graph:

1. On the x axis, put "Days". **The thing you change** (in this case time) is always put on the **x axis**. This is the independent variable. Space the lines along the axis equally. "Make an appropriate scale" by spacing the numbers on the graph so that all the data fits on the graph and it is easy to read. There must be an equal number of days between lines (see graph.) On the x axis, put one day between lines (every two lines) (scale on the x axis), then all the days in the data table fit on the graph and it is easy to read. Start counting the lines (example every two lines) **after** the axis (see graph).

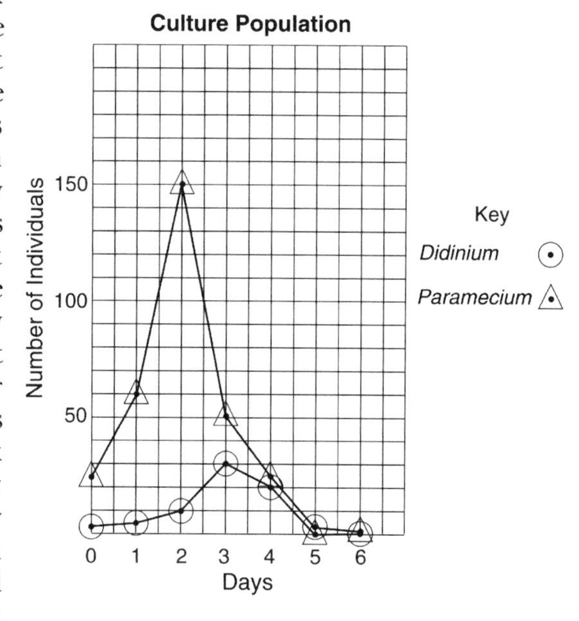

2. On the y axis, put "Number of Individuals". The **result** you get (number of individuals) is always put on the **y axis**. This is the dependent variable. Space the lines along the axis equally. There must be an equal number of individuals between lines (see graph). "Make an appropriate scale" by spacing the numbers on the graph so that all the data fits on the graph and it is easy to read.

   An easy way to figure out how to make the scale is as follows:

Look at the second column in the data table; you see you have a number of paramecium (0 to 150). Look at the y axis of the graph; it has 21 lines. To find the **number** of paramecium (or number of seconds in another example) that each line should show, divide $\frac{150 \ (paramecium)}{21 \ lines}$ = about $\frac{7\frac{1}{2} \ paramecium}{line}$. One line should show at least 7½ paramecium; therefore, **make each line** a **little bigger**, **easier number,** such as 10 paramecium.

On the y axis, put 50 individuals between lines (every five lines) (scale on the y axis), then all the individuals on the data table fit on the graph and it is easy to read. Start counting the lines (example every five lines) **after** the axis (see graph).

3. Plot the data for the number of *Didimium* on the graph. Surround each point with a small circle; draw a line that connects the points. Do not continue the line past the last data point. Make a key showing that ⊙ means *Didimium.*
4. Plot the data for the number of *Paramecium* on the graph. Surround each point with a small triangle; draw a line that connects the points. Do not continue the line past the last point. Make a key showing that ◬ means *Paramecium.*
5. Put a title on the graph which shows what the graph is about. Example: "Culture Population".

Problem 5 (at the end of Chapter 7):

Answer questions 1 through 3 based on the information below and on your knowledge of biology. The average level of carbon dioxide in the atmosphere has been measured for the past several decades. The data collected are shown in the table below:

Average CO_2 Levels in the Atmosphere

Year	**CO_2** (In parts per million)
1960	320
1970	332
1980	350
1990	361
2000	370

Questions 1 and 2: Using the information in the data table, construct a line graph on the grid, following the directions below.

Question 1: Mark an appropriate scale on each labeled axis.

Question 2: Plot the data on the grid. Surround each point with a small circle and connect the points. Example:

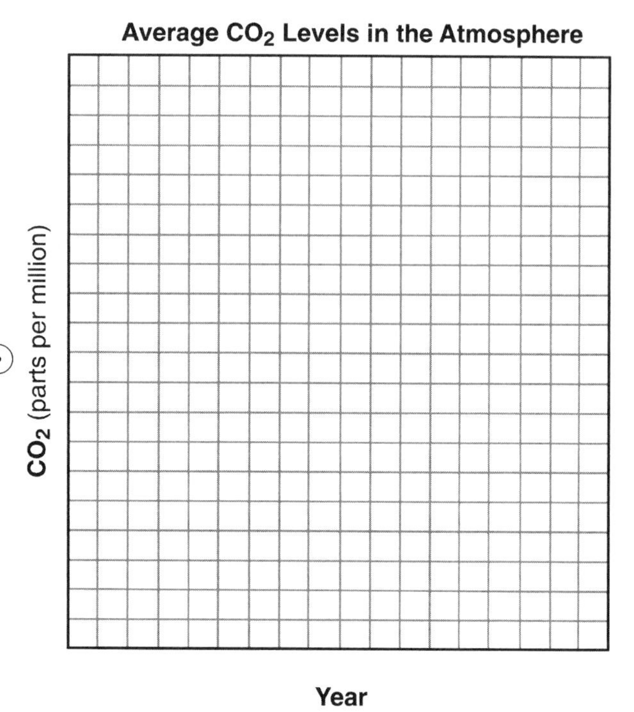

Question 3: Calculate the net change in CO_2 levels in parts per million (ppm) during the years 1960 through 2000.

Solution 1: Look at the line graph in the question. As you see, on the x axis is the thing you change (the independent variable), which is the year, such as 1960, 1970, etc. There must be an equal number of years between lines. "Make an appropriate scale" by spacing the numbers on the graph so that all of the data fits on the graph and it is easy to read. On the x axis, put 10

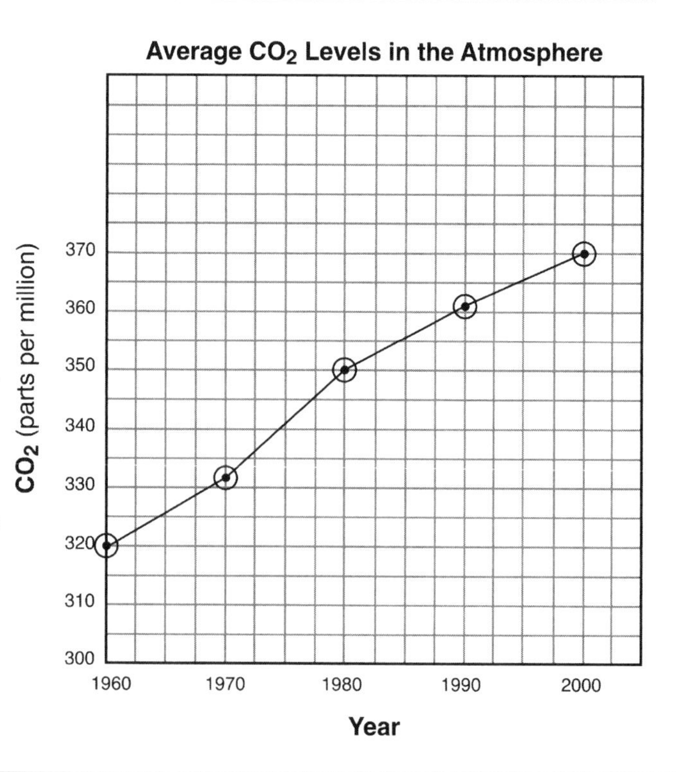

years between lines (every four lines)(scale on the x axis), then all the years on the data table fit on the graph and it is easy to read. Start counting the lines (example every four lines) **after** the axis (see graph). Look at the line graph again. On the y axis, put CO_2 (parts per million). The result you get the amount of CO_2 (carbon dioxide) in parts per million, is always put on the y axis. This is the dependent variable. There must be an equal number of parts per million of carbon dioxide between lines (every two lines). Space the lines along the axis equally. "Make an appropriate scale" by spacing the numbers on the graph so that all of the data fits on the graph and it is easy to read. On the y axis, put 10 parts per million of CO_2 between lines (every two lines) (scale on the y axis), then all the parts per million on the data table fit on the graph and it is easy to read. Start counting the lines (example every two lines) **after** the axis (see graph).

Solution 2: Plot the experimental data. Draw a circle around each point. Draw a line that connects the points. Do not continue the line past the last point (see graph at right).

Solution 3: Look at the data table or the graph.

In 2000, there were 370 parts per million (ppm) CO_2.

In 1960, there were 320 parts per million (ppm) CO_2.

Change in CO_2 from 1960-2000 is: 370 ppm − 320 ppm = 50 ppm (parts per million).

Now Do Homework Questions #50-82, pages 53-62.

Reading Line Graphs

Problem 1:

Students cut twenty rod-shaped pieces of potato of the same diameter (width) and length. Five pieces of potato were placed into each of four beakers containing different concentrations of sugar solutions. Each potato piece was measured again after 24 hours. The data table below shows the results of their experiment.

Change in Length

Concentration of Sugar Solution (grams per liter)	**Original Length of Potato Pieces** (mm)	**Average Length After 24 Hours** (mm)
0	50.0	52.0
5	50.0	44.0
8	50.0	43.5
10	50.0	42.5

Which graph best represents the information in the data table:

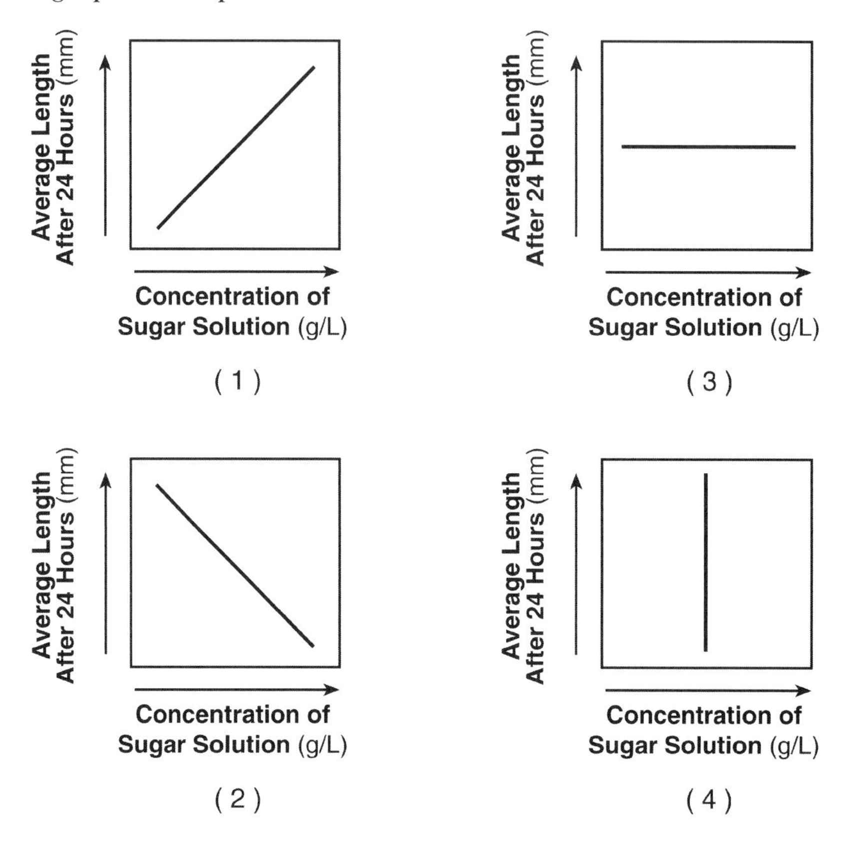

Solution: Look at the data table. You see as concentration of sugar solution (grams per liter, g/L) increases from 0 to 5 to 8 to 10 (as you go down the data table) the average length (in millimeters, mm) after 24 hours (third column on data table) decreases from 52.0 to 44.0 to 43.5 to 42.5 (as you go down the table).

Choice *2* is the correct answer. The graph shows as concentration increases (gets more) average length decreases (gets less).

Wrong choices:

Choice *1*. As concentration increases, average length after 24 hours increases.

Choice *3*. As concentration increases, average length after 24 hours remains the same.

Choice *4*. As concentration remains the same, average length after 24 hours increases.

Problem 2: (more advanced graphs)

These three line graphs show the speed of a runner (boy or girl running).

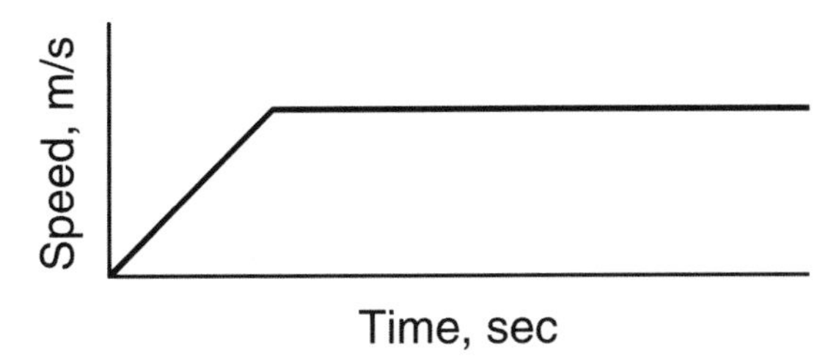

The boy's speed increased, then remained the same.

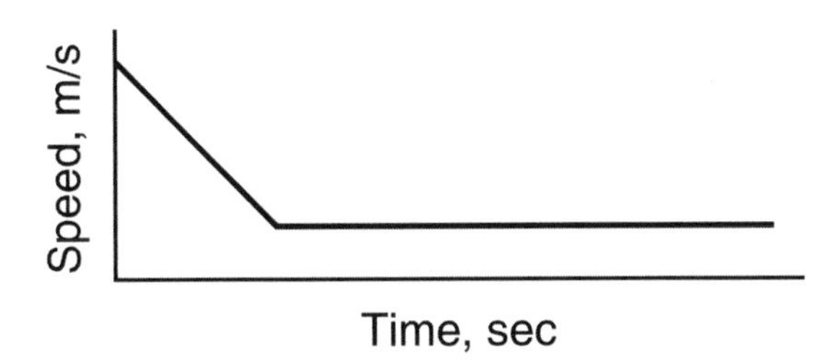

The boy's speed decreased, then remained the same.

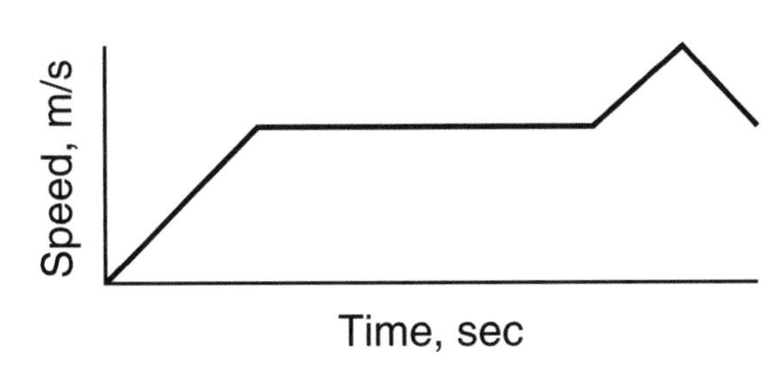

The boy's speed increased, then remained the same, then increased again, and then decreased.

Problem 3:(Review and study line graphs on enzyme activity and homeostasis). Review and study the four line graphs on enzyme activity in the middle of chapter 2.

Also review and study the line graph showing homeostasis (dynamic equilibrium) of blood sugar level with time in the middle of chapter 2.

Problem 4: (comparing graphs, from chapter 7)

A. Comparing the graph of carbon dioxide levels in the atmosphere (left) and the graph of atmospheric temperature (right), you can see the overall relationship that as carbon dioxide concentration increased from 1960 - 2000, the atmospheric temperature also increased.

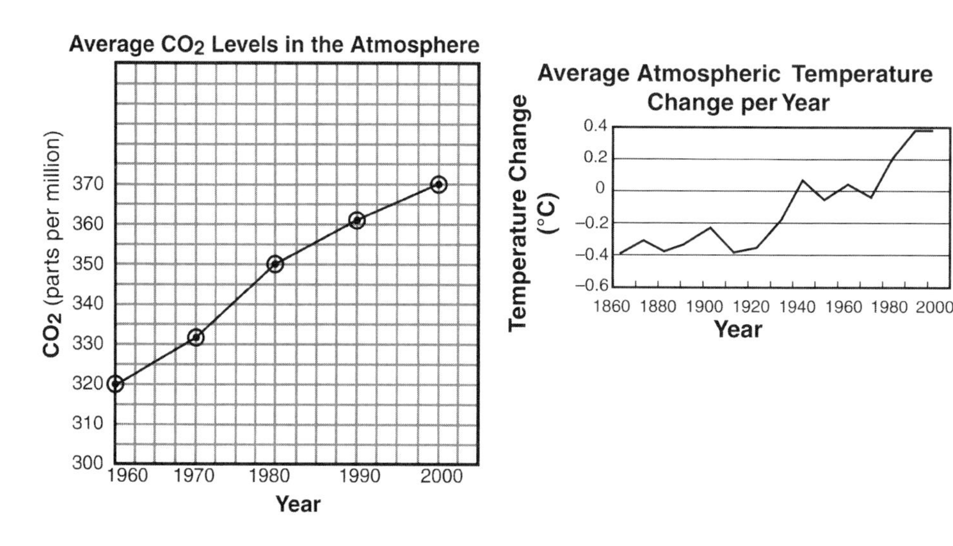

B. By comparing these two graphs, number of beavers and relative number of aspen trees, you see the relationship is as the number of beavers increases, the number of aspen trees decreases (because the beavers are chewing and destroying the trees).

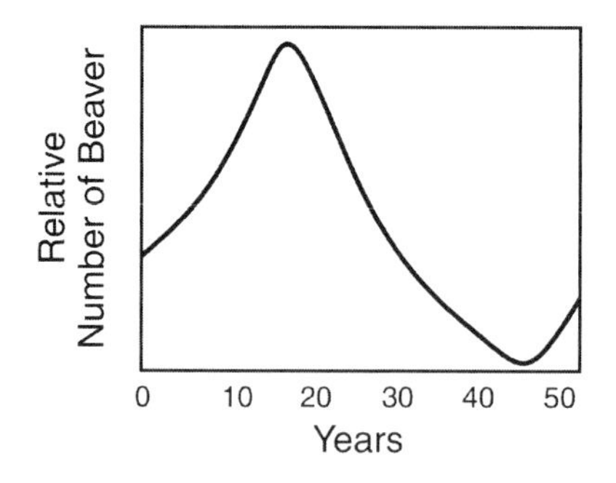

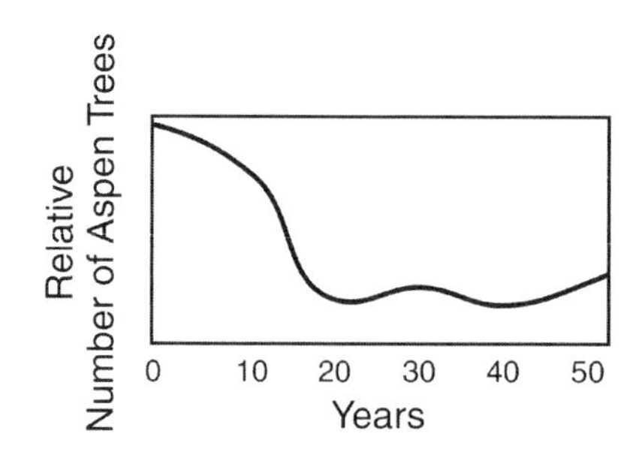

Now Do Homework Questions #83-94, pages 63-68.

Bar Graphs

Bar graphs, like line graphs, have the independent variable (the thing you change, example sitting, walking, or running) on the horizontal axis and the dependent variable (the result you get, example pulse rate) on the vertical axis.
In a bar graph, like a line graph, there is an equal number (in this example 20 (beats/minute)) between lines (0 to 20, 20 to 40, 40 to 60, 60 to 80, 80 to 100); see vertical axis on the bar graph.

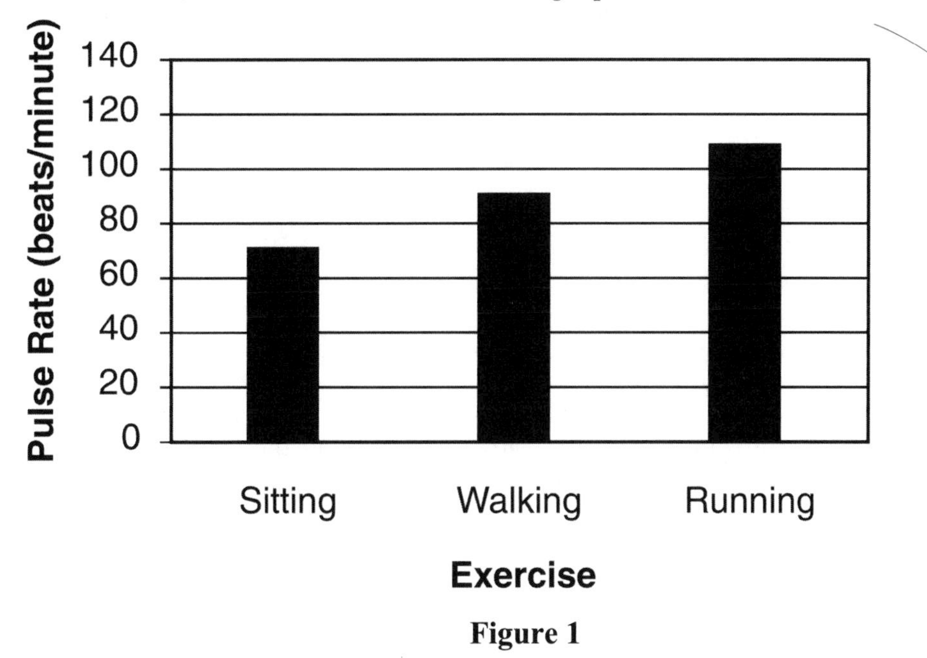

Figure 1

The difference between a bar graph and a line graph is that in a bar graph there are separate bars; there are spaces between each independent variable, such as sitting (one independent variable), walking (another independent variable) and running (a third independent variable).
In a line graph, there is a continuous line connecting the points.

Problem 1:
The **bar graph** in **Figure** 1 shows the pulse rate for **females** after sitting, walking, and running. (Sitting, walking, and running are the independent variables; pulse rate is the dependent variable).

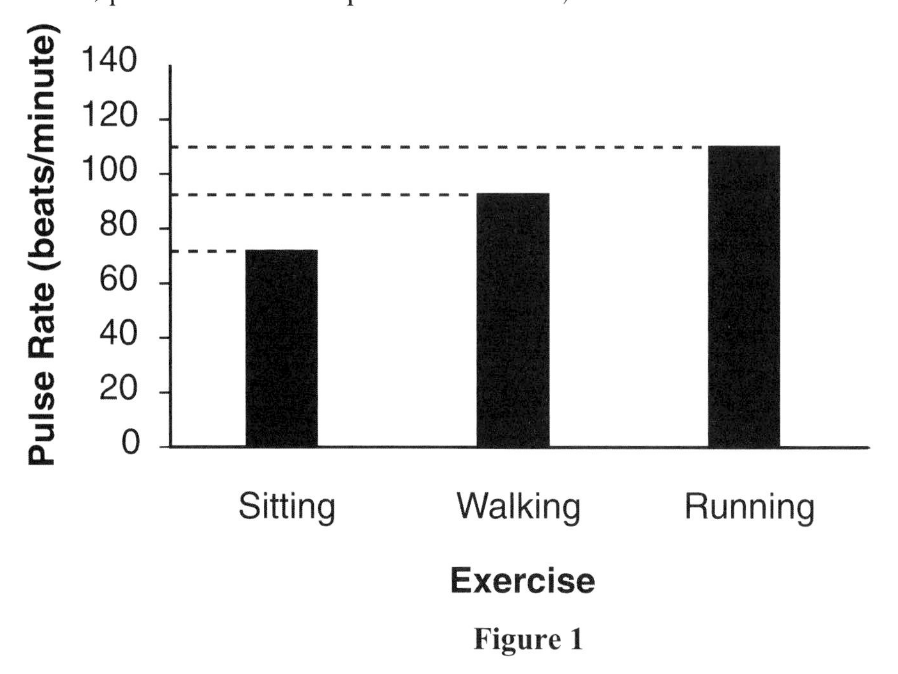

Figure 1

What is the pulse rate for females sitting? Look at the bar graph (figure 1 above). There are 20 beats per minute between each two numbers on the vertical axis (line). Look at the top of the bar for sitting and go across to the vertical axis (line); the dotted line is more than halfway between 60 beats per minute and 80 beats per minute, or about 72 beats per minute. The pulse rate for females sitting is about 72 beats per minute.

What is the pulse rate for females walking? Look at the bar graph (figure 1 above). Look at the top of the bar for walking and go across to the vertical line; the dotted line is more than halfway between 80 beats per minute and 100 beats per minute. The pulse rate for females walking is about 92 beats per minute. The pulse rate for females running is about 110 beats per minute, halfway between 100 and 120.

The **bar graph** in **Figure 2** below shows the pulse rate for both **males and females** sitting, walking, and running.

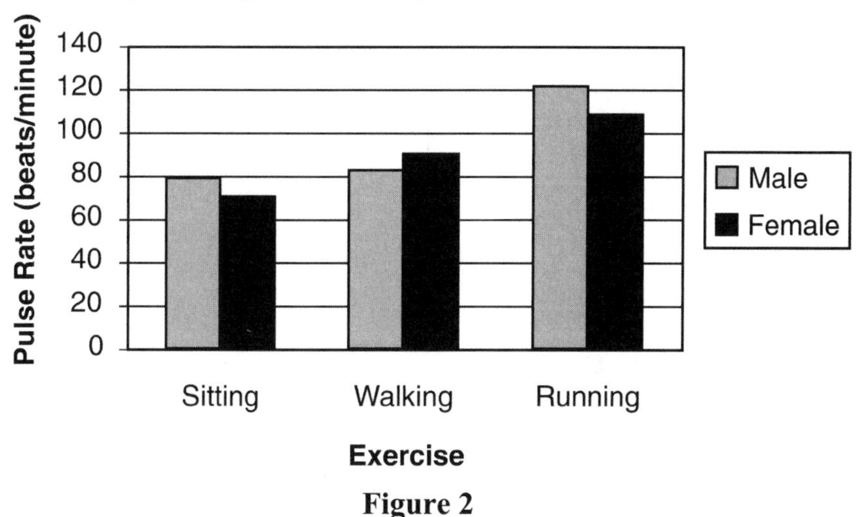

Figure 2

Look at the bar graph in figure 2. This is the male bar ▢ ;this is the female bar ■ . You can see the pulse rate for females sitting is about 72 beats per minute. The pulse rate for males sitting is about 80 beats per minute (the bar is on the line for 80); therefore, the pulse rate for males is higher than the pulse rate for females.
By just looking at the bar graphs, you can compare the pulse rate of males and females sitting. On the vertical axis is pulse rate. For sitting, the male has a taller (longer) bar than the female (shorter bar), which means the male has a higher pulse rate than the female.

Look again at the bar graph in figure 2. You can see the pulse rate for females walking is 92 beats per minute. The pulse rate for males walking is 84 beats per minute (the bar is a little less than 1/4 the way between 80 and 100); therefore, the pulse rate for females walking is higher than the pulse rate for males walking.
By just looking at the bar graphs, you can compare the pulse rate of males walking and females walking. On the vertical axis is pulse rate. For walking, the male has a shorter bar than the female (taller, longer bar), which means the male has a lower pulse rate than the female.

Problem 2: (from chapter 1)
There are two bar graphs below, one bar graph of a plant cell and one bar graph of an animal cell. Look at the graphs.

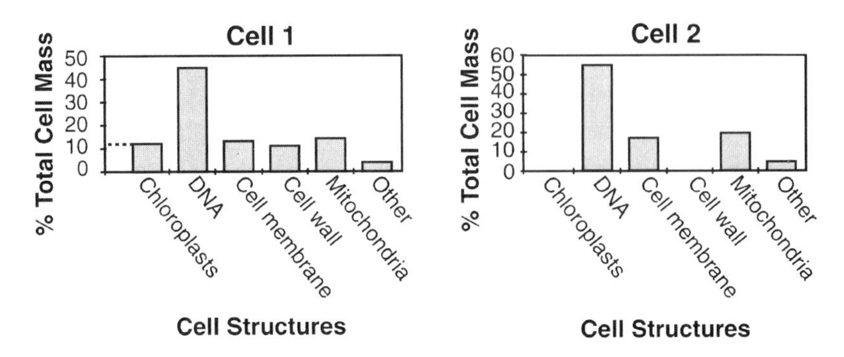

On the vertical axis is percent cell mass (example, DNA in cell 1 makes up 46% of cell mass). You can tell that **cell 1** is a **plant cell** because it has **chloroplasts** and a **cell wall**. Chloroplasts and cell wall are in plants (plant cells) and not in animals (animal cells). Look at the top of the bar for chloroplasts; the student draws a dotted line to the vertical axis (see cell 1). You see the dotted line is a little above 10% but less than 20%, therefore the chloroplasts are about 12% of the cell mass (material).
Look at cell 2. Cell 2 has **no** (zero) **chloroplasts** and **no cell wall** (there is no bar above the word chloroplasts and no bar above the word cell wall). **Cell 2** is an **animal cell**.

Drawing Bar Graphs

Make sure you have **equal spacing** on the **vertical axis.** Let's draw two bar graphs, one bar graph from problem 1, Figure 2, and the second bar graph from problem 2 on this page.

In problem 1 (shown at right), the equal spacing on the vertical axis (line) is 0 beats/minute, 20 beats/minute, 40, 60, 80, 100.

Pulse rate for female sitting equals 72 beats per minute, therefore, draw the height of the bar for female sitting, 72 beats per minute (see bar graph). Pulse rate for male sitting equals 80

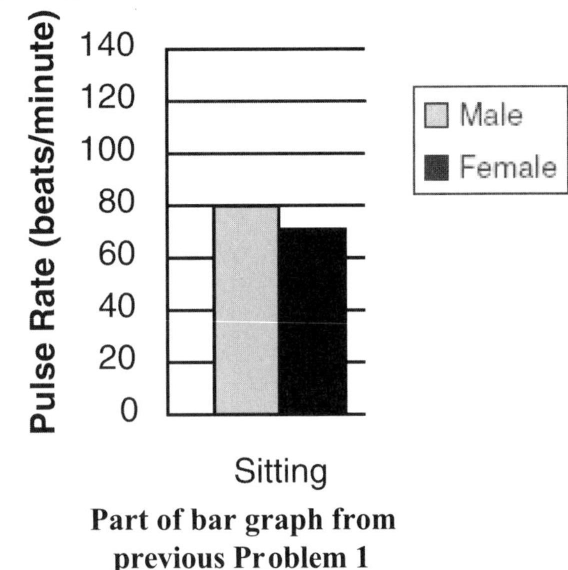

Part of bar graph from previous Problem 1

beats per minute, therefore, draw the height of the bar for male sitting, 80 beats per minute (see bar graph above).

In problem 2 (shown at right), the equal spacing on the vertical axis (line) is 0%, 10%, 20%, 30%, 40%, 50%.

Percent of cell mass for chloroplasts equals about 12 percent, therefore, draw the height of the bar for chloroplasts equal to 12 percent. Percent of cell mass for DNA equals 46 percent, therefore draw the height of the bar for DNA equal to 46 percent. Draw the bar for cell membrane equal to 13 percent, for cell wall equal to 10 percent, and for mitochondria equal to 14 percent.

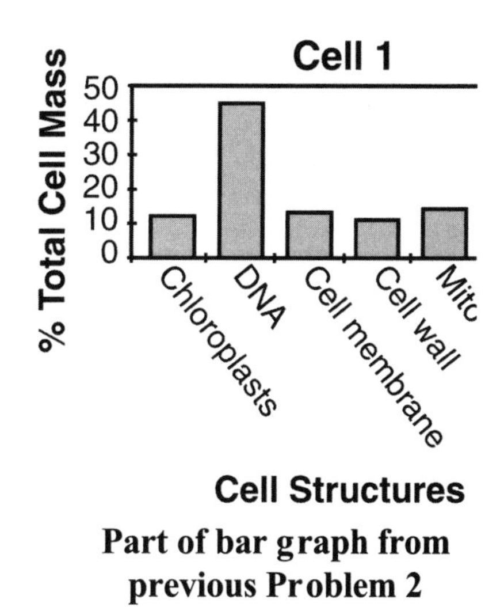

Part of bar graph from previous Problem 2

Problem 3:
Each year, a New York State power agency provides its customers with information about some of the fuel sources used in generating electricity. The table below applies to the period of 2002-2003.

Fuel Sources Used

Fuel Source	Percentage of Electricity Generated
hydro (water)	86
coal	5
nuclear	4
oil	1
solar	0

Using the information given, construct a bar graph on the grid, following the directions below.

1. Mark an appropriate scale on the axis labeled, "Percentage of Electricity Generated."

2. Construct vertical bars to represent the data. Shade in each bar.

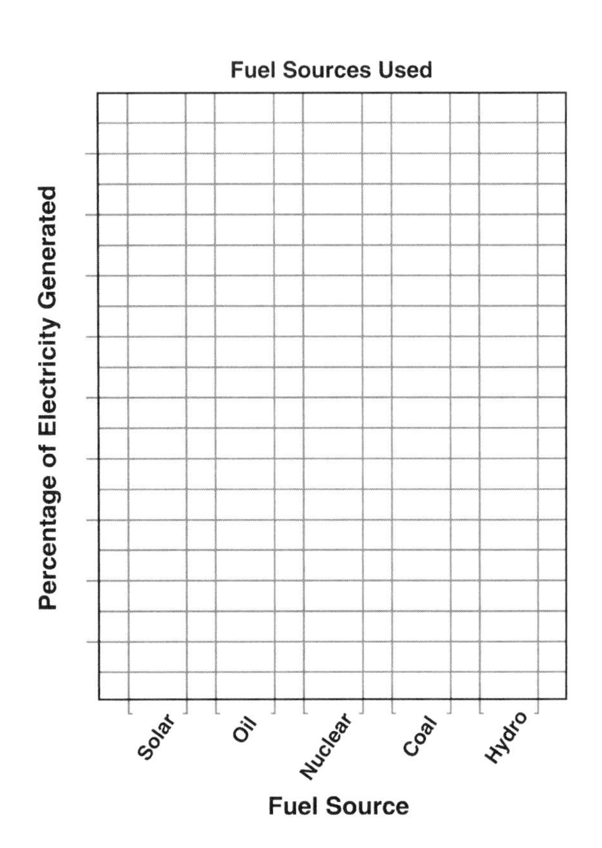

Solution:

1. This method might help you in figuring out how to make a scale. Look at the largest and smallest numbers for the item you are putting on the axis (in this example percentage of electricity generated, given in the data table). In the data table in this example, solar = 0%, and hydro (water) = 86%. On the graph, on the vertical axis, there must be space for at least 86%, or more. On the graph, on the vertical axis there are 20 lines. Find how much each line can be.

$$\frac{86 \textit{ percent}}{20 \textit{ lines}} = \frac{4.3 \textit{ percent}}{\textit{line}}$$

or

Proportion method:

$$\frac{86 \textit{ percent}}{20 \textit{ lines}} = \frac{x \textit{ percent}}{1 \textit{ line}}$$

Cross multiply:

$$20x = 86$$

$$x = 4.3\ \%/\text{line}$$

One line equals at least 4.3%, therefore, make each line a little bigger, a larger number such as 5% (two lines then = 10%).
Make sure that all the numbers fit on the graph. Make sure you have equal spacing on the vertical axis.

2. Draw the height of each bar (example height of coal) to equal the number on the data table (5 for coal). Look again at the data table. Draw the height of the bar for nuclear to equal the number on the data table, (4 for nuclear). Draw the height of the bar for hydro (water) to equal 86, draw the height of the bar for oil to equal 1 and draw the height of the bar for solar to equal 0.

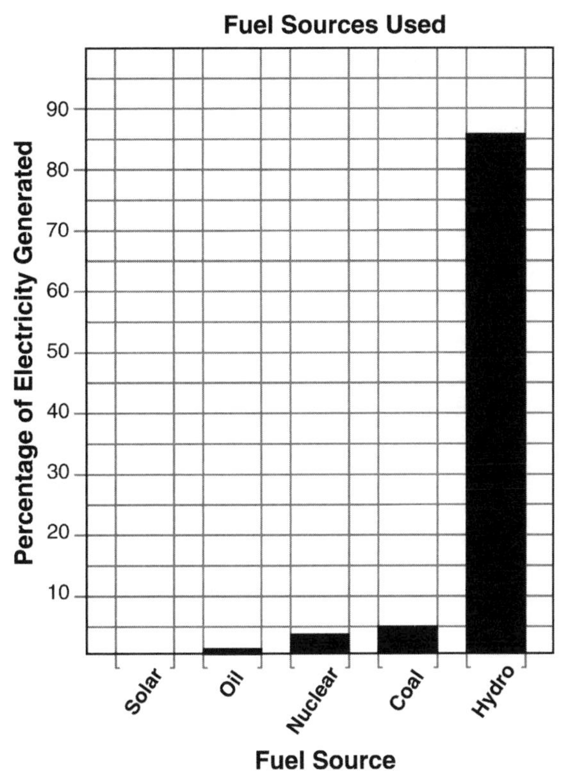

Try sample problems 4- 9 on bar graphs.

Problem 4:
The graph below shows the relative concentrations of different ions inside and outside of an animal cell.

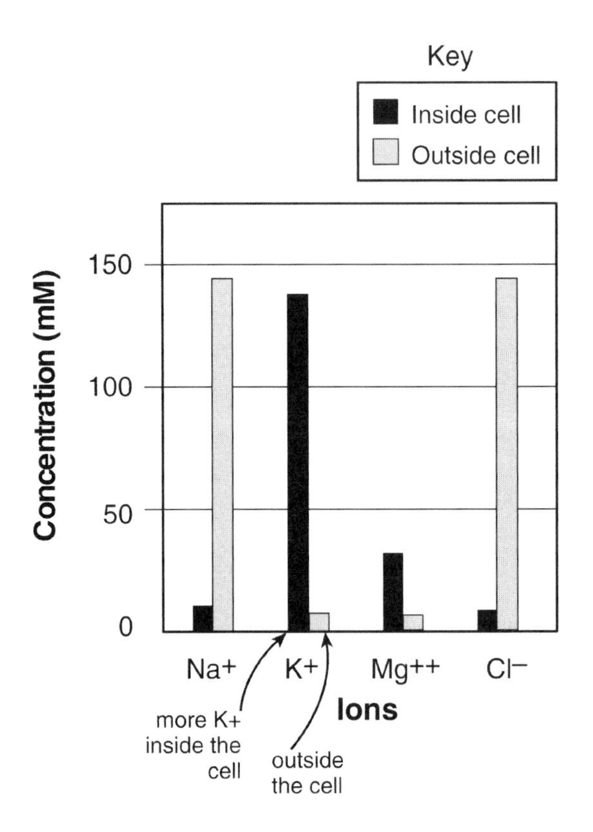

Which process is directly responsible for the net movement of K^+ and Mg^{++} into the animal cell?

(1) electrophoresis
(2) diffusion
(3) active transport
(4) circulation

Solution: Look at the bar graph. The key shows the black bar means inside the cell and the gray bar means outside he cell. Look at the two bars for K^+. The black bar for K^+ (inside the cell) is taller, which means there is **much more K^+ inside the cell.** In this example, K^+ went (moved) from lower concentration (outside the cell) to higher concentration (inside the cell), causing much more K^+ to be inside the cell. The only **process** that **moves materials** (example K^+ and also Mg^{++}) from **lower concentration** (in this example outside the cell) **to higher concentration** (in this example inside the cell) is **active transport.** Answer *3*

Problem 5: (from chapter 5)
Use the graph to answer the four questions below.

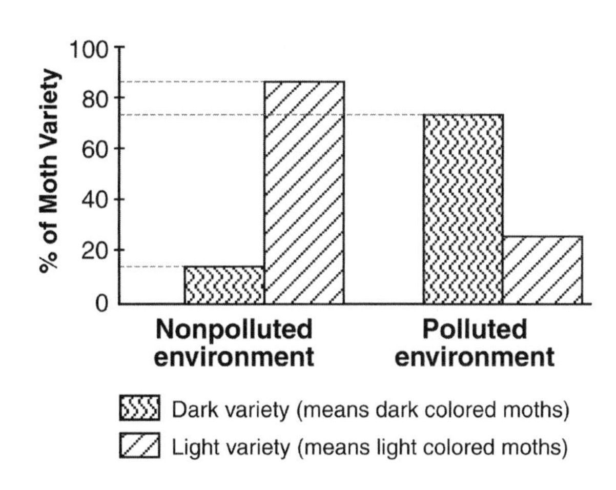

Background information: Environment (or environmental change) influences which color of moths will survive more in that environment. In England, before there were factories, when there was no soot and there was white lichen on the trees, white moths had the adaptive value (white color) of being hidden and more survived. Later, factories were built, which produced large amounts of black soot (polluted the environment), and more dark moths survived.

Question: What is the percentage of dark-colored moths in the nonpolluted environment?

Solution: Look at the bar graph above and at the right (part of the bar graph above). There is 20% between each two numbers on the vertical axis (line). In the non polluted environment, look at the top of the bar for the dark variety ▩, and go across to the vertical axis (line). You see the dotted line is more than halfway between 0% and 20%; with equal spacing between the numbers, it is about 16%.

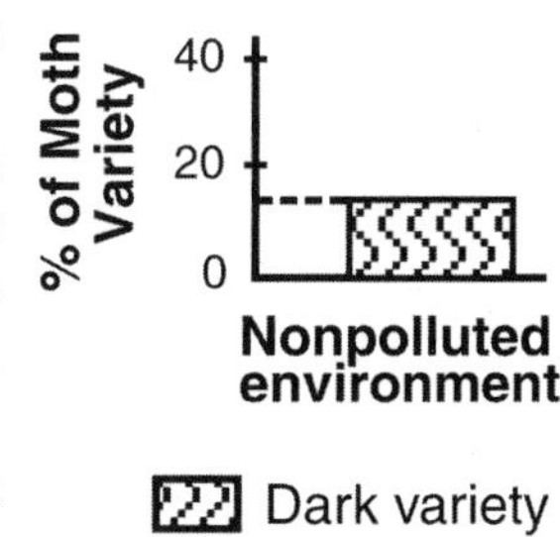

Question: What is the percentage oflight-colored moths in thenonpolluted environment?

Solution: Look at the bar graph above. There is 20% between each two numbers on the vertical axis (line). In the nonpolluted environment, look at the top of the bar for the light variety ▨ and go across to the vertical line. You see the dotted line is at a little more than 80% and less than 100%; with equal spacing between the numbers, it is about 84%.

Question: What is the percentage of light-colored moths in the polluted environment?

Solution: Look at the bar graph above. There is 20% between each two numbers on the vertical axis (line). In the polluted environment, look atthe top of the bar for the light variety ▨ and go across to the vertical line. You see it is at a little more than 20% and less than 40%; with equal spacing between 20% and 40%, it is about 24%.

Question: What is the percentage of dark-colored moths in the polluted environment?

Solution: Look at the bar graph. There is 20% between each two numbers on the vertical axis (line). In the polluted environment, look at the top of the bar for the dark variety ▩, and go across to the vertical line. You see the dotted line is more than halfway, about 3/4 of the way, between 60% and 80%; with equal spacing between the numbers, it is about 76%.

Problem 6: (from chapter 2)

Question: A student studied how the amount of oxygen affects ATP production in muscle cells. The data for amount X are shown in the graph.

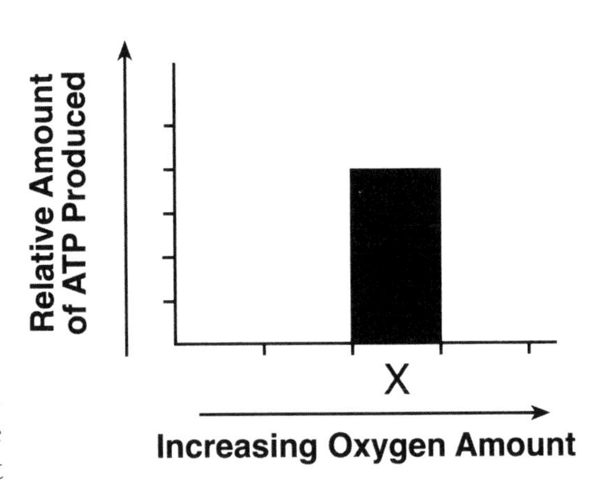

If the student supplies the muscle cells with *less* oxygen in a second trial of the investigation, a bar placed on the graph to represent the results of this trial would most likely be

(1) shorter than bar X and placed to the left of bar X
(2) shorter than bar X and placed to the right of bar X
(3) taller than bar X and placed to the left of bar X
(4) taller than bar X and placed to the right of bar X

Solution: If in a new trial, there was **less oxygen**, the new bar would be placed to the left of bar X , which is less oxygen, and the bar would be shorter because less ATP would be produced.

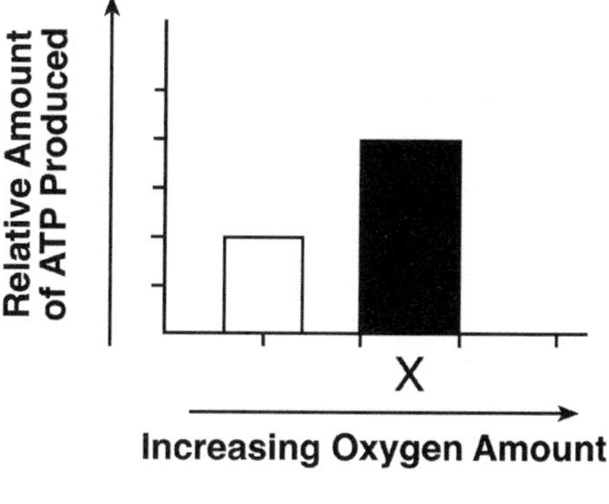

Answer *1*

Problem 7 (from chapter 2).
Question: Base your answer on the graph below.

Incidence of Three Human Diseases in Four Different Years

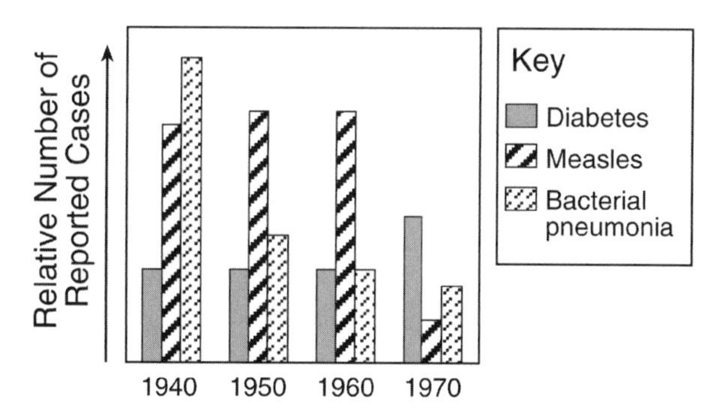

The greatest difference between the incidence of measles and the incidence of bacterial pneumonia occurred in
(1) 1940 (2) 1950 (3) 1960 (4) 1970

Solution: From the key, you see shows measles and shows pneumonia.
The taller (higher) the bar, the more people had that disease. The shorter the bar, the fewer people had the disease.

In 1940, there was a small difference between the height of the bars for measles and pneumonia.

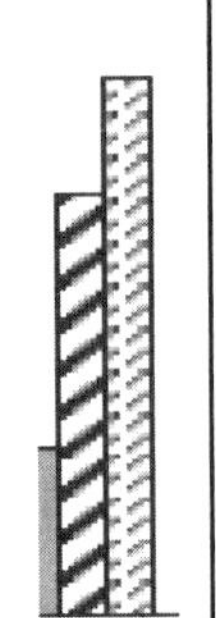

In 1960, there was the biggest difference (greatest difference) between the height of the bars for measles and pneumonia.

Answer *3*

Problem 8: (from chapter 4).

Question: The diagram below illustrates some of the changes that occur during gamete formation.

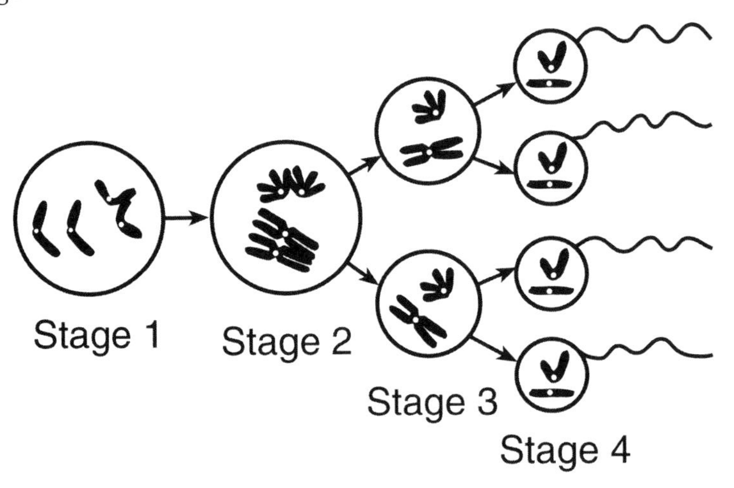

Which graph best represents the changes in the amount of DNA in one of the cells at each stage?

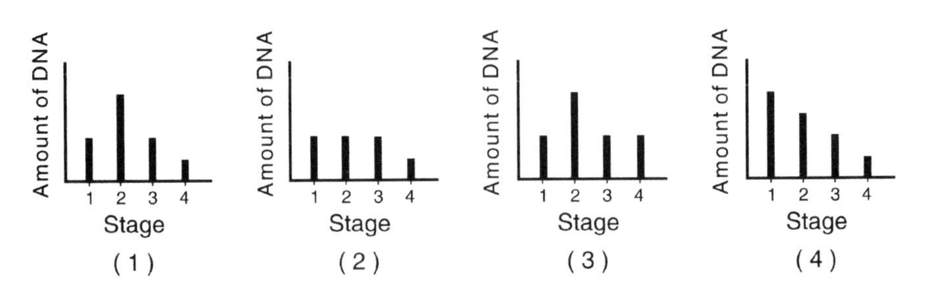

Solution: Look at the top diagram, showing stages 1-4 of gamete formation.

or = one chromosome.

Stage 1 Cell forms **four chromosomes**.
Stage 2 Cell has double the number of chromosomes as stage 1, which is **eight chromosomes.**
Stage 3 Cell in stage 2 divides, forming two cells. Each cell in stage 3 has the same number of chromosomes as stage 1 (**four chromosomes).**
Stage 4 Each cell in stage 3 divides again. Each cell in stage 4 has half the number of chromosomes as stage 1, which is **two chromosomes.**

Look at the graphs. Choose the bar graph that shows the changes in amount of DNA (chromosomes) in one cell at each stage. On the vertical axis is amount of DNA. Chromosomes are made of DNA.

Look at choice (1) in the question. Choice (1) is the correct answer.

Stage 1 on the bar graph shows the amount of DNA in stage 1 of gamete formation; stage 1 has four chromosomes (DNA that will form four chromosomes).

Stage 2 on the bar graph shows the amount of DNA in stage 2 of gamete formation; the bar in stage 2 is twice as long as in stage 1, because the bar represents eight chromosomes.

Stage 3 on the bar graph shows the amount of DNA in stage 3 of gamete formation; the bar in stage 3 is the same height as stage 1, because the bar represents four chromosomes (DNA that will form four chromosomes).

Stage 4 is half the height of stage 1, because the bar represents two chromosomes (DNA that will form two chromosomes).

The bar graph in choice (1) shows the changes in the amount of DNA in stages 1-4 of gamete formation (meiosis). Answer *1*

Wrong choices:
Choice (2) shows the first three stages have the same amount of DNA (same number of chromosomes), which is not correct.
Choice (3) shows that stages 1, 3, and 4 all have the same amount of DNA (same number of chromosomes), which is not correct.
Choice (4) shows that stage 1 has the most DNA (most chromosomes), which is not correct, and stage 3 has half the DNA (half the chromosomes) of stage 1, which is not correct.

Problem 9: (from chapter 5)
Question: The graph below shows the percent of variation for a given trait in four different populations of the same species. The populations inhabit similar environments.

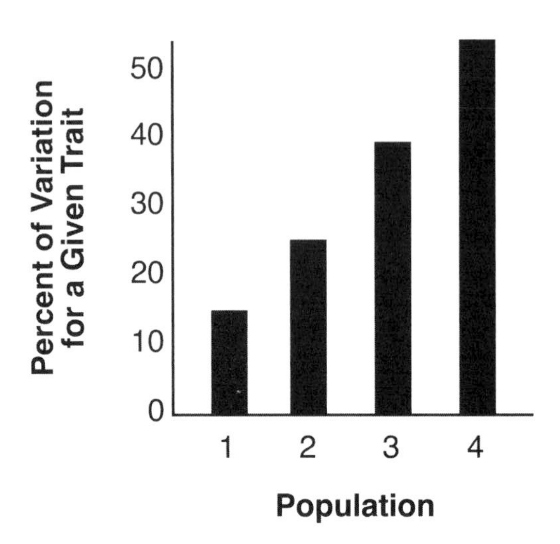

In which population will the greatest number of individuals most likely survive if a significant environmental change related to this trait occurs?
(1) 1 (2) 2 (3) 3 (4) 4

Solution: Background information: You learned species with organisms with **few variations** might not survive (or would become extinct) because none of the organisms might be able to protect themselves against the changing environment (such as flooding, or drought (very little rainfall), or extremely hot temperature). Species with organisms with **many different variations** (many different traits, and many different combinations of traits) will more likely survive and be able to reproduce in a changing environment (example flooding, or drought (very little rainfall), or freezing temperature). A **species** or a **population** (all members of a species living in one area) with **many different variations survives** (does not become extinct).

Look at the bar graph. On the vertical axis is percent of variation (how much variation there is).
Population 1 has the shortest bar, which means population 1 has the smallest amount of variation.
Population 2 has a taller (longer) bar which means population 2 has more variation.
Population 3 has an even taller (longer) bar, which means population 3 has even more variation.
Population 4 has the tallest (longest) bar, which means it has the most variation.
Answer: Choice *4*. When there is a big environmental change, the population (example population 4) with the most variation is most likely to survive.

Review of reading percentages on a bar graph: There is 10% between each two numbers on the vertical scale. Drawing a dotted line from the top of each bar to the vertical axis may make it easier to see the percent of variation.
For population 1, the bar is less than halfway between 10% and 20%; with equal spacing between the numbers, the variation in population 1 is about 13%.
For population 2, the bar is halfway between 20% and 30%; with equal spacing between the numbers, the variation in population 2 is about 25%.
For population 3, the bar is most of the way from 30% to 40%; with equal spacing between the numbers, the variation in population 3 is about 38%.

Now Do Homework Questions #95-108, pages 69-73.

Now Let's Try a Few Homework Questions

To assign fewer questions, you may want to skip the questions with underlined numbers.

1. A science researcher is reviewing another scientist's experiment and conclusion. The reviewer would most likely consider the experiment invalid if
   (1) the sample size produced a great deal of data
   (2) other individuals are able to duplicate the results
   (3) it contains conclusions not explained by the evidence given
   (4) the hypothesis was not supported by the data obtained

2. Which statement most accurately describes scientific inquiry?
   (1) It ignores information from other sources.
   (2) It does not allow scientists to judge the reliability of their sources.
   (3) It should never involve ethical decisions about the application of scientific knowledge.
   (4) It may lead to explanations that combine data with what people already know about their surroundings.

3. Which statement best describes the term *theory* as used in the gene-chromosome theory?
   (1) A theory is never revised as new scientific evidence is presented.
   (2) A theory is an assumption made by scientists and implies a lack of certainty.
   (3) A theory refers to a scientific explanation that is strongly supported by a variety of experimental data.
   (4) A theory is a hypothesis that has been supported by one experiment performed by two or more scientists.

4. In 1910, Thomas Morgan discovered a certain pattern of inheritance in fruit flies known as sex linkage. This discovery extended the ideas of inheritance that Gregor Mendel had discovered while working with garden peas in 1865. Which principle of scientific inquiry does this illustrate?
   (1) A control group must be part of a valid experiment.
   (2) Scientific explanations can be modified as new evidence is found.
   (3) The same experiment must be repeated many times to validate the results.
   (4) Values can be used to make ethical decisions about scientific discovery.

5. In one variety of corn, the kernels turn red when exposed to sunlight. In the absence of sunlight, the kernels remain yellow. Based on this information, it can be concluded that the color of these corn kernels is due to the
   (1) process of selective breeding
   (2) rate of photosynthesis
   (3) effect of environment on gene expression
   (4) composition of the soil

6. Researchers performing a well-designed experiment should base their conclusions on
   (1) the hypothesis of the experiment
   (2) data from repeated trials of the experiment
   (3) a small sample size to insure a reliable outcome of the experiment
   (4) results predicted before performing the experiment

7. Which source would provide the most reliable information for use in a research project investigating the effects of antibiotics on disease-causing bacteria?
   (1) the local news section of a newspaper from 1993
   (2) a news program on national television about antigens produced by various plants
   (3) a current professional science journal article on the control of pathogens
   (4) an article in a weekly news magazine about reproduction in pathogens

8. The analysis of data gathered during a particular experiment is necessary in order to
   (1) formulate a hypothesis for that experiment
   (2) develop a research plan for that experiment
   (3) design a control for that experiment
   (4) draw a valid conclusion for that experiment

9. A great deal of information can now be obtained about the future health of people by examining the genetic makeup of their cells. There are concerns that this information could be used to deny an individual health insurance or employment. These concerns best illustrate that
   (1) scientific explanations depend upon evidence collected from a single source
   (2) scientific inquiry involves the collection of information from a large number of sources
   (3) acquiring too much knowledge in human genetics will discourage future research in that area
   (4) while science provides knowledge, values are essential to making ethical decisions using this knowledge

10. A student hypothesized that lettuce seeds would not sprout (germinate) unless they were exposed to darkness. The student planted 10 lettuce seeds under a layer of soil and scattered 10 lettuce seeds on top of the soil. The data collected are shown in the table below.

Data Table

Seed Treatment	Number of Seeds Germinated
Planted under soil	9
Scattered on top of soil	8

One way to improve the validity of these results would be to
   (1) conclude that darkness is necessary for lettuce seed germination
   (2) conclude that light is necessary for lettuce seed germination
   (3) revise the hypothesis
   (4) repeat the experiment

11. Which statement best describes a controlled experiment?
   (1) It eliminates the need for dependent variables.
   (2) It shows the effect of a dependent variable on an independent variable.
   (3) It avoids the use of variables.
   (4) It tests the effect of a single independent variable.

12. The removal of nearly all the predators from an ecosystem would most likely result in
   (1) an increase in the number of carnivore species
   (2) a decrease in new predators migrating into the ecosystem
   (3) a decrease in the size of decomposers
   (4) an increase in the number of herbivores

13. A biologist used the Internet to contact scientists around the world to obtain information about declining amphibian populations. He was able to gather data on 936 populations of amphibians, consisting of 157 species from 37 countries. Results showed that the overall numbers of amphibians dropped 15% a year from 1960 to 1966 and continued to decline about 2% a year through 1997. What is the importance of collecting an extensive amount of data such as this?
   (1) Researchers will now be certain that the decline in the amphibian populations is due to pesticides.
   (2) The data collected will prove that all animal populations around the world are threatened.

(3) Results from all parts of the world will be found to be identical.
(4) The quantity of data will lead to a better understanding of the extent of the problem.

14. The first trial of a controlled experiment allows a scientist to isolate and test
(1) a logical conclusion
(2) a variety of information
(3) a single variable
(4) several variables

15. A science researcher is reviewing another scientist's experiment and conclusion. The reviewer would most likely consider the experiment invalid if
(1) the sample size produced a great deal of data
(2) other individuals are able to duplicate the results
(3) it contains conclusions not explained by the evidence given
(4) the hypothesis was not supported by the data obtained

16. A scientist conducted an experiment in which he fed mice large amounts of the amino acid cysteine. He observed that this amino acid protected mouse chromosomes from damage by toxic chemicals. The scientist then claimed that cysteine, added to the diet of all animals, will protect their chromosomes from damage. State whether or not this is a valid claim. Support your answer.

17. Students were asked to determine if they could squeeze a clothespin more times in a minute after resting than after exercising. An experiment that accurately tests this question should include all of the following except
(1) a hypothesis on which to base the design of the experiment
(2) a large number of students
(3) two sets of clothespins, one that is easy to open and one that is more difficult to open
(4) a control group and an experimental group with equal numbers of students of approximately the same age

18. You have been assigned to design an experiment to determine the effects of light on the growth of tomato plants. In your experimental design be sure to:

- state one hypothesis to be tested
- identify the independent variable in the experiment
- describe the type of data to be collected

19. Which statement about the use of independent variables in controlled experiments is correct?
(1) A different independent variable must be used each time an experiment is repeated.
(2) The independent variables must involve time.
(3) Only one independent variable is used for each experiment.
(4) The independent variables state the problem being tested.

20. Many plants can affect the growth of other plants near them. This can occur when one plant produces a chemical that affects another plant.

Design an experiment to determine if a solution containing ground-up goldenrod plants has an effect on the growth of radish seedlings. In your experimental design be sure to:

- state a hypothesis to be tested
- describe how the experimental group will be treated differently from the control group
- explain why the number of seedlings used for the experiment should be large
- identify the type of data that will be collected
- describe experimental results that would support your hypothesis

21. Information concerning nests built in the same tree by two different bird species over a ten-year period is shown in the table below.

Distance of Nest Above Ground (m)	Total Number of Nests Built by Two Different Species	
	A	B
less than 1	5	0
1–5	10	0
6–10	5	0
over 10	0	20

What inference best describes these two bird species?

(1) They most likely do not compete for nesting sites because they occupy different niches.
(2) They do not compete for nesting sites because they have the same reproductive behavior.
(3) They compete for nesting sites because they build the same type of nest.
(4) They compete for nesting sites because they nest in the same tree at the same time.

22. An experimental setup is shown below.

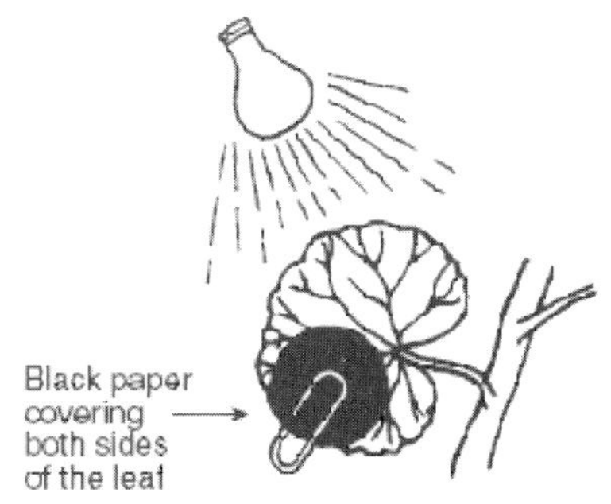

Which hypothesis would most likely be tested using this setup?

(1) Light is needed for the process of reproduction.
(2) Glucose is not synthesized by plants in the dark.
(3) Protein synthesis takes place in leaves.
(4) Plants need fertilizers for proper growth.

23. A student performed an experiment to determine if treating 500 tomato plants with an auxin (a plant growth hormone) will make them grow faster. The results are shown in the table below.

Days	Average Stem Height (cm)
1	10
5	13
10	19
15	26
20	32
25	40

Explain why the student can *not* draw a valid conclusion from these results.

24. The photograph below shows a pill bug. Pill bugs are small animals frequently found in wooded areas near decomposing organic material.

Describe some parts of an experiment to determine the preference of pill bugs for light or darkness. In your answer be sure to:

- state a hypothesis
- identify the independent variable in the experiment
- identify *two* conditions that should be kept the same in all experimental setups
- state *one* example of experimental data that would support your hypothesis

25. A certain plant has white flower petals and it usually grows in soil that is slightly basic. Sometimes the plant produces flowers with red petals. A company that sells the plant wants to know if soil pH affects the color of the petals in this plant. Design a controlled experiment to determine if soil pH affects petal color. In your experimental design be sure to:

- state the hypothesis to be tested in the experiment
- state one way the control group will be treated differently from the experimental group
- identify *two* factors that must be kept the same in both the control group and the experimental group
- identify the dependent variable in the experiment
- state *one* result of the experiment that would support the hypothesis

26. A scientist is planning to carry out an experiment on the effect of heat on the function of a certain enzyme. Which would not be an appropriate first step?
    (1) doing research in a library
    (2) having discussions with other scientists
    (3) completing a data table of expected results
    (4) using what is already known about the enzyme

27. The drugs usually used to treat high blood pressure do not affect blood vessels in the lungs. *Bosentan* is a new drug being studied as a treatment for high blood pressure in the lungs. In an experiment, patients treated with *bosentan* showed an improvement in the distance they could walk without fatigue within 12 weeks.

    Design an experiment to test the effectiveness of *bosentan* as a drug to treat high blood pressure in the lungs. In your answer be sure to:

- state the hypothesis your experiment will test
- state how the control group will be treated differently from the experimental group

- state *two* factors that must be kept the same in both the experimental and control groups
- state the type of data that should be collected to determine if the hypothesis is supported

28. The analysis of data gathered during a particular experiment is necessary in order to
    (1) formulate a hypothesis for that experiment
    (2) develop a research plan for that experiment
    (3) design a control for that experiment
    (4) draw a valid conclusion for that experiment

29. An experiment was carried out to determine which mouthwash was most effective against bacteria commonly found in the mouth. Four paper discs were each dipped into a different brand of mouthwash. The discs were then placed onto the surface of a culture plate that contained food, moisture, and bacteria commonly found in the mouth. The diagram below shows the growth of bacteria on the plate after 24 hours.

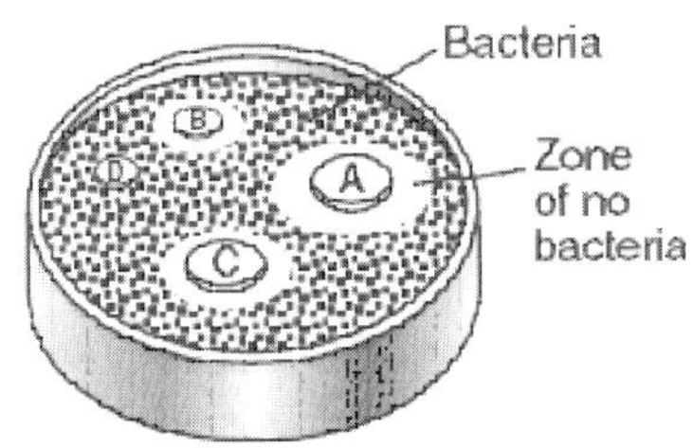

Which change in procedure would have improved the experiment?
    (1) using a smaller plate with less food and moisture
    (2) using bacteria from many habitats other than the mouth
    (3) using the same size paper discs for each mouthwash
    (4) using the same type of mouthwash on each disc

30. The concentration of salt in water affects the hatching of brine shrimp eggs. Brine shrimp eggs will develop and hatch at room temperature in glass containers of salt solution. Describe a controlled experiment using three experimental groups that could be used to determine the best concentration of salt solution in which to hatch brine shrimp eggs. Your answer must include at least:

- a description of how the control group and each of the three experimental groups will be different
- *two* conditions that must be kept constant in the control group and the experimental groups
- data that should be collected
- *one* example of experimental results that would indicate the best concentration of salt solution in which to hatch brine shrimp eggs

31. A scientist wants to determine the best conditions for hatching brine shrimp eggs. In a laboratory, brine shrimp hatch at room temperature in glass containers of salt water. The concentration of salt in the water is known to affect how many brine shrimp eggs will hatch.

    Design an experiment to determine which of three saltwater concentrations (2%, 4%, or 6%) is best for hatching brine shrimp eggs. In your experimental design, be sure to:

- state how many containers to use in the experiment, and describe what would be added to each container in addition to the eggs
- state *two* factors that must be kept constant in all the containers

- state what data must be collected during this experiment
- state *one* way to organize the data so that they will be easy to analyze
- describe a result that would indicate the best salt solution for hatching brine shrimp eggs

Base your answers to the next three questions on the information and chart below and on your knowledge of biology.

It has been hypothesized that a chemical known as BW prevents colds. To test this hypothesis, 20,000 volunteers were divided into four groups. Each volunteer took a white pill every morning for one year. The contents of the pill taken by the members of each group are shown in the chart below.

Group	Number of Volunteers	Contents of Pill	% Developing Colds
1	5,000	5 grams of sugar	20
2	5,000	5 grams of sugar 1 gram of BW	19
3	5,000	5 grams of sugar 3 grams of BW	21
4	5,000	5 grams of sugar 9 grams of BW	15

32. Which factor most likely had the greatest influence on these experimental results?
    (1) color of the pills
    (2) amount of sugar added
    (3) number of volunteers in each group
    (4) health history of the volunteers

33. Which statement is a valid inference based on the results?
    (1) Sugar reduced the number of colds.
    (2) Sugar increased the number of colds.
    (3) BW is always effective in the prevention of colds.
    (4) BW may not be effective in the prevention of colds.

34. Which group served as the control in this investigation?
    (1) 1 (2) 2 (3) 3 (4) 4

Base your answers to the next two questions on the information and data table below and on your knowledge of biology

Two students collected data on their pulse rates while performing different activities. Their average results are shown in the data table below.

Data Table

Activity	Average Pulse Rate (beats/min)
sitting quietly	70
walking	98
running	120

35. State the relationship between activity and pulse rate.

36. State *one* way that this investigation could be improved.

37. You have been assigned to design an experiment to determine the effects of light on the growth of tomato plants. In your experimental design be sure to:

- state *one* hypothesis to be tested
- identify the independent variable in the experiment
- describe the type of data to be collected

38. Base your answers to the next question on the diagram below and on your knowledge of biology. The diagram shows the results of a technique used to analyze DNA.

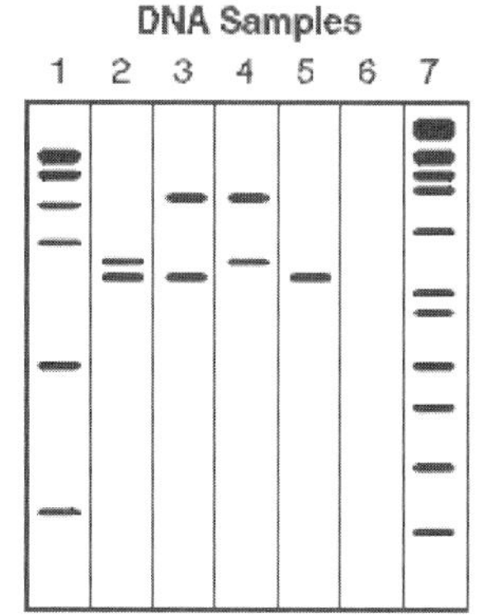

State *one* specific way the results of this laboratory technique could be used.

39. A scientist was investigating why a particular tree species grows only in a specific environment. To determine physical conditions the tree species needs to survive, an appropriate study should include
    (1) the identification of organisms in the food web in that environment
    (2) an analysis of the arrangement of the leaves on the trees
    (3) the identification of all tree species in the area
    (4) an analysis of the soil around the tree

40. Many plants can affect the growth of other plants near them. This can occur when one plant produces a chemical that affects another plant.

    Design an experiment to determine if a solution containing ground-up goldenrod plants has an effect on the growth of radish seedlings. In your experimental design be sure to:

- state a hypothesis to be tested
- describe how the experimental group will be treated differently from the control group
- explain why the number of seedlings used for the experiment should be large
- identify the type of data that will be collected
- describe experimental results that would support your hypothesis

41. Base your answers to the next question on the information and data table below and on your knowledge of biology.

    A student studied the location of single-celled photosynthetic organisms in a lake for a period of several weeks. The depth at which these organisms were found at different times of the day varied greatly. Some of the data collected are shown in the table below.

Data Table

Light Conditions at Different Times of the Day	Average Depth of Photosynthetic Organisms (cm)
full light	150
moderate light	15
no light	10

A valid inference based on these data is that
(1) most photosynthetic organisms live below a depth of 150 centimeters
(2) oxygen production increases as photosynthetic organisms move deeper in the lake
(3) photosynthetic organisms respond to changing light levels
(4) photosynthetic organisms move up and down to increase their rate of carbon dioxide production

42. Base your answers to this question on the information and data table below and on your knowledge of biology.

A number of bean seeds planted at the same time produced plants that were later divided into two groups, *A* and *B*. Each plant in group *A* was treated with the same concentration of gibberellic acid (a plant hormone). The plants in group *B* were not treated with gibberellic acid. All other growth conditions were kept constant. The height of each plant was measured on 5 consecutive days, and the average height of each group was recorded in the data table below.

Data Table

	Average Plant Height (cm)				
	Day 1	Day 2	Day 3	Day 4	Day 5
Group A	5	7	10	13	15
Group B	5	6	6.5	7	7.5

State a valid conclusion that can be drawn concerning the effect of gibberellic acid on bean plant growth.

43. Base your answers to the next question on the passage below and on your knowledge of biology.

Research indicates that many plants prevent the growth of other plants in their habitat by releasing natural herbicides (chemicals that kill plants). These substances are known as allelochemicals and include substances such as quinine, caffeine, and digitalis. Experiments have confirmed that chemicals in the bark and roots of black walnut trees are toxic, and when released into the soil they limit the growth of crop plants such as tomatoes, potatoes, and apples. Allelochemicals can alter growth and enzyme action, injure the outer cover of a seed so the seed dies, or stimulate seed growth at inappropriate times of the year. Studies on allelochemical effects help explain the observation that almost nothing grows under a black walnut tree even though light and moisture levels are adequate for growth.

What evidence from the data table shows that a salt-marsh plant is sensitive to its environment?

44. Some interactions in a desert community are shown in the diagram below.

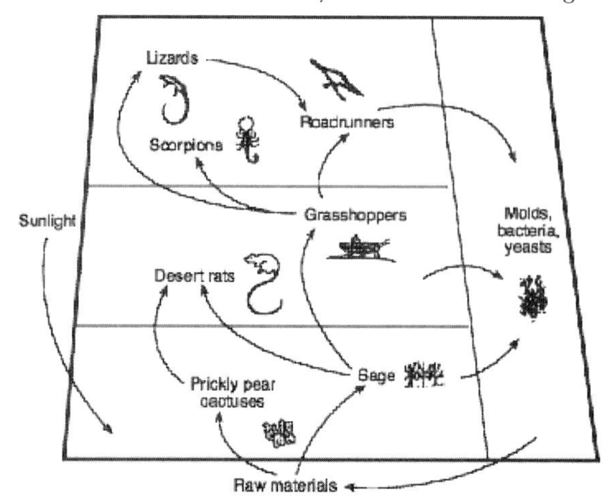

Which statement is a valid inference based on the diagram?

(1) Certain organisms may compete for vital resources.
(2) All these organisms rely on energy from decomposers.
(3) Organisms synthesize energy.
(4) All organisms occupy the same niche.

45. The diagram below shows the effect of spraying a pesticide on a population of insects over three generations.

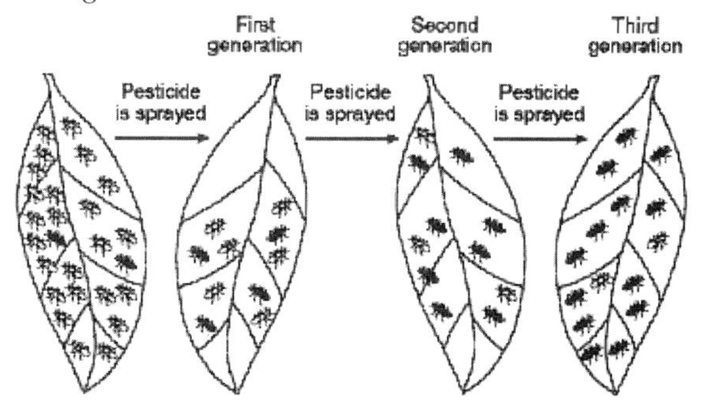

Which concept is represented in the diagram?

(1) survival of the fittest
(2) dynamic equilibrium
(3) succession
(4) extinction

46. The graph below shows the growth of a population of bacteria over a period of 80 hours.

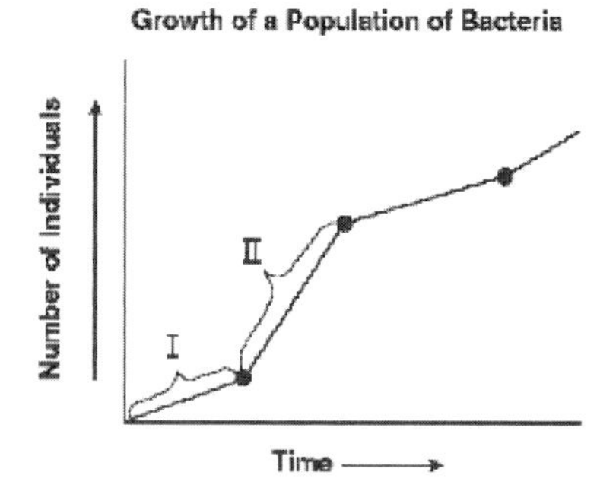

Which statement best describes section *II* of the graph?

(1) The population has reached the carrying capacity of the environment.
(2) The rate of reproduction is slower than in section *I*.
(3) The population is greater than the carrying capacity of the environment.
(4) The rate of reproduction exceeds the death rate.

47. Which procedure would most likely provide valid results in a test to determine if drug *A* would be effective in treating cancer in white mice?

(1) injecting 1 mL of drug A into 100 white mice with cancer
(2) injecting 1 mL of drug A into 100 white mice with cancer and 0.5 mL of drug *X* into 100 white mice without cancer
(3) injecting 1 mL of drug *A* into 100 white mice with cancer and 0.5 mL of drug *X* into another group of 100 white mice with cancer
(4) injecting 1 mL of drug *A* into 100 white mice with cancer and 1 mL of distilled water into another group of 100 white mice with cancer

48. In 1995, during an Ebola virus outbreak, approximately 80% of the infected individuals died. Which statement is an inference that could be made based on this information?

(1) The individuals who survived were able to produce antibodies against the Ebola virus.
(2) The individuals who survived were not exposed to the Ebola antigens.
(3) Eighty percent of the population had a natural immunity to the Ebola virus.
(4) Eighty percent of the population was infected with a viral antigen.

49. A biologist collected the data shown in the table below.

Data Table

Type of Organism	Number of Organisms in a Field		
	May	July	September
grasshoppers	100	500	150
birds	25	100	10
spiders	75	200	50

Which statement is supported by the data in the table?

(1) Populations do not vary from month to month.
(2) The populations are highest in September.
(3) The grasshoppers increased in length in July.
(4) Seasonal variations may affect populations.

Base your answers to the next four questions on the information below and on your knowledge of biology.

Insecticides are used by farmers to destroy crop-eating insects. Recently, scientists tested several insecticides to see if they caused damage to chromosomes. Six groups of about 200 cells each were examined to determine the extent of chromosome damage after each group was exposed to a different concentration of one of two insecticides. The results are shown in the data table below.

Cell Damage After Exposure to Insecticide

Insecticide	Insecticide Concentration (ppm)	Number of Cells with Damaged Chromosomes
Methyl parathion	0.01	7
	0.10	15
	0.20	30
Malathion	0.01	3
	0.10	4
	0.20	11

Directions for the next three questions: *Using the information in the data table, construct a line graph on the grid provided, following the directions below.*

Cell Damage After Exposure to Insecticide

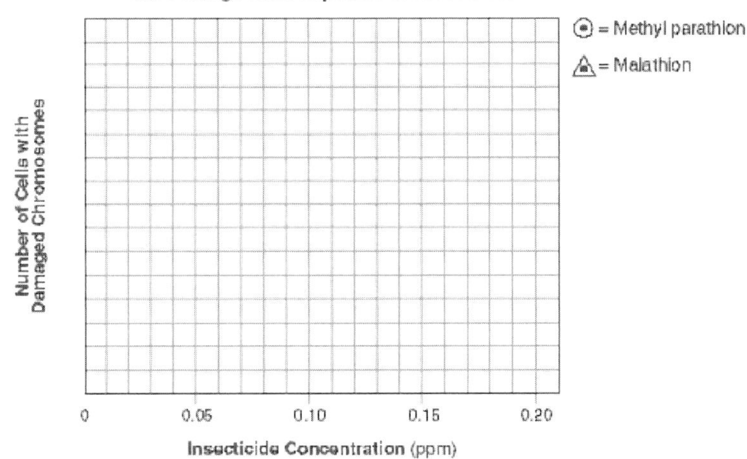

50. Mark an appropriate scale on the axis labeled, "Number of Cells with Damaged Chromosomes."

51. Plot the data for methyl parathion on the grid. Surround each point with a small circle and connect the points.

    Example:

52. Plot the data for malathion on the grid. Surround each point with a small triangle and connect the points.

    Example:

53. Which insecticide has a more damaging effect on chromosomes? Support your answer.

Base your answers to the next four questions on the information and data table below and on your knowledge of biology.

The effect of temperature on the action of pepsin, a protein-digesting enzyme present in stomach fluid, was tested. In this investigation, 20 milliliters of stomach fluid and 10 grams of protein were placed in each of five test tubes. The tubes were then kept at different temperatures. After 24 hours, the contents of each tube were tested to determine the amount of protein that had been digested. The results are shown in the table below.

Protein Digestion at Different Temperatures

Tube #	Temperature (°C)	Amount of Protein Digested (grams)
1	5	0.5
2	10	1.0
3	20	4.0
4	37	9.5
5	85	0.0

54. The dependent variable in this investigation is the
    (1) size of the test tube
    (2) time of digestion
    (3) amount of stomach fluid
    (4) amount of protein digested

55. ***Directions for the next two questions:*** *Using the information in the data table, construct a line graph on the grid provided, following the directions below.*

    Mark an appropriate scale on each axis.

56. Plot the data on the grid. Surround each point with a small circle and connect the points.

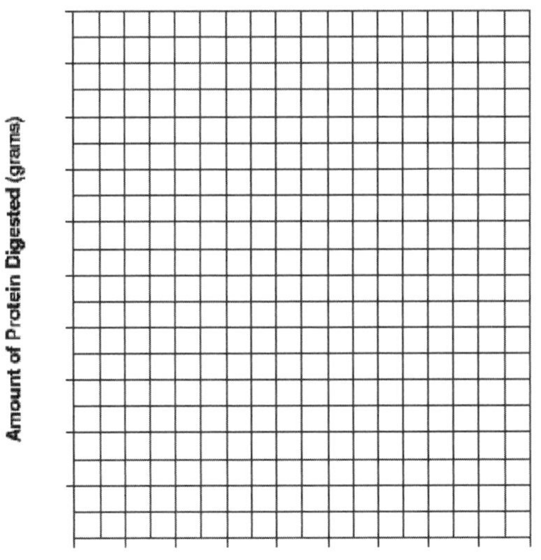

Example:

57. If a sixth test tube identical to the other tubes was kept at a temperature of 30°C for 24 hours, the amount of protein digested would most likely be
    (1) less than 1.0 gram
    (2) between 1.0 and 4.0 grams
    (3) between 4.0 and 9.0 grams
    (4) more than 9.0 grams

Base your answers to the next two questions on the information and data table below and on your knowledge of biology.

The results of blood tests for two individuals are shown in the data table below. The blood glucose level before breakfast is normally 80–90 mg/100 mL of blood. A blood glucose level above 110 mg/100 mL of blood indicates a failure in a feedback mechanism.

Injection of chemical *X*, a chemical normally produced in the body, may be required to correct this problem.

Data Table

Time	Blood Glucose (mg/100 mL)	
	Individual 1	**Individual 2**
7:00 a.m.	90	150
7:30 a.m.	120	180
8:00 a.m.	140	220
8:30 a.m.	110	250
9:00 a.m.	90	240
9:30 a.m.	85	230
10:00 a.m.	90	210
10:30 a.m.	85	190
11:00 a.m.	90	170

Directions for the next two questions: *Using the information in the data table, construct a line graph on the grid provided below, following the directions below.*

58. Mark an appropriate scale on each labeled axis.

59. Plot the blood glucose levels for the individual who will most likely need injections of chemical *X*. Surround each point with a small circle and connect the points.

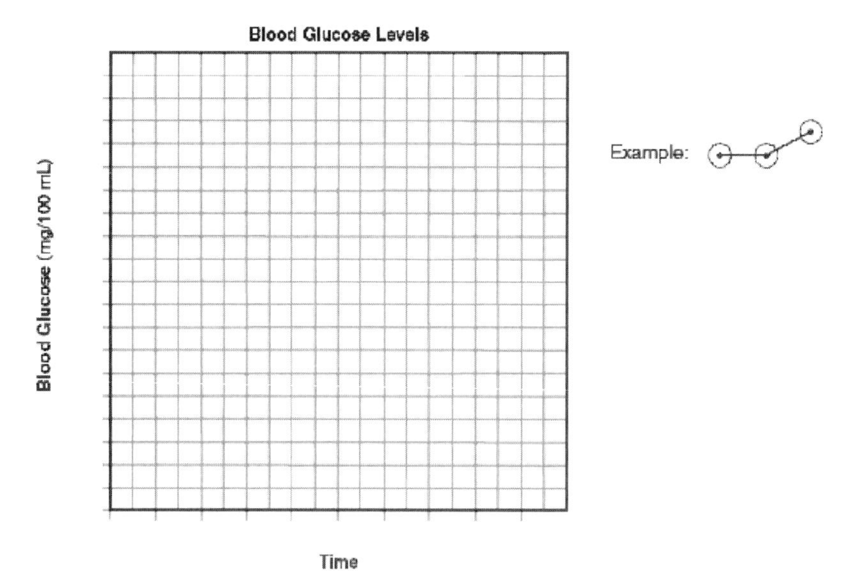

***Hint**: Review middle of Chapter 2.*

Base your answers to the next five questions on the passage and data table below and on your knowledge of biology.

The amount of oxygen gas dissolved in water is important to the organisms that live in a river. The amount of dissolved oxygen varies with changes in both physical factors and biological processes. The temperature of the water is one physical factor affecting dissolved oxygen levels as shown in the data table below. The amount of dissolved oxygen is expressed in parts per million (ppm).

Dissolved Oxygen Levels at Various Temperatures

Water Temperature (°C)	Level of Dissolved Oxygen (ppm)
1	14
10	11
15	10
20	9
25	8
30	7

Directions for the next two questions: *Using the information given, construct a line graph on the grid provided, following the directions below.*

60. Mark an appropriate scale on each labeled axis.

61. Plot the data for dissolved oxygen on the grid. Surround each point with a small circle and connect the points.

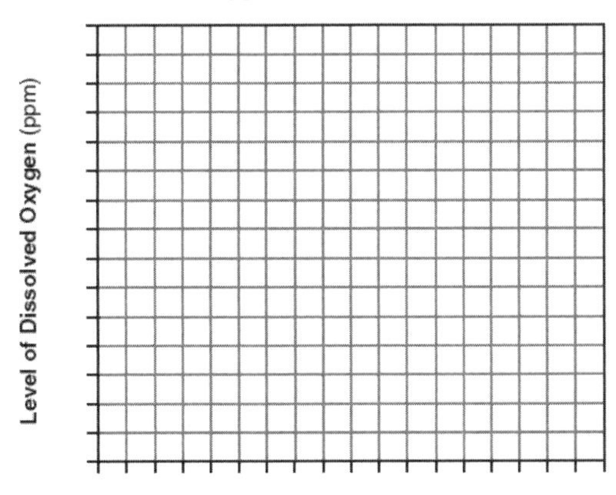

62. If the trend continues as shown in the data, what would the dissolved oxygen level most likely be if the temperature of the water was 35°C?

63. State the relationship between the level of dissolved oxygen and water temperature.

64. Identify one physical or biological process taking place within the river, other than temperature change, that would affect the level of dissolved oxygen and state whether this process would increase or decrease the level of dissolved oxygen.

Protein Digestion at Different Temperatures

Tube #	Temperature (°C)	Amount of Protein Digested (grams)
1	5	0.5
2	10	1.0
3	20	4.0
4	37	9.5
5	85	0.0

Base your answers to the next five questions on the information below and on your knowledge of biology.

Each year, a New York State power agency provides its customers with information about some of the fuel sources used in generating electricity. The table below applies to the period of 2002–2003.

Fuel Sources Used

Fuel Source	Percentage of Electricity Generated
hydro (water)	86
coal	5
nuclear	4
oil	1
solar	0

Directions for the next two questions: *Using the information given, construct a bar graph on the grid provided at right, following the directions below.*

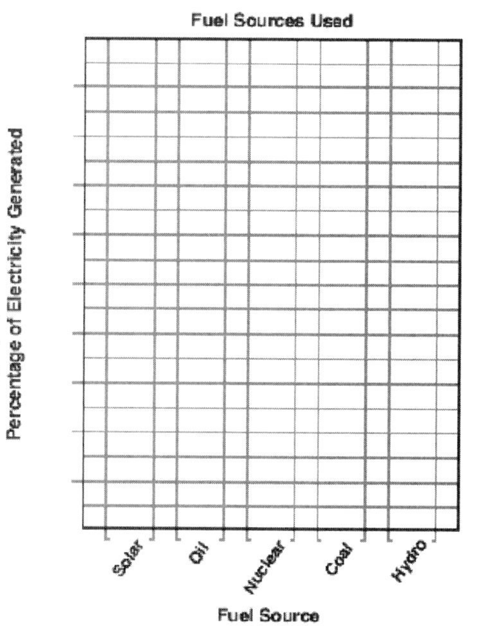

65. Mark an appropriate scale on the axis labeled "Percentage of Electricity Generated."

66. Construct vertical bars to represent the data. Shade in *each* bar.

67. Identify *one* fuel source in the table that is considered a fossil fuel.

68. Identify *one* fuel source in the table that is classified as a renewable resource.

69. State *one* specific environmental problem that can result from burning coal to generate electricity.

Base your answers to the next five questions on the information and data table below and on your knowledge of biology.

Biologists investigated the effect of the presence of aluminum ions on root tips of a variety of wheat. They removed 2-mm sections of the tips of roots. Half of the root tips were placed in a nutrient solution with aluminum ions, while the other half were placed in an identical nutrient solution without aluminum ions. The length of the root tips, in millimeters, was measured every hour for seven hours. The results are shown in the data table below.

Data Table

Time (hr)	Length of Root Tips in Solution With Aluminum Ions (mm)	Length of Root Tips in Solution Without Aluminum Ions (mm)
0	2.0	2.0
1	2.1	2.2
2	2.2	2.4
3	2.4	2.8
4	2.6	2.9
5	2.7	3.2
6	2.8	3.7
7	2.8	3.9

Directions for the next three questions: *Using the information in the data table, construct a line graph on the grid provided, following the directions below.*

70. Mark an appropriate scale on each labeled axis.

71. Plot the data for root tips in the solution with aluminum ions on the grid. Surround each point with a small circle and connect the points.

Example:

72. Plot the data for root tips in the solution without aluminum ions on the grid. Surround each point with a small triangle and connect the points.

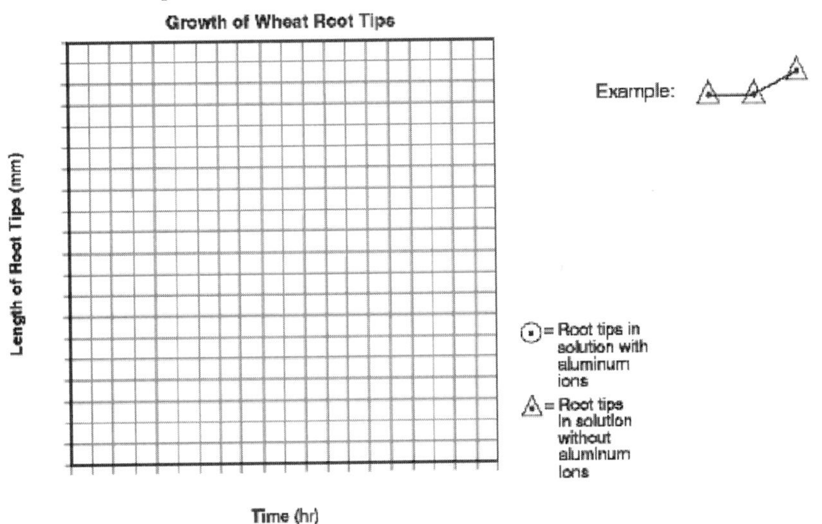

73. The aluminum ions most likely affected
   (1) photosynthetic rate
   (2) the union of gametes
   (3) mitotic cell division
   (4) starch absorption from the soil

74. Describe the effect of aluminum ions on the growth of the root tips of wheat.

Base your answers to the next five questions on the information and data table below and on your knowledge of biology.

The table shows data collected on the pH level of an Adirondack lake from 1980 to 1996.

Lake pH Level

Year	pH Level
1980	6.7
1984	6.3
1986	6.4
1988	6.2
1990	5.9
1992	5.6
1994	5.4
1996	5.1

***Directions**: Using the information in the data table, construct a line graph on the grid provided, following the directions below.*

75. Label the axes.

76. Mark an appropriate scale on the y-axis. The scale has been started for you.

77. Plot the data from the data table. Surround each point with a small circle and connect the points.

Example: 

Lake pH Level from 1980 to 1996

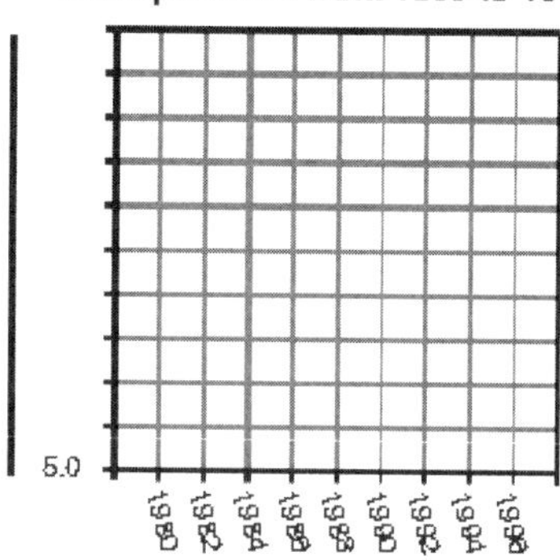

78. Describe the trend in pH level in the lake over this 16-year period.

79. Identify *one* factor that should have been kept constant each time water samples were collected from the lake.

Base your answers to the next two questions on the information below and on your knowledge of biology.

In a test for diabetes, blood samples were taken from an individual every 4 hours for 24 hours. The glucose concentrations were recorded and are shown in the data table below.

Blood Glucose Level Over Time

Time (h)	Blood Glucose Concentration (mg/dL)
0	100
4	110
8	128
12	82
16	92
20	130
24	104

80. State *one* likely cause of the change in blood glucose concentration between hour 16 and hour 20.

81. ***Directions:*** *Using the information given, construct a line graph on the grid provided, following the directions below.*

Blood Glucose Concentration Over Time

Blood Glucose Concentration (mg/dL)

0 2 4 6 8 10 12 14 16 18 20 22 24

Time (h)

Mark an appropriate scale on the axis labeled "Blood Glucose Concentration (mg/dL)."

82. An experiment was set up to test the effect of light intensity on the rate of photosynthesis, as shown in the diagram below.

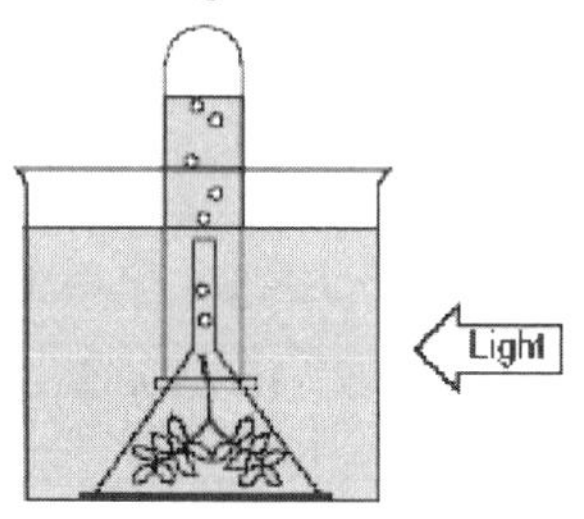

Data were collected by counting gas bubbles released in a 5-minute period when the light source was placed at various distances from the experimental setup. The data are shown in the table below.

Data Table

Distance From Light (cm)	Bubbles in 5-Minute Period
15	27
23	20
30	13
45	6

The number of bubbles released when the light source is at a distance of 38 centimeters would most likely be closest to

(1) 6 (2) 10 (3) 13 (4) 22

83. The relative amount of oxygen in the atmosphere of Earth over millions of years is shown in the graph below.

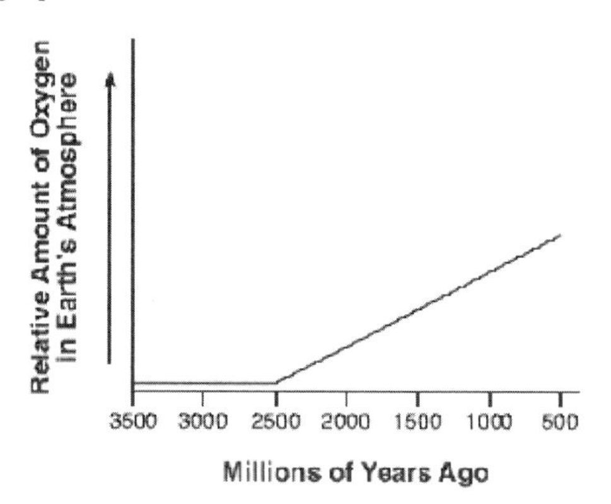

At what point in the history of Earth did autotrophs most likely first appear?

(1) 3500 million years ago
(2) 2500 million years ago
(3) 1500 million years ago
(4) 500 million years ago

***Hint**: Review autotrophs (beginning of Chap. 2).*

84. The graphs below show the changes in the relative concentrations of two gases in the air surrounding a group of mice.

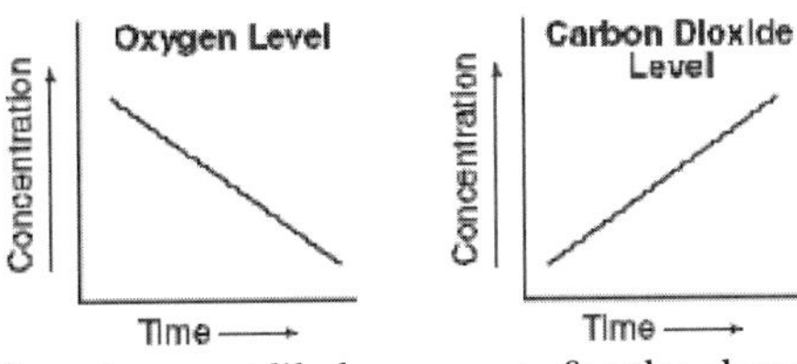

Which process in the mice most likely accounts for the changes shown?

(1) active transport
(2) evaporation
(3) respiration
(4) photosynthesis

Hint: *Review Chapter 2.*

85. The graph below shows the populations of two species of ants. Ants of species *2* have a thicker outer covering than the ants of species *1*. The outer covering of an insect helps prevent excessive evaporation of water.

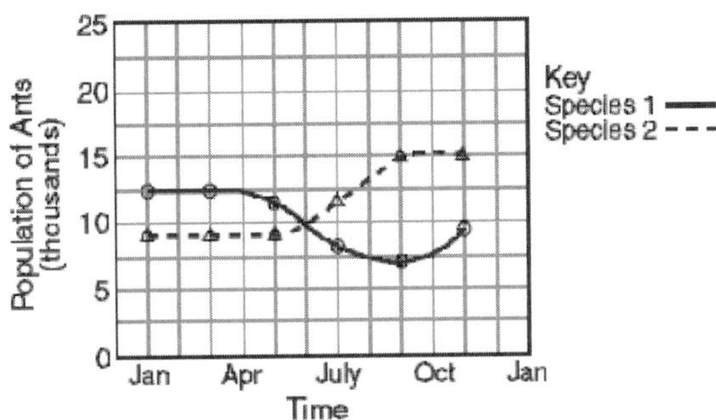

Which statement would best explain the population changes shown in the graph?

(1) The food sources for species *1* increased while the food sources for species *2* decreased from January through November.
(2) Disease killed off species *1* beginning in May.
(3) The weather was hotter and dryer than normal from April through September.
(4) Mutations occurred from April through September in both species, resulting in both species becoming better adapted to the environment.

Hint: *Review Evolution in a changing environment, middle of Chapter 5 (evolution).*

86. The graph below shows the results of an experiment in which a container of oxygen-using bacteria and strands of a green alga were exposed to light of different colors.

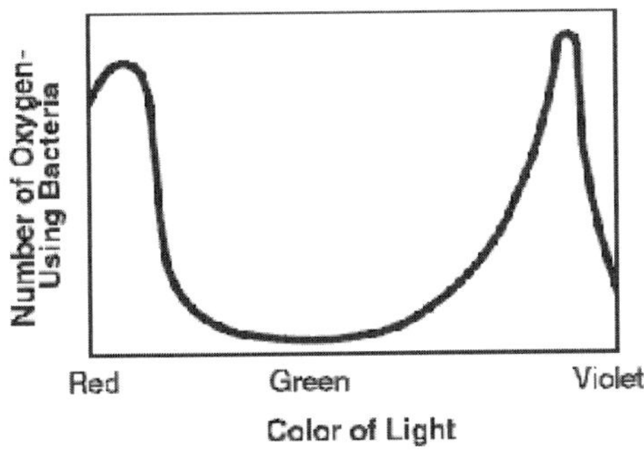

Which statement best explains the results of this experiment?

(1) The rate of photosynthesis is affected by variations in the light.
(2) In all environments light is a vital resource.
(3) The activities of bacteria and algae are not related.
(4) Uneven numbers and types of species can upset ecosystem stability.

87. The graph below provides information about the reproductive rates of four species of bacteria, *A*, *B*, *C*, and *D*, at different temperatures.

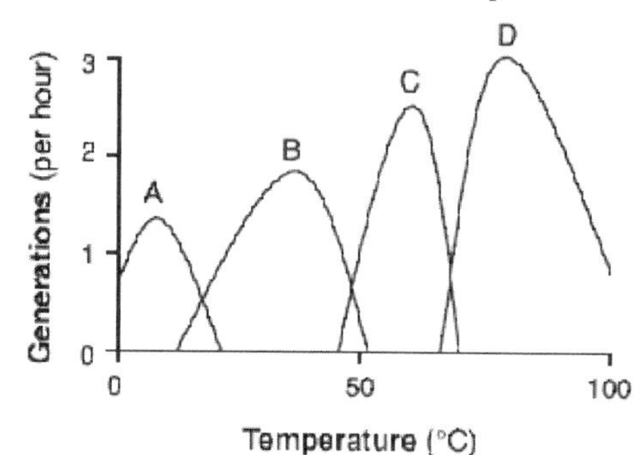

Which statement is a valid conclusion based on the information in the graph?

(1) Changes in temperature cause bacteria to adapt to form new species.
(2) Increasing temperatures speed up bacterial reproduction.
(3) Bacteria can survive only at temperatures between 0°C and 100°C.
(4) Individual species reproduce within a specific range of temperatures.

88. On which day did the population represented in the graph below reach the carrying capacity of the ecosystem?

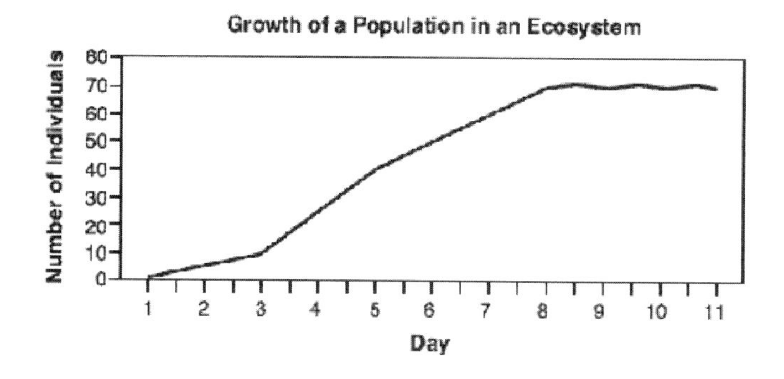

(1) day 11 (2) day 8 (3) day 3 (4) day 5

Hint: *Review carrying capacity, beginning of Chapter 6.*

89. Students conducting a study on an insect population placed 25 insects of the same size in a box. The amount of food, water, and shelter available to the insects was kept constant. Each month, students removed and counted the number of insects present, recorded the total, and returned the insects to the box. The graph below shows the number of insects in the box over a 12-month period.

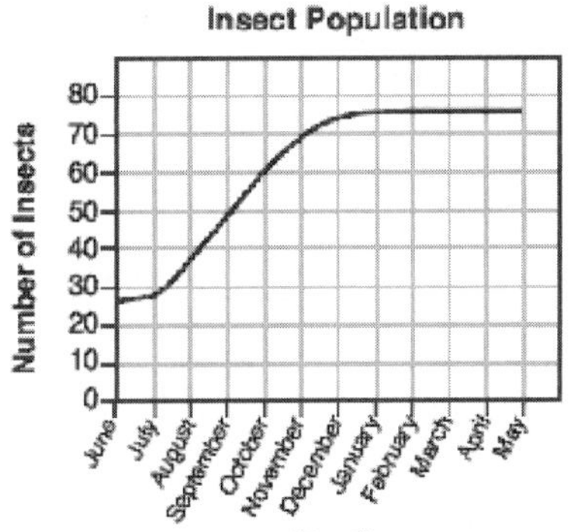

What inference can be made regarding this insect population?

(1) All the insects in the box are the same age.
(2) The insects hibernated from January to April.
(3) The population has carnivorous members.
(4) The population reached carrying capacity by January.

90. The graph below shows the growth of a population of bacteria over a period of 80 hours.

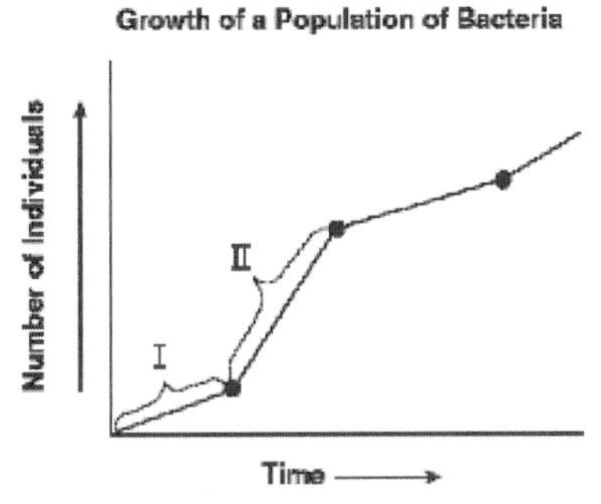

Which statement best describes section *II* of the graph?

(1) The population has reached the carrying capacity of the environment.
(2) The rate of reproduction is slower than in section *I*.
(3) The population is greater than the carrying capacity of the environment.
(4) The rate of reproduction exceeds the death rate.

91. The graph below represents the growth of a population of flies in a jar.

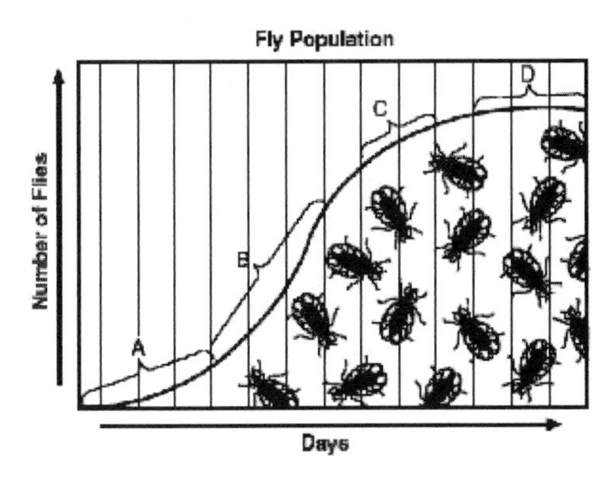

Which letter indicates the part of the graph that represents the carrying capacity of the environment in the jar?

(1) *A* (2) *B* (3) *C* (4) *D*

92. Compounds containing phosphorus that are dumped into the environment can upset ecosystems because phosphorus acts as a fertilizer. The graph below shows measurements of phosphorus concentrations taken during the month of June at two sites from 1991 to 1997.

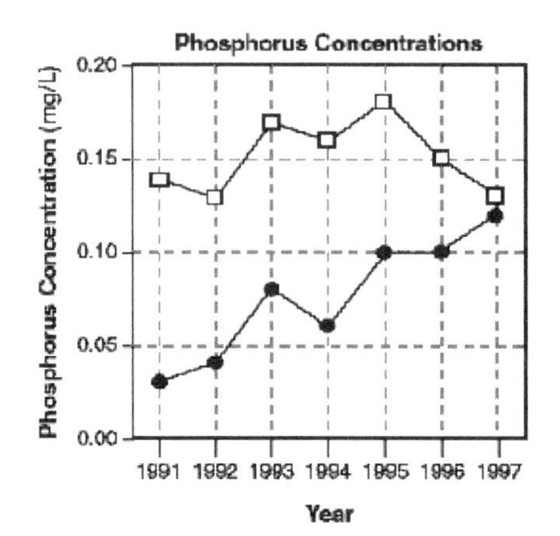

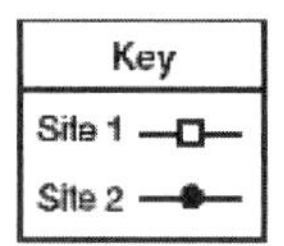

Which statement represents a valid inference based on information in the graph?

(1) There was no decrease in the amount of compounds containing phosphorus dumped at site *2* during the period from 1991 to 1997.
(2) Pollution controls may have been put into operation at site *1* in 1995.
(3) There was most likely no vegetation present near site *2* from 1993 to 1994.
(4) There was a greater variation in phosphorous concentration at site *1* than there was at site *2*.

93. As part of an experiment, the heart rate of a person at rest was measured every hour for 7 hours. The data are shown in the table below.

Data Table

Hour	Heart Rate (beats/min)
1	72
2	63
3	61
4	61
5	60
6	63
7	68

Which graphed line best represents this data?

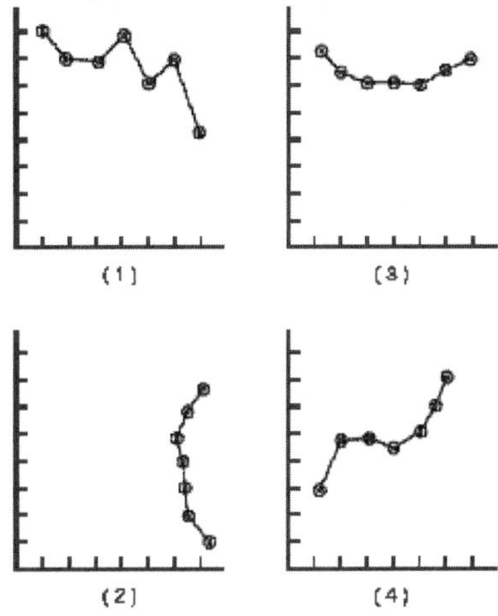

94. Base your answers to the next question on the information below and on your knowledge of biology.

Signs of a Changing Planet

While the changing climate endangers some species, a little global warming suits many shallow-water squid and octopuses just fine. Slightly higher ocean temperatures have been shown to boost the growth of these cephalopods, whose digestive enzymes speed up when warm. The tentacled creatures are also quick to colonize new territory as conditions become more favorable. Humboldt squid, which usually range from Southern California to South America, have been spotted as far north as Alaska. Deep-sea squid may not, however, adapt as readily.

Sierra Magazine, March/April 2005

Which graph most accurately shows the interaction between water temperature and digestive enzyme action in the shallow-water squid?

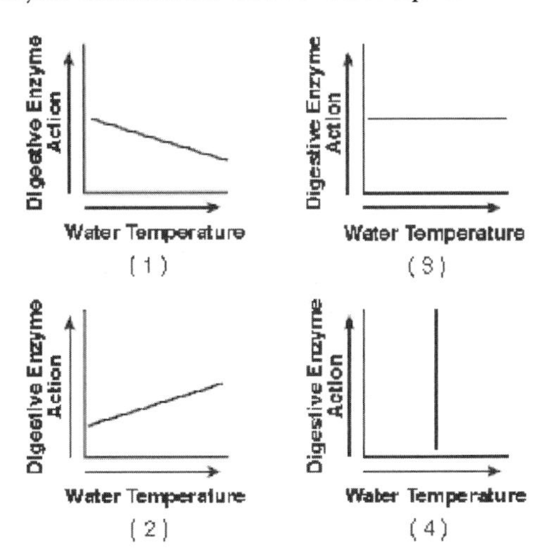

Base your answers to the next four questions on the information and data table below and on your knowledge of biology.

Tooth decay occurs when bacteria living in the mouth produce an acid that dissolves tooth enamel (the outer, protective covering of a tooth).

The Effect of Sugar Intake on Tooth Decay

World Regions	Average Sugar Intake per Person (kg/year)	Average Number of Teeth with Decay per Person
Americas	40	3.0
Africa	18	1.7
Southeast Asia	14	1.6
Europe	36	2.6

Directions for the next three questions: *Using the information in the data table, construct a bar graph on the grid provided, following the directions below.*

95. Mark an appropriate scale on the axis labeled "Average Sugar Intake per Person."

96. Construct vertical bars in the bracketed area for each world region to represent the "Average Sugar Intake per Person." Place the bars on the left side of each bracketed region and shade the bars as shown below. (The bar for Americas has been done for you.)

97. Construct vertical bars in the bracketed area for each world region to represent the "Average Number of Teeth with Decay per Person." Place the bars on the right side of each bracketed region and shade in each bar as shown below.

98. Which statement is a valid conclusion regarding tooth decay?
    (1) As sugar intake increases, the acidity in the mouth decreases, reducing tooth decay.
    (2) As sugar intake increases, tooth decay increases in Europe and the Americas, but not in Africa and Southeast Asia.
    (3) The greater the sugar intake, the greater the average number of decayed teeth.
    (4) The greater the sugar intake, the faster a tooth decays.

Base your answers to the next two questions on the information below and on your knowledge of biology.

A student read a magazine article that claimed people who exercise for 30 minutes are able to solve more math problems than if they had not exercised. The student convinced four of his friends to test this claim. First, he gave them 15 minutes to do 50 math problems. The number each person solved is shown in the trial 1 graph. Next, all four of the students exercised for 30 minutes. At the end of the 30 minutes, they were given another 50 math problems of equal difficulty for the same amount of time. The number of math problems each student solved is shown in the trial 2 graph.

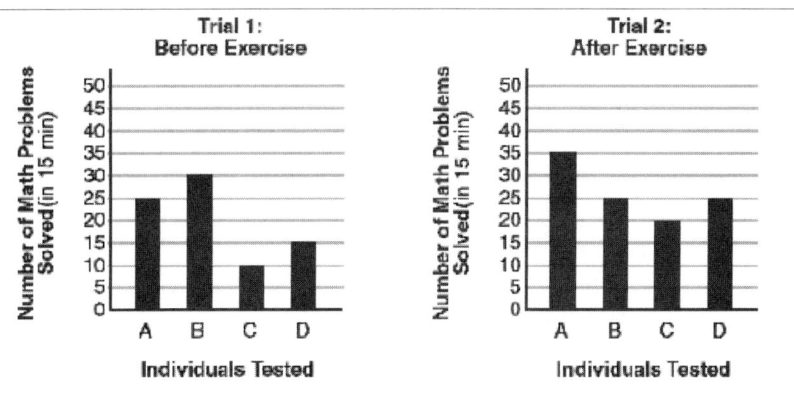

99. Explain why exercise could influence the ability of a student to solve math problems.

100. State whether or not exercising for 30 minutes improved the ability of students to solve math problems. Support your answer using data from the graphs.

101. A graph of the population growth of two different species is shown below.

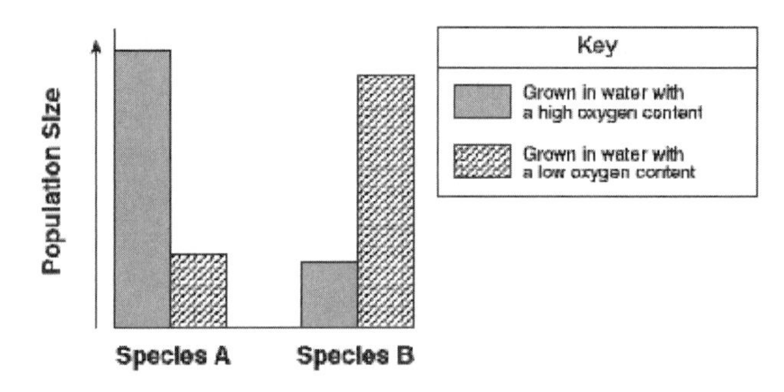

Which conclusion can be drawn from information in the graph?

(1) Oxygen concentration affects population sizes of different species in the same manner.

(2) Species *A* requires a high oxygen concentration for maximum population growth.

(3) Species *B* requires a high oxygen concentration to stimulate population growth.

(4) Low oxygen concentration does not limit the population size of either species observed.

102. Base your answers to the next question on the graph below and on your knowledge of biology. The graph illustrates a single species of bacteria grown at various pH levels.

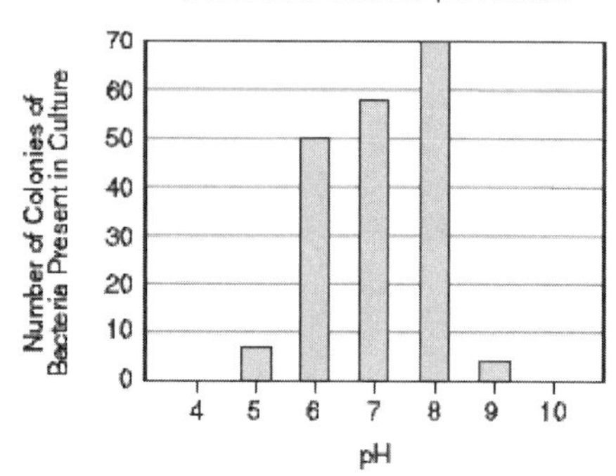

Which statement is supported by data from this graph?

(1) All species of bacteria can grow well at pH 7.
(2) This type of bacterium would grow well at pH 7.5.
(3) This type of bacterium would grow well at pH 2.
(4) Other types of bacteria can grow well at pH 4.

103. An energy pyramid is shown below.

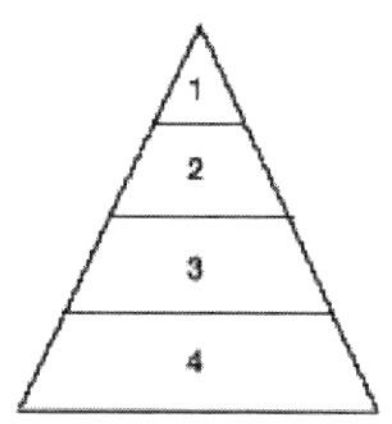

Which graph best represents the relative energy content of the levels of this pyramid?

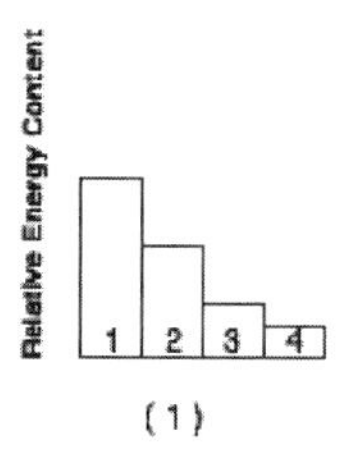

(1)

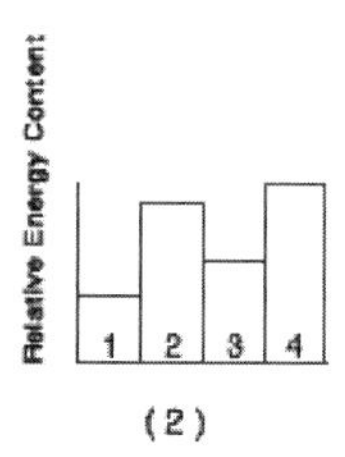

(2)

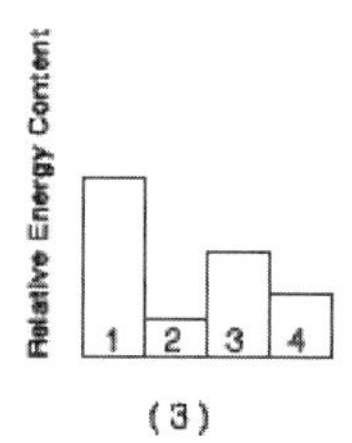

(3)

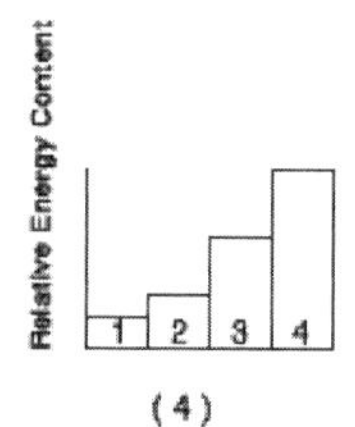

(4)

104. The graph below represents the amount of available energy at successive nutrition levels in a particular food web.

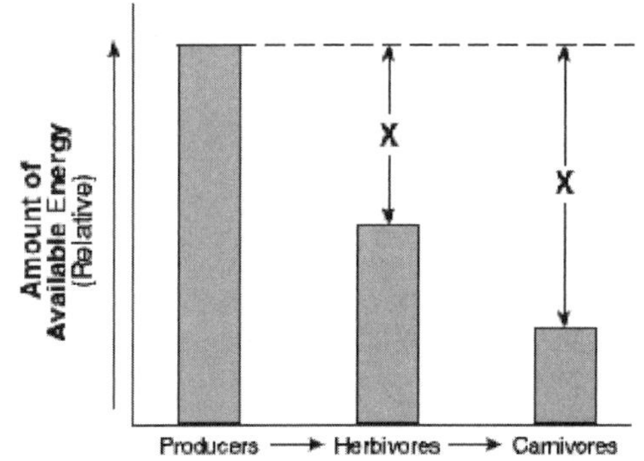

The X's in the diagram represent the amount of energy that was most likely

(1) changed into inorganic compounds
(2) retained indefinitely by the herbivores
(3) recycled back to the producers
(4) lost as heat to the environment

Base your answers to the next two questions on the information below and on your knowledge of biology.

Each year, a New York State power agency provides its customers with information about some of the fuel sources used in generating electricity. The table below applies to the period of 2002–2003.

Fuel Sources Used

Fuel Source	Percentage of Electricity Generated
hydro (water)	86
coal	5
nuclear	4
oil	1
solar	0

***Directions for the next two questions**: Using the information given, construct a bar graph on the grid provided, following the directions below.*

105. Mark an appropriate scale on the axis labeled "Percentage of Electricity Generated."

106. Construct vertical bars to represent the data. Shade in *each* bar.

Base your answers to the next two questions on the histograms below and on your knowledge of biology.

Students in a class recorded their resting pulse rates and their pulse rates immediately after strenuous activity. The data obtained are shown in the histograms below.

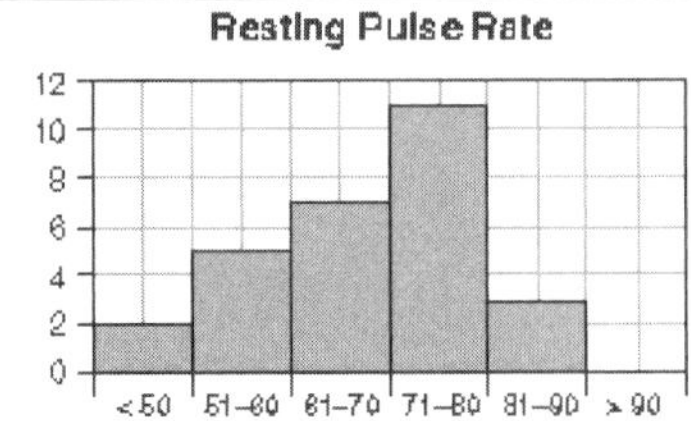

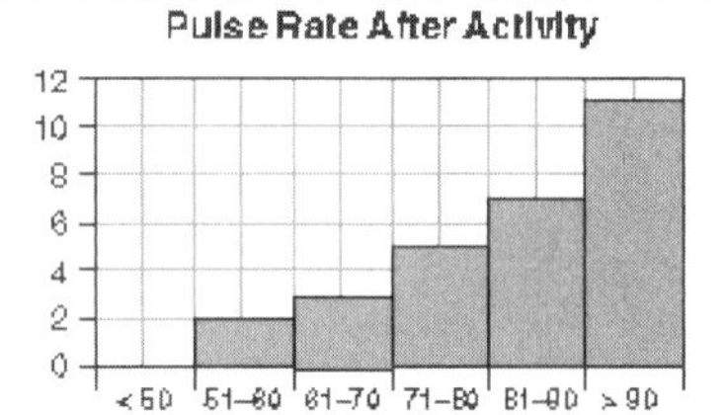

107. An appropriate label for the y-axis in each histogram would be
   (1) Number of Students
   (2) Average Number of Heartbeats
   (3) Time (min)
   (4) Amount of Exercise

108. According to the data, compared to the average resting pulse rate, the average pulse rate immediately after strenuous activity generally
   (1) decreased
   (2) increased
   (3) remained the same
   (4) decreased and leveled off

CHAPTER 9: LABORATORY

This chapter will also help in answering **Part D** of the Regents exam.

Safety Rules

Follow safety rules in the laboratory:

1. Do not start working in the laboratory until your teacher tells you what to do.
2. Keep sleeves and hair away from bunsen burners.
3. Do not bring food into the lab. Food can get contaminated.
4. When you heat a test tube, point the top of the test tube away from you.
5. Don't heat a test tube that is stoppered.
6. Heat flammable liquids (such as alcohol) in a hot water bath (a beaker of water on a hot plate, not on a flame).
7. Don't put electrical equipment near sinks (water).
8. Don't touch hot equipment.
9. Don't use glassware that is chipped or broken. If glassware is broken, tell your teacher.
10. Notify your teacher immediately of any accidents.
11. If a chemical spills on your skin or clothing, wash it off immediately and tell the teacher.
12. Do not pour chemicals back from your container into the original bottle (stock bottle); do not exchange stoppers between bottles.
13. Do not use dissection instruments (examples scalpel, forceps) until your teacher tells you how to use them.
14. Wear safety goggles when dissecting (cutting plants or dead animals to see the inside structures such as organs, etc.), using chemicals, or heating liquids.
15. Follow the teacher's instructions on how to dispose of (get rid of) chemicals.

Know where the safety equipment (fire extinguisher, fire blanket, safety shower, eye wash station, gas shutoff) is located and how to use it.

Select and Use Correct Measuring Instruments

1. **Use** a **metric ruler** to **measure length**.
The ruler is divided into centimeters (see 10 cm ruler below).

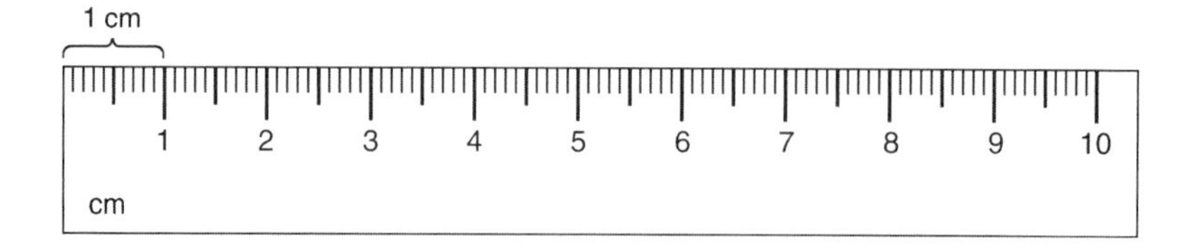

Each cm is divided into 10 parts (10 lines) see figure above.

1 centimeter (cm) = 10 millimeters (mm),
therefore, 2 cm = 20 mm (2 cm x 10 = 20 mm)
2.5 cm = 25 mm (2.5 cm x 10 = 25 mm)
3 cm = 30 mm
9 cm = 90 mm.

Reading a metric ruler:

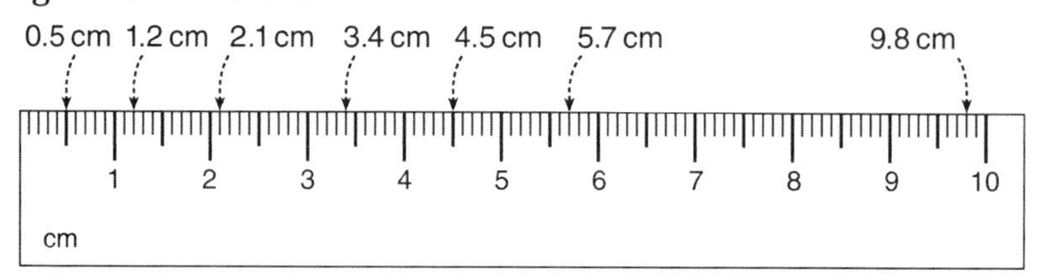

Example 1: Measure the leaf:

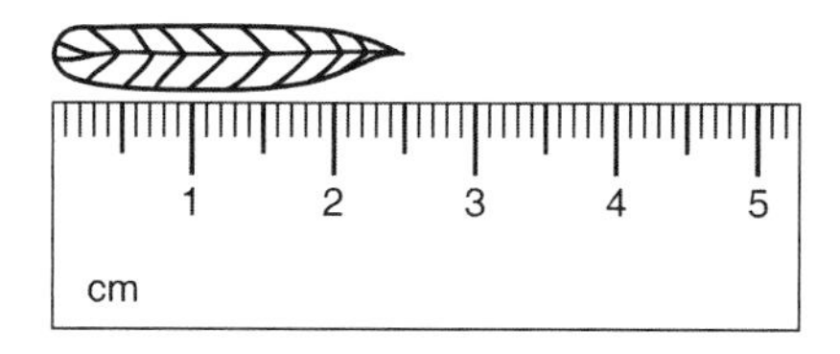

The leaf is 2.5 cm (from 0 to 2.5 cm on the ruler)
1 cm = 10 mm, therefore 2.5 cm leaf = 25 mm

Example 2: How many centimeters is the leaf in this figure?

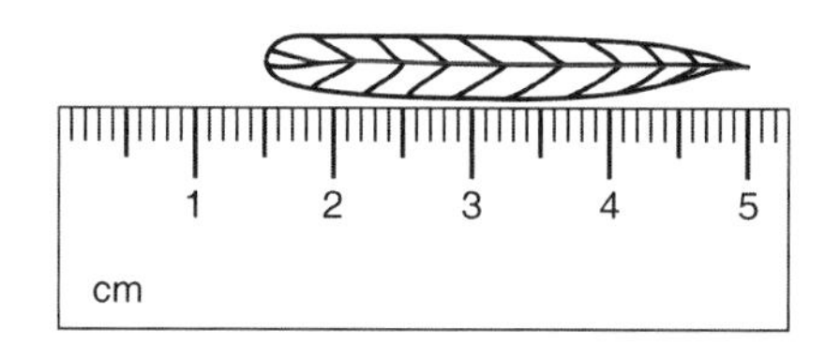

The leaf is 3.5 cm (from1.5 cm to 5.0 cm on the ruler)
1 cm = 10 mm, therefore 3.5 cm leaf = 35 mm

2. **Use** the **thermometer** to **measure temperature**. Temperature is measured in degrees Celsius (°C). Freezing point of water = 0°C. The boiling point of water = 100°C. Human body temperature =37°C.

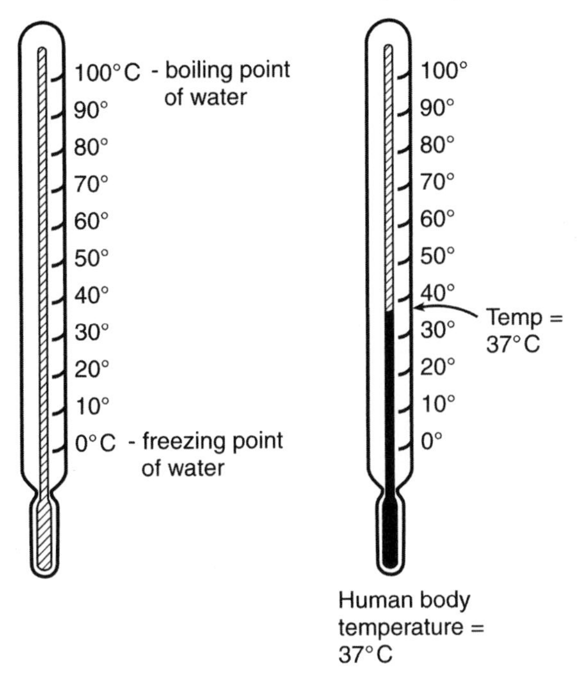

Thermometers

3. **Use** a **triple beam balance** or **electronic balance** to measure **mass**. Mass is how much material (matter) is in an object.

Triple beam balance: Look at Figure 1. The triple beam balance has a pan and three beams. One beam is divided into hundreds (100, 200, 300, 400, 500 grams).

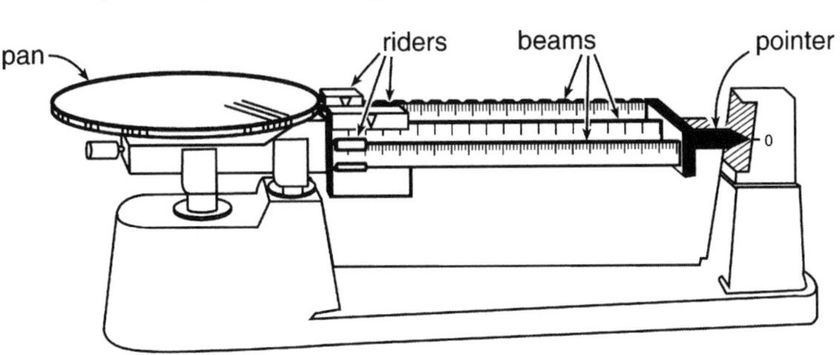

Triple-beam balance

Figure 1

Another beam is divided into tens (10, 20 30, 40, 50, 60, 70, 80, 90. 100 grams). The front beam is divided into ones (1, 2, 3, 4, 5, 6, 7, 8, 9, 10 grams). Before you put the object on the pan, the riders on each of the beams should be at zero and the pointer should be at zero.

Look at Figure 2. To find the mass of an object, put the object on the pan (the pointer now will move away from zero). Move the riders so the pointer will again be at zero. Add up the masses shown by the three riders (example 300 grams + 40 grams + 2.0 grams = 342.0 grams).

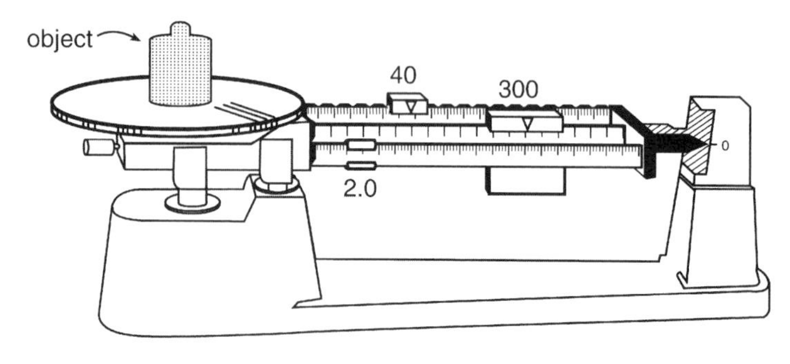

Mass of object is 342 grams

Figure 2

Electronic balance: You can also use an electronic balance to measure mass. Set the electronic balance to zero. Place an object on the pan of the balance. Read the number (example 124) shown on the electronic balance.

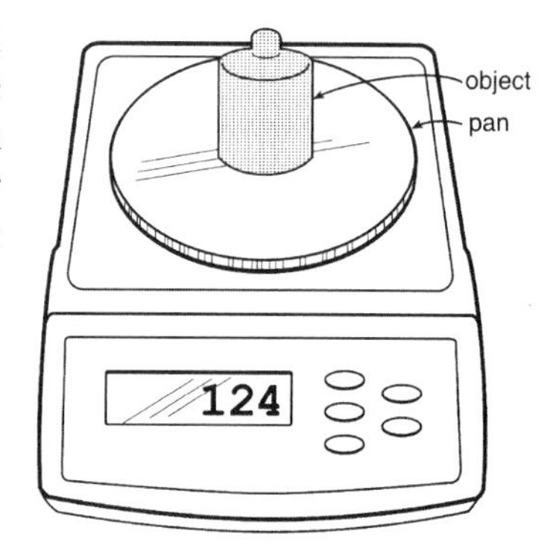

Electronic Balance

4. **Use** a **graduated cylinder** to **measure** volume (how much space it takes up). Volume is measured in mL (milliliters) and L (liters). Look at the graduated cylinder at right. This graduated cylinder has liquid (such as water) in it. The arrow points to the 5 mL line. In this graduated cylinder, there are 5 lines (4 small lines and 1 large line) up to 5 mL, therefore each line = 1 mL.

To find the volume of liquid, read the bottom of the meniscus (curve). There is 5 mL + 5 mL + 1 mL = 11mL. The volume of the liquid (example water) = 11 mL.

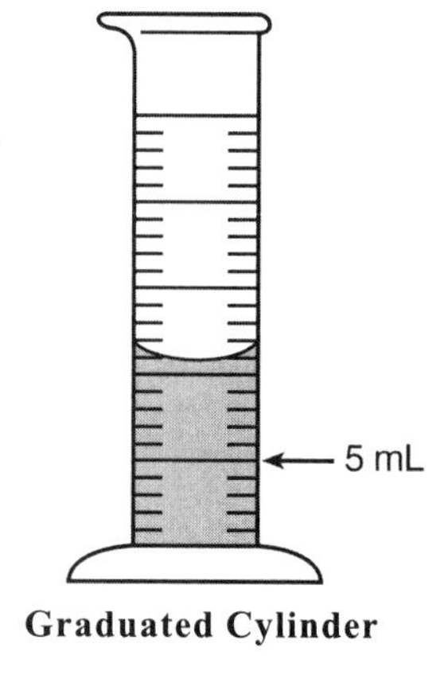

Graduated Cylinder

This graduated cylinder is different than the previous one; there are five lines from 0 to 10 mL, therefore each line = 2 mL. Read the bottom of the meniscus (curve). There is 10 mL + 10 mL + 6 mL = 26 mL. The volume of the liquid (example water) = 26 mL.

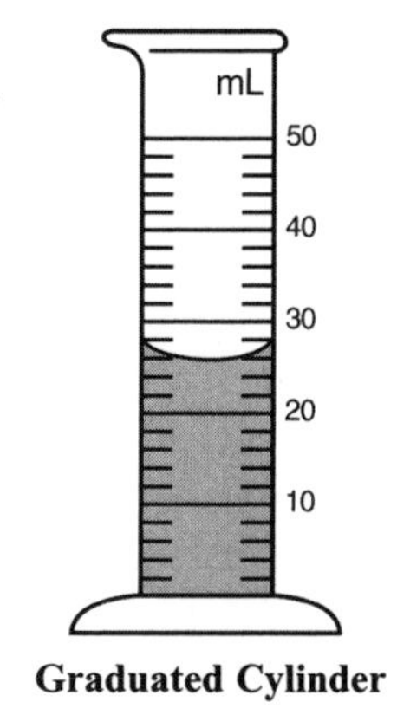

Graduated Cylinder

Question: How much water should be removed from the graduated cylinder to leave 5 milliliters of water in the cylinder?

1. 6 mL 2. 7 mL
3. 11 mL 4. 12 mL

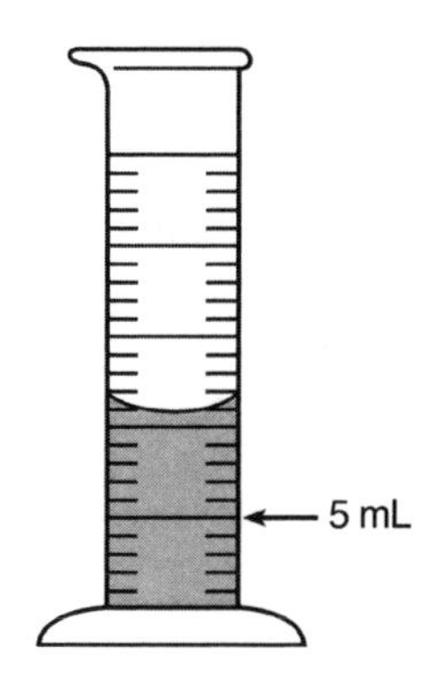

Solution: This graduated cylinder has water in it. The arrow points to the 5 mL line. In this graduated cylinder, there are 5 lines (4 small lines and 1 large line) up to 5 mL, therefore each line = 1 mL.

First find the volume of water by reading the bottom of the meniscus (curve). There is 5 mL + 5 mL + 1 mL = 11mL. The volume of the water in the graduated cylinder = 11 mL.

Then, you must **remove 6 mL** of water to have **5 mL left,**

11 mL − **6 mL** = 5 mL . Answer *1*

Now Do Homework Questions #1-6, pages 35-36

MICROSCOPE

When you look at objects (example plant cells, ameba) under a microscope, the microscope makes **objects look bigger (magnifies them)**. With a microscope, you can see objects you cannot see with your eyes alone or you can say with your naked eye (example animal cells, plant cells, or parts of a cell such as nucleus, etc). The microscope also helps you see clearly objects that are next to each other.

Two types of **microscopes** are: the **stereoscope** and the **compound light microscope**.

Stereoscope (also called dissecting microscope): Look at the stereoscope at right. The stereoscope has two ocular lenses (eyepieces), one lens for each eye, and one or more objective lenses.
(Note: The lens near the eye is called the eyepiece (ocular lens) and the lenses at the other end are called the objectives (objective lenses)).
The stereoscope only magnifies a little (gives low magnification), but gives a three-dimensional image. Light is reflected from (bounces off) the object (example grasshopper). The stereoscope can be used to examine an insect (example grasshopper) or a flower, etc.

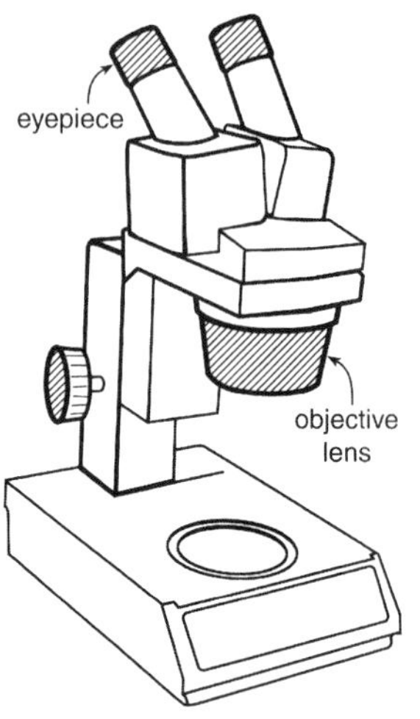

Stereoscope (Dissecting Microscope)

Compound microscope: The compound microscope generally has one ocular lens (eyepiece) and two or more objective lenses. Look at the figure below of the compound microscope.

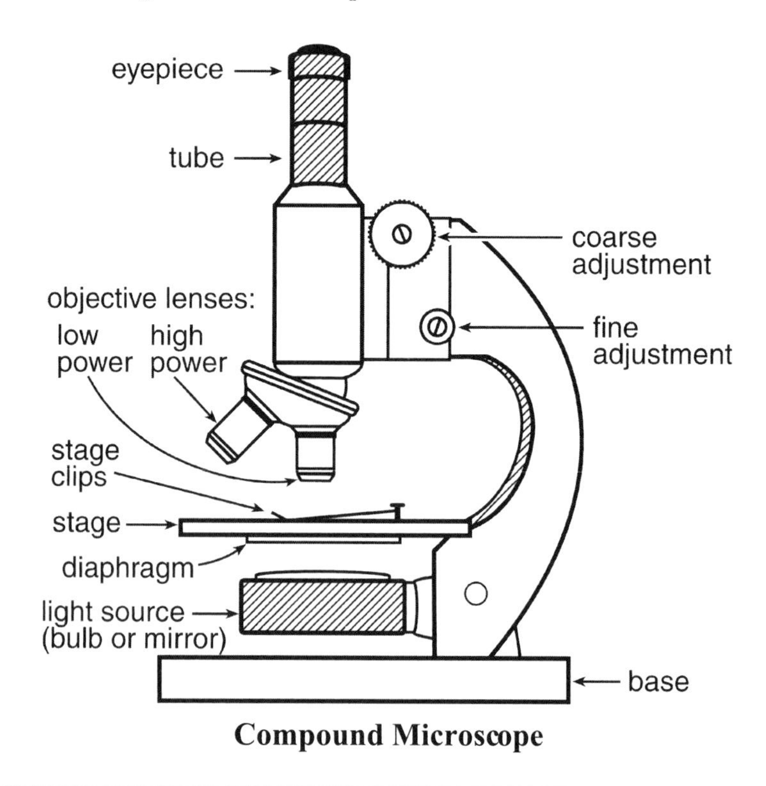

Compound Microscope

Parts of the microscope:
Look at the diagram of the compound microscope.

1. **Eyepiece** (ocular lens) is the **lens** on the **top** of the microscope that you look through.
2. **Objective lenses** are two (or three) **lenses** at the **other end** of the tube. The objective lens (low power) magnifies the object the least. The objective lens (high power) magnifies the object the most. These lenses are close to the object (specimen, example plant cell) that you are looking at.
3. **Tube connects** the **eyepiece and** the **objectives.**
4. **Stage** is a flat surface where you **put the slide** (on the slide is the specimen, example plant cell).
   Stage clips hold the slide in place.
5. **Diaphragm regulates** the **amount** of **light** entering the microscope.
6. **Light** source can be a **mirror** or a **light bulb.**
   Light goes upward from the light source (the bulb or mirror), through the specimen on the slide (the specimen must be very thin) to the objectives and then to the tube, to the eyepiece and to your eye (see diagram of compound microscope).
7. **Coarse adjustment** is the **larger knob** on the microscope. When you move the knob, it moves the tube up or down by large amounts. Under **low power, use** the **coarse adjustment to focus.** Never use the coarse adjustment under high power because you can break the slide.
8. **Fine adjustment** is the **small knob** on the microscope. The **fine adjustment sharpens** the **image** (makes what you see clearer) under **low power.** Under **high power,** use only the fine adjustment (not the coarse adjustment) **to focus.** With the fine adjustment, you can see **different layers** of a specimen (example plant cell).

Magnification: The microscope magnifies objects and makes them look bigger. Microscopes usually have two or three objective lenses (examples 4x, 10x, 40x or 64x). If the **eyepiece (lens) is 10x**, it makes the specimen (example plant cell or animal cell) look **ten times** its **actual size**; if the **objective (lens) is 40x**, it makes the specimen look **40 times** its **actual size**.
1. A microscope has a **10x eyepiece and a 40x** objective. To find **total magnification** (how much bigger the object looks under the microscope) multiply the eyepiece (10x) times the objective (40x) and you get 400x (total magnification). The microscope magnifies the object (specimen) 400 times its actual size (total magnification = 400x).

Eyepiece	X	Objective	=	Total Magnification
10x	X	40x	=	400x

2. If the **eyepiece is 10x** and the **objective is 10x**, find **total magnification:**

Eyepiece	X	Objective	=	Total Magnification
10x	X	10x	=	100x

The microscope magnifies the object 100 times.

Note: If a microscope has two objectives, 10x objective and 40x objective, the 10x is the low power (magnifies the object the least), the 40x is the high power (magnifies the object the most).

Slide: A slide is made of glass or plastic; the specimen (example onion cells, animal cells, etc.) is placed on the slide and covered with a glass or plastic square called a cover slip. A slide can be either a prepared slide made by a company or a wet mount slide, which will be explained later in the chapter.

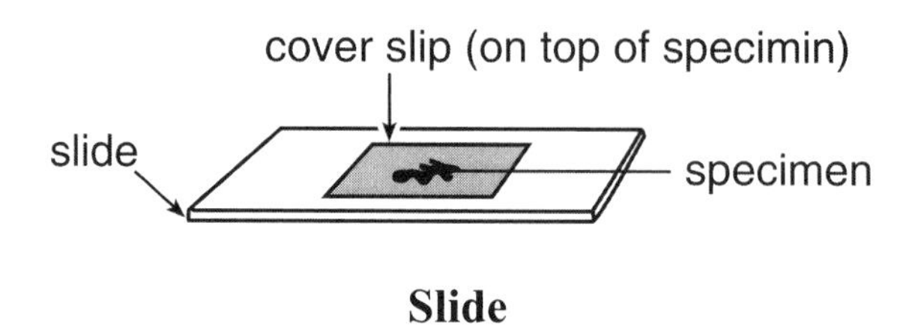

Slide

Focusing:
Look at the figure of the microscope which you had before.
1. Put your slide on the stage of the microscope. Use stage clips to keep the slide in place. First use the low power objective (example 10x) when using the microscope.
2. Use (turn) the coarse adjustment to lower the tube with the low power objective (example 10x) very close to the slide without breaking the slide (when lowering the low power objective, look at both the low power objective and the slide to make sure you do not break the slide). Then look through the eyepiece of the microscope and turn the coarse adjustment (which raises the objectives and the tube) to focus until you can see what is on the slide.
3. Use (turn) the fine adjustment to focus clearer (to see clearer).
4. To see the object (example plant cells) on the slide bigger, use (turn) the high power objective (example 40x). The high power objective is longer than the low power objective, so make sure the high power objective does not touch and break the slide. Under high power, only use the fine adjustment to focus (to see clearer); never use the coarse adjustment to focus, it might break the slide.
When you look under high power, the size of the specimen (object) is bigger, therefore you see less of the object or a smaller number of cells (the field of view (the area you see when looking under the microscope) is smaller).

Understand how to use the microscope:

1. Look at the **letter e** under the microscope. What you see when you look through the eyepiece is the image. The image is upside down and backward, therefore you see the letter e as shown at right. (upside down and backward).

2. On the slide, if the **organism** or **cells** you want to see are **moving** away **toward** the **upper right** of the field of view (area you can see under the microscope), the organism or cells could move out of the field of view. **Move** the **slide** in the opposite direction (**down** and **to** the **left**) to keep the object in the area you can see.

organism moves away
move slide
field of view
(area you can see)
organism (living thing)

3. Under **low power,** the **diameter** of the **field of view** of this microscope (as shown in the diagram) is **about 3.1 mm.** When you look at the widest part of the field of view (of what you can see), there are about three cells in a row across the width, therefore each cell is about 1 mm (3.1 mm/3 cells = about 1 mm/cell). 1 mm = 1000 micrometers, therefore you can say each cell = 1 mm or 1000 micrometers.

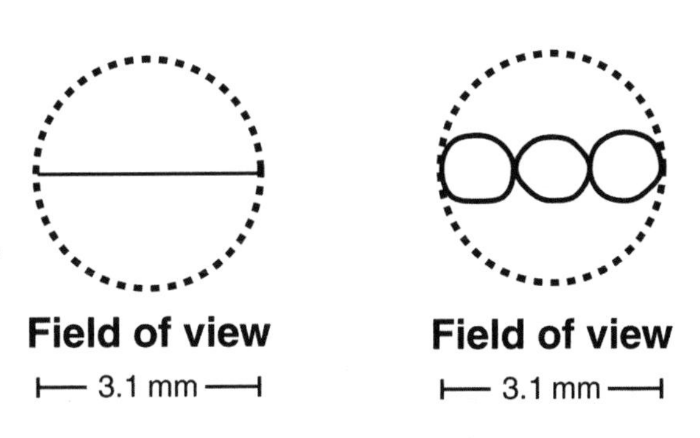

Note: Under high power, each cell looks bigger, therefore you see a smaller number of cells under the microscope.

Example: A plant cell in a microscope field of view is shown in the diagram. The diameter of the field of view is 4000µm. As you can see from the diagram, the width (w) of the plant cell is about 1/5 the field of view, which means the **plant cell** is 1/5 x 4000 µm = **800 µm.**

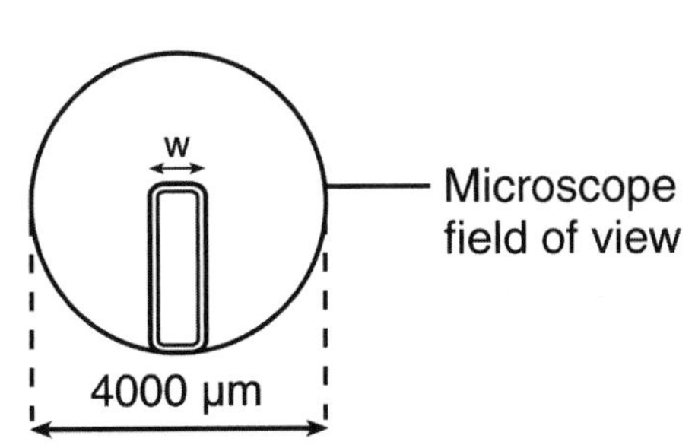

Question: The diagrams show four different one-celled organisms (shaded) in the field of view of the same microscope using different magnifications. Which illustration shows the largest one-celled organism?

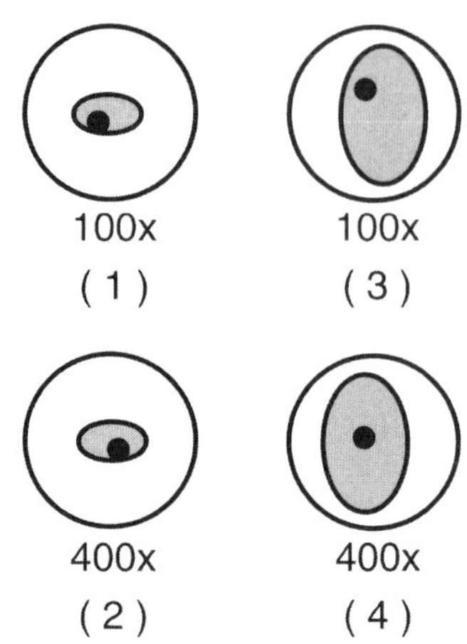

Solution: 100x means the microscope magnifies it 100 times. With the microscope, you can see the organism 100 times bigger than it actually is, which means the organism is 1/100 the size of what you see.
400x means the microscope magnifies it 400 times. With the microscope, you can see the organism 400 times bigger than it actually is, which means the organism is 1/400 the size. of what you see.

Look at the pictures. Choice 1 and choice 3 are both at 100x (organism magnified 100 times). You see the picture in choice 3 is bigger than the picture in choice 1, therefore the organism in **choice 3** is also **bigger.** Choice 2 and choice 4 are both at 400x (organism magnified 400 times). You see the picture in choice 4 is bigger, therefore the organism in **choice 4** is also **bigger**.

The pictures in choice 3 and choice 4 look about the same size. In choice 3, the organism is 1/100 the size of what you see; in choice 4, the organism is 1/400 the size of what you see, therefore, in choice 3 the organism is bigger. Answer *3*

Identifying and comparing cells and parts of cells

Using a **compound light microscope,** you can **see cells** (example plant and animal cells) and some **parts of a cell** (example nucleus, cell membrane). See plant and animal cells below.

Nucleus- dark stained round structure in the cell.

Cytoplasm- living material that fills up the space inside the cell membrane. With a compound microscope, you can see some organelles (parts of cells example nucleus) inside the cytoplasm.

Cell membrane- surrounds the cytoplasm. A cell membrane surrounds an animal cell. In a plant cell, the cell membrane is between the cytoplasm and the cell wall.

Cell wall- outside the cell membrane in plant cells; supports the cell.

Chloroplasts- green oval structures in a plant cell (mostly in leaves).

Vacuoles- clear spaces in the cytoplasm. Plant cells have large vacuoles; animal cells have smaller vacuoles.

Chromosomes- thread-like structures seen in cells that are dividing (in mitosis or meiosis).

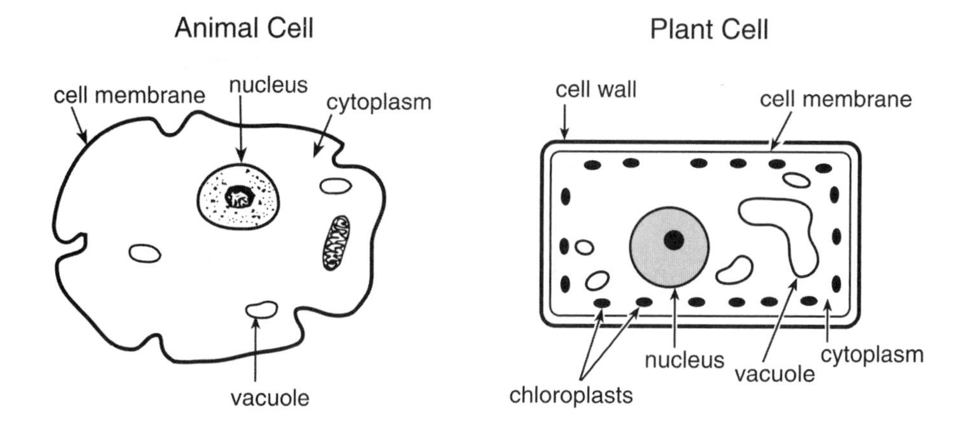

Making wet-mount slides: When you look under a microscope, the specimen (example plant or animal cells) must be thin. The problem is that if the specimen is thin, it can easily dry out. Therefore, we make a **wet-mount slide**:

1. Put a small drop of water in the center of the glass slide (you can use a dropper or a pipette).

2. Put the object (specimen, example plant cell) that you want to see in the drop of water.

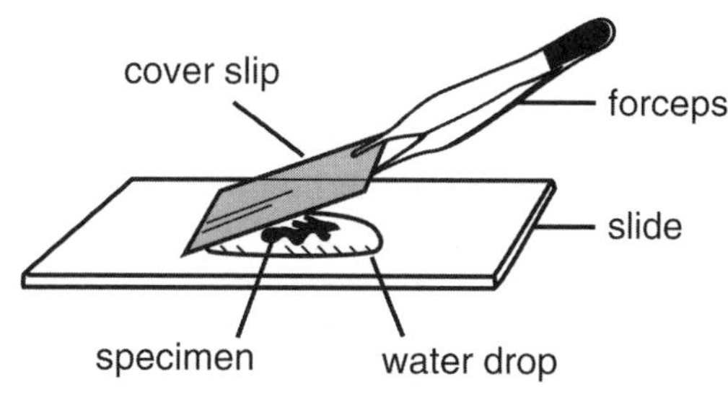

Wet Mount Slide

3. Take a cover slip. You can use forceps to position (see where you should put) the cover slip and lower the cover slip over the object on the slide slowly (to avoid air bubbles).

Staining techniques: Stains are used to make it easier to see different parts of the cell. Certain parts of the cell (example nucleus) turn darker when a stain is added.

Staining a wet mount slide: Add the stain (examples, iodine, methylene blue) to the wet mount slide. Put one drop of stain on the edge of the cover slip. Put a small piece of paper towel on the other edge of the cover slip. The paper towel absorbs the water from under the cover slip, which causes the stain to go under the cover slip. You can even stain the wet mount slide while it is on the microscope.

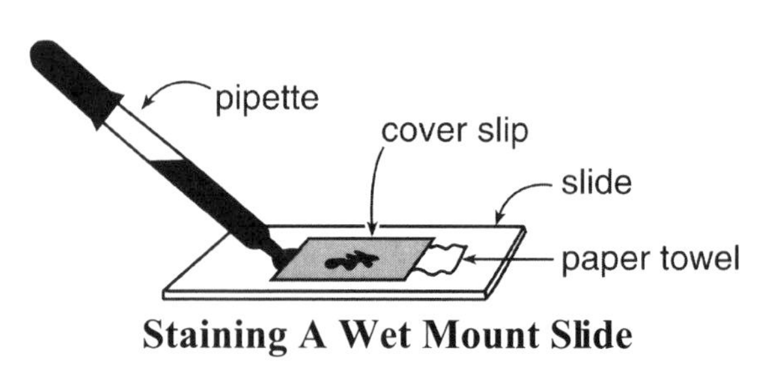

Staining A Wet Mount Slide

Now Do Homework Questions #7-14, pages 36-38

Indicators

Indicators are used to tell if a substance (example starch or sugar) is present (is there).

Lugol's solution or iodine (starch indicator solution), which has an amber color, turns blue-black if starch is present (is there). If Lugol's solution (iodine, starch indicator solution) turns blue black, we know starch is present.

Benedict's solution turns from blue to green, to yellow, to orange, then to brick red if a simple sugar is present.

Bromthymol blue is blue in a base and yellow in acid. When bromthymol blue turns yellow, we know it is an acid. In respiration, carbon dioxide is given off. Carbon dioxide in water forms a weak acid, which turns bromthymol blue yellow.

pH paper tells whether a liquid is acid, base, or neutral. Dip the pH paper into the liquid; see what color the pH paper turns. Then match the color of this paper with the same color on a scale that shows the color of pH paper at different pH's.

Diffusion

If there is more concentration of a dissolved substance (example, dissolved sugar) outside the cell and a lower concentration (less) of the dissolved substance (dissolved sugar) inside the cell, some dissolved sugar

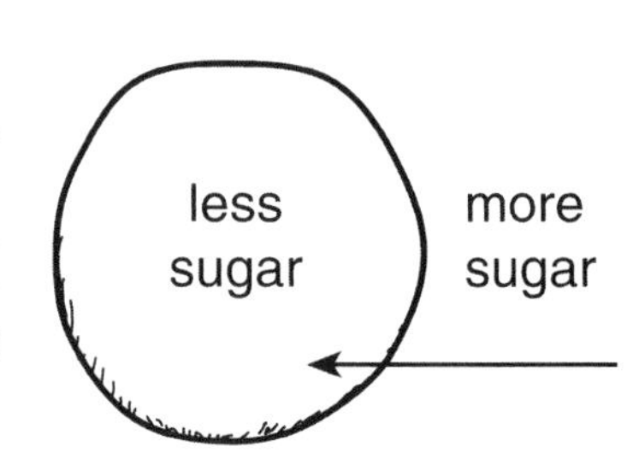

from outside will go into the cell so that there is the same concentration (amount) of sugar inside and outside the cell.

If there is more concentration of a dissolved substance (example dissolved sugar) inside the cell and a lower concentration (less) of a dissolved substance (dissolved sugar) outside the cell, some of the dissolved sugar will go across the membrane out of the cell so that there is the same concentration (amount) of dissolved sugar inside and outside the cell.

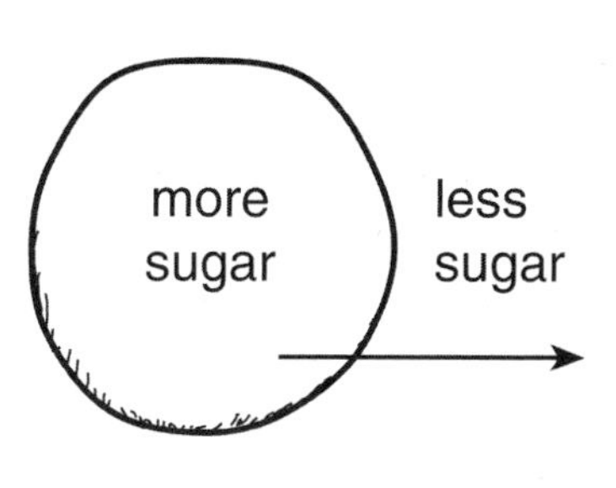

Diffusion: molecules go from an **area of higher concentration** (more concentration) across a membrane **to** an **area of lower concentration.**

Diffusion of water is called **osmosis.** When **water goes from higher concentration** of **water to a lower concentration** of **water** it is called **osmosis.**

Examples of osmosis:

Example 1: cell has 94% water, surrounding area has 96% water. Water goes from higher concentration of water across a membrane to lower concentration of water, therefore, water will enter the cell until the concentration of water inside and outside the cell is the same.

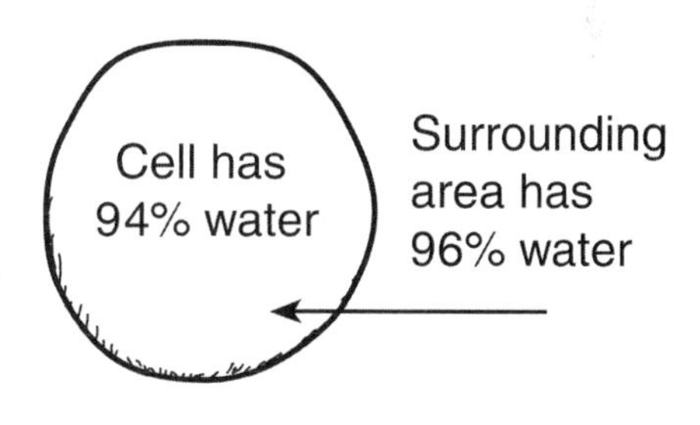

Example 2: cell has 97% water, surrounding area has 92% water. Water goes from higher concentration of water (97% water inside the cell) to outside the cell (92% water) until the concentration of water inside and outside the cell is the same.

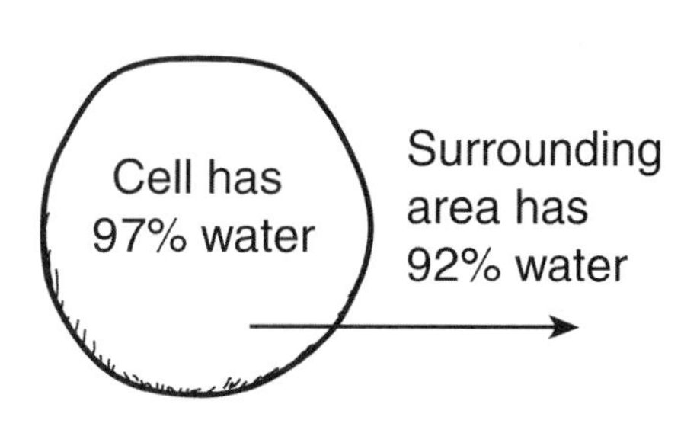

Note: A **3% salt solution** (salt in water) means it has 3% salt and **97% water**. Percent of salt (3%) and percent of water (97%) must equal 100%. An **8% sugar solution** (sugar in water) means it has 8% sugar and **92%** water. Percent of sugar (8%) and percent of water (92%) must equal 100%.

Diffusion and osmosis (diffusion of water) are called passive transport, because no energy is required for diffusion.

Example 1: Red onion cell in salt water (see Figure 1 below).
Look at Figure 2A below. **Rinse** the **onion cell with** distilled water (**pure water**, 100% water, which has no salt), which means there is a higher concentration of water outside the cell than inside the cell. You learned water goes from higher concentration of water (which is now outside the cell) to lower concentration of water (inside the cell), therefore water will enter the cell and cause the cell with its cell membrane to swell, get bigger, until the concentration of water outside and inside the cell is the same (see Figure 2B).

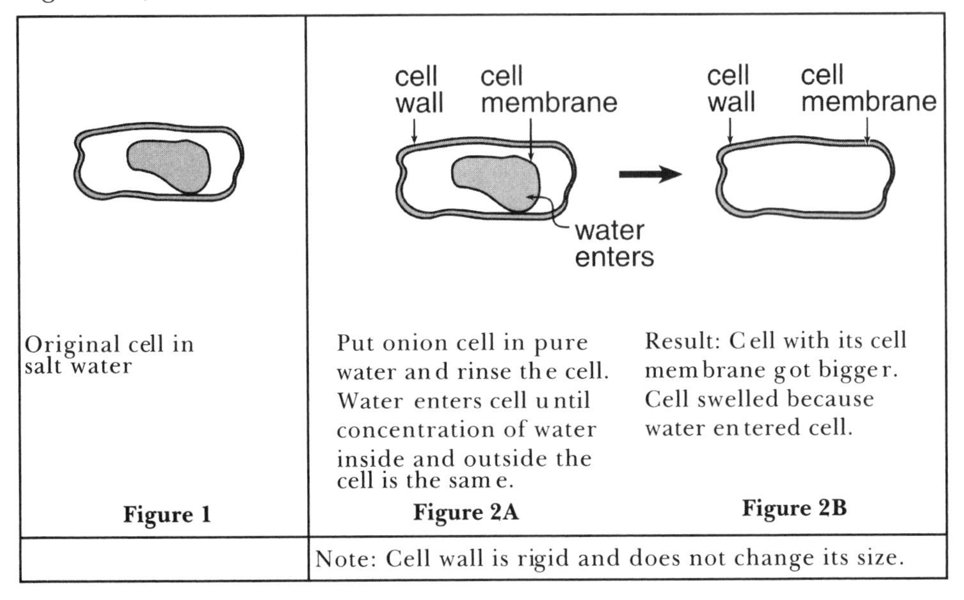

Original cell in salt water

Figure 1

Put onion cell in pure water and rinse the cell. Water enters cell until concentration of water inside and outside the cell is the same.

Figure 2A

Result: Cell with its cell membrane got bigger. Cell swelled because water entered cell.

Figure 2B

Note: Cell wall is rigid and does not change its size.

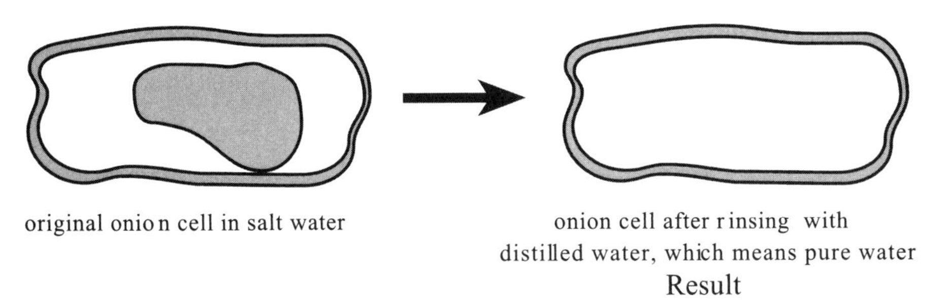

original onion cell in salt water

onion cell after rinsing with distilled water, which means pure water

Result

In short:
If a cell was left in pure water, the water would keep entering the cell and the cell could burst.

Example 2: Cell A shown at right is a typical red onion cell in water on a slide viewed with a compound light microscope.

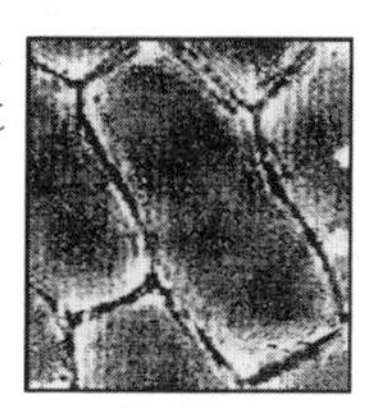

Cell A

Draw a diagram of how Cell A would most likely look after salt water has been added to the slide and label the cell membrane in your diagram.

Solution: Cell A is in water (see Figure 1 below).
Look at Figure 2A, which is the same picture as Figure 1, but is labeled. Now, put salt water on the slide (salt water is surrounding the cell). Since **salt** water has **some salt** in the **water**, obviously it has a lower concentration of water than pure water.
You learned water goes from higher concentration of water (inside the cell) to lower concentration of water (outside the cell), therefore water goes out of the cell and the **cell with its cell membrane shrinks (gets smaller)**. See figure 2B.

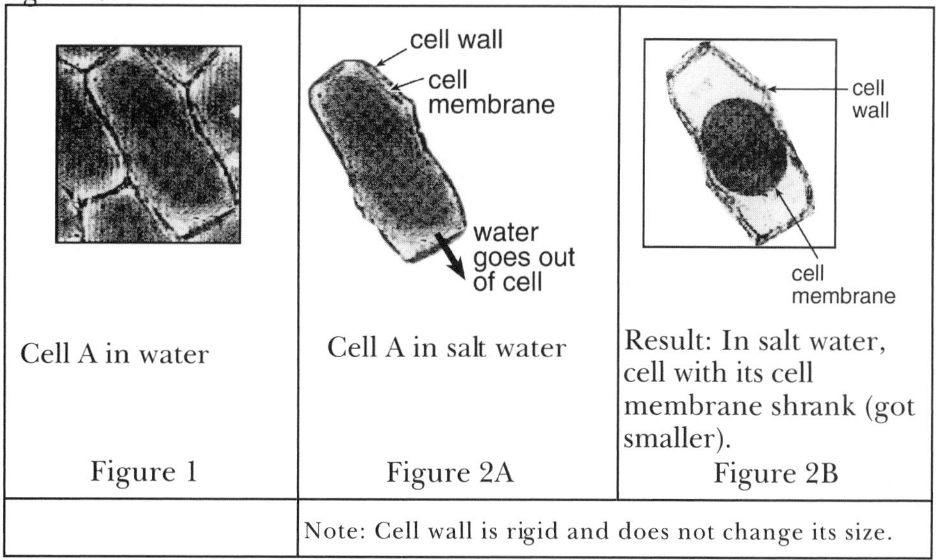

Cell A in water — Figure 1

Cell A in salt water — Figure 2A

Result: In salt water, cell with its cell membrane shrank (got smaller). — Figure 2B

Note: Cell wall is rigid and does not change its size.

The question asks to draw a diagram of how the cell would look in salt water. Draw the result, figure 2B.

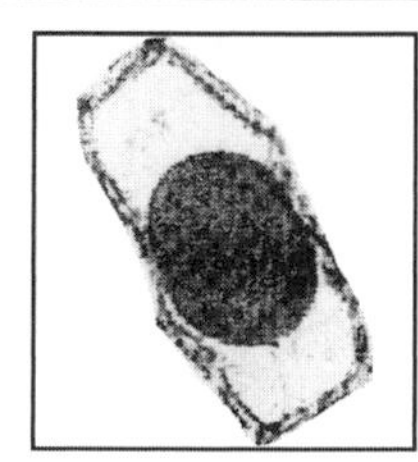

Similarly, just like in example 2 above, you see again that when a plant cell is put into salt water, the cell with its cell membrane shrinks (gets smaller).

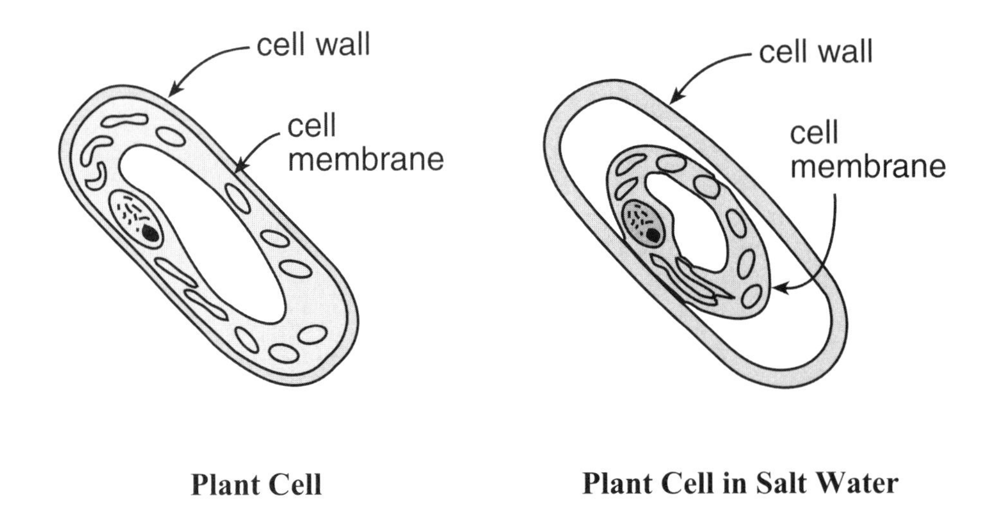

Plant Cell **Plant Cell in Salt Water**

Question: A wet mount of red onion cells as seen with a compound light microscope is shown below.

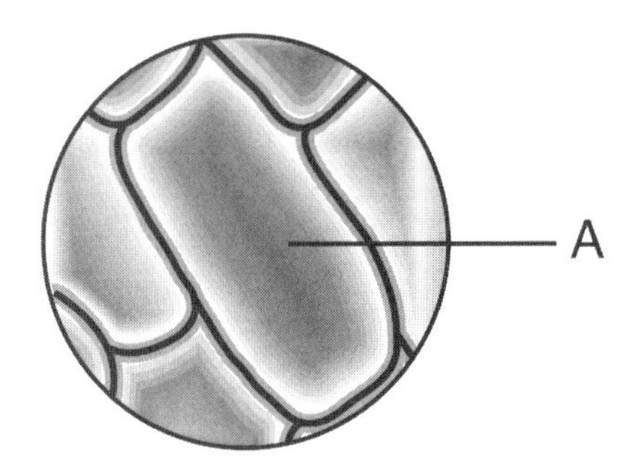

What technique would be used to add salt to a slide of onion cells?

Solution: Adding salt water to the slide is done just like adding a stain to a wet mount slide (see: Staining a wet mount slide in this chapter). Add the salt water to the wet mount slide. Put one drop of salt water on the edge of the cover slip. Put a small piece of paper towel on the other edge of the cover slip. The paper towel absorbs

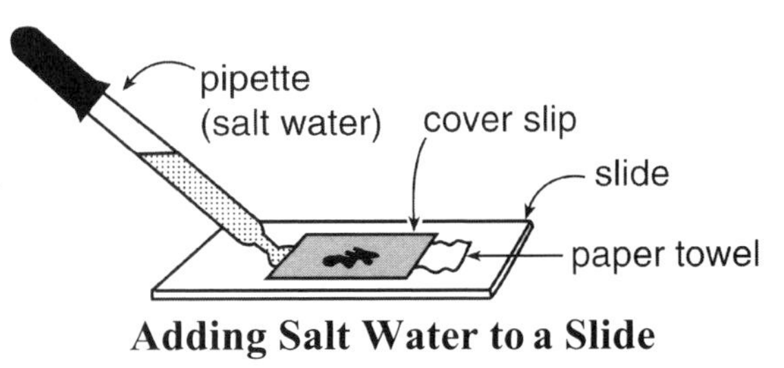

Adding Salt Water to a Slide

the water from under the cover slip, which causes the salt water to go under the cover slip.

Diffusion through a dialysis membrane in an artificial (model) cell

We can show diffusion or osmosis by **making an artificial cell (model cell)** using a bag made of dialysis tubing and placing it in a beaker of water. The bag represents a cell; the dialysis tubing (dialysis membrane) represents the cell membrane.

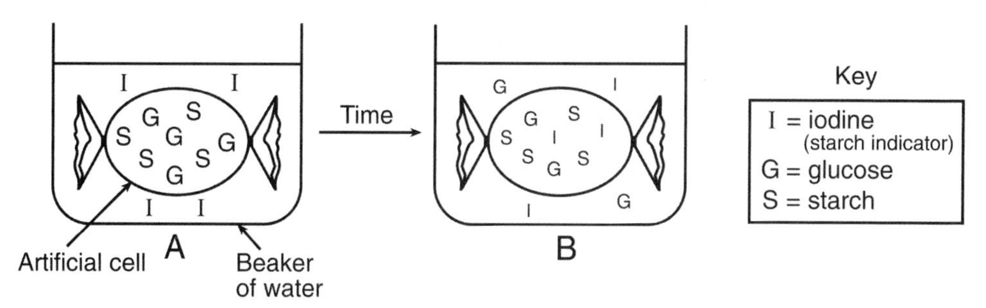

Diffusion Through a Membrane In an Artificial Cell
(The beaker and the artificial cell contain water)

Inside the bag (artificial cell), there is glucose, starch, and water. Outside the artificial cell is only water. In the figure above, glucose diffuses out of the artificial cell into the water in the beaker, going from the area of higher concentration (of glucose) to an area of lower concentration (of glucose). Glucose is a small molecule and will diffuse out of the bag (artificial cell) into the water in the beaker. Starch is a large molecule and will not diffuse out of the bag (artificial cell) into the water.

Iodine (starch indicator, amber colored) turns blue-black if starch is present. The iodine in the beaker of water did not change color because the **starch** (in the artificial cell) is a large molecule and did **not** diffuse (go) into the beaker of **water.**

Also, the iodine is a small molecule and will diffuse from outside the bag to inside the bag. The iodine inside the bag turns blue-black because there is starch inside the bag.

Glucose indicator solution (blue color), when heated, turns green, yellow, orange, or red if glucose is present. We took water from the beaker (around the artificial cell) and heated it in a test tube with glucose indicator solution. The glucose indicator solution turned green, yellow, orange, red, showing the presence of glucose. Glucose is a small molecule and diffuses out of the bag (artificial cell) into the water in the beaker.

Questions 1 and 2: The diagram shows the changes that occurred in a beaker after 30 minutes. The beaker contained water, food coloring, and a bag made from dialysis tubing membrane.

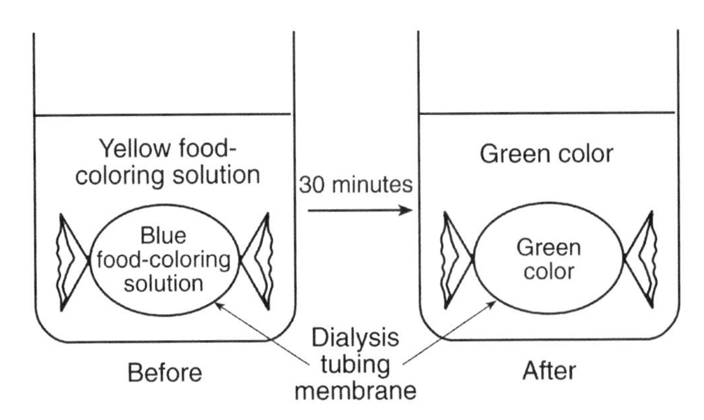

Question 1: When the colors yellow and blue are combined, they produce a green color. Which statement most likely describes the relative sizes of the yellow and blue food-coloring molecules in the diagram?

(1) the yellow food-coloring molecules are small, while the blue food-coloring molecules are large.

(2) the yellow food-coloring molecules are large, while the blue food-coloring molecules are small.

(3) both the yellow food-coloring molecules and the blue food-coloring molecules are large.

(4) both the yellow food-coloring molecules and the blue food-coloring molecules are small.

Question 2: Which statement best explains the changes shown?

(1) molecular movement was aided by the presence of specific carbohydrate molecules on the surface of the membrane.

(2) molecular movement was aided by the presence of specific enzyme molecules on the surface of the membrane.

(3) molecules moved across the membrane without additional energy being supplied.

(4) molecules moved across the membrane only when additional energy was supplied.

Solution 1: The water (both inside and outside the bag (made from dialysis tubing **membrane**)) became green (see diagram above at right). Blue color and yellow color mixed to form a green color.

Blue food color diffused from inside the bag to outside the bag and the **yellow food color diffused** from outside the bag to inside the bag **causing blue and yellow colors** to **mix** together and **form** a **green**

color (inside and outside the bag.). Therefore, the blue and yellow colors must be small molecules to diffuse (go) through a membrane (bag made from membrane) Large molecules cannot diffuse through a membrane. Answer *4*

Solution 2: Blue food color diffused across the membrane from inside the dialysis bag (high concentration of blue food color) to outside the bag (low concentration of blue color). Yellow food color diffused across the membrane from outside the dialysis bag (high concentration of yellow food color) to inside the bag (low concentration of yellow color). Diffusion is when molecules go from an area of high concentration to an area of low concentration across a membrane and does not need energy, Answer *3*

Darwin's Finches

Background information: A few finches (a type of bird) originally left the South American mainland and arrived at the isolated Galapagos Islands (the islands are 600 miles from the mainland and the islands are separated by water). There were different types of food for the finches on the different islands; some foods were plants, other food were animals. Charles Darwin saw that the finches on all the islands were very similar but the finches have different types of bills and eat different types of food.

Finch	Type of Bill	Type of Food
sharp-billed ground finch	crushing	plants
warbler finch	probing	animals

The finches (example large ground finch, woodpecker finch, and vegetarian finch) had evolved (evolution) into different species (types) of finch which have different types of bills and eat different types of food; these finches cannot interbreed with each other. On each island, those finches with adaptations that matched the food on that island survived and had offspring (babies) (natural selection); those finches not adapted to the food on that island died out.

The diagram below shows different species of finches (examples large ground finch, vegetarian finch, large tree finch), their bills, and the food they eat. The title of the diagram is Finch Diversity, which means different varieties (species) of finches.

Finch Diversity

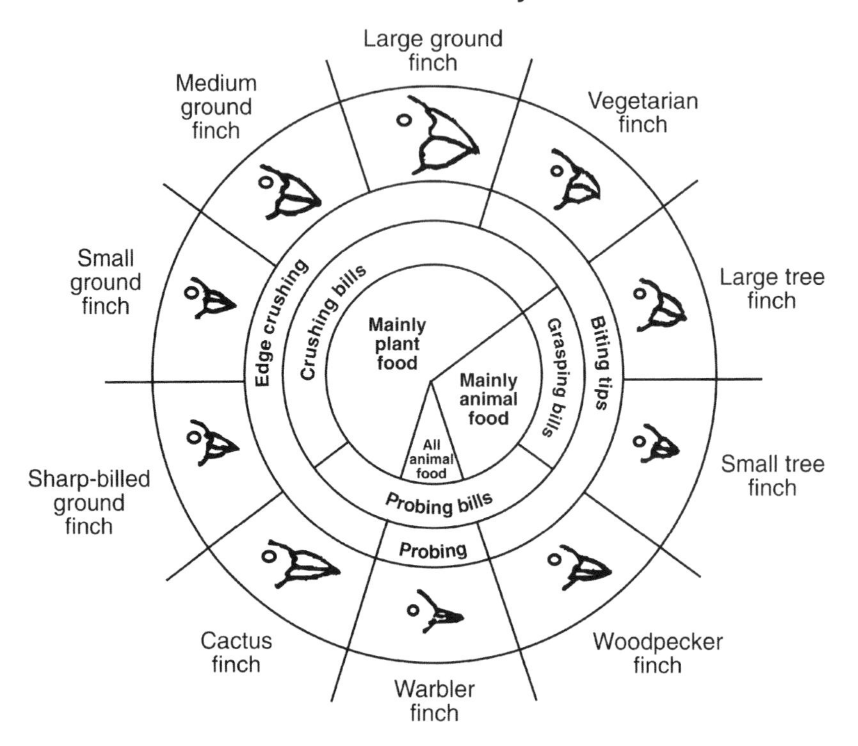

The outer circle shows drawings of the bills of each species of finch. The next two circles, inside the drawings of the bills, describe the type of bill each species has. The innermost circle shows the type of food each species eats. For example, the large tree finch and the small tree finch have grasping bills and biting tips and eat mainly animal food. Both species compete for the same food with the same type of bill.

Similarly, the sharp-billed ground finch, the small ground finch, the medium ground finch, and the large ground finch all have edge crushing and crushing bills and mainly eat plant food. These species compete for the same food with the same type of bill.

Question: In the *Beaks of Finches* laboratory activity, students were each assigned a tool to use to pick up seeds. In round one, students acting as birds used their assigned tools to pick up small seeds from their own large dishes (the environment) and place them in smaller dishes (their stomachs). The seeds collected by each student were counted. Some students were able to collect many seeds, while others collected just a few.

In round two, students again used their assigned tools to collect seeds. This time several students were picking up seeds from the same dish of seeds.

Question 1: Explain how this laboratory activity illustrates the process of natural selection.

Question 2: One factor that influences the evolution of a species that was *not* part of this laboratory activity is

(1) struggle for survival (3) competition
(2) variation (4) overproduction

Question 3: Identify *one* trait, other than beak characteristics, that could contribute to the ability of a finch to feed successfully.

Solution 1: The tools the students used represented (modeled) the beaks of the finches. The students with the better tools (meaning birds with better beaks) are able to get more food and survive better. This is natural selection.

Solution 2: The lab discussed variations in beaks, competition for seeds, and struggle for survival. The lab did not discuss overproduction. Overproduction means too many offspring are produced and therefore there is not enough food to support all of them.

Solution 3: The stronger birds can compete better for food. The birds with better vision (seeing better) can find the food more easily. Faster birds can get to the food more quickly.

Now Do Homework Questions #15-23, pages 38-41

Electrophoresis

Gel electrophoresis can analyze DNA, RNA, and protein (example from a person, dog, cat, etc.). The big **DNA** molecule (example from a person) is broken (cut) into **smaller pieces** by restriction **enzymes**.

The smaller pieces of DNA are put in a well at one end of the gel (see diagram). An electric current is put through the gel, causing **one end** to be **negative (−)** and the **other end** to be **positive (+)**. Because **DNA** is **negatively charged**, it **moves toward** the **positive** part of the **gel.** The **broken pieces** of **DNA separate** from each other **by size**, forming bands of DNA on the gel.

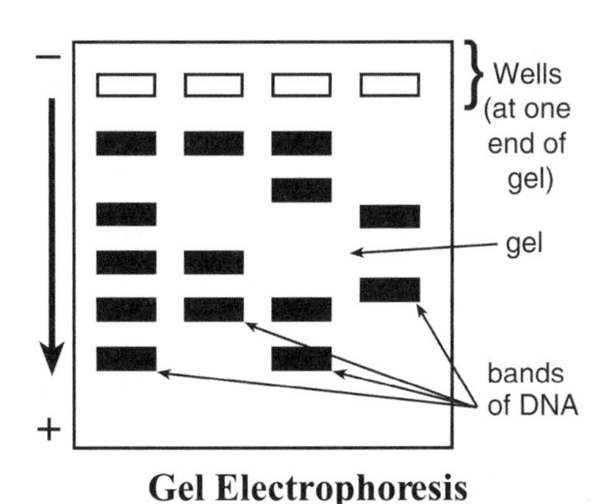

Gel Electrophoresis

Bigger pieces (fragments) of DNA move slower, therefore they go a smaller distance away from the well. Smaller pieces (fragments) of DNA move faster, therefore they go a bigger distance, further away from the well. Medium sized pieces of DNA are between the bigger and smaller pieces of DNA. As you can see, the technique of **gel electrophoresis separates DNA pieces** (fragments) by **size.** Different people have different band patterns in their DNA.

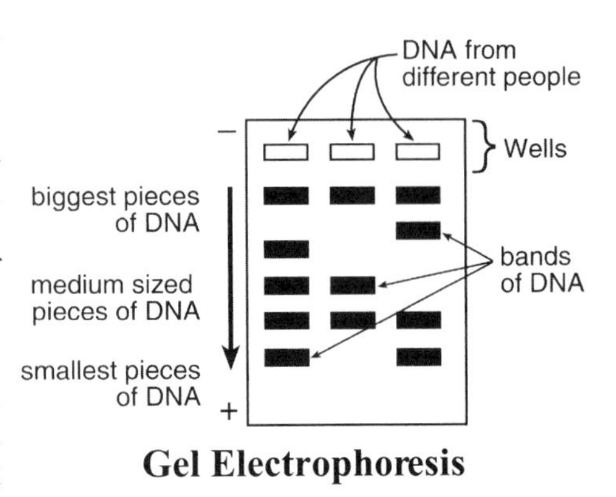

Gel Electrophoresis

Uses of Electrophoresis

1. Evolutionary relationships (to find out which organisms are closely related). Take DNA from an **animal.** Add an enzyme that will selectively break up the DNA into chunks (fragments) of DNA. Separate the DNA pieces (fragments) into larger and smaller pieces. The larger fragments of DNA are near the top; the smaller pieces are near the bottom.

Compare the DNA from a lion with the DNA from a cat and see how the DNA of a cat compares with a lion's DNA. If the lines (**bands**) in the **gel** of the **cat's DNA** are **in similar places** to the lines (**bands)** of the **lion's DNA**, you know **both animals have similar DNA's** and are **closely related.**

Using gel electrophoresis, you can see which organisms have more similar DNA (more similar chemical makeup) and therefore are more closely related. Just observing (looking at) organisms (example structure such as type of paws, or type of teeth, or vein pattern in leaves) is less reliable, not as good.

Look at the electrophoresis diagram. The lines (bands) in the gel of animal A's DNA are in similar places to the lines (bands) of the lion's DNA, therefore you know animal A and the lion have similar DNA's and are closely related. Animal B's bands (lines) are not similar to the lion's bands (lines), therefore you know they are not closely related.

Lion A B

Gel Electrophoresis

2. Solving crimes. A person was attacked. There may be blood or skin under the victim's nails, or at the crime scene. Or, there can be semen from the attacker.
Take the DNA from the blood, skin, or semen. Add an enzyme that will break up (cut) the DNA (at specific places in the sequence of bases) into fragments (pieces) of DNA. **Compare** the **bands from** the **DNA** of the blood or semen found on the victim (or at the crime scene) with the **DNA from** the suspected **criminal's blood** or skin sample. If the bands (lines) on the electrophoresis gel are the same, he is probably the criminal.

3. Paternity suits. **Take DNA** from the **baby** and **DNA from** possible **parents.** Using the same process (electrophoresis), **compare** the **DNA bands** (lines) of the **child** and the **possible parents**. If the DNA of the father matches (is similar to) the DNA of the child, he is the father.

4. Gene testing for diagnosis. The pattern of **DNA bands** can be used to **recognize** that a **person has** a **gene for** certain **hereditary diseases**.
In short, electrophoresis is used to separate pieces of DNA (see 1-4 above).

5. **Separating proteins**. Electrophoresis is also used to separate a mixture of proteins. The pattern of bands may show if a person has a hereditary disease.

Question: In preparation for an electrophoresis procedure, enzymes are added to DNA in order to

1. convert the DNA into gel
2. cut the DNA into fragments
3. change the color of the DNA
4. produce longer segments of DNA

Solution: Restriction enzymes cut the DNA into fragments (pieces).
Answer 2

Question: The chart below represents the results of gel electrophoresis of the DNA from the unknown plant species and the four known species.

Results of Gel Electrophoresis of DNA from Five Plant Species

Unknown Species	Species A	Species B	Species C	Species D
—			—	—
		—		
—			—	—
	—			
—			—	
		—		
		—		
—				
—	—		—	—
	—			

The unknown species is most closely related to which of the four known species? Support your answer.

Solution: The unknown species is most closely related to species C because the bands of DNA (pieces of DNA) from the unknown species are most similar to the bands of species C.

Now Do Homework Questions #24-46, pages 41-45

Chromatography

Chromatography is a way to separate different molecules (example different amino acids) in a mixture. Example: Put a drop from a mixture of amino acids near one end of the chromatography paper or filter paper. Put the end of the paper in a liquid (solvent) like alcohol. The mixture of amino acids must be above the liquid. (Do not put the amino acid mixture into the liquid because the amino acids would go off the paper into the liquid). The solvent alcohol moves up the paper, which has the amino acids on it. Different molecules (example different amino acids) in the mixture move different distances up the paper, separating the molecules (example amino acids) in the mixture.

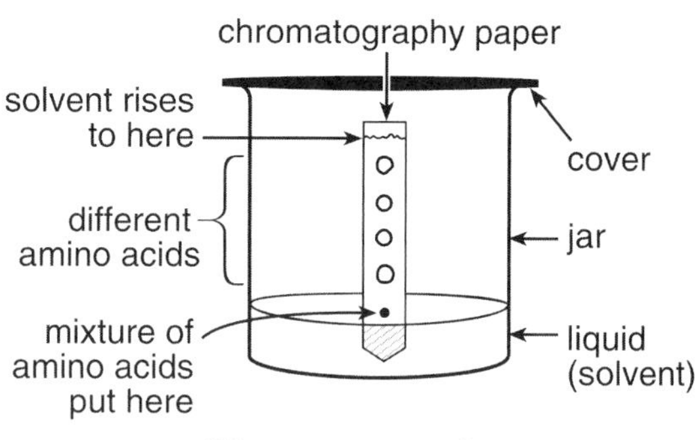

Chromatography

As you know, chromatography is used to separate chlorophyll molecules from other molecules in plant leaves. (The molecules in plant leaves are colored, therefore they are easy to see on chromatography paper).

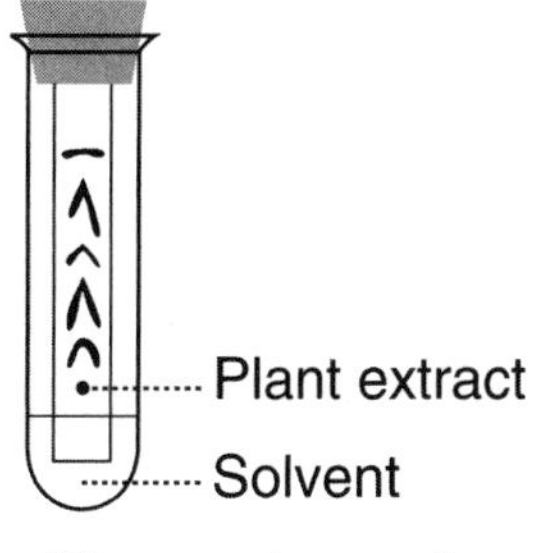

Chromatography

Chromatography is used to investigate evolutionary relationships. Leaves from a plant were ground and mixed with a solvent (liquid). The mixture of ground leaves and solvent was then filtered (poured through a filter paper, similar to a coffee filter); the liquid that goes through the filter paper is called the filtrate. Using a toothpick, some of the filtrate (in this example, liquid from the leaves) was placed at a spot near one end of a strip of chromatography paper (see figure).

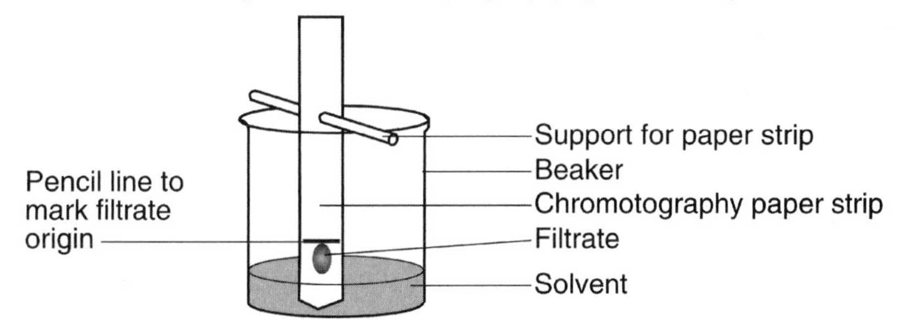

Liquid From Ground Leaves (Filtrate) Placed on Chromatography Paper

The exact procedure was repeated again for two more species of plants. Using a clean new toothpick, a little liquid (filtrate) from the second species was put on another chromatography paper. Using a clean new toothpick, a little liquid (filtrate) from the third species was put on a new chromatography paper. The three chromatography papers were put in three separate beakers; each beaker had the same solvent (this can be the same or a different solvent than the one that was added to the ground leaves).

Look at the figure below. The solvent (example alcohol) moves up the papers, separating the different molecules in each filtrate. Each chromatography paper will have a number of spots; each spot is a different molecule. Compare the spots on the three chromatography papers for the three species. The more similar the patterns of spots, the more closely related the species are.

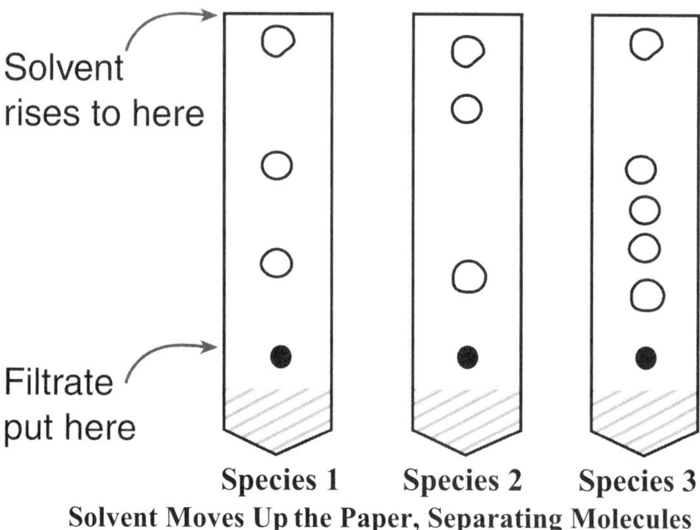

Solvent Moves Up the Paper, Separating Molecules

The spots on the chromatography papers of species one and two are similar which shows that species one and two are closely related. The spots on the chromatography papers one and three are different which shows that species one and three are not closely related.

Question: State one reason for using a new toothpick for the filtrate from each plant in the experiment above.

Solution: Using a new toothpick makes sure the filtrates (liquids) from the three species are kept separate, not mixed together.

Question: State one way the strips of chromatography paper would most likely be different from each other after being removed from the beakers.

Solution: Each strip would have a different pattern of spots.

Question: State how a comparison of these strips could show evolutionary relationships.

Solution: The spots on the chromatography papers of species one and two are similar which shows that species one and two are closely related. The spots on the chromatography papers one and three are different which shows that species one and three are not closely related.

Pulse Rate

The heart beats (contracts and relaxes) to pump blood. Every time the heart contracts, blood is forced into the arteries, making the arteries bulge (get a little wider). When an artery is close under the skin, we feel these bulges as a pulse.

Your **heart rate** (example 70 beats per minute) is **equal** to the **pulse** (example 70 per minute.) During exercise, heart rate and pulse increase.

When sitting your pulse might be 70 beats per minute, when walking, your pulse might increase to 80 beats per minute and when running your pulse might even increase more to 90 beats per minute.
Note: During exercise, the heart beats faster, therefore the blood moves faster and brings more oxygen and glucose to the cells of the body (example to muscle cells) and removes more wastes.

Experiment

In an investigation, 14 female (girl) students in a class determined their pulse rates after performing each of three different activities. The bar graph below shows the average pulse rate for females after sitting, walking, and running.
Look at the bar graph. The vertical axis shows pulse rate. The bar for sitting is shorter than the bar for walking, which means the pulse rate while sitting is less than the pulse rate when walking. The bar for walking is shorter than the bar for running, which means the pulse rate when walking is less than the pulse rate when running.

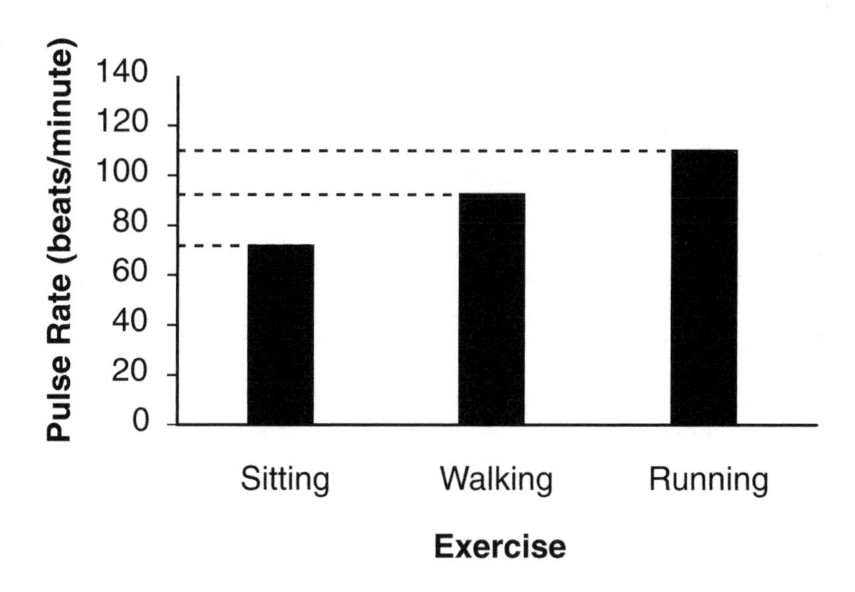

What is the pulse rate for females sitting? Look at the bar graph. There are 20 beats per minute between each two numbers on the vertical line. Look at the top of the bar for sitting and go across to the vertical line; the dotted line is more than halfway between 60 beats and 80 beats per minute, or about 72 beats per minute.
What is the pulse rate for females walking? Look at the bar graph. Look at the top of the bar for walking and go across to the vertical line; the dotted line is more than halfway between 80 beats and 100 beats per minute. The pulse rate for females walking is about 92 beats per minute. The pulse rate

for females running is about 110 beats/minute, halfway between 100 and 120.

Lab Activity on Pulse Rates

Students in a class recorded their resting pulse rates and their pulse rates immediately after strenuous activity. The data obtained are shown in the resting pulse rate histogram (graph) and in the pulse rate after activity histogram (graph).

Description of the resting pulse histogram (graph): Look at the resting pulse histogram (graph) below.

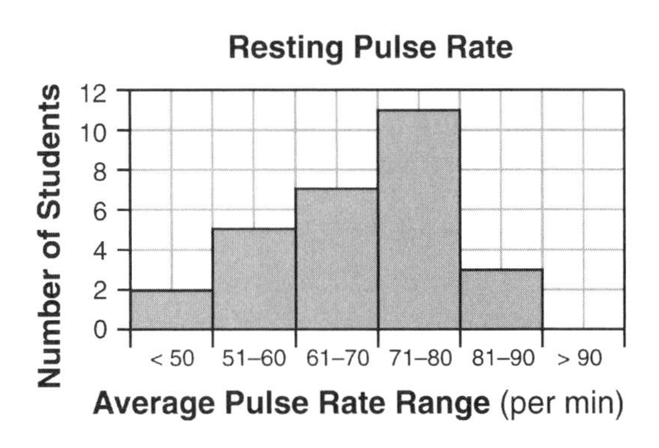

On the horizontal axis is average pulse rate range. The **pulse rate** from 0-90 is divided into groups, pulse rate less than (<) 50, 51-60, 61-70, 71-80, 81-90, and above (>) 90.

The vertical axis shows the number of students in each group (example two students (in the group) with a pulse less than 50, five students with a pulse between 51 and 60).

On a histogram, the vertical axis shows the number (example 2, 4, 10, etc.) in each group; each group is shown by one bar (example one group has a pulse of less than 50; that group has two students).

Look again at the **resting pulse rate histogram** (graph). Two students have a pulse rate of less than 50, five students have a pulse rate between 51 and 60, seven students have a pulse rate between 61-70, 11 students between 71-80, three students between 81-90, and zero students above 90. The total number of students in the class recording their pulse rates are 2 + 5 + 7 + 11 + 3 = 28 students.

Description of pulse rate after activity histogram (graph). Look at the histogram (graph) of pulse rate after activity (exercise) below.

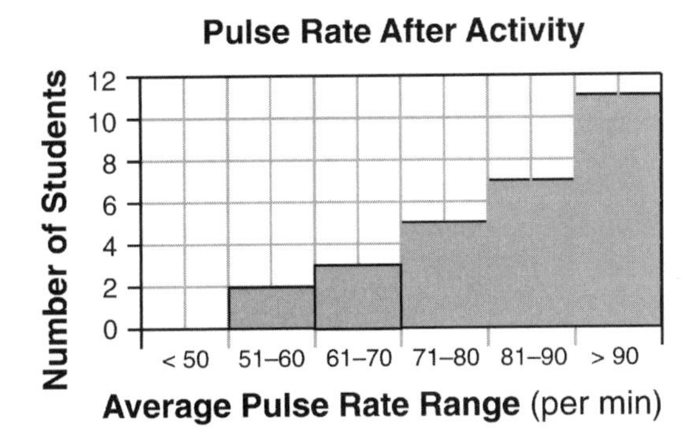

Two students have a pulse rate between 51-60, three students between 61-70, five students between 71-80, seven students between 81-90, and eleven students over 90. The same number of students (28 students) who recorded pulse rate after activity recorded resting pulse rate.

By looking at (comparing) the graph of the resting pulse rate and the graph of the pulse rate after activity (exercise), you see the average pulse rate generally increased. Before activity (exercise), zero students had pulse rates above 90, but after activity (exercise), eleven students had a pulse rate above 90. Before activity, only three students had a pulse rate above 81, but after activity, seven students (pulse rates of 81-90) and eleven students (pulse rates of 91 and above) see graph, which means 18 students, had a pulse rate above 81. Before activity, two students had a low pulse rate of below 50 per minute, but after activity, no student had such a low pulse rate; the lowest pulse rate after activity was between 51-60.

Now Do Homework Questions #47-54, pages 45-46

Classification with a dichotomous key

A dichotomous key is used to help identify species.
Here is shown a dichotomous key representing genera of birds. You want to identify bird X.

Bird W

Bird X

Bird Y

Bird Z

Dichotomous Key to Representative Birds	
1. a. The beak is relatively long and slender	*Certhidea*
b. The beak is relatively stout and heavy	go to 2
2. a. The bottom surface of the lower beak is flat and straight	*Geospiza*
b. The bottom surface of the lower beak is curved	go to 3
3. a. The lower edge of the upper beak has a distinct bend	*Camarhynchus*
b. The lower edge of the upper beak is mostly flat	*Platyspiza*

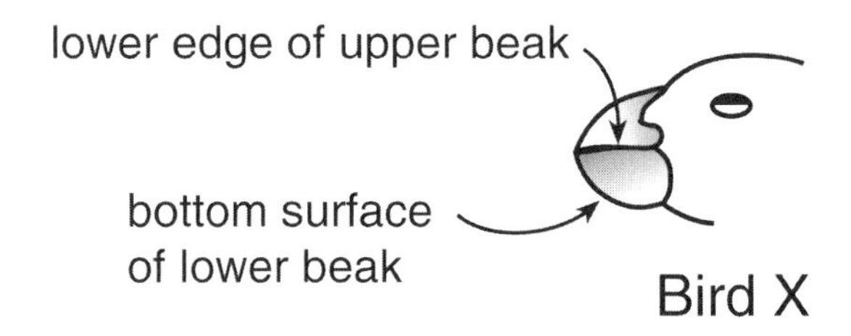

How to **use** a **dichotomous key** to **identify bird X:**

Look at #1 in the key:

1a. It says if the beak is relatively long and slender (thin), the bird is Certhidea.

1b. It says if the beak is relatively stout and heavy, go to 2.
Bird X has a stout (short) and heavy beak, therefore go to #2.

Look at #2 in the key:

2a. If the bottom surface of the lower beak is flat and straight, the bird is Geospiza.

2b. If the bottom surface of the lower beak is curved, go to 3.
In bird X, the bottom surface of the lower beak is curved, therefore go to 3.

Look at #3 in the key:

3a. If the lower edge of the upper beak has a distinct bend, the bird is Camarhynchus.

3b. If the lower edge of the upper beak is mostly flat, the bird is Platyspiza.
In bird X, the lower edge of the upper beak is mostly flat, therefore it is Platyspiza.

Note: In each number in the key, a and b have opposite types of the same trait. For example, 1a is beak long and slender; 1b is beak stout and heavy.

2a has the bottom surface of the lower beak flat and straight; 2b has the bottom surface of the lower beak curved.

Rule: when you want to identify an unknown organism (example plant or animal) look at the unknown organism (example plant or animal). Choose the type of trait in the key that describes the unknown organism. Always start at #1. Either you will find the name of the species or you will be told to go to another number (example, #2 or #3). Continue going down the key, following the directions, until you get the name of the species.

Now let's identify bird Y in the diagram above. Look at bird Y. Look at #1 in the dichotomous key above. 1a says that if the beak is relatively long and slender (thin), the bird is Certhidea. Bird Y has a relatively long and slender beak, therefore bird Y is Certhidea.

Let's see again how we use a dichotomous key to identify species.

The diagram represents five insect species, Species A through Species E.

Species E Species D

Question 1: A dichotomous key to these five species is shown below. Complete the missing information for sections 4a. and 4b. so that the key is complete for all five species.

Dichotomous Key

1. a. has small wings..................................go to 2
   b. has large wings..................................go to 3

2. a. has a single pair of wings....................Species A
   b. has a double pair of wings....................Species B

3. a. has a double pair of wings....................go to 4
   b. has a single pair of wings....................Species C

4. a. ________________________________ ......Species E
   b. ________________________________ ......Species D

Question 2: Use the key to identify the drawings of species A, B, and C. Place the letter of each species on the line below the drawing of the species.

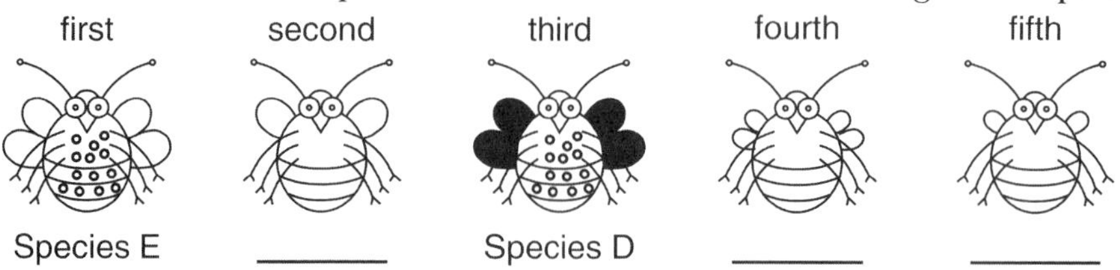

Solutions 1 and 2:
Look at #1 in the dichotomous key above.
Let's follow the direction of 1a.
1a. It says if it has small wings, go to 2. The fourth and fifth insects in the drawing in the question have small wings, therefore go to 2.

2a. It says (see key above) if it has a **single pair** of **wings**, it is **species A.** The last drawing in the question has a single pair of wings, therefore label the fifth insect species A.
2b. It says if it has a **double pair** of **wings**, it is **species B**. The next to the last drawing has a double pair of wings, therefore label the fourth insect species B.

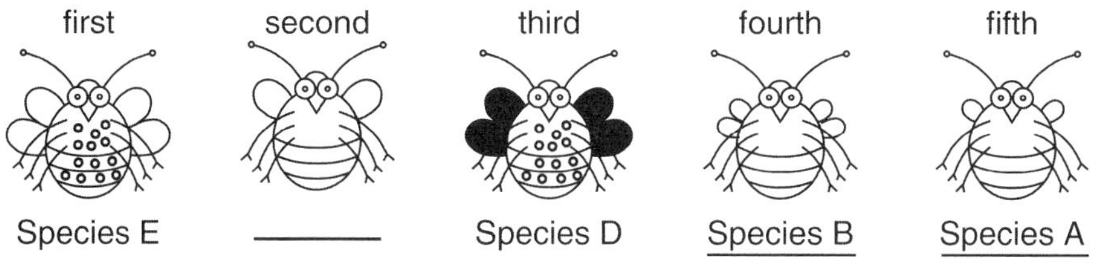

Let's now follow the direction of 1b in the dichotomous key.
1b. It says if it has large wings, go to 3. The first, second, and third insects in the drawing have large wings, therefore go to 3.

3a. It says if it has a **double pair** of **wings, go** to **4.**
3b. It says if it has a **single pair** of **wings**, it is **species C.** The second insect has a single pair of wings, therefore label the second insect species C.

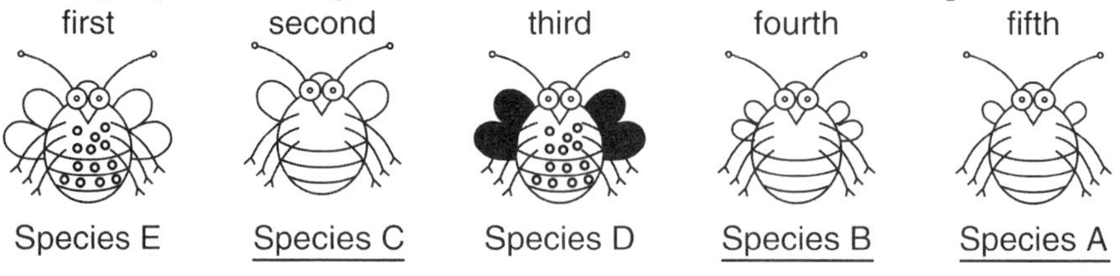

Let's follow the directions of 3a in the dichotomous key. It says if it has a double pair of wings, go to 4. The first and third insects in the drawings (species E and D) have a double pair of wings. The only thing different between E and D is the color of the wings, white wings for species E and black wings for species D. Therefore, on line 4a of the dichotomous key, write white wings for species E. On line 4b, write black wings for species D:

4. a. has white wings...Species E
   b. has black wings..Species D

Solution to question 1: Question 1 was to complete 4a and 4b. By using the dichotomous key, we figured out the answers to 4a and 4b (see 4a and 4b above).

4. a. has white wings...Species E
   b. has black wings..Species D

Now Do Homework Questions 55-56, page 47

Classification

There are different types of plants (examples, grass, oak trees) and different types of animals (examples, dogs, humans, birds). Let's discuss the animal kingdom. The **kingdom** has the most different types of animals. A **phylum** has fewer different types of animals. A **class** has even less, **genus** even less, and a **species** is only one type of animal.
Similarly in plants, the **kingdom** of plants has the most different types of plants, a **phylum** has fewer, a **class** even less, a **genus** even less, and a **species** is only one type of plant.

This chart shows the classification system. Classification is a way to group organisms by similarities (you do not have to memorize the names of the animals). Realize that all the members of a genus (see chart below) are more closely related (more similar to each other) than all members of a class, phylum, or kingdom.

	Kingdom animals	**Phylum** chordates	**Class** mammals	**Genus** canis	**Species** familiaris
Different types of animals:	most	fewer types	even fewer	even less	only one
	dog wolf human monkey frog snake butterfly	dog wolf human monkey frog snake **NO** ~~butterfly~~	dog wolf human monkey **NO** ~~frog~~ **NO** ~~snake~~ **NO** ~~butterfly~~	dog wolf **NO** ~~human~~ **NO** ~~monkey~~ **NO** ~~frog~~ **NO** ~~snake~~ **NO** ~~butterfly~~ In a genus, animals are most closely related.	dog

The animal kingdom has all animals (all the animals listed above and many more animals). The phylum chordates are animals that have a backbone, therefore you have fewer types of animals in the phylum. The class of mammals are only animals that have a backbone and produce milk, therefore you have even fewer different types of animals (see chart above). The genus canis has even fewer different types (see chart above). A species has only one type of animal.

In short, in a kingdom (example animal kingdom) there are many different types of animals (all types of animals). In a genus (example genus canis) there are relatively only a few different types of animals, such as dog and wolf (see chart above), therefore the animals in a genus are more closely related.

Now Do Homework Question #57, page 47

To answer Part D questions, also review:

Evolutionary trees	Chapter 5, pages 5:1-5:5
DNA controls protein synthesis	Chapter 3, pages 3:8-3:15

Now let's try a few Homework Questions

To assign fewer questions, you may want to skip the questions with underlined numbers.

1. What is the approximate length of the earthworm shown in the diagram below?

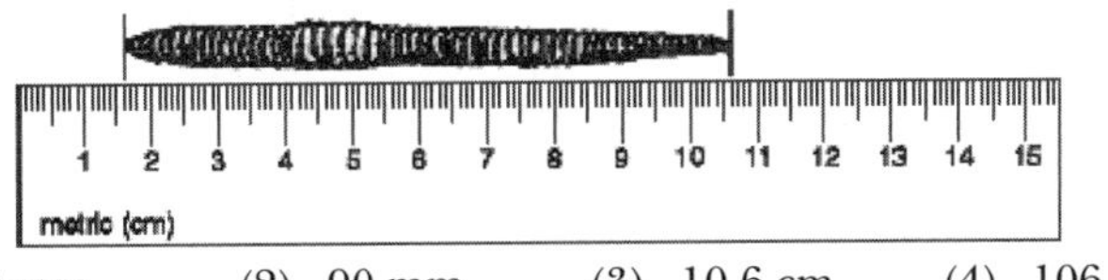

(1) 9 mm (2) 90 mm (3) 10.6 cm (4) 106 cm

2. The diagram below represents the measurement of two leaves.

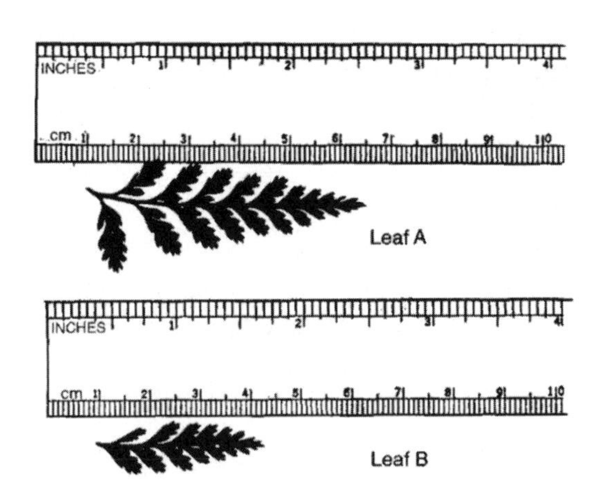

The difference between leaves *A* and *B* is closest to

(1) 20 mm (2) 20 cm (3) 0.65 mm (4) 1.6 μm

3. A student measured an earthworm using a metric ruler, as shown in the diagram below. What is the length of *A*?

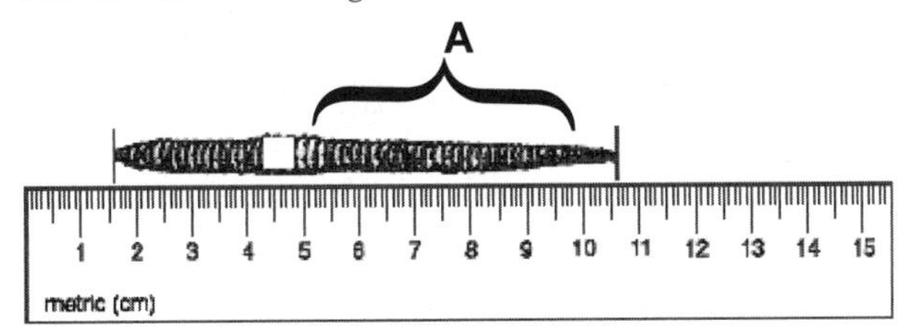

(1) 5.5 cm (2) 11.6 mm (3) 46 cm (4) 23 cm

4. The diagram below represents a thermometer.

The temperature reading on this thermometer would most likely indicate the temperature of

(1) of the human body on a very hot summer day
(2) a which water freezes
(3) at which water boils
(4) of a human with a very high fever

5. What is the mass of the object on the triple beam balance?

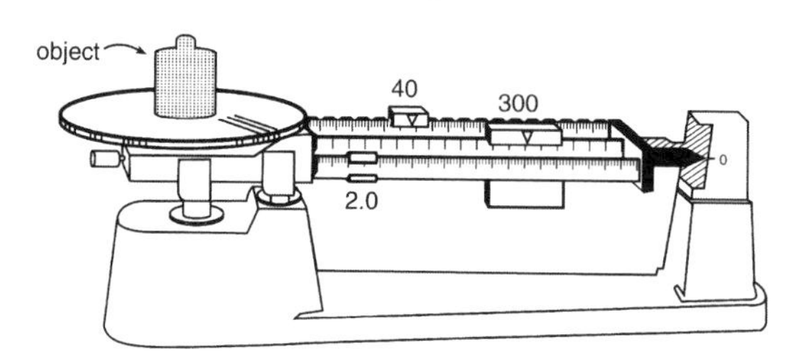

(1) 243 grams (2) 432 grams (3) 342 grams (4) 72 grams

---

6. Which graduated cylinder represents a volume of liquid closest to 15 milliliters?

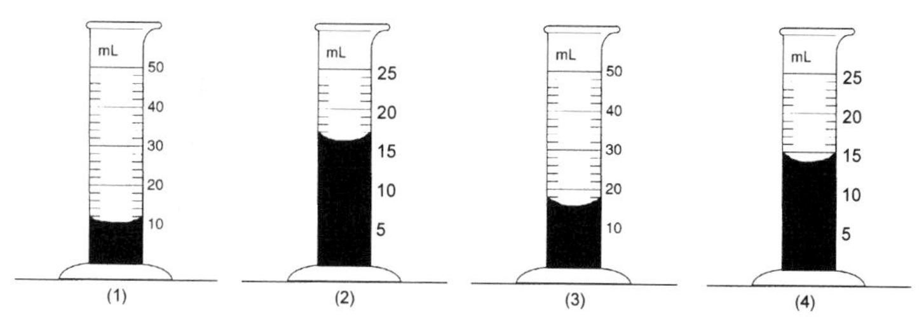

7. A plant cell in a microscopic field of view is represented below.

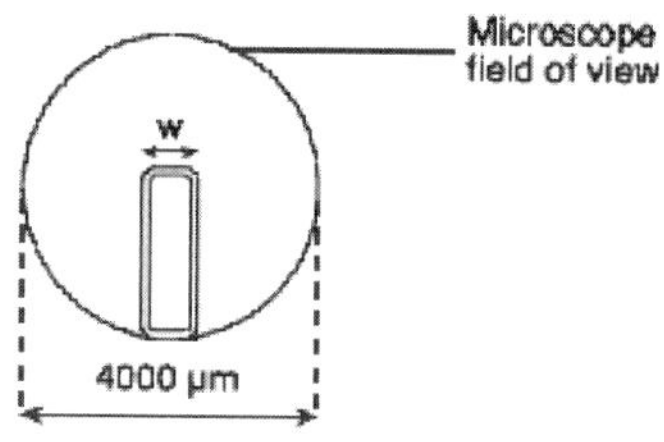

The width (w) of this plant cell is closest to

(1) 200 μm (2) 800 μm (3) 1200 µm (4) 1600 µm

8. Which activity might lead to damage of a microscope and specimen?

(1) cleaning the ocular and objectives with lens paper
(2) focusing with low power first before moving the high power into position

(3) using the coarse adjustment to focus the specimen under high power
(4) adjusting the diaphragm to obtain more light under high power

9. The diagram below shows how a coverslip should be lowered onto some single-celled organisms during the preparation of a wet mount.

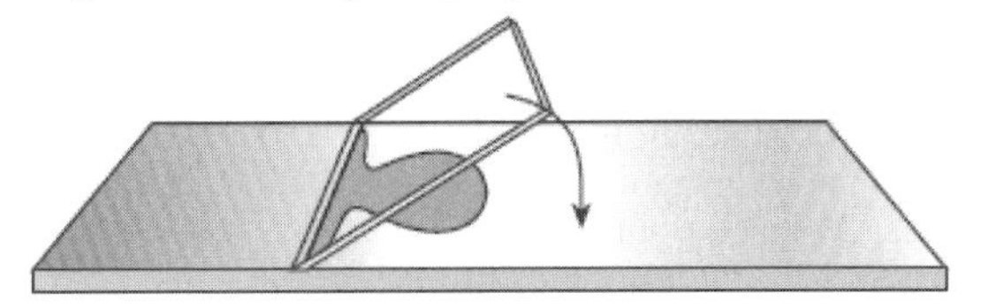

Why is this a preferred procedure?
(1) The coverslip will prevent the slide from breaking.
(2) The organisms will be more evenly distributed.
(3) The possibility of breaking the coverslip is reduced.
(4) The possibility of trapping air bubbles is reduced.

10. The diagrams below show four different one-celled organisms (shaded) in the field of view of the same microscope using different magnifications. Which illustration shows the largest one-celled organism?

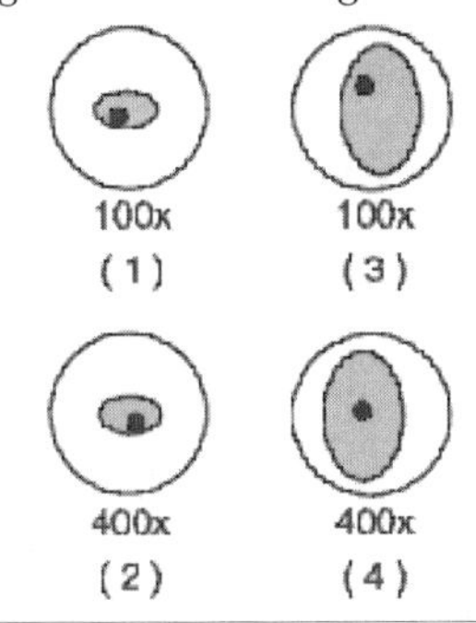

Base your answers to the next two questions on the information and data table below and on your knowledge of biology.

A student studied the location of single-celled photosynthetic organisms in a lake for a period of several weeks. The depth at which these organisms were found at different times of the day varied greatly. Some of the data collected are shown in the table below.

Data Table

Light Conditions at Different Times of the Day	Average Depth of Photosynthetic Organisms (cm)
full light	150
moderate light	15
no light	10

11. Which materials would the student most likely have used in this investigation?
(1) microscope, pipette, and slides with coverslips
(2) graduated cylinder, triple-beam balance, and chromatography paper
(3) thermometer, electric balance, and biological stains
(4) computer, pH paper, and gel electrophoresis apparatus

12. A student prepared a slide of pollen grains from a flower. First the pollen was viewed through the low-power objective lens and then, without moving the slide, viewed through the highpower objective lens of a compound light microscope.

Which statement best describes the relative number and appearance of the pollen grains observed using these two objectives?

(1) low power: 25 small pollen grains
high power: 100 large pollen grains
(2) low power: 100 small pollen grains
high power: 25 large pollen grains
(3) low power: 25 large pollen grains
high power: 100 small pollen grains
(4) low power: 100 large pollen grains
high power: 25 small pollen grains

---

13. While viewing a specimen under high power of a compound light microscope, a student noticed that the specimen was out of focus. Which part of the microscope should the student turn to obtain a clearer image under high power?

(1) eyepiece
(2) coarse adjustment
(3) fine adjustment
(4) nosepiece

14. Base your answers to the next question on the diagram below and on your knowledge of biology. The diagram illustrates what happens when a particular solution is added to a wet-mount slide containing red onion cells being observed using a compound light microscope.

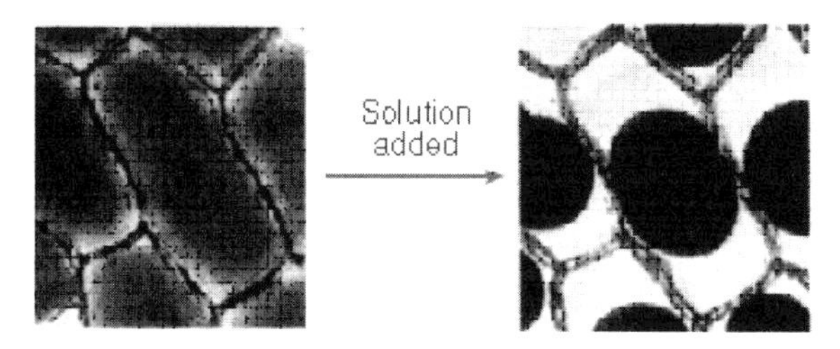

To observe the cells on this slide it is best to start out using the

(1) high-power objective and focus using the coarse adjustment, only
(2) low-power objective and focus using the fine adjustment, only
(3) high-power objective and focus using the fine adjustment
(4) low-power objective and focus using the coarse adjustment

15. A solution containing both starch and glucose was placed inside the model cell represented below. The model cell was then placed in a beaker containing distilled water.

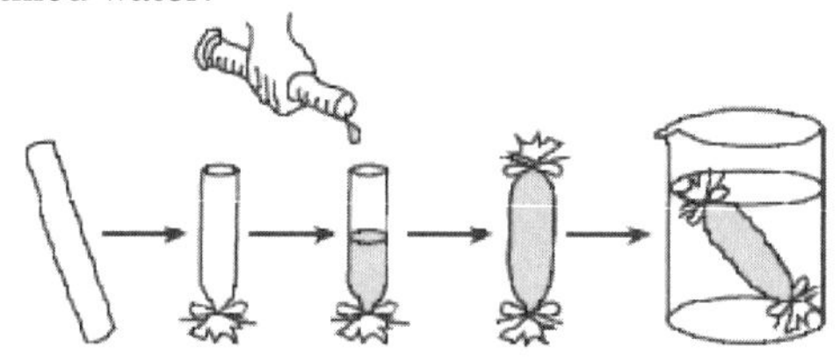

Identify *one* specific substance that should have been added to the distilled water so that observations regarding movement of starch could be made.

16. Phenolphthalein is a chemical that turns pink in the presence of a base. A student set up the demonstration shown in the diagram below.

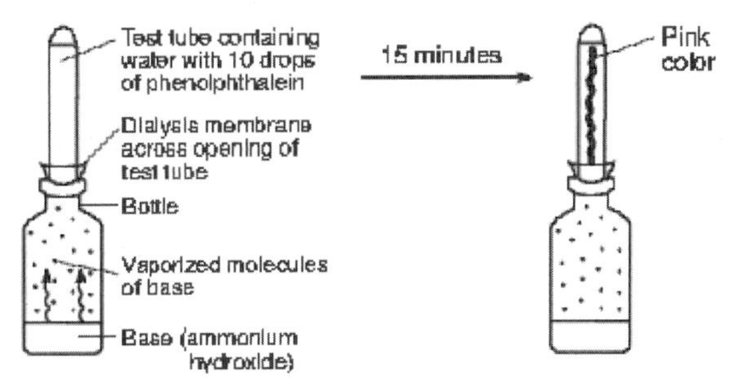

The appearance of the pink color was due to the movement of

(1) phenolphthalein molecules from low concentration to high concentration
(2) base molecules from high concentration through the membrane to low concentration
(3) water molecules through the membrane from high concentration to low concentration
(4) phenolphthalein molecules in the water from high concentration to low concentration

---

Base your answers to the next three questions on the experimental setup shown below.

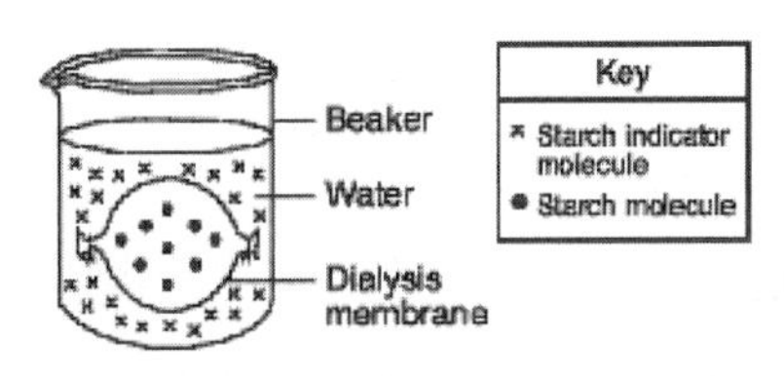

17. On the diagram provided, draw in the expected locations of the molecules after a period of one hour.

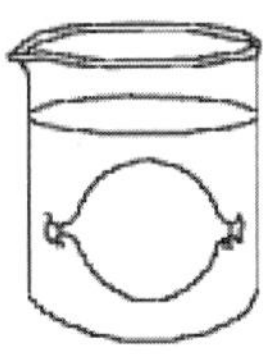

18. When starch indicator is used, what observation would indicate the presence of starch?

19. State *one* reason why some molecules can pass through a certain membrane, but other molecules can not.

---

20. Glucose indicator was added to a beaker of an unknown liquid. Starch indicator was added to a different beaker containing the same unknown liquid. The color of the indicator solutions before they were added to the beakers and the color of the contents of the beakers after adding the indicator solution are recorded in the chart below.

Beaker	Solution	Color of Indicator Solution Before Adding to Beaker	Color of Contents of Beaker After Adding Indicator Solution
1	unknown liquid + glucose indicator	blue	blue (after heating)
2	unknown liquid + starch indicator	amber	blue black

Which carbohydrate is present in the unknown liquid? Support your answer.

21. A laboratory setup of a model cell is shown in the diagram below.

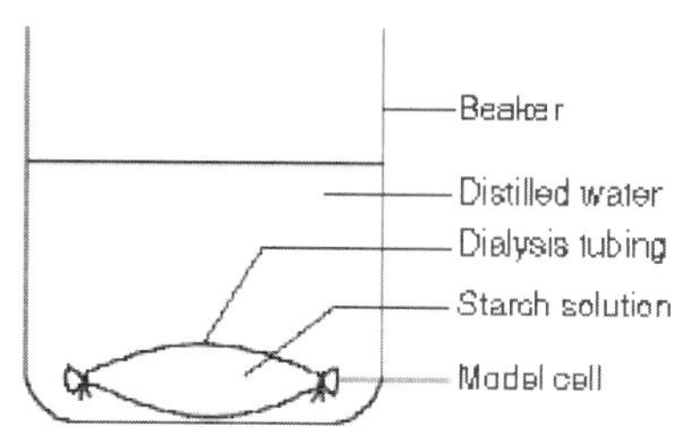

Which observation would most likely be made 24 hours later?

(1) The contents of the model cell have changed color.
(2) The diameter of the model cell has increased.
(3) The model cell has become smaller.
(4) The amount of distilled water in the beaker has increased.

Base your answers to the next two questions on the diagram below and on your knowledge of biology. The diagram illustrates what happens when a particular solution is added to a wet-mount slide containing red onion cells being observed using a compound light microscope.

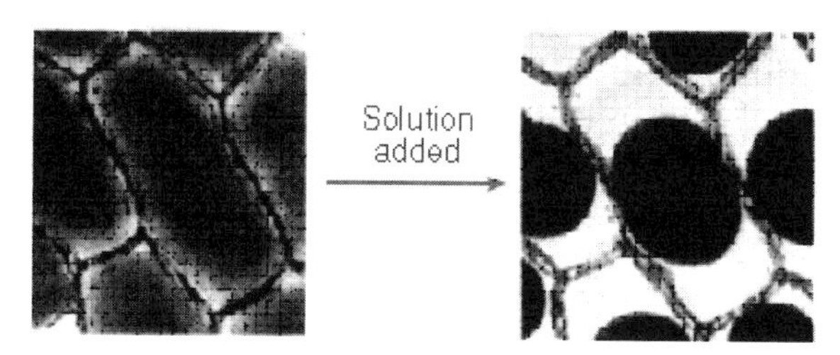

22. Identify a process that caused the change in the cells.

23. The diagram below represents the distribution of some molecules inside and outside of a cell over time.

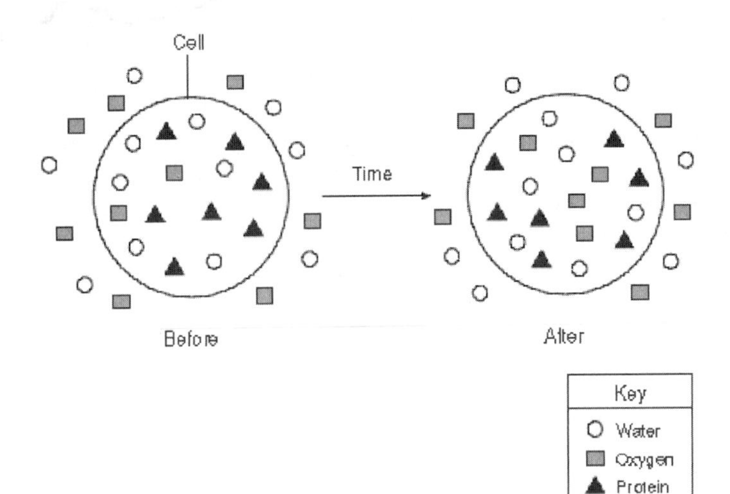

Which factor prevented the protein molecules (▲) from moving out of the cell?

(1) temperature
(2) pH
(3) molecule size
(4) molecule concentration

24. In preparation for an electrophoresis procedure, enzymes are added to DNA in order to

(1) convert the DNA into gel
(2) cut the DNA into fragments
(3) change the color of the DNA
(4) produce longer sections of DNA

25. Electrophoresis is a method of

(1) separating DNA fragments
(2) changing the genetic code of an organism
(3) indicating the presence of starch
(4) separating colored compounds on a strip of paper

Base your answers to the next four questions on the information and diagram below and on your knowledge of biology.

The four wells represented in the diagram were each injected with fragments that were prepared from DNA samples using identical techniques.

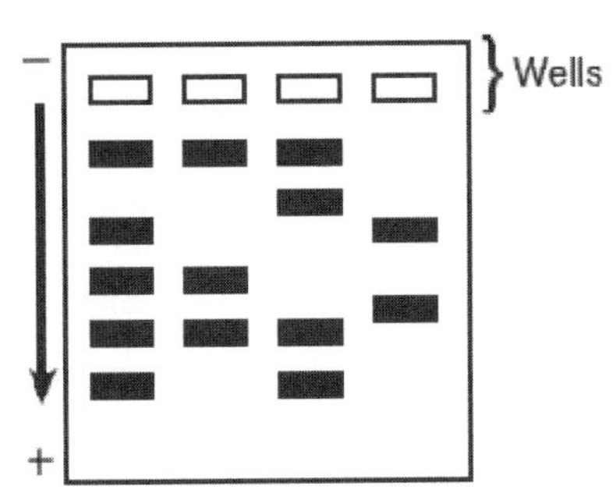

26. This laboratory procedure is known as

(1) cloning
(2) gel electrophoresis
(3) chromatography
(4) use of a dichotomous key

27. The arrow represents the direction of the movement of the DNA fragments. What is responsible for the movement of the DNA in this process?

28. The four samples of DNA were taken from four different individuals. Explain how this is evident from the results shown in the diagram.

29. Identify the substance that was used to treat the DNA to produce the fragments that were put into the wells.

Base your answers to the next three questions on the information and diagram below and on your knowledge of biology.

The DNA of three different species of birds was analyzed to help determine if there is an evolutionary relationship between these species. The diagram shows the results of this analysis.

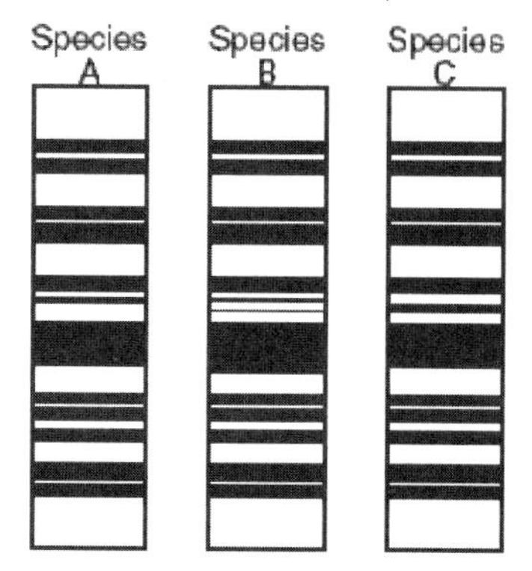

30. Identify the technique normally used to separate the DNA fragments to produce the patterns shown in the diagram.

31. The chart below contains amino acid sequences for part of a protein that is found in the feathers on each of these three species of birds.

Species	Amino Acid Sequence
A	Arg-Leu-Glu-Gly-His-His-Pro-Lys-Arg
B	Arg-Gly-Glu-Gly-His-His-Pro-Lys-Arg
C	Arg-Leu-Glu-Gly-His-His-Pro-Lys-Arg

State *one* way this data supports the inference that these three bird species may be closely related.

32. State *one* type of additional information that could be used to determine if these three species are closely related.

Base your answers to the next four questions on the information below and on your knowledge of biology.

To demonstrate techniques used in DNA analysis, a student was given two paper strip samples of DNA. The two DNA samples are shown below.

Sample 1: ATTCCGGTAATCCCGTAATGCCGGATAATACTCCGGTAATATC
Sample 2: ATTCCGGTAATCCCGTAATGCCGGATAATACTCCGGTAATATC

The student cut between the *C* and *G* in each of the shaded CCGG sequences in *sample 1* and between the *A*s in each of the shaded TAAT sequences in *sample 2*.

33. The action of what kind of molecules was being demonstrated when the DNA samples were cut?

34. Identify the technique that was being demonstrated when the fragments were arranged on the gel model.

35. The results of this type of DNA analysis are often used to help determine
   (1) the number of DNA molecules in an organism
   (2) if two species are closely related
   (3) the number of mRNA molecules in DNA
   (4) if two organisms contain carbohydrate molecules

36. State *one* way that the arrangement of the two samples on the gel model would differ.

Base your answers to the next three questions on the diagram below and on your knowledge of biology. The diagram shows the results of a technique used to analyze DNA.

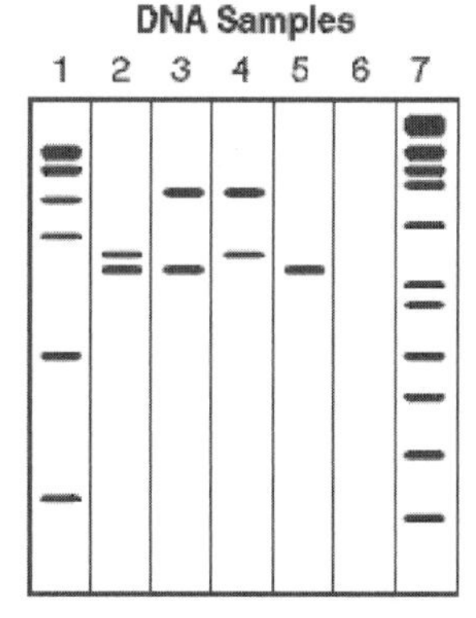

37. This technique used to analyze DNA directly results in
   (1) synthesizing large fragments of DNA
   (2) separating DNA fragments on the basis of size
   (3) producing genetically engineered DNA molecules
   (4) removing the larger DNA fragments from the samples

38. This laboratory technique is known as
   (1) gel electrophoresis
   (2) DNA replication
   (3) protein synthesis
   (4) genetic recombination

39. State *one* specific way the results of this laboratory technique could be used.

Base your answers to the next two questions on the information below and on your knowledge of biology.

Scientists found members of a plant species they did not recognize. They wanted to determine if the unknown species was related to one or more of four known species, *A*, *B*, *C*, and *D*.

The relationship between species can be determined most accurately by comparing the results of gel electrophoresis of the DNA from different species.

The chart below represents the results of gel electrophoresis of the DNA from the unknown plant species and the four known species.

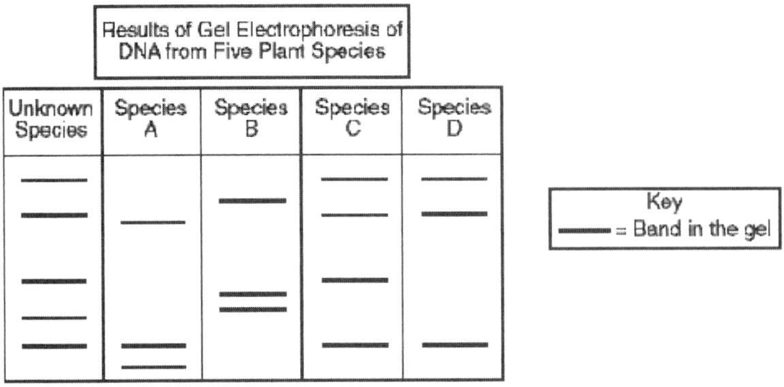

Results of Gel Electrophoresis of DNA from Five Plant Species

Unknown Species	Species A	Species B	Species C	Species D

Key
—— = Band in the gel

40. The unknown species is most closely related to which of the four known species? Support your answer.

41. Explain why comparing the DNA of the unknown and known plant species is probably a more accurate method of determining relationships than comparing only the physical characteristic you identified in the previous question.

Base your answers to the next three questions on the information below and on your knowledge of biology.

In an investigation, DNA samples from four organisms, *A*, *B*, *C*, and *D*, were cut into fragments. The number of bases in the resulting DNA fragments for each sample is shown below.

Data Table

Sample	Number of Bases in DNA Fragments
A	3, 9, 5, 14
B	8, 4, 12, 10
C	11, 7, 6, 8
D	4, 12, 8, 11

42. The diagram below represents the gel-like material through which the DNA fragments moved during gel electrophoresis. Draw lines to represent the position of the fragments from each DNA sample when electrophoresis is completed.

Wells →	Sample A	Sample B	Sample C	Sample D	– Negative pole
15					
14					
13					
12					
11					
Number of DNA bases 10					
9					
8					
7					
6					
5					
4					
3					
2					+ Positive pole
1					

43. Which *two* DNA samples are the most similar? Support your answer using data from this investigation.

44. State *one* specific use for the information obtained from the results of gel electrophoresis.

Base your answers to the next two questions on the diagram below that illustrates the results of a laboratory technique and on your knowledge of biology.

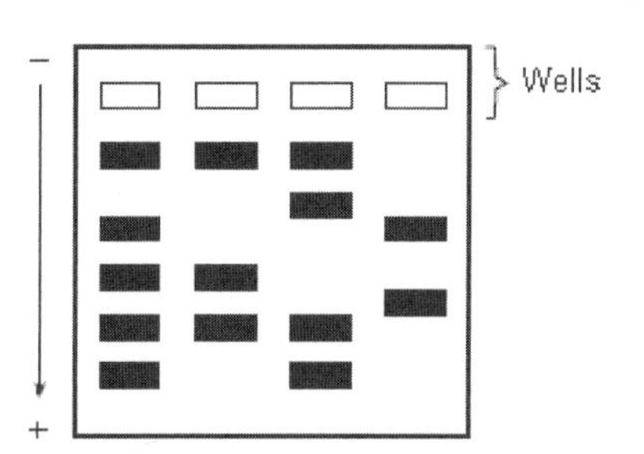

45. State *one* way the information obtained by this technique can be used.

46. The results of which laboratory technique are represented in the diagram?
(1) chromatography
(2) manipulation of genes
(3) genetic engineering
(4) gel electrophoresis

47. An experiment was done using chromatography paper to determine evolutionary relationships among different species. The spots on the chromatography papers of species *2* and *3* are similar; the spots on the chromatography papers of species *2* and *4* are different. What does a comparison of these strips tell us about evolutionary relationships between these three species?

48. Paper chromatography is a laboratory technique that is used to
(1) separate different molecules from one another
(2) stain cell organelles
(3) indicate the pH of a substance
(4) compare relative cell sizes

49. A laboratory technique is illustrated in the diagram below.

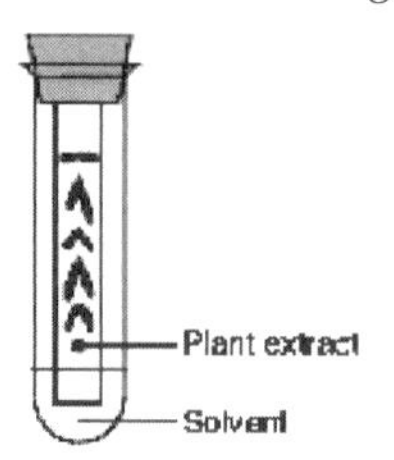

This technique is used to
- (1) determine volume
- (2) separate molecules in a mixture
- (3) measure length
- (4) analyze data from an experiment

Base your answers to the next three questions on the histograms below and on your knowledge of biology.

Students in a class recorded their resting pulse rates and their pulse rates immediately after strenuous activity. The data obtained are shown in the histograms below.

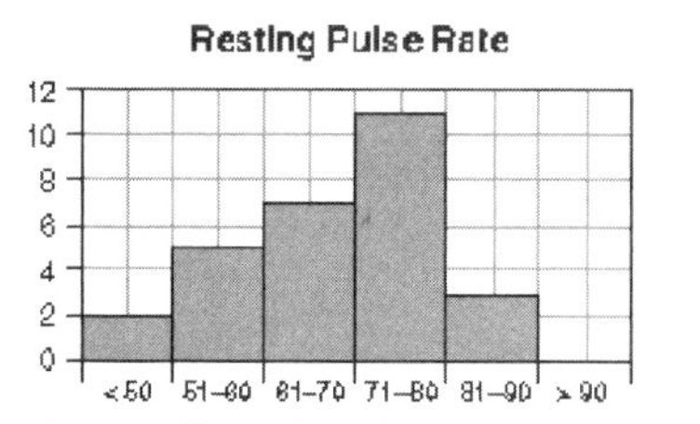

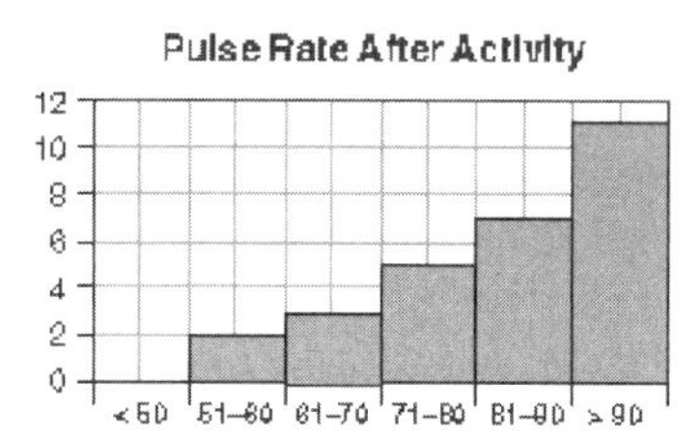

50. An appropriate label for the y-axis in each histogram would be
- (1) Number of Students
- (2) Average Number of Heartbeats
- (3) Time (min)
- (4) Amount of Exercise

51. According to the data, compared to the average resting pulse rate, the average pulse rate immediately after strenuous activity generally
- (1) decreased
- (2) increased
- (3) remained the same
- (4) decreased and leveled off

52. State *one* biological explanation for the fact that not all students had the same resting pulse rate.

53. How many students had a resting pulse rate between 61 and 70?
(1) 5 (2) 7 (3) 5, 6, 7 (4) 2, 5, 6, 7

54. How many students had a pulse rate after activity above 90 beats per minute?
(1) 0 (2) 7 (3) 10 (4) 11

Base your answers to the next two questions on the diagram below and on your knowledge of biology. The diagram represents six insect species.

Species E

Species F

55. A dichotomous key to these six species is shown below. Complete the missing information for sections 5.a. and 5.b. so that the key is complete for all six species.

DICHOTOMOUS KEY

1.	a.	has small wings	go to 2
	b.	has large wings..	go to 3
2.	a.	has a single pair of wings	Species A
	b.	has a double pair of wings	Species B
3.	a.	has a double pair of wings	go to 4
	b.	has a single pair of wings................................	Species C
4.	a.	has spots	go to 5
	b.	does not have spots......................................	Species D
5.	a.	______________________........................	Species E
	b.	______________________........................	Species F

56. Use the key to identify the drawings of species *A*, *B*, *C*, and *D*. Place the letter of each species on the line located below the drawing of the species.

DICHOTOMOUS KEY

1.	a.	has small wings	go to 2
	b.	has large wings..	go to 3
2.	a.	has a single pair of wings	Species U
	b.	has a double pair of wings	Species V
3.	a.	has a double pair of wings	go to 4
	b.	has a single pair of wings................................	Species W
4.	a.	has spots	go to 5
	b.	does not have spots......................................	Species Z
5.	a.	______________________........................	Species Y
	b.	______________________........................	Species X

Species ___ Species Y Species ___ Species X Species ___ Species ___

57. A classification system is shown in the table below.

Classification	Examples
Kingdom — animal	△, ○, ⏢, ☆, ▭, ◇, ξ, ▽
Phylum — chordata	△, ⏢, ξ, ☆, ▭
Genus — *Felis*	▭, ξ
Species — *domestica*	▭

This classification scheme indicates that ▭ is most closely related to

☆ (1) △ (2) ⏢ (3) ξ (4)

Appendix I: Glossary

A

abiotic factors: nonliving things in an ecosystem, examples water, oxygen, sunlight, minerals.

acid rain: industries burn fossil fuels, giving off sulfur and nitrogen compounds which combine with water vapor in the air to produce acids. These acids, with water, fall to Earth as rain or snow, which is called acid rain.

active transport: using energy to force materials across a membrane from low concentration to high concentration.

air pollution: harmful materials in the air; includes acid rain, smog, global warming, and ozone depletion.

allergy: body's immune system responds to a usually harmless substance.

artery: blood vessel that goes away from the heart.

asexual reproduction: producing offspring using only one parent; offspring are identical to the parent.

ATP: (adenosine triphosphate) is a usable form of energy; ATP is made in mitochondria and stored in cells.

autotroph: plant or algae that makes its own food by photosynthesis.

B

Benedict's solution: an indicator that tests for sugar. When Benedict's solution is heated with sugar, the color changes from blue to green to yellow, and then to red.

biodiversity: having many different species of plants and animals in an ecosystem (example forest, meadow).

biological control: using living organisms or materials from living organisms to control pests; does not use chemicals, therefore does not harm the environment..

biosphere: the part of the Earth where life exists.

biotechnology: using technology with biological science (biology). Examples are selective breeding and genetic engineering.

biotic factors: living things in an ecosystem, example geese, frogs, fish.

bromthymol blue: indicator that is blue in a base and yellow in an acid.

C

carbon dioxide-oxygen cycle: plants take in carbon dioxide for photosynthesis and give off oxygen; animals take in oxygen for respiration and give off carbon dioxide.

carnivore: animal that eats animals.

carrying capacity: the number of organisms of a species that can live or survive in an area.

cell: basic unit of structure and function in living things.

cell membrane: surrounds the cell; made mostly of fats (lipids) and protein; the cell membrane regulates which materials enter and leave the cell.

cell wall: only found in plant cells; cell wall is outside the cell membrane and supports the plant.

cellular respiration: glucose unites with oxygen, forming water and carbon dioxide, plus ATP (a usual form of energy; cellular respiration takes place in mitochondria.

CFC's: gases that destroy the ozone layer; they are used as the cooling liquid in refrigerators and air conditioners.

chloroplast: in cells of plant leaves and some one-celled organisms, example euglena; photosynthesis takes place in chloroplasts.

chromatography: a way to separate different molecules (example different amino acids) in a mixture. A drop of the mixture is put on filter paper and a solvent (liquid, example alcohol) moves up the paper. Different molecules (example different amino acids) in the mixture move different distances up the paper, separating the molecules.

chromosome: thread-like structure in the nucleus; chromosome has genes that have genetic information (DNA).

circulatory system:; made of heart, blood vessels, and blood; carries dissolved materials (examples glucose, amino acids, wastes, carbon dioxide).

classification: a way to group organisms by similarities.

climax community: a community that is stable and is not replaced by another community, example an oak-hickory forest.

cloning: a type of asexual reproduction; produces new organisms identical to the parent (genetically identical, same genes, same number of chromosomes).

commensalism: relationship between two organisms in which one benefits and the other one is not affected, example orchids gain by growing on trees without harming the tree.

community: all populations (all species) in a given area, example deer, squirrels, and oak trees in a forest.

competition: when organisms use or fight for the same limited resources they need to live.

consumers: (also called heterotrophs) eat plants or animals for food.

control: one group in an experiment that is identical to the experimental group, except that only the experimental group gets the thing being tested (example, takes a diet pill).

crossing over: exchanging part of one chromosome for part of another chromosome.

cytoplasm: jellylike substance inside a cell.

D

decomposer: organisms (example some bacteria and fungi) that eat wastes and dead organisms; they decompose (break down chemically) the wastes and dead materials into simple materials (example carbon dioxide and nitrogen compounds) and recycle materials to air and soil.

deforestation: destroying forests.

dependent variable: thing that is changed (example height of plants) because of the independent variable (example gibberellic acid).

diaphragm: regulates the amount of light entering a microscope.

dichotomous key: a key used to identify living things. At each step, the key divides the living things into two groups. Each group is then divided again into two as many times as needed until all the living things are identified.

diffusion: dissolved substances (example dissolved sugar) go across a membrane from area of high concentration to area of low concentration.

direct harvesting: destroying or removing a species from its habitat.

disease: disturbance (disruption, failure) in homeostasis (maintaining constant internal environment); diseases can be caused by pathogens (examples viruses, bacteria, fungi, parasites), organ malfunction, heredity, etc.

DNA: DNA (deoxyribonucleic acid) is a very large molecule in the nucleus that has the genetic code (genes, genetic information); made of sugar, phosphate, and nitrogen bases.

E

ecosystem: part of the environment, examples forest, lake, pond, or desert. An ecosystem is an area where living things interact with the living and nonliving things.

egestion: elimination; undigested food (wastes) leaves the body through an opening called the anus.

egg: female gamete (sex cell); has half the number of chromosomes as a regular cell in the body.

electrophoresis: method for separating proteins or pieces of DNA by using electric current. Electrophoresis can be used to find evolutionary relationships (which species are closely related), to establish paternity, and to identify criminals by matching DNA.

embryo: early stage of development of a new living thing.

endangered species: species that is in danger of becoming extinct, example blue whale.

energy flow: energy flows (goes) from the sun to photosynthetic organisms (example plants and algae), then to herbivores and then to carnivores. Energy flows from organisms (plants and animals) at all levels of the pyramid to decomposers.

energy pyramid: a diagram showing the flow of energy in an ecosystem. The base (producers) is widest, showing that they have the most energy; the upper layers are narrower, showing that there is less energy in herbivores, carnivores, and higher level carnivores.

environment: everything that surrounds an organism.

enzyme: a substance (protein) that speeds up chemical reactions.

evolution: change in species over time.

excretion: removing (getting rid of) cellular wastes (wastes from life processes) from the body.

extinction: when a species dies out.

eyepiece: lens on a microscope that you look through.

F

fertilization: conbining sperm and egg to form a zygote (fertilized egg).

fetus: when an embryo is more developed (after about the first two months in humans), it is a fetus.

fine adjustment: small knob on the microscope that sharpens the image (makes what you see clearer).

food chain: shows which organisms eat other organisms; shows how energy is transferred from sun through producers to consumers and decomposers.

food web: many food chains interconnected together.

fossil: remains or trace of an organism that was once living.

fossil fuel: fuel made from remains of plants and animals that lived millions of years ago. Examples: coal, petroleum, oil, natural gas.

G

gamete: sex cell (example sperm or egg); gamete has half the number of chromosomes as a regular cell in the body.

gene: piece of DNA with genetic information for one trait (there are many genes on each chromosome).

gene expression: genes that are used in a cell are "expressed;" genes not used in a particular type of cell are not "expressed."

genetic engineering: making new or improved varieties of organisms by inserting genes (DNA) from one organism into the DNA of another organism.

genetic recombination: combining genes from both parents to form a new living thing.

global warming: increase in the average temperature over the whole Earth, caused by increased amounts of carbon dioxide and other gases in the air.

graduated cylinder: instrument that measures volume of a liquid.

growth: an organism gets bigger; increase in number of cells in an organism.

H

habitat: where an animal lives, example fish live in a pond.

herbivore: animal that eats plants, examples cow, zebra.

heredity: passage of genetic information by genes from one generation to the next generation.

homeostasis: a constant internal environment in a living thing (also called steady state or dynamic equilibrium).

hormone: chemical produced by an endocrine gland that regulates body functions. Example: insulin from pancreas.

hypothesis: educated guess; a prediction of what will happen in an experiment.

I

immune system: protects the body from disease; uses white blood cells and antibodies.

imported species: species brought by humans to a new environment; imported species often become a problem because they have no natural enemies in the new environment.

independent variable: thing we change in an experiment (example temperature).

indicator: changes color to tell if a substance, example starch, sugar, acid, or base, is present.

inorganic molecule: does not have both carbon and hydrogen. Examples: water, salt.

invalid experiment: has conclusions not based on observations and data.

J

K

L

locomotion: moving from place to place.

M

meiosis: a cell dividing into two cells; each cell has half the original number of chromosomes. Gametes (sperm and egg) form by meiosis.

metabolism: all life functions (life processes) together.

microscope: instrument that uses lenses to magnify objects 40 to hundreds of times.

mitochondria: organelles (structures inside the cell) where cellular respiration takes place.

mitosis: a cell dividing into two identical cells, each with all the genetic information and the same number of chromosomes as the original cell.

mutation: change in the order of bases in a gene or chromosome.

mutualsim: relationship between two organisms in which both benefit, example nitrogen-fixing bacteria live on the roots of some plants; the bacteria get a place to live and the plants get nitrates to make proteins.

N

natural selection: organisms that are best adapted in a particular environment survive and pass on their characteristics to their offspring.

neurotransmitter: chemical made by a nerve cell that carries messages from one nerve cell to another nerve cell (or to a muscle or gland).

niche: an animal's habitat (where it lives) and its feeding style (what it eats and where it eats).

nitrogen bases: contain nitrogen. A (adenine), T (thymine), C (cytosine), and G (guanine) are nitrogen bases that carry genetic information in DNA, etc.

nitrogen cycle: plants use nitrates to make nitrogen compounds like proteins; decomposers break proteins, etc. into simple nitrogen compounds that plants can take in.

nonrenewable resource: resource (example fossil fuels, minerals) that cannot be replaced in a reasonable amount of time.

nuclear energy: energy produced when mass is changed to energy, such as in a fission reaction (one atom splits into two or more pieces, giving off a lot of energy).

nucleus: organelle (structure inside the cell) that contains DNA and is the control center of the cell. It stores genetic information and directs protein synthesis.

nutrition: taking in food, digestion, and elimination (getting rid of undigested food (wastes).

O

objective lenses: lenses at the bottom of the tube of the microscope that magnify the object.

organ: different types of tissues combine and work together to form an organ, which carries out a life function; examples of organs are heart and stomach.

organ system: different organs working together, example digestive system (mouth, stomach, intestines, pancreas, etc.).

organelle: a structure inside a cell, examples nucleus, chloroplasts.

organic molecule: has both carbon and hydrogen examples sugars, proteins.

osmosis: diffusion of water; water goes across (through) a membrane from area of high concentration to area of low concentration.

ovary: organ in female reproductive system; produces egg cells (female gametes) and the hormones estrogen and progesterone.

ozone depletion: lowering of the amount of ozone in the atmosphere (ozone protects living things from ultraviolet radiation) caused by some pollutants, such as CFC's.

P

parasite: organism that lives by harming another organism.

parasitism: relationship between two organisms in which one organism benefits and one organism is harmed (example tapeworm benefits and the animal it lives in is harmed).

pathogen: virus, bacteria, fungus, or parasite that causes disease.

pH: a scale that measures whether a substance is acid, base, or neutral.

pH paper: paper put in a liquid, changes color to tell whether a liquid is acid, base, or neutral; some pH papers show how acidic or how basic a liquid is.

photosynthesis: in the presence of sunlight, plants take in carbon dioxide and water, producing glucose and oxygen.

placenta: structure attached to the wall of the uterus where oxygen, food, and wastes are exchanged between mother and embryo.

population: all members of one species that live in one area, example all deer in a forest or all pigeons in New York City.

predator: animal that kills other animals for food.

producers: (also called autotrophs) make their own food, examples plants and algae.

pulse: each time the heart beats, there is a bulge in the arteries; we feel these bulges as a pulse. The pulse rate equals the heart rate.

Q

R

receptor: protein on a cell surface that responds to a chemical, example to a hormone).

recycle: using materials over again to make new products.

regulation: control and coordination of all life activities (life processes) by nerves and chemicals in the blood (hormones).

renewable resource: resource (example sun's energy, water, animals, and plants) that can be replaced in a reasonable amount of time.

replication: a DNA molecule making an identical copy of itself.

reproduction: producing offspring (children).

reproductive technology: using scientific methods to produce organisms with desired characteristics, example in vitro fertilization.

research plan: A research plan tells what you need to do to make an experiment. This involves getting background information, seeing what experiments have been done, making a hypothesis, and designing an experiment.

respiration: glucose unites with oxygen in cells, producing water, carbon dioxide, and ATP (a usable form of energy).

respiratory system: Consists of nose, trachea, bronchi (bronchial tubes), and lungs; respiratory system brings oxygen into the body and removes carbon dioxide.

ribosome: organelle (in the cytoplasm) that is the site of protein synthesis (place where proteins are made).

S

scavenger: animal that eats dead animals, example vulture.

science inquiry: the way scientists get their ideas to explain how the world works, by questioning, observing, inferring, experimenting, finding evidence, collecting and organizing data, drawing valid conclusions, and peer review.

selective breeding: mating organisms with different desirable traits (example large strawberries and sweet strawberries) to try to produce organisms with both desirable traits (large, sweet strawberries) in one organism.

sexual reproduction: producing offspring using two parents. Gametes (sperm and egg) from the two parents combine, giving offspring not identical to the parents.

slide: clear plastic or glass rectangle that a specimen is put on for examination under a microscope.

species: group of closely related organisms that can reproduce together.

sperm: male gamete (sex cell); has half the number of chromosomes as a regular cell in the body.

stomate: opening in a leaf, lets carbon dioxide in and oxygen out. Stomates also let water vapor go out.

succession: when one community of organisms is replaced by another community, until a stable climax community is reached.

synthesis: combining small molecules (example amino acids) to form large molecules (example protein).

T

testes: organ in male reproductive system; produces sperm (male gametes) and the hormone testosterone.

theory: group of ideas which explains many observations on one topic.

tissue: group of similar cells working together.

theory of evolution: change in species is caused by overproduction, limited resources, genetic variability, and natural selection.

thermal pollution: heating of the water in a lake by water that is used to cool machinery in a factory or power plant.

toxic wastes: poisonous substances that are dumped into the ground or into lakes and streams.

transport: taking materials (digested food from the digestive system, oxygen from the lungs, etc.) into the organism (living thing) and spreading (circulating) the material (digested food, oxygen, etc.) throughout the organism.

triple beam balance: instrument that measures mass.

U

uterus: organ in female reproductive system where a baby develops.

V

vaccine: made from weakened or dead pathogens; protects against a disease by stimulating the body to produce white blood cells or antibodies to fight that pathogen.

vacuole: storage sac inside the cytoplasm.

valid conclusion: based on observations and data.

valid experiment: an experiment that has condusions based on observations; experiment that has only one variable (one thing that is different between the experimental and control groups).

variation: difference among organisms of the same species (examples blue eyes and brown eyes).

vein: blood vessel that goes toward the heart.

W

water cycle: water evaporates from oceans, lakes, etc., forming water vapor; the water vapor rises, forming clouds. Water in clouds comes down as rain or snow, etc., and goes into lakes and oceans.

wet mount slide: slide used on a microscope where a specimen (example plant cell) is put on the slide and covered with liquid.

water pollution: harmful materials in the water; includes nitrogen and phosphorus pollution, sediment pollution, toxic wastes, and thermal pollution.

X

Y

Z

zygote: a fertilized egg.

Appendix II: Index

F

G

H

I

J

K

L

M

N

Z

APPENDIX III: NYS REGENTS EXAMS

DESCRIPTION OF THE LIVING ENVIRONMENT REGENTS

The Regents consists of Parts A, B, C. and D.

Part A has 30 multiple choice questions: 30 points.

Part B has multiple choice and short response questions.

Part C has extended constructed response questions-a series of interrelated questions based on topics and applications. Some of these questions may have several parts, which can include examples, how to design experiments, differences, and applications.

Part D consists of multiple choice and constructive response questions which are based on the New York State required laboratory activities for that year. Teachers should check with the New York State Education Department which four laboratory activities are required, because the required laboratory activities can vary from year to year.

The 85 points for Parts A, B, C, and D are scaled up to 100 points.

All questions on the Living Environment Regents must be answered.

STRATEGIES FOR TEST TAKING

1. Read each question carefully.

2. For a **lengthy question**, pick out the **main words** of the question to make it easier to find the answer.

3. If you do not understand the choices (which choice is correct), look at the question, think of how you would answer it, then see which choice is closest to your answer.

4. You must answer all questions. If you do not know the answer, take a guess. You don't lose points for wrong answers.

5. When you guess **eliminate the wrong choices**. This increases the chance of getting the right answer.

6. Stay with your first guess if you don't have a better one.

7 For questions which have models, graphs, data tables, or pictures:

Read every word in the model, graph, data table, or picture very carefully.

a. For graphs, look at the title and the labels on the x and y axes. See what the relationship is between the two variables.

b. For data tables, see the title and the headings on the top and sides of the columns. Look for the relationships shown in the table. You can be asked to draw a graph based on the data table.

c. For models, diagrams, and pictures, look carefully at all details in the model, picture, diagram, etc, and use **all** the information given to answer the question.

8 For longer constructed responses, each part of the question should be answered in a separate sentence or paragraph, so the one marking it knows where each answer is.

If a question has 3 or 4 parts, make sure you answer **all** the parts.

See what the question asks. Does the question ask for similarities, differences, or examples? Does the question ask why, explain why (give a reason) or how (by what means or what method)? Make sure you answer what the question asks for.

9 Go to the easy questions first. If there is a hard question, skip it and come back to it. Make a mark near the question to remember to come back for it.

10 Pace yourself. Make sure you have enough time to complete the exam.

11 Review the test.

12 Make sure you sign the declaration on the answer sheet.

REGENTS HIGH SCHOOL EXAMINATION

Friday, January 25, 2008
9:15 a.m. to 12:15 p.m., only

Part A

Answer all questions in this part. [30]

Directions (1–30): For *each* statement or question, write on your separate answer sheet the *number* of the word or expression that, of those given, best completes the statement or answers the question.

1 The transfer of genes from parents to their offspring is known as

(1) differentiation
(2) heredity
(3) immunity
(4) evolution

2 Damage to which structure will most directly disrupt water balance within a single-celled organism?

(1) ribosome
(2) cell membrane
(3) nucleus
(4) chloroplast

3 Two primary agents of cellular communication are

(1) chemicals made by blood cells and simple sugars
(2) hormones and carbohydrates
(3) enzymes and starches
(4) hormones and chemicals made by nerve cells

4 The function of most proteins depends primarily on the

(1) type and order of amino acids
(2) environment of the organism
(3) availability of starch molecules
(4) nutritional habits of the organism

5 Which procedure would most likely provide valid results in a test to determine if drug *A* would be effective in treating cancer in white mice?

(1) injecting 1 mL of drug *A* into 100 white mice with cancer
(2) injecting 1 mL of drug *A* into 100 white mice with cancer and 0.5 mL of drug *X* into 100 white mice without cancer
(3) injecting 1 mL of drug *A* into 100 white mice with cancer and 0.5 mL of drug *X* into another group of 100 white mice with cancer
(4) injecting 1 mL of drug *A* into 100 white mice with cancer and 1 mL of distilled water into another group of 100 white mice with cancer

6 The table below provides some information concerning organelles and organs.

Function	Organelle	Organ
gas exchange	cell membrane	lung
nutrition	food vacuole	stomach

Based on this information, which statement accurately compares organelles to organs?

(1) Functions are carried out more efficiently by organs than by organelles.
(2) Organs maintain homeostasis while organelles do not.
(3) Organelles carry out functions similar to those of organs.
(4) Organelles function in multicellular organisms while organs function in single-celled organisms.

7 In order to produce the first white marigold flower, growers began with the lightest yellow-flowered marigold plants. After crossing them, these plants produced seeds, which were planted, and only the offspring with very light-yellow flowers were used to produce the next generation. Repeating this process over many years, growers finally produced a marigold flower that is considered the first white variety of its species. This procedure is known as

(1) differentiation
(2) cloning
(3) gene insertion
(4) selective breeding

8 Chromosomes can be described as

(1) large molecules that have only one function
(2) folded chains of bonded glucose molecules
(3) reproductive cells composed of molecular bases
(4) coiled strands of genetic material

9 Some interactions in a desert community are shown in the diagram below.

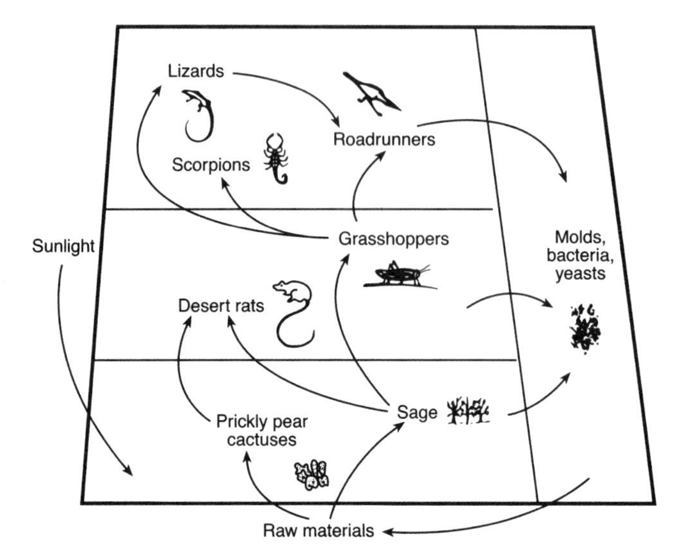

Which statement is a valid inference based on the diagram?

(1) Certain organisms may compete for vital resources.
(2) All these organisms rely on energy from decomposers.
(3) Organisms synthesize energy.
(4) All organisms occupy the same niche.

10 Which concept is best illustrated in the flowchart below?

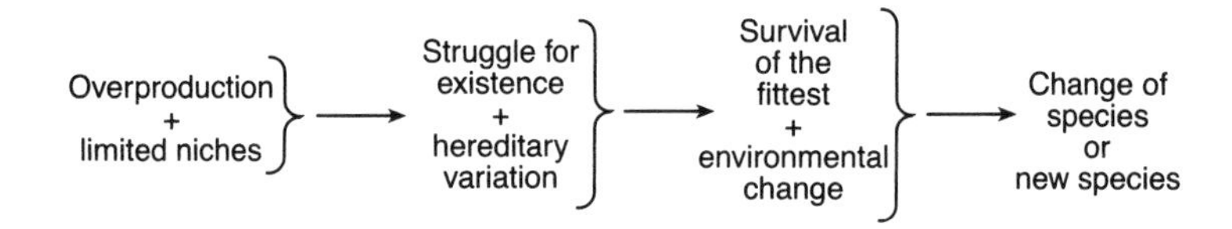

(1) natural selection
(2) genetic manipulation
(3) dynamic equilibrium
(4) material cycles

11 The headline "Improved Soybeans Produce Healthier Vegetable Oils" accompanies an article describing how a biotechnology company controls the types of lipids (fats) present in soybeans. The improved soybeans are most likely being developed by the process of

(1) natural selection
(2) asexual reproduction
(3) genetic engineering
(4) habitat modification

12 Which statement indicates one difference between the gene that codes for insulin and the gene that codes for testosterone in humans?

(1) The gene for insulin is replicated in vacuoles, while the gene for testosterone is replicated in mitochondria.
(2) The gene for insulin has a different sequence of molecular bases than the gene for testosterone.
(3) The gene for insulin is turned on in liver cells, but the gene for testosterone is not.
(4) The gene for insulin is a sequence of five different molecular bases while the gene for testosterone is a sequence of only four different molecular bases.

13 Cells that develop from a single zygote all contain identical DNA molecules. However, some of these cells will develop differently because

(1) different groups of cells containing the DNA may be exposed to different environmental conditions
(2) only the DNA in certain cells will replicate
(3) some of the DNA in some of the cells will be removed by chemical reactions
(4) DNA is functional in only 10% of the cells of the body

14 Which sequence represents the correct order of processes that result in the formation and development of an embryo?

(1) meiosis → fertilization → mitosis
(2) mitosis → fertilization → meiosis
(3) fertilization → meiosis → mitosis
(4) fertilization → mitosis → meiosis

15 The graph below shows the percent of variation for a given trait in four different populations of the same species. The populations inhabit similar environments.

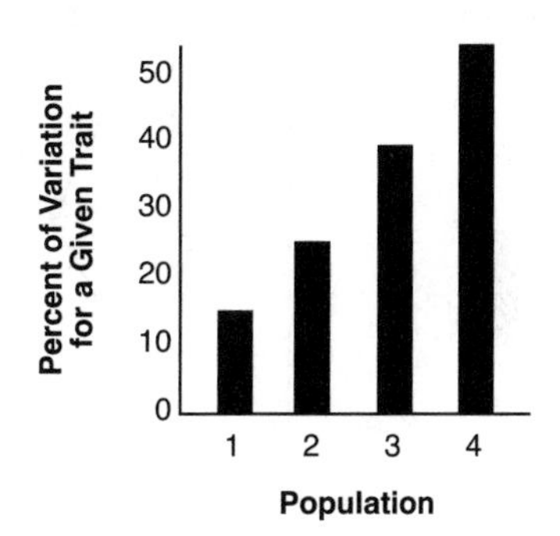

In which population will the greatest number of individuals most likely survive if a significant environmental change related to this trait occurs?

(1) 1
(2) 2
(3) 3
(4) 4

16 The sequence of events occurring in the life cycle of a bacterium is listed below.

(A) The bacterium copies its single chromosome.
(B) The copies of the chromosome attach to the cell membrane of the bacterium.
(C) As the cell grows, the two copies of the chromosome separate.
(D) The cell is separated by a wall into equal halves.
(E) Each new cell has one copy of the chromosome.

This sequence most closely resembles the process of

(1) recombination
(2) zygote formation
(3) mitotic cell division
(4) meiotic cell division

17 The diagram below illustrates possible evolutionary pathways of some species.

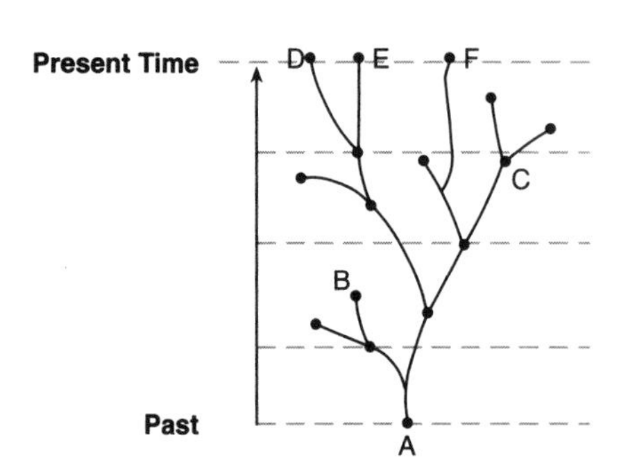

Which statement is a valid inference based on the information in the diagram?

(1) Species *A* is the common ancestor of all life on Earth.
(2) Species *D* is more closely related to species *E* than to species *F*.
(3) Species *B* is the ancestor of species *F*.
(4) Species *C* is the ancestor of species that exist at the present time.

18 The diagram below represents stages in the processes of reproduction and development in an animal.

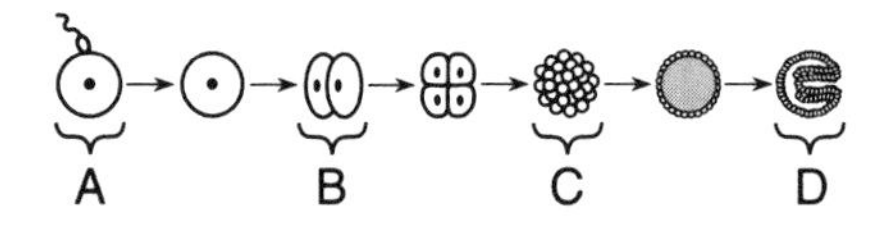

Cells containing only half of the genetic information characteristic of this species are found at

(1) *A* (3) *C*
(2) *B* (4) *D*

19 Which hormones most directly influence the uterus during pregnancy?

(1) testosterone and insulin
(2) progesterone and testosterone
(3) estrogen and insulin
(4) progesterone and estrogen

20 The diagram below represents the human female reproductive system.

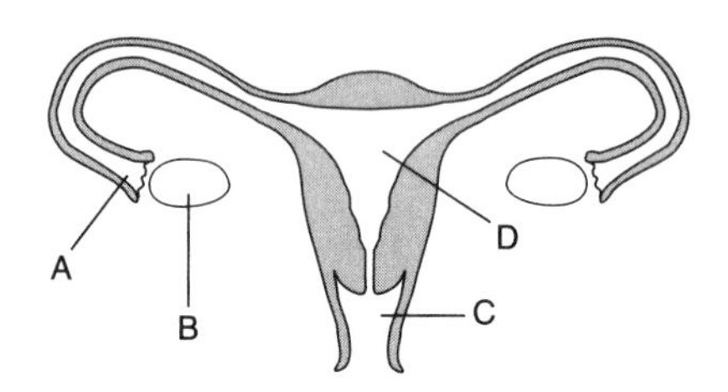

Exposure to radiation or certain chemicals could alter the genetic information in the gametes that form in structure

(1) *A* (3) *C*
(2) *B* (4) *D*

21 All life depends on the availability of usable energy. This energy is released when

(1) organisms convert solar energy into the chemical energy found in food molecules
(2) respiration occurs in the cells of producers and high-energy molecules enter the atmosphere
(3) cells carry out the process of respiration
(4) animal cells synthesize starch and carbon dioxide

22 The sweet taste of freshly picked corn is due to the high sugar content in the kernels. Enzyme action converts about 50% of the sugar to starch within one day after picking. To preserve its sweetness, the freshly picked corn is immersed in boiling water for a few minutes, and then cooled.

Which statement most likely explains why the boiled corn kernels remain sweet?

(1) Boiling destroys sugar molecules so they cannot be converted to starch.
(2) Boiling kills a fungus on the corn that is needed to convert sugar to starch.
(3) Boiling activates the enzyme that converts amino acids to sugar.
(4) Boiling deactivates the enzyme responsible for converting sugar to starch.

23 One biotic factor that affects consumers in an ocean ecosystem is

(1) number of autotrophs
(2) temperature variation
(3) salt content
(4) pH of water

24 Which component of a stable ecosystem can *not* be recycled?

(1) oxygen
(2) water
(3) energy
(4) nitrogen

25 A food web is represented in the diagram below.

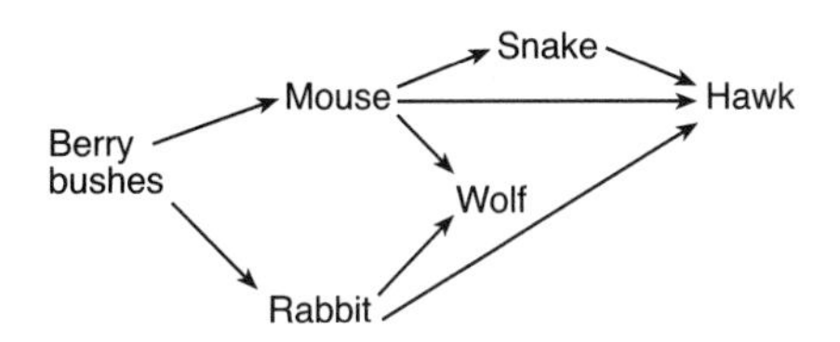

Which population in this food web would most likely be *negatively* affected by an increase in the mouse population?

(1) snake
(2) rabbit
(3) wolf
(4) hawk

26 Years after the lava from an erupting volcano destroyed an area, grasses started to grow in that area. The grasses were gradually replaced by shrubs, evergreen trees, and finally, by a forest that remained for several hundred years. This entire process is an example of

(1) feedback
(2) ecological succession
(3) plant preservation
(4) deforestation

27 Increased industrialization will most likely

(1) decrease available habitats
(2) increase environmental carrying capacity for native species
(3) increase the stability of ecosystems
(4) decrease global warming

28 A five-year study was carried out on a population of algae in a lake. The study found that the algae population was steadily decreasing in size. Over the five-year period this decrease most likely led to

(1) a decrease in the amount of nitrogen released into the atmosphere
(2) an increase in the amount of oxygen present in the lake
(3) an increase in the amount of water vapor present in the atmosphere
(4) a decrease in the amount of oxygen released into the lake

29 Which result of technological advancement has a positive effect on the environment?

(1) development of new models of computers each year, with disposal of the old computers in landfills
(2) development of new models of cars that travel fewer miles per gallon of gasoline
(3) development of equipment that uses solar energy to charge batteries
(4) development of equipment to speed up the process of cutting down trees

30 The diagram below represents a biological process.

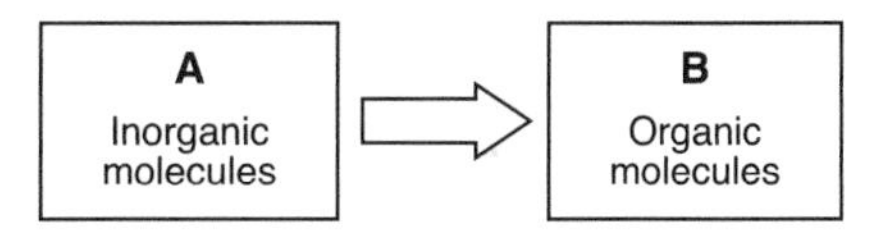

Which set of molecules is best represented by letters *A* and *B*?

(1) *A*: oxygen and water
B: glucose
(2) *A*: glucose
B: carbon dioxide and water
(3) *A*: carbon dioxide and water
B: glucose
(4) *A*: glucose
B: oxygen and water

Part B–1

Answer all questions in this part. [12]

Directions (31–42): For *each* statement or question, write on the separate answer sheet the *number* of the word or expression that, of those given, best completes the statement or answers the question.

31 A biologist used the Internet to contact scientists around the world to obtain information about declining amphibian populations. He was able to gather data on 936 populations of amphibians, consisting of 157 species from 37 countries. Results showed that the overall numbers of amphibians dropped 15% a year from 1960 to 1966 and continued to decline about 2% a year through 1997.

What is the importance of collecting an extensive amount of data such as this?

(1) Researchers will now be certain that the decline in the amphibian populations is due to pesticides.
(2) The data collected will prove that all animal populations around the world are threatened.
(3) Results from all parts of the world will be found to be identical.
(4) The quantity of data will lead to a better understanding of the extent of the problem.

32 The first trial of a controlled experiment allows a scientist to isolate and test

(1) a logical conclusion
(2) a variety of information
(3) a single variable
(4) several variables

33 A student studied how the amount of oxygen affects ATP production in muscle cells. The data for amount *X* are shown in the graph below.

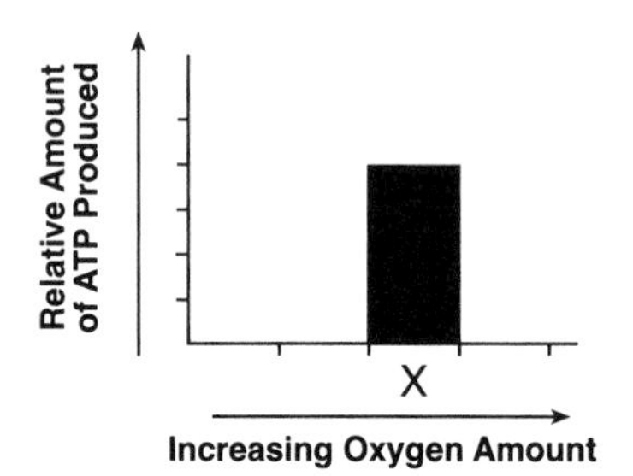

If the student supplies the muscle cells with *less* oxygen in a second trial of the investigation, a bar placed on the graph to represent the results of this trial would most likely be

(1) shorter than bar *X* and placed to the left of bar *X*
(2) shorter than bar *X* and placed to the right of bar *X*
(3) taller than bar *X* and placed to the left of bar *X*
(4) taller than bar *X* and placed to the right of bar *X*

34 Some steps involved in DNA replication and protein synthesis are summarized in the table below.

Step A	DNA is copied and each new cell gets a full copy.
Step B	Information copied from DNA moves to the cytoplasm.
Step C	Proteins are assembled at the ribosomes.
Step D	Proteins fold and begin functioning.

In which step would a mutation lead directly to the formation of an altered gene?

(1) *A* (3) *C*
(2) *B* (4) *D*

35 Species *A*, *B*, *C*, and *D* are all different heterotrophs involved in the same food chain in an ecosystem. The chart below shows the population of each species at the same time on a summer day.

Species	Population
A	847
B	116
C	85
D	6

Which statement best describes one of these species of heterotrophs?

(1) Species *A* is the most numerous because it can make its own food.
(2) Species *B* probably feeds on species *D*.
(3) Species *C* and *B* interbred to produce species *A*.
(4) Species *D* is most likely the top predator in the food chain.

36 Students conducting a study on an insect population placed 25 insects of the same size in a box. The amount of food, water, and shelter available to the insects was kept constant. Each month, students removed and counted the number of insects present, recorded the total, and returned the insects to the box. The graph below shows the number of insects in the box over a 12-month period.

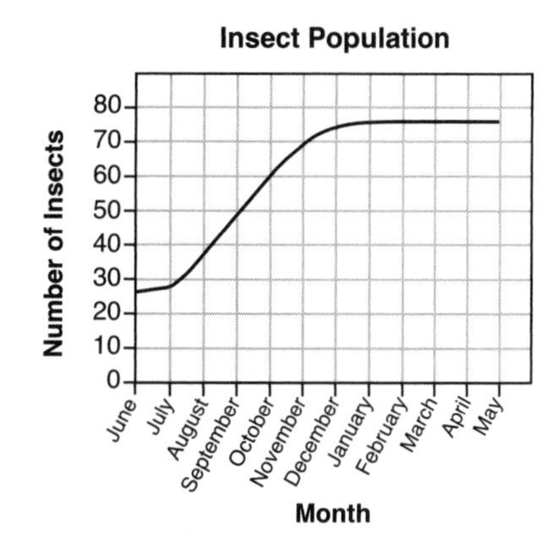

What inference can be made regarding this insect population?

(1) All the insects in the box are the same age.
(2) The insects hibernated from January to April.
(3) The population has carnivorous members.
(4) The population reached carrying capacity by January.

37 The relative amount of oxygen in the atmosphere of Earth over millions of years is shown in the graph below.

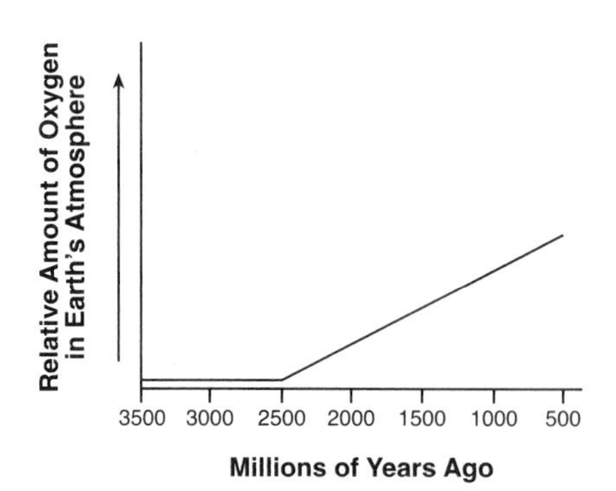

At what point in the history of Earth did autotrophs most likely first appear?

(1) 3500 million years ago
(2) 2500 million years ago
(3) 1500 million years ago
(4) 500 million years ago

38 A biologist collected the data shown in the table below.

Data Table

Type of Organism	Number of Organisms in a Field		
	May	July	September
grasshoppers	100	500	150
birds	25	100	10
spiders	75	200	50

Which statement is supported by the data in the table?

(1) Populations do not vary from month to month.
(2) The populations are highest in September.
(3) The grasshoppers increased in length in July.
(4) Seasonal variations may affect populations.

Base your answers to questions 39 and 40 on the diagram below, which represents possible relationships between animals in the family tree of the modern horse, and on your knowledge of biology.

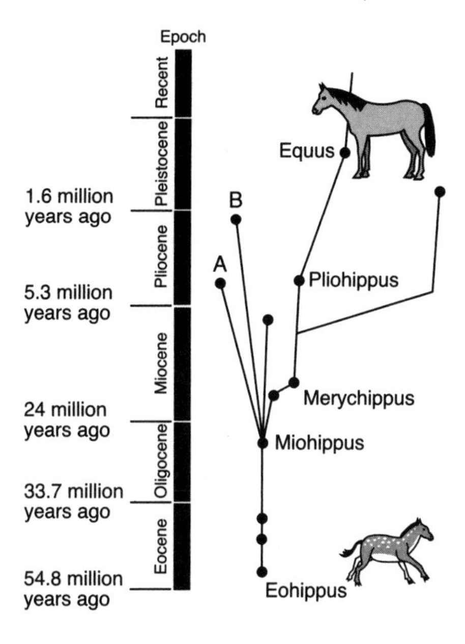

39 One possible conclusion that can be drawn regarding ancestral horses *A* and *B* is that

(1) *A* was better adapted to changes that occurred during the Pliocene Epoch than was *B*
(2) the areas that *B* migrated to contained fewer varieties of producers than did the areas that *A* migrated to
(3) competition between *A* and *B* led to the extinction of *Pliohippus*
(4) the adaptive characteristics present in both *A* and *B* were insufficient for survival

40 *Miohippus* has been classified as a browser (an animal that feeds on shrubs and trees) while *Merychippus* has been classified as a grazer (an animal that feeds on grasses). One valid inference that can be made regarding the evolution of modern horses based on this information is that

(1) *Eohippus* inhabited grassland areas throughout the world
(2) *Pliohippus* had teeth adapted for grazing
(3) *Equus* evolved as a result of the migration of *Pliohippus* into forested areas due to increased competition
(4) ecological succession led to changes in tooth structure during the Eocene Epoch

---

Base your answers to questions 41 and 42 on the diagram below and on your knowledge of biology.

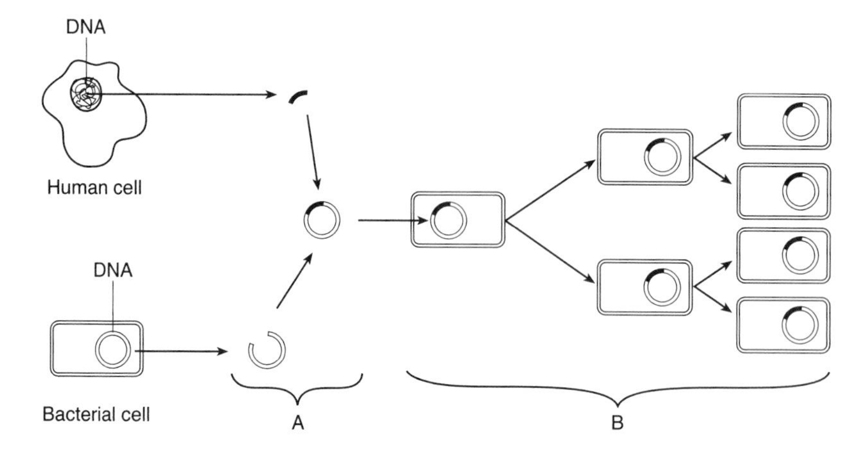

41 In the procedure indicated by letter *A*, DNA segments from humans and bacteria are joined by the action of

(1) starch molecules
(2) simple sugars
(3) enzymes
(4) hormones

42 Which process is indicated by letter *B*?

(1) natural selection
(2) asexual reproduction
(3) sexual reproduction
(4) gene deletion

Part B–2

Answer all questions in this part. [13]

Directions (43–55): For those questions that are followed by four choices, circle the *number* of the choice that, of those given, best completes the statement or answers the question. For all other questions in this part, follow the directions given in the question and record your answers in the spaces provided.

For Teacher Use Only

43 Select *one* of the paired items below and describe how the first item in the pair regulates the second item for the maintenance of homeostasis. [1]

insulin—blood sugar level
CO_2 in blood—breathing rate
activity of guard cells—water loss from a leaf

____

____ 43 ☐

44 Explain how harmful substances in the blood of a pregnant female can enter a fetus even though the blood vessels of the mother and fetus are *not* directly connected. [1]

____

____ 44 ☐

45 Identify *one* farming practice that could be a source of environmental pollution. [1]

____

____ 45 ☐

Base your answers to questions 46 through 49 on the passage below and on your knowledge of biology.

For Teacher Use Only

When humans perspire, water, urea, and salts containing sodium are removed from the blood. Drinking water during extended periods of physical exercise replenishes the water but not the sodium. This increase in water dilutes the blood and may result in the concentration of sodium dropping low enough to cause a condition known as hyponatremia.

Symptoms of hyponatremia include headache, nausea, and lack of coordination. Left untreated, it can lead to coma and even death. The body has a variety of feedback mechanisms that assist in regulating water and sodium concentrations in the blood. The kidneys play a major role in these mechanisms, as they filter the blood and produce urine.

46 The best way to reduce the symptoms of hyponatremia would be to

(1) drink more water

(2) eat chocolate

(3) eat salty foods

(4) drink cranberry juice

46 ☐

47 Many runners pour water on their bodies during a race. Explain how this action helps to maintain homeostasis. [1]

____

____

47 ☐

48 How would running in a marathon on a warm day most likely affect urine production? Support your answer. [1]

____

____

48 ☐

49 Many people today drink sport drinks containing large amounts of sodium. Describe *one* possible effect this might have on a person who is *not* very active. [1]

____

____

49 ☐

50 Data from two different cells are shown in the graphs below.

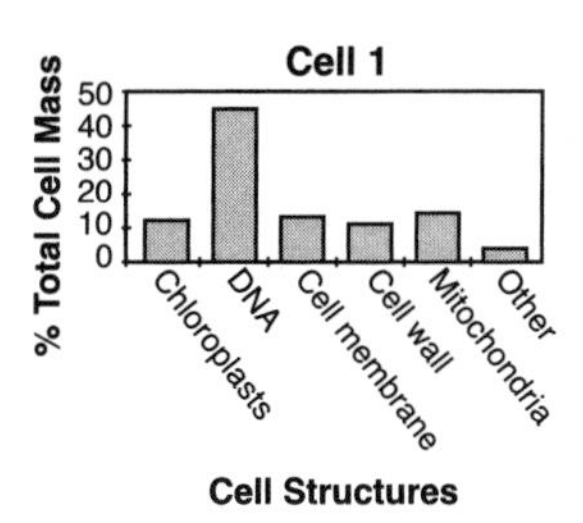

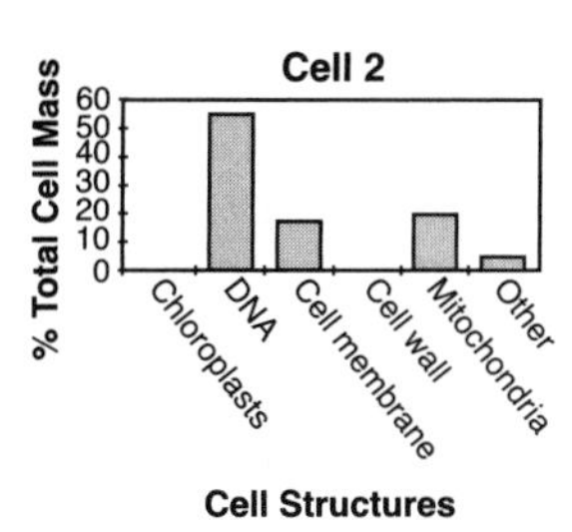

Which cell is most likely a plant cell? Support your answer. [1]

____

____

For Teacher Use Only

50 ☐

Base your answers to questions 51 through 55 on the information below and on your knowledge of biology. The average level of carbon dioxide in the atmosphere has been measured for the past several decades. The data collected are shown in the table below.

Average CO_2 Levels in the Atmosphere

Year	CO_2 (in parts per million)
1960	320
1970	332
1980	350
1990	361
2000	370

Directions (51 and 52): Using the information in the data table, construct a line graph on the grid on the next page, following the directions below.

51 Mark an appropriate scale on each labeled axis. [1]

52 Plot the data on the grid. Surround each point with a small circle and connect the points. [1]

Example: 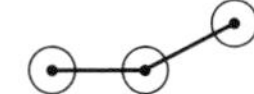

For Teacher Use Only

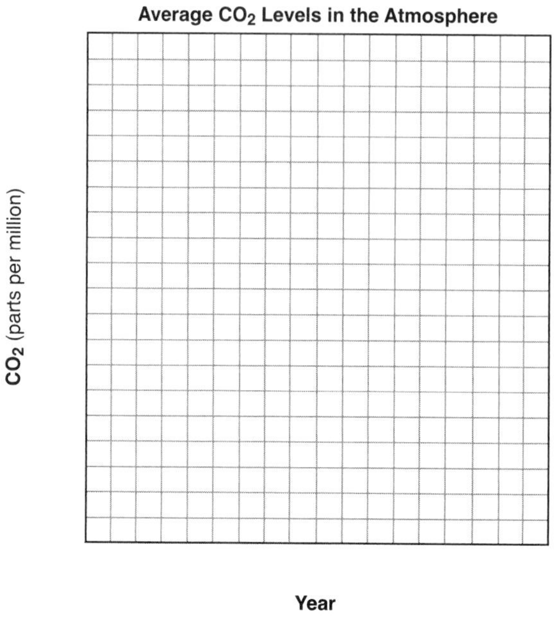

Average CO_2 Levels in the Atmosphere

CO_2 (parts per million)

Year

51

52

53 Identify *one* specific human activity that could be responsible for the change in carbon dioxide levels from 1960 to 2000. [1]

____

____

53

54 State *one* possible *negative* effect this change in CO_2 level has had on the environment of Earth. [1]

____

____

54

55 Calculate the net change in CO_2 level in parts per million (ppm) during the years 1960 through 2000. [1]

__________ **ppm**

55

Part C

Answer all questions in this part. [17]

Directions (56–65): Record your answers in the spaces provided in this examination booklet.

For Teacher Use Only

56 Smallpox is a disease caused by a specific virus, while the common cold can be caused by over 100 different viruses. Explain why it is possible to develop a vaccine to prevent smallpox, but it is difficult to develop a vaccine to prevent the common cold. In your answer be sure to:

- identify the substance in a vaccine that makes the vaccine effective [1]
- explain the relationship between a vaccine and white blood cell activity [1]
- explain why the response of the immune system to a vaccine is specific [1]
- state *one* reason why it would be difficult to develop a vaccine to be used against the common cold [1]

56 ☐

Base your answers to questions 57 through 59 on the information below and on your knowledge of biology.

Untreated organic wastes were accidentally discharged into a river from a sewage treatment plant. The graph below shows the dissolved oxygen content of water samples taken from the river at specific distances downstream from the plant, both before, and then three days after the discharge occurred.

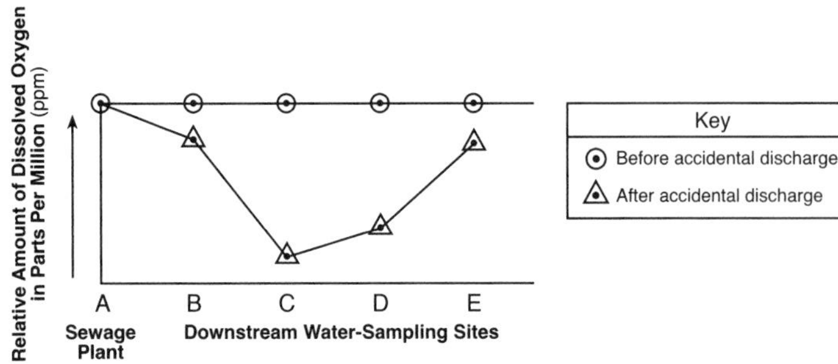

57 State why this accident would be expected to benefit the decomposers in the river below the sewage plant. [1]

____

____

58 Explain why an energy-releasing process occurring in the mitochondria of the decomposer organisms is most likely responsible for the change indicated by the data shown at sampling site *C* in the graph. [1]

____

____

59 State *one* reason why the statement below is correct.

"The effects of the accidental discharge are not expected to last for a long time." [1]

____

____

For Teacher Use Only

57 ☐

58 ☐

59 ☐

60 The photograph below shows a pill bug. Pill bugs are small animals frequently found in wooded areas near decomposing organic material.

For Teacher Use Only

Describe some parts of an experiment to determine the preference of pill bugs for light or darkness. In your answer be sure to:

- state a hypothesis [1]
- identify the independent variable in the experiment [1]
- identify *two* conditions that should be kept the same in all experimental setups [1]
- state *one* example of experimental data that would support your hypothesis [1]

60 ☐

Base your answers to questions 61 through 64 on the information below and on your knowledge of biology.

For Teacher Use Only

In recent years, the striped bass population in Chesapeake Bay has been decreasing. This is due, in part, to events known as "fish kills," a large die-off of fish. Fish kills occur when oxygen-consuming processes in the aquatic ecosystem require more oxygen than the plants in the ecosystem produce, thereby reducing the amount of dissolved oxygen available to the fish.

One proposed explanation for the increased fish kills in recent years is that human activities have increased the amount of sediment suspended in the water of Chesapeake Bay, largely due to increased erosion into its tributary streams. The sediment acts as a filter for sunlight, which causes a decrease in the intensity of the sunlight that reaches the aquatic plants in the Chesapeake Bay ecosystem.

61 Identify *one* abiotic factor in the Chesapeake Bay ecosystem involved in the fish kills. [1]

____

61

62 Identify the process carried out by organisms that uses oxygen and contributes to the fish kills. [1]

____

62

63 State *one* way humans have contributed to the *decrease* of the striped bass population in Chesapeake Bay. [1]

____

____

63

64 State how a *decrease* in the amount of light may be responsible for fish kills in the Chesapeake Bay area. [1]

____

____

____

64

65 Over the past few decades, many oil companies have discovered oil below the seafloor near the coasts of many states. Some states, however, refuse to permit offshore oil drilling, fearing it might damage the environment.

Discuss both sides of this issue. In your answer, be sure to:

- state *one* way in which offshore oil drilling might have a long-term *negative* effect on the environment [1]
- state *one* way in which offshore oil drilling could benefit society [1]

____

____

____

____

____

____

For Teacher Use Only

65 ☐

Part D

Answer all questions in this part. [13]

Directions (66–75): For those questions that are followed by four choices, circle the *number* of the choice, that, of those given, best completes the statement or answers the question. For all other questions in this part, follow the directions given in the question and record your answers in the spaces provided.

For Teacher Use Only

66 Researchers discovered four different species of finches on one of the Galapagos Islands. DNA analysis showed that these four species, shown in the illustration below, are closely related even though they vary in beak shape and size. It is thought that they share a common ancestor.

Which factor most likely influenced these differences in beak size and shape?

(1) Birds with poorly adapted beaks changed their beaks to get food.

(2) Birds with yellow beaks were able to hide from predators.

(3) Birds with successful beak adaptations obtained food and survived to have offspring.

(4) Birds with large, sharp beaks become dominant.

66 ☐

67 Relationships between plant species may most accurately be determined by comparing the

(1) habitats in which they live

(2) structure of guard cells

(3) base sequences of DNA

(4) shape of their leaves

67 ☐

Base your answers to questions 68 through 70 on the information below and on your knowledge of biology.

Cytochrome c is an enzyme located in the mitochondria of many types of cells. The number of differences in the amino acid sequences of Cytochrome c from different species are compared to human Cytochrome c in the data table below.

Differences in Amino Acid Sequences

Organism	Number of Differences in Cytochrome c Compared to Humans
tuna	21
mold	48
moth	31
dog	11
horse	12
chicken	13
monkey	1

68 Of the organisms listed below, which one has a DNA code for Cytochrome c that is most similar to that of a human?

(1) tuna

(2) chicken

(3) moth

(4) dog

69 The fact that all of these organisms contain Cytochrome c could lead to the inference that

(1) Cytochrome c is essential for the reproduction of all organisms

(2) these organisms have all evolved from an ancestor that produced Cytochrome c

(3) mutations in genes that code for Cytochrome c always occur during DNA replication

(4) only heterotrophs make Cytochrome c

For Teacher Use Only

68 ☐

69 ☐

70 Cytochrome c is most likely a

(1) protein molecule

(2) material containing genes

(3) carbohydrate that is absorbed by cells

(4) component of the membrane around the cell

For Teacher Use Only

70 ☐

71 The data table below compares blood flow in various human body structures, both at rest and during strenuous exercise.

Structure	**Blood Flow at Rest** (mL/min)	**Blood Flow During Strenuous Exercise** (mL/min)
heart	250	750
skeletal muscle	1200	12,500
digestive organs	1400	600

Select *one* structure from the data table and write its name in the space below. Explain *one* way that the change in the rate of blood flow in this structure helps maintain homeostasis during exercise. [1]

Structure: ______________________________

____

____

71 ☐

For Teacher Use Only

Base your answers to questions 72 and 73 on the information and table below and on your knowledge of biology.

A model of a cell is prepared and placed in a beaker of fluid as shown in the diagram below. The letters *A*, *B*, and *C* represent substances in the initial experimental setup.

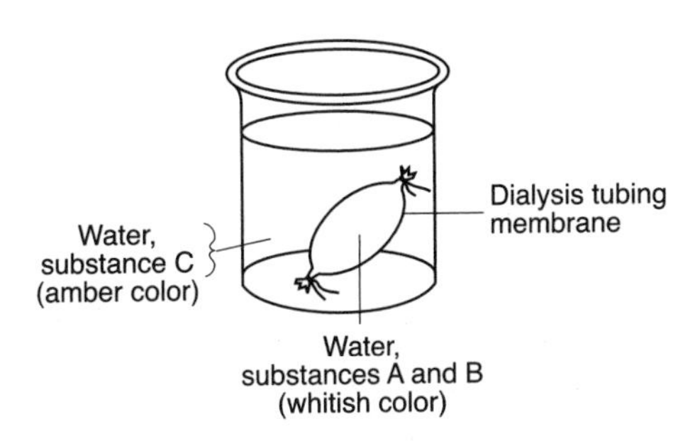

The table below summarizes the content and appearance of the cell model and beaker after 20 minutes.

Results After 20 Minutes

	Outside of Cell Model	Inside of Cell Model
Substances	water, A, C	water, A, B, C
Color	amber	blue black

72 Complete the table below to summarize a change in location of substance *C* in the experimental setup. [3]

Name of Substance C	Direction of Movement of Substance C	Reason for the Movement of Substance C

72 ☐

73 Identify substance *B* and explain why it did *not* move out of the model cell. [2]

Substance: ______________________________

____

____

____

For Teacher Use Only

73 [ ]

74 Species of finches are represented in the diagram below.

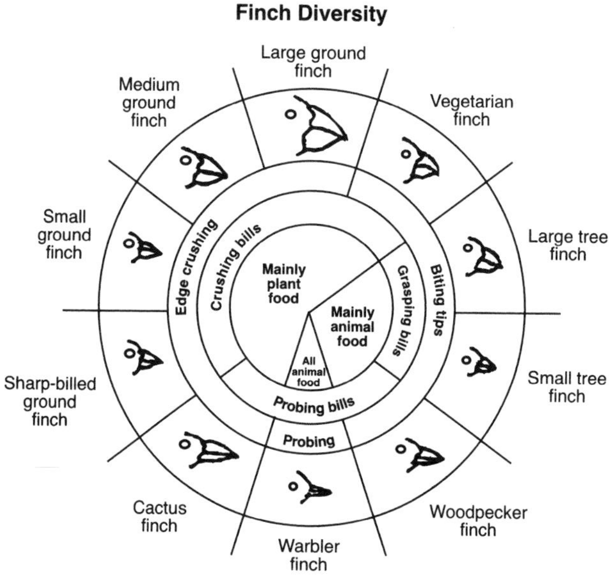

Finch Diversity

State the name of *one* species of finch from the diagram that is most likely to compete with the small tree finch if they lived on the same island. Support your answer with an explanation. [1]

Species: ______________________________

____

____

____

74 [ ]

75 Electrophoresis is a method of

(1) separating DNA fragments

(2) changing the genetic code of an organism

(3) indicating the presence of starch

(4) separating colored compounds on a strip of paper

For Teacher Use Only

75

Tear Here

The University of the State of New York

REGENTS HIGH SCHOOL EXAMINATION

LIVING ENVIRONMENT

Friday, January 25, 2008 — 9:15 a.m. to 12:15 p.m., only

ANSWER SHEET

Student .... Sex: ☐ Female ☐ Male

Teacher ....

School .... Grade ....

Part	Maximum Score	Student's Score
A	30	
B–1	12	
B–2	13	
C	17	
D	13	
Total Raw Score (maximum Raw Score: 85)		
Final Score (from conversion chart)		

Raters' Initials

Rater 1 .... **Rater 2** ....

Record your answers to Part A and Part B–1 on this answer sheet.

Part A

1	11	21 ....
2	12	22 ....
3	13	23 ....
4	14	24 ....
5	15	25 ....
6	16	26 ....
7	17	27 ....
8	18	28 ....
9	19	29 ....
10	20	30 ....

Part A Score

Part B–1

31	37 ....
32	38 ....
33	39 ....
34	40 ....
35	41 ....
36	42 ....

Part B–1 Score

The declaration below must be signed when you have completed the examination.

I do hereby affirm, at the close of this examination, that I had no unlawful knowledge of the questions or answers prior to the examination and that I have neither given nor received assistance in answering any of the questions during the examination.

Signature

Tear Here

REGENTS HIGH SCHOOL EXAMINATION

Tuesday, June 24, 2008
9:15 a.m. to 12:15 p.m., only

Part A

Answer all questions in this part. [30]

Directions (1–30): For *each* statement or question, write on your separate answer sheet the *number* of the word or expression that, of those given, best completes the statement or answers the question.

1 The chart below contains both autotrophic and heterotrophic organisms.

A	owl	cat	shark
B	mouse	corn	dog
C	squirrel	bluebird	alga

Organisms that carry out only heterotrophic nutrition are found in

(1) row *A*, only
(2) row *B*, only
(3) rows *A* and *B*
(4) rows *A* and *C*

2 A stable pond ecosystem would *not* contain

(1) materials being cycled
(2) oxygen
(3) decomposers
(4) more consumers than producers

3 Although all of the cells of a human develop from one fertilized egg, the human is born with many different types of cells. Which statement best explains this observation?

(1) Developing cells may express different parts of their identical genetic instructions.
(2) Mutations occur during development as a result of environmental conditions.
(3) All cells have different genetic material.
(4) Some cells develop before other cells.

4 Humans require organ systems to carry out life processes. Single-celled organisms do not have organ systems and yet they are able to carry out life processes. This is because

(1) human organ systems lack the organelles found in single-celled organisms
(2) a human cell is more efficient than the cell of a single-celled organism
(3) it is not necessary for single-celled organisms to maintain homeostasis
(4) organelles present in single-celled organisms act in a manner similar to organ systems

5 Certain poisons are toxic to organisms because they interfere with the function of enzymes in mitochondria. This results directly in the inability of the cell to

(1) store information
(2) build proteins
(3) release energy from nutrients
(4) dispose of metabolic wastes

6 At warm temperatures, a certain bread mold can often be seen growing on bread as a dark-colored mass. The same bread mold growing on bread in a cooler environment is red in color. Which statement most accurately describes why this change in the color of the bread mold occurs?

(1) Gene expression can be modified by interactions with the environment.
(2) Every organism has a different set of coded instructions.
(3) The DNA was altered in response to an environmental condition.
(4) There is no replication of genetic material in the cooler environment.

7 Asexually reproducing organisms pass on hereditary information as

(1) sequences of A, T, C, and G
(2) chains of complex amino acids
(3) folded protein molecules
(4) simple inorganic sugars

8 Species of bacteria can evolve more quickly than species of mammals because bacteria have

(1) less competition
(2) more chromosomes
(3) lower mutation rates
(4) higher rates of reproduction

9 The diagram below represents the synthesis of a portion of a complex molecule in an organism.

Building blocks → Product

Which row in the chart could be used to identify the building blocks and product in the diagram?

Row	Building Blocks	Product
(1)	starch molecules	glucose
(2)	amino acid molecules	part of protein
(3)	sugar molecules	ATP
(4)	DNA molecules	part of starch

10 Which diagram best represents the relative locations of the structures in the list below?

A–chromosome
B–nucleus
C–cell
D–gene

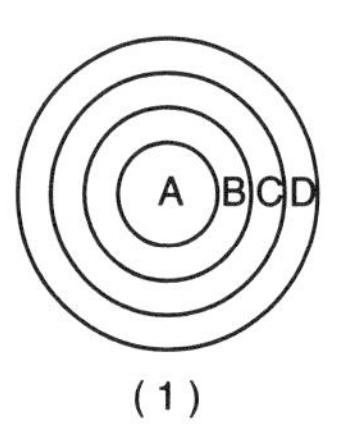

(1)

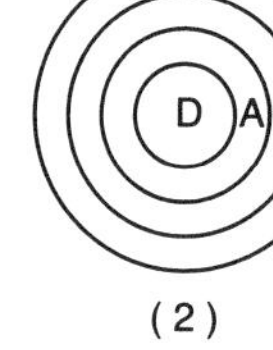

(2)

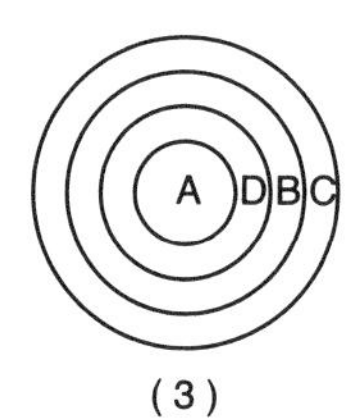

(3)

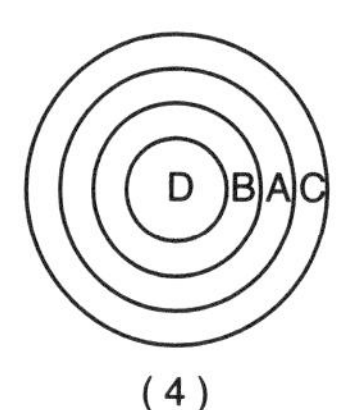

(4)

11 Which nuclear process is represented below?

A DNA molecule untwists. → The two strands of DNA separate. → Molecular bases pair up. → Two identical DNA molecules are produced.

(1) recombination
(2) fertilization
(3) replication
(4) mutation

12 For centuries, certain animals have been crossed to produce offspring that have desirable qualities. Dogs have been mated to produce Labradors, beagles, and poodles. All of these dogs look and behave very differently from one another. This technique of producing organisms with specific qualities is known as

(1) gene replication
(2) natural selection
(3) random mutation
(4) selective breeding

13 Certain insects resemble the bark of the trees on which they live. Which statement provides a possible biological explanation for this resemblance?

(1) The insects needed camouflage so they developed protective coloration.
(2) Natural selection played a role in the development of this protective coloration.
(3) The lack of mutations resulted in the protective coloration.
(4) The trees caused mutations in the insects that resulted in protective coloration.

14 When is extinction of a species most likely to occur?

(1) when environmental conditions remain the same and the proportion of individuals within the species that lack adaptive traits increases
(2) when environmental conditions remain the same and the proportion of individuals within the species that possess adaptive traits increases
(3) when environmental conditions change and the adaptive traits of the species favor the survival and reproduction of some of its members
(4) when environmental conditions change and the members of the species lack adaptive traits to survive and reproduce

15 In what way are photosynthesis and cellular respiration similar?

(1) They both occur in chloroplasts.
(2) They both require sunlight.
(3) They both involve organic and inorganic molecules.
(4) They both require oxygen and produce carbon dioxide.

16 Which process will increase variations that could be inherited?

(1) mitotic cell division
(2) active transport
(3) recombination of genes
(4) synthesis of proteins

17 Some cells involved in the process of reproduction are represented in the diagram below.

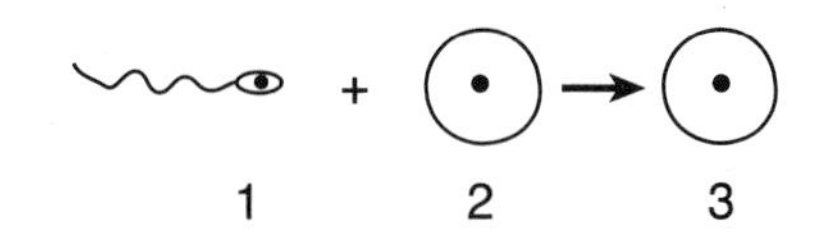

The process of meiosis formed

(1) cell 1, only
(2) cells 1 and 2
(3) cell 3, only
(4) cells 2 and 3

18 Kangaroos are mammals that lack a placenta. Therefore, they must have an alternate way of supplying the developing embryo with

(1) nutrients
(2) carbon dioxide
(3) enzymes
(4) genetic information

19 Which substance is the most direct source of the energy that an animal cell uses for the synthesis of materials?

(1) ATP
(2) glucose
(3) DNA
(4) starch

20 To increase chances for a successful organ transplant, the person receiving the organ should be given special medications. The purpose of these medications is to

(1) increase the immune response in the person receiving the transplant
(2) decrease the immune response in the person receiving the transplant
(3) decrease mutations in the person receiving the transplant
(4) increase mutations in the person receiving the transplant

21 The diagram below represents the cloning of a carrot plant.

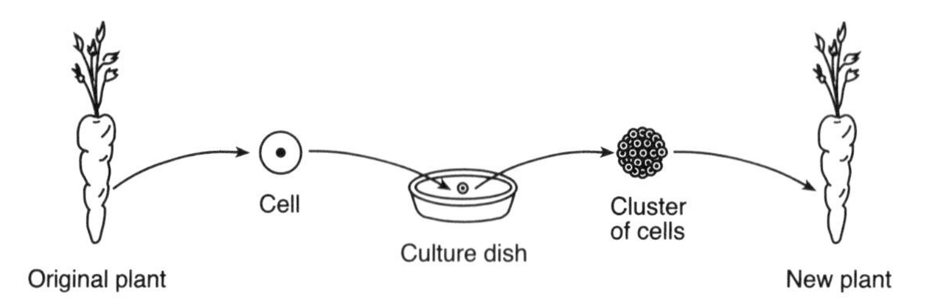

Compared to each cell of the original carrot plant, each cell of the new plant will have

(1) the same number of chromosomes and the same types of genes
(2) the same number of chromosomes, but different types of genes
(3) half the number of chromosomes and the same types of genes
(4) half the number of chromosomes, but different types of genes

22 The development of an embryo is represented in the diagram below.

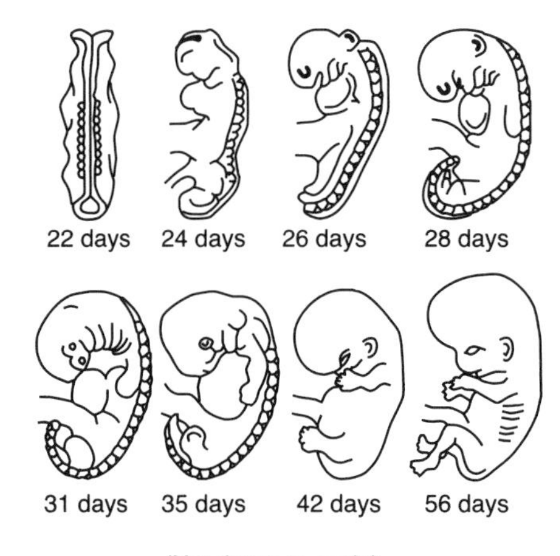

(Not drawn to scale)

These changes in the form of the embryo are a direct result of

(1) uncontrolled cell division and mutations
(2) differentiation and growth
(3) antibodies and antigens inherited from the father
(4) meiosis and fertilization

23 The diagram below represents an event that occurs in the blood.

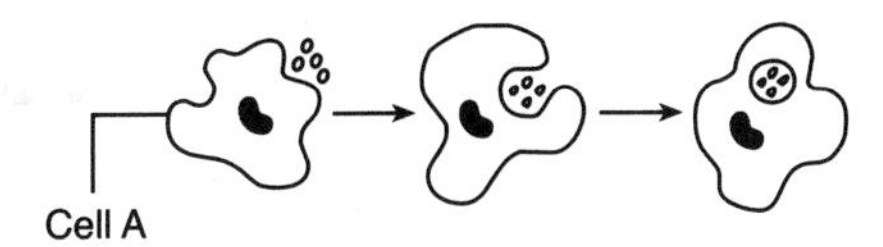

Which statement best describes this event?

(1) Cell *A* is a white blood cell releasing antigens to destroy bacteria.
(2) Cell *A* is a cancer cell produced by the immune system and it is helping to prevent disease.
(3) Cell *A* is a white blood cell engulfing disease-causing organisms.
(4) Cell *A* is protecting bacteria so they can reproduce without being destroyed by predators.

24 In an ecosystem, the growth and survival of organisms are dependent on the availability of the energy from the Sun. This energy is available to organisms in the ecosystem because

(1) producers have the ability to store energy from light in organic molecules
(2) consumers have the ability to transfer chemical energy stored in bonds to plants
(3) all organisms in a food web have the ability to use light energy
(4) all organisms in a food web feed on autotrophs

25 Which factor has the greatest influence on the type of ecosystem that will form in a particular geographic area?

(1) genetic variations in the animals
(2) climate conditions
(3) number of carnivores
(4) percentage of nitrogen gas in the atmosphere

26 Farming reduces the natural biodiversity of an area, yet farms are necessary to feed the world's human population. This situation is an example of

(1) poor land use
(2) a trade-off
(3) conservation
(4) a technological fix

27 A food chain is represented below.

Grass → Cricket → Frog → Owl

This food chain contains

(1) 4 consumers and no producers
(2) 1 predator, 1 parasite, and 2 producers
(3) 2 carnivores and 2 herbivores
(4) 2 predators, 1 herbivore, and 1 producer

28 A volcanic eruption destroyed a forest, covering the soil with volcanic ash. For many years, only small plants could grow. Slowly, soil formed in which shrubs and trees could grow. These changes are an example of

(1) manipulation of genes
(2) evolution of a species
(3) ecological succession
(4) equilibrium

29 A major reason that humans can have such a significant impact on an ecological community is that humans

(1) can modify their environment through technology
(2) reproduce faster than most other species
(3) are able to increase the amount of finite resources available
(4) remove large amounts of carbon dioxide from the air

30 Rabbits are herbivores that are not native to Australia. Their numbers have increased steadily since being introduced into Australia by European settlers. One likely reason the rabbit population was able to grow so large is that the rabbits

(1) were able to prey on native herbivores
(2) reproduced more slowly than the native animals
(3) successfully competed with native herbivores for food
(4) could interbreed with the native animals

Part B–1

Answer all questions in this part. [12]

Directions (31–42): For *each* statement or question, write on the separate answer sheet the *number* of the word or expression that, of those given, best completes the statement or answers the question.

31 Which laboratory procedure is represented in the diagram below?

(1) placing a coverslip over a specimen
(2) removing a coverslip from a slide
(3) adding stain to a slide without removing the coverslip
(4) reducing the size of air bubbles under a coverslip

32 In the United States, there has been relatively little experimentation involving the insertion of genes from other species into human DNA. One reason for the lack of these experiments is that

(1) the subunits of human DNA are different from the DNA subunits of other species
(2) there are many ethical questions to be answered before inserting foreign genes into human DNA
(3) inserting foreign DNA into human DNA would require using techniques completely different from those used to insert foreign DNA into the DNA of other mammals
(4) human DNA always promotes human survival, so there is no need to alter it

33 The development of an experimental research plan should *not* include a

(1) list of safety precautions for the experiment
(2) list of equipment needed for conducting the experiment
(3) procedure for the use of technologies needed for the experiment
(4) conclusion based on data expected to be collected in the experiment

34 A student performed an experiment to demonstrate that a plant needs chlorophyll for photosynthesis. He used plants that had green leaves with white areas. After exposing the plants to sunlight, he removed a leaf from each plant and processed the leaves to remove the chlorophyll. He then tested each leaf for the presence of starch. Starch was found in the area of the leaf that was green, and no starch was found in the area of the leaf that was white. He concluded that chlorophyll is necessary for photosynthesis.

Which statement represents an assumption the student had to make in order to draw this conclusion?

(1) Starch is synthesized from the glucose produced in the green areas of the leaf.
(2) Starch is converted to chlorophyll in the green areas of the leaf.
(3) The white areas of the leaf do not have cells.
(4) The green areas of the leaf are heterotrophic.

35 The diagram below represents an interaction between parts of an organism.

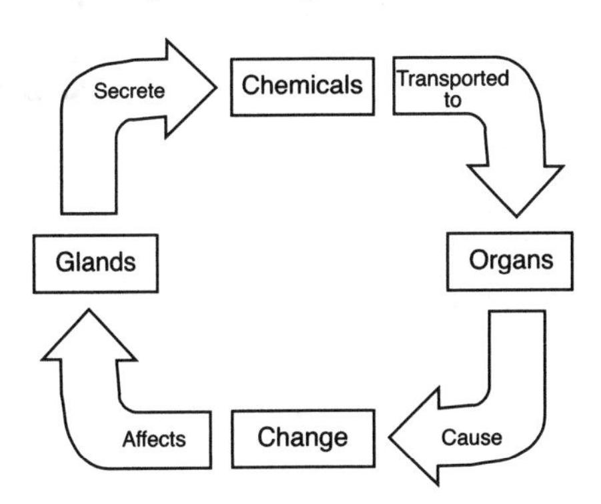

The term *chemicals* in this diagram represents

(1) starch molecules
(2) DNA molecules
(3) hormone molecules
(4) receptor molecules

36 The diagram below represents two cells, X and Y.

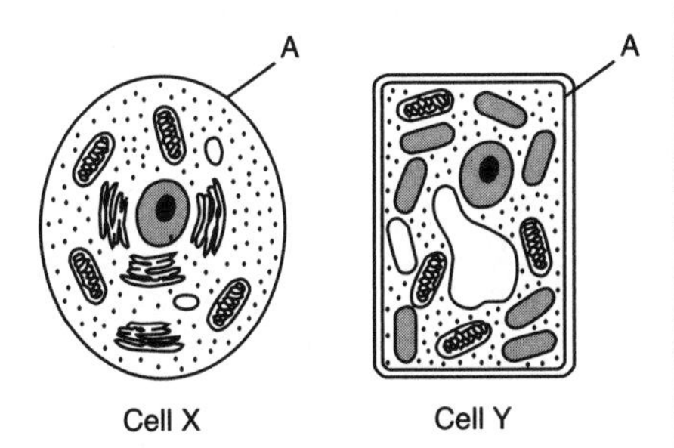

Which statement is correct concerning the structure labeled A?

(1) It aids in the removal of metabolic wastes in both cell X and cell Y.
(2) It is involved in cell communication in cell X, but not in cell Y.
(3) It prevents the absorption of CO_2 in cell X and O_2 in cell Y.
(4) It represents the cell wall in cell X and the cell membrane in cell Y.

37 The graph below provides information about the reproductive rates of four species of bacteria, A, B, C, and D, at different temperatures.

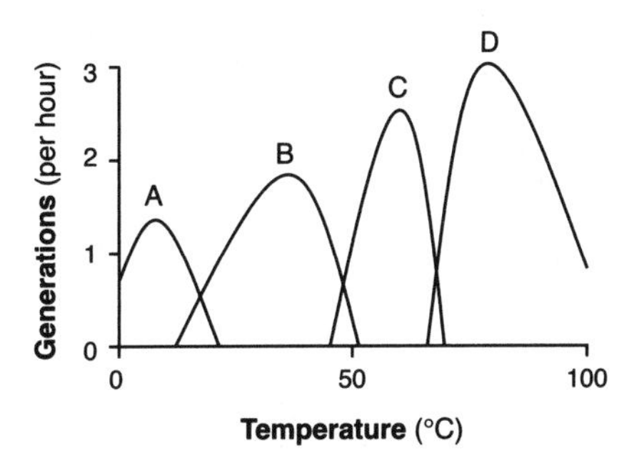

Which statement is a valid conclusion based on the information in the graph?

(1) Changes in temperature cause bacteria to adapt to form new species.
(2) Increasing temperatures speed up bacterial reproduction.
(3) Bacteria can survive only at temperatures between 0°C and 100°C.
(4) Individual species reproduce within a specific range of temperatures.

38 The diagram below shows some of the steps in protein synthesis.

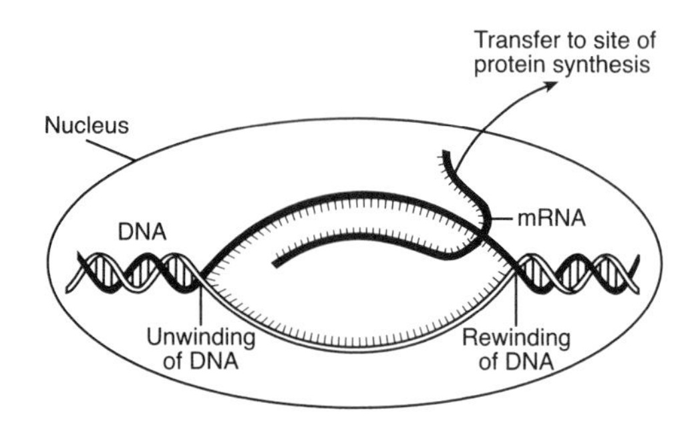

The section of DNA being used to make the strand of mRNA is known as a

(1) carbohydrate
(2) gene
(3) ribosome
(4) chromosome

39 An energy pyramid is shown below.

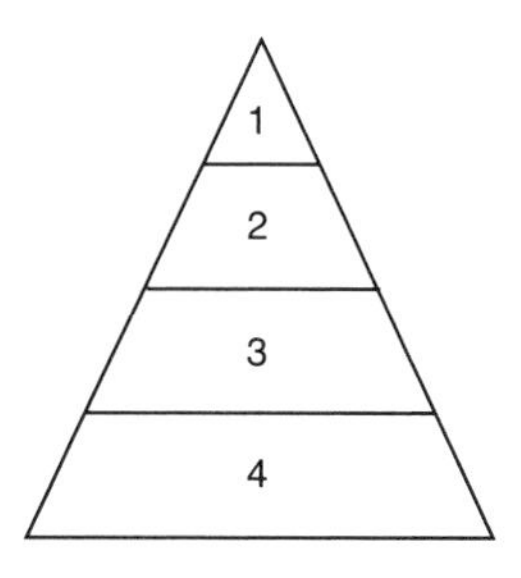

Which graph best represents the relative energy content of the levels of this pyramid?

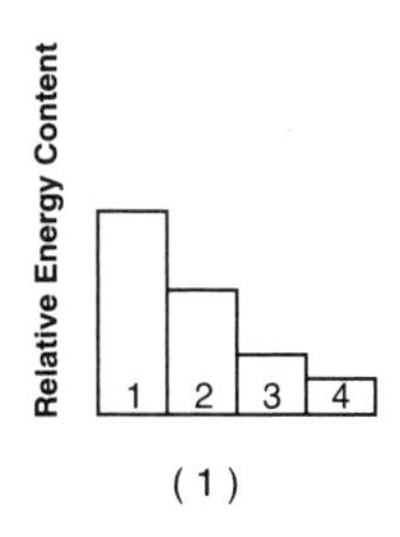

(1)

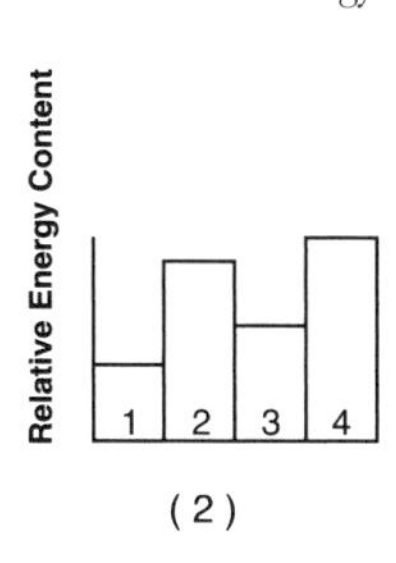

(2)

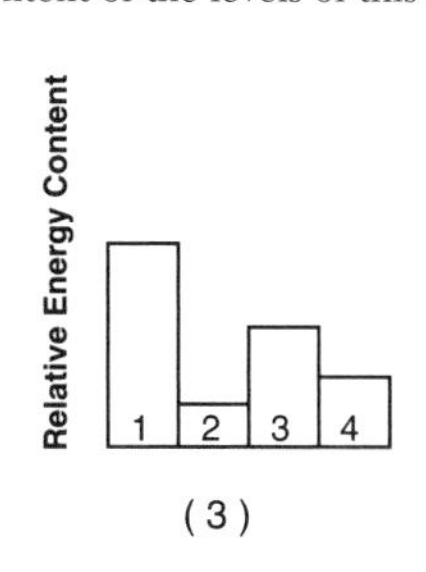

(3)

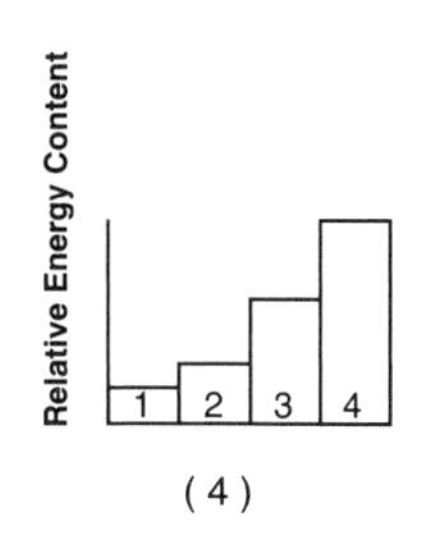

(4)

40 The diagram below represents four different species of bacteria.

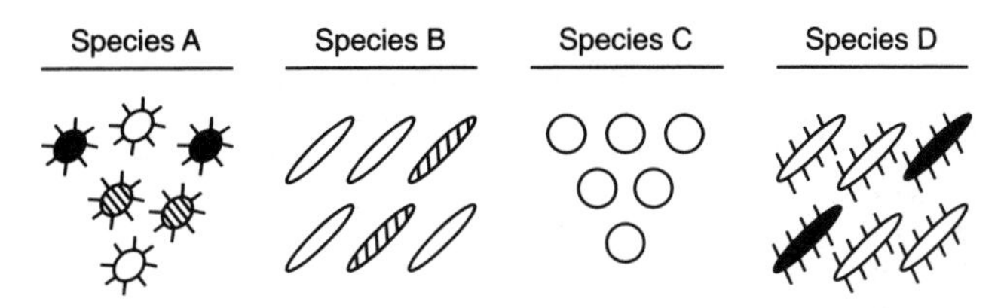

Which statement is correct concerning the chances of survival for these species if there is a change in the environment?

(1) Species *A* has the best chance of survival because it has the most genetic diversity.
(2) Species *C* has the best chance of survival because it has no gene mutations.
(3) Neither species *B* nor species *D* will survive because they compete for the same resources.
(4) None of the species will survive because bacteria reproduce asexually.

41 The diagram below represents possible evolutionary relationships between groups of organisms.

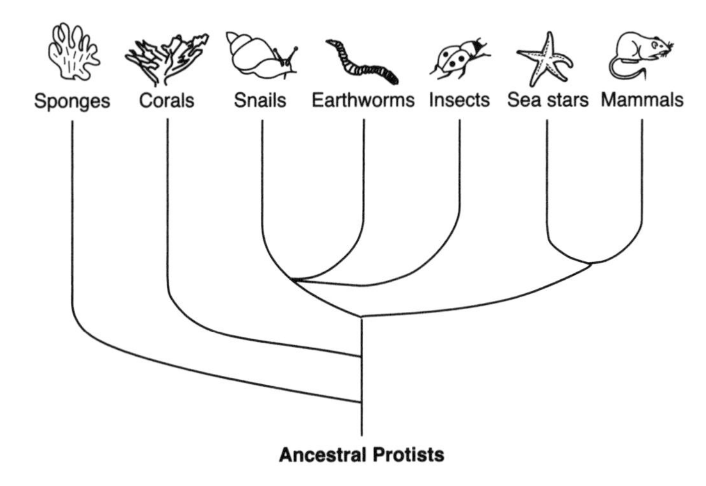

Which statement is a valid conclusion that can be drawn from the diagram?

(1) Snails appeared on Earth before corals.
(2) Sponges were the last new species to appear on Earth.
(3) Earthworms and sea stars have a common ancestor.
(4) Insects are more complex than mammals.

42 On which day did the population represented in the graph below reach the carrying capacity of the ecosystem?

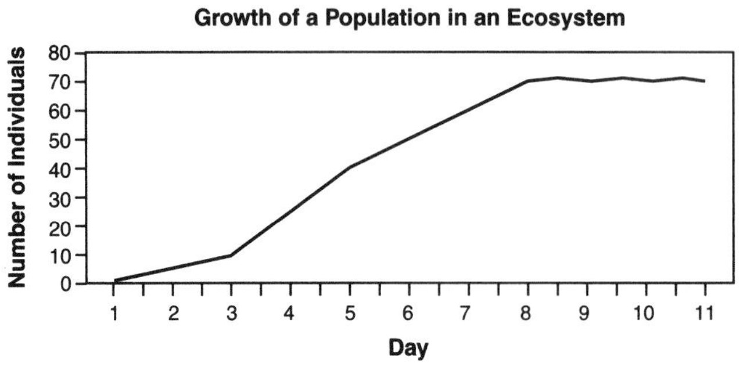

(1) day 11
(2) day 8
(3) day 3
(4) day 5

Part B–2

Answer all questions in this part. [13]

Directions (43–55): For those questions that are followed by four choices, circle the *number* of the choice that, of those given, best completes the statement or answers the question. For all other questions in this part, follow the directions given in the question and record your answers in the spaces provided.

For Teacher Use Only

Base your answers to questions 43 through 47 on the information below and on your knowledge of biology.

Each year, a New York State power agency provides its customers with information about some of the fuel sources used in generating electricity. The table below applies to the period of 2002–2003.

Fuel Sources Used

Fuel Source	Percentage of Electricity Generated
hydro (water)	86
coal	5
nuclear	4
oil	1
solar	0

Directions (43 and 44): Using the information given, construct a bar graph *on the grid on the next page*, following the directions below.

43 Mark an appropriate scale on the axis labeled "Percentage of Electricity Generated." [1]

44 Construct vertical bars to represent the data. Shade in *each* bar. [1]

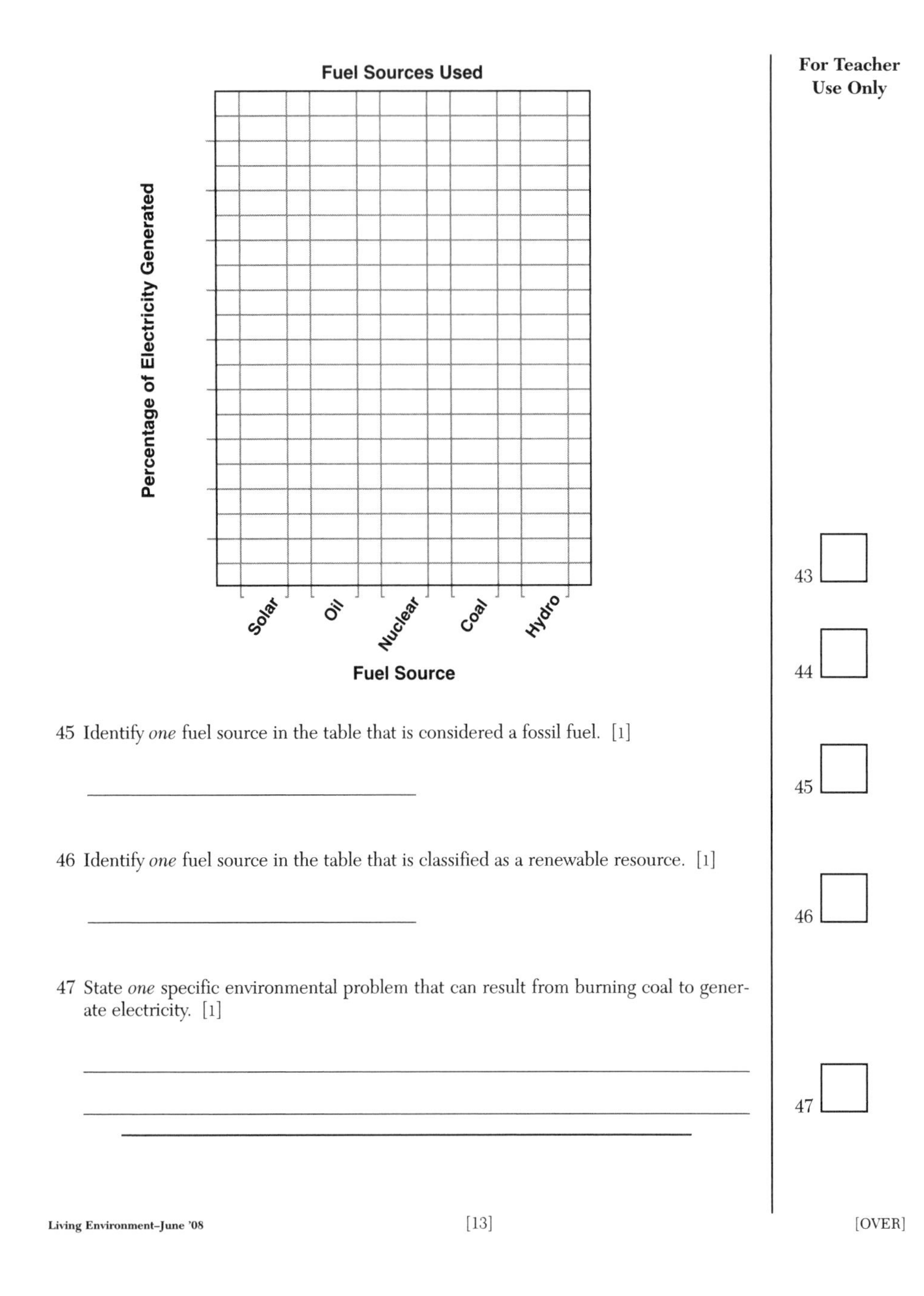

For Teacher Use Only

43

44

45 Identify *one* fuel source in the table that is considered a fossil fuel. [1]

____

45

46 Identify *one* fuel source in the table that is classified as a renewable resource. [1]

____

46

47 State *one* specific environmental problem that can result from burning coal to generate electricity. [1]

____

____

47

Base your answers to questions 48 and 49 on the diagram below that shows some interactions between several organisms located in a meadow environment and on your knowledge of biology.

For Teacher Use Only

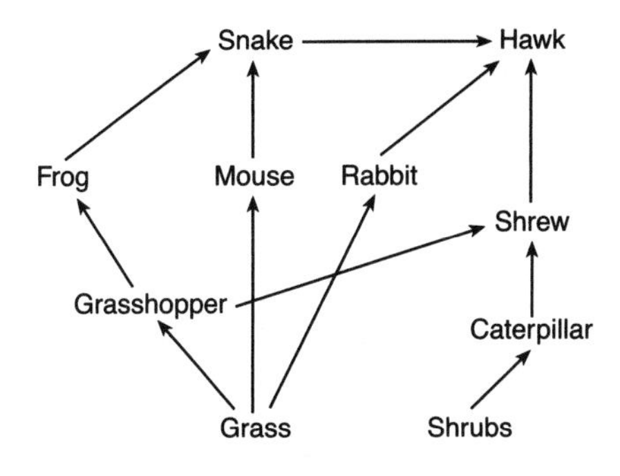

48 A rapid *decrease* in the frog population results in a change in the hawk population. State how the hawk population may change. Support your answer. [1]

____

____

48 ☐

49 Identify *one* cell structure found in a producer in this meadow ecosystem that is *not* found in the carnivores. [1]

____

49 ☐

50 Individuals of some species, such as earthworms, have both male and female sex organs. In many cases, however, these individuals do not fertilize their own eggs.

State *one* genetic advantage of an earthworm mating with another earthworm for the production of offspring. [1]

____

____

50 ☐

Base your answers to questions 51 and 52 on the diagram below and on your knowledge of biology. The diagram represents six insect species.

Species E   Species F

51 A dichotomous key to these six species is shown below. Complete the missing information for sections 5.*a*. and 5.*b*. so that the key is complete for all *six* species. [1]

Dichotomous Key

1. a. has small wings ....go to 2
   b. has large wings..go to 3

2. a. has a single pair of wings ....Species A
   b. has a double pair of wings ....Species B

3. a. has a double pair of wings ....go to 4
   b. has a single pair of wings..........................Species C

4. a. has spots ....go to 5
   b. does not have spots ....Species D

5. a. ______________________________ ....Species E

   b. ______________________________ ....Species F

51 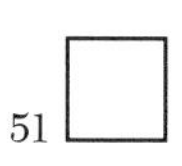

52 Use the key to identify the drawings of species *A*, *B*, *C*, and *D*. Place the letter of *each* species on the line located below the drawing of the species. [1]

Species ___   Species E   Species ___   Species F   Species ___   Species ___

52 ☐

For Teacher Use Only

Base your answers to questions 53 through 55 on the information below and on your knowledge of biology.

For Teacher Use Only

Proteins on the surface of a human cell and on a bird influenza virus are represented in the diagram below.

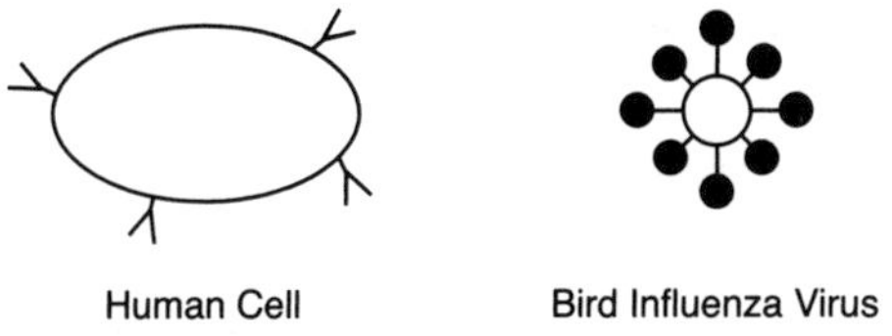

Human Cell

Bird Influenza Virus

53 In the space below, draw a change in the bird influenza virus that would allow it to infect this human cell. [1]

53

54 Explain how this change in the virus could come about. [1]

___

___

54

55 Identify the relationship that exists between a virus and a human when the virus infects the human. [1]

___

___

55

Part C

Answer all questions in this part. [17]

Directions (56–67): Record your answers in the spaces provided in this examination booklet.

For Teacher Use Only

Base your answers to questions 56 and 57 on the information below and on your knowledge of biology.

Insulin is a hormone that has an important role in the maintenance of homeostasis in humans.

56 Identify the structure in the human body that is the usual source of insulin. [1]

____

56 ☐

57 Identify a substance in the blood, other than insulin, that could change in concentration and indicate a person is not secreting insulin in normal amounts. [1]

____

____

57 ☐

Base your answers to questions 58 and 59 on the information below and on your knowledge of biology.

The hedgehog, a small mammal native to Africa and Europe, has been introduced to the United States as an exotic pet species. Scientists have found that hedgehogs can transfer pathogens to humans and domestic animals. Foot-and-mouth viruses, *Salmonella*, and certain fungi are known pathogens carried by hedgehogs. As more and more of these exotic animals are brought into this country, the risk of infection increases in the human population.

58 State *one negative* effect of importing exotic species to the United States. [1]

____

____

58 ☐

59 State *one* way the human immune system might respond to an invading pathogen associated with handling a hedgehog. [1]

____

____

____

59 ☐

Base your answers to questions 60 through 62 on the information below and on your knowledge of biology.

For Teacher Use Only

The last known wolf native to the Adirondack Mountains of New York State was killed over a century ago. Several environmental groups have recently proposed reintroducing the wolf to the Adirondacks. These groups claim there is sufficient prey to support a wolf population in this area. These prey include beaver, deer, and moose. Opponents of this proposal state that the Adirondacks already have a dominant predator, the Eastern coyote.

60 State *one* effect the reintroduction of the wolf may have on the coyote population within the Adirondacks. Explain why it would have this effect. [1]

____

____ 60

61 Explain why the coyote is considered a limiting factor in the Adirondack Mountains. [1]

____

____ 61

62 State *one* ecological reason why some individuals might support the reintroduction of wolves to the Adirondacks. [1]

____

____ 62

For Teacher Use Only

63 You have been assigned to design an experiment to determine the effects of light on the growth of tomato plants. In your experimental design be sure to:

- state *one* hypothesis to be tested [1]
- identify the independent variable in the experiment [1]
- describe the type of data to be collected [1]

63

64 In some land plants, guard cells are found only on the lower surfaces of the leaves. In some water plants, guard cells are found only on the upper surfaces of the leaves. Explain how guard cells in both land and water plants help maintain homeostasis. In your answer be sure to:

- identify *one* function regulated by the guard cells in leaves [1]
- explain how guard cells carry out this function [1]
- give *one* possible evolutionary advantage of the position of the guard cells on the leaves of land plants [1]

64

For Teacher Use Only

Base your answers to questions 65 and 66 on the information below and on your knowledge of biology.

Scientists are increasingly concerned about the possible effects of damage to the ozone layer.

65 Damage to the ozone layer has resulted in mutations in skin cells that lead to cancer. Will the mutations that caused the skin cancers be passed on to offspring? Support your answer. [1]

____

____

65 ☐

66 State *two* specific ways in which an ocean ecosystem will change (other than fewer photosynthetic organisms) if populations of photosynthetic organisms die off as a result of damage to the ozone layer. [2]

____

____

____

____

____

66 ☐

67 Lawn wastes, such as grass clippings and leaves, were once collected with household trash and dumped into landfills. Identify *one* way that this practice was harmful to the environment. [1]

____

____

67 ☐

Part D

Answer all questions in this part. [13]

Directions (68–80): For those questions that are followed by four choices, circle the *number* of the choice, that, of those given, best completes the statement or answers the question. For all other questions in this part, follow the directions given in the question and record your answers in the spaces provided.

For Teacher Use Only

68 In preparation for an electrophoresis procedure, enzymes are added to DNA in order to

(1) convert the DNA into gel

(2) cut the DNA into fragments

(3) change the color of the DNA

(4) produce longer sections of DNA

68 ☐

69 Paper chromatography is a laboratory technique that is used to

(1) separate different molecules from one another

(2) stain cell organelles

(3) indicate the pH of a substance

(4) compare relative cell sizes

69 ☐

70 A marathon runner frequently experiences muscle cramps while running. If he stops running and rests, the cramps eventually go away. The cramping in the muscles most likely results from

(1) lack of adequate oxygen supply to the muscle

(2) the runner running too slowly

(3) the runner warming up before running

(4) increased glucose production in the muscle

70 ☐

Base your answers to questions 71 through 73 on the information below and on your knowledge of biology.

For Teacher Use Only

A series of investigations was performed on four different plant species. The results of these investigations are recorded in the data table below.

Characteristics of Four Plant Species

Plant Species	Seeds	Leaves	Pattern of Vascular Bundles (structures in stem)	Type of Chlorophyll Present
A	round/small	needle-like	scattered bundles	chlorophyll a and b
B	long/pointed	needle-like	circular bundles	chlorophyll a and c
C	round/small	needle-like	scattered bundles	chlorophyll a and b
D	round/small	needle-like	scattered bundles	chlorophyll b

71 Based on these data, which *two* plant species appear to be most closely related? Support your answer. [1]

Plant species ____A____ and ____C____

____

____

71 [ ]

72 What additional information could be gathered to support your answer to question 71? [1]

____

____

72 [ ]

73 State *one* reason why scientists might want to know if two plant species are closely related. [1]

____

____

73 [ ]

For Teacher Use Only

Base your answers to questions 74 and 75 on the data table below and on your knowledge of biology.

Dietary Preferences of Finches

Species of Finch	Preferred Foods
A	nuts and seeds
B	worms and insects
C	fruits and seeds
D	insects and seeds
E	nuts and seeds

74 Based on its preferred food, species *B* would be classified as a

(1) decomposer

(2) producer

(3) carnivore

(4) parasite

74 ☐

75 Which two species would most likely be able to live in the same habitat without competing with each other for food?

(1) *A* and *C*

(2) *B* and *C*

(3) *B* and *D*

(4) *C* and *E*

75 ☐

Base your answers to questions 76 and 77 on the experimental setup shown below.

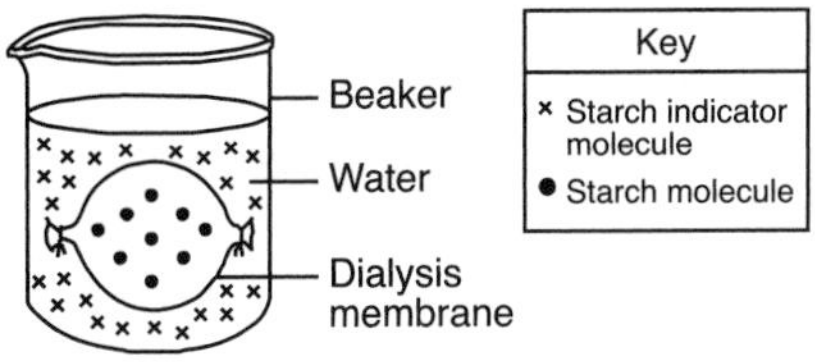

76 On the diagram below, draw in the expected locations of the molecules after a period of one hour. [1]

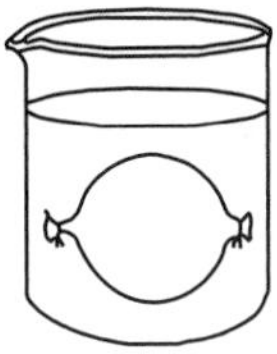

77 When starch indicator is used, what observation would indicate the presence of starch? [1]

____

____

78 State *one* reason why some molecules can pass through a certain membrane, but other molecules can *not*. [1]

____

____

For Teacher Use Only

76 ☐

77 ☐

78 ☐

79 A plant cell in a microscopic field of view is represented below.

For Teacher Use Only

The width (w) of this plant cell is closest to

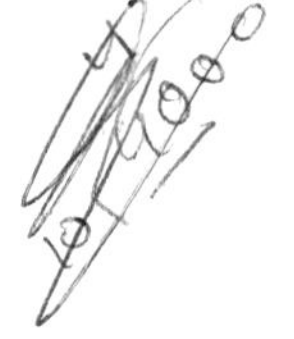

(1) 200 μm

(2) 800 μm

(3) 1200 μm

(4) 1600 μm

79

80 The diagram below represents a plant cell in tap water as seen with a compound light microscope.

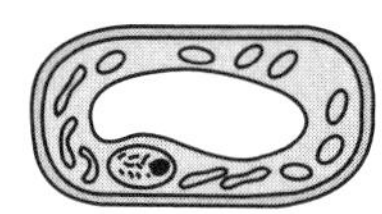

Which diagram best represents the appearance of the cell after it has been placed in a 15% salt solution for two minutes?

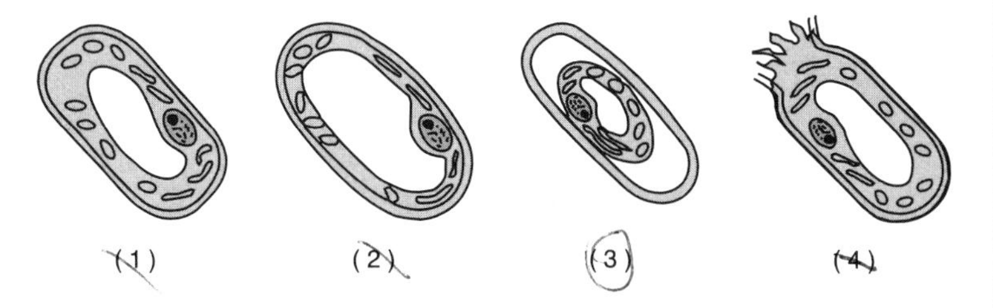

80

The University of the State of New York

REGENTS HIGH SCHOOL EXAMINATION

LIVING ENVIRONMENT

Tuesday, June 24, 2008 — 9:15 a.m. to 12:15 p.m., only

ANSWER SHEET

Student .... Sex: ☐ Female ☐ Male

Teacher ....

School .... Grade ....

Part	Maximum Score	Student's Score
A	30	
B–1	12	
B–2	13	
C	17	
D	13	
Total Raw Score (maximum Raw Score: 85)		
Final Score (from conversion chart)		

Raters' Initials

Rater 1 .... **Rater 2** ....

Record your answers to Part A and Part B–1 on this answer sheet.

Part A

1	11	21 ....
2	12	22 ....
3	13	23 ....
4	14	24 ....
5	15	25 ....
6	16	26 ....
7	17	27 ....
8	18	28 ....
9	19	29 ....
10	20	30 ....

Part A Score

Part B–1

31	37 ....
32	38 ....
33	39 ....
34	40 ....
35	41 ....
36	42 ....

Part B–1 Score

The declaration below must be signed when you have completed the examination.

I do hereby affirm, at the close of this examination, that I had no unlawful knowledge of the questions or answers prior to the examination and that I have neither given nor received assistance in answering any of the questions during the examination.

____
Signature

REGENTS HIGH SCHOOL EXAMINATION **Part A** **Wednesday,** August 13, 2008 12:30 to 3:30p.m., only

Answer all questions in this part. [30]

Directions (1–30): For *each* statement or question, write on your separate answer sheet the *number* of the word or expression that, of those given, best completes the statement or answers the question.

1 Scientists studying ocean organisms are discovering new and unusual species. Which observation could be used to determine that an ocean organism carries out autotrophic nutrition?

(1) Chloroplasts are visible inside the cells.
(2) Digestive organs are visible upon dissection.
(3) The organism lives close to the surface.
(4) The organism synthesizes enzymes to digest food.

2 Abiotic factors that characterize a forest ecosystem include

(1) light and biodiversity
(2) temperature and amount of available water
(3) types of producers and decomposers
(4) pH and number of heterotrophs

3 One season, there was a shortage of producers in a food web. As a result, the number of deer and wolves decreased. The reason that both the deer and wolf populations declined is that

(1) producers are not as important as consumers in a food web
(2) more consumers than producers are needed to support the food web
(3) organisms in this food web are interdependent
(4) populations tend to stay constant in a food web

4 Which statement best describes a population of organisms if cloning is the only method used to reproduce this population?

(1) The population would be more likely to adapt to a changing environment.
(2) There would be little chance for variation within the population.
(3) The population would evolve rapidly.
(4) The mutation rate in the population would be rapid.

5 An organelle that releases energy for metabolic activity in a nerve cell is the

(1) chloroplast (3) mitochondrion
(2) ribosome (4) vacuole

6 A student notices that fruit flies with the curly-wing trait develop straight wings if kept at a temperature of 16°C, but develop curly wings if kept at 25°C. The best explanation for this observation is that

(1) wing shape is controlled by behavior
(2) wing shape is influenced by light intensity
(3) gene expression can be modified by interactions with the environment
(4) gene mutations for wing shape can occur at high temperatures

7 In all organisms, the coded instructions for specifying the characteristics of the organism are directly determined by the arrangement of the

(1) twenty kinds of amino acids in each protein
(2) twenty-three pairs of genes on each chromosome
(3) strands of simple sugars in certain carbohydrate molecules
(4) four types of molecular bases in the genes

8 Which sequence shows a *decreasing* level of complexity?

(1) organs → organism → cells → tissues
(2) organism → cells → organs → tissues
(3) cells → tissues → organs → organism
(4) organism → organs → tissues → cells

9 Which row in the chart below contains an event that is paired with an appropriate response in the human body?

Row	Event	Response
(1)	a virus enters the bloodstream	increased production of antibodies
(2)	fertilization of an egg	increased levels of testosterone
(3)	dehydration due to increased sweating	increased urine output
(4)	a drop in the rate of digestion	increased respiration rate

10 The diagram below represents a genetic procedure.

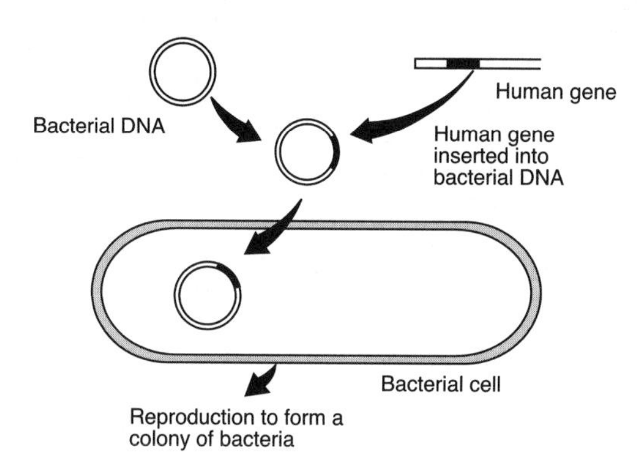

Which statement best describes the outcome of this procedure?

(1) Bacterial cells will destroy defective human genetic material.
(2) Bacterial cells may form a multicellular embryo.
(3) The inserted human DNA will change harmful bacteria to harmless ones.
(4) The inserted human DNA may direct the synthesis of human proteins.

11 The diagram below represents the genetic contents of cells before and after a specific reproductive process.

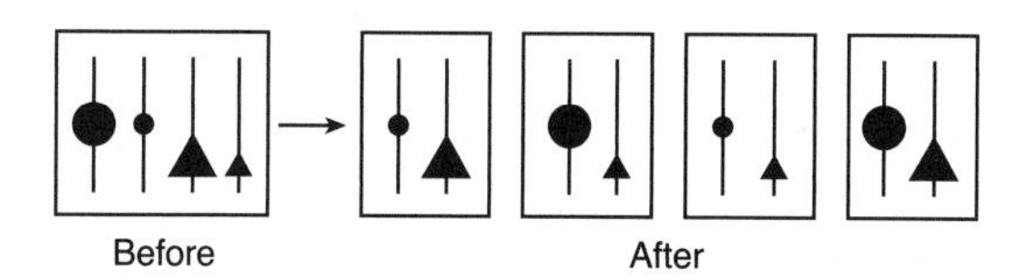

This process is considered a mechanism of evolution because it

(1) decreases the chance for new combinations of inheritable traits in a species
(2) decreases the probability that genes can be passed on to other body cells
(3) increases the chance for variations in offspring
(4) increases the number of offspring an organism can produce

12 One *disadvantage* of a genetic mutation in a human skin cell is that it

(1) may result in the production of a defective protein
(2) may alter the sequence of simple sugars in insulin molecules
(3) can lead to a lower mutation rate in the offspring of the human
(4) can alter the rate of all the metabolic processes in the human

13 The DNA of a human cell can be cut and rearranged by using

(1) a scalpel
(2) electrophoresis
(3) hormones
(4) enzymes

14 Much of the carbon dioxide produced by green plants is *not* excreted as a metabolic waste because it

(1) can be used for photosynthesis
(2) is too large to pass through cell membranes
(3) is needed for cellular respiration
(4) can be used for the synthesis of proteins

15 In several species of birds, the males show off their bright colors and long feathers. The dull-colored females usually pick the brightest colored males for mates. Male offspring inherit their father's bright colors and long feathers. Compared to earlier generations, future generations of these birds will be expected to have a greater proportion of

(1) bright-colored females
(2) dull-colored females
(3) dull-colored males
(4) bright-colored males

16 To determine evolutionary relationships between organisms, a comparison would most likely be made between all of the characteristics below *except*

(1) methods of reproduction
(2) number of their ATP molecules
(3) sequences in their DNA molecules
(4) structure of protein molecules present

17 The females of certain species of turtles will sneak into a nest of alligator eggs to lay their own eggs and then leave, never to return. When the baby turtles hatch, they automatically hide from the mother alligator guarding the nest and go to the nearest body of water when it is safe to do so. Which statement best explains the behavior of these baby turtles?

(1) More of the turtles' ancestors who acted in this way survived to reproduce, passing this behavioral trait to their offspring.
(2) The baby turtles are genetically identical, so they behave the same way.
(3) Turtles are not capable of evolving, so they repeat the same behaviors generation after generation.
(4) The baby turtles' ancestors who learned to behave this way taught the behaviors to their offspring

18 A pattern of reproduction and growth in a one-celled organism is shown below.

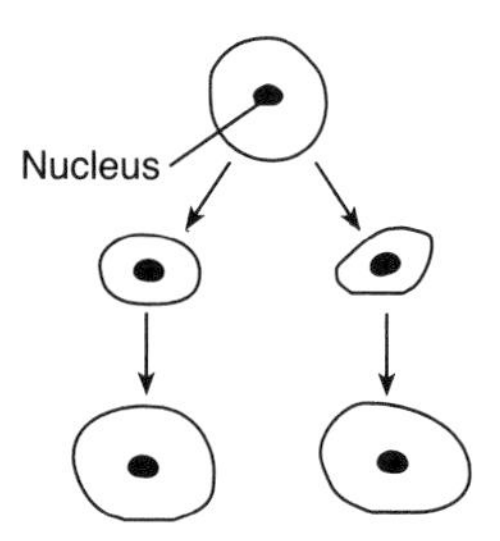

Which statement best describes this pattern of reproduction?

(1) All genetic material comes from one parent.
(2) Only some of the genetic material comes from one parent.
(3) The size of the parent determines the amount of genetic material.
(4) The size of the parent determines the source of the genetic material.

19 A technique used to reproduce plants is shown in the diagram below.

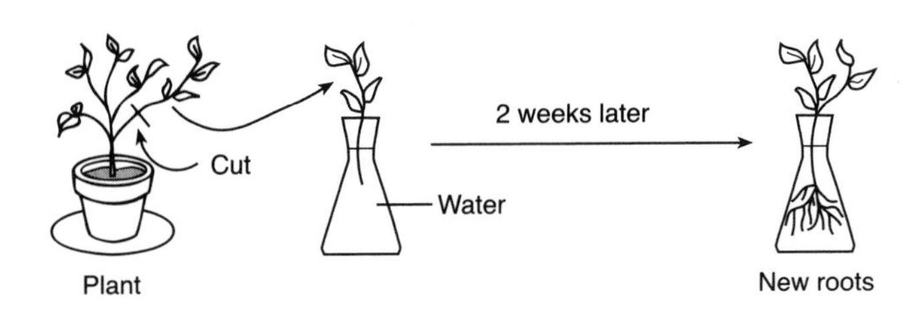

This technique is a form of

(1) sexual reproduction
(2) asexual reproduction
(3) gamete production
(4) gene manipulation

20 The diagram below represents a biochemical process.

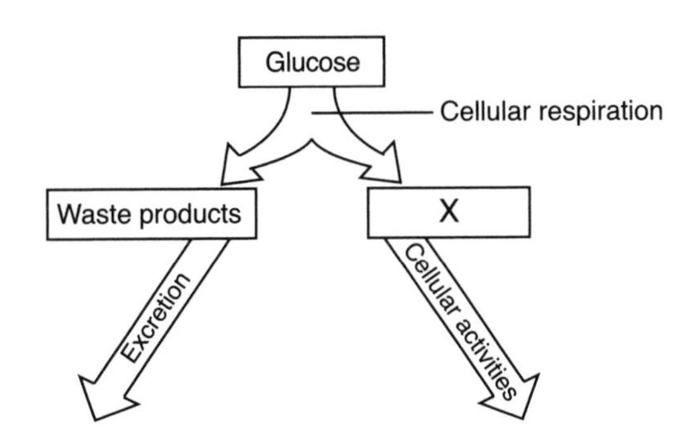

Which molecule is represented by *X*?

(1) DNA
(2) starch
(3) protein
(4) ATP

21 The diagram below represents early stages of embryo development.

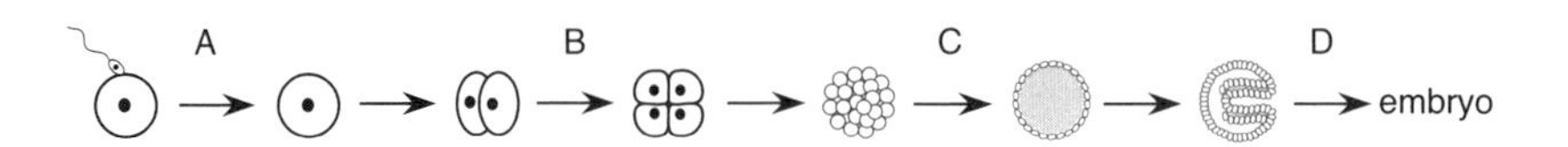

The greatest amount of differentiation for organ formation most likely occurs at arrow

(1) *A*
(2) *B*
(3) *C*
(4) *D*

22 The diagram below shows a cell in the human body engulfing a bacterial cell.

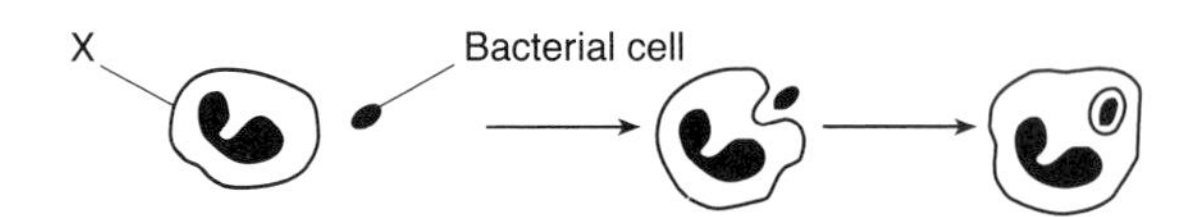

The cell labeled *X* is most likely a

(1) red blood cell
(2) white blood cell
(3) liver cell
(4) nerve cell

23 The graph below shows the levels of glucose and insulin in the blood of a human over a period of time.

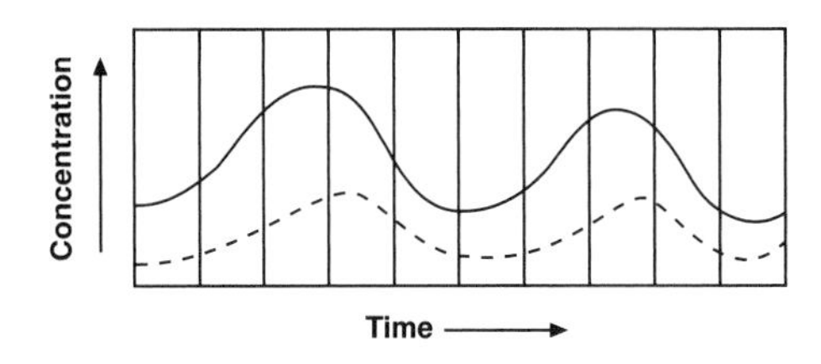

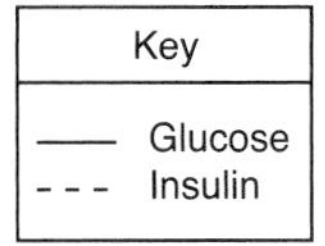

This graph represents

(1) an allergic reaction
(2) an antigen-antibody reaction
(3) maintenance of homeostasis
(4) autotrophic nutrition

24 Which concept is best represented in the diagram shown below?

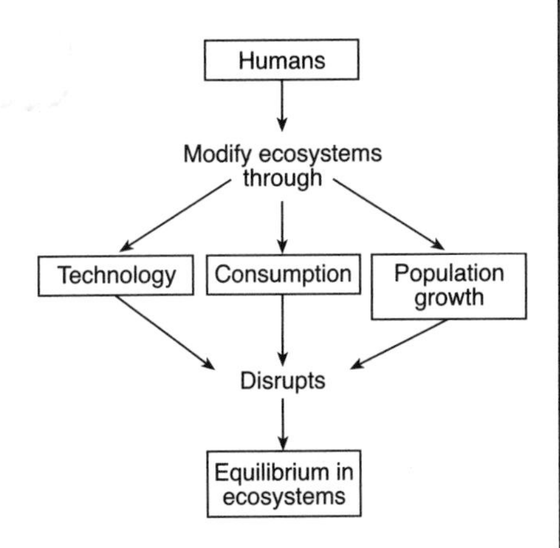

(1) Human actions are a threat to equilibrium in ecosystems.
(2) Equilibrium in ecosystems requires that humans modify ecosystems.
(3) Equilibrium in ecosystems directly affects how humans modify ecosystems.
(4) Human population growth is the primary reason for equilibrium in ecosystems.

25 One possible reason for the rise in the average air temperature at Earth's surface is that

(1) decomposers are being destroyed
(2) deforestation has increased the levels of oxygen in the atmosphere
(3) industrialization has increased the amount of carbon dioxide in the air
(4) growing crops is depleting the ozone shield

26 The size of a frog population in a pond remains fairly constant over a period of several years because of

(1) decreasing competition
(2) environmental carrying capacity
(3) excessive dissolved oxygen
(4) the depth of water

27 Plants such as the Venus flytrap produce chemical compounds that break down insects into substances that are usable by the plant. The chemical compounds that break down the insects are most likely

(1) fats
(2) minerals
(3) biological catalysts
(4) complex carbohydrates

28 In December 2004, a tsunami (giant wave) destroyed many of the marine organisms along the coast of the Indian Ocean. What can be expected to happen to the ecosystem that was most severely hit by the tsunami?

(1) The ecosystem will change until a new stable community is established.
(2) Succession will continue in the ecosystem until one species of marine organism is established.
(3) Ecological succession will no longer occur in this marine ecosystem.
(4) The organisms in the ecosystem will become extinct.

29 Many homeowners who used to collect, bag, and discard grass clippings are now using mulching lawnmowers, which cut up the clippings into very fine pieces and deposit them on the soil. The use of mulching lawnmowers contributes most directly to

(1) increasing the diversity of life
(2) recycling of nutrients
(3) the control of pathogens
(4) the production of new species

30 Deforestation of areas considered to be rich sources of genetic material could limit future agricultural and medical advances due to

(1) the improved quality of the atmosphere
(2) the maintenance of dynamic equilibrium
(3) an increase in the rate of evolutionary change
(4) the loss of biodiversity

Part B–1

Answer all questions in this part. [12]

Directions (31–42): For *each* statement or question, write on the separate answer sheet the *number* of the word or expression that, of those given, best completes the statement or answers the question.

31 In 1883, Thomas Engelmann, a German botanist, exposed a strand of algae to different wavelengths of light. Engelmann used bacteria that concentrate near an oxygen source to determine which sections of the algae were releasing the most O_2. The results are shown below.

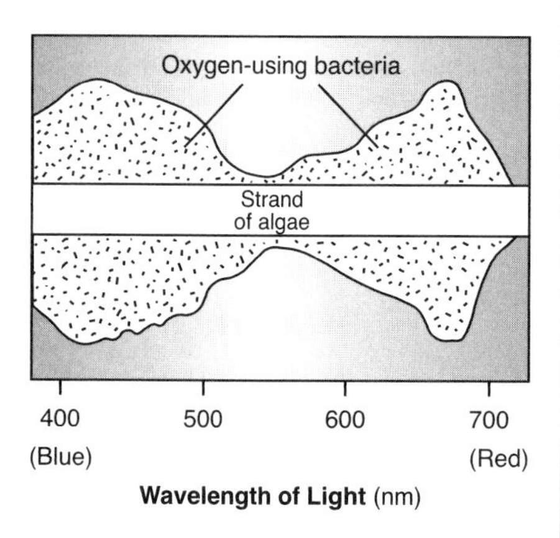

Which statement is a valid inference based on this information?

(1) Oxygen production decreases as the wavelength of light increases from 550 to 650 nm.
(2) Respiration rate in the bacteria is greatest at 550 nm.
(3) Photosynthetic rate in the algae is greatest in blue light.
(4) The algae absorb the greatest amount of oxygen in red light.

32 The graph below represents a predator-prey relationship.

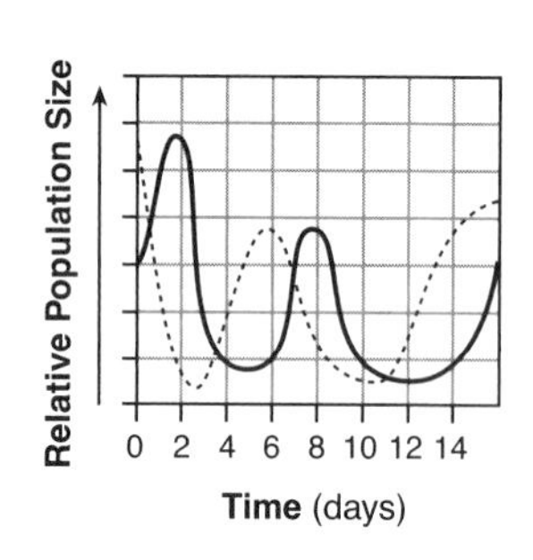

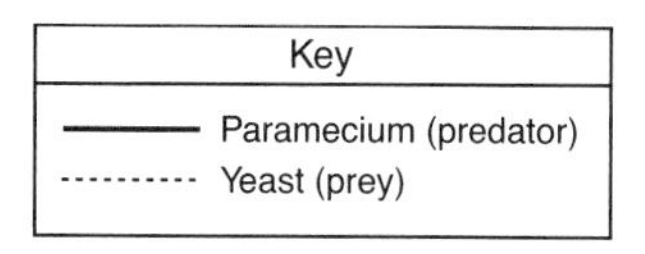

What is the most probable reason for the increasing predator population from day 5 to day 7?

(1) an increasing food supply from day 5 to day 6
(2) a predator population equal in size to the prey population from day 5 to day 6
(3) the decreasing prey population from day 1 to day 2
(4) the extinction of the yeast on day 3

33 A single cell and a multicellular organism are represented below.

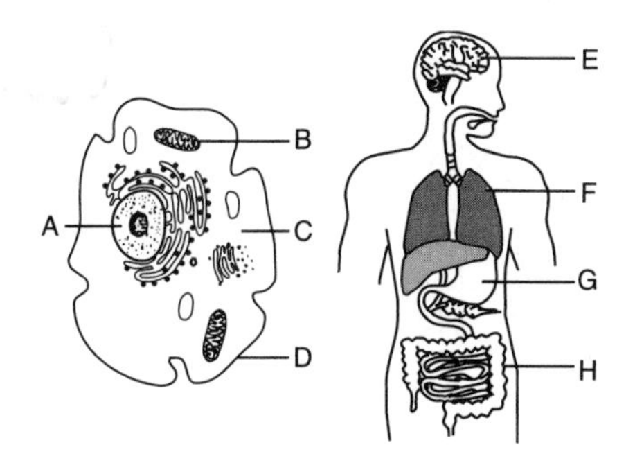

Which structures are correctly paired with their primary function?

(1) *A* and *G*—transmission of nerve impulses
(2) *B* and *E*—photosynthesis
(3) *C* and *H*—digestion of food
(4) *D* and *F*—gas exchange

34 The graph below indicates the size of a fish population over a period of time.

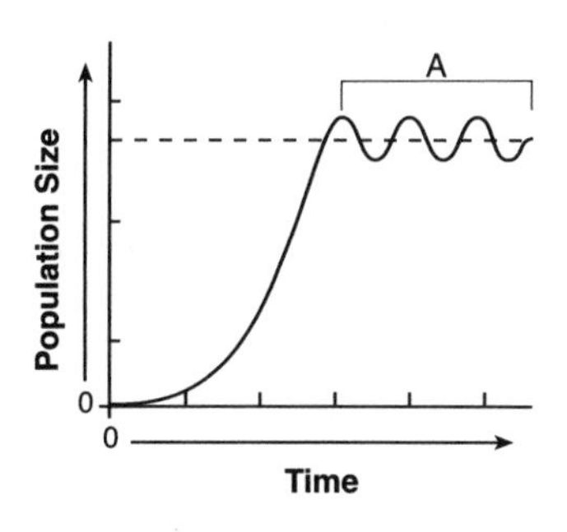

The section of the graph labeled *A* represents

(1) biodiversity within the species
(2) nutritional relationships of the species
(3) a population becoming extinct
(4) a population at equilibrium

35 The data table below shows the presence or absence of DNA in four different cell organelles.

Data Table

Organelle	DNA
cell membrane	absent
cell wall	absent
mitochondrion	present
nucleus	present

Information in the table suggests that DNA functions

(1) within cytoplasm and outside of the cell membrane
(2) both inside and outside of the nucleus
(3) only within energy-releasing structures
(4) within cell vacuoles

Base your answers to questions 36 and 37 on the diagram below, which represents stages in the digestion of a starch, and on your knowledge of biology.

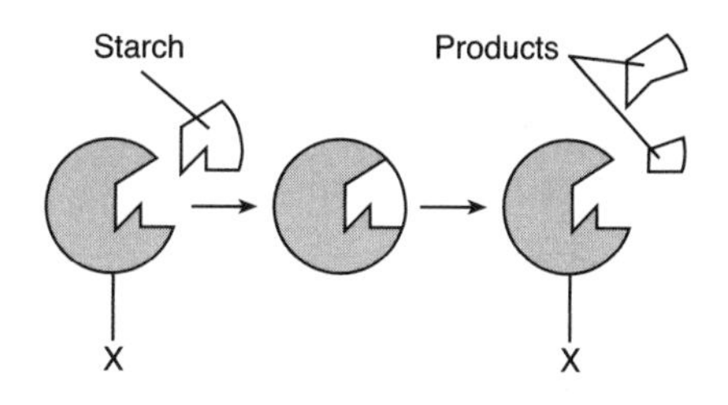

36 The products would most likely contain

(1) simple sugars
(2) fats
(3) amino acids
(4) minerals

37 The structure labeled *X* most likely represents

(1) an antibody
(2) a receptor molecule
(3) an enzyme
(4) a hormone

Base your answers to questions 38 through 40 on the diagram below and on your knowledge of biology. Each arrow in the diagram represents a different hormone released by the pituitary gland that stimulates the gland indicated in the diagram. All structures are present in the same organism.

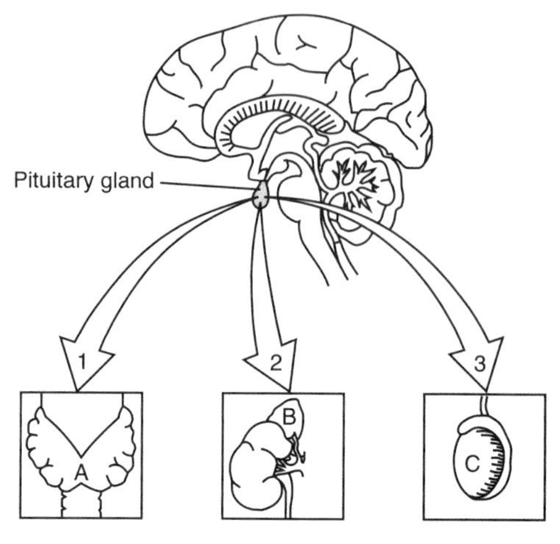

(Not drawn to scale)

38 The pituitary gland may release hormone 2 when blood pressure drops. Hormone 2 causes gland *B* to release a different hormone that raises blood pressure which, in turn, stops the secretion of hormone 2. The interaction of these hormones is an example of

(1) DNA base substitution
(2) manipulation of genetic instructions
(3) a feedback mechanism
(4) an antigen-antibody reaction

39 What would most likely occur if the interaction is blocked between the pituitary and gland *C*, the site of meiosis in males?

(1) The level of progesterone would start to increase.
(2) The pituitary would produce another hormone to replace hormone 3.
(3) Gland *A* would begin to interact with hormone 3 to maintain homeostasis.
(4) The level of testosterone may start to decrease.

40 Why does hormone 1 influence the action of gland *A* but *not* gland *B* or *C*?

(1) Every activity in gland *A* is different from the activities in glands *B* and *C*.
(2) The cells of glands *B* and *C* contain different receptors than the cells of gland *A*.
(3) Each gland contains cells that have different base sequences in their DNA.
(4) The distance a chemical can travel is influenced by both pH and temperature.

Base your answers to questions 41 and 42 on the information and diagram below and on your knowledge of biology.

A small water plant (elodea) was placed in bright sunlight for five hours as indicated below. Bubbles of oxygen gas were observed being released from the plant.

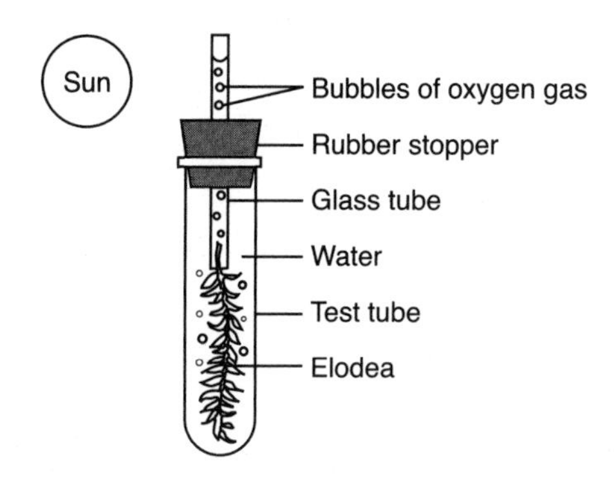

41 Since oxygen gas is being released, it can be inferred that the plant is

(1) producing glucose
(2) making protein
(3) releasing energy from water
(4) carrying on active transport

42 What substance did the plant most likely absorb from the water for the process that produces the oxygen gas?

(1) dissolved nitrogen
(2) carbon dioxide
(3) an enzyme
(4) a hormone

Part B–2

Answer all questions in this part. [13]

Directions (43–55): For those questions that are followed by four choices, circle the *number* preceding the choice that, of those given, best completes the statement or answers the question. For all other questions in this part, follow the directions given in the question and record your answers in the spaces provided.

For Teacher Use Only

Base your answers to questions 43 through 45 on the information below and on your knowledge of biology.

Human reproduction is influenced by many different factors.

43 Identify *one* reproductive hormone and state the role it plays in reproduction. [1]

____

____ 43 ☐

44 Identify the structure in the uterus where the exchange of material between the mother and the developing fetus takes place. [1]

____ 44 ☐

45 Identify *one* harmful substance that can pass through this structure and describe the *negative* effect it can have on the fetus. [1]

____

____ 45 ☐

---

46 The flow of materials through ecosystems involves the interactions of many processes and organisms. State how decomposers aid in the flow of materials in an ecosystem. [1]

____

____ 46 ☐

Base your answers to questions 47 through 49 on the information below and on your knowledge of biology.

For Teacher Use Only

Honeybees have a very cooperative way of living. Scout bees find food, return to the hive, and do the "waggle dance" to communicate the location of the food source to other bees in the hive. The waggle, represented by the wavy line in the diagram below, indicates the direction of the food source, while the speed of the dance indicates the distance to the food. Different species of honeybees use the same basic dance pattern in slightly different ways as shown in the table below.

Number of Waggle Runs in 15 Seconds		Distance to Food (feet)
Giant Honeybee	Indian Honeybee	
10.6	10.5	50
9.6	8.3	200
6.7	4.4	1000
4.8	2.8	2000

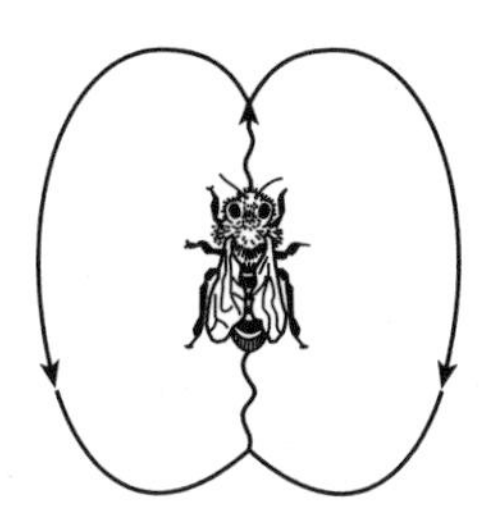

47 State the relationship between the distance to the food source and the number of waggle runs in 15 seconds. [1]

____

____

47 ☐

48 Explain how waggle-dance behavior increases the reproductive success of the bees. [1]

____

____

48 ☐

49 The number of waggle runs in 15 seconds for each of these species is most likely due to

(1) behavioral adaptation as a result of natural selection

(2) replacement of one species by another as a result of succession

(3) alterations in gene structure as a result of diet

(4) learned behaviors inherited as a result of asexual reproduction

49 ☐

Base your answers to questions 50 through 54 on the information and data table below and on your knowledge of biology.

For Teacher Use Only

The table shows data collected on the pH level of an Adirondack lake from 1980 to 1996.

Lake pH Level

Year	pH Level
1980	6.7
1984	6.3
1986	6.4
1988	6.2
1990	5.9
1992	5.6
1994	5.4
1996	5.1

Directions (50–54): Using the information in the data table, construct a line graph on the grid *on the next page*, following the directions below.

50 Label the axes. [1]

51 Mark an appropriate scale on the y-axis. The scale has been started for you. [1]

52 Plot the data from the data table. Surround each point with a small circle and connect the points. [1]

Example: 

For Teacher Use Only

Lake pH Level from 1980 to 1996

5.0

1980 1982 1984 1986 1988 1990 1992 1994 1996

50 ☐

51 ☐

52 ☐

53 Describe the trend in pH level in the lake over this 16-year period. [1]

53 ☐

54 Identify *one* factor that should have been kept constant each time water samples were collected from the lake. [1]

54 ☐

55 Two cultures, each containing a different species of bacteria, were exposed to the same antibiotic. Explain how, after exposure to this antibiotic, the population of one species of bacteria could increase while the population of the other species of bacteria decreased or was eliminated. [1]

55 ☐

Part C

Answer all questions in this part. [17]

Directions (56–71): Record your answers in the spaces provided in this examination booklet.

Base your answers to questions 56 through 58 on the information below and on your knowledge of biology.

For Teacher Use Only

Throughout the world, in nearly every ecosystem, there are animal and plant species present that were introduced into the ecosystem by humans or transported to the ecosystem as a result of human activities. Some examples are listed in the chart below.

Examples of Introduced Species

Organism	New Location
purple loosestrife (plant)	wetlands in New York State
zebra mussel	Great Lakes
brown tree snake	Guam

56 State *one* reason why an introduced species might be very successful in a new environment. [1]

____

____ 56 ☐

57 Identify *one* action the government could take to prevent the introduction of additional new species. [1]

____

____ 57 ☐

58 Identify *one* introduced organism and write its name in the space below. Describe *one* way in which this organism has altered an ecosystem in the new location. [1]

Organism: ________________________

____

____ 58 ☐

Base your answers to questions 59 and 60 on the information and food web below and on your knowledge of biology.

The organisms in the food web below live near large cattle ranches. Over many years, mountain lions occasionally killed a few cattle. One year, a few ranchers hunted and killed many mountain lions to prevent future loss of their cattle. Later, ranchers noticed that animals from this food web were eating large amounts of grain from their fields.

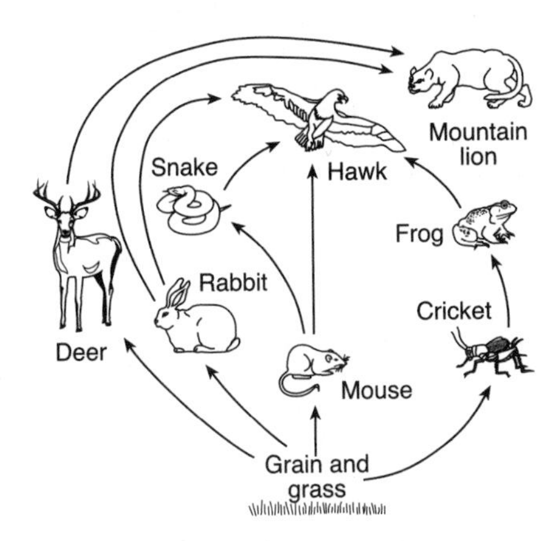

59 Identify *two* specific populations that most likely increased in number after the mountain lion population *decreased*. Support your answer. [2]

____

____

____

____

60 Explain how killing many mountain lions affected other ranchers in the community. [1]

____

____

For Teacher Use Only

59 ☐

60 ☐

Base your answers to questions 61 through 64 on the passage below and on your knowledge of biology. The letters indicate paragraphs.

For Teacher Use Only

Yellow Fever

Paragraph *A*

A team of doctors was sent to Havana, Cuba, to study a yellow fever epidemic. The doctors wanted to find out how the pathogenic microbe that causes yellow fever is transferred from those who are sick to those who are well. Some people thought that the disease was spread by having contact with a person who had the disease or even through contact with clothing or bedding that they had used.

Paragraph *B*

It was known that yellow fever occurred more frequently in swampy environments than in environments that were dry. Consequently, some people thought that the disease was due to contact with the atmosphere of the swamps. A respected doctor in Havana was convinced that a particular species of mosquito, *Aedes calopus,* spread the disease.

Paragraph *C*

The team of doctors carried out several experiments and collected data. They built poorly ventilated houses in which American soldiers volunteered to sleep on bedding used by individuals who had recently died of yellow fever in local hospitals. The soldiers also wore the nightshirts of those who had died. The houses were fumigated to kill all mosquitoes and the doors and windows of the houses were screened. None of the soldiers living in these houses contracted the disease, though the experiment was tried repeatedly.

Paragraph *D*

In another experiment, the team built houses that were tightly sealed. The doors and windows were screened. The insides of the houses were divided into two parts by mosquito netting. One part of the house contained a species of mosquito, *Aedes calopus,* that had been allowed to bite yellow fever patients in the hospital. There were no mosquitoes in the other part of the house. A group of volunteers lived in each part of the house. A number of those who lived in the part of the house with the mosquitoes became infected; none of those in the other part of the house did.

Paragraph *E*

Putting these facts together with other evidence, the team concluded that *Aedes calopus* spread the disease. The validity of this conclusion then had to be tested. All newly reported cases of yellow fever were promptly taken to well-screened hospitals and their houses were fumigated to kill any mosquitoes. The breeding places of the mosquitoes in and around Havana were drained or covered with a film of oil to kill mosquito larvae. Native fish species known to feed on mosquito larvae were introduced into streams and ponds. The number of yellow fever cases steadily declined until Havana was essentially free of the epidemic.

61 State the problem the team of doctors was trying to solve. [1]

For Teacher Use Only

61 ☐

62 State *one* hypothesis from paragraph *A* that was tested by one of the experiments. [1]

62 ☐

63 Describe the control that should have been set up for the experiment described in paragraph *C*. [1]

63 ☐

64 Explain why the use of native fish (described in paragraph *E*), rather than the use of pesticides, is less likely to have a *negative* impact on the environment. [1]

64 ☐

Base your answers to questions 65 through 67 on the information below and on your knowledge of biology.

For Teacher Use Only

Vaccines play an important role in the ability of the body to resist certain diseases.

65 Describe the contents of a vaccine. [1]

65

66 Identify the system in the body that is most directly affected by a vaccination. [1]

66

67 Explain how a vaccination results in the long-term ability of the body to resist disease. [1]

67

Base your answers to questions 68 and 69 on the information below and on your knowledge of biology.

A factory in Florida had dumped toxic waste into the soil for 40 years. Since the company is no longer in business, government officials removed the toxic soil and piled it up into large mounds until they can finish evaluating how to treat the waste.

68 State *one* way these toxins could move from the soil into local ecosystems, such as nearby lakes and ponds. [1]

68

69 State *one* way these toxins might affect local ecosystems. [1]

69

Base your answers to questions 70 and 71 on the diagram below and on your knowledge of biology. The diagram shows some of the gases that, along with their sources, contribute to four major problems associated with air pollution.

For Teacher Use Only

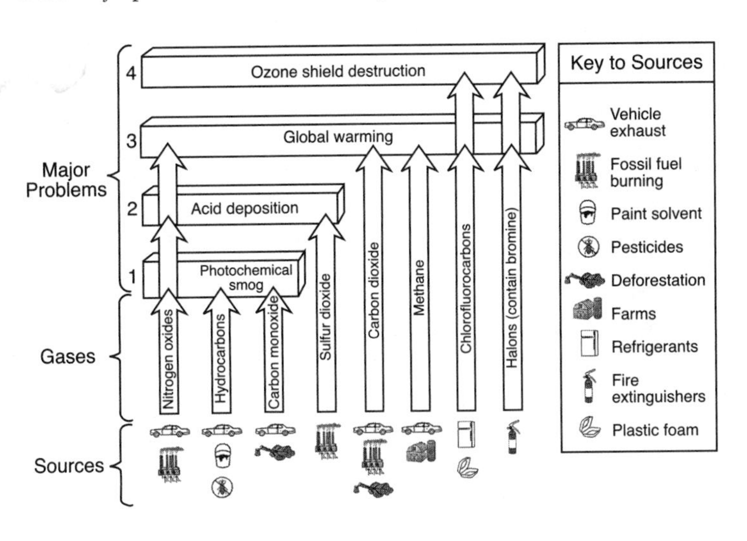

70 Select *one* of the four major problems from the diagram and record the number of the problem on the line below. Identify a gas that contributes to the problem you selected and state *one* way in which the amount of this gas can be reduced. [1]

Problem number: ____________

Gas: ______________________________

____

____

70 ☐

71 Explain why damage to the ozone shield is considered a threat to many organisms. [1]

____

____

____

71 ☐

Part D

Answer all questions in this part. [13]

Directions (72–84): For those questions that are followed by four choices, circle the *number* of the choice, that, of those given, best completes the statement or answers the question. For all other questions in this part, follow the directions given in the question and record your answers in the spaces provided.

For Teacher Use Only

72 A laboratory technique is illustrated in the diagram below.

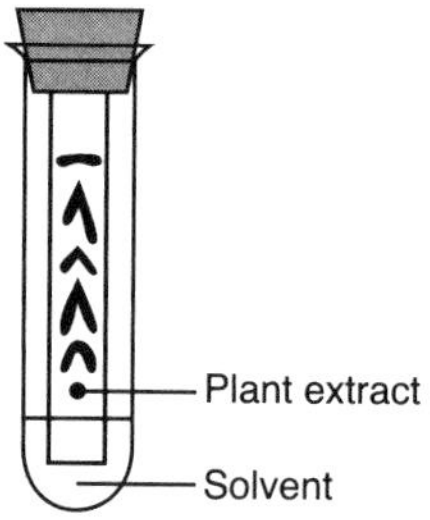

This technique is used to

(1) determine volume

(2) separate molecules in a mixture

(3) measure length

(4) analyze data from an experiment

72 ☐

73 As part of an experiment, the heart rate of a person at rest was measured every hour for 7 hours. The data are shown in the table below.

Data Table

Hour	Heart Rate (beats/min)
1	72
2	63
3	61
4	61
5	60
6	63
7	68

Which graphed line best represents this data?

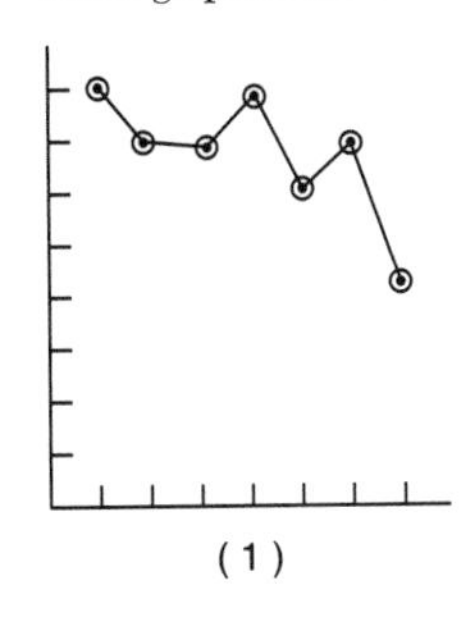

(1)

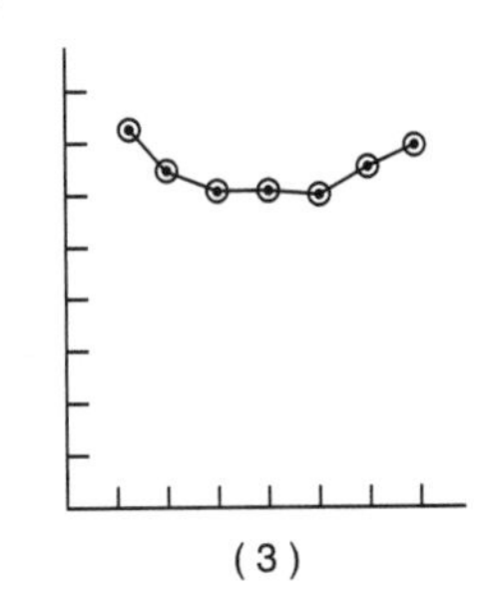

(3)

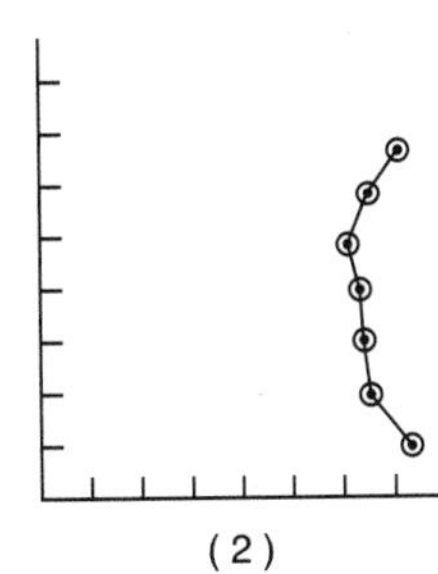

(2)

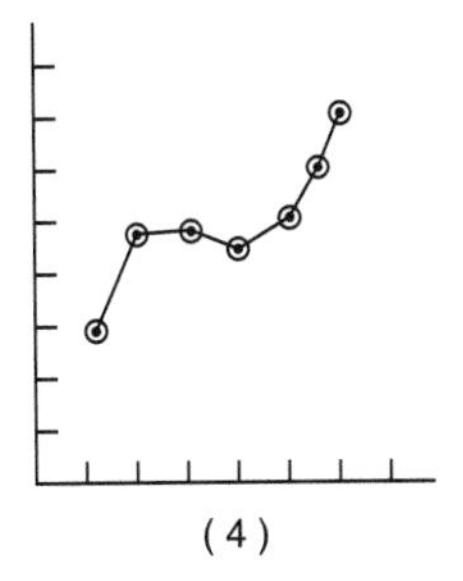

(4)

For Teacher Use Only

73 ☐

Base your answers to questions 74 through 77 on the Universal Genetic Code Chart below and on your knowledge of biology. Some DNA, RNA, and amino acid information from the analysis of a gene present in five different species is shown in the chart on the next page.

For Teacher Use Only

Universal Genetic Code Chart

Messenger RNA Codons and Amino Acids for Which They Code

First base	Second base: U	Second base: C	Second base: A	Second base: G	Third base
U	UUU, UUC: **PHE** UUA, UUG: **LEU**	UCU, UCC, UCA, UCG: **SER**	UAU, UAC: **TYR** UAA, UAG: **STOP**	UGU, UGC: **CYS** UGA: **STOP** UGG: **TRP**	U C A G
C	CUU, CUC, CUA, CUG: **LEU**	CCU, CCC, CCA, CCG: **PRO**	CAU, CAC: **HIS** CAA, CAG: **GLN**	CGU, CGC, CGA, CGG: **ARG**	U C A G
A	AUU, AUC, AUA: **ILE** AUG: **MET or START**	ACU, ACC, ACA, ACG: **THR**	AAU, AAC: **ASN** AAA, AAG: **LYS**	AGU, AGC: **SER** AGA, AGG: **ARG**	U C A G
G	GUU, GUC, GUA, GUG: **VAL**	GCU, GCC, GCA, GCG: **ALA**	GAU, GAC: **ASP** GAA, GAG: **GLU**	GGU, GGC, GGA, GGG: **GLY**	U C A G

74 Using the Universal Genetic Code Chart, fill in the missing amino acids in the amino acid sequence for species *A* in the chart *on the next page.* [1]

75 Using the information given, fill in the missing mRNA bases in the mRNA strand for species *B* in the chart *on the next page.* [1]

76 Using the information given, fill in the missing DNA bases in the DNA strand for species *C* in the chart *on the next page.* [1]

Species A	DNA strand:	TAC	CGA	CCT	TCA
	mRNA strand:	AUG	GCU	GGA	AGU
	Amino acid sequence:	____	____	____	____
Species B	DNA strand:	TAC	TTT	GCA	GGA
	mRNA strand:	____	____	____	____
	Amino acid sequence:	MET	LYS	ARG	PRO
Species C	DNA strand:	____	____	____	____
	mRNA strand:	AUG	UUU	UGU	CCC
	Amino acid sequence:	MET	PHE	CYS	PRO
Species D	DNA strand:	TAC	GTA	GTT	GCA
	mRNA strand:	AUG	CAU	CAA	CGU
	Amino acid sequence:	MET	HIS	GLN	ARG
Species E	DNA strand:	TAC	TTC	GCG	GGT
	mRNA strand:	AUG	AAG	CGC	CCA
	Amino acid sequence	MET	LYS	ARG	PRO

For Teacher Use Only

74 ☐

75 ☐

76 ☐

77 According to the information, which *two* species are most closely related? Support your answer. [1]

Species:________ and ________

____

____

____

77 ☐

Base your answers to questions 78 and 79 on the information below and on your knowledge of biology. The diagram below represents the relationship between beak structure and food in several species of finches in the Galapagos Islands.

For Teacher Use Only

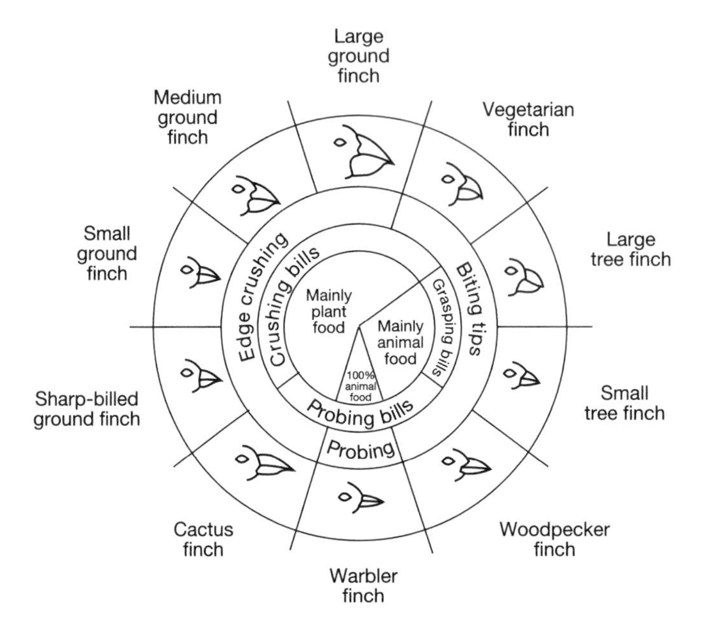

From: *Galapagos: A Natural History Guide*

Variations in Beaks of Galapagos Islands Finches

78 Which factor most directly influenced the evolution of the diverse types of beaks of these finches?

(1) predation by humans

(2) available food sources

(3) oceanic storms

(4) lack of available niches

78 ☐

79 State *one* reason why the large tree finch and the large ground finch are able to coexist on the same island. [1]

____

____

____

For Teacher Use Only

79 ☐

80 Phenolphthalein is a chemical that turns pink in the presence of a base. A student set up the demonstration shown in the diagram below.

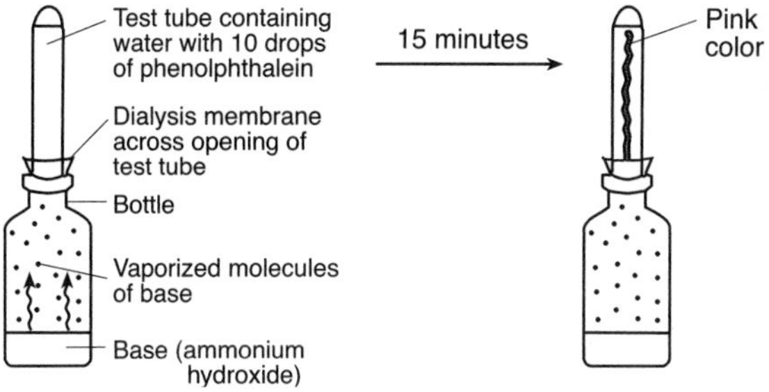

The appearance of the pink color was due to the movement of

(1) phenolphthalein molecules from low concentration to high concentration

(2) base molecules from high concentration through the membrane to low concentration

(3) water molecules through the membrane from high concentration to low concentration

(4) phenolphthalein molecules in the water from high concentration to low concentration

80 ☐

For Teacher Use Only

Base your answers to questions 81 and 82 on the information and data table below and on your knowledge of biology.

A student cut three identical slices from a potato. She determined the mass of each slice. She then placed them in labeled beakers and added a different solution to each beaker. After 30 minutes, she removed each potato slice from its solution, removed the excess liquid with a paper towel, and determined the mass of each slice. The change in mass was calculated and the results are shown in the data table below.

Change in Mass of Potato in Different Solutions

Beaker	Solution	Change in Mass
1	distilled water	gained 4.0 grams
2	6% salt solution	lost 0.4 gram
3	16% salt solution	lost 4.7 grams

81 Identify the process that is responsible for the change in mass of each of the three slices. [1]

____ 81 ☐

82 Explain why the potato slice in beaker 1 increased in mass. [1]

____

____ 82 ☐

83 Which activity might lead to damage of a microscope and specimen?

(1) cleaning the ocular and objectives with lens paper

(2) focusing with low power first before moving the high power into position

(3) using the coarse adjustment to focus the specimen under high power

(4) adjusting the diaphragm to obtain more light under high power

83 ☐

84 A solution containing both starch and glucose was placed inside the model cell represented below. The model cell was then placed in a beaker containing distilled water.

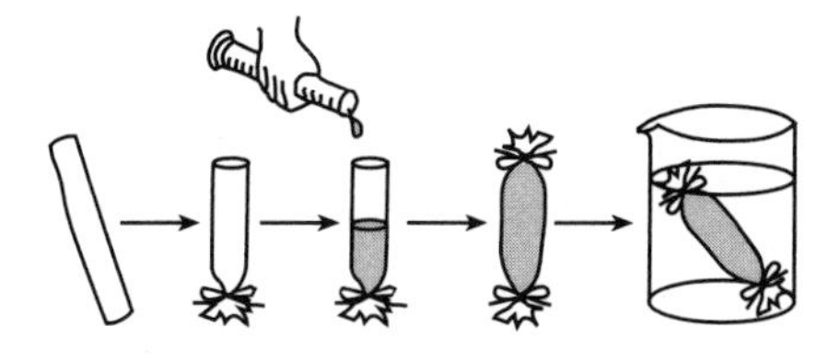

Identify *one* specific substance that should have been added to the distilled water so that observations regarding movement of starch could be made. [1]

____

____

For Teacher Use Only

84 [ ]

Tear Here

The University of the State of New York

REGENTS HIGH SCHOOL EXAMINATION

LIVING ENVIRONMENT

Wednesday, August 13, 2008 — 12:30 to 3:30 p.m., only

ANSWER SHEET

Student .... Sex: ☐ Female ☐ Male

Teacher ....

School .... Grade ....

Part	Maximum Score	Student's Score
A	30	
B–1	12	
B–2	13	
C	17	
D	13	
Total Raw Score (maximum Raw Score: 85)		☐
Final Score (from conversion chart)		☐

Raters' Initials

Rater 1 .... **Rater 2** ....

Record your answers to Part A and Part B–1 on this answer sheet.

Part A

1	11	21 ....
2	12	22 ....
3	13	23 ....
4	14	24 ....
5	15	25 ....
6	16	26 ....
7	17	27 ....
8	18	28 ....
9	19	29 ....
10	20	30 ....

Part A Score ☐

Part B–1

31	37 ....
32	38 ....
33	39 ....
34	40 ....
35	41 ....
36	42 ....

Part B–1 Score ☐

Tear Here

The declaration below must be signed when you have completed the examination.

I do hereby affirm, at the close of this examination, that I had no unlawful knowledge of the questions or answers prior to the examination and that I have neither given nor received assistance in answering any of the questions during the examination.

____

Signature

REGENTS HIGH SCHOOL EXAMINATION

Part A

Tuesday, January 27, 2009
9:15a.m. to 12:15p.m., only

Answer all questions in this part. [30]

Directions (1–30): For *each* statement or question, write on your separate answer sheet the *number* of the word or expression that, of those given, best completes the statement or answers the question.

1 Scientists in the United States, Europe, and Africa have now suggested that the hippopotamus is a relative of the whale. Earlier studies placed the hippo as a close relative of wild pigs, but recent studies have discovered stronger evidence for the connection to whales. This information suggests that

(1) genetic engineering was involved in the earlier theories
(2) structural evidence is the best evolutionary factor to consider
(3) natural selection does not occur in hippopotamuses
(4) scientific explanations are tentative and subject to change

2 A stable ecosystem would *not* contain

(1) materials being cycled
(2) consumers without producers
(3) decomposers
(4) a constant source of energy

3 A human liver cell and a human skin cell in the same person have the same genetic sequences. However, these cells are different because the liver cell

(1) has more dominant traits than the skin cell
(2) can reproduce but the skin cell cannot
(3) carries out respiration but the skin cell does not
(4) uses different genes than the skin cell

4 Abiotic factors that could affect the stability of an ecosystem could include

(1) hurricanes, packs of wolves, and temperature
(2) blizzards, heat waves, and swarms of grasshoppers
(3) droughts, floods, and heat waves
(4) species of fish, number of decomposers, and supply of algae

5 Many viruses infect only a certain type of cell because they bind to certain

(1) other viruses on the surface of the cell
(2) mitochondria in the cell
(3) hormones in the cell
(4) receptor sites on the surface of the cell

6 The respiratory system includes a layer of cells in the air passages that clean the air before it gets to the lungs. This layer of cells is best classified as

(1) a tissue
(2) an organ
(3) an organelle
(4) an organ system

7 The diagram below represents a typical energy pyramid.

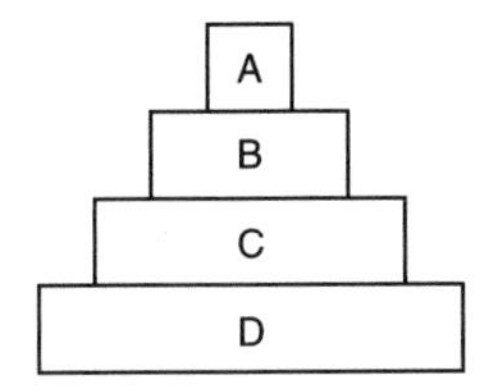

Which level in the pyramid includes autotrophs?

(1) *A*
(2) *B*
(3) *C*
(4) *D*

8 Mustard gas removes guanine (G) from DNA. For developing embryos, exposure to mustard gas can cause serious deformities because guanine

(1) stores the building blocks of proteins
(2) supports the structure of ribosomes
(3) produces energy for genetic transfer
(4) is part of the genetic code

9 The diagram below represents a food web.

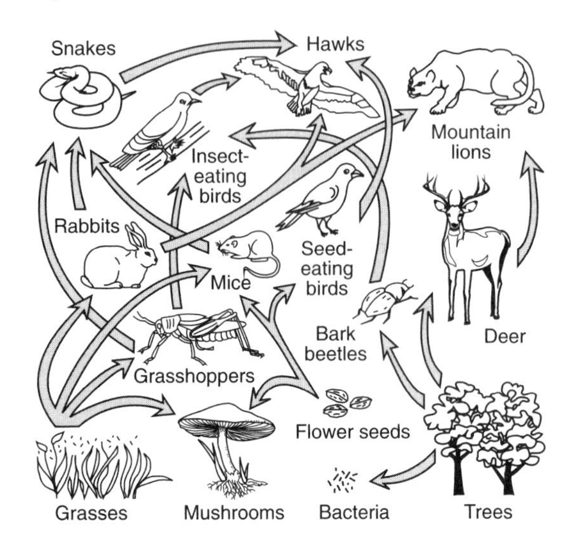

Which organisms are correctly paired with their nutritional roles?

(1) hawk—decomposer; insect-eating bird—parasite
(2) mouse—autotroph; flower seed—heterotroph
(3) mountain lion—predator; bark beetle—herbivore
(4) grasshopper—carnivore; grass—autotroph

10 Which process usually results in offspring that exhibit new genetic variations?

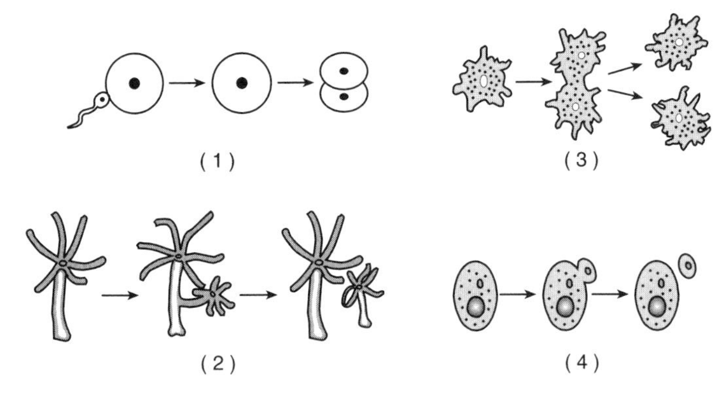

11 Which observation could best be used to indicate an evolutionary relationship between two species?

(1) They have similar base sequences.
(2) They have similar fur color.
(3) They inhabit the same geographic regions.
(4) They occupy the same niche.

12 A species in a changing environment would have the best chance of survival as a result of a mutation that has a

(1) high adaptive value and occurs in its skin cells
(2) low adaptive value and occurs in its skin cells
(3) high adaptive value and occurs in its gametes
(4) low adaptive value and occurs in its gametes

13 In an area of Indonesia where the ocean floor is littered with empty coconut shells, a species of octopus has been filmed "walking" on two of its eight tentacles. The remaining six tentacles are wrapped around its body. Scientists suspect that, with its tentacles arranged this way, the octopus resembles a rolling coconut. Local predators, including sharks, seem not to notice the octopus as often when it behaves in this manner. This unique method of locomotion has lasted over many generations due to

(1) competition between octopuses and their predators
(2) ecological succession in marine habitats
(3) the process of natural selection
(4) selective breeding of this octopus species

14 Which statement concerning production of offspring is correct?

(1) Production of offspring is necessary for a species to survive, but it is not necessary for an individual to survive.
(2) An organism can reproduce without performing any of the other life processes.
(3) Production of offspring is necessary for an individual organism to survive, while the other life processes are important for a species to survive.
(4) Reproduction is a process that requires gametes in all species.

15 Limited resources contribute to evolutionary change in animals by increasing

(1) genetic variation within the population
(2) competition between members of the species
(3) the carrying capacity for the species
(4) the rate of photosynthesis in the population

16 Some chemical interactions in a human are shown in the graph below.

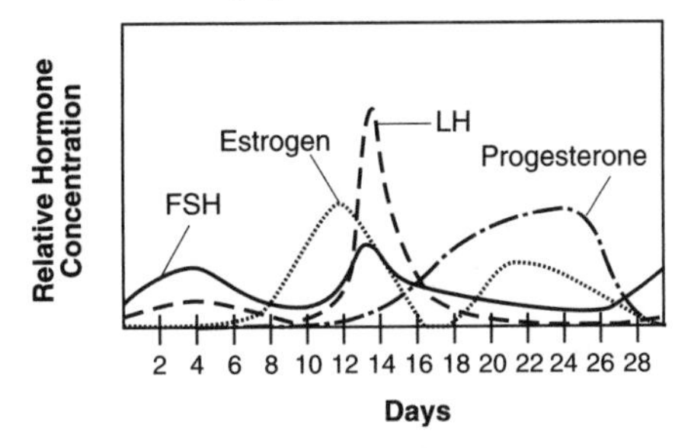

This graph represents hormones and events in the

(1) process of fetal growth and development
(2) process of meiotic cell division during sperm development
(3) reproductive cycle of males
(4) reproductive cycle of females

17 German measles is a disease that can harm an embryo if the mother is infected in the early stages of pregnancy because the virus that causes German measles is able to

(1) be absorbed by the embryo from the mother's milk
(2) be transported to the embryo in red blood cells
(3) pass across the placenta
(4) infect the eggs

18 In lakes in New York State that are exposed to acid rain, fish populations are declining. This is primarily due to changes in which lake condition?

(1) size
(2) temperature
(3) pH
(4) location

19 The diagram below represents a system in the human body.

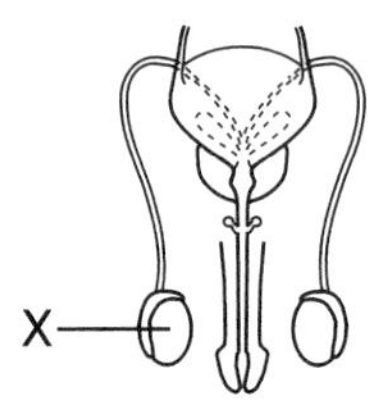

The primary function of structure *X* is to

(1) produce energy needed for sperm to move
(2) provide food for the sperm to carry to the egg
(3) produce and store urine
(4) form gametes that may be involved in fertilization

20 The diagram below represents an autotrophic cell.

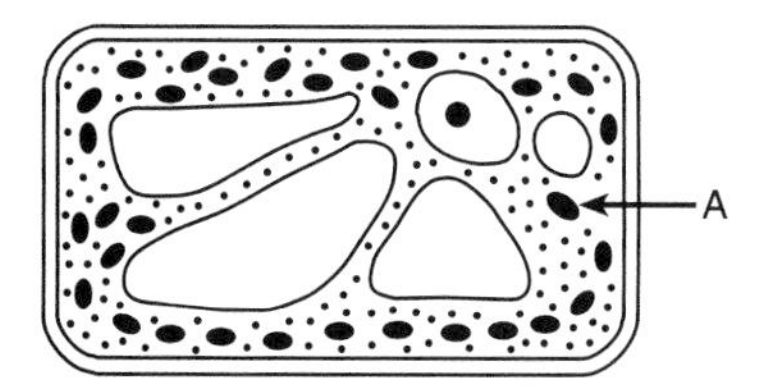

For the process of autotrophic nutrition, the arrow labeled *A* would most likely represent the direction of movement of

(1) carbon dioxide, water, and solar energy
(2) oxygen, glucose, and solar energy
(3) carbon dioxide, oxygen, and heat energy
(4) glucose, water, and heat energy

21 Which statement describes starches, fats, proteins, and DNA?

(1) They are used to store genetic information.
(2) They are complex molecules made from smaller molecules.
(3) They are used to assemble larger inorganic materials.
(4) They are simple molecules used as energy sources.

22 In 1995, during an Ebola virus outbreak, approximately 80% of the infected individuals died. Which statement is an inference that could be made based on this information?

(1) The individuals who survived were able to produce antibodies against the Ebola virus.
(2) The individuals who survived were not exposed to the Ebola antigens.
(3) Eighty percent of the population had a natural immunity to the Ebola virus.
(4) Eighty percent of the population was infected with a viral antigen.

23 In some people, substances such as peanuts, eggs, and milk cause an immune response. This response to usually harmless substances is most similar to the

(1) action of the heart as the intensity of exercise increases
(2) mechanism that regulates the activity of guard cells
(3) action of white blood cells when certain bacteria enter the body
(4) mechanism that maintains the proper level of antibiotics in the blood

24 The ivory-billed woodpecker, long thought to be extinct, was recently reported to be living in a southern swamp area. The most ecologically appropriate way to ensure the natural survival of this population of birds is to

(1) feed them daily with corn and other types of grain
(2) destroy their natural enemies and predators
(3) move the population of birds to a zoo
(4) limit human activities in the habitat of the bird

25 Millions of acres of tropical rain forest are being destroyed each year. Which change would most likely occur over time if the burning and clearing of these forests were stopped?

(1) an increase in the amount of atmospheric pollution produced
(2) a decrease in the source of new medicines
(3) an increase in the amount of oxygen released into the atmosphere
(4) a decrease in the number of species

26 The diagram below represents a biological process taking place in an area of New York State unaffected by natural disasters.

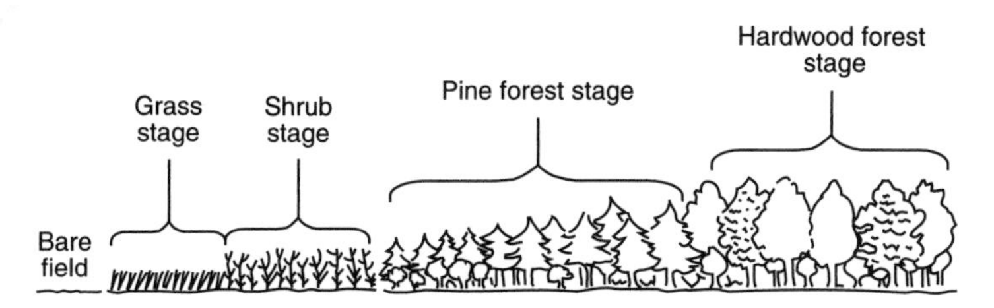

Which statement correctly describes a stage in this process?

(1) The grass stage is the most stable stage and exists for thousands of years.
(2) The shrub stage modifies the ecosystem, making it more suitable for the pine forest.
(3) The pine forest stage has no biodiversity and the least competition.
(4) The hardwood forest stage will be replaced by a pine forest.

27 Which sequence of natural events is likely to lead to ecosystem stability?

(1) sexual reproduction → genetic variation → biodiversity → ecosystem stability
(2) asexual reproduction → genetic variation → cloning → ecosystem stability
(3) genetic variation → asexual reproduction → biodiversity → ecosystem stability
(4) genetic variation → sexual reproduction → cloning → ecosystem stability

28 The Susquehanna River, which runs through the states of New York, Pennsylvania, and Maryland, received the designation "America's Most Endangered River" in 2005. One of the river's problems results from the large number of sewage overflow sites that are found along the course of the river. These sewage overflow sites are a direct result of an increase in

(1) global warming
(2) human population
(3) recycling programs
(4) atmospheric changes

29 Many farmers plant corn, and then harvest the entire plant at the end of the growing season. One *negative* effect of this action is that

(1) soil minerals used by corn plants are not recycled
(2) corn plants remove acidic compounds from the air all season long
(3) corn plants may replace renewable sources of energy
(4) large quantities of water are produced by corn plants

30 Which human activity is correctly paired with its likely future consequence?

(1) overfishing in the Atlantic — increase in supply of flounder and salmon as food for people
(2) development of electric cars or hybrid vehicles — increased rate of global warming
(3) use of fossil fuels — depletion of underground coal, oil, and natural gas supplies
(4) genetically engineering animals — less food available to feed the world's population

Part B–1

Answer all questions in this part. [12]

Directions (31–42): For *each* statement or question, write on the separate answer sheet the *number* of the word or expression that, of those given, best completes the statement or answers the question.

31 An experiment was set up to test the effect of light intensity on the rate of photosynthesis, as shown in the diagram below.

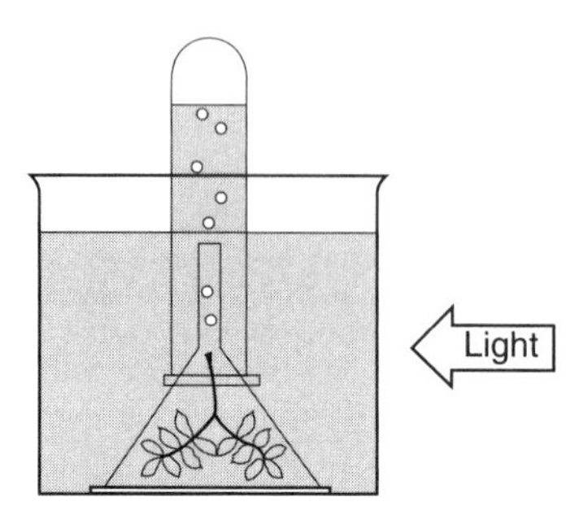

Data were collected by counting gas bubbles released in a 5-minute period when the light source was placed at various distances from the experimental setup. The data are shown in the table below.

Data Table

Distance From Light (cm)	Bubbles in 5-Minute Period
15	27
23	20
30	13
45	6

The number of bubbles released when the light source is at a distance of 38 centimeters would most likely be closest to

(1) 6
(2) 10
(3) 13
(4) 22

32 Which diagram represents the relative sizes of the structures listed below?

Structures

A	gene
B	cell
C	chromosome
D	nucleus

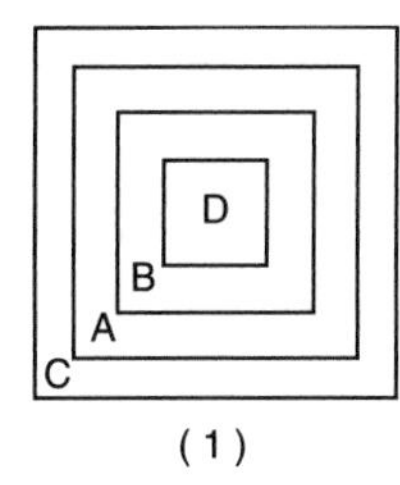

(1)

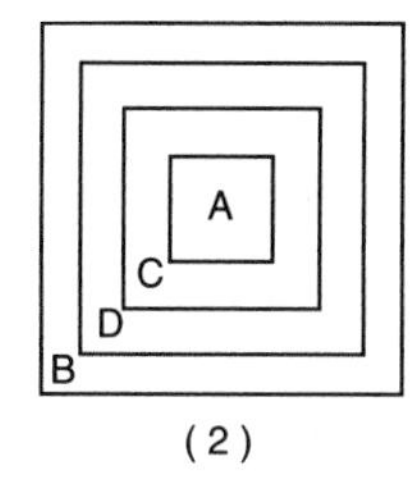

(2)

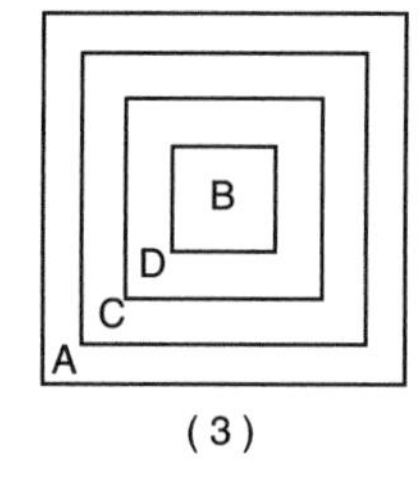

(3)

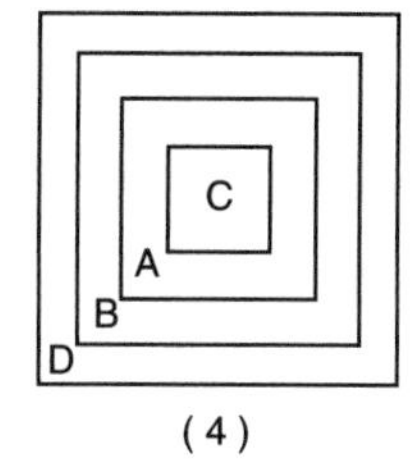

(4)

Base your answers to questions 33 through 35 on the diagram below, which represents some stages in the development of an embryo, and on your knowledge of biology.

A → B → C → D → E → F → G $\xrightarrow{X}$ Embryo

33 This entire sequence (*A* through embryo) started with

(1) the periodic shedding of a thickened uterine lining
(2) mitotic cell division in a testis
(3) meiotic cell division in the placenta
(4) the process of fertilization

34 If cell *A* has 46 chromosomes, how many chromosomes will most likely be found in each cell of stage *G*?

(1) 23
(2) 46
(3) 69
(4) 92

35 The arrow labeled *X* represents the process of

(1) meiosis
(2) recombination
(3) differentiation
(4) cloning

36 Which statement about the use of independent variables in controlled experiments is correct?

(1) A different independent variable must be used each time an experiment is repeated.
(2) The independent variables must involve time.
(3) Only one independent variable is used for each experiment.
(4) The independent variables state the problem being tested.

37 A scientist was investigating why a particular tree species grows only in a specific environment. To determine physical conditions the tree species needs to survive, an appropriate study should include

(1) the identification of organisms in the food web in that environment
(2) an analysis of the arrangement of the leaves on the trees
(3) the identification of all tree species in the area
(4) an analysis of the soil around the tree

38 The process illustrated in the sequence below occurs constantly in the biosphere.

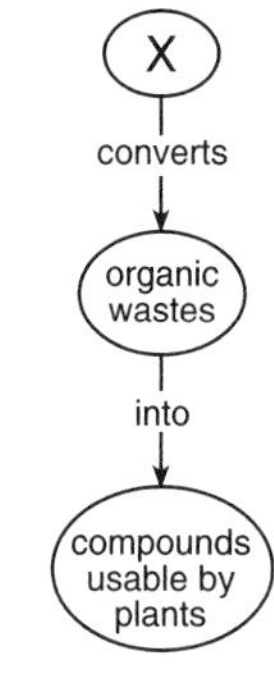

Which type of organism is most likely represented by *X*?

(1) decomposer
(2) producer
(3) herbivore
(4) carnivore

39 The direct source of ATP for the development of a fetus is

(1) a series of chemical activities that take place in the mitochondria of fetal cells
(2) a series of chemical activities that take place in the mitochondria of the uterine cells
(3) the transport of nutrients by the cytoplasm of the stomach cells of the mother
(4) the transport of nutrients by the cytoplasm of the stomach cells of the fetus

40 A sample of bacteria was added to a culture dish containing a food supply. The dish was kept in an incubator for two weeks, where temperature and other conditions that favored bacterial growth were kept constant. The graph below shows changes that occurred in the bacterial population over the two weeks.

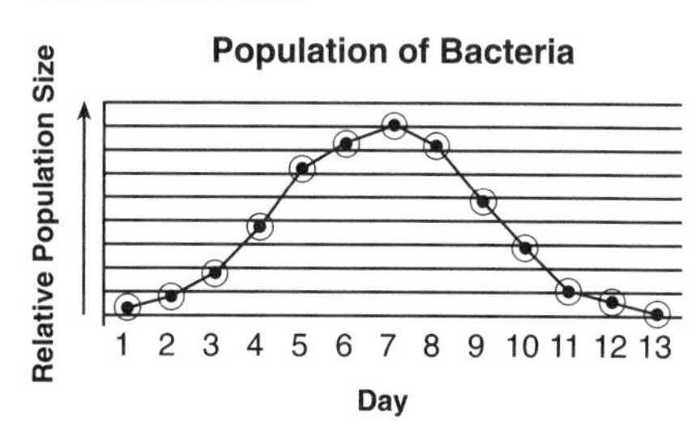

Which statement provides the best explanation for some of the changes observed?

(1) The bacteria were unable to reproduce until day 8.
(2) The bacteria consumed all of the available food.
(3) The culture dish contained an antibiotic for the first five days.
(4) The temperature increased and the bacteria died.

41 The diagram below represents a process involved in reproduction in some organisms.

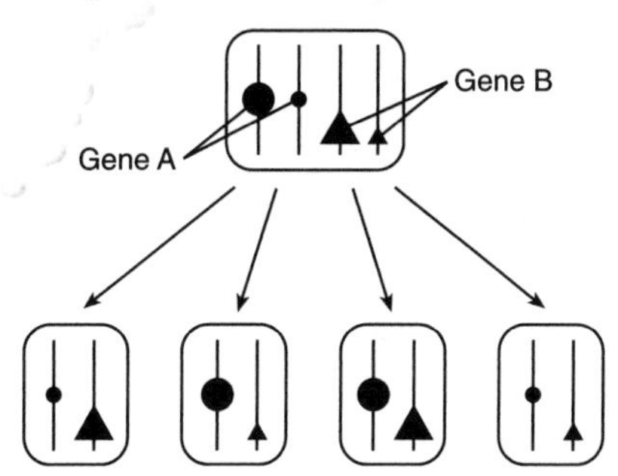

This process is considered a mechanism of evolution because

(1) mitosis produces new combinations of inheritable traits
(2) it increases the chances of DNA alterations in the parent
(3) it is a source of variation in the offspring produced
(4) meiosis prevents recombination of lethal mutations

42 The graph below shows the changes in the size of a fish population over a period of time.

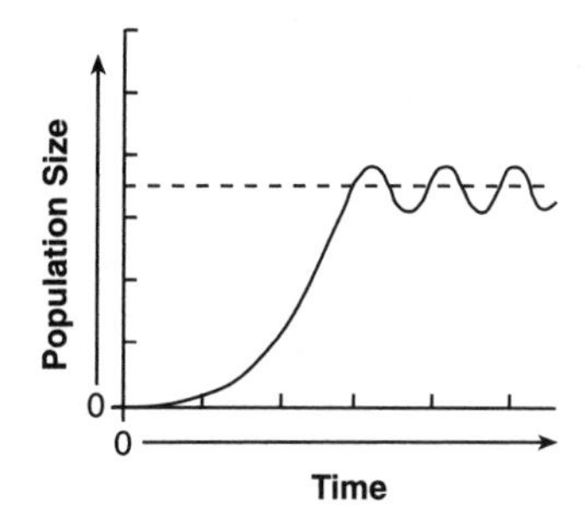

The dashed line on the graph represents the

(1) carrying capacity of the environment
(2) life span of the species
(3) level at which extinction is reached
(4) level of maximum biodiversity of the species

Part B–2

Answer all questions in this part. [13]

Directions (43–54): For those questions that are followed by four choices, circle the *number* preceding the choice that, of those given, best completes the statement or answers the question. For all other questions in this part, follow the directions given in the question and record your answers in the spaces provided.

For Teacher Use Only

43 Complete the chart below by identifying *two* cell structures involved in protein synthesis and stating how *each* structure functions in protein synthesis. [2]

Cell Structure	Function in Protein Synthesis

43 ☐

44 The diagram below represents stages in the digestion of an organic compound.

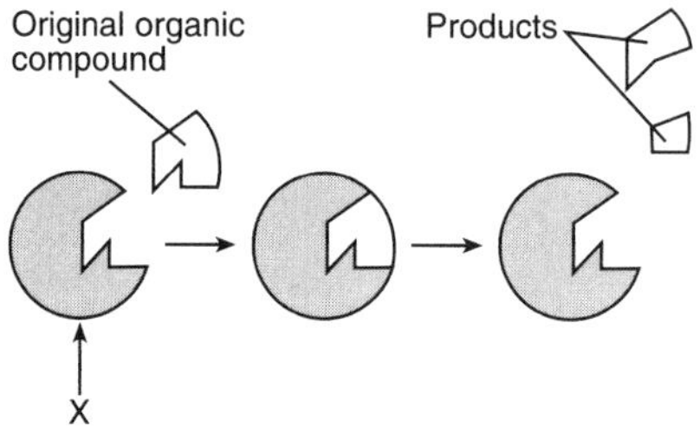

Explain why substance *X* would *not* be likely to digest a different organic compound. [1]

____

____

44 ☐

Base your answers to questions 45 through 47 on the passage below and on your knowledge of biology.

For Teacher Use Only

Overstaying Their Welcome: Cane Toads in Australia

Everyone in Australia is in agreement that the cane toads have got to go. The problem is getting rid of them. Cane toads, properly known as *Bufo marinus*, are the most notorious of what are called invasive species in Australia and beyond. But unlike other species of the same classification, cane toads were intentionally introduced into Australia. The country simply got much more and much worse than it bargained for.

Before 1935, Australia did not have any toad species of its own. What the country did have, however, was a major beetle problem. Two species of beetles in particular, French's Cane Beetle and the Greyback Cane Beetle, were in the process of decimating [destroying] the north-eastern state of Queensland's sugar cane crops. The beetle's larvae were eating the roots of the sugar cane and stunting, if not killing, the plants. The anticipated solution to this quickly escalating problem came in the form of the cane toad. After first hearing about the amphibians in 1933 at a conference in the Caribbean, growers successfully lobbied to have the cane toads imported to battle and hopefully destroy the beetles and save the crops....

The plan backfired completely and absolutely. As it turns out, cane toads do not jump very high, only about two feet actually, so they did not eat the beetles that for the most part lived in the upper stalks of cane plants. Instead of going after the beetles, as the growers had planned, the cane toads began going after everything else in sight—insects, bird's eggs and even native frogs. And because the toads are poisonous, they began to kill would-be predators. The toll on native species has been immense....

Source: Tina Butler, mongabay.com, April 17, 2005

45 State *one* reason why the cane toads were imported to Australia. [1]

____

____

45 ☐

46 Identify *one* adaptation of cane toads that made them successful in their new environment. [1]

____

____

46 ☐

47 State *one* specific example of how the introduction of the cane toads threatened biodiversity in Australia. [1]

____

____

47 ☐

Base your answers to questions 48 and 49 on the information below and on your knowledge of biology.

Signs of a Changing Planet

While the changing climate endangers some species, a little global warming suits many shallow-water squid and octopuses just fine. Slightly higher ocean temperatures have been shown to boost the growth of these cephalopods, whose digestive enzymes speed up when warm. The tentacled creatures are also quick to colonize new territory as conditions become more favorable. Humboldt squid, which usually range from Southern California to South America, have been spotted as far north as Alaska. Deep-sea squid may not, however, adapt as readily.

Sierra Magazine, March/April 2005

For Teacher Use Only

48 Which graph most accurately shows the interaction between water temperature and digestive enzyme action in the shallow-water squid?

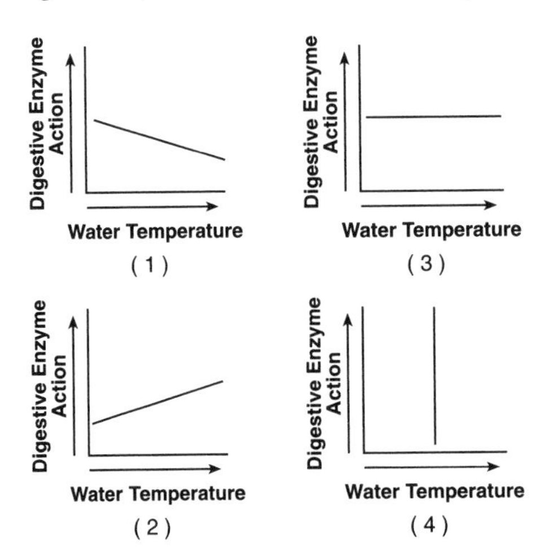

48 ☐

49 Although warming of the ocean may favor the migration of these squid into new territory, there may be biotic factors that make it difficult for these squid to live there. Identify *one* of these biotic factors, and explain why this factor would make it difficult for these squid to live in the new territory. [1]

____

____

____

49 ☐

Base your answers to questions 50 through 54 on the information below and on your knowledge of biology.

For Teacher Use Only

In a test for diabetes, blood samples were taken from an individual every 4 hours for 24 hours. The glucose concentrations were recorded and are shown in the data table below.

Blood Glucose Level Over Time

Time (h)	Blood Glucose Concentration (mg/dL)
0	100
4	110
8	128
12	82
16	92
20	130
24	104

50 State *one* likely cause of the change in blood glucose concentration between hour 16 and hour 20. [1]

____

____

50 ☐

Directions (51 and 52): Using the information given, construct a line graph on the grid *on the next page,* following the directions below.

51 Mark an appropriate scale on the axis labeled "Blood Glucose Concentration (mg/dL)." [1]

52 Plot the data from the data table. Surround each point with a small circle and connect the points. [1]

Example: 

For Teacher Use Only

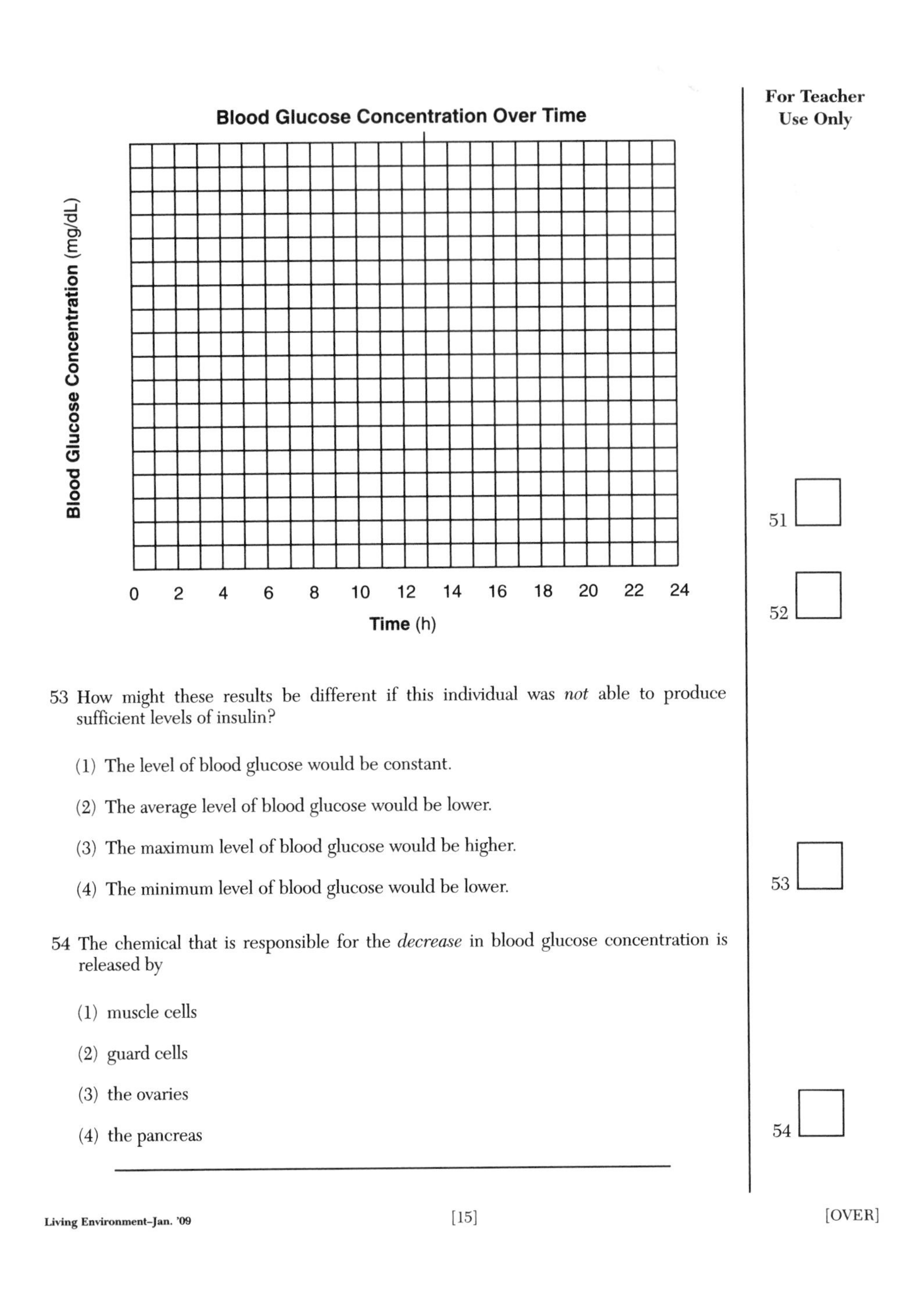

51

52

53 How might these results be different if this individual was *not* able to produce sufficient levels of insulin?

(1) The level of blood glucose would be constant.

(2) The average level of blood glucose would be lower.

(3) The maximum level of blood glucose would be higher.

(4) The minimum level of blood glucose would be lower.

53

54 The chemical that is responsible for the *decrease* in blood glucose concentration is released by

(1) muscle cells

(2) guard cells

(3) the ovaries

(4) the pancreas

54

Part C

Answer all questions in this part. [17]

Directions (55–67): Record your answers in the spaces provided in this examination booklet.

For Teacher Use Only

55 Many plants can affect the growth of other plants near them. This can occur when one plant produces a chemical that affects another plant.

Design an experiment to determine if a solution containing ground-up goldenrod plants has an effect on the growth of radish seedlings. In your experimental design be sure to:

- state a hypothesis to be tested [1]
- describe how the experimental group will be treated differently from the control group [1]
- explain why the number of seedlings used for the experiment should be large [1]
- identify the type of data that will be collected [1]
- describe experimental results that would support your hypothesis [1]

____

____

____

____

____

____

____

____

____

____

____

____

____

55 ☐

Base your answers to questions 56 and 57 on the information below and on your knowledge of biology.

A biologist at an agriculture laboratory is asked to develop a better quality blueberry plant. He is given plants that produce unusually large blueberries and plants that produce very sweet blueberries.

56 Describe *one* way the biologist could use these blueberry plants to develop a plant with blueberries that are both large and sweet. [1]

____

____

56 ☐

57 The biologist is successful in producing the new plant. State *one* method that can be used to produce many identical blueberry plants of this new type. [1]

____

____

57 ☐

Base your answers to questions 58 and 59 on the information below and on your knowledge of biology.

Two adaptations of the monarch butterfly that aid in its survival are the production of a certain chemical and a distinctive coloration that other animals can easily recognize. When a monarch butterfly is eaten, the presence of the chemical results in a bad taste to the predator. Although the viceroy butterfly does not contain the chemical that tastes bad to a predator, it does resemble the monarch in size, shape, and coloration.

58 Explain how the combination of this chemical and the distinctive coloration aid in the survival of the monarch butterfly. [1]

____

____

58 ☐

59 How do the characteristics of the viceroy butterfly aid in its survival? [1]

____

____

59 ☐

For Teacher Use Only

Base your answers to questions 60 through 62 on the information below and on your knowledge of biology.

For Teacher Use Only

Food is often treated to lower the risk of disease and spoilage, as shown in the chart below.

Food Preservation Methods

Method	Description of Method	Example of Food Treated With This Method
canning	heating at 115°C for 30 minutes	green beans
freezing	storing between –10°C and –18°C for extended time	meat, fish, poultry
salting	soaking in a salt solution for several days or weeks	pickles, sauerkraut

60 Identify *one* type of organism that is controlled by these food preservation methods. [1]

____

60 ☐

61 State *one* way extremely high temperatures can affect biological catalysts found in these organisms. [1]

____

____

61 ☐

62 Explain why high salt concentrations can kill organisms. [1]

____

____

62 ☐

63 An industry releases small amounts of a chemical pollutant into a nearby river each day. The chemical is absorbed by the microscopic water plants in the river. It causes the plants no apparent harm. Explain how this small amount of the chemical in the microscopic plants could enter the food chain and endanger the lives of birds that live nearby and feed on the fish from the river each day. [1]

For Teacher Use Only

63

Base your answers to questions 64 through 67 on the information below and on your knowledge of biology.

> Carbon, like many other elements, is maintained in ecosystems through a natural cycle. Human activities have been disrupting the carbon cycle.

64 Identify *one* process involved in recycling carbon dioxide within ecosystems. [1]

64

65 State *one* reason why the amount of carbon dioxide in the atmosphere has increased in the last 100 years. [1]

65

66 Identify *one* effect this increase in carbon dioxide could have on the environment. [1]

66

67 Describe *one* way individuals can help slow down or reverse the increase in carbon dioxide. [1]

67

Part D

Answer all questions in this part. [13]

Directions (68–80): For those questions that are followed by four choices, circle the *number* of the choice, that, of those given, best completes the statement or answers the question. For all other questions in this part, follow the directions given in the question and record your answers in the spaces provided.

Base your answer to question 68 on the chart below and on your knowledge of biology.

For Teacher Use Only

Species	Sequence of Four Amino Acids Found in the Same Part of the Hemoglobin Molecule of Species
human	Lys–Glu–His–Phe
horse	Arg–Lys–His–Lys
gorilla	Lys–Glu–His–Lys
chimpanzee	Lys–Glu–His–Phe
zebra	Arg–Lys–His–Arg

68 Which evolutionary tree best represents the information in the chart?

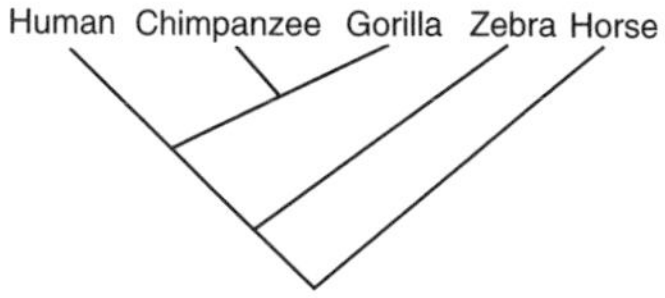

Human Chimpanzee Gorilla Zebra Horse

(1)

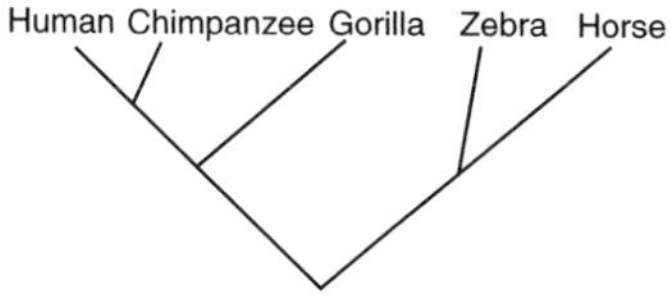

Human Chimpanzee Gorilla Zebra Horse

(3)

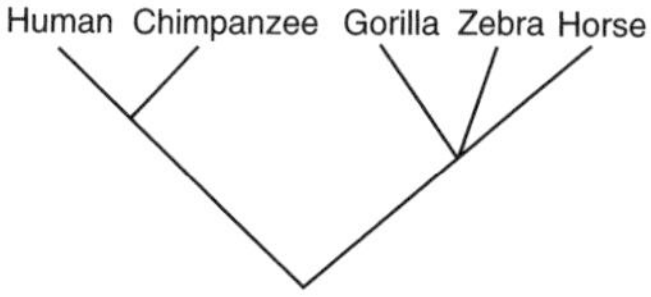

Human Chimpanzee Gorilla Zebra Horse

(2)

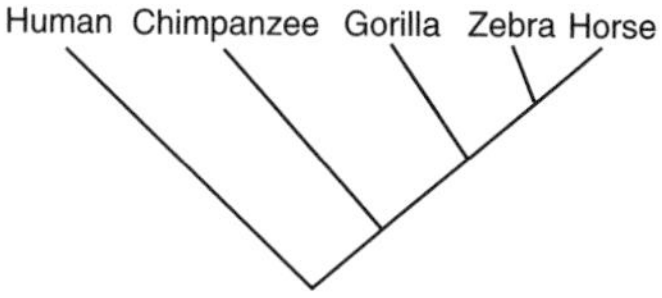

Human Chimpanzee Gorilla Zebra Horse

(4)

68

Base your answers to questions 69 and 70 on the diagram below that illustrates the results of a laboratory technique and on your knowledge of biology.

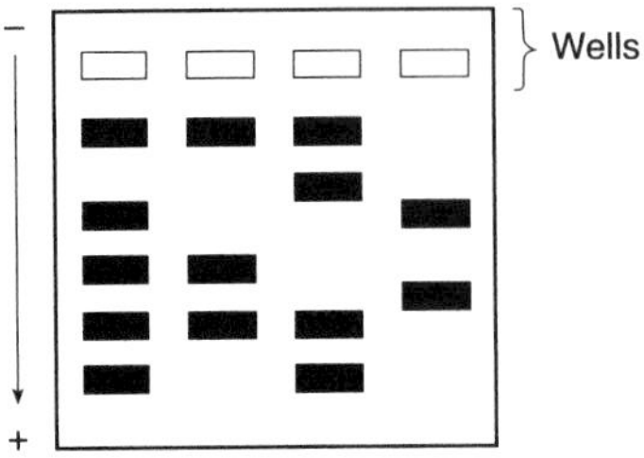

For Teacher Use Only

−

Wells

+

69 State *one* way the information obtained by this technique can be used. [1]

____

____

69

70 The results of which laboratory technique are represented in the diagram?

(1) chromatography

(2) manipulation of genes

(3) genetic engineering

(4) gel electrophoresis

70

Base your answers to questions 71 through 73 on the histograms below and on your knowledge of biology.

For Teacher Use Only

Students in a class recorded their resting pulse rates and their pulse rates immediately after strenuous activity. The data obtained are shown in the histograms below.

Resting Pulse Rate

12
10
8
6
4
2
0
< 50 51–60 61–70 71–80 81–90 > 90

Average Pulse Rate Range (per min)

Pulse Rate After Activity

12
10
8
6
4
2
0
< 50 51–60 61–70 71–80 81–90 > 90

Average Pulse Rate Range (per min)

71 An appropriate label for the *y*-axis in each histogram would be

(1) Number of Students

(2) Average Number of Heartbeats

(3) Time (min)

(4) Amount of Exercise

71 ☐

72 According to the data, compared to the average resting pulse rate, the average pulse rate immediately after strenuous activity generally

(1) decreased

(2) increased

(3) remained the same

(4) decreased and leveled off

72 ☐

73 State *one* biological explanation for the fact that *not* all students had the same resting pulse rate. [1]

____

____

____

73 ☐

Base your answers to questions 74 and 75 on the diagram below and on your knowledge of biology.

For Teacher Use Only

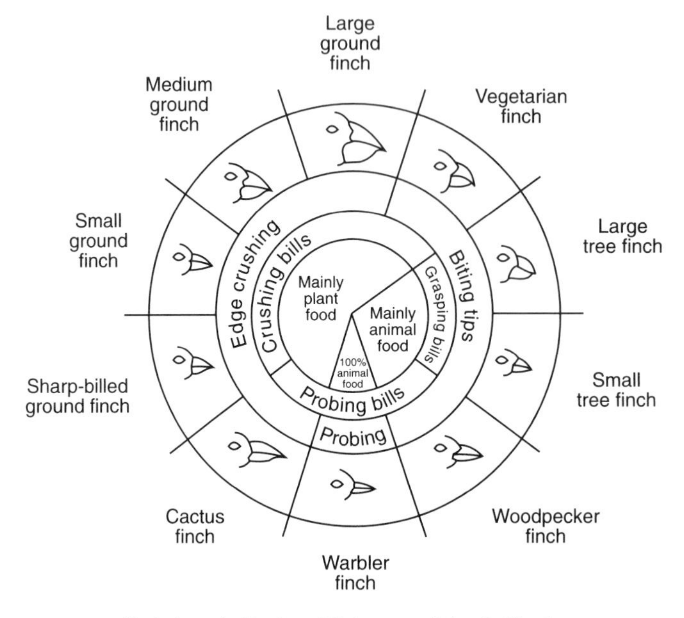

Variations in Beaks of Galapagos Islands Finches

74 The only finch that is completely carnivorous has a beak adapted for

(1) probing, only

(2) probing and edge crushing

(3) probing and biting

(4) biting and edge crushing

74 ☐

75 Which two finches would compete the *least* for food?

(1) small ground finch and large ground finch

(2) large ground finch and sharp-billed ground finch

(3) small tree finch and medium ground finch

(4) vegetarian finch and small ground finch

75 ☐

76 Glucose indicator was added to a beaker of an unknown liquid. Starch indicator was added to a different beaker containing the same unknown liquid. The color of the indicator solutions before they were added to the beakers and the color of the contents of the beakers after adding the indicator solution are recorded in the chart below.

For Teacher Use Only

Beaker	Solution	Color of Indicator Solution Before Adding to Beaker	Color of Contents of Beaker After Adding Indicator Solution
1	unknown liquid + glucose indicator	blue	blue (after heating)
2	unknown liquid + starch indicator	amber	blue black

Which carbohydrate is present in the unknown liquid? Support your answer. [1]

____

____

76 ☐

77 A laboratory setup of a model cell is shown in the diagram below.

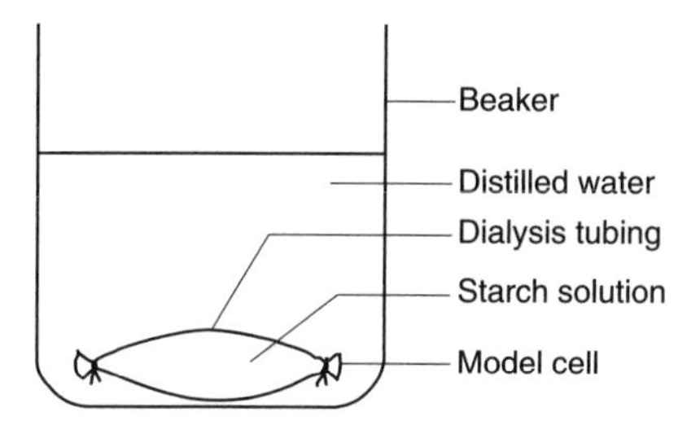

Which observation would most likely be made 24 hours later?

(1) The contents of the model cell have changed color.

(2) The diameter of the model cell has increased.

(3) The model cell has become smaller.

(4) The amount of distilled water in the beaker has increased.

77 ☐

Base your answers to questions 78 and 79 on the diagram below and on your knowledge of biology. The diagram illustrates what happens when a particular solution is added to a wet-mount slide containing red onion cells being observed using a compound light microscope.

For Teacher Use Only

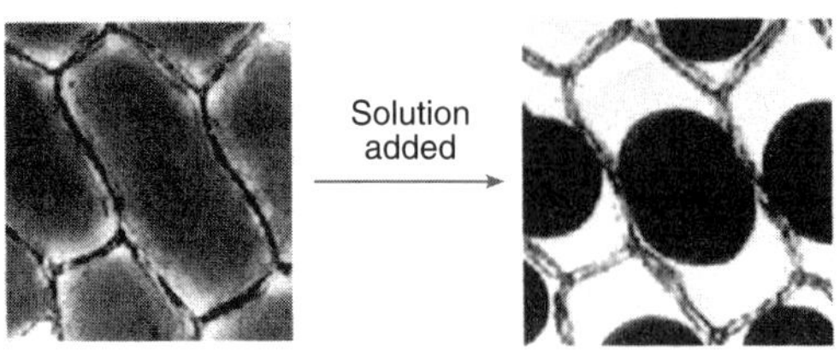

78 Identify a process that caused the change in the cells. [1]

____

78 ☐

79 To observe the cells on this slide it is best to start out using the

(1) high-power objective and focus using the coarse adjustment, only

(2) low-power objective and focus using the fine adjustment, only

(3) high-power objective and focus using the fine adjustment

(4) low-power objective and focus using the coarse adjustment

79 ☐

80 The diagram below represents the distribution of some molecules inside and outside of a cell over time.

For Teacher Use Only

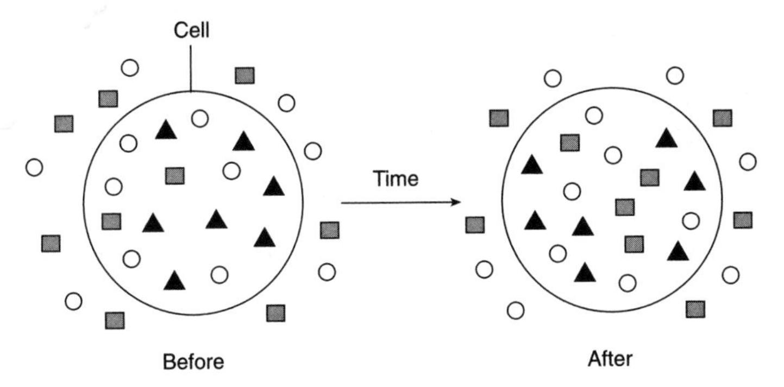

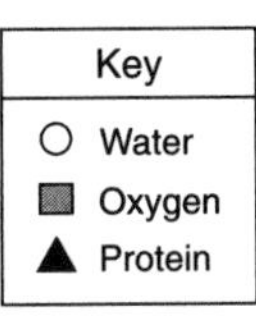

Which factor prevented the protein molecules (▲) from moving out of the cell?

(1) temperature

(2) pH

(3) molecule size

(4) molecule concentration

80 ☐

Tear Here

The University of the State of New York

REGENTS HIGH SCHOOL EXAMINATION

LIVING ENVIRONMENT

Tuesday, January 27, 2009 — 9:15 a.m. to 12:15 p.m., only

ANSWER SHEET

Student .... Sex: ☐ Female ☐ Male

Teacher ....

School .... Grade ....

Part	Maximum Score	Student's Score
A	30	
B–1	12	
B–2	13	
C	17	
D	13	
Total Raw Score (maximum Raw Score: 85)		
Final Score (from conversion chart)		

Raters' Initials

Rater 1 .... Rater 2 ....

Record your answers to Part A and Part B–1 on this answer sheet.

Part A

1	11	21 ....
2	12	22 ....
3	13	23 ....
4	14	24 ....
5	15	25 ....
6	16	26 ....
7	17	27 ....
8	18	28 ....
9	19	29 ....
10	20	30 ....

Part A Score

Part B–1

31	37 ....
32	38 ....
33	39 ....
34	40 ....
35	41 ....
36	42 ....

Part B–1 Score

Tear Here

The declaration below must be signed when you have completed the examination.

I do hereby affirm, at the close of this examination, that I had no unlawful knowledge of the questions or answers prior to the examination and that I have neither given nor received assistance in answering any of the questions during the examination.

Signature

REGENTS HIGH SCHOOL EXAMINATION **Part A** **Thursday,** June 18, 2009
1:15 to 4:15p.m., only

Answer all questions in this part. [30]

Directions (1–30): For *each* statement or question, write on your separate answer sheet the *number* of the word or expression that, of those given, best completes the statement or answers the question.

1 Which statement best describes one of the stages represented in the diagram below?

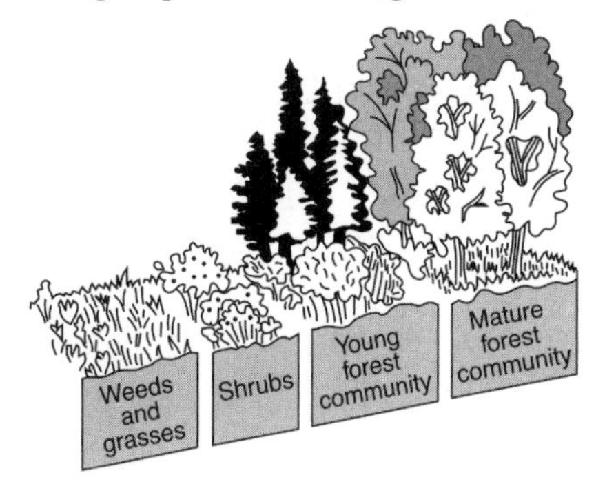

(1) The mature forest will most likely be stable over a long period of time.
(2) If all the weeds and grasses are destroyed, the number of carnivores will increase.
(3) As the population of the shrubs increases, it will be held in check by the mature forest community.
(4) The young forest community will invade and take over the mature forest community.

2 Which organ system in humans is most directly involved in the transport of oxygen?

(1) digestive
(2) nervous
(3) excretory
(4) circulatory

3 Which cell structure contains information needed for protein synthesis?

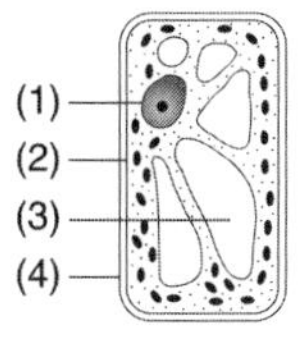

4 The human liver contains many specialized cells that secrete bile. Only these cells produce bile because

(1) different cells use different parts of the genetic information they contain
(2) cells can eliminate the genetic codes that they do not need
(3) all other cells in the body lack the genes needed for the production of bile
(4) these cells mutated during embryonic development

5 Although identical twins inherit exact copies of the same genes, the twins may look and act differently from each other because

(1) a mutation took place in the gametes that produced the twins
(2) the expression of genes may be modified by environmental factors
(3) the expression of genes may be different in males and females
(4) a mutation took place in the zygote that produced the twins

6 Which hormone does *not* directly regulate human reproductive cycles?

(1) testosterone
(2) estrogen
(3) insulin
(4) progesterone

7 Owls periodically expel a mass of undigested material known as a pellet. A student obtained several owl pellets from the same location and examined the animal remains in the pellets. He then recorded the number of different prey animal remains in the pellets. The student was most likely studying the

(1) evolution of the owl
(2) social structure of the local owl population
(3) role of the owl in the local ecosystem
(4) life cycle of the owl

8 Which sequence best represents the relationship between DNA and the traits of an organism?

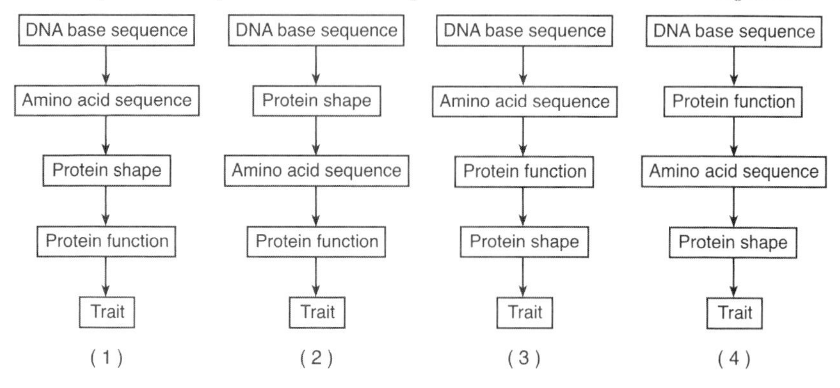

9 A sequence of events associated with ecosystem stability is represented below.

sexual reproduction → genetic variation → biodiversity → ecosystem stability

The arrows in this sequence should be read as

(1) leads to
(2) reduces
(3) prevents
(4) simplifies

10 In some people, the lack of a particular enzyme causes a disease. Scientists are attempting to use bacteria to produce this enzyme for the treatment of people with the disease. Which row in the chart below best describes the sequence of steps the scientists would most likely follow?

Row	Step A	Step B	Step C	Step D
(1)	identify the gene	insert the gene into a bacterium	remove the gene	extract the enzyme
(2)	insert the gene into a bacterium	identify the gene	remove the gene	extract the enzyme
(3)	identify the gene	remove the gene	insert the gene into a bacterium	extract the enzyme
(4)	remove the gene	extract the enzyme	identify the gene	insert the gene into a bacterium

11 What will most likely occur as a result of changes in the frequency of a gene in a particular population?

(1) ecological succession
(2) biological evolution
(3) global warming
(4) resource depletion

12 The puppies shown in the photograph below are all from the same litter.

The differences seen within this group of puppies are most likely due to

(1) overproduction and selective breeding
(2) mutations and elimination of genes
(3) evolution and asexual reproduction
(4) sorting and recombination of genes

13 Carbon dioxide makes up less than 1 percent of Earth's atmosphere, and oxygen makes up about 20 percent. These percentages are maintained most directly by

(1) respiration and photosynthesis
(2) the ozone shield
(3) synthesis and digestion
(4) energy recycling in ecosystems

14 Which sequence represents the order of some events in human development?

(1) zygote → sperm → tissues → egg
(2) fetus → tissues → zygote → egg
(3) zygote → tissues → organs → fetus
(4) sperm → zygote → organs → tissues

15 A variety of plant produces small white fruit. A stem was removed from this organism and planted in a garden. If this stem grows into a new plant, it would most likely produce

(1) large red fruit, only
(2) large pink fruit, only
(3) small white fruit, only
(4) small red and small white fruit on the same plant

16 A mutation that can be inherited by offspring would result from

(1) random breakage of chromosomes in the nucleus of liver cells
(2) a base substitution in gametes during meiosis
(3) abnormal lung cells produced by toxins in smoke
(4) ultraviolet radiation damage to skin cells

17 The diagram below represents a process that occurs in organisms.

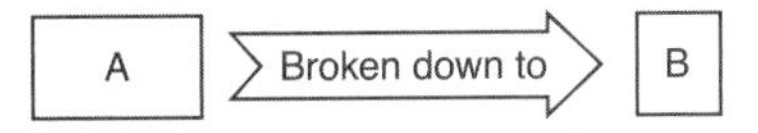

Which row in the chart indicates what *A* and *B* in the boxes could represent?

Row	A	B
(1)	starch	proteins
(2)	starch	amino acids
(3)	protein	amino acids
(4)	protein	simple sugars

18 Some organs of the human body are represented in the diagram below.

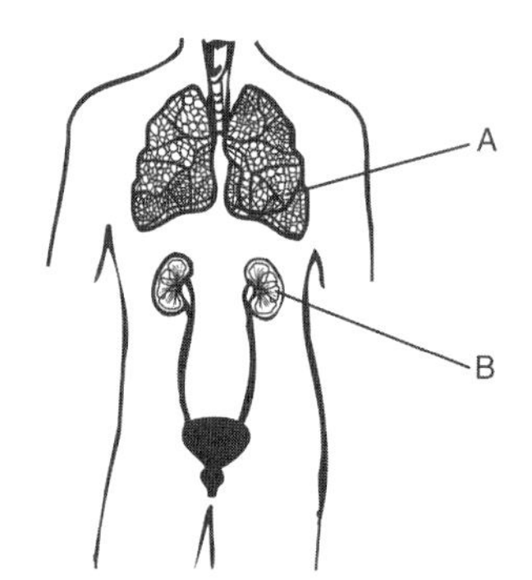

Which statement best describes the functions of these organs?

(1) *B* pumps blood to *A* for gas exchange.
(2) *A* and *B* both produce carbon dioxide, which provides nutrients for other body parts.
(3) *A* releases antibodies in response to an infection in *B*.
(4) The removal of wastes from both *A* and *B* involves the use of energy from ATP.

19 *Salmonella* bacteria can cause humans to have stomach cramps, vomiting, diarrhea, and fever. The effect these bacteria have on humans indicates that *Salmonella* bacteria are

(1) predators
(2) pathogenic organisms
(3) parasitic fungi
(4) decomposers

20 The virus that causes AIDS is damaging to the body because it

(1) targets cells that fight invading microbes
(2) attacks specific red blood cells
(3) causes an abnormally high insulin level
(4) prevents the normal transmission of nerve impulses

21 In the leaf of a plant, guard cells help to

(1) destroy atmospheric pollutants when they enter the plant
(2) regulate oxygen and carbon dioxide levels
(3) transport excess glucose to the roots
(4) block harmful ultraviolet rays that can disrupt chlorophyll production

22 An antibiotic is effective in killing 95% of a population of bacteria that reproduce by the process shown below.

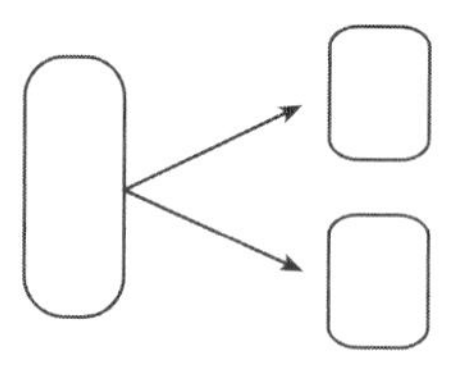

Which statement best describes future generations of these bacteria?

(1) They will be produced by asexual reproduction and will be more resistant to the antibiotic.
(2) They will be produced by sexual reproduction and will be more resistant to the antibiotic.
(3) They will be produced by asexual reproduction and will be just as susceptible to the antibiotic.
(4) They will be produced by sexual reproduction and will be just as susceptible to the antibiotic.

23 The size of plant populations can be influenced by the

(1) molecular structure of available oxygen
(2) size of the cells of decomposers
(3) number of chemical bonds in a glucose molecule
(4) type of minerals present in the soil

24 Competition between two species occurs when

(1) mold grows on a tree that has fallen in the forest
(2) chipmunks and squirrels eat sunflower seeds in a garden
(3) a crow feeds on the remains of a rabbit killed on the road
(4) a lion stalks, kills, and eats an antelope

25 A food chain is illustrated below.

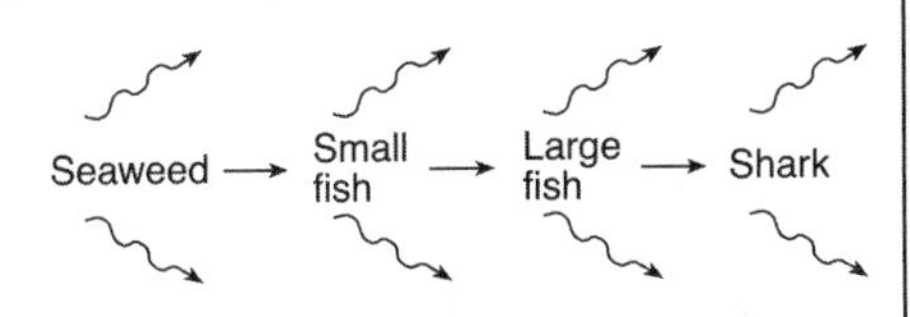

The arrows represented as ⤳ most likely indicate

(1) energy released into the environment as heat
(2) oxygen produced by respiration
(3) the absorption of energy that has been synthesized
(4) the transport of glucose away from the organism

26 If several species of carnivores are removed from an ecosystem, the most likely effect on the ecosystem will be

(1) an increase in the kinds of autotrophs
(2) a decrease in the number of abiotic factors
(3) a decrease in stability among populations
(4) an increase in the rate of succession

27 Some people make compost piles consisting of weeds and other plant materials. When the compost has decomposed, it can be used as fertilizer. The production and use of compost is an example of

(1) the introduction of natural predators
(2) the use of fossil fuels
(3) the deforestation of an area
(4) the recycling of nutrients

28 Which statement best describes a chromosome?

(1) It is a gene that has thousands of different forms.
(2) It has genetic information contained in DNA.
(3) It is a reproductive cell that influences more than one trait.
(4) It contains hundreds of genetically identical DNA molecules.

29 The graph below shows how the level of carbon dioxide in the atmosphere has changed over the last 150,000 years.

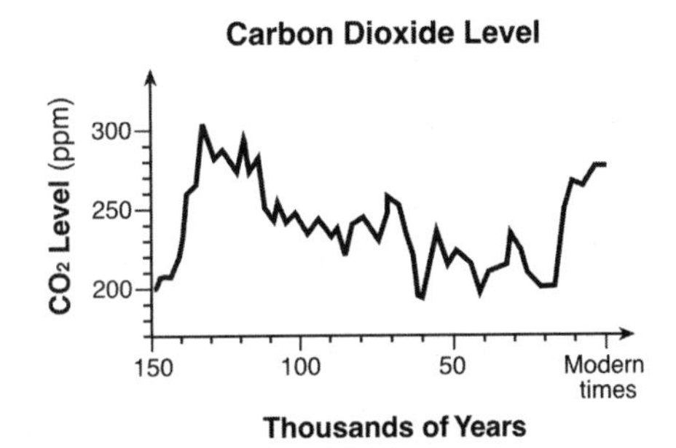

Which environmental factor has been most recently affected by these changes in carbon dioxide level?

(1) light intensity
(2) types of decomposers
(3) size of consumers
(4) atmospheric temperature

30 One reason why people should be aware of the impact of their actions on the environment is that

(1) ecosystems are never able to recover once they have been adversely affected
(2) the depletion of finite resources cannot be reversed
(3) there is a decreased need for new technology
(4) there is a decreased need for substances produced by natural processes

Part B–1

Answer all questions in this part. [11]

Directions (31–41): For *each* statement or question, write on the separate answer sheet the *number* of the word or expression that, of those given, best completes the statement or answers the question.

31 The diagram below represents the process used in 1996 to clone the first mammal, a sheep named Dolly.

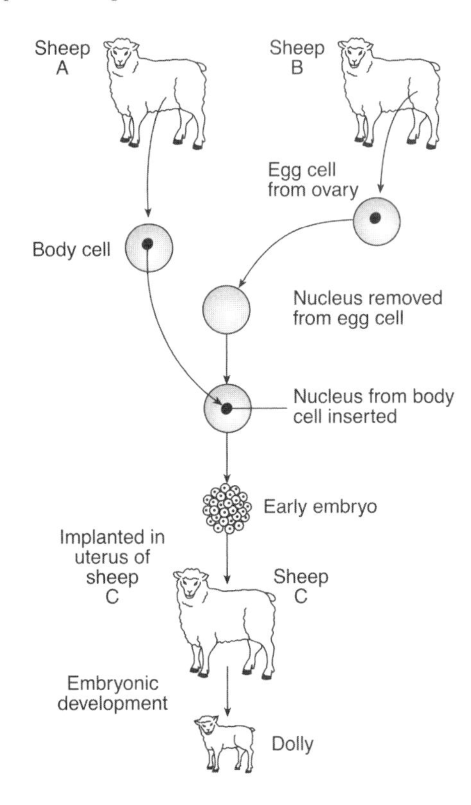

Which statement concerning Dolly is correct?

(1) Gametes from sheep *A* and sheep *B* were united to produce Dolly.
(2) The chromosome makeup of Dolly is identical to that of sheep *A*.
(3) Both Dolly and sheep *C* have identical DNA.
(4) Dolly contains genes from sheep *B* and sheep *C*.

32 The diagram below represents a cell.

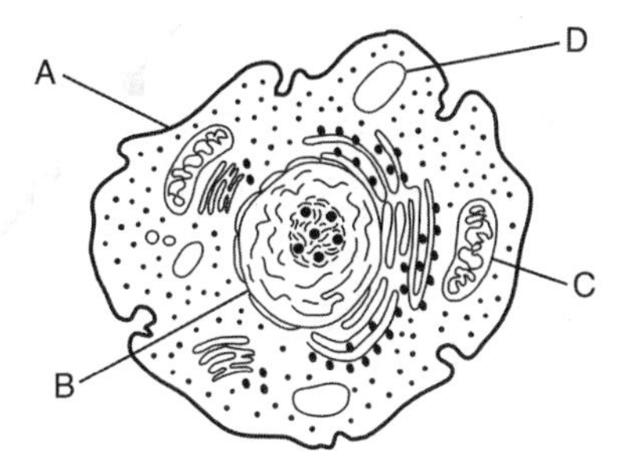

Which statement concerning ATP and activity within the cell is correct?

(1) The absorption of ATP occurs at structure *A*.
(2) The synthesis of ATP occurs within structure *B*.
(3) ATP is produced most efficiently by structure *C*.
(4) The template for ATP is found in structure *D*.

33 The diagram below illustrates some functions of the pituitary gland. The pituitary gland secretes substances that, in turn, cause other glands to secrete different substances.

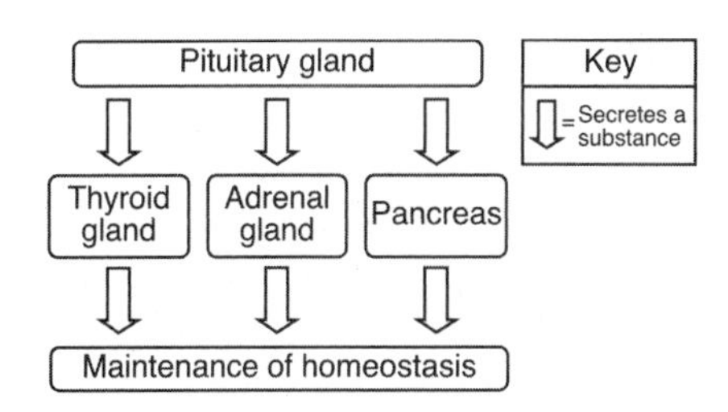

Which statement best describes events shown in the diagram?

(1) Secretions provide the energy needed for metabolism.
(2) The raw materials for the synthesis of secretions come from nitrogen.
(3) The secretions of all glands speed blood circulation in the body.
(4) Secretions help the body to respond to changes from the normal state

34 A pond ecosystem is shown in the diagram below.

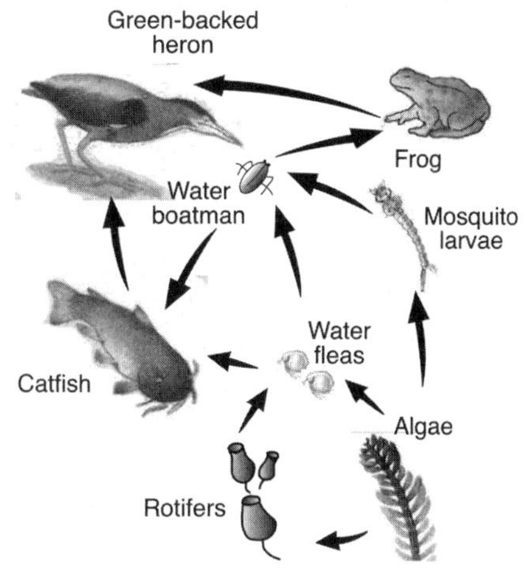

(Not drawn to scale)

Which statement describes an interaction that helps maintain the dynamic equilibrium of this ecosystem?

(1) The frogs make energy available to this ecosystem through the process of photosynthesis.
(2) The algae directly provide food for both the rotifers and the catfish.
(3) The green-backed heron provides energy for the mosquito larvae.
(4) The catfish population helps control the populations of water boatman and water fleas.

35 The diagram below represents a portion of a cell membrane.

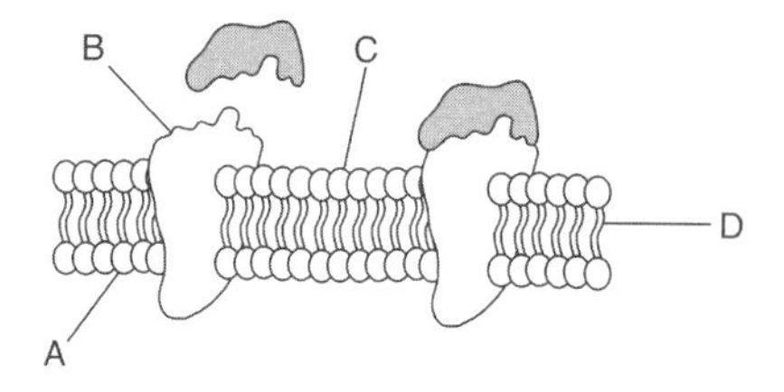

Which structure may function in the recognition of chemical signals?

(1) *A* (3) *C*
(2) *B* (4) *D*

36 Some evolutionary pathways are represented in the diagram below.

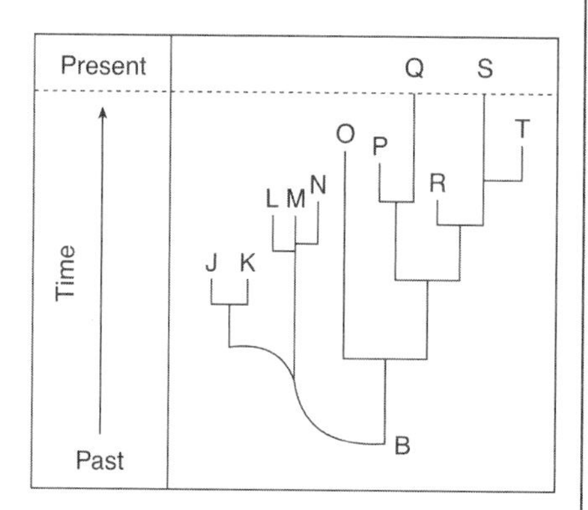

An inference that can be made from information in the diagram is that

(1) many of the descendants of organism *B* became extinct
(2) organism *B* was probably much larger than any of the other organisms represented
(3) most of the descendants of organism *B* successfully adapted to their environment and have survived to the present time
(4) the letters above organism *B* represent members of a single large population with much biodiversity

37 Which species in the chart below is most likely to have the fastest rate of evolution?

Species	Reproductive Rate	Environment
A	slow	stable
B	slow	changing
C	fast	stable
D	fast	changing

(1) *A* (3) *C*
(2) *B* (4) *D*

Base your answers to questions 38 and 39 on the diagram below that represents an energy pyramid in a meadow ecosystem and on your knowledge of biology.

38 Which species would have the largest amount of available energy in this ecosystem?

(1) *A* (3) *C*
(2) *B* (4) *E*

39 Which two organisms are carnivores?

(1) *A* and *B* (3) *B* and *D*
(2) *A* and *E* (4) *C* and *E*

40 The kit fox and red fox species are closely related. The kit fox lives in the desert, while the red fox inhabits forests. Ear size and fur color are two differences that can be observed between the species. An illustration of these two species is shown below.

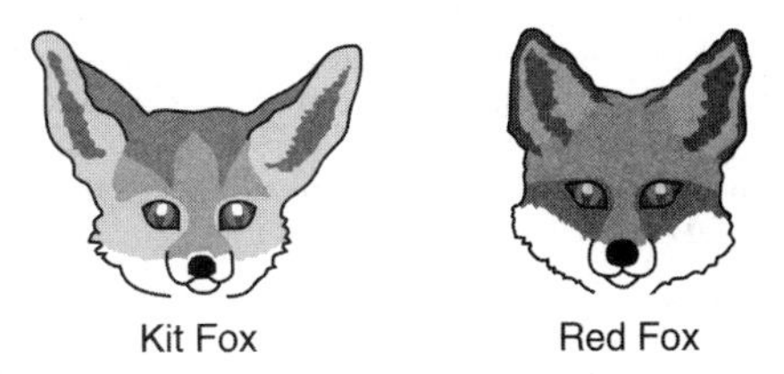

Which statement best explains how the differences between these two species came about?

(1) Different adaptations developed because the kit fox preferred hotter environments than the red fox.
(2) As the foxes adapted to different environments, differences in appearance evolved.
(3) The foxes evolved differently to prevent overpopulation of the forest habitat.
(4) The foxes evolved differently because their ancestors were trying to avoid competition.

41 An ecosystem is represented below.

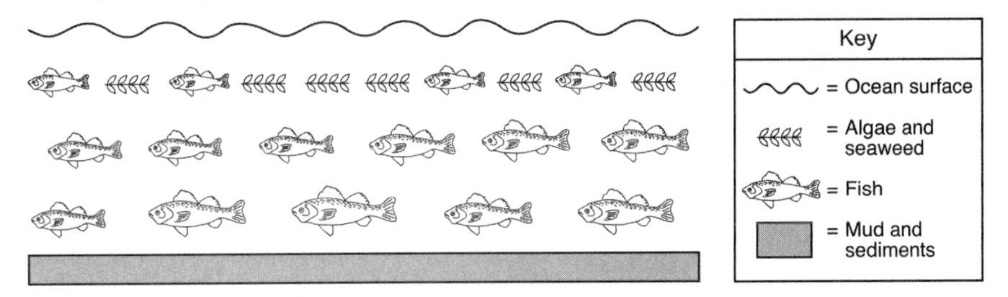

The organisms represented as [algae symbol] are found in the area shown due to which factor?

(1) pH
(2) sediment
(3) light intensity
(4) colder temperature

Part B–2

Answer all questions in this part. [14]

Directions (42–51): For those questions that are followed by four choices, circle the *number* preceding the choice that, of those given, best completes the statement or answers the question. For all other questions in this part, follow the directions given in the question and record your answers in the spaces provided.

For Teacher Use Only

42 The graphs below show dissolved oxygen content, sewage waste content, and fish populations in a lake between 1950 and 1970.

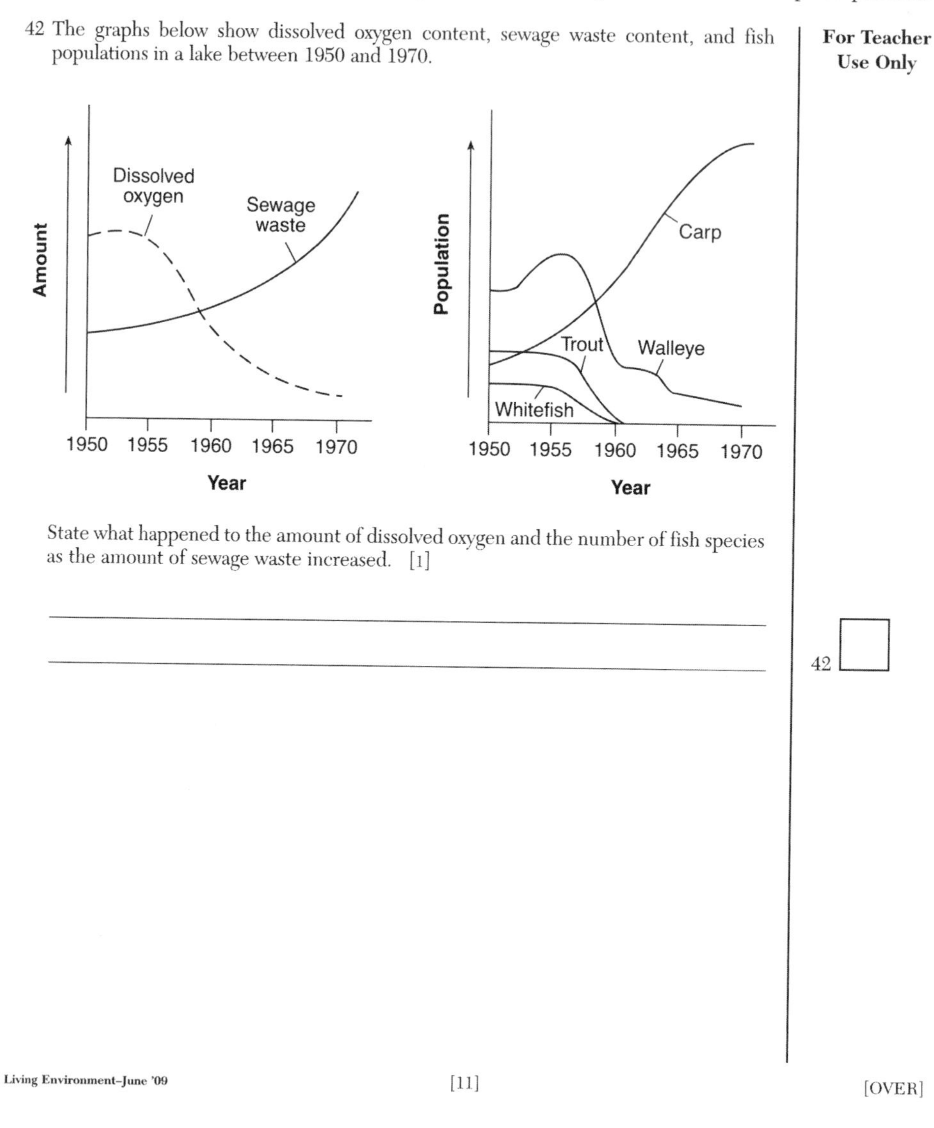

State what happened to the amount of dissolved oxygen and the number of fish species as the amount of sewage waste increased. [1]

____

____

42 ☐

Base your answers to questions 43 through 46 on the information below and on your knowledge of biology.

For Teacher Use Only

Yeast cells carry out the process of cellular respiration as shown in the equation below.

$$C_6H_{12}O_6 \rightarrow 2C_2H_5OH + 2CO_2$$

glucose → ethyl alcohol + carbon dioxide

An investigation was carried out to determine the effect of temperature on the rate of cellular respiration in yeast. Five experimental groups, each containing five fermentation tubes, were set up. The fermentation tubes all contained the same amounts of water, glucose, and yeast. Each group of five tubes was placed in a water bath at a different temperature. After 30 minutes, the amount of gas produced (*D*) in each fermentation tube was measured in milliliters. The average for each group was calculated. A sample setup and the data collected are shown below.

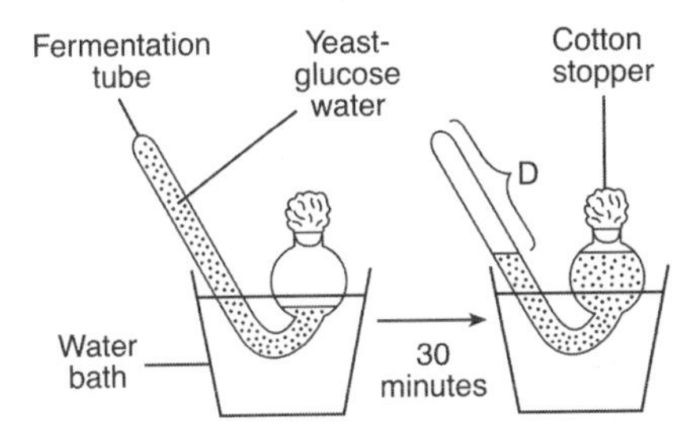

Average Amount of Gas Produced (D) After 30 Minutes at Various Temperatures

Group	Temperature (°C)	D (mL)
1	5	0
2	20	5
3	40	12
4	60	6
5	80	3

Directions (43 and 44): Using the information in the data table, construct a line graph on the grid below, following the directions below.

For Teacher Use Only

43 Mark an appropriate scale on each labeled axis. [1]

44 Plot the data from the data table. Surround each point with a small circle, and connect the points. [1]

Example:

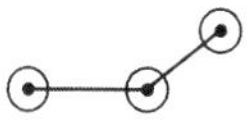

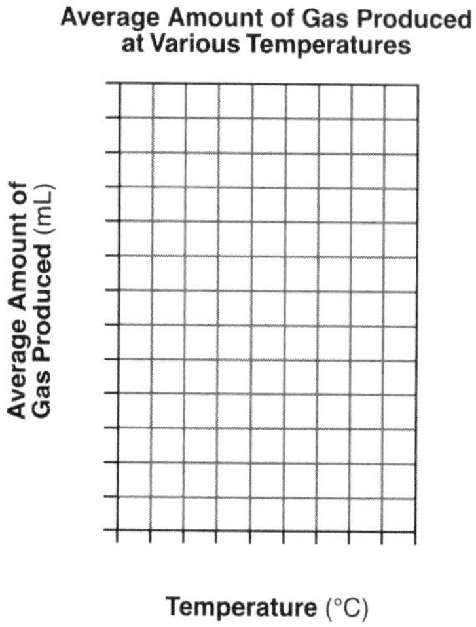

Average Amount of Gas Produced at Various Temperatures

Average Amount of Gas Produced (mL)

Temperature (°C)

43

44

45 The maximum rate of cellular respiration in yeast occurred at which temperature?

(1) 5°C

(2) 20°C

(3) 40°C

(4) 60°C

45

46 Compared to the other tubes at the end of 30 minutes, the tubes in group 3 contained the

(1) smallest amount of CO_2

(2) smallest amount of glucose

(3) smallest amount of ethyl alcohol

(4) same amounts of glucose, ethyl alcohol, and CO_2

46

Base your answers to questions 47 through 49 on the information below and on your knowledge of biology.

For Teacher Use Only

An ecologist made some observations in a forest ecosystem over a period of several days. Some of the data collected are shown in the table below.

Observations in a Forest Environment

Date	Observed Feeding Relationships	Ecosystem Observations
6/2	• white-tailed deer feeding on maple tree leaves • woodpecker feeding on insects • salamander feeding on insects	• 2 cm of rain in 24 hours
6/5	• fungus growing on a maple tree • insects feeding on oak trees	• several types of sedimentary rock are in the forest
6/8	• woodpecker feeding on insects • red-tailed hawk feeding on chipmunk	• air contains 20.9% oxygen
6/11	• chipmunk feeding on insects • insect feeding on maple tree leaves • chipmunk feeding on a small salamander	• soil contains phosphorous

47 On the diagram below, complete the food web by placing the names of *all* the organisms in the correct locations. [1]

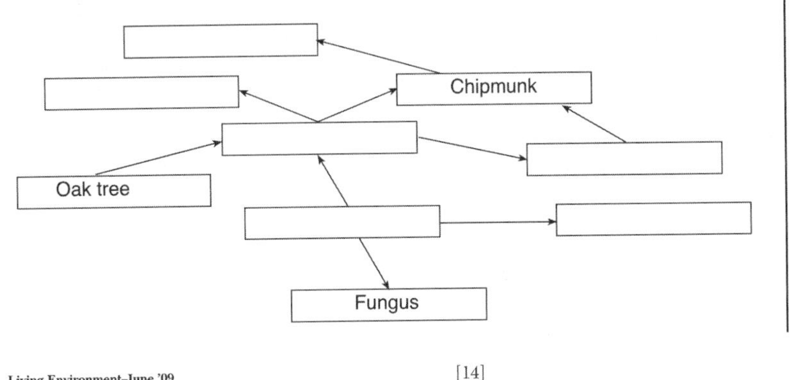

47 ☐

For Teacher Use Only

48 Identify *one* producer recorded by the ecologist in the data table. [1]

____

48 ☐

49 Which statement describes how one biotic factor of the forest uses one of the abiotic factors listed in the data table?

(1) Trees absorb water as a raw material for photosynthesis.

(2) Insects eat and digest the leaves of trees.

(3) Erosion of sedimentary rock adds phosphorous to the soil.

(4) Fungi release oxygen from the trees back into the air.

49 ☐

50 Fill in all of the blanks in parts 2 and 3 of the dichotomous key below, so that it contains information that could be used to identify the four animals shown below. [2]

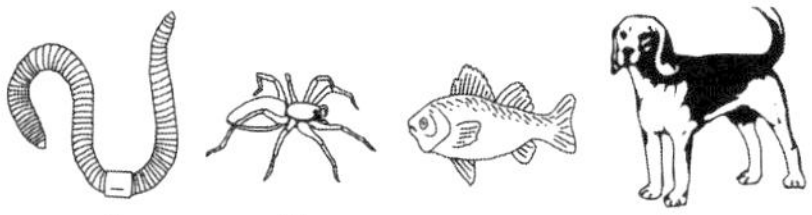

I II III IV

Dichotomous Key

1. a. Legs present.. Go to 2
   b. Legs not present.. Go to 3

	Characteristic		**Organism**
2.	a. ______________________	....	______________________
	b. ______________________	....	______________________
3.	a. ______________________	....	______________________
	b. ______________________	....	______________________

50 ☐

51 The human female reproductive system is represented in the diagram below.

For Teacher Use Only

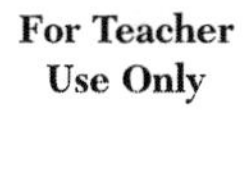

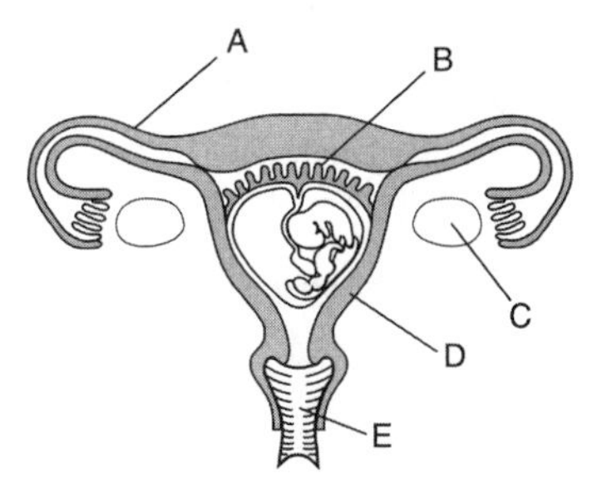

Complete boxes 1 through 4 in the chart below using the information from the diagram. [4]

Name of Structure	Letter on Diagram	Function of Structure
1 ________________	2 _______	produces gametes
uterus	D	3 ______________________
4 ________________	B	transports oxygen directly to the embryo

51 ☐

Part C

Answer all questions in this part. [17]

Directions (52–59): Record your answers in the spaces provided in this examination booklet.

For Teacher Use Only

52 Humans have many interactions with the environment. Briefly describe how human activities can affect the environment of organisms living 50 years from now. In your answer, be sure to:

- identify *one* human activity that could release chemicals harmful to the environment [1]
- identify the chemical released by the activity [1]
- state *one* effect the release of this chemical would most likely have on future ecosystems [1]
- state *one* way in which humans can reduce the production of this chemical to lessen its effect on future ecosystems [1]

52

For Teacher Use Only

53 Plants respond to their environment in many different ways. Design an experiment to test the effects of *one* environmental factor, chosen from the list below, on plant growth.

Acidity of precipitation
Temperature
Amount of water

In your answer, be sure to:

- identify the environmental factor you chose
- state *one* hypothesis the experiment would test [1]
- state how the control group would be treated differently from the experimental group [1]
- state *two* factors that must be kept the same in both the experimental and control groups [1]
- identify the independent variable in the experiment [1]
- label the columns on the data table below for the collection of data in your experiment [1]

Environmental factor: ______________________________

____

____

____

____

____

____

____

____

____

____

Data Table

53 ☐

Base your answer to question 54 on the article below and on your knowledge of biology.

For Teacher Use Only

Power plan calls for windmills off beach

The Associated Press

Several dozen windmills taller than the Statue of Liberty will crop up off Long Island — the first source of off-shore wind power outside of Europe, officials said.

The Long Island Power Authority [LIPA] expects to choose a company to build and operate between 35 and 40 windmills in the Atlantic Ocean off Jones Beach, The New York Times reported Sunday [May 2, 2004]. Cost and completion date are unknown.

Energy generated by the windmills would constitute about 2 percent of LIPA's total power use. They are expected to produce 100 to 140 megawatts, enough to power 30,000 homes....

But some Long Island residents oppose the windmills, which they fear will create noise, interfere with fishing, and mar ocean views....

Source: "Democrat and Chronicle", Rochester, NY 5/3/04

54 State *two* ways that the use of windmills to produce energy would be beneficial to the environment. [2]

(1) ____

(2) ____

____

54 ☐

55 Importing a foreign species, either intentionally or by accident, can alter the balance of an ecosystem. State *one* specific example of an imported species that has altered the balance of an ecosystem and explain how it has disrupted the balance in that ecosystem. [2]

____

____

____

55 ☐

Base your answers to questions 56 through 59 on the passage below and on your knowledge of biology.

For Teacher Use Only

Avian (Bird) Flu

Avian flu virus H5N1 has been a major concern recently. Most humans have not been exposed to this strain of the virus, so they have not produced the necessary protective substances. A vaccine has been developed and is being made in large quantities. However, much more time is needed to manufacture enough vaccine to protect most of the human population of the world.

Most flu virus strains affect the upper respiratory tract, resulting in a runny nose and sore throat. However, the H5N1 virus seems to go deeper into the lungs and causes severe pneumonia, which may be fatal for people infected by this virus.

So far, this virus has not been known to spread directly from one human to another. As long as H5N1 does not change to another strain that can be transferred from one human to another, a worldwide epidemic of the virus probably will not occur.

56 State *one* difference between the effect on the human body of the usual forms of flu virus and the effect of H5N1. [1]

____

____

56 ☐

57 Identify the type of substance produced by the human body that protects against antigens, such as the flu virus. [1]

____

57 ☐

58 State what is in a vaccine that makes the vaccine effective. [1]

____

58 ☐

59 Identify *one* event that could result in the virus changing to a form able to spread from human to human. [1]

____

59 ☐

Part D

Answer all questions in this part. [13]

Directions (60–72): For those questions that are followed by four choices, circle the *number* of the choice, that, of those given, best completes the statement or answers the question. For all other questions in this part, follow the directions given in the question and record your answers in the spaces provided.

For Teacher Use Only

60 The data table below shows the number of amino acid differences in the hemoglobin molecules of several species compared with amino acids in the hemoglobin of humans.

Amino Acid Differences

Species	Number of Amino Acid Differences
human	0
frog	67
pig	10
gorilla	1
horse	26

Based on the information in the data table, write the names of the organisms from the table in their correct positions on the evolutionary tree below. [1]

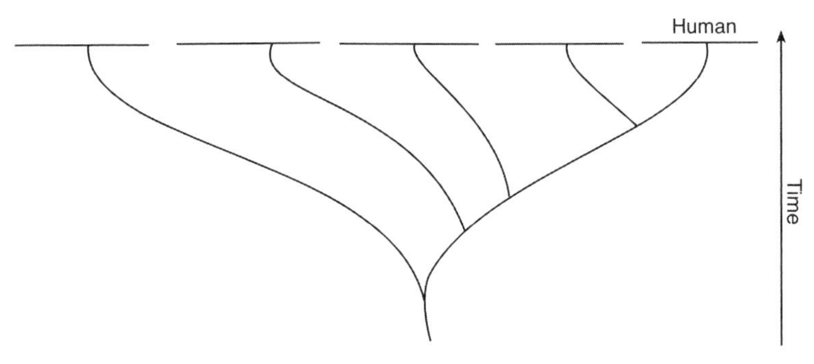

60

61 Explain why comparing the vein patterns of several leaves is a less reliable means of determining the evolutionary relationship between two plants than using gel electrophoresis. [1]

____

____

61

Base your answer to question 62 on the information and diagram below and on your knowledge of biology.

An enzyme and soluble starch were added to a test tube of water and kept at room temperature for 24 hours. Then, 10 drops of glucose indicator solution were added to the test tube, and the test tube was heated in a hot water bath for 2 minutes.

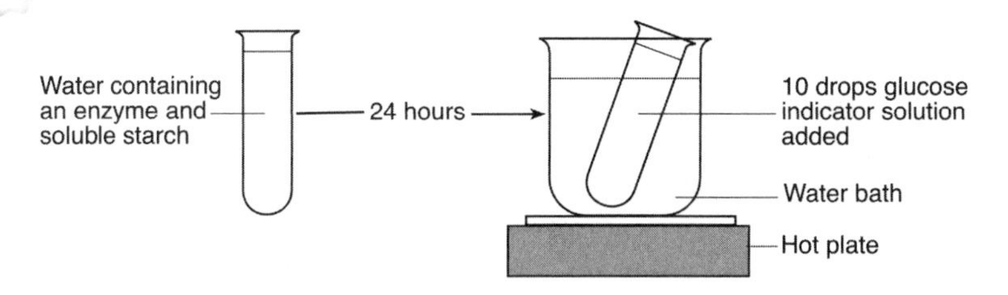

62 The test was performed in order to

(1) measure the quantity of fat that is converted to starch

(2) determine if digestion took place

(3) evaporate the water from the test tube

(4) cause the enzyme to bond to the water

63 A chromatography setup is shown below.

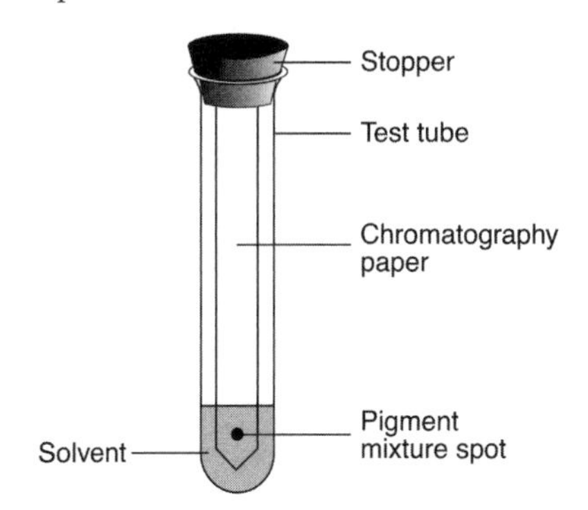

Identify *one* error in the setup. [1]

The Pigment mixture spot is below the solvent

For Teacher Use Only

62 ☐

63 ☐

Base your answers to questions 64 through 66 on the information and data table below and on your knowledge of biology.

For Teacher Use Only

During a laboratory activity, a group of students obtained the data shown below.

Pulse Rate Before and After Exercise

Student Tested	Pulse Rate at Rest (beats/min)	Pulse Rate After Exercise (beats/min)
A	70	97
B	74	106
C	83	120
D	60	91
E	78	122
Group Average		107

64 Which procedure would increase the validity of the conclusions drawn from the results of this experiment?

(1) increasing the number of times the activity is repeated

(2) changing the temperature in the room

(3) decreasing the number of students participating in the activity

(4) eliminating the rest period before the resting pulse rate is taken

64 ☐

65 Calculate the group average for the resting pulse rate. [1]

__________ **beats/min**

65 ☐

66 A change in pulse rate is related to other changes in the body. Write the name of *one* organ that is affected when a person runs a mile and describe *one* change that occurs in this organ. [1]

Organ: ____________________

____

____

____

66 ☐

For Teacher Use Only

Base your answers to questions 67 through 69 on the information and diagram below and on your knowledge of biology.

A wet mount of red onion cells as seen with a compound light microscope is shown below.

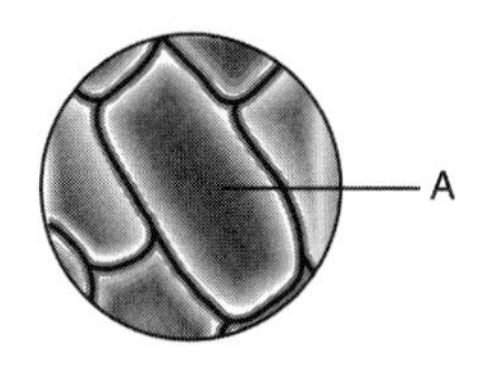

67 Which diagram best illustrates the technique that would most likely be used to add salt to these cells?

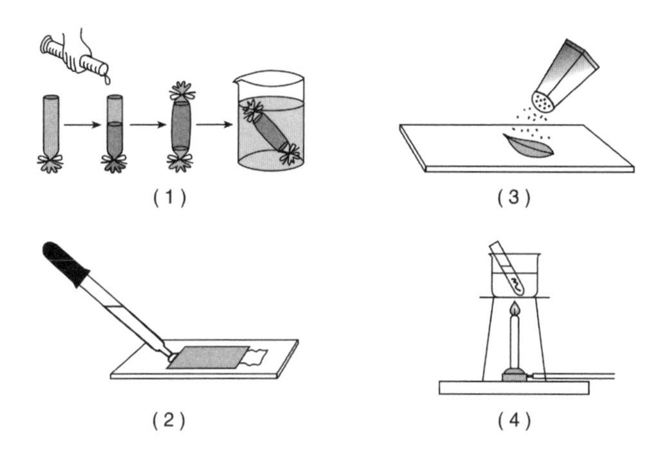

67 ☐

68 In the space below, sketch what cell *A* would look like after the addition of the salt. [1]

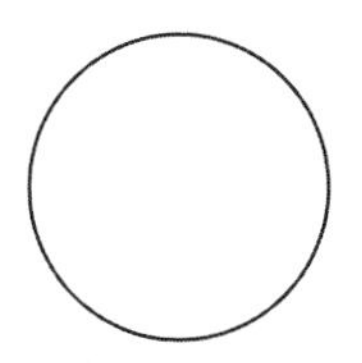

68 ☐

For Teacher Use Only

69 Which substance would most likely be used to return the cells to their original condition?

(1) starch indicator

(2) dialysis tubing

(3) glucose indicator solution

(4) distilled water

69 ☐

---

70 DNA electrophoresis is used to study evolutionary relationships of species. The diagram below shows the results of DNA electrophoresis for four different animal species.

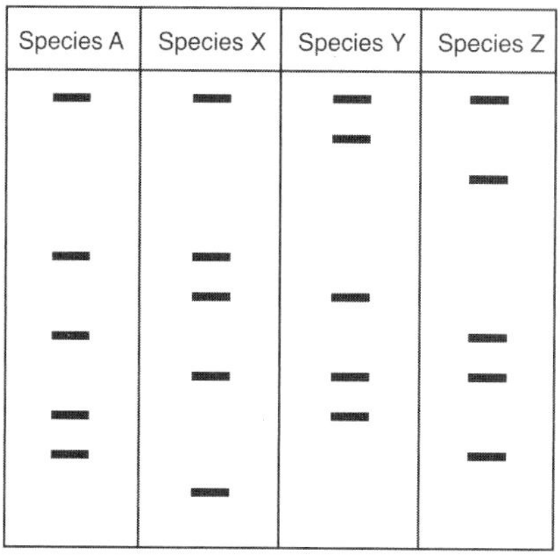

Species A Species X Species Y Species Z

Which species has the most DNA in common with species *A*?

(1) *X* and *Y*, only

(2) *Y*, only

(3) *Z*, only

(4) *X*, *Y*, and *Z*

70 ☐

Base your answers to questions 71 and 72 on the diagram below that shows variations in the beaks of finches in the Galapagos Islands and on your knowledge of biology.

For Teacher Use Only

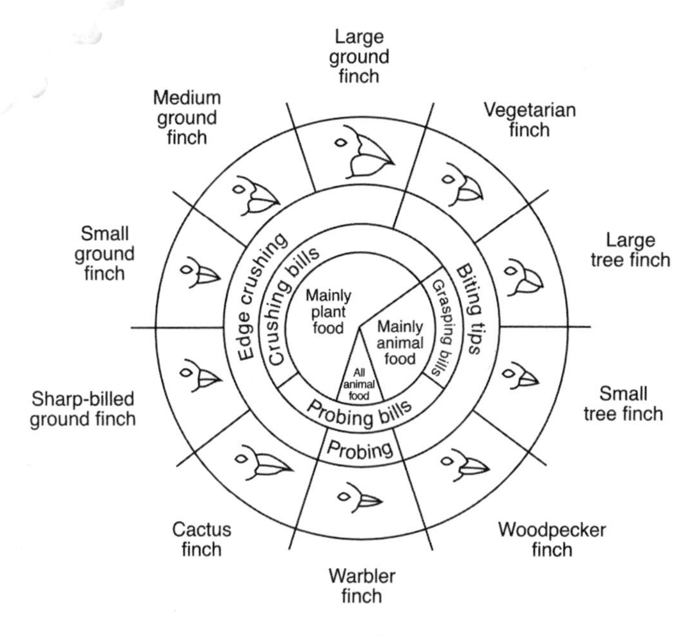

From: *Galapagos: A Natural History Guide*

71 The diversity of species seen on the Galapagos Islands is mostly due to

(1) gene manipulation by scientists

(2) gene changes resulting from mitotic cell division

(3) natural selection

(4) selective breeding

71 ☐

72 State *one* reason why large ground finches and large tree finches can coexist on the same island. [1]

____

____

____

72 ☐

Tear Here

The University of the State of New York

REGENTS HIGH SCHOOL EXAMINATION

LIVING ENVIRONMENT

Thursday, June 18, 2009 — 1:15 to 4:15 p.m., only

ANSWER SHEET

Student .... Sex: ☐ Female ☐ Male

Teacher ....

School .... Grade ....

Part	Maximum Score	Student's Score
A	30	
B–1	11	
B–2	14	
C	17	
D	13	
Total Raw Score (maximum Raw Score: 85)		
Final Score (from conversion chart)		

Raters' Initials

Rater 1 .... Rater 2 ....

Record your answers to Part A and Part B–1 on this answer sheet.

Part A

1	11	21 ....
2	12	22 ....
3	13	23 ....
4	14	24 ....
5	15	25 ....
6	16	26 ....
7	17	27 ....
8	18	28 ....
9	19	29 ....
10	20	30 ....

Part A Score ☐

Part B–1

31	37 ....
32	38 ....
33	39 ....
34	40 ....
35	41 ....
36	

Part B–1 Score ☐

Tear Here

The declaration below must be signed when you have completed the examination.

I do hereby affirm, at the close of this examination, that I had no unlawful knowledge of the questions or answers prior to the examination and that I have neither given nor received assistance in answering any of the questions during the examination.

____
Signature

NOTES

NOTES

NOTES

NOTES

NOTES

NOTES